Alloy Phase Stability

NATO ASI Series

Advanced Science Institutes Series

A Series presenting the results of activities sponsored by the NATO Science Committee, which aims at the dissemination of advanced scientific and technological knowledge, with a view to strengthening links between scientific communities.

The Series is published by an international board of publishers in conjunction with the NATO Scientific Affairs Division

A Life Sciences **B Physics**	Plenum Publishing Corporation London and New York
C Mathematical and Physical Sciences **D Behavioural and Social Sciences** **E Applied Sciences**	Kluwer Academic Publishers Dordrecht, Boston and London
F Computer and Systems Sciences **G Ecological Sciences** **H Cell Biology**	Springer-Verlag Berlin, Heidelberg, New York, London, Paris and Tokyo

Series E: Applied Sciences - Vol. 163

Alloy Phase Stability

edited by

G. M. Stocks

Metals and Ceramics Division,
Oak Ridge National Laboratory,
Oak Ridge, Tennessee, U.S.A.

and

A. Gonis

Division of Chemistry and Materials Science,
Lawrence Livermore National Laboratory,
Livermore, U.S.A.

Kluwer Academic Publishers

Dordrecht / Boston / London

Published in cooperation with NATO Scientific Affairs Division

Proceedings of the NATO Advanced Study Institute on
Alloy Phase Stability
Maleme, Crete, Greece
June 13–17, 1987

Library of Congress Cataloging in Publication Data

NATO Advanced Study Institute on Alloy Phase Stability (1987 : Maleme, Crete)
Alloy phase stability : proceedings of the NATO Advanced Study Institute on Alloy Phase Stability, Maleme, Crete, Greece, June 13-27, 1987 / [edited by] G.M. Stocks and A. Gonis.
p. cm. -- (NATO ASI series. Series E, Applied sciences ; vol. 163)
"Published in cooperation with the NATO Scientific Affairs Division."
Includes bibliographies and index.
1. Order-disorder in alloys--Congresses. 2. Phase rule and equilibrium--Congresses. I. Stocks, G. M., 1943- . II. Gonis, Antonios, 1945- . III. North Atlantic Treaty Organization. Scientific Affairs Division. IV. Title. V. Series: NATO ASI series. Series E, Applied sciences ; no. 163.
TN690.N327 1987
669'.94--dc19 89-2802

ISBN-13: 978-94-010-6901-4 e-ISBN-13: 978-94-009-0915-1
DOI: 10.1007/978-94-009-0915-1

Published by Kluwer Academic Publishers,
P.O. Box 17, 3300 AA Dordrecht, The Netherlands.

Kluwer Academic Publishers incorporates the publishing programmes of
D. Reidel, Martinus Nijhoff, Dr W. Junk and MTP Press.

Sold and distributed in the U.S.A. and Canada
by Kluwer Academic Publishers,
101 Philip Drive, Norwell, MA 02061, U.S.A.

In all other countries, sold and distributed
by Kluwer Academic Publishers Group,
P.O. Box 322, 3300 AH Dordrecht, The Netherlands.

CONTENTS

SECTION 5. Electronic Theories of Phase Stability: First Principles Theory

SECTION 6. The Effects of Strain and Macroscopic Defects on Phase Stability

SECTION 7. General Topics

*Invited lectures

PREFACE

One of the ultimate goals of materials research is to develop a fundamental and predictive understanding of the physical and metallurgical properties of metals and alloys. Such an understanding can then be used in the design of materials having novel properties or combinations of properties designed to meet specific engineering applications. The development of new and useful alloy systems and the elucidation of their properties are the domain of metallurgy. Traditionally, the search for new alloy systems has been conducted largely on a trial and error basis, guided by the skill and intuition of the metallurgist, large volumes of experimental data, the principles of 19th century thermodynamics and ad hoc semi-phenomenological models. Recently, the situation has begun to change. For the first time, it is possible to understand the underlying mechanisms that control the formation of alloys and determine their properties. Today theory can begin to offer guidance in predicting the properties of alloys and in developing new alloy systems.

Historically, attempts directed toward understanding phase stability and phase transitions have proceeded along distinct and seemingly diverse lines. Roughly, we can divide these approaches into the following broad categories.

1. Experimental determination of phase diagrams and related properties,
2. Thermodynamic/statistical mechanical approaches based on semi-phenomenological models, and
3. Ab initio quantum mechanical methods.

Metallurgists have traditionally concentrated their efforts in categories 1 and 2, while theoretical physicists have been preoccupied with 2 and 3. The latter group is further divided into scientists pursuing a more phenomenological approach and those involved in developing completely parameter-free theories.

On the experimental front, new techniques based on synchrotron radiation, photoemission, positron annihilation, and electron microscopy are being used to provide a deeper insight into alloy phase diagrams, short-range order, and the approach to phase transitions than has been hitherto possible. The semi-empirical thermodynamic approach of the CALPHAD school has now progressed to the stage where it can produce phase diagrams for complex, multicomponent systems. Recent studies of the anisotropic next-nearest-neighbor interaction (ANNNI) model have revealed the richness of the phases which can result from even quite simple models of the interaction between atoms. Advances in the calculation of the electronic structure and total energy of both ordered and disordered alloys make it possible to calculate their ground state properties ab initio. These latter advances now offer the possibility, when integrated with modern statistical mechanical techniques, of providing a parameter-free understanding of alloy phase stability.

The need to bridge the gaps which exist between the various scientific groups has been felt for a long time. Recently, a number of scientific meetings have taken place for this purpose, and it is evident that scientists from the various disciplines are beginning to speak a common language. Thus, a broad scientific gathering on the general subject of alloy phase stability and its relation to subjects as diverse as alloy design and microscopic theory seemed warranted.

The NATO Advanced Study Institute (ASI) on Alloy Phase Stability which was held June 14-27, 1987 in Maleme-Chania, Greece, was conceived to provide a forum for furthering the interaction and dialogue between the various disciplines, and to promote communication between the students new to the various subjects and the "experts." Judging from the enthusiastic response of the lecturers and participants, the high attendance of the lectures (despite the beckoning weather and sea outside), and the comments upon completion of the ASI, we gather that the primary aims of the meeting were accomplished. The lectures, delivered by outstanding researchers in their respective fields, were lively and informative and were often followed by animated, indeed at times heated, discussions.

The present volume contains the lecture notes of the invited lecturers at the ASI as well as a number of papers contributed by the participants. The authors have made a valiant attempt to make their presentation simple enough to be understandable by the informed nonspecialist without seriously compromising the scientific content.

In order to emphasize the interdisciplinary nature of the ASI and to present the material as clearly as possible, the book is divided into seven sections corresponding to the various fields represented at the meeting. This division is, admittedly, somewhat arbitrary, and the editors bear full responsibility for distributing the articles in this particular manner. Although no effort has been made to cross-reference the articles, reading the articles makes it rather clear that the authors were cognizant of the scope of the contributions of other lecturers and participants. We hope that this helps toward a feeling of overall unity. The individual articles are testimony to the great progress that has been made in recent years along a broad front of endeavors that contributes to the general understanding of metallurgy, and the physics of alloy phase stability.

The editors are indebted to the NATO Scientific Affairs Division which was the main sponsor of the ASI. They are particularly grateful to Dr. C. Sinclair, Director of the ASI Program, for his support and assistance in bringing this rather complex task to fruition.

The directors also wish to express their gratitude for the cosponsorship of the meeting by the Office of Naval Research and the U.S. Department of Energy through the Metals and Ceramics Division of the Oak Ridge National Laboratory and the Chemistry and Materials Sciences Division of the Lawrence Livermore National Laboratory. These institutions provided financial and, in the case of the Oak Ridge National Laboratory, technical support that was essential to the success of this endeavor.

The editors also wish to thank the members of the organizing committee, R. W. Cahn, J. S. Faulkner, F. Gautier, and A. R. Miedema, for a number of helpful suggestions and criticism which greatly helped in shaping the format of the summer school.

It is with particular pleasure that we acknowledge the extraordinarily efficient assistance of Ms. Judy Turner during the rather hectic two years that the summer school was in preparation. Not only were forms and other incidentals handled quickly and accurately, but the final manuscript was processed to completion by her capable hands. Of equal importance to the success of the ASI was the competent handling of the conference onsite by Ms. Maria Besieri. If the entire proceedings gave the impression of faultless organization, it is only because of her untiring dedication and the efficiency with which potentially catastrophic situations were handled. To both of them, we express our heartfelt gratitude.

Last, but certainly not least, the editors acknowledge the organizational assistance and competent handling of the reservations by the staff of the Chandris Hotel. Mrs. Rae Derke, the Chandris Representative in New York, Mr. K. Apostolopoulos, the Reservations Director at Maleme, and Mr. Z. Mourikas, Director of the Chandris Hotel, are due special mention and thanks.

A. Gonis
G. M. Stocks
Directors, NATO Advanced Study
Institute on Alloy Phase Stability

This project was sponsored by the

North Atlantic Treaty Organization (NATO)
through the Scientific Affairs Division

and was cosponsored by the

U. S. Department of Energy through the Oak Ridge National Laboratory
Lawrence Livermore National Laboratory
U. S. Office of Naval Research

ALLOY PHASE STABILITY: OPENING REMARKS

S. C. Moss
Department of Physics
University of Houston
Houston, Texas 77004

The problems of phase stability in alloys involve subjects that fall quite naturally between metallurgy and metal physics. By metallurgy we mean, for the most part, the more traditional aspects of the structure-property relations in metals and alloys and their manipulation by heat treatment, although a principal purpose of this Advanced Study Institute (ASI) is to demonstrate the continuing success that physics is having in dealing with metallurgical complexity. By metal physics, we mean the application of quantum theory, classical mechanics and elasticity, and statistical mechanics to metallic alloys. The dichotomy, actually, has always been somewhat artificial, especially as many of the most insightful advances in metallurgy have come from chemists, physicists and crystallographers who have chosen to work on alloys and their phase relations and properties. It could nonetheless have been said some years ago that if one studied the Fermi surface of pure copper one was a physicist whereas if one wanted to understand the alloying effects of Zn on, for example, the hardness of Cu, one was a physical metallurgist. [And if one wished to design a heat treatment to optimize the properties of an alloy, ferrous or otherwise, one was certainly a metallurgist!]

To return to CuZn alloys, for a description of the β-brass order-disorer transition from the bcc to CsCl structure, the Ising antiferromagnet of physics is a uniquely suitable analogue because order-disorder in β-brass is a continuous second order phase transition. The first order γ-α transition in iron, on the other hand, is often referred to by metallurgists as a "phase transformation", the distinction being that the latter can be sluggish, hysteretic and disruptive with complex crystallographic mechanisms. Ordered β-brass, for example, transforms martensitically on cooling to a more complex orthorhombic structure in a first order fashion with a transition start temperature that changes by hundreds of degrees over a few percent in concentration of Zn. The electronic and lattice dynamical basis for this phase change, how it is "nucleated", and why there is such a striking concentration dependence of the transition temperature remain of continuing interest as examples of current problems in alloy phase stability.

The disciplines that will inevitably be touched on in these proceedings include: thermodynamics and statistical mechanics, kinetics, elasticity theory, crystallography, electronic structure and lattice dynamics, and the theory and observation of defects and defect structures. Among the ultimate objectives are: calculation of phase diagrams, understanding of first-order phase transitions (transformations), prediction of complex structures and finally, as always, the elucidation of at least some of the simpler aspects of structure-property relations. This is, of course, a rather forbidding list and the ASI will be fortunate

G. M. Stocks and A. Gonis (eds.), Alloy Phase Stability, 1–4.

to come to terms with one or two. But it is important to keep in mind that alloys pose for the physicist a great range of interesting problems that only recently have become tractable and are being successfully addressed.

In discussing phase stability we often make reference to mechanical analogues such as a brick on a table: lying flat, it is stable, on its end it is metastable while on an edge it is unstable because any small fluctuation will lower its potential energy. But an object is only metastable if it is so with respect to all infinitesimal changes. Alloy stability by analogy must be viewed with respect to all paths. It may be an important issue because so many of the states that we regard as both metastable and useful are often quenched-in configurations that represent intermediate steps on the path to equilibrium (i.e. tempered steel and other aged alloys). Such issues while perhaps outside the nominal scope of this ASI are nevertheless worthy of mention. For example, the calculation of equilibrium phase diagrams often reveals metastable competing phases that may actually be more readily accessed on cooling than the equilibrium structures.

By way of illustration, Cu-Ni provides one of the simplest alloy phase diagrams in which there are simple liquidus and solidus lines with no evidence in the solid state of anything but a single phase fcc solution. The atom sizes are close but not identical and, of course, Ni is magnetic while Cu is not. Because they are neighboring elements in the periodic table, the electron and X-ray diffraction patterns of the alloys are not very sensitive to local order fluctuations. Null-matrix neutron scattering studies of Cu-Ni alloys, however, have shown a clear tendency to phase-separate with a predicted critical temperature at ~ 420°C. The influence of this clustering tendency on the magnetic and optical properties of the alloys is quite significant even though the metastable solid solution never truly phase-separates for kinetic reasons (too low a spinodal temperature). For example, the small enhancement of Ni neighbors of a Ni atom, over the random value, produces exchange-enhanced giant polarization clouds and a superparamagnetic response of the solid solution.

The concentration fluctuations in Cu-Ni alloys are coherent in that they modulate an average lattice whose Bragg peaks are sharp. Because of the relative lattice parameter difference, $\Delta a/a \simeq 0.025$, these concentration fluctuations will cost elastic energy which in turn adds a term to the free energy that is otherwise absent. The contribution of strain energy to phases under stress (mutual or otherwise) is in general very important in the consideration of stability, as Cahn first showed some years ago. It leads to phase diagrams and phase relations for stressed solids that may differ appreciably from the usual unstressed case which is represented by the published phase diagram (although the phase boundaries on experimental phase diagrams may often be in error by virtue of these coherency/incoherency effects, i.e., they are inadvertently not equilibrium incoherent stress-free diagrams).

Incoherent phases, as suggested above, do not share a common lattice. Their interfaces therefore introduce a surface free energy, in place of the elastic contribution, which is usually negligible when the phases are macroscopic. It is when incoherent phases evolve out of coherent phase fluctuations that this trade-off is most critical (as with artificial multilayers). For example, the coherent structural fluctuations that appear as precursors above the structural phase transitions discussed earlier lead directly to new phases which are crystallographically related to, but incoherent with, the high temperature phases. The traditional description of such first order transitions by Bain strains, shears and "shuffles", while often useful and insightful, may conceal the underlying lattice dynamical basis and, ultimately, the physis of the structural change.

Central to the question of mechanism in metallurgical structural transitions has been the origin of the precursor fluctuations or embryos. These embryos can be either intrinsic or extrinsic. By extrinic we mean defect-initiated and these phase changes thereby provide us with an excellent example of the influence of defects on phase stability. As Yamada has emphasized, a modulated lattice relaxation about a defect or stress center can produce a martensitic embryo if a "soft phonon" of the lattice couples directly to the defect stress field. The origin of the soft phonon then becomes an important issue that can be resolved through electronic structure calculations. Intrinsic sources of embryos are of two basic types. The first, proposed by Krivoglaz, posits a localized heterogeneous state due to the response function of the alloy at $2k_F$ (k_F = Fermi wavevector). The real-space consequence of the enhanced screening (with lower effective dimensionality or strong electron-lattice coupling) will be a uniform distribution of "phased" embryos much as Tanner and others have observed in high resolution electron microscopy above the martensitic transition. A second source of intrinsic embryo formation that does not depend upon the Fermi surface and screening, and is therefore relevant to non-metals as well, is the presence of strong amharmonic coupling together with large elastic anisotropy to produce, at a "soft-mode" wavevector, a modulated displacment correlation function. As Krumhansl and co-workers emphasize, it is the elastic softness at a particular wavevector combined with the pronounced anisotropy that acts to create wave-packet-like embryos.

The issue of stability is clearly involved in the way in which one phase transforms into another - the path it chooses - although kinetics and mechanism determine how (and if) it will in fact transform. Stability arguments, of course, also govern what phase will form from among the likely candidates, the balance being as always between energetic and entropic considerations. The bcc phase is stabilized at high temperature, for instance, by vibrational entropy via anharmonic phonon-phonon coupling. We may also, however, consider more complex phases among which number the great variety of incommensurate modulated structures with long periods. For these structures we have the formalism of the ANNNI model, as described in the lectures by Selke, which quite successfully accounts for a host of subtle transitions from one (incommensurate) period to another; van Tendeloo and Loiseau will present beautiful high resolution electron microscopic images of sequences of these phases. The origin of the competing long-range interactions responsible for the ANNNI phases must for metals remain electronic and oscillatory. For example, if we are discussing an fcc lattice with nearest and next nearest neighbor $\{100\}$ planar interactions, J_1 and J_2, the actual interatomic pairwise interactions must be appreciable out to the <u>eighth</u> neighbor. This suggests that band structure effects and screening will be important. Similarly for the Laves phases, Bruinsma and Zangwill proposed a long-range oscillatory (Friedel) potential as the stabilizing influence for the remarkable stacking sequences observed by Komura and co-workers. There must, therefore, be an indirect electronic origin of the phenomenological ANNNI interactions in metals via screening. When the structures involve pure anti-phasing it is hard to think of another mechanism. Where, however, stacking is involved, transverse phonons and electron-phonon (or phonon-phonon) interactions may also play a role.

The Laves phases bring us to a related question in alloy theory, namely the stability (and prediction) of the tetrahedrally close packed structures, or Frank-Kasper phases, to which Laves phases belong. For example, it is interesting to ask what stabilizes the FeCr σ-phase over the bcc phase from which it forms. This may perhaps seem like too difficult a question for the current state of alloy theory, but it is the sort of issue which Pettifor and Liu have been considering and which must somehow be decomposed into the stability of smaller, more local, units. It also leads us to quasicrystals which have been observed as both metastable and stable phases in alloys where the competing crystalline phases are invariably complex, and Frank-Kasper-like, with (distorted) icosahedra as building blocks. The quasicrystals are, in a sense, Frank-Kasper phases with infinite unit cells and their stability must therefore involve rather subtle local considerations. Finally, extending these ideas to the liquid state, it seems now as if supercooled liquid alloys, out of which the quasicrystals nucleate, may form metallic glass structures that are directly derivable from the quasicrystal state through the introduction of defects. This establishes a kind of conceptual sequence: the local order in complex phases, and the demands of close packing, determine the complexity of the structural units that form either out of the melt (or glass) or from competing crystal phases where entropy may play a role. The complex stacking in these structures is then finally decided by long-range competing interactions.

Our very brief discussion has touched upon several of the issues that will be treated in these proceedings, namely: nucleation and spinodal decomposition, the structure of defects and interfaces, the influence of stress on phase relations and morphologies in solids, the nature of long-period structures, and, overall, the electronic structure of alloys. The last topic has advanced enormously over the past several years and will be central to many presentations. The successes of Ducastelle, Gonis and Freeman, Stocks and Gyorffy and de Fontaine and coworkers, not just in developing successful band theories but in predicting structures and phase relations through the combination of band theory and statistical mechanics, represent some of the most elegant and promising developments in alloy theory. And Pettifor's phenomenological treatment of structural themes meshes well with the band theory and leads to far-ranging predictions and trends. One can therefore look forward to these lectures and to the concomitant growth of the entire enterprise of alloy physics.

SECTION 1

ALLOY DESIGN

PHASE STABILITY AND ALLOY DESIGN OF ORDERED INTERMETALLICS*

C. T. LIU
METALS AND CERAMICS DIVISION, OAK RIDGE NATIONAL LABORATORY, OAK RIDGE, TN 37831-6115

1. INTRODUCTION

This paper summarizes recent research on phase stability and grain boundary structure in ordered intermetallics. Emphasis will be placed on control of ordered crystal structure and grain-boundary bonding behavior for the purpose of alloy design of ductile intermetallic alloys for structural use at elevated temperatures.

Ordered intermetallic alloys constitute a unique class of metallic materials which form long-range ordered crystal structures below their melting points (T_m) or critical ordering temperatures (T_c). The various atomic species in these alloys tend to occupy specific sublattice sites and form superlattice structures. In ordered lattices, dislocations travel in pairs or groups, referred to as superlattice dislocations, and their motion is thus subject to certain constraints, particularly at elevated temperatures (1–4). In general, the strength of ordered intermetallics does not degrade rapidly with increasing temperature. In many cases, these alloys exhibit yield stresses that increase with test temperature (1–6) rather than decrease, as is common for conventional or disordered alloys. Long-range order produces stronger bonding and closer packing between atoms. This restricts atom mobility and generally leads to slower diffusion processes and better creep resistance in ordered lattices. Ordered intermetallics such as aluminides and silicides are usually resistant to oxidation and corrosion because of their ability to form compact, adherent oxide surface films that protect the base metal from excessive attack (7).

Ordered intermetallic alloys, however, generally exhibit low ductility and brittle fracture (5,6, 8–14), which severely restrict their use as structural materials. Many intermetallics are so brittle that they simply cannot be fabricated into useful structural components. Even when fabricated, their low fracture toughness severely limits their use in structural applications. The brittleness in intermetallics can be related to one of the three major causes: (1) Low crystal symmetry and insufficient number of slip systems, (2) poor cleavage strength, and (3) intrinsic and extrinsic grain-boundary weakness.

Many ordered intermetallics that crystallize in low-crystal symmetries (e.q. DO_{22}, $L1_0$, etc.) simply do not offer enough slip systems to permit extensive plastic deformation. Some intermetallics with sufficient slip systems, however, fail prematurely because of poor atomic bonding across certain crystallographic planes (i.e. poor cleavage strength). In some

*Research sponsored by the Division of Materials Sciences, U.S. Department of Energy under contract DE-AC05-84OR21400 with Martin Marietta Energy Systems, Inc.

G. M. Stocks and A. Gonis (eds.), Alloy Phase Stability, 7–21.

cases, adequate strength and deformation modes exist, yet the materials are brittle because of easy crack propagation along grain boundaries. The nickel aluminide Ni_3Al is a classic example of such behavior. Single crystals of Ni_3Al are highly ductile but polycrystals are extremely brittle (15,16,17).

During the past several years, significant progress has been made on improving ductility and toughness of ordered intermetallics through macroalloying and microalloying processes. Macroalloying is employed to control bulk crystal structure and phase stability through major alloy additions (>1 at. %). Microalloying is used to control grain-boundary structure and chemistry through minor alloying additions, often in the parts-per-million range. This paper focuses on recent progress on experimental research on control of ordered crystal structure and grain-boundary strength in several alloy systems, including Ni_3V-Co_3V-Fe_3V, Cu_3Ti-Ni_3Ti-Co_3Ti-Fe_3Ti, Ni_3V-Ni_3Al, Al_3R (R = rare-earth elements), Fe_3Al-FeAl and Ni_3X (X = Fe, Mn, Al, Ga, Si and Ge).

2. CONTROL OF ORDERED CRYSTAL STRUCTURES

Bulk materials of many ordered intermetallics are brittle because of low-symmetry crystal structures that have limited numbers of slip systems. The ductility of these alloys can be substantially improved by control of ordered crystal structures — that is, changing the crystal structure from low symmetry (such as ordered hexagonal structure) to high symmetry (such as ordered cubic structures) through macroalloying. This section focuses on the control of ordered crystal structure by systematically adjusting electron concentration (e/a), atom size ratio (R_A/R_B), or electronegativity in A_3B intermetallics.

2.1. Electron concentration and ordered crystal structures in close-packed A_3B intermetallics

Many alloys of the general composition A_3B exist in ordered crystal structures (18–23) based on close-packed ordered layers, as shown in Fig. 1. These structures are built from the regular stacking of these layers. There are two basic types of close-packed ordered layers (18), designated as triangular (T) type and rectangular (R) type, as illustrated

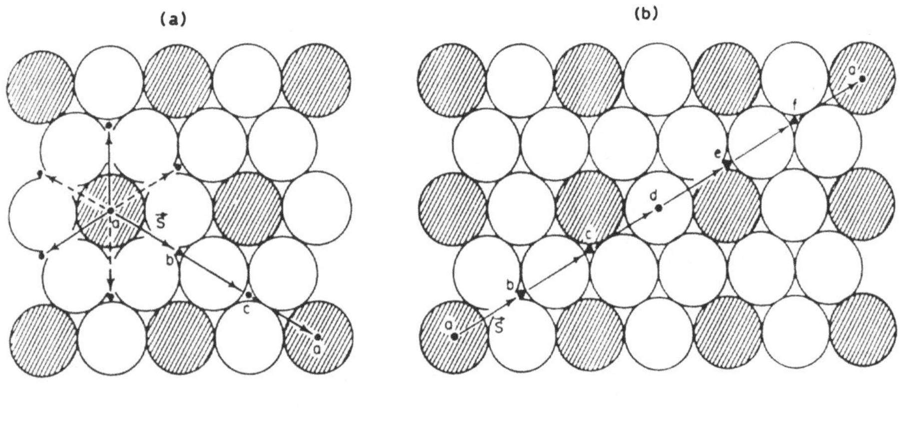

FIGURE 1. Atomic structure on close-packed ordered planes; (a) triangular (T)-type ordered layer and (b) rectangular (R)-type ordered layer (18).

in Fig. 1(a) and (b), respectively. Stacking of the T layers gives ordered structures of cubic or hexagonal symmetry, depending on the stacking sequence. Some transition structures are complicated, with unit cells extending over 15 layers (23). Ordered structures formed by stacking of R layers generally have a tetragonal symmetry.

Systematic studies of close-packed ordered A_3B structures have revealed that the stacking character in pseudobinary alloys can be correlated with alloying variables, such as atomic size and electron concentration. Beck and Dwight (18,21) first correlated the stacking character of the T layers with the e/a ratio in alloys. A layer is characterized as hexagonal if the layers adjacent to it have the same stacking position. For example, with a stacking sequence ABA, the middle layer B is designated as hexagonal. On the other hand, the B layer is a cubic layer in a stacking sequence ABC, where its adjacent layers A and C have different stacking positions. With an increase in e/a, the ordered structure changes from predominantly cubic to predominantly hexagonal stacking. Further increase in e/a leads to a change in the basic layer structure from T to R type.

The electron concentration correlation provides a useful guidance in the control of ordered crystal structures of A_3B intermetallics containing major transition elements. Sinha (19) and Liu and Inouye (24,25) found that the ordered crystal structure in Ni_3V-Co_3V-Fe_3V alloys can be correlated with e/a. The electron concentration in Co_3V can be increased by partial replacement of Co (e/a = 9) with Ni (e/a = 10): $(Ni,Co)_3V$. With increase of e/a, the hexagonality of the ordered structure increases systematically from 33.3 to finally 100 percent (Fig. 2). Further increase in e/a above 8.54 produces a change in the basic layer structure from T type to R type (Fig. 1), and stacking of the R layers gives a tetragonal ordered structure similar to Ni_3V (DO_{22}). On the other hand, e/a in Co_3V can be reduced by partial substitution of Fe (e/a = 8) for Co:

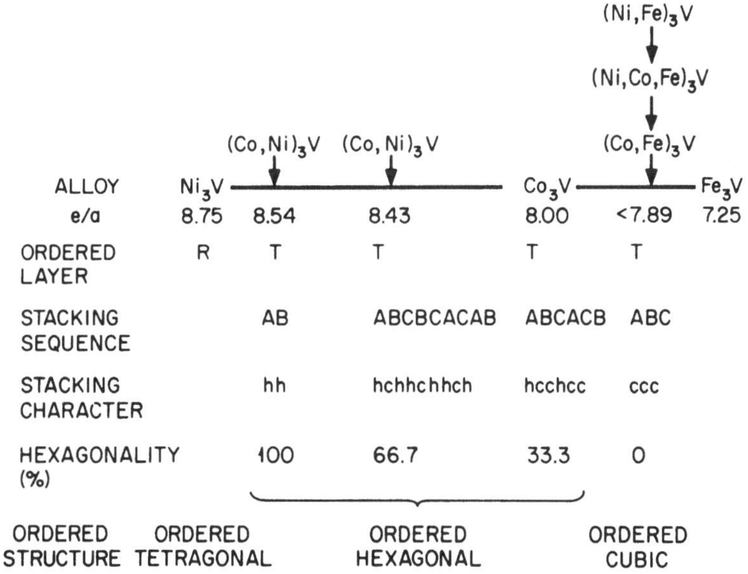

FIGURE 2. Effect of electron concentration (e/a) on the stability of ordered crystal structures in Ni_3V-Co_3V-Fe_3V alloys (ref. 25).

$(Co,Fe)_3V$. With an e/a below 7.89 the $L1_2$ ordered cubic structure having the stacking sequence ABC (ccc) is stabilized in $(Co,Fe)_3V$ alloys. A similar phase relationship had been observed in the Cu_3Ti-Ni_3Ti-Co_3Ti-Fe_3Ti system, where e/a is taken as 11, 10, 9, 8, and 4 for Cu, Ni, Co, Fe, and Ti elements, respectively. All these observations agree well with the prediction from the e/a correlation.

Control of e/a provides considerable scope for alloy design. For instance, the cobalt content in $(Co,Fe)_3V$ can be reduced by replacing cobalt with an equal amount of an equiatomic mixture of nickel and iron. This scheme alters the alloy composition but not e/a. As indicated in Fig. 2, the ordered cubic structure remains stable in $(Ni,Co,Fe)_3V$ alloys as long as the electron concentration falls roughly in the same range as $(Co,Fe)_3V$. All cobalt atoms can be eventually replaced by nickel and iron atoms, resulting in ordered cubic alloys of the composition $(Ni,Fe)_3V$. With this scheme, the $L1_2$ structure is stabilized in $(Ni,Fe)_3V$ alloys without cobalt, an expensive strategic element. However, other properties such as the critical ordering temperature, T_c, are influenced by the alloy composition at a constant electron concentration.

Sinha (26) suggested that the electron effect may originate from the interaction of the electron concentration-dependent Fermi surface with the corresponding Brillouin zone. Recently, Pei and Massalski (27) have found that free electron approach fails to predict the ordered crystal structures in the Ni_3V-Co_3V-Fe_3V alloy system. A full understanding of the e/a correlation requires a study of the electronic structure using first-principles calculations.

The importance of the ordered cubic structure is shown by the room-temperature ductility of several ordered hexagonal and cubic alloys (25) (Fig. 3). Ordered cubic alloys ($L1_2$) of the compositions of $(Fe,Co)_3V$, $(Fe,Co,Ni)_3V$, and $(Fe,Ni)_3V$ are all ductile, with a tensile elongation of 40 percent or higher. However, the hexagonally ordered alloys Co_3V and $(Ni,Co)_3V$ are brittle with less than 1 percent elongation at room temperature. The deformation behavior of ordered cubic alloys is similar to that of fcc materials having twelve {111} slip systems. The brittleness of ordered hexagonal alloys is presumably related to the limited number of

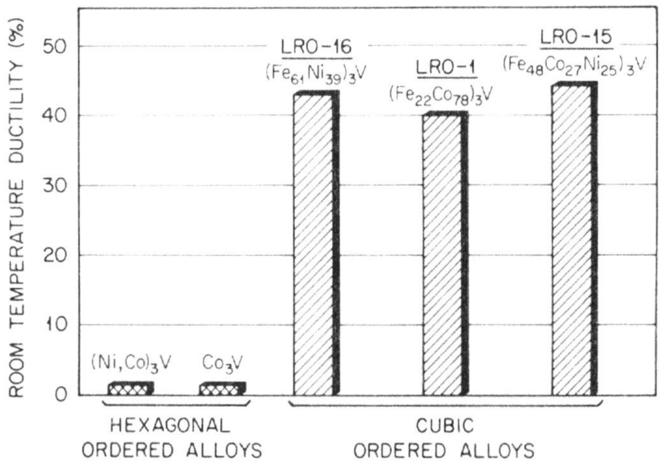

FIGURE 3. Comparisons of room-temperature tensile elongations of cubic and hexagonal ordered alloys (ref. 25).

slip systems available in ordered hexagonal crystal structures. The hexagonal alloys have ductilities too low to permit easy fabrication, while the cubic alloys have excellent fabricability at both room and elevated temperatures. The ability to control ordered crystal structure and ductility by alloying represents a major advance in alloy design of intermetallics.

The yield strengths, σ_y, of the cubic ordered alloys $(Fe,Co,Ni)_3V$ are shown in Fig. 4 as a function of temperature (25). The distinct feature of the plot is that the yield strengths of these ordered alloys increase with temperature, rather than decrease, as do those of conventional disordered alloys such as Hastelloy-X and type 316 stainless steel. The σ_y values for these ordered alloys increase substantially with temperature above 400°C, and reach a maximum around T_c. Because of the increase, the ordered alloys become much stronger than disordered solid-solution alloys at elevated temperatures. The strengths of the ordered alloys are lower than those of precipitation-strengthened superalloys such as Inconel 718 and Waspaloy, at lower temperatures, but comparable to or higher than them near T_c. For instance, in LRO-1 alloy σ_y reaches 520 MNm^{-2} at 900°C, a value markedly higher than those for Inconel 718 (300 MNm^{-2}) and Waspaloy (400 MNm^{-2}). The strength of the ordered LRO alloys decreases sharply above T_c, apparently because of the loss of long-range order. The increase in σ_y with temperature has been successfully explained by the cross-slip-pinning model proposed by Takeuchi and Kuramoto (1) and modified by Paidar, Pope, and Vitek (28), and Yoo (29).

2.2. Atomic size ratio and ordered crystal structures

VanVucht and Buschow (22,23) noted a general correlation between the stacking character of each layer and the radius ratio of A and B atoms, R_A/R_B. As the ratio of R_A/R_B decreases in A_3B ordered alloys, the stacking character changes from purely cubic, through different ordered mixtures of cubic and hexagonal layers, to purely hexagonal. The change in the

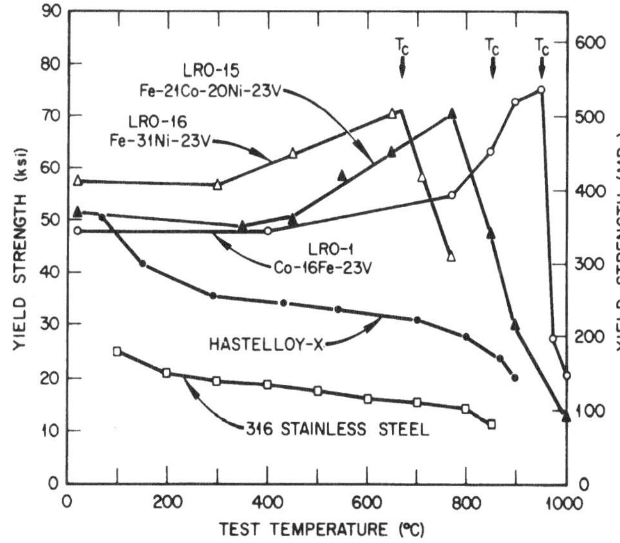

FIGURE 4. Variation of yield strength with test temperature for cubic ordered alloys and commercial solid-solution strengthened alloys Hastelloy-X and type 316 stainless steel (ref. 25).

stacking character can be rationalized from consideration of space packing density. Clustering of smaller A atoms in the layer lowers the symmetry of space group from cubic to hexagonal, but admits a better packing when the larger B atoms of both adjoining layers profit from the open spaces. For instance, a change in A_3B structure (with $R_A/R_B = 0.80$) from a cubic to hexagonal symmetry results in an increase in fraction of space filled from 0.64 to 0.69, based on a hard-sphere model (22).

In the classic study of VanVucht and Buschow (22), the atomic radius of rare-earth elements (R) was correlated well with the stacking character of trialuminides (Al_3R), as shown in Fig. 5. With the decrease in atomic size, the hexagonal stacking decreases step by step from 100 percent (purely hexagonal ordered structure) to 0 percent (purely cubic ordered structure, $L1_2$ type). VanVucht (23) further extended the atom size correlation to other pseudobinary alloy systems.

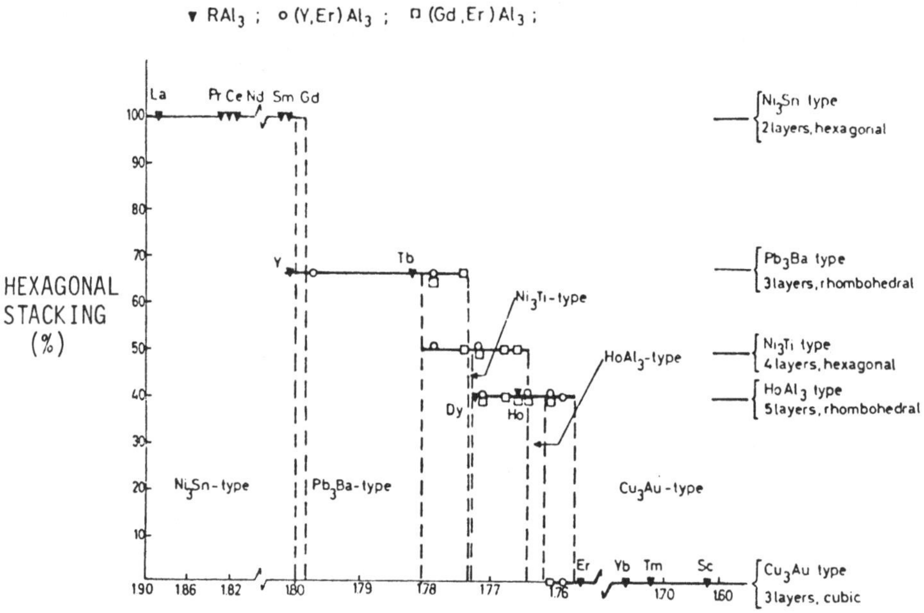

FIGURE 5. Decrease of hexagonal stacking character with the decrease in atom size of rare-earth element (R_A) in trialuminides (ref. 22).

2.3. Ordered structures in Ni_3V-Ni_3Al

The ordered crystal structure in the pseudobinary alloy system Ni_3V-Ni_3Al was studied recently by DasGupta et al. (30). The table on the next page summarizes alloy variables and ordered structures for Ni_3V and Ni_3Al. Because of the distinct difference in both layer structure and crystal structure between Ni_3V and Ni_3Al, we initially expected to see a series of transition structures formed in these pseudobinary alloys. However, except for the end structures, no evidence for formation of complex structures with close-packed ordered A_3B layers was found. The absence of such transition structures is interpreted as evidence of their metastable nature (Fig. 6), in view of the high stability of Ni_3Al and Ni_3V. Recently, the first-principles calculations by Freeman et al. (31) confirm the high stability of $L1_2$-Ni_3Al and DO_{22}-Ni_3V.

	Ni_3V	Ni_3Al
Ordered layer	R	T
Stacking sequence	ABC	ABC
Ordered structure	Tetragonal DO_{22}	Cubic $L1_2$
e/a	8.75	8.25
R_{Ni}/R_X	0.91	0.87

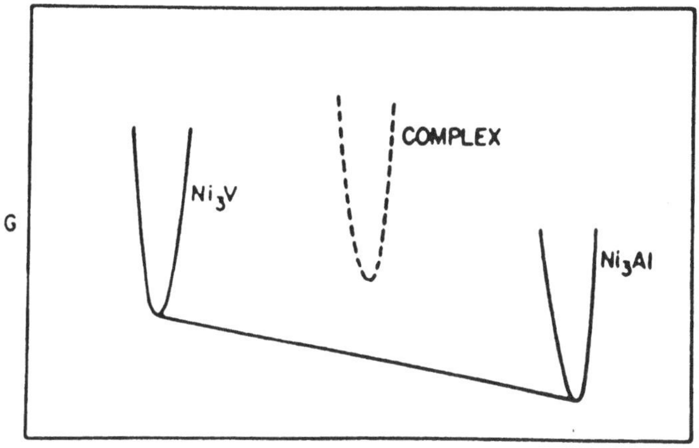

FIGURE 6. Schematic diagram of Gibbs free energy G vs composition for Ni_3Al, Ni_3V, and a possible phase with complex transition structure. The transition-structure phase is rendered metastable due to the high stabilities of both Ni_3Al and Ni_3V (ref. 30).

2.4. Structural maps and control of ordered crystal structure

Recently, Pettifor (32—34) has developed structural maps for ordered intermetallics. An example of the AB_3 structure is shown in Fig. 7. The maps were constructed in the way that a string running through a modified periodic table puts all the elements in sequential order, given by the Mendeleev number. This number, in principle, includes alloy effects from atom size, electronegativity and valence electron. A detailed description of the structural maps is given in Pettifor's paper in this proceedings (32). The structural maps have successfully grouped each ordered structure into the "domain" regions as shown in Fig. 7. Furthermore, the maps successfully predict the "instability" of compositions located near the structure domain boundary.

The structural maps provide a very useful guidance in control of ordered crystal structures in intermetallics. Pettifor (32) has started to extend the structural maps to include ternary alloy systems. At present, much of the experimental work has just been initiated to verify the structural maps for pseudobinary alloy systems and to design ordered intermetallics with controlled crystal structures.

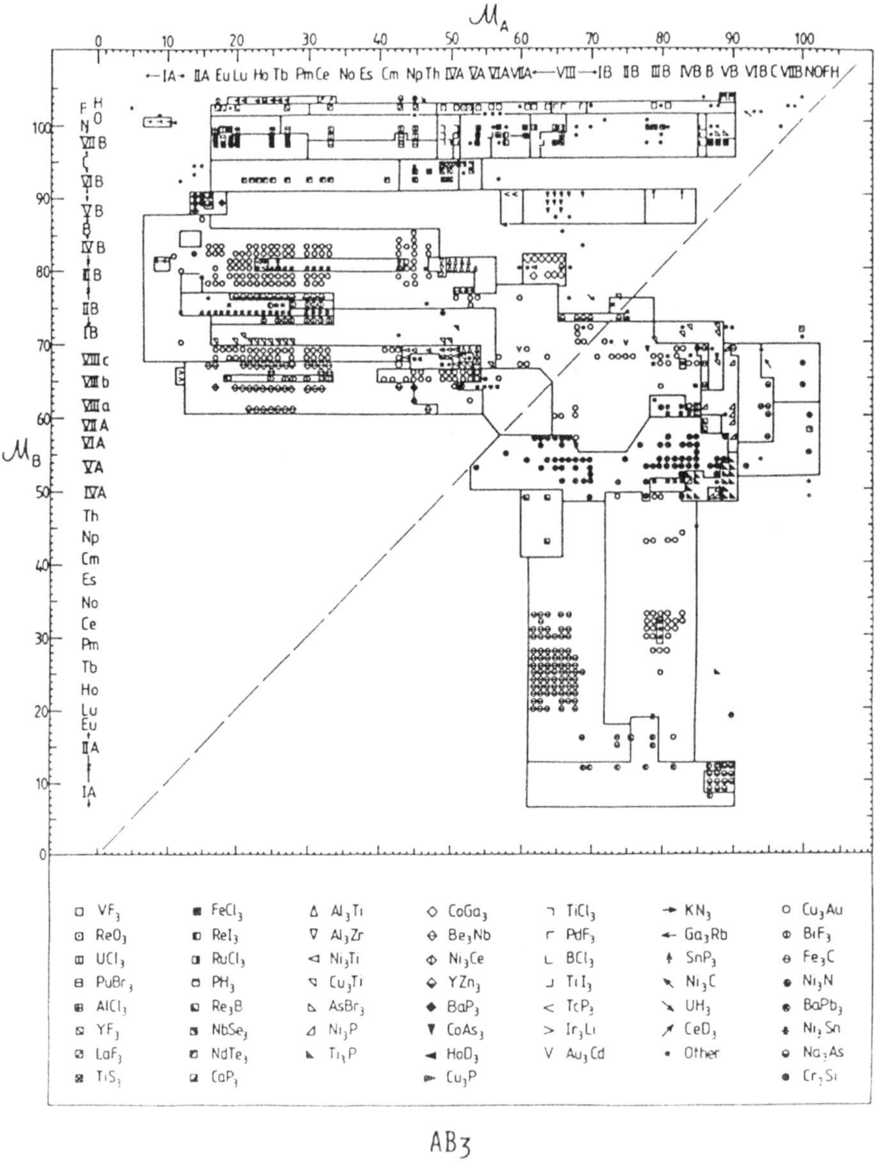

FIGURE 7. The AB_3 structural map developed recently by Pettifor (32).

3. CONTROL OF CLEAVAGE STRENGTH

It is known that many bcc ordered intermetallics (16,35,36) such as NiAl, FeAl, and Fe_3Al exhibit brittle cleavage fracture and low ductility at ambient temperatures. The brittleness is apparently associated with weak atomic bonding across certain crystallographic planes. However, not much work has been done on correlating cleavage strength with electronic structure in these intermetallics. Recently, Porter, Oliver, and Schneibel (37) have observed brittle cleavage fracture in titanium and zirconium trialuminides with the $L1_2$ ordered structure, even though these intermetallics are relative soft with microhardness at a level of 200 DPH.

High-purity Fe_3Al alloys showed completely cleavage fracture with a low ductility (38). McKamey et al. (39) have recently observed that the ductility of Fe_3Al alloys can be substantially improved by alloying with moderate amounts of chromium. The increase in ductility is accompanied by a change in fracture mode from cleavage to mixed fracture including grain-boundary separation. This observation suggests the possibility of enhancing cleavage strength and reducing cleavage fracture by alloying with chromium additions. The detailed role of chromium in affecting electronic structure and atomic bonding among iron and aluminum atoms is not understood; however, this work demonstrates the possibility of control of atomic bonding and cleavage fracture by alloying additions.

4. CONTROL OF GRAIN-BOUNDARY STRUCTURE AND CHEMISTRY

Many ordered intermetallics exhibit severe brittleness that originates at grain boundaries. A classic example is Ni_3Al. Single crystals of Ni_3Al are highly ductile, whereas polycrystals are very brittle at ambient temperatures even though there are 12 independent slip systems (16,17,40–42). In most metals and alloys, intergranular brittleness is associated with strong segregation of harmful impurities (e.g., S, P) to grain boundaries (43), causing embrittlement. Studies of fracture in high-purity polycrystalline Ni_3Al and Ni_3Si using Auger electron spectroscopy (AES), however, revealed brittle intergranular fracture without appreciable segregation of impurities at grain boundaries (15,44–47). The grain boundary is, therefore, considered to be intrinsically weak in these alloys. Note that the grain boundaries in Ni_3Al can be further embrittled by segregation of impurities. Sulfur was identified as a trace element that segregates to and embrittles grain boundaries in impure Ni_3Al (48).

In an attempt to understand the nature of intrinsic grain-boundary weakness, Takasugi and Izumi (49) initiated a systematic study of the effect of metallurgical, mechanical and chemical factors on grain-boundary cohesion in $L1_2$ ordered A_3B alloys. They found that the valency difference (ΔZ) between A and B atoms is the dominant factor controlling the grain-boundary cohesive strength, and that the tendency for grain-boundary fracture increases with increasing ΔZ. They also considered the importance of the atomic size difference and postulate that a better correlation can be obtained by a combined consideration of both electron valency and atom size differences. Their correlation appears to correctly rank the grain-boundary cohesive strength of $L1_2$ ordered nickel-based alloys in the order $Ni_3Fe > Ni_3Mn > Ni_3Al > Ni_3Ga > Ni_3Si > Ni_3Ge$, which is in a good agreement with experimental data tabulated in Table 1 by Taub et al. (50).

For the past several years, substantial progress has been made in improving the grain-boundary strength and ductility of Ni_3Al, Ni_3Si, and Ni_3Ga. Boron has been found to be most effective in ductilizing these alloys (15,50–52). Figure 8 is a plot of room-temperature tensile ductility as a function of boron addition in a hypostoichiometric Ni_3Al alloy

TABLE 1. Valency-size effect-electronegativity correlation with ductility in the $L1_2$ Ni_3X alloys (50)

X Species	Valency Difference (Δz [19])	Lattice Dilation ($a-a_{Ni}/a_{Ni}$)	Electronegativity Difference (Pauling's)	Undoped Alloy	Boron-Doped Alloy
Fe	0.2	+1.0%	−0.08	T	-
Mn	0.9	+2.2%	−0.36	T	-
Al	3.0	+1.5%	−0.30	I	T
Ga	3.0	+1.6%	−0.10	I	T
Si	4.0	−0.04%	−0.01	I	M
Ge	4.0	+1.5%	+0.10	I	I

T = transgranular, I = intergranular, M = mixed mode.

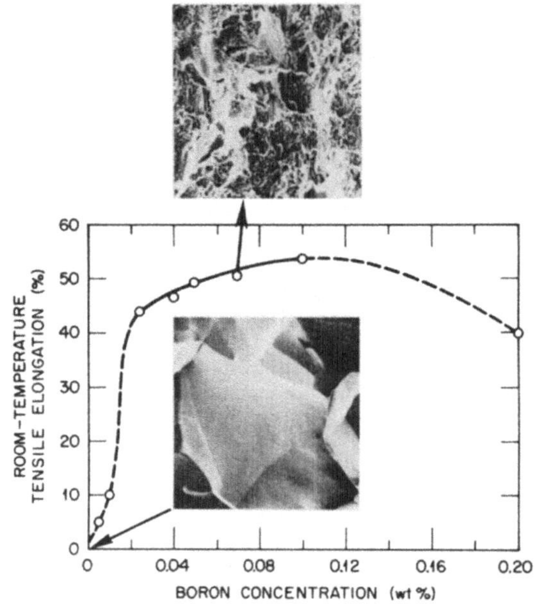

FIGURE 8. Effect of boron additions on tensile elongation and fracture behavior of Ni_3Al (24 at. % Al) tested at room temperature.

(Ni-24 at. % Al). Microalloying with boron sharply increases the ductility and completely suppresses brittle intergranular fracture. This striking effect of boron on the ductility of Ni_3Al, first discovered by Aoki and Izumi (51) occurs over a wide range of boron concentration where boron is in solid solution [the equilibrium solubility of boron ≃ 1.5 at. % (44)]. Liu et al. (44) and White et al. (53) have observed an unusual segregation behavior of boron in Ni_3Al, using Auger electron spectroscopy (AES). Boron tends to segregate strongly to grain boundaries in Ni_3Al but not to free surfaces. According to the solute segregation theory developed by Rice (54) based on thermodynamic considerations, the unusual segregation behavior of boron leads to substantially increased grain-boundary cohesive energy. Recently, theoretical calculations based on first-principles cluster calculations (55), cluster calculation of density of states (56), and embedded atom calculations (57), have all indicated the beneficial effect of boron on the cohesive strength of Ni_3Al grain boundaries.

Recently, atom probe (58) and scanning-transmission electron microscopic (STEM) (59) studies have shown a strong cosegregation of nickel and boron atoms to grain boundaries, possibly leading to disordered Ni_3Al grain boundaries. The disordered boundaries are expected to be more ductile, because many dislocation interactions are permissible when chemical ordering is not required to be maintained there (60). Thus, additional beneficial effect of boron is expected to come from the enhancement of plastic flow near grain boundaries.

Taub et al. (50,61) have studied grain-boundary fracture in boron-doped and undoped binary and pseudobinary intermetallic alloys based on Ni_3X (X = Al, Ga, Si or Ge), prepared by a melt-spinning technique. They have found that both bend ductility and fracture behavior can be better correlated with the electronegativity consideration rather than the valency difference proposed by Takasagi and Izumi (49,62). As shown in Fig. 9, the valency model requires two parameters, while the electronegativity model requires only a single parameter to successfully correlate the data. In addition to the data fitting, the electronegativity model provides a better understanding of atomic bonding. As compared with Al, Ga and Si, Ge atoms are more electronegative with respect to Ni atoms; consequently, Ge has a greater tendency to pull electron charge out of

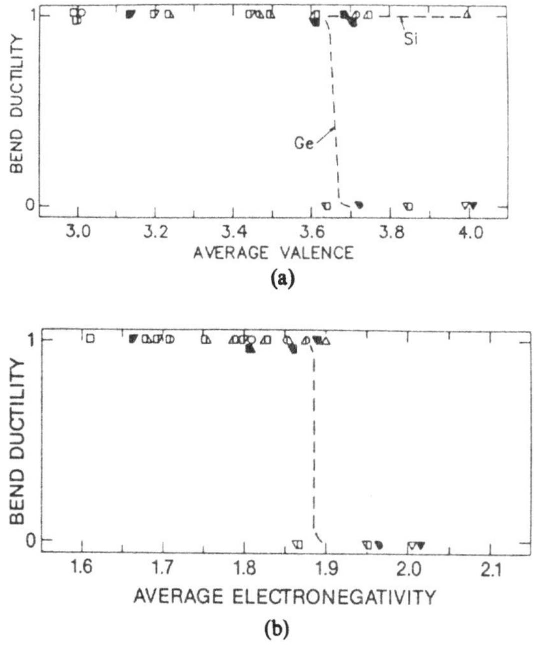

FIGURE 9. The bend ductility of Ni_3X alloys as a function of (a) average valence and (b) averaged electronegativity of the X species (ref. 57).

Ni-Ni bonds, thereby further reducing the cohesive strength and promoting intergranular fracture in Ni_3Ge. Thus, grain boundaries in Ni_3Ge are substantially weaker than those in Ni_3Al, hence boron is ineffective in ductilizing Ni_3Ge and $Ni_3(Al,Ga)$ alloys containing high levels of germanium [e.g., Ge > 15 at. % in $Ni_3(Ge,Al)$].

The most interesting feature in connection with the study of boron in Ni_3Al is the discovery of the alloy stoichiometric effect (44). Alloy stoichiometry was found to have a large effect on the ductility and fracture behavior of boron-doped Ni_3Al. Boron is most effective in improving the ductility and suppressing intergranular fracture in Ni_3Al alloys containing <24 at. % Al. As the aluminum concentration is increased, the ductility decreases sharply (Fig. 10), and the failure mode changes from transgranular to mixed mode and then to mainly intergranular fracture. Auger studies (44) of freshly fractured surfaces of boron-doped samples indicate that the intensity of boron segregation decreases significantly and the grain-boundary aluminum concentration increases moderately with increasing bulk aluminum concentration (Fig. 11). These results simply suggest that deviations from alloy stoichiometry influence grain-boundary chemistry, which, in turn, affects grain-boundary cohesion and the overall ductility of nickel aluminides. The boundaries remain brittle when the amount of boron segregation is insufficient. In line with this thought, Choudhury et al. (63) have observed that Ni_3Al is more resistant to intergranular fracture when a higher level of boron is present at the boundaries.

It has been clearly demonstrated that boron is effective in suppressing brittle grain-boundary fracture and increasing the ductility of a number of $L1_2$ ordered intermetallics. The beneficial effect of boron has also been observed in other ordered intermetallics, including B2-ordered FeAl (40% Al) (35) and DO_{22}-ordered $TiAl_3$ alloys (64), although the effect of boron is most pronounced in the $L1_2$ intermetallics.

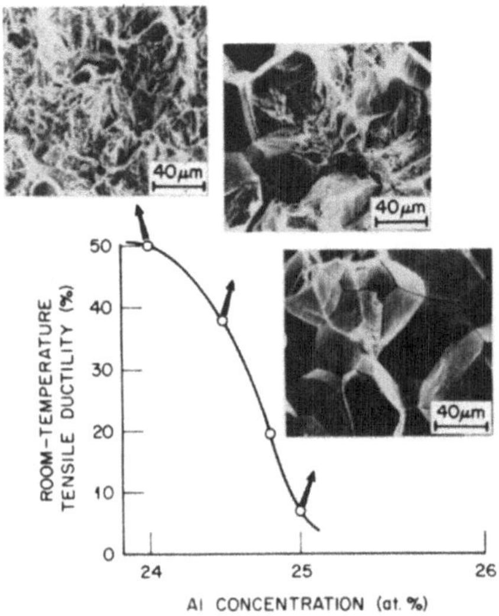

FIGURE 10. Effect of aluminum concentration on room-temperature ductility and fracture behavior of Ni_3Al doped with 0.1 at. % B.

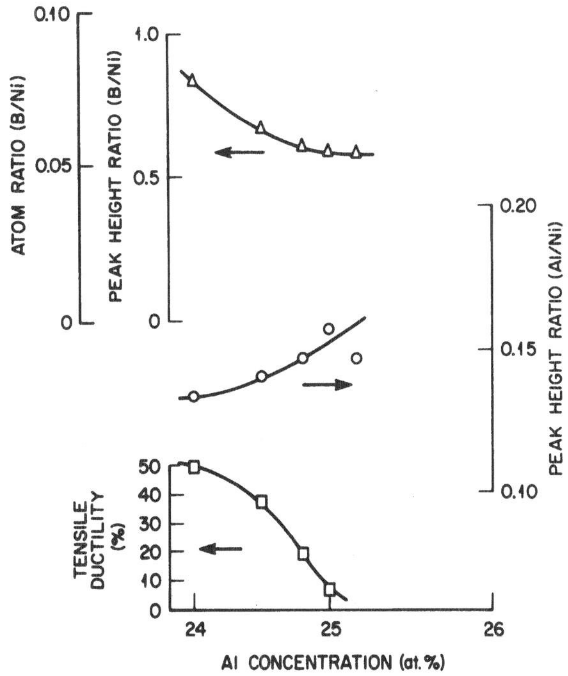

FIGURE 11. Effect of aluminum concentration on room-temperature ductility and boron and aluminum concentrations at grain boundaries (ref. 44).

5. CONCLUSTIONS

Ordered intermetallics have many attractive high-temperature properties; however, low ductility and brittle fracture limit their use for structural applications. The embrittlement in intermetallics is mainly caused by low-crystal symmetry and insufficient number of slip systems, poor cleavage strength, and intrinsic and extrinsic grain-boundary weakness. Experimental studies of ordered phase stability and grain-boundary structure have led to substantial improvement in the ductility and fracture toughness of several ordered intermetallics, including $(Co,Fe)_3V$, $(Ni,Fe)_3V$, $(Co,Ni,Fe)_3V$, Fe_3Al, Ni_3Al, Ni_3Ga, and Ni_3Si. At present, the control of ordered structures and grain-boundary properties is mainly based on some empirical and/or phenomenological relationship. The first-principles calculations have advanced to the stage that they are able to help us to understand the electron structure and phase stablity and to design ductile ordered intermetallic alloys for structural use at elevated temperatures.

6. ACKNOWLEDGMENTS

The author would like to thank Connie Dowker for preparation of the manuscript.

REFERENCES

1. Takeuchi S and Kuramoto E: Acta Metall. 21, 415 (1973).
2. Thornton PH, Davies RG, and Johnston TL: Metall. Trans. 1, 207 (1970).
3. Copley SM and Kear BH: Trans. Metall. Soc. AIME 239, 977 (1967).
4. Pope, DP: Philos. Mag. 25, 917 (1972).
5. Koch CC, Liu CT, and Stoloff NS(ed): High-Temperature Ordered Intermetallic Alloys, Materials Research Society, Pittsburgh, PA.
6. Stoloff NS, Koch CC, Liu CT, and Izumi O(ed): Ordered Intermetallic Alloys II, Materials Research Society, Pittsburgh, PA, 1987.
7. Aitken EA: Intermetallic Compounds, ed. J. H. Westbrook, Wiley, New York, 1967, pp. 491–516.
8. Stoloff, NS and Davies RG: Prog. Mater. Sci. 13(1), 1 (1966).
9. Kear, BH, Sims, CT, Stoloff, NS, and Westbrook JH(ed): Ordered Alloys—Structural Applications and Physical Metallurgy, Proceedings of the 3rd Bolton Landing Conference, September 1969, Claitor's Baton Rouge, La., 1970.
10. Westbrook JH(ed): Mechanical Properties of Intermetallic Compounds, Wiley, New York, 1959.
11. Westbrook, JH(ed): Intermetallic Compounds, Wiley, New York, 1967.
12. Tanner LE, et al.: Mechanical Behavior of Intermetallic Compounds. Report AST-TDR62-1087, Manlabs, Inc., Cambridge, Mass., 1963–1964, parts 1–3.
13. Lipsitt HA, Schechtman D, and Schafrik RE: Metall. Trans. A 11A, 1369 (1980).
14. Aoki K and Izumi O: Acta Metall. 27, 807 (1979).
15. Liu CT and Koch CC: Proceedings of a Public Workshop on Trends in Critical Materials Requirements for Steels of the Future: Conservation and Substitution Technology for Chromium NBSIR-83-2679-2, National Bureau of Standards, Washington, D.C., June 1983.
16. Grala EM: Mechanical Properties of Intermetallic Compounds, ed. J. H. Westbrook, Wiley, New York, 1960, pp. 358–404.
17. Aoki K and Izumi O: Trans. Jpn. Inst. Met. 19, 203 (1978).
18. Beck PA: Adv. X-ray Anal. 12, 1 (1969).
19. Sinha AK: Trans. Metall. Soc. AIME 245, 911 (1969).
20. Saito S: Acta Crystallogr. 12, 500 (1959).
21. Dwight AE and Beck PA: Trans. Metall. Soc. AIME 215, 976 (1959).
22. VanVucht JHN and Buschow KH: J. Less-Common Met. 10, 98 (1965).
23. VanVucht JHN: J. Less-Common Met. 11, 308 (1966).
24. Liu CT and Inouye H: Metall. Trans. A 10A, 1515 (1979).
25. Liu CT: Int. Metall. Rev. 29, 168 (1984).
26. Sinha AK: Prog. Mater. Sci. 15(2), 81 (1972).
27. Pei SY and Massalski TB: Carnegie Mellon University, private communication, 1987.
28. Paidar V, Pope DP, and Vitek V: Acta Metall. 32, 435–48 (1984).
29. Yoo MH: Scr. Metall. 20, 915 (1986).
30. DasGupta A, Horton JA, and Liu CT: High-Temperature Alloys: Theory and Design, ed. J. O. Stiegler, AIME Publication, 1984, pp. 115–124.
31. Freeman AJ: Structural Stability of Intermetallic Compounds: A Computational Metallurgical Approach, in this proceedings.
32. Pettifor DG: Quantum Mechanics in Alloy Design, in this proceedings.
33. Pettifor DG: New Scientist 110(1510), 48 (1986).
34. Pettifor DG: J. Phys. C19, 285 (1986).
35. Crimp MA and Vedula K: Mater. Sci. Eng. 78, 193 (1986).
36. Vedula K and Stephens JR: MRS Proc. 81, ed. N. S. Stoloff, C. C. Koch, C. T. Liu, and O. Izumi, MRS Publication, 1987, pp. 381–92.

37. Porter WD, Oliver WC, and Schneibel JH: Oak Ridge National Laboratory, private communication, 1987.
38. Mendiratta MG, Ehlers SK, Chatterjee DK, and Lipsitt HA, Proceedings of the Third Conference on Rapid Solidification, NBS Publication, 1983, p. 240.
39. McKamey CG and Liu CT: Oak Ridge National Laboratory, unpublished results, 1987.
40. Aoki K and Izumi O: Nippon Kinzoku Gakkaishi **41**, 170 (1977).
41. Moskovich, R: J. Mater. Sci. **13**, 1901 (1978).
42. Seybolt AV and Westbrook JH: Acta Metall. **12**, 449 (1964).
43. Stein, DF and Heldt LA: Interfacial Segregation, ed. W. C. Johnson and J. M. Blakely, ASM, Metals Park, Ohio, pp. 239–260.
44. Liu CT, White CL, and Horton JA: Acta Metall. **33**, 213–219 (1985).
45. Takasugi T, George EP, Pope DP, and Izumi O: Scr. Metall. **19**, 551–556 (1985).
46. Ogura T, Hanada S, Masumoto T, and Izumi O: Metall. Trans. A **16A**, 441–443 (1985).
47. Oliver WC: Oak Ridge National Laboratory, private communication.
48. White CL and Stein DF: Metall. Trans. A **9A**, 13 (1978).
49. Takasugi, T. and Izumi O: Acta Metall. **33**, 1247–1258 (1985).
50. Taub AI, Briant CL, Huang SC, Chang KM, and Jackson MR: Scr. Metall. **20**, 129–134 (1986).
51. Aoki K and Izumi O: Nippon Kinzoku Gakkaishi **43**, 1190 (1979).
52. Taub AI, Huang SC, and Chang KM: Metall. Trans. A **15A**, 399 (1984).
53. White CL, Padgett RA, Liu CT, and Yalisove SM: (1984).
54. Rice JR: The Effect of Hydrogen on the Behavior of Metals, AIME publication, New York, New York, pp. 455–465.
55. Painter GS and Averill FW: Phys. Rev. Lett. **58**, 234 (1987).
56. Eberhart ME and Vvedinsky DD: Phys. Rev. Lett. **58**, 61 (1987).
57. Chen SP and Srolovitz DJ: private communication, 1987.
58. Brenner SS: University of Pittsburgh, unpublished results, 1987.
59. Baker I and Schulson EM: Dartmouth College, unpublished results, 1987.
60. King AH and Yoo MH: High-Temperature Ordered Intermetallic Alloys II, Materials Research Society, Pittsburgh, PA, p. 99.
61. Taub AI and Briant CL: Acta Metall. **35**, 1597–1603 (1987).
62. Takasugi T, Izumi O, and Masahashi N: Acta Metall. **33**, 1259 (1985).
63. Choudhury A, White CL, and Brooks CR: Scr. Metall. **20**, 1060–66 (1986).
64. Yamaguchi M: Kyoto University, Japan, private communication, 1987.

MECHANICAL PROPERTIES AND PHASE STABILITY OF $L1_2$ Ni_3Al TERNARY COMPOUNDS

Yoshinao MISHIMA and Tomoo SUZUKI
Research Laboratory of Precision Machinery and Electronics,
Tokyo Institute of Technology, Midori-ku, Yokohama 227, Japan

1.INTRODUCTION

$L1_2$ Ni_3Al is well known for its anomalous positive temperature dependence of strength. The compounds based on Ni_3Al have been a major strengthener in commercial nickel-base superalloys, but for this reason it is anticipated to utilize them as a single phase high temperature material. In order to design an alloy compound for the purpose, it is necessary to make comprehensive understanding on the effect of ternary additions on the temperature dependence of strength in both poly- and single-crystalline compounds.

The cross slip mechanism, known as Kear-Wilsdorf mechanism, has been believed to provide the anomalous positive temperature dependence of strength in the compound(1,2). The driving force for the cross slip is a large anisotropy in anti-phase boundary(APB) energies between (001) and (111) planes, by which the cross slip of screw dislocation is thermally activated from its glide (111) plane to non-glide (001) plane to increase plastic flow stress. This fact implies that the alloying elements to reduce (001) APB energy would enhance the mechanical anomaly of Ni_3Al. It is important to understand such situation to be closely related to the phase stablity of $L1_2$ structure against other geometrically closed packed(GCP) phases, in particular for this case against DO_{22} phase. DO_{22} ordered structure is based on fcc and it is constructed by periodically placing APBs on (001) planes of $L1_2$. Therefore it is expected that the reduction in the phase stability of $L1_2$ against DO_{22} would enhance the mechanical anomaly. The present authors have already shown that the relative magnitude of the mechanical anomaly among many $L1_2$ ordered alloys is successfully interpreted by two parameters to alter the stability of $L1_2$ against DO_{22}, namely the electron-atom ratio of the alloy and the atomic radius ratio of the component elements(3-6).

The present paper deals with the effect of ternary additions on the mechanical anomaly in Ni_3Al and an effort is described on elucidating the role of alloying elements systematically according to their group and period and to the way they affect the stability of $L1_2$ ordered crystal structure.

2. EFFECT OF VALENCE OF TERNARY ELEMENT ON MECHANICAL ANOMALY

A typical example of how ternary additions alter the temperature dependence of 0.2% compressive flow stress in Ni_3Al is shown in Fig. 1 for the additions of niobium. The compressive flow stress was measured using a 3x3x6 mm^3 specimen on an Instron-type testing machine. Details of such experiments have been described in earlier work by the present authors(7,8). Then according to a schematic representation as in Fig. 2, activation energy for the thermally activated cross slip to provide the mechanical anomaly can be obtained by analyzing the flow stress component

G. M. Stocks and A. Gonis (eds.), Alloy Phase Stability, 23–27.

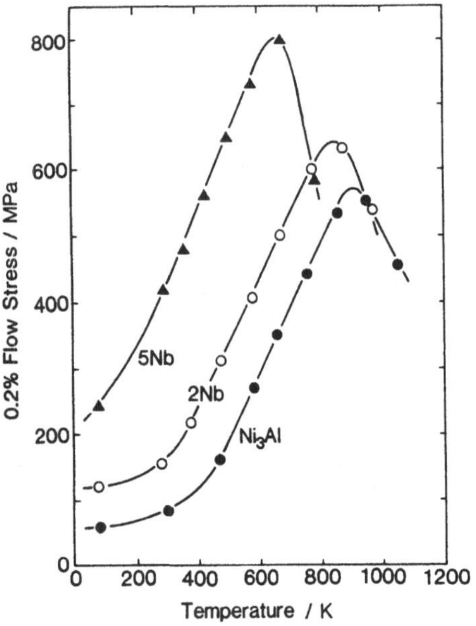

FIGURE 1 Temperature dependence of 0.2% flow stress in Ni_3Al with additions of niobium.

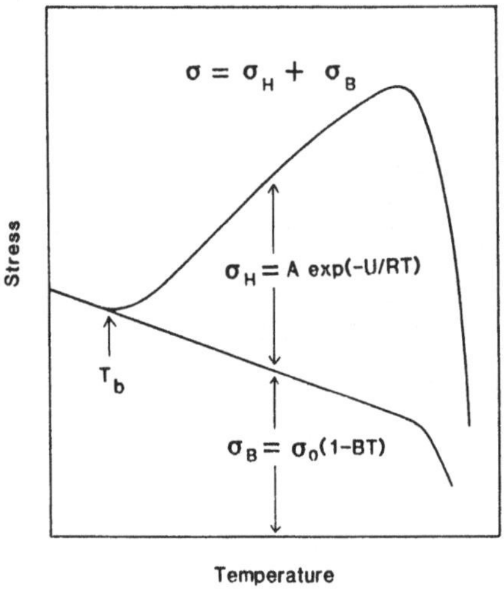

FIGURE 2 A schematic representation of flow stress-temperature curve showing anomalous positive temperature dependence of strength.

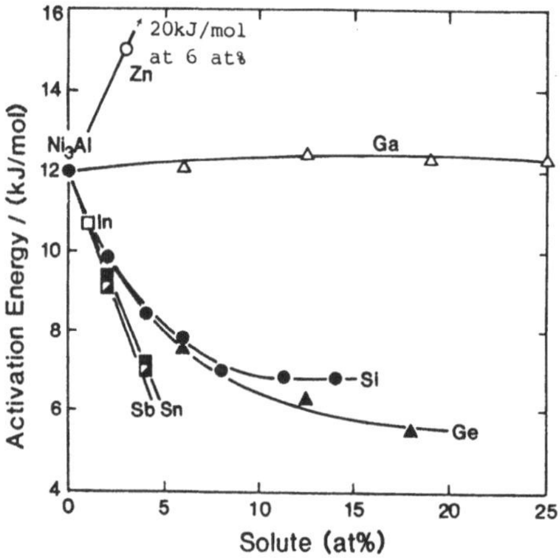

FIGURE 3 Change in activation energy with solute concentration in Ni_3Al by additions of B-subgroup elements.

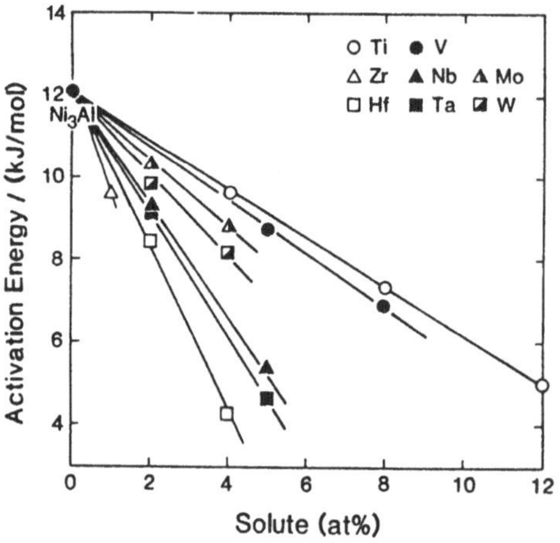

FIGURE 4 Change in activation energy with solute concentration in Ni_3Al by additions of transition metal elements.

denoted as σ_H. In Fig. 2, the component σ_B is the flow stress determined by negative temperature dependence of shear modulus so that the total flow stress is expressed as ($\sigma_H + \sigma_B$) and T_b is a temperature at which onset of the mechanical anomaly becomes visible. Then the concentration dependence of the activation energy, U, is shown in Fig. 3 for the additions of B-subgroup elements and in Fig. 4 of transition metal elements. The concentration dependence of activation energy is expressed by straight lines, though limited to dilute solutions for some B-subgroup elements, and the slope is always negative except in the case of zinc. From such results we can define dU/dc which is the change in activation energy per one atomic percent of solute addition.

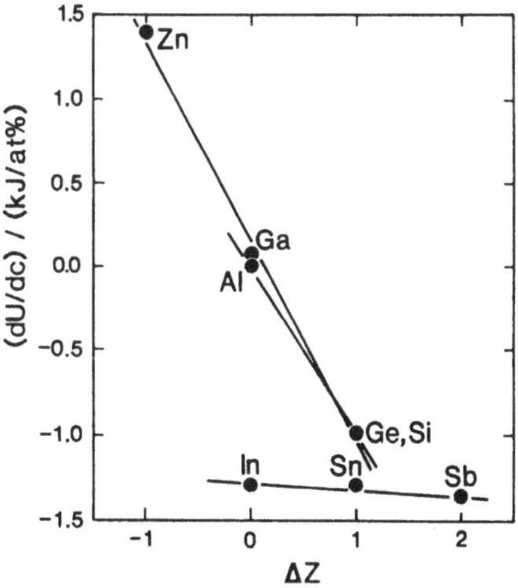

FIGURE 5 Relation between rate of activation energy change, dU/dc, and valence difference between solute and aluminum.

For the additions of B-subgroup elements, the effect of valence on the magnitude of dU/dc can be clearly seen. In Fig. 5 is shown such an effect where larger the valence is and larger the period is, negative larger the value of dU/dc. Knowing that a negative larger value for U provides higher magnitude of the mechanical anomaly, an element having larger group and period numbers is favourable for ternary additions. Note that additions of such elements to Ni_3Al are to increase the e ectron-atom ratio of the alloy.

3. PARAMETERS AFFECTING MECHANICAL ANOMALY

In discussing the phase stability of $L1_2$ to DO_{22} ordered crystal structures as affected by the ternary additions, we have to take into account an effect arising from introducing the third foreign atoms of a different atomic size. According to a systematic investigations on the substitution behavior of ternary elements in Ni_3Al by Ochiai et al.(9), all the elements appearing in Figs. 3 and 4 have been shown to substitute aluminum in the compound and the change in lattice parameters due to such substitutions has also been measured by the present authors(10). It was shown through their work that the lattice parameters of Ni_3Al linearly change in its homogeneous solid solutions and thereby da/dc, the change in lattice parameter per one atomic percent of solute, has been assigned for each element.

Figure 6 shows the relation between dU/dc and da/dc thus introduced for both B-subgroup and transition metal elements. The figure enables us to take a glance on both valence effect and the size effects on the magnitude of the mechanical anomaly in the compound. As shown by bold curves the relation is systematically established for B-subgroup elements of the same group. It is now clear that at a constant valence, dU/dc is negative larger with larger da/dc. Also at a constant da/dc, dU/dc is also found negative larger with larger valence. For transition metals, for which definition of valence is vague, the apparent valence could be assigned by analyzing the equi-valence contour establishied for B-subgroup elements. The best reasonable estimation of valence turns out to be 3.5 for titanium, 3.7 for vanadium and molybdenum followed by an order such as Zr>Hf>(Ta, Nb)> 5. Although the physical meaning of the apparent valence is left unknown, such an approach does provide useful information on the role of alloying elements to alter the mechanical properties of Ni_3Al.

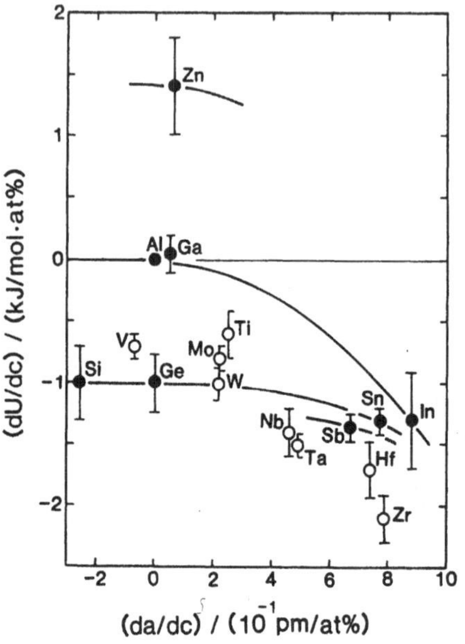

FIGURE 6 Relation between rate of activation energy change, dU/dc, and that of lattice parameter change, da/dc, in ternary Ni_3Al with additions of transition metal and B-subgroup elements.

4. PHASE STABILITY AND MAGNITUDE OF MECHANICAL ANOMALY

The assignment of apparent valence to the transition metal elements is of importance when we discuss the relative magnitude of the mechanical anomaly in Ni_3Al in terms of the phase stability of $L1_2$ ordered crystal structure against other GCP phases. According to notations to represent particular GCP phases described previously(11,12), $L1_2$ is expressed as cT where c denotes cubic stacking and T triangle network of minority component on the closed packed plane. The increase in electron-atom ratio, being achieved by additions of B-subgroup elements with larger valence, results in increasing tendency to shift from c to h(hexagonal) stacking and from T to R(rectangular) network(13,14). Relevant ordered crystal structures being resulted from such shift would be DO_{24}(chT), DO_{19}(hT), and DO_{22}(cR) and, therefore, the additions of elements having a larger valence can be considered to reduce the stability of $L1_2$ against these phases. As is shown in Fig. 5 this results in making the value of activation energy, U, a negative larger and thereby enhancing the mechanical anomaly. Reducing phase stability of $L1_2$ in such a way should alter stacking fault and/or APB energies in the crystal structure. The thermally activated cross slip to provide the mechanical anomaly is driven by an large anisotropy in APB energies between (111) and (001) and, as already described above, DO_{22} is a crystal structure having (001) APB on every other (001) planes of $L1_2$ crystal structure. The phase stability of $L1_2$ against DO_{22}, therefore, would be most important in determining the magnitude of the mechanical anomaly in Ni_3Al.

The other important parameter to affect the magnitude of the mechanical anomaly, as found in Fig. 6, is the rate of lattice parameter change, da/dc, encountered by ternary additions to Ni_3Al. The parameter is directly related to apparent atomic diameter of a ternary element in Ni_3Al which here is defined as an value obtained by an extraporation of the linear variation in lattice parameter shown in Fig. 6 to 100% of solute concentration. It has been shown that increasing R_B/R_A ratio in an A_3B compounds results in an increase in hexagonality(15,16) and therefore in

Ni_3Al additions of ternary elements with a larger apparent atomic diameter would also cause the same effect as what being arised from increasing electron-atom ratio. The effect of the atomic radius ratio can be best illustrated in A_3B transition metal compounds where appearance of GCP phases with various combinations of c, h, T and R components has successfully been interpreted together with the effect of electron-atom ratio(3-6). For the present case, Fig. 6 does provide a clear evidence of such combined effects the magnitude of the mechanical anomaly in Ni_3Al.

5.CONCLUSION

The role of alloying elements on the anomalous positive temperature dependence of strength is summarized herein with particular emphasis on how the ternary additions alter the stability of $L1_2$ crystal structure and consequently stacking fault and/or APB energies. The relevance between the characteristic mechanical anomaly and the stability of the phase seems to be successfully demonstrated by analyses of the experimental results of a systematic investigations. The results presented here would be useful in carrying out a more effective alloy design of nickel-base superalloys or Ni_3Al itself and are at the same time valuable for the theoretical treatment of phase stability by providing real-world information to be plugged in to prove its consistency.

REFERENCES

1.B.H.Kear and H.G.F.Wilsdorf, Trans. TMS-AIME, 224(1962), 382.
2.S.Takeuchi and E.Kuramoto, Acta Met., 21(1973), 415.
3.D.M.Wee and T.Suzuki, Trans. Japan Inst. Metals, 20(1979), 634.
4.D.M.Wee, O.Noguchi, Y.Oya and T.Suzuki, ibid., 21(1980), 237.
5.T.Suzuki and Y.Oya, J. Mater. Sci., 16(1982), 2737.
6.Y.Mishima, Y.Oya and T.Suzuki, High-Temperature Ordered Intermetallic Alloys, Ed. by C.C.Koch, C.T.Liu and N.S.Stoloff, Materials Research Society, Pennsylvania, (1985), p.263.
7.S.Ochiai, Y.Mishima, M.Yodogawa and T.Suzuki, Trans. Japan Inst. Metals, 27(1986), 32.
8.Y.Mishima, S.Ochiai, M.Yodogawa and T.Suzuki, ibid., 27(1986), 41.
9.S.Ochiai, Y.Oya and T.Suzuki, Acta Met., 32(1984), 289.
10.Y.Mishima, S.Ochiai and T.Suzuki, ibid., 33(1985), 1161.
11.H.J.Beattie, Jr., Intermetallic Compounds, Ed. by J.H.Westbrook, John Wiley and Sons Inc., New York, (1967), p.144.
12.W.B.Pearson, The Crystal Chemistry and Physics of Metals and Alloys, John Wiley and Sons Inc., New York, (1972), p.1.
13.A.K.Sinha, Trans. TMS-AIME, 245(1969), 237 and 911.
14.V.Sadagopan and B.C.Giessen, J. Metals, 17(1965), 1058.
15.J.H.N. van Vucht and K.H.J.Buschow, ibid., 10(1965), 98.
16.J.H.N. van Vucht, J. Less Common Met., 11(1966), 308.

ORDERED B2 ALUMINIDES

K. VEDULA

Department of Materials Science and Engineering
Case Western Reserve University
Cleveland, Ohio 44106

1. INTRODUCTION

B2 aluminides are potential high temperature structural materials because of their low densities and excellent oxidation resistance (1). B2 NiAl and FeAl are of particular interest because of lack of strategic elements in these alloys. The phase diagrams containing these two ordered structures are illustrated in Figures 1 and 2. The B2 crystal structure is illustrated in Figure 3 and is seen to extend over a wide range of aluminum contents for both FeAl and NiAl. NiAl has a melting point of 1640°C which gives it a temperature advantage over conventional superalloys.

These compounds, however, suffer from the problem of low tensile ductilities at room temperature. This short paper is intended to provide some insights into this low temperature ductility problem.

2. SINGLE CRYSTAL DEFORMATION

Single crystals of FeAl and NiAl are ductile at room temperature. The details of the operative slip systems have been studied to some extent and the behavior is shown to be a sensitive function of orientation and composition (2-15). Of particular importance is the difference in behavior of FeAl and NiAl, both having the same crystal structure.

In the B2 crystal structure, slip of <111> dislocations requires the formation of wrong types of bonds (i.e. Ni-Ni and Al-Al bonds instead of the lower energy Ni-Al bonds). Slip of <100> dislocations does not require the formation of such types of bonds. Which of these two slip directions is active in a crystal is determined by the magnitude of the energy for the formation of wrong types of bonds, since from geometric considerations the preferred slip vector for BCC crystals is <111>.

The major difference between FeAl and NiAl is that the bond energy between Fe and Al is less than that between Ni and Al. As a consequence, wrong bonds can form more readily in FeAl then in NiAl. The slip direction in FeAl is, hence, generally <111>, whereas it is <100> in NiAl. These have been observed experimentally in single crystals of FeAl and NiAl.

3. POLYCRYSTALLINE DEFORMATION

Deformation of polycrystals requires consideration of the number of independent slip systems. According to Von Mises criterion, a minimum of 5 independent slip systems is required for strain compatibility across grain boundaries in a polycrystal. FeAl, with <111> slip, possesses the required number of slip systems, whereas NiAl, with <100> slip does not possess the required number. Polycrystals of FeAl are, therefore, more likely to be ductile at room temperature compared with those of NiAl.

G. M. Stocks and A. Gonis (eds.), Alloy Phase Stability, 29–32.

An additional factor in polycrystalline deformation in tension is the mechanism of fracture, which can be either intergranular or transgranular. This depends upon the relative magnitudes of intergranular fracture stress and cleavage fracture stress. If either of these fracture stresses is lower than the stress needed to activate 5 independent slip systems, brittle fracture is likely to occur.

In the case of NiAl, transgranular cleavage fracture normally occurs before bulk plastic yielding as a consequence of difficulty of <111> slip. Hence, polycrystalline ductility at room temperature is typically not obtained.

On the other hand, in the case of FeAl, the ease of <111> slip makes several slip systems operate and when the fracture strength is higher than the yield strength, room temperature ductility is observed (16,17). This is, in fact, the case for lower aluminum contents within the B2 phase field (e.g. Fe-40 at% Al). At higher Al contents, the yield strength increases, presumably due to an increase in antiphase boundary energy causing the fracture stress to be lower than the yield strength, thereby resulting in a loss of ductility. Fe-50 at% Al is, hence, brittle at room temperature in tension.

The above discussion on the deformation of polycrystalline NiAl and FeAl shows that the stress to activate appropriate slip systems, as well as the stress to fracture grain boundaries and the cleavage fracture stress, are all important factors controlling ductility of these intermetallic compounds. In order to modify the behavior of such compounds through alloying additions, it is necessary to be able to predict the effect of these additions on bonding energy, on stress to cause different kinds of slip, and on fracture stress of the intergranular, as well as the cleavage type.

Such a predictive capability must be developed through suitable interactions between theoretical and experimental materials scientists. For example, in the case of NiAl, it would be useful to identify elements which could decrease the bond energy of Ni-Al in the compound. This decrease in bond energy may encourage easier <111> slip. It would also be necessary to predict the effect of the same alloying additions on fracture stress as well, in order to ensure that yielding occurs before fracture.

Theoretical modeling using techniques such as total energy calculations, embedded atom methods and cluster variation methods would be extremely beneficial in this regard. Even more empirical approaches using approximate pseudo-potential theories may have a useful role to play in providing qualitative information regarding alloying additions.

In this manner more scientific approaches towards alloy development based on intermetallic compounds may be made possible.

REFERENCES

1. K. Vedula and J. R. Stephens, Mat. Res. Soc., Symp. Proc. Vol. 81, 1987, p. 381.
2. T. Yamagata and H. Yoshida, Mat. Sci. and Engg. Vol. 12, p. 95, 1973.
3. R. C. Crawford, Phil. Mag. Vol. 33, p. 529, 1976.
4. Y. Umakoshi and M. Yamaguchi, Phil. Mag. A, Vol. 41, p. 573, 1980.
5. Y. Umakoshi and M. Yamaguchi, Phil. Mag. A, Vol. 44, p. 711, 1981.
6. I. L. F. Ray, R. C. Crawford and D. J. H. Cockayne, Phil. Mag. Vol. 21, p. 1027, 1970.
7. M. A. Crimp, Ph.D. Thesis and M. A. Crimp and K. Vedula, unpublished papers.

8. M. J. Marcinkowski and N. Brown, Acta Met. Vol. 9, p. 764, 1961.
9. T. Yamagata, Trans. JIM, Vol. 18, p. 715, 1977.
10. M. Mendiratta, H. M. Kim and H. A. Lipsitt, Met. Trans. 15A, p. 395, 1984.
11. A. Ball and R. E. Smallman, Acta Met. Vol. 14, p. 1349, 1966.
12. R. J. Wasilewski, S. R. Butler and J. E. Hanlon, Trans. Met. Soc. AIME, Vol. 239, p. 1357, 1967.
13. R. T. Pascoe and C. W. A. Newey, Met. Sci. J. Vol. 2, p. 138, 1968.
14. R. T. Pascoe and C. W. A. Newey, Phys. Stat. Sol. Vol. 29, p. 357, 1968.
15. M. H. Loretto and R. J. Wasilewski, Phil. Mag. Vol. 23, p. 1311, 1971.
16. M. A. Crimp, K. Vedula and D. J. Gaydosh, Mat. Res. Soc. Symp. Proc. Vol. 81, 1987.
17. M. A. Crimp and K. Vedula, Mat. Sci. and Eng., Vol. 78, p. 193, 1986.

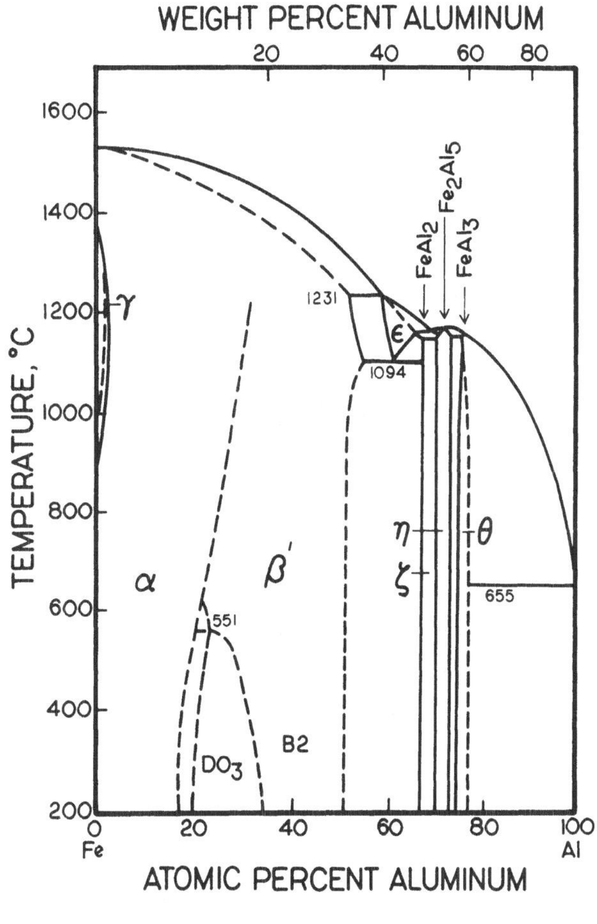

Figure 1. Phase diagram for Fe-Al

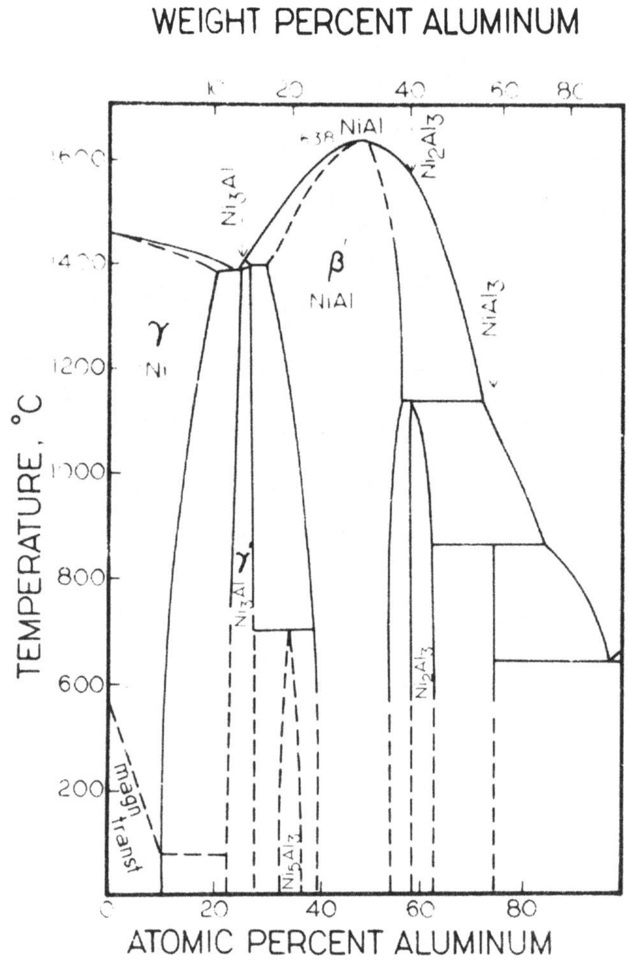

Figure 2. Phase diagram for Ni-Al

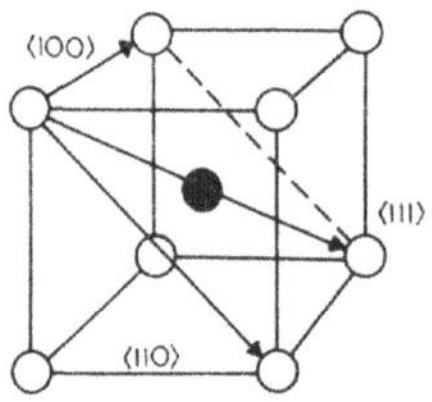

Figure 3. The B2 crystal structure

SECTION 2

EXPERIMENTAL PROBES OF ATOMIC AND ELECTRONIC STRUCTURE

PHOTOEMISSION AND THE STUDY OF ORDER-DISORDER TRANSFORMATIONS

R.G. JORDAN*+ and P.J. DURHAM+

*Department of Physics, University of Birmingham, Birmingham B15 2TT, UK.
+SERC Daresbury Laboratory, Warrington WA4 4AD, UK.

This article is based on a series of invited lectures delivered at the NATO Advanced Study Institute on Alloy Phase Stability held at Maleme, Crete, June 14-27, 1987.

INTRODUCTION

There can be little doubt that the introduction of angle-resolving techniques in photoelectron spectroscopy [1] in the 1970's heralded a new era in the study of metallic solids. Although at that time certain fundamental microscopic properties, such as the phonon spectrum, were reasonably well known for a wide range of materials, there was little detailed information about the electronic energy bands except in some ordered systems at the Fermi level, principally through measurements of the de Haas-van Alphen effect. This was an unfortunate situation for it had been long recognised that the physical and metallurgical properties of metals and alloys - their strength, structure, phase stability, conductivity, magnetism, etc - are related directly to the underlying electronic structure. To understand these properties fundamentally - in order to be able to manipulate and modify them to advantage perhaps - one requires a realistic and parameter-free, i.e., first principles, description of the electron states. However, it is essential that the results of any electronic structure calculation are tested and verified in detail experimentally. Most spectroscopies such as optical, soft x-ray and angle-integrated photoemission effectively average over the Brillouin zone and reveal only the gross features of the density of states. The development of angle-resolved UV photoelectron spectroscopy represented a major breakthrough, therefore, because many of the features in experimental spectra can be understood in terms of wave-vector ($\underline{k}$) conserving direct transitions. By measuring the kinetic energies of the photoelectrons in a particular direction for various photon energies, say, one obtains data which are related to the spectral density of electrons as a function of energy and $\underline{k}$ (i.e., the spectral function). Thus, in principle, theory can be confronted in a most direct and profound way. Within the past few years a complementary technique - $\underline{k}$-resolved inverse photoemission [2] - has been developed and used to probe the unoccupied regions of the electronic structure.

During the past decade or so there has been major progress in alloy theory and one is beginning to see where first principles calculations can have an impact on the design and optimisation of alloys for specific applications [3]. The 'modern' theory of alloys - based on the coherent potential approximation (CPA) - has proved to be adequate for first principles calculations of electron states and the development of the statistical mechanics of compositional and magnetic configurations on the basis of the electronic structure. Also, it provides a sound conceptual framework within which one can interpret the results of searching experiments such as angle-resolved UV photoemission. The recent advances in experimental and theoretical methods, therefore, provide the foundation for re-investigating many of the central

G. M. Stocks and A. Gonis (eds.), Alloy Phase Stability, 35–74.

issues of metal-physics from a new, fully electronic point of view. Potentially, such work can lead to new and deeper insights into the basic principles governing stability and the phase diagram, transport properties, itinerant ferromagnetism, and so on.

Historically, these developments ran very much hand-in-hand. The two papers upon which the experimental and theoretical progress is based were both published in 1970. There was the first serious use of the CPA to calculate the electron states in a real alloy system using a tight-binding (TB) model [4] and the angle-integrated UV photoemission studies of Cu-Ni alloys by Seib and Spicer [5]; the photoemission measurements were especially significant because they proved beyond all doubt that the rigid band model was totally inappropriate, since individual Cu- and Ni-related features could be identified in the spectra. Stocks, Williams and Faulkner [6] improved the initial TB-CPA calculations by using better potentials and by considering the complete range of compositions. In fact, it was these later calculations which first demonstrated that the CPA was able to reproduce the experimental observations. Further XPS [7] and higher resolution UV photoemission measurements [8] also compared very well with the (improved) TB-CPA calculations. However, since the experiments involved polycrystals and angle-integrated detection contact with theory could only be made via the density of states. Nevertheless, it was clear that the CPA gave a reliable description of the split-band behaviour, in complete contrast to the rigid band model or the virtual crystal approximation. The effort to investigate the electron states in alloys in more quantitative detail, e.g., as a function of E and $\underline{k}$, succeeded in 1977 with the first study of a single crystal Cu-Ni alloy using angle-resolved UV photoelectron spectroscopy [9], the results of which compared very favourably with the much more refined and detailed CPA calculations using KKR band theory methods - the so-called KKRCPA [10]. (Since angle-resolved spectra contain much more selective information than angle-integrated and XPS measurements they provide a critical examination of the predictions of the KKRCPA.) The success of these studies of Cu-Ni alloys provided the impetus to investigate both experimentally and theoretically a range of other systems and more complex issues including, for instance, other binary alloys, e.g., Ag-Pd [11] and Cu-Pd [12], 3-component systems, e.g. the spin glass Ag-Mn [13], and, more recently, magnetism at finite temperatures [14]. The current standing and very wide applicability of the CPA as a theoretical tool for studying the properties of alloys can be gauged by the various topics discussed at this Summer School meeting and through articles published as part of the NATO ASI series [15].

As indicated above, a realistic description of the electronic spectral density is crucial for a fundamental understanding of the properties of alloys. In Fig.1 we show just how important it really is. The 1-e Bloch spectral function can now be calculated reasonably straightforwardly using the KKRCPA (including charge self-consistency and spin- polarisation). As we shall show it can be probed - more or less directly - by angle-resolved UV photoelectron spectroscopy; the experimental complications such as many-body effects, the effect of the surface, segregation of the constituents, etc., can be realistically taken into account. In principle, therefore, by combining theory with experiment one can establish a reliable picture of the spectral function. (Some other experiments such as 2-D angular correlation of (positron) annihilation radiation (ACAR) measurements, also provide a test but in a less direct way). From the spectral function one can calculate the sum over the 1-e eigenvalues - the structure dependent part of the total electronic energy - and determine a range of ground state properties such as the structure, lattice spacing, bulk modulus, magnetic moment, etc; clearly, such information will have a vital role to play in the design of alloys for

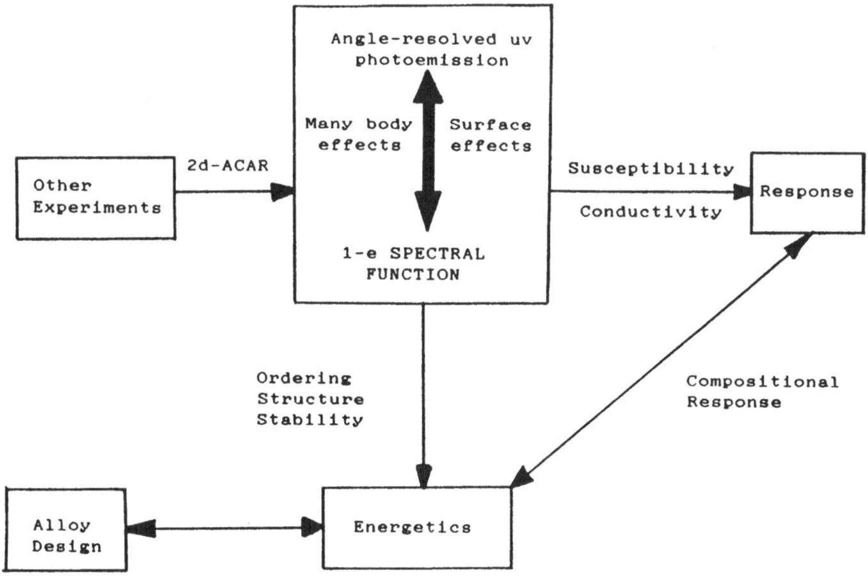

FIGURE 1. Block diagram indicating the central role of the spectral function.

specific applications. Also, one can determine the response of the electron system to external perturbations, i.e., the generalised susceptibility, the electrical conductivity, etc. Furthermore, techniques have been developed which permit the calculation of concentration correlation functions on the basis of the spectral function; these theories have application in first principles studies of, for instance, short-range ordering and Guinnier-Preston zones.

We will concentrate our discussion, therefore, on the spectral function. In section I we describe the photoemission technique and how experimental measurements are related to the 1-e spectral density. In section II we outline the basics of the calculation of the Bloch spectral function. In order to interpret quantitatively the richly structured experimental spectra in terms of the electronic structure, they must be compared with proper photocurrent calculations rather than with the 1-e (bulk) spectral density itself. Accordingly, in section III, we show how the photocurrent can be calculated on the same theoretical footing as the spectral function and we compare a selection of experimental and theoretical results to demonstrate the effectiveness and potential of our joint approach. Ordering in alloys is a scientifically interesting and technologically important phenomenon and the statistical mechanics of order-disorder phase transitions has been studied for decades. What is less clear is the electronic origin of the interatomic forces, which from the point of view of the general problem of alloy phase stability, is crucial. In section IV we give some results from our recent studies of the transformations in CsCℓ-bcc (CuZn and AgZn) and L12-fcc (Cu_3Au) systems, and we conclude by indicating the current direction of our work in this area.

Note on Units

Throughout this article all equations are expressed in atomic units (a.u.): $\hbar=m=e=1$. Thus lengths are gven as multiples of the Bohr radius (a_0=0.529177 Å), and energies are given in Hartrees (1H=27.211608 eV).

SECTION I

1. Photoemission measurements

In principle, angle-resolved UV photoemission measurements are relatively straightforward. In a typical experiment UV photons (with energy ω) are incident (at an angle ϕ) upon the clean and crystallographically ordered surface of a single crystal specimen maintained in ultra-high vacuum ($<10^{-9}$ mbar), Fig.2(a). If the photon energy is sufficiently high (>5 eV, say) electrons are emitted through the surface of the specimen. These photoelectrons propagating at angles θ,ψ, are counted and energy analysed as a function of $\omega,\phi,\theta,\psi,\underline{e}$ (where $\underline{e}$ is the polarisation direction of the light) and, in the most up-to-date experiments, $\underline{\sigma}$ (the electron spin). In a typical spectrometer the photon source is fixed and the specimen and analyser can be moved independently so that a range of ϕ,θ and ψ can be realised; in all the experimental data reported here, however, $\psi = 0$. The source may be an inert gas discharge lamp, producing unpolarised radiation at the energies of the resonance lines of HeI (21.2 eV), NeI (16.8 eV) or ArI (11.7 eV), or tuned and linearly polarised radiation from a storage ring, in the range $20 \text{ eV} < \omega < 100 \text{ eV}$, say, viz. station 6.2 at the Synchrotron Radiation Source (SRS), Daresbury Laboratory. The irradiated area is $\sim 1 \text{ mm}^2$ and the

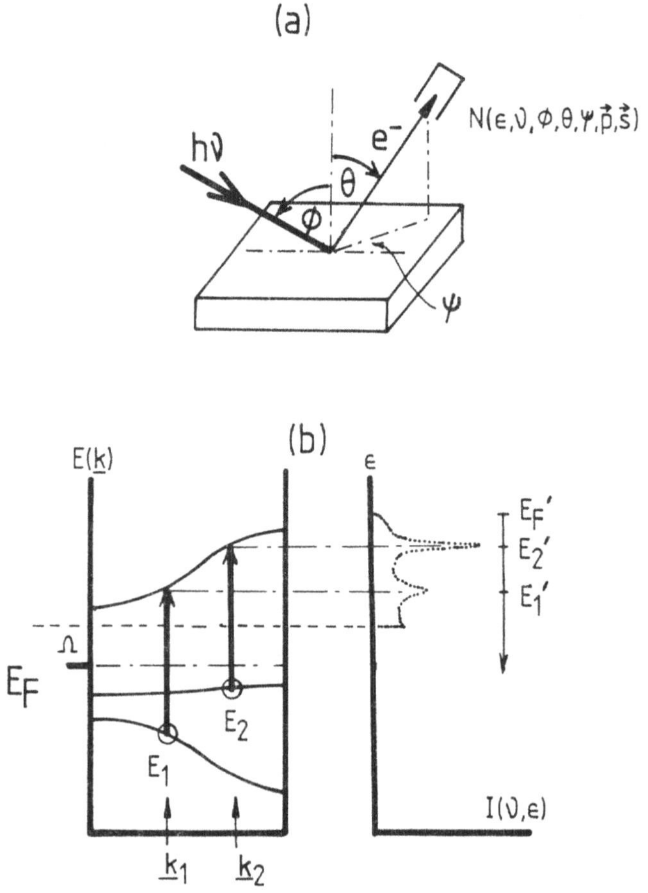

FIGURE 2. Schematic diagram of photoemission.

angular and energy resolutions are, typically, ±2° and 50-100 meV, respectively, when a hemispherical electron energy analyser is used. Clean and ordered surfaces are not always easily achieved - often considerable detailed study is required, particularly in the case of alloys. Preparation usually involves repeated Ar^+ bombardment and annealing cycles with characterisation by low energy electron diffraction (LEED) and Auger electron spectroscopy (AES).

2. Simple model of photoemission

Some insight into photoemission and the potential of photoelectron spectroscopy can be realised through a highly simplified description involving a typical band structure, as shown in Fig.2(b). Following interaction with a photon of energy ω, an electron with an initial energy $E_1(\underline{k})$ undergoes a direct transition (i.e., vertical in the reduced zone scheme; the momentum of a UV photon is negligible on the scale of the Brillouin zone) to an unoccupied state $E_2(\underline{k})$ above the vacuum level E_V. Transitions are confined to points in $\underline{k}$-space such that

$$E_2(\underline{k}) - E_1(\underline{k}) - \omega = 0.$$

When crossing the surface the parallel component of the wave-vector is conserved, modulo a surface reciprocal lattice vector, i.e., $\underline{K}_\parallel = \underline{k}_\parallel + \underline{G}_\parallel$. Thus, if the detector is set to accept photoelectrons with kinetic energy ε emitted at an angle θ with respect to the surface normal, the following relationships can be deduced:

$$|\underline{K}_\parallel| = \sqrt{2\varepsilon}\ \mathrm{Sin}\theta \qquad (I.1)$$

and for the initial state the binding energy is

$$E_F - E_1 = \omega - \Omega - \varepsilon \qquad (I.2)$$

where Ω is the work function of the surface. If energy is scanned in the detector a response like that shown schematically on the right hand side of Fig.2(b) is obtained. (Many-body and resolution effects have been added in order to simulate a real spectrum.) The Fermi level can be identified in spectra and so a knowledge of Ω is not necessary since

$$E_F - E_1 = E_F' - E_1' .$$

It is usual, therefore, for photoelectron spectra to be shown as a function of binding energy. Clearly, as ω is varied the peak positions in the spectra may change according to the dispersion of the initial states. A particularly simple and useful geometry is normal emission (i.e., $\underline{k}_\parallel = 0$) since the initial states will then lie along a fixed and known direction in $\underline{k}$-space; the particular direction depends on the orientation of the specimen surface. For example, normal emission from the (100) surface of an fcc crystal involves initial states along the ΓX (Δ) direction.

It is not possible to map the band structure $(E,\underline{k})$ of a material directly using angle-resolved photoelectron spectroscopy since one has no direct knowledge of the wave-vector $\underline{k}$ for the transition(s); only the component parallel to the surface, from eqn.(I.1). However, two approaches have been used to deduce values of $\underline{k}$:

- The first involves normal emission and approximating the final states by a free-electron parabola (or a semi-empirical band) if the critical points are known. The value of k can be determined, therefore, from the measurement of ε [1,16].

- The second is known as the triangulation method. The angular dependence of emission from two or more different crystal orientations is measured in the same emission plane. If the energy-coincidence criterion is applied to identify the same transition from the different surfaces a unique value for k can be deduced [17].

Despite the limitations and assumptions involved in this simplified single-particle approach, in the case of particular transition metals, for example, experimentally determined and calculated (1-e) bands are often remarkably similar. However, it should be recognised that this procedure is an approximation and is only applicable under certain circumstances, as we shall demonstrate below. In other cases the approximation is poor or even breaks down completely. The various factors which complicate this elementary picture and should be taken into account for a more realistic and quantitative approach include:

- The effects of correlation and fluctuation. Photoemission involves an excited state of the system and there are many-body in teractions involving, for example, the hole and the Fermi sea. The net result is that the features in spectra given by eqn.(I.2) are broadened and often shifted from the ground state values.
- The system is semi-infinite. The presence of the surface can modify considerably the electronic structure in the relatively narrow spatial region probed in photoemission. Highly localised (surface) states often occur in bulk band gaps, for instance, which result in additional features in photoelectron spectra.
- Not all transitions are observed. The intensity of spectral features depends on the transition probability involving the polarisation of the radiation and the symmetries of the initial and final states.

It is not altogether appropriate, therefore, to simply compare experimentally determined spectra with 1-e (groundstate) band structure calculations for the infinite solid without some consideration of the above points. A rather more sophisticated approach is required in order to provide more meaningful comparisons. We shall now consider the photoemission process in more detail and expand our discussion of the various complicating factors.

3. More detailed model of photoemission

The effect of the absorption of the photon is to create an electron-hole pair. While the photoelectron is on its way out the surrounding electron cloud responds to the perturbation in a number of ways. The actual importance of these many-electron effects depends on the system [18]. For instance, when the photoelectron and hole separate they form an electric dipole and for small separations the surrounding electrons will respond to the dipolar field. As the separation of the electron-hole pair increases the surrounding electrons will notice the individual charges of the electron and hole and will respond in order to screen out these charges. These response processes are complicated but an insight into the problem can be obtained through a 1-electron picture.

In a real system a given electron moves in the instantaneous field due to the others and so its motion is correlated with that of the ensemble. The most straightforward approach, first advanced by Hartree, is to consider the interaction between one electron and the average charge distribution of the other electrons. Nearly all modern self-consistent-field band structure calculations are carried out within the framework of density functional theory (DFT), usually adopting the local density approximation. DFT rests on a theorem which shows that the <u>exact</u> total energy of the ground state can be obtained by minimising a certain unique functional of the charge density, and this is usually achieved by solving an effective 1-electron Schrödinger equation resembling that of the Hartree method, but with exchange and correlation included. It is the implementation of DFT in accurate calculations which has led to the present high level of understanding electron structure. It was clear from the outset, however, that DFT has very little to say about electronic excitation energies, despite its status as a formally exact theory for the ground state total energy. In using DFT eigenvalues or energy bands to interpret photoemission (and other excitation processes), we are therefore neglecting all final state interactions, relaxation, decay, etc. [52]. On the other hand, one can still use DFT to calculate excited states, not as an exact theory but as a physically motivated approximation scheme. As we shall see, in many systems this approach works rather well; in others it fails. In what follows, when we use a single-particle picture, we shall always have DFT in mind. We shall refer to the DFT energy bands as 1-electron bands or ground state bands more or less interchangeably.

The 1-electron approximation for the photocurrent into the detector can be written in the golden rule form [19]

$$I(\omega,\varepsilon) = 2\pi \sum_{n,\underline{k}} \left|\langle\Psi(\varepsilon,\underline{k})\left|\Delta\right|\Psi(E_n^0,\underline{k})\rangle\right|^2 \delta(\varepsilon - E_n^0(\underline{k}) - \omega) \tag{I.3}$$

where the electron-photon interaction is $\Delta = -\alpha\, \underline{A}.\underline{p} = i\alpha\, \underline{A}.\nabla v/\omega$ (α being the fine structure constant, $\alpha^{-1} = 137.03604$, $\underline{A}$ the photon vector potential, p the electron momentum, and v the potential in which the electron moves), and ε is the kinetic energy of the photoelectron; we put $E = \varepsilon - \omega$ and write

$$\delta(E - E_n^0(\underline{k})) = A_n^0(\underline{k},E)$$

and

$$\begin{aligned} A^0(\underline{k},E) &= \Sigma_n\, A_n^0(\underline{k},E) \\ &= \Sigma_n\, \delta(E - E_n^0(\underline{k})) \end{aligned} \tag{I.4}$$

i.e., a sum of δ-function peaks - the ground state 1-e spectral function - whose loci in $E,\underline{k}$ space trace the band structure. Thus, eqns.(I.3) and (I.4) imply that in an ordered system the photoelectron spectrum will consist of a set of sharp peaks - depending on the matrix elements - at energies corresponding to the eigenvalues of the system. In the case of a disordered alloy, the picture is more complex since the spectral function is no longer a set of δ-function peaks [10]. Instead, it comprises broadened features whose widths vary in a complicated way as a function of E and $\underline{k}$ as we shall show later. Nevertheless, the approach is still valid.

It is easy to understand, therefore, why photoemission is such a powerful probe of the electronic structure in both ordered and disordered systems; to zero-order the photocurrent gives a direct picture of the spectral function modulated by matrix element effects. In fact, the result is valid beyond the 1-electron picture. In the presence of an interaction between the electron-hole pair and the surrounding electron cloud the photoelectron spectrum will describe the 1-e eigenvalue spectrum but with shifted and broadened main lines, etc., as we shall now demonstrate.

(a) Many-body effects. Following the creation of the hole and the photoelectron many-body interactions will arise and the system will begin to relax. The key quantity for describing this behaviour is the self-energy

$S_n(\underline{k},E)$ which represents a generalised correction to the 1-e energy $E_n^0(\underline{k})$ - for simplicity we will continue with an ordered system. In the non-interacting case the spectral function (eqn.(I.4)) can be expressed in terms of the 1-e Green function $G_n^0(\underline{k},E)$, i.e.,

$$A_n^0(\underline{k},E) = -\frac{1}{\pi}\,\mathrm{Im}\,G_n^0(\underline{k},E) \qquad (I.5)$$

where

$$G_n^0(\underline{k},E) = \frac{1}{E - E_n^0(\underline{k}) + i\mu}\,. \qquad (I.6)$$

[Here the limit $\mu \rightarrow 0^+$ is to be taken, when $A_n^0(\underline{k},E) = \delta(E-E_n^0(\underline{k}))$.]
If we switch on the interaction the modified Green function becomes

$$G_n = G_n^0 + G_n^0 S_n G_n = \frac{1}{E - E_n^0(\underline{k}) - S_n(\underline{k},E)}\,. \qquad (I.7)$$

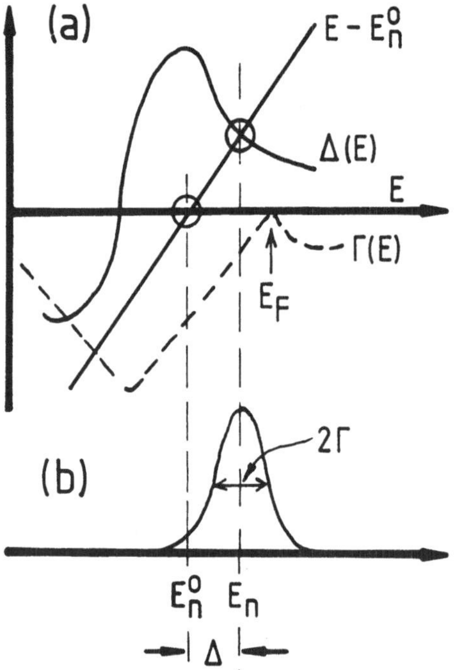

FIGURE 3. (a) Typical variation of the real (Δ) and imaginary (Γ) parts of the self-energy (assumed to be independent of $\underline{k}$). (b) The effect on a photoemission features.

The self-energy will comprise real ($\Delta_n(\underline{k},E)$) and imaginary ($\Gamma_n(\underline{k},E)$) parts and so the spectral function for the interacting (i.e., excited) system becomes

$$A_n(\underline{k},E) = \frac{1}{\pi}\;\frac{|\Gamma_n(\underline{k},E)|}{[E - E_n^0(\underline{k}) - \Delta_n(\underline{k},E)]^2 + [\Gamma_n(\underline{k},E)]^2}\,. \qquad (I.8)$$

Thus, its structure - and consequently the photoemission spectrum - depends on the form of the self-energy. A typical variation of Δ_n and Γ_n with energy in the valence band region of a metal is shown schematically in Fig.3(a).

Providing Γ_n is not too large eqn.(I.8) will have resonance maxima when

$$E - E_n^0(\underline{k}) - \Delta_n(\underline{k},E) = 0 .$$

The solutions of this Dyson-type equation are given by the intersection of the straight line $E - E_n^0(\underline{k})$ and the curve for $\Delta_n(\underline{k},E)$, see Fig.3(b). Thus, in addition to the instrumental resolution, the photoelectron spectrum will comprise broadened peaks at E_n (shifted by Δ_n from E_n^0) with a FWHM $\sim 2|\Gamma_n|$. If the real part of the self-energy varies rather slowly near resonance there is a single solution which picks up most of the spectral strength according to

$$\int E\, A_n(\underline{k},E)dE = E_n^0(\underline{k}) .$$

Therefore, the main features of the 1-e ground state spectral function will remain distinct, although broadened and shifted, making comparisons between photoelectron spectra and band structure calculations possible. If, on the other hand, S_n exhibits very strong dispersions the quasi-particle picture of a well-defined hole breaks down since the spectral strength may be spread over a number of excitations; in this regime comparisons become very difficult.

As far as photoemission measurements are concerned the difference between the excited (N-1) and N electron systems depends on the localisation of the hole. Qualitatively, if the hole state is extended - as in the valence band - the modification to the potential is small and the photoelectron spectrum, in general, will agree with predictions based on the ground state spectral function. In contrast, photoemission from core levels or localised valence band states (e.g., and 4f levels in rare earths) will be subject to strong many-body effects.

The d-bands in transition and noble metals provide some qualitative insights into the complexities of self-energy corrections. The d-states are extended (although their mobilities are considerably less than states with sp character) so one might expect the band picture to be satisfactory. However, in the presence of intra-band fluctuation and correlation effects, many-body corrections may be apparent. In Ni, for instance, the 3d-band is not filled and so there is a high density of states in the vicinity of E_F. Rather than propagating freely a hole in the 3d-valence band couples to localised two-hole-one-electron excitations resulting from intra 3d-band fluctuation. This leads to large variations in the self-energy and to shifts (compared with the calculated 1-e bands) and satellite peaks in photoelectron spectra. The shifts are positive (see Fig.3(b)) and lead to a narrowing of the band [20]. In Cu, the top of the 3d-band is ~2 eV below E_F and the bandwidth is ~3 eV. Since the 3d-band is full and there is a low density of states in the vicinity of E_F, there is no phase-space for intra-3d-band fluctuation, and localisation does not occur. The hole propagates and the lifetime broadening effect is relatively small. In Zn the 3d-levels form a narrow band, ~1 eV wide, with some dispersion. However, in view of this small bandwidth a 3d-hole wave packet will move very slowly and will be screened almost as effectively as a localised hole. It is unlikely, therefore, that a 1-e band structure calculation will place the 3d-levels at the same energy as that indicated by photoemission - indeed, a discrepancy of several eV's is apparent [21] - although there will be only minor lifetime broadening.

The lifetimes of holes in the valence band, therefore, vary considerably from material to material. Close to the Fermi energy the variation of the inverse hole lifetime is [22]

$$|\Gamma^h| \propto (E - E_F)^2$$

with typical values in the middle of the band of ~0.5 eV. Since the characteristic decay process for holes in the valence band is via an Auger decay mechanism, i.e., the creation of further electron-hole pairs, the inverse lifetimes in insulators are typically much smaller than in metals, because zero density of states above E_F inhibits the Auger process.

As far as the photoelectron is concerned the most important many-body effect is the strong inelastic damping due to electron-electron interactions. As a result the final state bands become broadened making minor band-shape details less important. Above the characteristic excitation energies of plasmons (~10-20 eV) the inverse lifetime ($|\Gamma^e|$) varies only slowly and is almost material independent (typically 4 eV). For a free electron final state band the mean free path is related to the lifetime as follows

$$\lambda^e = \frac{\sqrt{2E}}{2|\Gamma|^e}$$

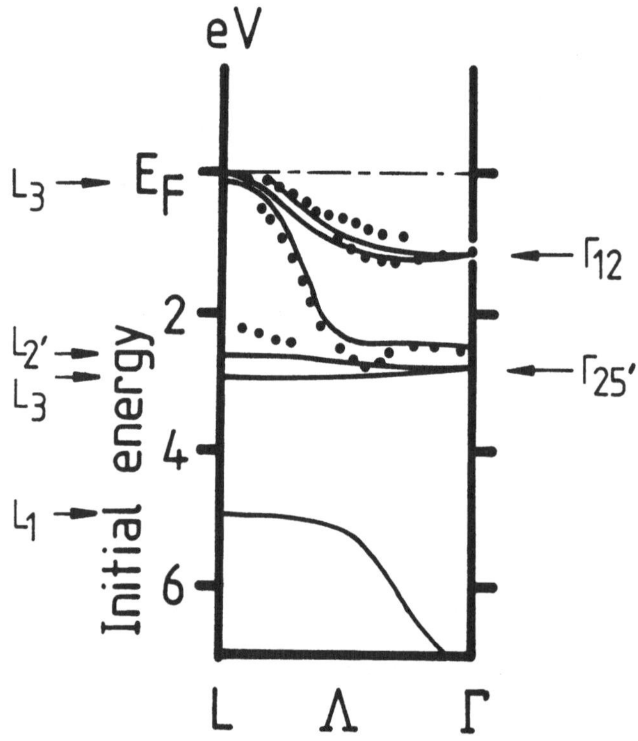

FIGURE 4. The calculated and 'experimental' band structure of Pd (from ref.[16]).

giving the widely-quoted 'characteristic mean free path curve'. At lower energies (<30 eV) Γ^e, and hence λ^e, vary rapidly.

As indicated previously, for some metals the bands deduced from photoemission measurements bear a close qualitative resemblance to those calculated. However, there are often quantitative discrepancies associated with the self-energy effects, see fig.4 (from [16]). Despite their complicated nature there have been some significant advances recently in dealing quantitatively with the many-body effects in photoemission. For example, Hybertsen and Louie [23] made a parameter-free calculation of quasi-particle energies in Si using the GW approximation for the self-energy and realistic energy bands. The results for the band gaps and band dispersions were in excellent agreement with experiments. A clear discussion of this approach to quasi-particle energies and their relationship to density functional energy bands is given by Godby et al.[52]. There is every reason to expect quantitative self-energy calculations, not necessarily involving the GW approximation, to be applied to other systems in the near future.

(b) Surface effects

We have commented above that a photoelectron spectrum often closely resembles the spectral function calculated for the infinite (or bulk) crystal. One might be surprised by this observation since the escape depths for UV photoexcitation (~10-100 eV) are typically 0.5-1.5 nm so that emission is predominantly from states at the surface. What it does demonstrate is the necessity to use a wave-mechanical approach to the process [19]. For instance, using the 3-step model of photoemission - (1) excitation, (2) transport towards the surface, (3) transmission through the surface - the electrons moving towards the surface will have a group velocity given by

$$V_g = \nabla_{\underline{k}}(E) = \partial E/\partial k \qquad \text{(in 1-dimension)}$$

Since the density of (final) states is

$$N_f(E) \propto (1/\pi)(\partial k/\partial E)$$

the rate at which electrons with energy E arrive at the surface is

$$\propto V_g N_f(E) = 1/\pi$$

i.e., constant, implying that all band structure effects are absent in spectra! This obvious contradiction points to the need to consider the process in terms of matching the Bloch waves in the solid and plane waves in the vacuum. A Bloch state in the crystal can be expressed as

$$\Psi(\underline{k}) = \Sigma_{\underline{G}}\, a_G(\underline{k}) \left| \underline{k} + \underline{G} \right> . \qquad (I.9)$$

For electrons travelling through the surface the $\underline{k}_{\parallel}$ conservation condition means that particular plane wave components of eqn.(I.9) are selected. Thus, the relative intensities of peaks in photoelectron spectra will depend on the strengths of the Fourier coefficients $a_G(\underline{k})$. The surface, then, acts like a filter [19].

With large systems the introduction of a surface does not affect appreciably the energy eigenvalues of the Bloch states - only by an amount of order 1/N where N is the number of atoms normal to the surface. However, the interference of incoming and reflected waves can lead to densities of states at the surface which are different from the bulk.

In addition to the rapid change in the potential at the surface (which from eqn.(I.3) will give rise to photoemission because of the ∇V term), it is well established that atom positions and layer spacings may be different from the bulk values. As a result the local density of electron states may be changed drastically in the surface region. Large densities of states may occur at the surface which decay towards the interior of the crystal; such states are called surface states. There are two types; Tamm and Shockley surface states. The former are derived from single bands and result from the changes of the parameters near the surface. If the potential in the surface region is less attractive than the bulk, say, the uppermost state may split off out of the band region and become localised. (There is the obvious analogy to core level shifts at the surface due to the modified local potential in the surface layer(s)). Shockley states are due only to the termination of the crystal and are located in certain band gaps. In band gaps the Bloch solutions have complex wave-vectors and so are not reasonable states for an infinite system. However, if at the surface such a solution, which decays towards the interior, can be matched to an exponential tail in the vacuum, then a Shockley-type surface state occurs. Such matching is only possible at particular energies under certain circumstances; for instance, if the states at the bottom and the top of the gap have p- and s-character, respectively. The most studied examples of Shockley-type surface states occur in the s-p hybridisation gap near E_F at the L point in the (fcc) noble metals and some (fcc) noble metal-based random alloys, see section III.

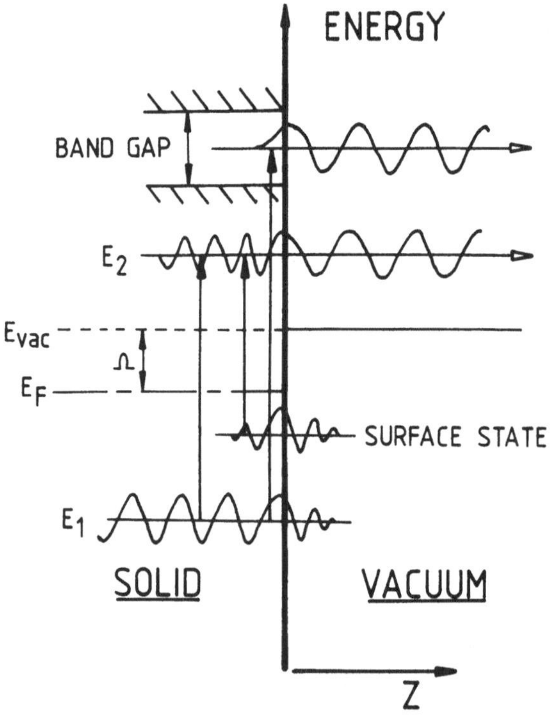

FIGURE 5. Schematic diagram of electron wavefunctions at the solid/vacuum interface and possible transitions (adapted from ref.[19]).

The idea of matching plane waves in the vacuum with the Bloch waves in the solid allows us to illustrate the components in a photoelectron spectrum in a particular elegant and simple way. In Fig.5 we show how several possible contributions to the photocurrent can arise. We represent the electrons arriving at the detector by a 'time-reversed' incoming plane wave. If the final state energy falls within an allowed energy band of the solid it may match the plane wave and produce a photocurrent; the initial state could be either an occupied Bloch state or an occupied surface state. Even if the final state energy falls in a band gap a photocurrent can be produced. A plane wave in the vacuum will have an exponential tail into the solid (~ few layer spacings) which is sufficient for photoexcitation and emission to occur at the surface. This effect is called surface photoemission. The implication is that there are always final states which can couple to initial states.

The modification of the potential in the surface region can also affect the core level binding energies, resulting in surface core level shifts. Such shifts can be +ve or -ve depending on the precise nature of the surface potential. In transition metals, for instance, as a result of a reduction of the d-bandwidth at the surface, if the d-band is less than half-filled there is a tendency to shift to higher binding energy (i.e., +ve), whereas if the band is more than half-filled the shift tends to be -ve. We will give an example in section III.

(c) Selection rules

We have alluded to certain selection rules already in angle-resolved photoemission, namely the conservation of $\underline{k}$ (modulo a reciprocal lattice vector) and the conservation of energy. (We note that the k-conservation may be relaxed for excitations occurring close to the surface where translational invariance in the normal direction is broken. If the surface is ordered, $k_{\parallel}$ is always conserved, of course.) We have shown how they combine so that photoemission measurements provide information about the $\underline{k}$-dependence of the electronic structure. There are, in addition, the matrix element effects which modulate the intensity. Since from eqn.(I.3)

$$\Delta = -\alpha \, \underline{A}\cdot\underline{p} = -\alpha \, A_0 \underline{e}\cdot\underline{p}$$

where $\underline{e}$ is the unit polarisation vector of the light, a polarisation selection rule is realised which involves parity conservation of the relevant initial and final state wavefunctions in an appropriate mirror plane. If, for example, $\underline{e}\cdot\underline{p}$ changes sign on reflection the matrix element can only couple states with different parities with respect to the emission plane in order to preserve reflection invariance. Thus, the polarisation of the light can play an important role in identifying the character (symmetry) of the initial states [24].

4. Experimental Spectra

In fig.6(a) we show a photoelectron spectrum from Y(0001) at normal emission [25] in order to illustrate some of the points raised above. In Fig.6(b) we have drawn the region of the calculated (1-e) band structure probed by these measurements. According to a simplified approach we would expect only four features in the photoelectron spectrum for ω = 40 eV corresponding to emission from the 4p(3/2) and 4p(1/2) core levels (~20 eV below E_F) and from the two Δ_1 valence bands. Clearly, the experimental spectrum is much more complicated. Without justifying our reasons here (see ref.[25]) we believe that (i) peak a arises from a Tamm-type surface state, (ii) peak b is due to emission from the states at the top of the uppermost

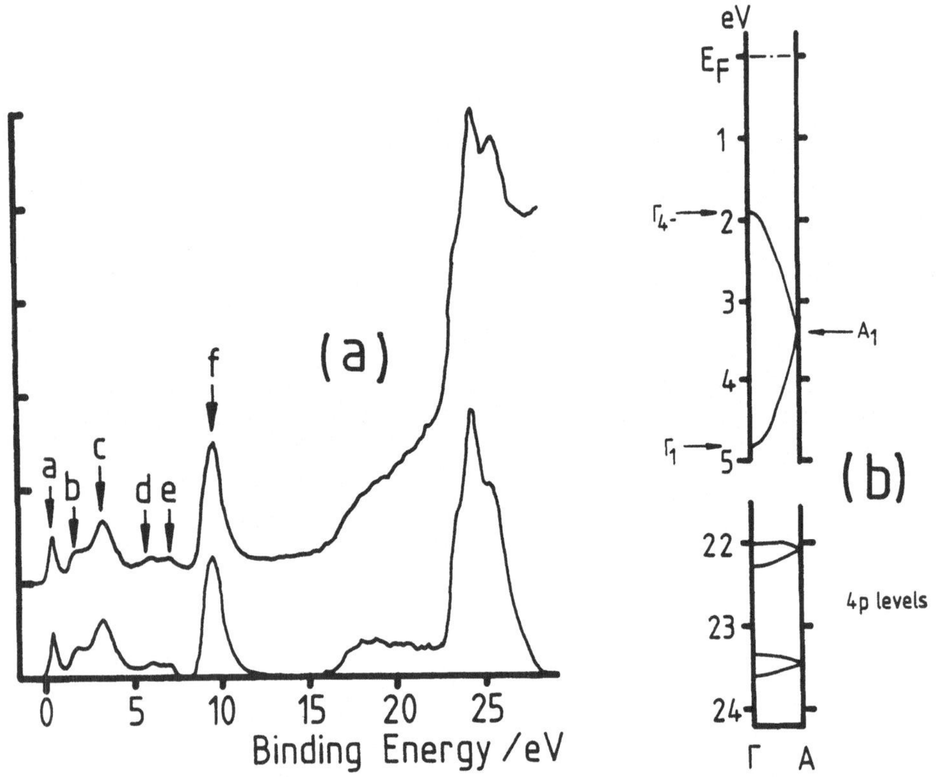

FIGURE 6. Experimental photoemission spectra from Y(0001). Normal emission, photon energy = 40 eV, angle of incidence 30°, also occupied bands in Y along the ΓA direction calculated using the LMTO-ASA method (see ref.[25].

Δ_1 band at the Γ point, (iii) peaks c, e and f are probably due to many-body effects, (iv) peak d may be due to a small amount of impurities on the surface of the crystal, (v) the structure in the 18-20 eV range is due to Auger electrons and (vi) the complex at 24-25 eV is due to emission from the spin-orbit split 4p levels. We believe that the shoulder on the leading edge is due to emission from the 4p(3/2) levels of the bulk atoms, the peak near 24 eV arises from an overlap of the (shifted) 4p(3/2) levels of the surface atoms and the bulk 4p(1/2) levels, and the peak at 25 eV is the surface 4p(1/2) contribution. Although the binding energy of peak b agrees within experimental error with that calculated using the LMTO method, indicating only very small self-energy effects, the 4p levels on the other hand are highly localised and we can anticipate disagreement between the measured and calculated binding energies. There is, in fact, a discrepancy of some 1.5 eV.

The example above shows how intimately the 'surface' and 'bulk' contributions to the photoelectron spectrum are connected. However, in spite of the apparent complexities angle-resolved UV photoelectron spectroscopy is the most powerful probe of the electronic structure in metallic solids. In contrast to Fermi surface investigations, which provide information only at E_F, and optical studies, which provide some sort of integrated picture of the bands, photoemission measurements can yield details of the occupied part of the spectral function, thus providing the severest test to any band structure calculation. To quote John Pendry [22] "...interpretation of photoemission spectra is not simply a matter of chasing a few peaks around in momentum space..." - i.e., peakology - "...The interpretation is much more involved...". As we shall show in section III real and important insights into the electronic structure in metals and alloys can be gained through a proper examination of experimental spectra using first principles calculations of the photocurrent in which the surface, the matrix element effects, the hole and photoelectron lifetimes, surface segregation in alloys, etc., are taken into account.

SECTION II

1. The spectral function in a disordered system

In the previous section we showed that, with certain important provisos, angle-resolved UV photoelectron spectroscopy gives very direct information on the bulk spectral function which describes the distribution E and $\underline{k}$ of the electronic excitations of the system. Due to the electron- electron interaction, the excitation spectrum is essentially a many-body object and is conveniently written as the imaginary part of the Green function. It is to be distinguished from the 1-e spectrum represented by conventional energy band calculations; as we showed in section I the latter must be corrected by a so-called self-energy whose evaluation is, in general, a difficult problem of many-body theory. Often, however, the E and $\underline{k}$ dependence of the self-energy is small and the band picture is not altered too much; this is the case in most of the systems we shall be considering. In this section, therefore, we will describe in more detail the 1-e spectral function for disordered alloys, the supposition being that many-body corrections are much the same as in ordered systems.

In a random alloy the language of band structure theory is inappropriate because of the lack of translational symmetry. We must find, therefore, an alternative way of describing the electron states. To date, KKRCPA method [10,26] has proved the most useful for this purpose, and the spectral functions we shall show in this article are calculated exclusively by this technique. However, we can note some generic features of the spectral function which remain no matter how it may be calculated. In an ordered system the Green function has the following symmetry under translation through a lattice vector $\underline{R}_n$

$$G(\underline{r},\underline{r}',E) = G(\underline{r} + \underline{R}_n,\underline{r}' + \underline{R}_n,E) . \qquad (II.1)$$

The Green function, in its full $\underline{r}$ and $\underline{r}'$ dependence, contains information about the shape of the wavefunctions within each atomic cell as well as the site-to-site phase changes determined by the Bloch $\underline{k}$-vector. The former is less useful and we can integrate it out by defining a Bloch transformed Green function

$$G_B(\underline{k},E) = \Sigma_n \int_{cell} d\underline{r}\ G(\underline{r},\underline{r}' + \underline{R}_n,E) \exp(i\underline{k}.\underline{R}_n) . \qquad (II.2)$$

The Bloch spectral function derived from this quantity

$$A_B(\underline{k},E) = -\frac{1}{\pi} \text{Im } G_B(\underline{k},E)$$

is the quantity referred to in eqn.(I.4), i.e.,

$$A_B(\underline{k},E) = \Sigma_n \delta(E - E_n(\underline{k}))$$

where $\underline{k}$ is in the first Brillouin zone. A disordered system violates (II.1), of course, but a substitutionally disordered system does possess translational symmetry on average, i.e.,

$$\langle G(\underline{r},\underline{r}',E)\rangle = \langle G(\underline{r} + \underline{R}_n,\underline{r}' + \underline{R}_n,E)\rangle$$

where $\langle\ldots\rangle$ denotes an average over all configurations of the alloy. Thus, a configurationally averaged Bloch spectral function

$$\bar{A}_B(\underline{k},E) = -\frac{1}{\pi}\ \mathrm{Im}\ \langle G_B(\underline{k},E)\rangle$$

can be rigorously defined for a substitutional system. This is the natural way to describe the average $\underline{k}$-dependence of the electron states. Note that this is true whether or not the alloy has short range order, atomic size effects, etc; the statement is simply that the average lattice is periodic. Thus, the spectral function has a physical significance extending beyond any particular calculational scheme with its attendant simplifications.

The central difficulty, therefore, is to determine the configurationally averaged Green function from which all one particle observables, such as the spectral function, may be calculated. Multiple scattering theory provides a very convenient framework within which to evaluate the configurational averages [26]. Moreover, the same techniques can be used to construct accurate expressions for theoretical photoemission intensities - and other spectroscopic observables - as we show in section III. The Green function for any system of non-overlapping muffin-tin potentials is given by

$$G(\underline{r},\underline{r}',E) = \sum_{LL'} Z_L^{(i)}(\underline{r}_i,E)\tau_{LL'}^{ij}(E)Z_{L'}^{(j)}(\underline{r}_j',E) - \delta_{ij}\sum_L Z_L^{(i)}(\underline{r}_i,E)J_L^{(i)}(\underline{r}_i',E) \tag{II.4}$$

where $\underline{r} = \underline{r}_i + \underline{R}_i$, $\underline{r}' = \underline{r}_j' + \underline{R}_j$ ($\underline{R}_i$ being the lattice vector of site i), and $r_i < r_j' < S$ (the muffin-tin radius). $Z_L^{(i)}(\underline{r},E)$ and $J_L^{(i)}(\underline{r},E)$ are regular and irregular solutions, with angular momentum $L = (\ell,m)$ and energy E, of the Schrödinger equation for muffin-tin i only. $\tau_{LL'}^{ij}(E)$ is an element of the so-called scattering path matrix - it gives the amplitude with which a spherical wave in channel L' and energy E incident on site j is elastically scattered into a spherical wave in channel L emerging from site i. If it is assumed that the alloy is perfectly random and that all atoms are located precisely on the ideal lattice points of the average periodic lattice, then the KKRCPA formalism allows one to express the configurational average of (II.4) in terms of the τ-matrix of an ordered lattice of effective scatterers, a quantity which can be calculated by the standard techniques of KKR band theory. Using this KKRCPA average Green function in (II.2) and (II.3) we find the following expression for the average Bloch spectral function [26]

$$\bar{A}_B(\underline{k},e) = -\frac{1}{\pi}\ \mathrm{Im}\sum_{LL'}\left\{F_{LL'}^{CC}\ \tau_{LL'}^{C}(\underline{k},E) + (F_{LL'}^{C} - F_{LL'}^{CC})\ \tau_{L'L}^{C,00}(E)\right\} \tag{II.5}$$

where

$$\tau_{LL'}^{C}(\underline{k},E) = \frac{1}{N}\sum_{ij}\ \tau_{LL'}^{C,ij}(E)\ \exp i\underline{k}.(\underline{R}_i - \underline{R}_j)$$

$$F^{c}_{LL'} = \sum_{\alpha,L''} C_\alpha \left\{ \int_{cell} d\underline{r}\; Z^{(\alpha)}_{L}(\underline{r},E)\; Z^{(\alpha)}_{L''}(\underline{r},E) \right\} D^{(\alpha)}_{L''L'}(E)$$

$$F^{cc}_{LL'} = \sum_{\substack{\alpha L_1 \\ \beta L_2}} C_\alpha D^{(\alpha)}_{LL'} \left\{ \int_{cell} d\underline{r}\; Z^{(\alpha)}_{L_1}(\underline{r},E)\; Z^{(\beta)}_{L_2}(\underline{r},E) \right\} C_\beta D^{(\beta)}_{L_2L'}$$

$$D^{(\alpha)}_{LL'}(E) = \left[\left(1 + \tau^{c,oo}(t_\alpha^{-1} - t_c^{-1})\right)^{-1} \right]_{LL'}$$

Here $\tau^{c,ij}_{LL'}$ is the scattering path matrix of the effective ordered, coherent lattice, C_α is the concentration of α-type atoms, $t_{\alpha,L}$ is the t-matrix of an α-type muffin-tin and $t_{c,L}$ is the effective single-site t-matrix occupying each site of the coherent lattice. An example of a calculation on a Cu-40%Ni alloy using this formalism is shown in Fig.7. Leaving for the moment a full discussion of these calculations it is clear that, while disorder scattering certainly broadens out the delta function structure of the pure components, there still remains a set of peaks which are well separated on the scale of their width. It is easy to follow the dispersion of these peaks with $\underline{k}$ and, in this sense, it is possible to speak of energy bands, Fermi surfaces, and so on, in random alloys. In the introduction we indicated the relationship between the spectral function and the energetics of phase stability and

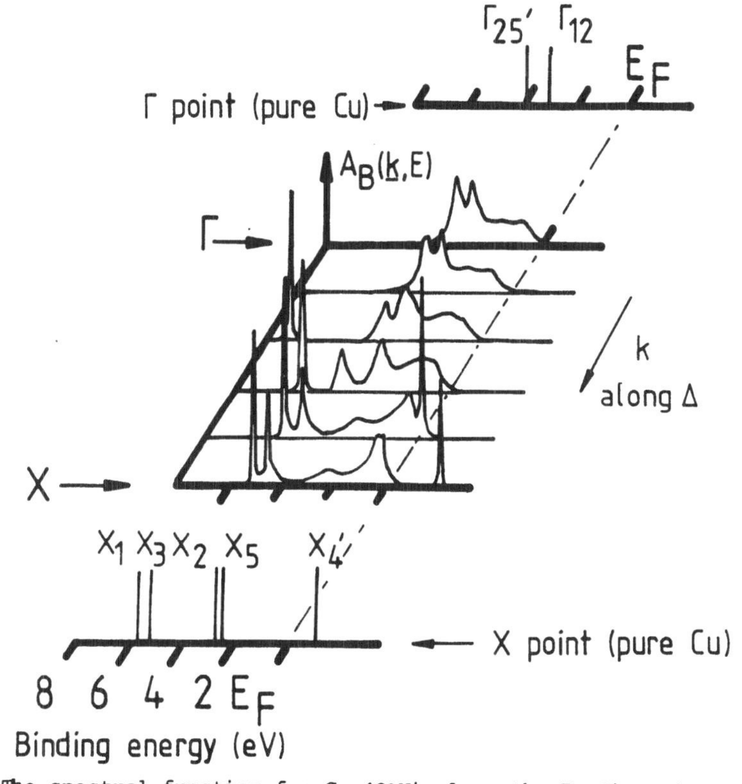

FIGURE 7. The spectral function for Cu-40%Ni along the ΓX direction calculated using the KKRCPA. Also shown are the corresponding features for pure Cu at the Γ and X points.

concentration fluctuations, to the electronic response of alloys to external perturbations, and to all of their electronic properties. An impressive illustration of this is given by the calculation by Stocks and Butler of the resistivity of Ag-Pd alloys [27] in which the precise form of the spectral function played a key role in determining the concentration dependence.

It is a central and typical result of KKRCPA calculations that the average spectral densities in random alloys are as richly structured in E and $\underline{k}$ as Fig.7 would suggest. Because, as we have argued, it contains so much physics, it is vital that our description of the spectral function be of known accuracy. The primary role of our photoemission studies is to test this description with as much quantitative rigour as possible.

SECTION III

1. Photoemission from random alloys

We can now examine in detail what angle-resolved UV photoelectron spectroscopy can tell us about the average spectral density in alloys. In the previous section we noted the connection between the measured photocurrent and the spectral density of occupied states. We have already pointed out the various factors intrinsic to the photoemission process which necessarily complicate the interpretation and seriously limit the accuracy of a naive bulk direct transition model. We begin, therefore, by describing how one can evaluate the configurationally averaged photocurrent from first principles methods which are on the same footing as bulk KKRCPA calculations of the spectral function.

Our model consists of a semi-infinite array of muffin-tin potentials separated from the vacuum by a surface barrier potential. The height of the latter is equal to the sum of the Fermi energy and the work function and its shape is taken simply to be square i.e., independent of coordinates parallel to the surface and a step function in the normal direction, see Fig.8. Experience shows that this is a reasonable and acceptable approximation except, perhaps, for quantitative details of electron states at the surface. (However, as we shall show later, even here the model works rather well [28]). The electronic structure of this model system can be calculated

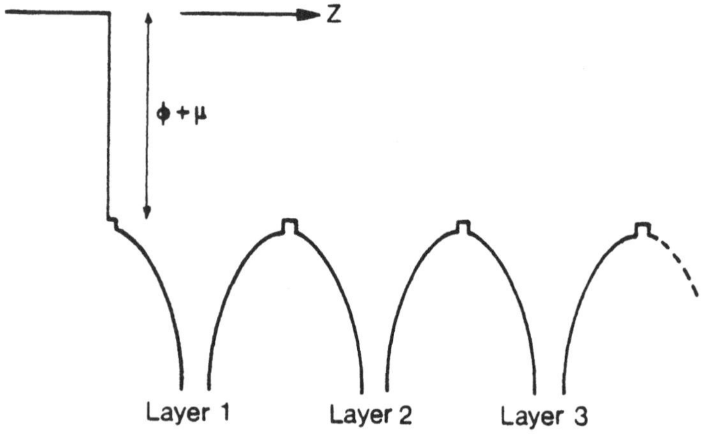

FIGURE 8. The form of the potential used in the photocurrent calculations. $\phi + \mu$ is the work function plus the Fermi energy measured from the muffin-tin zero.

accurately by means of layer multiple-scattering techniques, as can the angle-resolved photocurrent which it emits. The first such calculations were made on ordered metals - using the so-called PEOVER1 code [29]. The approach has been generalised to ordered compounds - the NEWPOOL code [30] - and, by combining KKRCPA ideas with layer multiple-scattering techniques, to substitutionally disordered alloys [31]. In several studies these calculations have proved to give good agreement with experimental photoemission data on ordered and disordered materials; we give some examples in the next subsection.

The photocurrent emitted from a surface is given by an expression of the form [31]

$$I(\underline{k}_{\parallel},E,\omega) = -\frac{1}{\pi} \text{ Im} \sum_{\substack{i,j,L,\\ L_1,L_2,L}} M^{(i)*}_{LL_1}(\underline{k}_{\parallel})\ \tau^{ij}_{L_1L_2}(E)\ M^{(j)}_{L_2L'}(\underline{k}_{\parallel}) \qquad \text{(III.1)}$$

in which the electron-photon matrix $M^{(i)}$ elements link a single scatterer wavefunction on site i with the so-called time-reversed LEED function, a wavefunction formed from a linear combination of Bloch states of energy $E+\omega$ inside the crystal matching smoothly at the surface on to that plane wave which takes the electron into the detector [32]. In a disordered system the configurational averaging procedure associated with the KKRCPA replaces the matrix elements for each type of atom by an effective average matrix element and the τ-matrix by τ^c. (There is another incoherent term in the average photocurrent but it carries no $\underline{k}$-dependence and is frequently small [33]). Both the LEED function and the scattering path matrix τ have to be calculated in the semi-infinite geometry of the (ideal) surface. Thus, although the same key ingredient, the τ-matrix, appears in eqn.(II.5) for the spectral function and eqn.(III.1) for the angle-resolved photocurrent, the two expressions are not as closely related as they seem. If we were to ignore the surface and replace the LEED function by a plane wave, the photocurrent would resemble the spectral function multiplied by simple electron-photon matrix elements (see eqns.(I.3) and (I.4). Of course, this would be quite inadmissable in general but it does suggest that those features of photoemission spectra which are least sensitive to the surface should reflect the behaviour of the many 'band mapping' measurements on ordered systems, see for example ref.[16], and by our own work on disordered Cu-Ni alloys [33-35].

2. Comparisons with experiment

At this point it is appropriate to illustrate the quality of the agreement between experiment and theory that can be achieved. We shall take three examples from our recent researches; (a) a study of the $\underline{k}$-dependence of the disorder-broadening in a Cu-Ni alloy, (b) the photon energy dependence of the emission from bulk states in Y and the existence of a surface 4p-core level shift, and (c) investigations of Shockley-type surface states in a number of noble metal-based random alloys.

(a) The calculated spectral function for Cu-40%Ni along the ΓX (Δ) direction is shown in Fig.7. The Ni-related 3d-states lie about 0.5 eV below E_F. At the Γ point they appear as a broad feature because of considerable disorder-broadening. However, the feature sharpens and moves slightly towards E_F as the X point is approached, where it is possible to resolve the Ni-related X_2 and X_5 states. According to our discussion in section I, such detail should be observable in an angle-resolved UV photoemission experiment. In a direct transition picture, varying the photon energy at normal

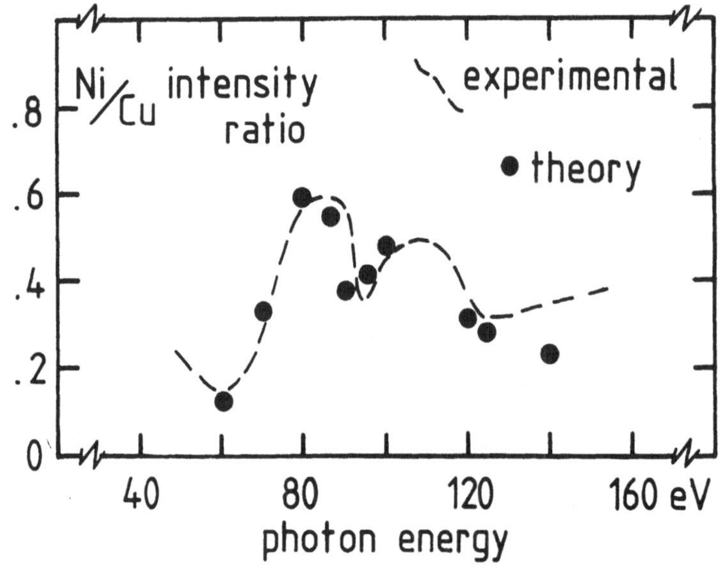

FIGURE 9. Comparison between the measured and calculated ratios of the intensities of the Ni and Cu features in photoemission spectra from the (100) surface of a Cu-40%Ni alloy as a function of photon energy.

emission from a (100) surface of a single crystal is equivalent to probing the spectral function along the Δ direction (and back) at a rate determined by the dispersion of the initial and final states. In such an experiment [35] the changes in the structure of the Ni-related states can be seen in photoelectron spectra through variations of the ratio of the intensities of the Ni and Cu peaks (at ~0.5 eV and ~2.5 eV binding energies, respectively, with photon energy, see fig.9. (The range 20 eV < ω < 140 eV corresponds to scanning in $\underline{k}$ from near X → Γ → X → Γ, etc). We find excellent agreement between the measurements and photocurrent calculations, not only in the positions of the maxima and minima but in amplitude also. (In the calculation, an alloy of uniform composition was assumed with a surface layer of pure Cu). The minima at 60 eV and 125 eV correspond to emission from near the Γ point, the maximum near 90 eV to emission from the X point. The dip in the experimental data close to 90 eV, which is reproduced in the calculations, is due to a final state band gap at the X point some 90 eV above E_F. In addition, the measurements confirm the small change in binding energy of the Ni-related feature referred to above. Thus, we conclude that the KKRCPA describes the $\underline{k}$-dependent disorder-broadening very well, a crucial effect in any realistic alloy theory. This experiment also provides a dramatic demonstration of the combined power of angle-resolved photoemission measurements and proper theoretical analysis to test a key effect in electronic structure calculations for alloys.

In Fig.10 we show photoelectron spectra with different polarisations. States with Δ_1 character couple strongly to p-polarisation and we can see this effect on a normal emission spectrum with ω = 96 eV. When the photon incidence angle is changed from 20° to 60° the p-component increases and a peak appears near 5 eV. We associate this feature with emission from the Cu-related X_1-states which can be seen in the spectral function, Fig.7.

(b) In section I we showed that photoelectron spectra from Y(0001) were much more complicated than expected from a single particle excitation picture.

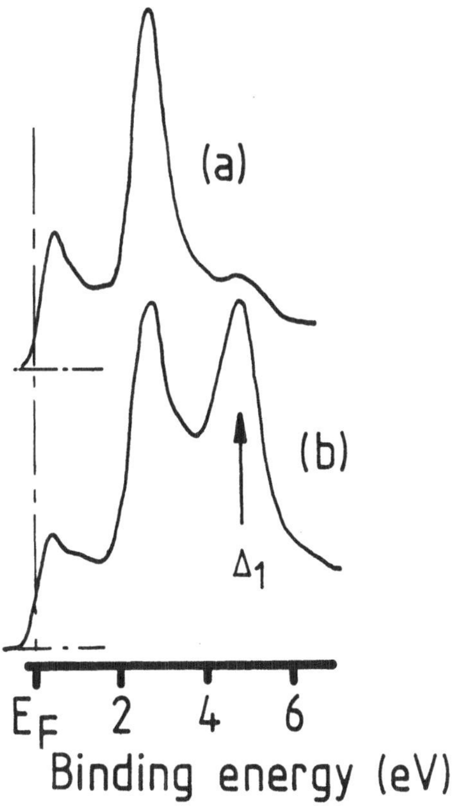

FIGURE 10. Showing the effect of polarisation on the intensities of photoemission features from the (100) surface of a Cu-40%Ni alloy, ω = 96 eV, normal emission. (a) Angle of incidence is 20°. (b) Angle of incidence is 60°. There is an increase in the p-component of the radiation in the latter configuration.

Nevertheless, certain features can be reproduced well and explained in terms of the 1-e bands [25]. In particular, the photon energy dependence of the intensity of peak b in Fig.6(a) agrees well with that calculated using the NEWPOOL photocurrent code and potentials from a SCF-LMTO-ASA calculation, as shown in Fig.11. Peak b, which shows very little dispersion with photon energy, is produced by surface photoemission (see section I.3(b)) from the high density of states at the top of the uppermost occupied band along ΓA. The increased intensity near ω = 30 eV results from quasi-direct transitions to unoccupied Bloch final states near the Γ point. The calculated binding energy of the Γ_{4-} state at the top of the band is 1.9 eV; experimentally it occurs at 1.7 ± 0.2 eV. The quality of the agreement, with no adjustable parmeters, indicates that the (1-e) LMTO band structure calculation is quite reasonable, even well above E_F.

Another feature in the Y spectrum which can be reproduced in the calculations is the ~0.8 eV shift to increased binding energy of the 4p core level due to the modified surface potential. We simulate this change by rigidly shifting the SCF-LMTO-ASA potential in the 1st layer by -1 eV and the results we achieve are shown in Fig. 12. (Note: the photocurrent code does not take into account spin-orbit splitting and so only a single 4p core level

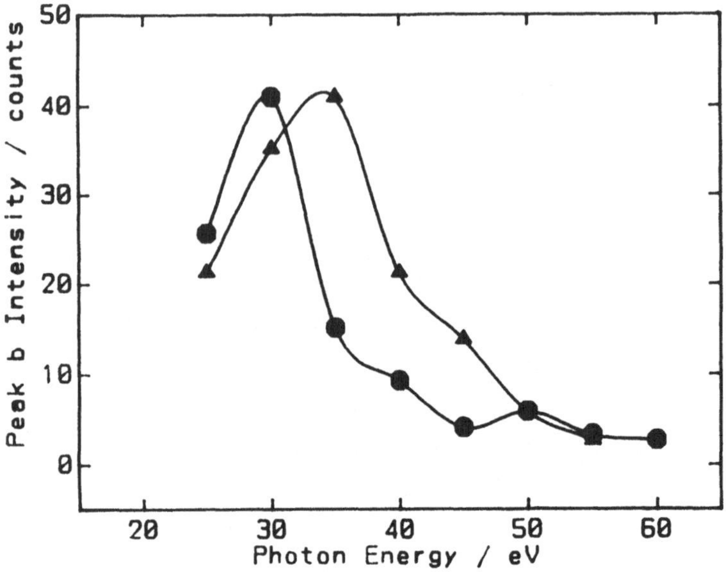

FIGURE 11. A comparison of the experimental (●) with the calculated (▲) intensity variation of peak b shown in Fig.6(a). The experimental data are flux-normalised with the background subtracted.

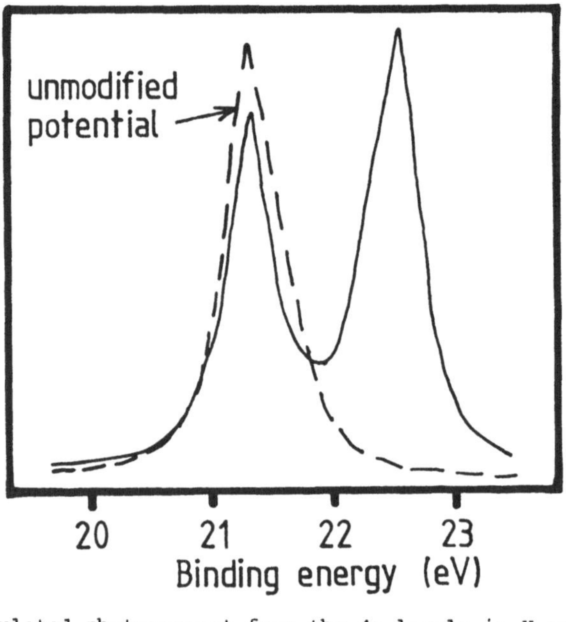

FIGURE 12. Calculated photocurrent from the 4p levels in Y using two different potential functions.

peak appears in the unmodified spectrum). The relative intensities of the calculated 'surface' and 'bulk' peaks depends on the choice of the inverse lifetime of the final state; we used a value of -3.8 eV corresponding to a mean free path (or escape depth) of ~0.4 nm which is quite reasonable.

(c) The Shockley-type surface states on the (111) surfaces of noble metal-based random alloys provide a most interesting topic for study. We have shown, for instance, that they can provide information about the Fermi surface which may not be accessible with other experimental techniques [36-38]. The Shockley surface states we are concerned with lie in the s-p hybridisation gap at the L point of the fcc alloys which spans E_F. We showed that in Cu-Ni alloys emission from an occupied surface state - derived from the Cu-related '$L_{2'}$' state - appears just below E_F in photoelectron spectra from the (111) surface at normal emission and that changes in binding energy are related directly to changes in the E_F - '$L_{2'}$' gap. Furthermore, since the surface state is bounded in k-space by the uppermost bulk band around the L point we showed that its occupied extent in $k_\parallel$-space is a measure of the neck radius on the Fermi surface. (The dispersion is measured by changing the angle of emission θ; the value of $k_\parallel$ can be calculated from eqn.(I.1)). In the case of Cu-Ni alloys the change in neck radius with composition deduced in this way agree with that calculated using KKRCPA, as shown in Fig.13.

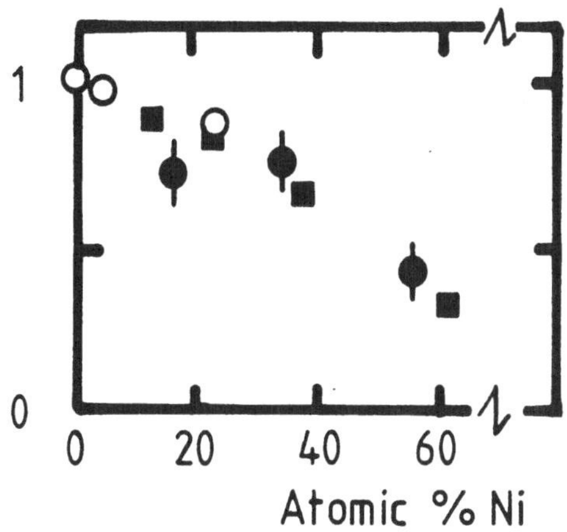

FIGURE 13. Normalised values of the $k_\parallel$-extent of the occupied Shockley-type surface states and the neck radii in Cu-Ni alloys as a function of composition. (o) $k_\parallel$ extents, ([]) neck radii calculated by B.E.A. Gordon, W.M. Temmerman and B.L. Gyorffy, J. Phys. F: Metal Phys. 11, 821 (1981) and (•) neck radii measured by M. Hasegawa, T. Suzuki and M. Mirabayashi, J. Phys. Soc. Japan 37, 85 (1974).

The KKRCPA calculations for Ag-Mn alloys suggested that as Mn is added to Ag, the Ag-related states at the top of the sp (Λ_1) band near E_F are lowered in energy through interaction with the exchange-split Mn states just above E_F leading to a possible increase in the neck radius [39]. Photoemission measurements of the binding energies of the Shockley surface states on the (111) surface showed that there is an increase in the E_F - '$L_{2'}$' gap. Furthermore, the measurements of their occupied extents in $k_\parallel$-space indicated that the neck radius increases with Mn concentration [36]. In fact, the observation of this strictly non-rigid-band-like behaviour was the first confirmation that an exchange-split Mn band occurs just above E_F.

In Figs.14(a) and (b) we compare the experimental and theoretical photoelectron spectra from the (111) surface of a Cu-Pd alloy [28]; the Shockley surface state is labelled S. No particular significance should be attached to the differences in either the intensity or binding energy of S. (The in-

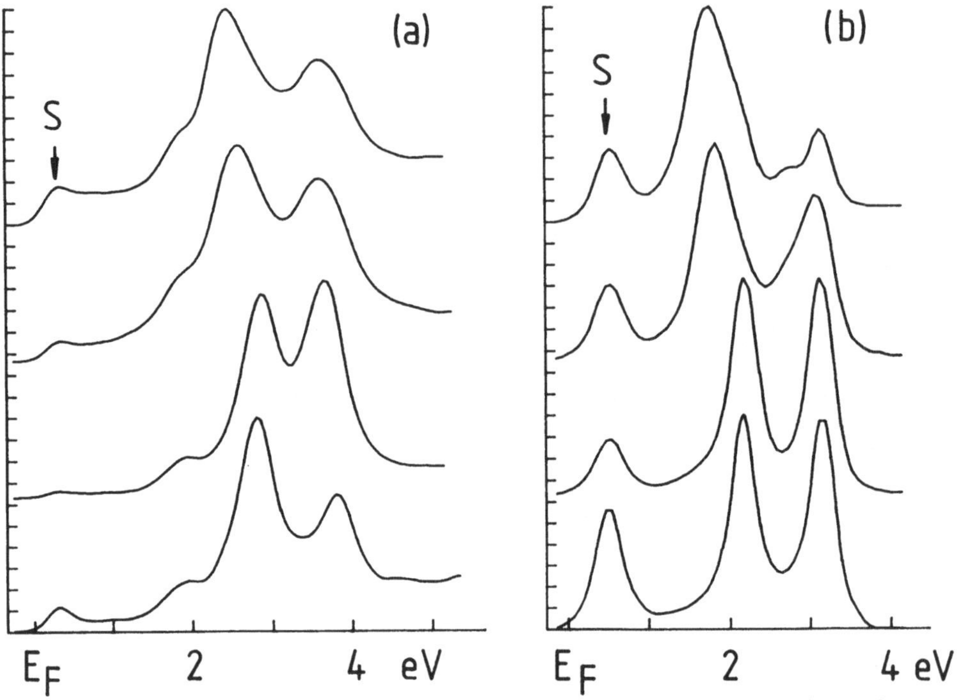

FIGURE 14. A comparison of the (a) experimental and (b) theoretical photoemission spectra from the (111) surface of a Cu-6.8% Pd alloy at various photon energies: 20 eV, 35 eV, 55 eV and 65 eV reading upwards. The angle of incidence = 45°, normal emission. In (b) the Fermi energy is positioned 0.620 Ryd above the muffin-tin zero given in ref.[12]. S is a Shockley-type surface state.

tensity in experimental spectra depends on parameters like surface crystallographic order, roughness, cleanliness, etc which may be difficult to quantify. In addition, the fact that the calculated photocurrent is enhanced is not surprising in view of the idealised model of the surface barrier with ∇V appearing in the matrix elements). In the calculation the energy of S is sensitively dependent on the position of the surface barrier, due to the matching conditions. (In fact, this is a test for Shockley surface states). We chose arbitrarily to place the barrier in contact with the outer layer of muffin-tins; moving it closer to the surface layer of atoms reduces the binding energy.

The photon energy dependence of the intensity of S is reproduced by the calculations as shown in Fig.15. This dependence arises through changes in the overlap between the initial and final state wavefunctions. In the experiment we are probing the surface state wavefunction and since the general shape of the curve agrees with the calculation - in particular the minimum near 40 eV and the enhancement near 65 eV - we believe we have established a good description of the surface state wavefunction.

The potential function used in this calculation was obtained from a bulk SCF-KKRCPA calculation. This may account for the discrepancies in the calculated peak positions and the lack of a feature split-off from the top of the Cu-related d-band complex near 2 eV which is visible in experimental spectra. For more quantitative comparisons input potentials to the calcula-

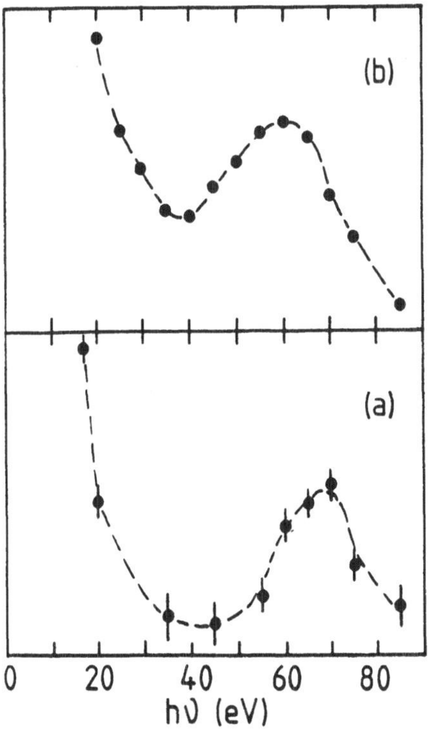

FIGURE 15. (a) The experimentally determined and (b) theoretically calculated variation of the photocurrent from the Shockley-type surface state on the (111) surface of a Cu-6.8%Pd alloy.

tions should be fully self-consistent with respect to the surface. However, we believe the results are reliable in this case because:

- The surface state is derived from the very sharp, bulk Cu- related $L_{2'}$ state which has a considerable coherence length. It will be only slightly perturbed by the surface (section I.3(b).
- The surface state wavefunction has purely p-character at the L point. The scattering in the p-channel is very small and so the surface state is free electron-like and not particularly sensitive to the surface potential. (Note: this would not be true of Tamm surface states).

What we have tried to demonstrate in this section is that photoemission measurements, when combined with a proper first principles analysis, are an appropriate probe of the surface and bulk electronic structure in metals and alloys. What this implies is that, in principle, we have the means to examine electronic structures in $\underline{k}$-point by $\underline{k}$-point detail. Since measurements can be made over a range of temperature we have the opportunity to investigate many of the crucial and complex issues in metal and alloy physics directly in electronic terms. One such topic is the order-disorder transformation which we now consider.

SECTION IV

In the preceeding sections our discussion has focussed essentially on the intimate connection between photoelectron spectra and the electronic structure. However, measurements of this type are of no immediate interest to the materials scientist who, for instance, is attempting to design alloys.

In this section we will deal with a specific example of our joint experimental and theoretical approach to understand a well-known metallurgical property exhibited by certain alloys with possible important technical application, namely ordering [40]. We will deal with examples from the two most common order-disorder systems, (a) CsCℓ-bcc and (b) L12-fcc.

Order-disorder phenomena come about because of an imbalance between the interatomic forces. If these forces are modelled in some simple way, for example as a set of pair potentials, then statistical mechanics of arbitrary complexity can be practised upon the resulting Hamiltonian yielding a variety of results from the phase diagram to critical phenomena. Alternatively, since interatomic forces are determined by the electrons the energetics of ordering can be expressed in terms of the total electronic energies of the ordered and disordered phases. This is now a practicable possibility through the application of density functional methods. However, ordering energies are small (~ 0.1 eV) and an accurate calculation of the total energy E(S) as a function of the long range order parameter S is a very stiff test of any theory of disorder. As a result little work along these lines has been attempted.

Both approaches have drawbacks. The first lacks predictive power simply because it is not a first principles approach but a phenomenological description. Furthermore, models based on pair potentials, in particular, have no fundamental justification in terms of the quantum mechanics of the electrons (except in weak scattering systems whose energy can be written accurately as a second order perturbation expression in pseudopotentials) and are wrong in general. The second approach involving total electronic energy calculations, has a rather "brute force" character, giving little idea of the underlying mechanisms at work (such as the role of the Fermi surface, etc).

Now the primary requirement for any electronic theory of phase stability is that it must describe accurately the electronic structure of each phase. This immediately suggests an approach directed, in the ordering problem, to understanding the spectral function in the ordered and disordered states. It is clear that angle-resolved photoelectron spectroscopy is the obvious tool for this, given that a reliable means of analysing the spectra for both phases is available. In the next two subsections we shall demonstrate how photoelectron spectroscopy can be used to follow the electronic spectral density of alloys across order-disorder transitions.

(i) bcc-CsCℓ systems

50/50 alloys which transform from a high temperature substitutionally disordered bcc phase to a low temperature ordered CsCℓ (or B2) phase form, perhaps, the simplest class of ordering systems. The phase transition is second order - the long range order parameter S, measuring the difference in concentration on the two sublattices, varies continuously from 1 at T=0 to 0 at the transition temperature T_0 - and mean field theory describes the phase diagram reasonably well. It is natural, therefore, to begin a study of electrons in partially ordered systems by considering this canonical case and we will discuss two such alloys. It should be noted, however, that their class is not large; many alloys which crystallise with the ordered CsCℓ-structure are so stable that they melt before disordering (e.g., TiFe).

(a) CuZn

β-brass is the most widely known ordering alloy, undergoing a bcc-CsCℓ transition at 740 K. From the theoretical point of view it is relatively simple, being non-magnetic and, to a good approximation, non-relativistic. (Previous calculations have concentrated on Cu-rich alloys and achieved

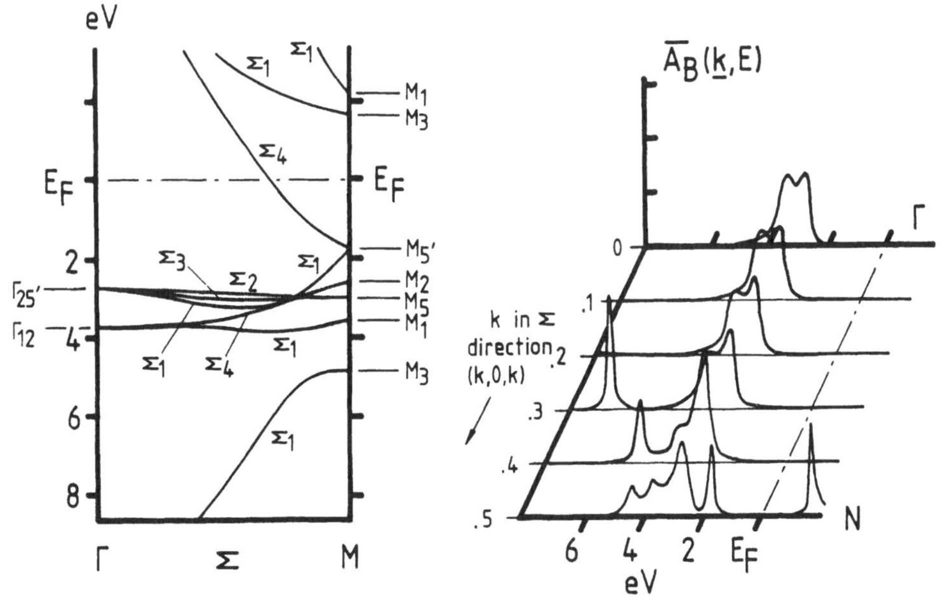

FIGURE 16. (a) The calculated band structure for ordered CuZn along the ΓM (Σ) direction. (b) The calculated spectral function for a disordered Cu-50%Zn alloy along the ΓN (Σ) direction.

excellent results for the concentration dependence of the low temperature specific heat coefficient, electrical resistivity, thermopower [41] and lattice constant [42]). We have carried out KKR calculations of the ordered CsCℓ phase and KKRCPA calculations of the disordered phase of the 50/50 alloy together with the corresponding photocurrents, calculated as described in section III. For our present purposes, the most important results of the study can be seen in Figs.16(a) and (b) showing the electronic structure of each phase in the Σ direction. The system is an extreme split-band alloy; in fact, the Zn d-bands lie at negative energies and so the figure shows only the Cu d-states together with states of s-p symmetry.

The topology of the dispersion of the Cu d-levels in the disordered phase is typical of a bcc system - in fact, it resembles 'bcc Cu' with a reduced overall band width. In contrast, the Cu d-states in the ordered phase are almost identical to those of 'simple cubic Cu' with the lattice parameter of CuZn, since at the Cu d-band energies the d-scattering on the Zn sites is very small. There are clear and obvious differences, then, in the electronic spectral density of the ordered and disordered phases. The question which arises is 'will such differences be observable in an experiment?'.

The question is answered in Figs.17(a) and (b). They show the calculated photocurrent for normal emission from the (110) faces and, in direct transition language, correspond to scanning the initial states along the Σ direction. In all calculations the polar angle of incidence was 45° and two polarisations were investigated, s-polarisation, in which the vector potential lies in the plane of the surface and p-polarisation, its orthogonal partner, in which there are components of the vector potential normal and

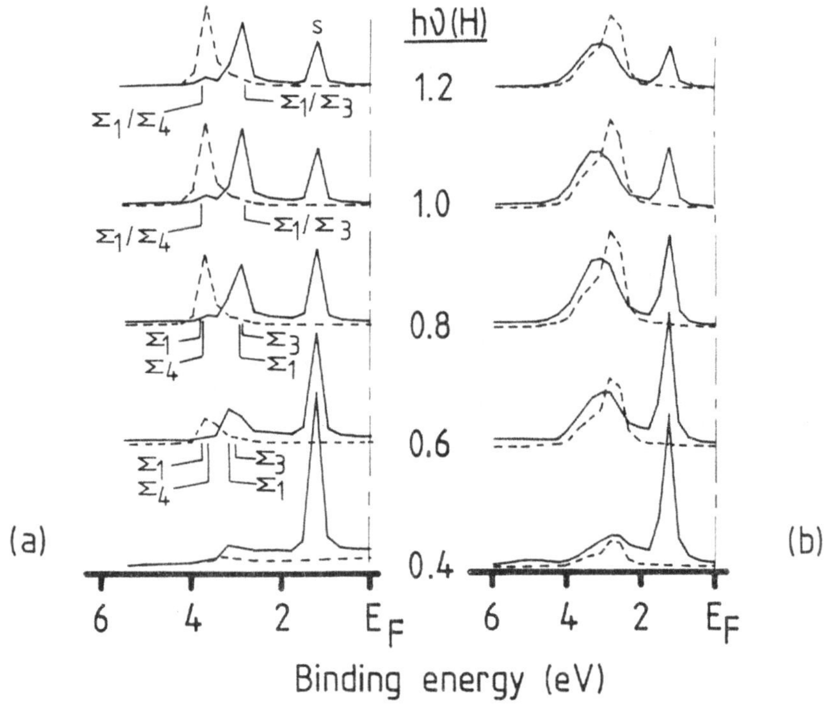

FIGURE 17. Calculated photocurrents emitted normally from the (110) surfaces of (a) ordered and (b) disordered β-brass, for different photon energies. The polar angle of incidence is 45° and the azimuth is zero. The solid curves are for p-polarisation, the dashed curves are for s-polarisation. S is a surface state.

parallel to the surface plane. The calculation of the occupied states was carried out using an imaginary part to the energy of -0.8 eV whilst for the final states the imaginary part was -4 eV; these values are characteristic of the broadening effect seen in photoemission from transition metals. Again the Zn d-bands lie below our muffin-tin zero potential and are not shown in the photocurrent calculations.

The spectra for the ordered phase, shown in Fig.17(a) can be readily understood by reference to the band structure of Fig.16(a), noting that direct transitions would occur near Γ for ω=1.2H (~33 eV) and at k~2/3 of the ΓN distance for ω=0.4H (~11 eV), assuming a free-electron final state band. For the particular photon angles we have used initial states of Σ_4 symmetry can be excited in s-polarisation [24], while p-polarisation excites Σ_1 and Σ_3 type initial states. The positions of these states predicted by the direct transition model are shown in the figure, and it can be seen that this somewhat crude interpretation is actually helpful in the present case. the spectra for s-polarisation are particularly simple and follow the dispersion of the Σ_4 band quite closely. The most prominent feature of the spectra for p-polarisation is the peak just below E_F for all photon energies. It does not originate from the bulk band structure; it arises from a surface state lying in the Σ_1 gap where a free-electron Fermi surface would push through the Brillouin zone face. The remaining features in the p-polarised spectra correspond closely to Σ_1 and Σ_3 bulk bands. The presence of the lower parts of the s-p band are faintly visible for the lower photon energies, but they give rise to extremely broad peaks compared with the d-bands. This is a generic feature of photoemission spectra and is always observed in experiments [43].

The spectra for the disordered phase show a strong dependence on polarisation also. Assuming, as the theory does, that the photoemission occurs from a sufficiently large region of the specimen for self-averaging to occur, the spectra must reflect the electronic properties of the average lattice. Since a substitutional alloy possesses translational symmetry on average - a property which allows a $\underline{k}$-space description of the electron states and, therefore, underlies all use of the average Bloch spectral function - it is legitimate to classify the average spectral density using the same symmetries as the corresponding ordered system. In this sense we can apply the above polarisation-dependent selection rules to the following discussion of the results for the disordered phase. Firstly, we see that the surface state is still present. This is natural - it is rather free-electron-like and should be little affected by the state of order (see also section III). However, this is not true for the Cu d-band features. While the disorder-broadening is apparent, the most striking effect is the polarisation dependence, which is opposite to that of the ordered case. It seems that the Σ_4 type states, for example, are shifted by about 1 eV to higher energies in the disordered phase and show signs of being split into two components. This effect, which would be observable in an experiment, can be related simply to the change in symmetry of the system. At the Γ point, see the ω=1.2H spectrum, the Σ_4 states merge with the e_g or Γ_{12} states. In a bcc (or fcc) system the e_g orbitals, being directed along the cube axes, point towards next nearest neighburs which means that the e_g levels lie higher in energy than the t_{2g} levels. Such behaviour can be observed in the Bloch spectral function for the disordered phase, Fig.16(b). However, in the ordered phase the Cu atoms occupy a simple cubic sublattice and so the e_g orbitals point towards nearest neighbour Cu atoms; the Zn d-orbitals lie at very much lower energies and have little influence on the Cu d-bands. Hence, the e_g states are lower in energy than the t_{2g} states as can be seen in the band structure, Fig.16(a), and this is the origin of the shift in the position of the Σ_4 related photocurrent peak. In fact, the Σ_4 states in the ordered phase behave rather like the Σ_3 states in the disordered phase and vice versa. It turns out that one can pick up emission from the Σ_3 initial states in s-polarisation if the azimuthal incidence angle of the photons is rotated by 90° [24] and this is illustrated in Figs.18(a) and (b). The predicted positions of direct transitions are marked on the spectra for the ordered phase and it is clear that the peaks follow the dispersion of the two bulk bands rather closely. The positions of the two peaks are reversed in the disordered phase, as the above argument suggests. Clearly, therefore, there are rather strong ordering effects on photoemission spectra which are related directly, in this case, to changes in the bulk spectral density. Note that the length of the Σ line is the same in both phases; no folding back occurs in this direction.

(b) AgZn

Although CuZn is the most obvious alloy for experimental study, angle-resolved UV photoemission measurements at temperatures near T_0 are impossible due to the high vapour pressure of Zn. AgZn (containing a few percent of Au) also shows a bcc-CsCℓ transition but at a lower temperature (T_0~600 K), see Fig.19. (The Au substitutes randomly on the Ag sites and stabilises the

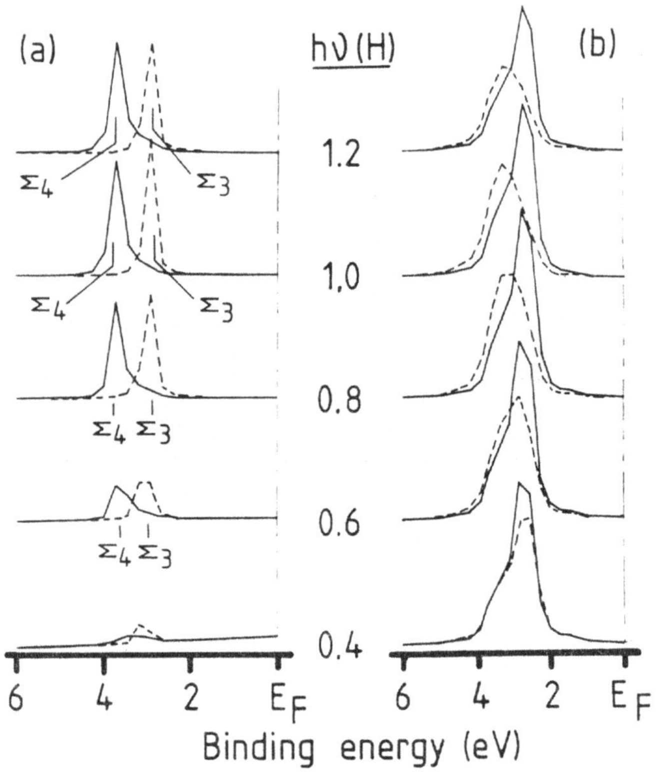

FIGURE 18. Calculated photocurrents emitted normally from the (110) surface of (a) ordered and (b) disordered β-brass, for different photon energies with an angle of incidence = 45° and s-polarisation. The solid curves correspond to an azimuth of zero, the dashed curves correspond to an aximuth of 90°.

CsCℓ structure. Without the Au substitution a complex hexagonal phase results). We felt, therefore, that it was possible to contemplate photoemission studies of this closely related, isoelectronic alloy over a range of temperatures where the long range order parameter varies considerably from unity. However, despite the lower transition temperature, we were still not able to carry out measurements much above room temperature. We were forced, therefore, to concentrate on measurements of the ordered phase. In addition, since we could not produce surfaces exhibiting sharp LEED patterns - indicating a small coherence length of the (ordered) domains on the surface - we decided to use x-ray photoelectron spectroscopy and to compare these with calculated spectra in order to study the electronic structure of <u>ordered</u> AgZn. (XPS spectra do not provide the detail contained in angle-resolved measurements, but they are closely related to the density of states [12], viz.

$$I_{XPS}(E) \propto \Sigma_{A,\ell} \left|M_{\ell\ell'}(E,\omega)\right|^2 n_{\ell,A}(E) \qquad \text{(IV.1)}$$

where $n_{\ell,A}(E)$ is the local density of states for angular momentum ℓ and atom type A, and $M_{\ell\ell'}$ is an electron-photon matrix element).

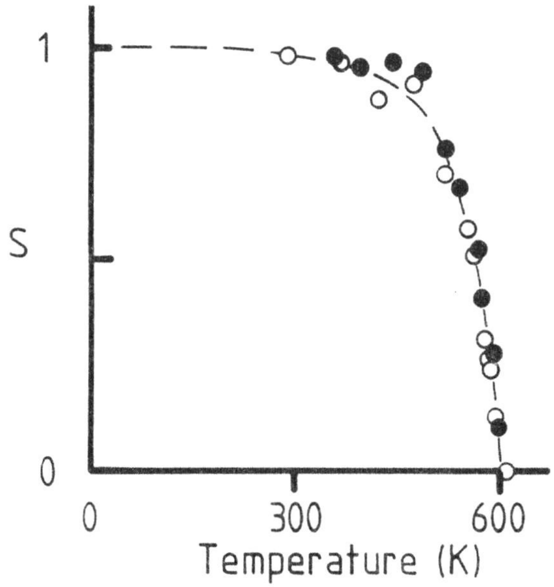

FIGURE 19. The long range order parameter for AgZn(+3% Au) measured by J.I. Langford and R.G. Jordan (unpublished).

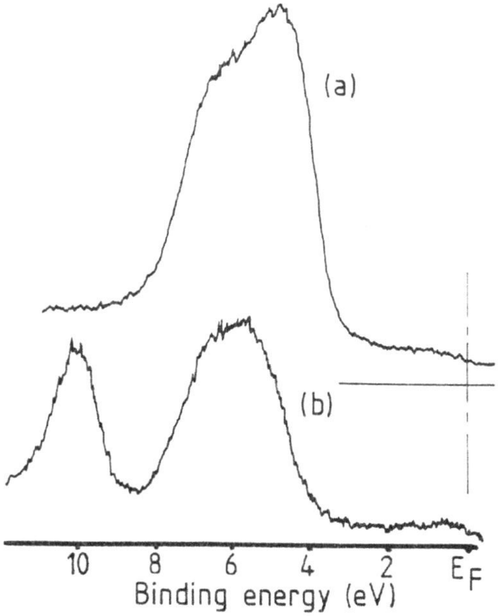

FIGURE 20. Experimental room temperature XPS valence band data for (a) pure Ag and (b) ordered AgZn(+3% Au), Aℓ-Kα radiation.

In fig.20 we show experimental XPS spectra (with hν=1486.6 eV) from the valence band region in AgZn(+3% Au) obtained at room temperature [44]. In comparison with the XPS spectra from pure Ag and Zn we can establish the

following observations:
- The shape of the Ag-related d-band peak in the alloy is substantially different from that for pure Ag. Its binding energy is increased by ~1 eV compared with pure Ag and there is a ~0.7 eV reduction in the FWHM.
- The Zn-related d-bands in the alloy occur at the same energy as in the pure metal. However, the FWHM is ~0.3 eV smaller in the alloy compared with pure Zn.

Furthermore, we find also that, within the detection limit of the analyser, the Ag 3d(5/2) and 3d(3/2), the Zn 2p(3/2), 2p(1/2), 3p(3/2), 3p(1/2) and 3s core levels for the alloy are not shifted from their pure metal values. However, we find a reproducible shift of several tenths of eV's to increased binding energy in the case of the Au 4f(7/2) core level.

In order to understand the ordered phase we have made self-consistent field LMTO-ASA calculations for AgZn with the CsCℓ-structure as well as pure Ag (fcc) and Zn (hcp) for comparison. The densities of states, see Fig.21, show the expected split-band behaviour with the Zn d-bands about 3 eV below the Ag d-bands. Thus, AgZn resembles CuZn but with a smaller separation between the d-bands. There is essentially no shift of the d-bands relative to the pure metals indicating little charge transfer. However, the d-band

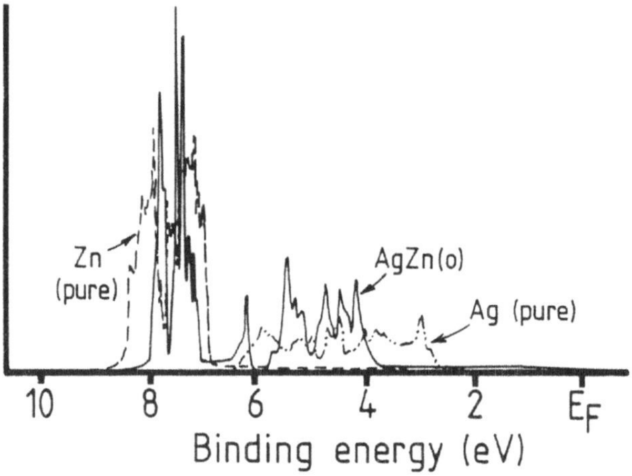

FIGURE 21. Densities of states of pure Ag, pure Zn and ordered AgZn calculated using the SCF-LMTO-ASA.

widths are substantially different, because in split-band alloys the widths of the d-bands are dominated by overlap between d-orbitals of the same species; in the CsCℓ-structure the nearest neigbours are unlike atoms and so the relevant overlaps are small. (An interesting feature of Fig.21 is the peak 6 eV below E_F which is caused by strong hybridisation between d-states on Ag sites and s-states on Zn sites. In the language of scattering theory it is a sympton of the s-wave scattering which is much stronger on Zn atoms than Ag atoms. It does not appear in CuZn because the Cu d-states are at higher energies).

We have calculated the XPS spectra according to eqn.(IV.1) (folded with a Lorentzian broadening function of width 0.02H) for ordered AgZn, pure Ag and Zn as shown in Fig.22. The calculated spectra clearly reflect the density of states features described in the previous paragraph. The influence of the electron-photon matrix elements can be seen in the relative heights of the Ag- and Zn-related peaks (c.f. the density of states in Fig.21), and in the skewed shape of the Ag-related peak. All the experimental observations

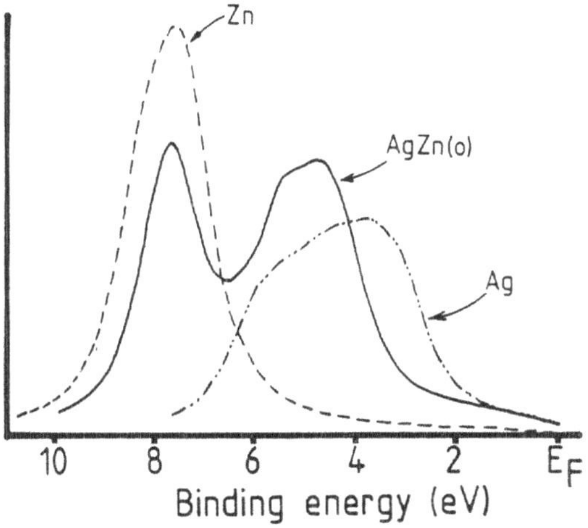

FIGURE 22. Calculated XPS valence band spectra for pure Ag, pure Zn and ordered AgZn, Aℓ-Kα radiation.

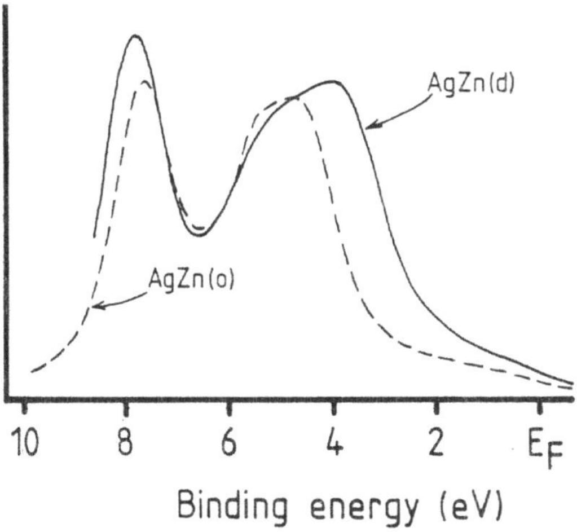

FIGURE 23. Calculated XPS valence band spectra for ordered and disordered AgZn, Aℓ-Kα radiation.

noted previousy are reproduced in these spectra; the lack of any measurable shifts of the Ag and Zn core levels indicates little charge transfer. We note, however, that the Zn d-band appears somewhat lower in energy (~2.5 eV) compared with the calculated LMTO value due to self-energy effects (see section I). We conclude that our picture of the electronic structure in ordered AgZn, at least, is satisfactory.

Although we are not able to probe the disordered spectral function experimentally we have, nevertheless, calculated the density of states using the KKRCPA. We used the same self-consistent potential functions derived in the LMTO calculations. As expected, the split-band behaviour persists but the band widths are greater since in the random bcc structure an atom has a 50% chance of finding a similar atom as a nearest neighbour. The corresponding calculated XPS spectrum is shown in fig.23 and is rather different from that of the ordered phase. The apparent shift of the d-band is associated with a strong narrowing in the ordered case; the centres of the Ag d-bands are essentially at the same energies in both phases, as are those of Zn.

(ii) fcc-L12 system

As we indicated above because of the practical difficulties we have made only limited experimental progress on studies of the bcc-CsCℓ systems. However, we have carried out a photoemission study of Cu_3Au which undergoes a first order L12-fcc (order-disorder) transition at T_0~660 K. It is well suited to investigation by photoelectron spectroscopy as the temperature required for a dynamic study is easily accessible in an experiment and as we shall show, it is possible, therefore, to make direct observations of the changes in the electronic structure across the transition. Certain L12 compounds exhibit high strength, ductility and fabricability which change abruptly at the ordering temperature [40]. Thus, apart from the intrinsic scientific interest in understanding the electronic driving forces responsible for order-disorder there is the potential for designing ordered intermetallic alloys suitable for high temperature structural uses, for example.

In a recent study of angle-resolved photoemission from the (100) surface of Cu_3Au [45] our aim was to locate regions in E and $\underline{k}$ where the effects of the order-disorder transition are most evident in the spectra. Since we were restricted to a fixed photon energy (ω=21.2 eV) in these initial experiments, we varied the emission angle θ in order to probe different initial states (see eqn.(I.1)). We found that the most striking changes occurred in the range $20° \leq \theta \leq 35°$. Several spectral features varied with temperature in such a way that they must be associated with ordering. For example, in ϕFg.24(a) we show spectra with θ=25° taken over a range of temperature; in Fig.24(b) we show a set from pure Cu under the same conditions. In addition to the (split) predominantly Cu d-band features at 2-3 eV we identify several sharp peaks A, B, C and D which are associated with alloying. D is essentially independent of temperature but peaks A, B and C are present only in the ordered phase.

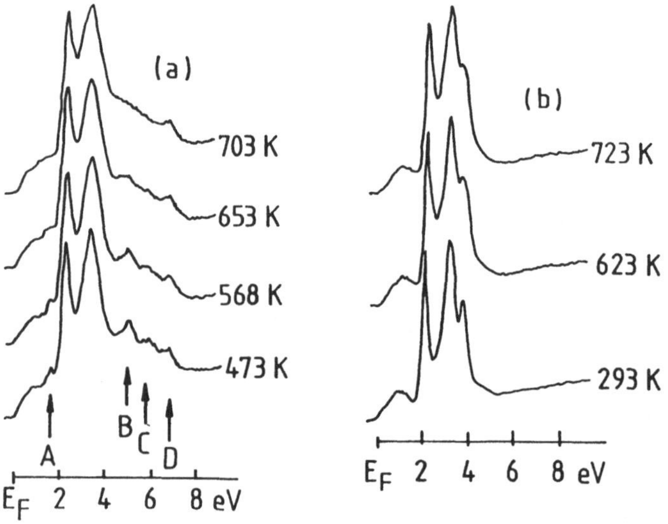

FIGURE 24. Angle-resolved photoemission spectra from the (100) surfaces of (a) Cu_3Au and (b) Cu at various temperatures. The emission angle is 25° in the ΓXLK plane, HeI radiation (21.2 eV) incident at 20°.

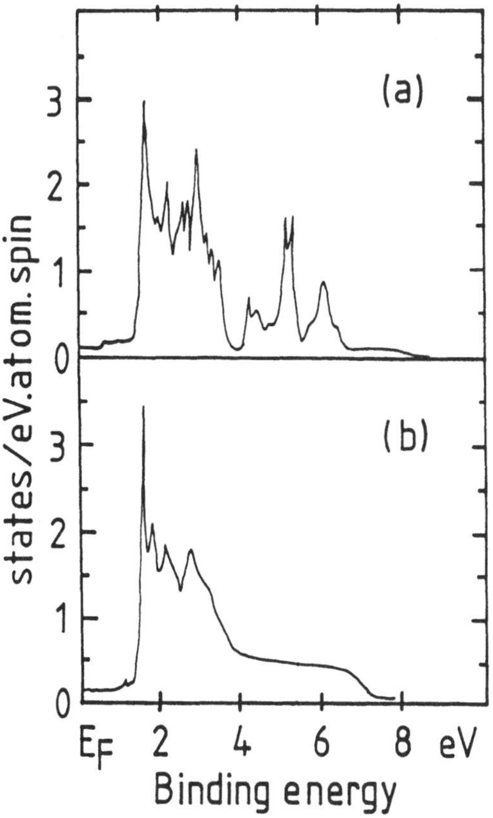

FIGURE 25. Calculated densities of states for a Cu-25% Au alloy: (a) is a self-consistent semi-relativistic LMTO-ASA calculation for ordered Cu_3Au; (b) is a semi-relativistic KKRCPA calculation for the random alloy using the same potential functions.

In order to attempt to understand these observations we carried out some calculations of the densities of states of ordered and disordered Cr_3Au using the same potential functions, which we show in figs.25(a) and (b). Between E_F and 4 eV the shapes are similar but below 4 eV the states in the random alloy are smeared out. The site-decomposed densities of states show that between E_F and 4 eV the Cu and Au states form a common band whereas below 4 eV the Au d-bands are split-off. The calculation suggests that the latter states are more sensitive to the atomic re-arrangement resulting from the change in the long range order parameter. Note that peaks B and C occur in this range.

A common feature of KKRCPA calculations for random alloys is that the disorder-broadening varies markedly with E and $\underline{k}$; viz. sections II and III. It is possible, therefore, that the behaviour of A, B and C is related to some highly non-uniform broadening of the band reflecting the incomplete long range order. On the other hand, when the symmetry of the fcc phase is broken, new features associated with the L12 phases may appear in the spectral density, their strength increasing with the degree of order.

Since these data were collected off-normal it is not clear which region of $\underline{k}$-space is being probed. However, it is apparent that the changes in spectra as a function of temperature are $\underline{k}$-specific. For instance, in Fig.26 we show a series of photoemission measurements above and below T_0 for varius photon energies taken at normal emission from the (100) face. These data correspond to scanning the spectral density along the ΓX (Δ) direction in the fcc Brillouin zone. At certain photon energies (ω ~ 25-30 eV) only minor changes occur in spectra, at others (ω ~ 40-50 eV) there are substantial differences. At ω = 40 eV, for example, in the ordered state the feature at 2-4 eV binding energy is broader than the corresponding peak in the disordered state. We have not completed our analysis of these measurements but we conjecture that at these latter photon energies initial states near the Γ point are being probed. When the alloy orders new zone boundaries are created half-way along the (fcc) ΓX direction and so we anticipate some changes at Γ when the bands at X are 'folded-back' to Γ.

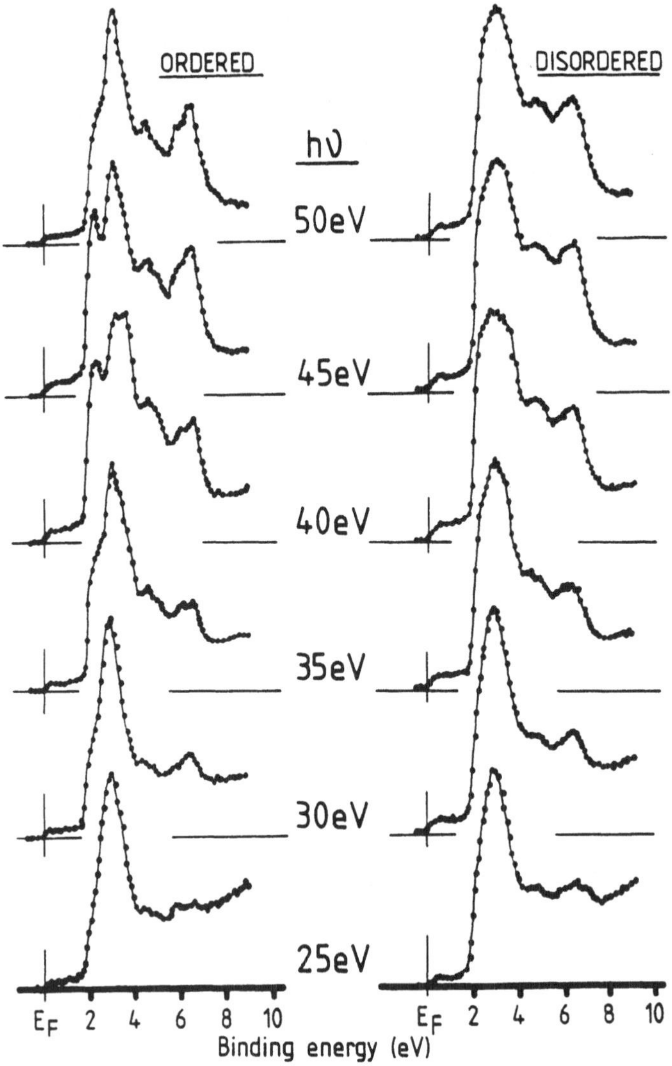

FIGURE 26. Angle-resolved photoemission spectra from the (100) surface of ordered and disordered Cu_3Au for different photon energies. Normal emission, angle of incidence = 30°.

(iii) Photoelectron spectra and long range order

What do these measurements at different temperatures actually mean? In the partially ordered state one of the four interpenetrating simple cubic sublattices has an Au concentration greater than 25%, i.e.,

$$C(Au) = 0.25(1 + 3S)$$

while the other three are correspondingly Au deficient, i.e.,

$$C'(Au) = 0.25(1 - S).$$

When S=1 the spectral density reduces to the bands of the ordered state and when S=0, to those of the random alloy. For $0 < S < 1$ the spectral density will interpolate between these limits in a complex way. However, S is a simple function of temperature. The temperature variation of A-D are shown in Fig.27 and we conclude that the strong temperature dependence of A, B and C and the weak effects on D shows that the three former peaks must be associated with ordering. Indeed, their temperature dependence reminds one irresistably of the variation of the long range order parameter. (Actually, the intensities of A-C do not decrease as abruptly as the long range order parameter at T_0; however, it is not obvious that a quantity such as photoemission should be discontinuous at a first-order transition; since it measures the electronic spectral density not the positions of atoms. In fact, the situation as far as photoemission is concerned is rather more complicated for although the bulk ordering of Cu_3Au is a classic example of a discontin-

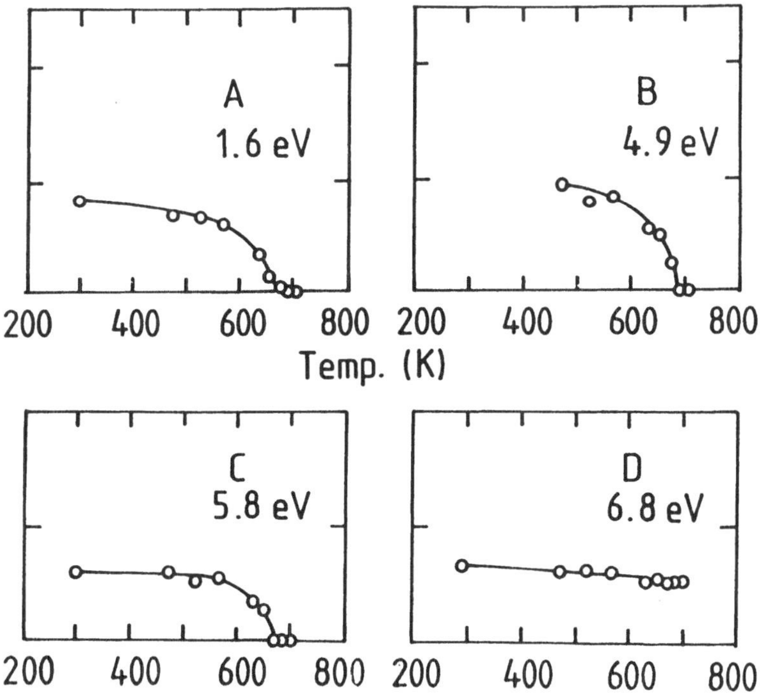

FIGURE 27. Temperature variation of peaks A, B, C and D from Fig.24.

uous transition, a transition occurs at the surface at approximately at the same temperature but with a modified form. There have been a number of studies of the surface transition by LEED [46,47,48] which suggest that it is continuous or weakly first-order [49]. In addition, Ne scattering experiments provide rather surprising evidence for 2-dimensional disordering at the surface since the compositions of the 1st and 2nd layers (i.e., 50% Cu/50% Au and 100% Cu, respectively) appear to remain unchanged at the ordering temperature [50]. Nevertheless, we stress that the effects shown in Fig.27 must arise predominantly from the ordering.)

The message that emerges with great clarity from the above discussion is that, whilst inhomogeneous random alloys angle-resolved photoelectron spectroscopy measures the Bloch spectral function, in partially ordered systems the temperature dependence of spectra reflects the dependence of the electronic spectral density on the long range order parameter; it measures, therefore, a generalised spectral function $A_B(\underline{k},E,S)$. This spectral function is directly related to the (structure-sensitive) 1-e part of the total energy

$$E(S) = \int dk \int E\, A_B(\underline{k},E,S)dE$$

from which the thermodynamics of ordering can be derived. However, the microscopic picture embodied in the spectral function also contains within it a general description of the electronic mechanisms behind the ordering process. Clearly, it subsumes 'simple' density of states arguments, Fermi surface mechanisms (including nesting at special wavevectors), and so on. It is an understanding of these mechanisms which will have an impact on alloy design and the formulation of new materials for specific applications.

EPILOGUE

To date most calculations have been carried out for either the completely random or completely ordered systems. However, 'real' alloys have a tendency to cluster or to order. (Recently, there has been some very interesting work on the driving forces for clustering and short-range order based on the KKRCPA [26]). As discussed in section IV, when an alloy develops long-range order the symmetry of the average lattice is broken and the system must be described in terms of two or more sublattices. At intermediate temperatures, where $0 < S < 1$, the system consists of disordered sublattices of different concentrations. Our current work in this area involves the calculation of the spectral function of such a partially ordered system, using a multi-sublattice KKRCPA calculation of the type developed by Temmerman and Pindor [51]. To our knowledge, this technique has not been applied to binary alloys in a state of partial long-range order. In addition, we are making more detailed relativistic calculations of the spectral function and photocurrents for Cu_3Au in an attempt to clarify the picture as regards our experimental spectra. In particular, it might be instructive to investigate the existence of order-sensitive surface states; the temperature dependence of the corresponding photoemission feature may be different from those of bulk direct transitions because the surface order parameter differs from that of the bulk.

As we stated in the introduction, the major goal of our studies is to try to understand the physical properties of alloys. The explanation and agreement we seek between photoemission experiments and calculations is a powerful a way of establishing the accuracy of the theoretical methods. Once a reliable spectral function has been established, for example, we can approach some of the more complex electrical and metallurgical problems of 'real' alloys. We believe, therefore, that theoretical and experimental studies of the spectral function offer one of the most attractive routes from 'fundamental' research through to the practice of the materials science of alloys. This new direction promises to maintain the central role of photoemission in the future development of the electronic theory of alloys, as solid state physics and materials science draw closer together.

ACKNOWLEDGEMENTS

We thank the SERC for financial support of most of this work and the provision of synchrotron radiation facilities at the Daresbury Laboratory. We are grateful to Drs. W.M. Temmerman, G.M. Stocks and B.L. Gyorffy for many collaborations and discussions on alloys.

REFERENCES

1. F.J. Himpsel, Adv. Phys. 32, 1 (1983).
2. N.V. Smith, Vacuum 33, 803 (1983): for an example of a study of an alloy see R.G. Jordan, W. Drube, D. Straub and F.J. Himpsel, Phys. Rev. B33, 5280 (1986).
3. See articles in "High Temperature Alloys: Theory and Design" ed. J.O. Stiegler, (Metallurgy Soc. of AIME - Pennsylvania, 1984).
4. S. Kirkpatrick, B. Velicky and H. Ehrenreich, Phys. Rev. B1, 3250 (1970).
5. D.H. Seib and W.E. Spicer, Phys. Rev. B2, 1676 (1970).
6. G.M. Stocks, R.W. Williams and J.S. Faulkner, Phys. Rev. B4, 4390 (1971).
7. S. Hufner, G.K. Wertheim and J.H. Wernick, Phys. Rev. B10, 3411 (1973).
8. N.J. Shevchik and C.M. Penchina, Phys. Stat. Sol.(b) 70, 619 (1975).
9. B.L. Gyorffy, G.M. Stocks, W.M. Temmerman, R.G. Jordan, D.R. Lloyd, C.M. Quinn and N.V. Richardson, Solid St. Commun. 23, 637 (1977).
10. J.S. Faulkner, Prog. Mat. Sci. 27, 1 (1982).
11. M. Pessa, H. Asonen, M. Lindroos, A. Pindor, B.L. Gyorffy and W.M. Temmerman, J. Phys. F: Metal Phys. 11, L33 (1981).
12. H. Winter, P.J. Durham, W.M. Temmerman and G.M. Stocks, Phys. Rev. B33, 2370 (1986).
13. R.G. Jordan, Physica Scripta T13, 22 (1986).
14. J. Staunton, B.L. Gyorffy, D.D. Johnson, F.J. Pinski and G.M. Stocks - this meeting; E. Kisker in "Polarised Electrons in Surface Physics" ed. R. Feder, (World Scientific Pub. Co. - Singapore 1985) p513.
15. See for example, "Electronic Structure of Complex Systems", NATO ASI Series, Ser.B, Vol.113 (Plenum - New York, 1982).
16. D.E. Eastman and F.J. Himpsel in "Physics of Transition Metals 1980", IOP Conf. Ser.55. (IOP - Bristol 1981) p115.
17. See for example: M. Pessa, M. Lindroos, H. Asonen and N.V. Smith, Phys. Rev. B25, 738 (1982).
18. G. Wendin "Breakdown of the One Electron Pictures in Photoelectron Spectroscopy", Structure and Bonding Vol.45. (Springer Verlag - Berlin 1981).
19. See for example R. Willis and B. Feuerbacher in "Photoemission and the Electronic Properties of Surfaces" eds. B. Feuerbacher, B. Fitton and R.F. Willis. (John Wiley - New York 1978) p281.
20. See for example: G. Treglia, M.C. Desjonqueres, F. Ducastelle and D. Spanjaard in "Physics of Transition Metals 1980", IOP Conf. Ser.55. (IOP - Bristol 1981) p45.
21. F.J. Himpsel, D.E. Eastman, E.E. Koch and A.R. Williams, Phys. Rev. B22, 4604 (1980).

22. J.B. Pendry in "Photoemission and the Electronic Perperties of Surfaces" eds. B. Feuerbacher, B. Fitton and R.F. Willis. (John Wiley - New York 1978) p87.
23. M.S. Hybertsen and S.G. Louie, Phys. Rev. Lett. 55, 1418 (1985).
24. J. Hermanson, Solid St. Commun. 22, 9 (1977): N.V. Richardson, D.R. Lloyd and C.M. Quinn, J. Electron Spectrosc. & Rel. Phenom. 15, 177 (1979).
25. S.D. Barrett and R.G. Jordan, Z. Phys. B66, 375 (1987).
26. G.M. Stocks and H. Winter in ref.[15].
27. G.M. Stocks and W. Butler, Phys. Rev. Lett. 48, 55 (1982).
28. R.G. Jordan, G.S. Sohal and P.J. Durham, J. Phys. F: Metal Phys. 16, L135 (1986).
29. J.F.L. Hopkinson, J.B. Pendry and D.J. Titterington, Comp. Phys. Commun. 19, 69 (1980).
30. C.G. Larsson, Surf. Sci. 152/153, 213 (1985).
31. P.J. Durham, J. Phys. F: Metal Phys. 11, 2475 (1981).
32. P.J. Feibelman and D.E. Eastman, Phys. Rev. B10, 4932 (1974).
33. N.K. Allen, P.J. Durham, B.L. Gyorffy and R.G. Jordan, J. Phys. F: Metal Phys. 13, 223 (1983).
34. N.K. Allen, R.G. Jordan, P.J. Durham, W.M. Temmerman and B.L. Gyorffy, Vacuum 31, 455 (1981).
35. P.J. Durham, R.G. Jordan, G.S. Sohal and L.T. Wille, Phys. Rev. Lett. 53, 2038 (1984).
36. R.G. Jordan, Vacuum 33, 827 (1983).
37. L.T. Wille, P.J. Durham and R.G. Jordan, Solid St. Commun. 49, 617 (1984).
38. R.G. Jordan and P.J. Durham, Solid St. Commun. 49, 622 (1984).
39. M.C. Muñoz, B.L. Gyorffy and K. Verhuyck, J. Phys. F: Metal Phys. 13, 1847 (1983).
40. C.T. Liu in ref.[3], p289.
41. J.C. Swihart, W.H. Butler, G.M. Stocks, D.M. Nicholson and R.C. Ward, Phys. Rev. Lett. 57, 1181 (1986).
42. D.D. Johnson, D.M. Nicholson, F.J. Pinski, B.L. Gyorffy and G.M Stocks, Phys. Rev. Lett. 56, 2088 (1986).
43. P.J. Durham and N. Kar, Surf. Sci. 111, L648 (1981).
44. R.G. Jordan, D.M. Zehner, N.J. Harrison and P.J. Durham - unpublished.
45. R.G. Jordan, G.S. Sohal, B.L. Gyorffy, P.J. Durham, W.M. Temmerman and P. Weinberger, J. Phys. F: Metal Phys. 15, L135 (1985).
46. V.S. Sundaram, R.S. Alben and W.D. Robertson, Surf. Sci. 46, 653 (1974).
47. K.D. Jamison, D.M. Lind, F.B. Dunning and G.K. Walters, Surf. Sci. 154, 451 (1985).
48. S.F. Alvarado, M. Campagna, A. Fattah and W. Nelhoff, Z. Phys. B66, 103 (1987).
49. E.G. McRae and R.A. Malic, Surf. Sci. 148, 551 (1984).
50. T. Buck, G.H. Wheatley and L. Marchut, Phys. Rev. Lett. 51, 43 (1983).
51. W.M. Temmerman and A.J. Pindor, J. Phys. F: Metal Phys. 13, 1869 (1983).
52. R.W. Godby, M. Schlüter and L.J. Sham, Phys. Rev. B in press (1987).

ELECTRON MICROSCOPY OF ORDERING IN ALLOYS

G. VAN TENDELOO
University of Antwerp, RUCA, Groenenborgerlaan 171, B-2020 Antwerp, Belgium

1. INTRODUCTION

The aim of this contribution is certainly not to provide a general course in electron microscopy, other and more elaborated works can be consulted for this purpose, e.g. [1,2]. We will only try to clarify these aspects of electron microscopy which are useful in the study of ordered alloys or in the evolution towards ordering. First we will study the symmetry aspects of the ordered alloy and predict the number of orientation as well as translation variants of the ordered phase in the disordered matrix. Due to the usual lowering of symmetry upon ordering a number of orientation and translation interfaces will result in ordered alloys. Their number and internal relationship can be predicted based on simple symmetry rules. The two dimensional defects related to ordering can be identified from electron microscope observations and sometimes produce specific electron diffraction effects. High resolution microscopy combined with computer simulated images allow to obtain atomic scale information of ordering related defects. Important is that we can show that the minority atoms in a large range of alloys can be systematically imaged as bright dots; this provides us with an easy code to interpret sometimes complex high resolution images.

In a second part (chapters 6 and 7) the ordering process from the short range ordered state towards the long range ordered state will be followed with the help of both electron diffraction and high resolution electron microscopy.

Since electron microscopy would be rather dull when only applied to fictive materials such as jellium we will use the $D1_a$ type ordered alloys with composition A_4B as a kind of leading thread through the story. Materials based on a $D1_a$ ordering are e.g. Ni_4Mo, Ni_4W, Au_4Mn, Au_4Cr, Au_4V. Their structure is tetragonal and based on a face centered cubic lattice which is also the structure of the disordered high temperature phase. The relation between the base vectors of the ordered phase ($\bar{A}_i$) and those of the disordered phase ($\bar{a}_i$) is given by :

$$\begin{bmatrix}\bar{A}_1\\ \bar{A}_2\\ \bar{A}_3\end{bmatrix} = \begin{bmatrix}3/2 & 1/2 & 0\\ -1/2 & 3/2 & 0\\ 0 & 0 & 1\end{bmatrix}\begin{bmatrix}\bar{a}_1\\ \bar{a}_2\\ \bar{a}_3\end{bmatrix}$$

This relationship is obvious from the structure representation of fig. 1 which is projected along $\bar{A}_3 \equiv \bar{a}_3$. The tetragonal unit cell is body centered and contains 8 A-atoms and 2 B-atoms. Lattice parameters vary slightly depending on the alloy; for Au_4Mn they are :

$$|\bar{A}_1| = 0.645 \text{ nm}$$

$$|\bar{A}_3| = 0.403 \text{ nm}$$

G. M. Stocks and A. Gonis (eds.), Alloy Phase Stability, 75–100.

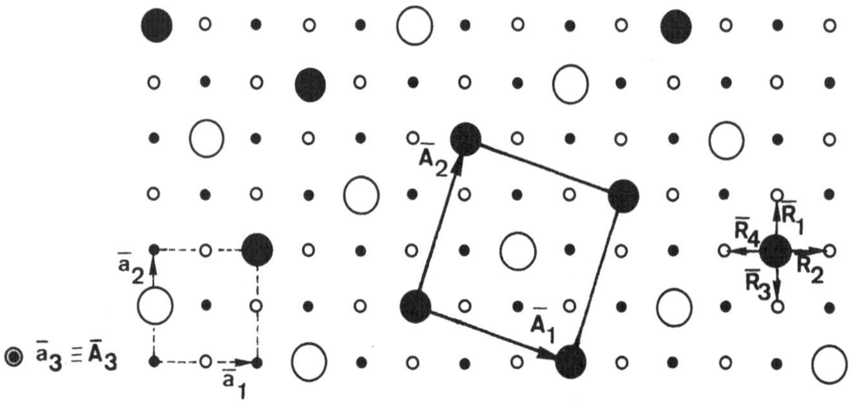

FIGURE 1. Structure of $D1_a$ ordered alloys projected along the $\bar{A}_3$ axis. The minority atoms of this A_4B structure are represented as large circles; open and closed circles refer to the level z = 0 or z = 1/2.
The four possible displacement vectors for antiphase boundaries are indicated as $\bar{R}_1$ through $\bar{R}_4$.

The ordering process of $D1_a$ type alloys is particularly well documented mainly because it is one of the few systems where the long range order (LRO) reflections appear at positions different from the short range order (SRO) positions. A more complete bibliography can be found in [3].

2. SYMMETRY ASPECTS OF ORDERING

Landau theory of phase transitions points out that for a second order transition the space group of the product phase (ordered phase) is induced by an irreducible representation of the space group of the parent phase (disordered phase). However also a first order transition will transform a single parent phase into a number of domains of the product phase (variants) which are strictly equivalent. Domain formation is clearly a consequence of the fact that on going through the order-disorder transformation the different variants of the ordered phase have equal probability of being generated if their structures are related by a symmetry operation of the disordered phase which is not a symmetry operation of the ordered phase. We will in this part try to determine the number of such variants as well as their internal relationship. We will follow here a very simple group theoretical approach put forward some time ago in [4] which however is valid for the overwhelming part of the order-disorder transitions. Later the theory has been extended and generalised [5,6,7].

We will assume that we can separately treat the translation symmetry and the orientation symmetry, which will allow us to consider the point group instead of the space group when deducing the number of and relation between orientation variants. We further assume that we have a pure order-disorder transition i.e. small changes in lattice parameter as a result of ordering will be neglected. They are unimportant as long as they do not further lower the symmetry of the ordered structure. However if the slight deformations do cause the domain fragmentation it is easy to take care of these; see e.g. [4,6].

The ordering is usually accompanied by a decrease in symmetry in such a way that the pointgroup of the ordered structure H (of order q) is a subgroup of the pointgroup G (of order p) of the disordered phase. If the symmetry of the lattice differs from the symmetry of the structure we can label them respectively H_ℓ and G_ℓ such that :

$$H \subset H_\ell$$

and

$$G \subset G_\ell$$

The lattice translations of the parent phase are represented by the group T and those of the product phase by $T^{(0)}$.
The structure of the different ordered variants will be denoted V_j and their pointgroup H_j. Different H_j can clearly only differ in orientation of their elements. It is clear that the structures (V_j) as well as the pointgroups are related by operations of G which are not elements of H.

if $\quad g\,V_i = V_j \qquad g \in G;\ g \notin H_i$

then $\quad H_j = g\,H_i g^{-1}$

i.e. the pointgroups of different variants are conjugate in G.
Moreover if $g\,V_i = V_j$ <u>all</u> operations that transform V_i into V_j are given by $g\,H_i$.
We can then construct the set of operations which transforms a given variant V_i into all other variants and we obtain the following sets :

$$g_1H_i,\ g_2H_i,\ \ldots\ ,\ g_jH_i,\ \ldots,\ g_nH_i$$

(with g_1 being the identity).
Since :
(a) all elements in each set belongs to G
(b) two sets have no element in common (such element would transform V_i into two different variants V_j and V_k which is clearly impossible.
We can write G as :

$$G = g_1H_i + g_2H_i + \ldots + g_nH_i$$

where $g_j \quad G \in \{g_j\} \cap H_i = \{E\}$.
Which immediately allows to conclude :
1) the number of orientation variants is given by n=p/q
2) a set of all operations that generates all variants can be obtained by taking one operation from each coset in the development of G into cosets of H.

The number of variant generating sets is very large (q^n); some of these sets however will form a group [a variant generating group (VGG)]. It is easy to show that if a variant generating group V exist it should obey the following conditions :

1) order V = p/q
2) $H \cap V = \{E\}$
3) $H \cdot V = G$ where $\cdot$ indicates the weak direct product introduced by Melvin [8]

The number of translation variants is the number of different ways the superstructure can be built in a parallel orientation within a given orientation variant i.e. the number of translation vectors τ such that :

$$\tau \in T \quad \text{but} \quad \tau \notin T^{(0)}$$

If we write the transformation between the primitive base ($\bar{a}_i$) of the disordered structure and the primitive base of the ordered structure ($\bar{A}_i$) as

$$\begin{bmatrix} \bar{A}_1 \\ \bar{A}_2 \\ \bar{A}_3 \end{bmatrix} = M \begin{bmatrix} \bar{a}_1 \\ \bar{a}_2 \\ \bar{a}_3 \end{bmatrix}$$

the volumes of the primitive unit cell of both phases are related as :

$$V^{(o)} = |M| \ V$$

The number of translation variants t can therefore simply be written as :

$$t = |M|$$

If the unit cells have respective multiplicities m and $m^{(o)}$ the number of translation variants becomes :

$$t = \frac{m}{m^{(o)}} |M|$$

Applied to the example of $D1_a$ ordering from a face centered cubic matrix we have :

$G = 4/m\ \bar{3}\ 2/m$	of order	$p = 48$	for the high temperature fcc phase
$H = 4/m$	of order	$q = 8$	for the ordered $D1_a$ alloy

The number of different orientation variants n becomes p/q = 48/8 = 6 and a variant generating group V of order 6, with no elements in common with 4/m can be chosen as : $V_1 = 3m$ or $V_2 = 32$

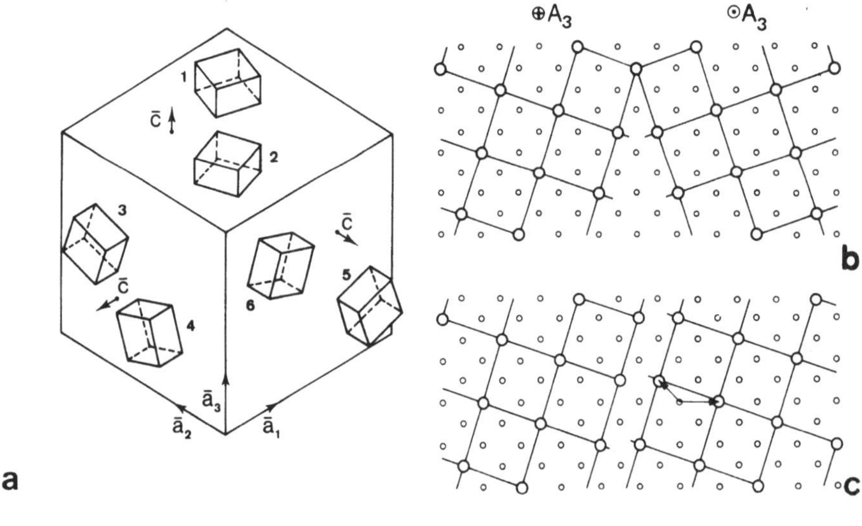

FIGURE 2. (a) Schematic representation of the six different ordered $D1_a$ variants with respect to the cubic phase.
(b) Structure of a twin boundary in $D1_a$ alloys having a parallel $\bar{A}_3$-axis (only the configuration in one (001) plane is represented).
(c) Structure of an antiphase boundary with displacement vector $\bar{R}_3$.

Since the structure of both the disordered and the ordered phase is very simple this has an easy intuitive interpretation. The $\bar{A}_3$-axis of the ordered structure can be chosen along any of the three cube directions (they are related by a rotation of $2\pi/3$ around the threefold axis along [111]) and for each choice of the $\bar{A}_3$ axis two possibilities exist for the $\bar{A}_1$ and $\bar{A}_2$ axis (they are related by a (110) mirror plane or a π-rotation along the [110]), resulting in six different orientation variants for the ordered $D1_a$. These six orientation variants are reproduced schematically in fig. 2a while fig. 2b shows a projection along the tetragonal axis for the two orientation variants having a common $\bar{A}_3$-axis.
The number of translation variants :

$$t = \frac{m}{m^{(o)}} |M|$$

then becomes

$$t = \frac{4}{2} \cdot \frac{5}{2} = 5$$

Between these five different translation variants from different translation interfaces (or antiphase boundaries) exist. The displacement vectors of these four antiphase boundaries are indicated by $\bar{R}_1$ to $\bar{R}_4$ in fig. 1a, while the atomic configuration projected along the tetragonal axis over a translation defect defined by $\bar{R}_3$ is reproduced in fig. 2c.

3. THE ELECTRON MICROSCOPY OF ORDERED ALLOYS

The working principle of the electron microscope differs in no way from its optical analogue. The specimen is irradiated by a monochromatic electron beam and a rotated and enlarged image is produced by the objective lens.

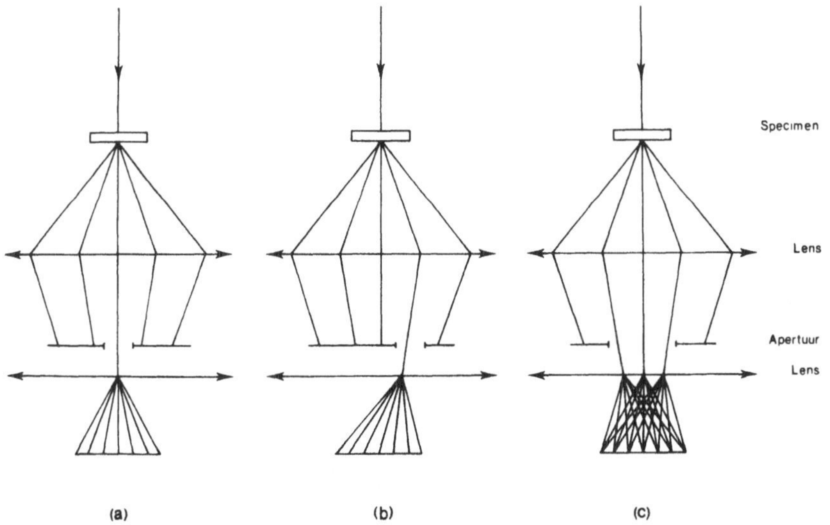

FIGURE 3. Different modes of image formation in a transmission electron microscope.
(a) bright field
(b) dark field
(c) structure imaging in HREM.

This image is further projected on the fluorescent screen by means of a combined system of intermediate and projector lenses. Together with the enlarged image a selected area diffraction pattern can be obtained; therefore the intermediate lenses are focussed onto the back focal plane of the objective lens. All strongly inclined electron beams that are severely affected by the spherical aberration are eliminated by an aperture placed in the back focal plane of the objective lens. Doing so one restricts the beams contributing to the final image and by changing the aperture size or the position of the aperture with respect to the transmitted beam different information can be obtained. When only selecting out the transmitted beam a bright field image is obtained (fig. 3a), by selecting out one of the diffracted beams a dark field image results (fig. 3b). When the image is formed by a large number of beams, structure or lattice imaging can be expected, exhibiting a contrast depending on the number and the type of the selected reflections. This technique can be applied in bright field as well as in dark field. A schematic representation of these different modes of image formation is given in figure 3.
Using the different modes in electron microscopy and combining information from the direct space (image) with information from reciprocal space (diffraction pattern) the different orientation and translation interfaces resulting from an ordering reaction can be identified and characterized.

Translation defects [stacking faults or antiphase boundaries (APB's)] which can be characterized by a constant displacement vector $\bar{R}$ do not produce any diffraction effect (apart from a faint streaking perpendicular to the defect plane). Only when such interfaces become periodic or pseudo periodic extra satellites appear associated with the periodicity introduced. This effect however will be discussed in more detail when long period superstructures will be considered. In the direct image translation interfaces can be recognized by their typical fringe contrast while the background contrast on both sides of the defect is the same (see fig. 4a).

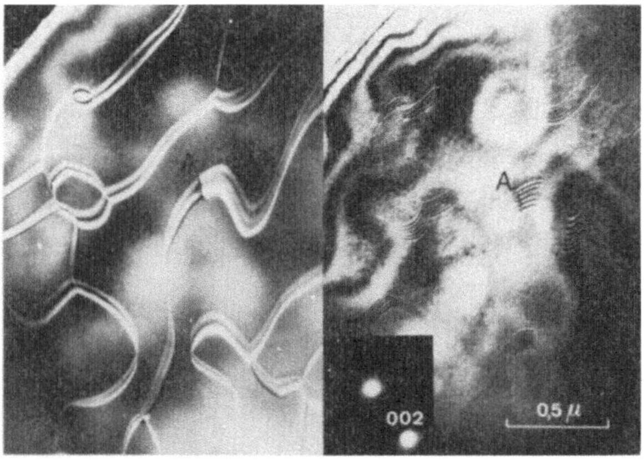

FIGURE 4. Antiphase boundaries in Ni_4Mo.
(a) When the interfaces are "in contrast"
(b) When the interfaces are "out of contrast" for a two beam situation (000, 002); the diffraction pattern is shown as an inset.

When a "two-beam" situation is realised (i.e. apart from the central beam only one strongly diffracted beam is present) the bright field image of the fringe pattern is symmetrical (see fig. 4a) while the corresponding dark field pattern is asymmetrical. The displacement vector $\bar{R}$ can be determined in an extremely elegant way. When different "two-beam" situations are realised, the antiphase boundary will be invisible (in extinction) for some of the two beam situations. One can prove (see e.g. [2] p. 107) that when a translation defect is out of contrast in a two beam situation, corresponding with a reciprocal vector $\bar{g}_i$, the dot product of $\bar{g}_i$ with the translation vector $\bar{R}$ equals an integer n, i.e.

$$\bar{g}_i \cdot \bar{R} = n$$

From three independent extinctions the displacement vector $\bar{R}$ can easily be determined. In fig. 4b the antiphase boundaries in Ni_4Mo with displacement vector $\bar{R}=[uvw]$ are out of contrast for a reflection $[002]^*$ which means that e.g. :

$$2w = n \quad \text{i.e.} \quad w = o \quad \text{or} \quad w = 1/2$$

The faint remaining contrast in fig. 4b which shows far more fringes than the same defects in fig. 4a is termed residual contrast and may result or from an incomplete two-beam situation or from the fact that the actual displacement vector $\bar{R}$ is not exactly a lattice vector of the disordered phase $\bar{R}_0$ but slightly deviates from it due to the ordering :

$$\text{i.e.} \quad \bar{R} = \bar{R}_0 + \bar{\varepsilon}$$

From the behaviour of the residual contrast for different $\bar{g}$ vectors the direction and magnitude of $\bar{\varepsilon}$ can be determined see e.g. [9].

As most common orientation defects one discerns twin boundaries (rotation twins or mirror twins) and inversion boundaries. A recent review of Boulesteix explicitely treats the image formation of twin boundaries and the diffraction effects associated with it [10]. In the direct image twin

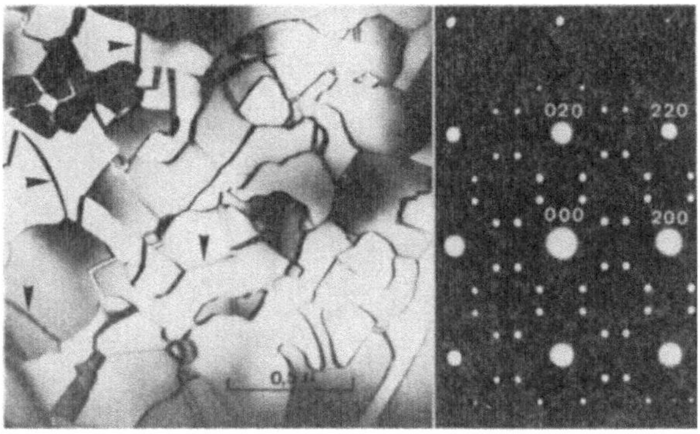

FIGURE 5. (a) Twin boundaries in Ni_4Mo to be recognized by a difference in background contrast. Antiphase boundaries (e.g. those indicated by an arrow) are present as well.
(b) Electron diffraction pattern taken over the area of (a) along the common $\bar{A}_3$-axis.

boundaries can be readily discerned from translation interfaces because of the difference in background intensity on both sides of the twin boundary. Fig. 5 shows twin boundaries with a parallel $\bar{A}_3$-axis in Ni_4Mo imaged under an arbitrary orientation. The corresponding atomic configuration is shown in fig. 3b. When the twin interface coincides with the mirror plane the twin is called coherent. The mirror plane can be determined from the electron diffraction pattern because it corresponds to a common plane of both crystal parts and therefore in the electron diffraction pattern the systematic row of diffraction spots is unsplit, while other reflections are split depending on their distance from the common row. This is illustrated in fig. 6 for twin formation in ZrO_2. High resolution image and diffraction pattern are oriented correctly with respect to each other; the unsplit [110]* row is perpendicular to the (110) twin-mirror plane. The amount of splitting also allows to deduce the displacement field $\bar{R}(z)$ where z is the distance away from the unsplit [110]* row.

Inversion boundaries can of course only occur in non-centrosymmetrical crystals an have a rather peculiar contrast behaviour. From the fringe contrast they cannot be discerned from antiphase boundaries : a symmetrical fringe pattern in bright field and an asymmetrical pattern in dark field. The background contrast on both sides however can be different, but only in the dark field mode (see fig. 7). This effect is based on the violation of Friedel's law [11]. In the framework of two-beam theory for non-centrosymmetrical crystals, contrast at inversion boundaries will only be found if the corresponding Fourier coefficient of the real and the imaginary part of the lattice potential have different phases. The largest intensity difference between inversion domains is found in multiple beam situations (i.e. more than one diffracted beam is strongly excited). Of course if only reflections operate which belong to a zone along which the projection is centrosymmetric, the crystal behaves as such and the contrast between the two domains disappears. The diffraction pattern is clearly not affected by the presence

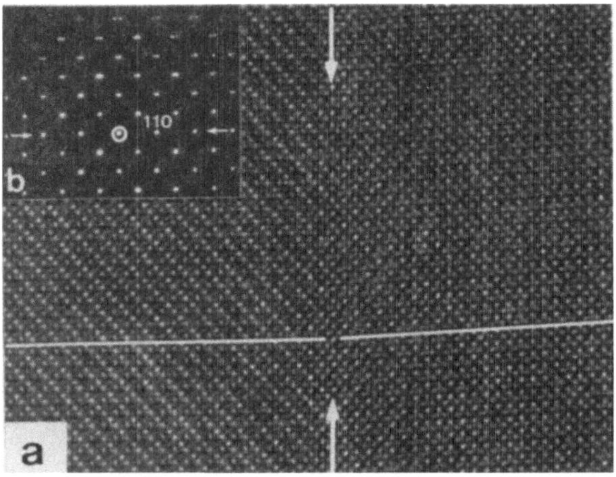

FIGURE 6. (a) High resolution image of a (110) twin boundary in the monoclinic ZrO_2; the unit cell is indicated in both variants.
(b) Corresponding electron diffraction pattern; note the spot splitting parallel to the [110]* row.

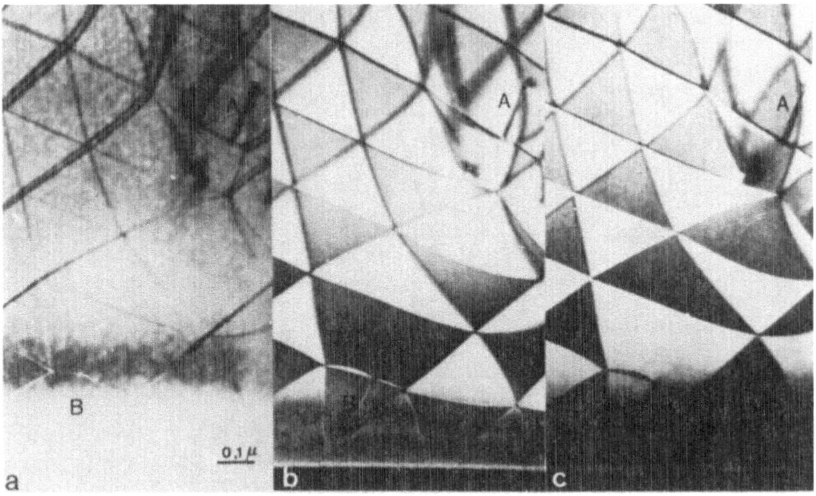

FIGURE 7. Inversion boundaries in the χ-phase.
(a) In bright field no background contrast is ever observed
(b)(c) in dark field inversion domains may produce pronounced contrast differences. The contrast can be inversed (compare b and c) by a slight change of orientation.

of an inversion boundary since the lattice is the same in both domains. Similar to antiphase boundaries or twins, inversion boundaries may also occur periodically as in γ-brass structures [12] or in the χ-phase [13]. In that case the diffraction pattern will again reflect the extra periodicity.

4. HIGH RESOLUTION ELECTRON MICROSCOPY (HREM)

In spide of the short wavelength of the accelerated electrons (λ=3.7 pm for 100 kV electrons) the resolution of present day high resolution electron microscopes is "only" 0.14 nm. The reason of this limitation is related to the use of electromagnetic lenses which are never perfect and invariably suffer from spherical and chromatic aberration. Further effects reducing the resolution are the presence of an objective aperture (used to select in the diffraction pattern the beams of interest), the beam divergence and the material under investigation. Resolutions quoted are always obtained for extremely thin samples which are perfectly stable under the electron beam. In practice a number of materials do not resist to the high energy electrons and desintegrate, change phase, disorder or render amorphous. Alloys fortunately are mostly stable under the electron beam (except if the voltage used exceeds the threshold value for atom displacement). The main problem for alloys is certainly the specimen preparation. In order to be suitable for HREM alloys should be as thin as possible and certainly below ~30 nm. Mostly the specimens are prepared by electropolishing techniques using a specific electrolyte at the right voltage and an adapted flow rate see eg. [14]. For some alloys, mainly those containing noble metals, the preparation can impose serious restrictions on the resolution.

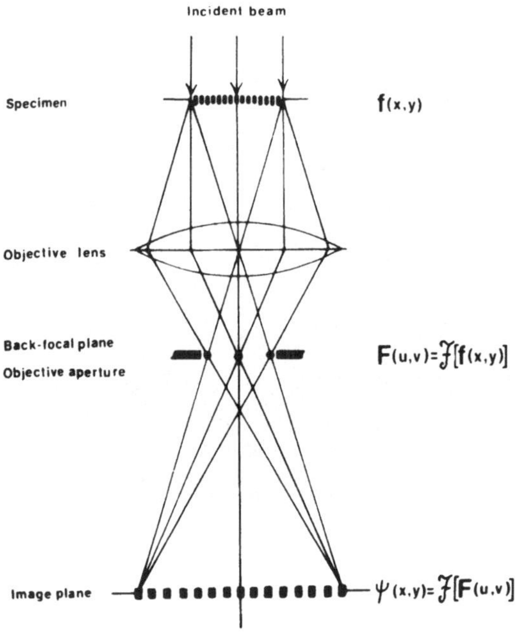

FIGURE 8. Schematic representation of the image formation in transmission electron microscopy.

A good reference work mainly on the theoretical background of the high resolution microscopy technique has been recently written by J. Spence [15] but here we will follow the ideas and notations as proposed by Van Dyck [16].

The image formation in an ideal electron microscope is extremely simple (see fig. 8). A very thin sample can always be described as a two dimensional transmission function f(x,y). The exit plane of the crystal acts as a Huygens source for spherical waves. The objective lens will focus each diffracted beam into a point of the back focal plane. The amplitude of the diffracted beams is simply the fourier transform F(u,v) of the transmission function. If the crystal is periodic the diffraction pattern consists of sharp spots; an amorphous material will give rise to a continuous diffraction pattern. In a second stage of the image formation the back focal plane acts again as a set of Huygens sources; spherical waves will through a lens, interfere and the amplitude in the image plane is given by an inverse fourier transform $\psi(x,y)$ which reconstructs in an enlarged way the original object function f(x,y). In such an ideal microscope the resolution would only be limited by the wavelength and the electron beam divergence (as it is the case in an optical microscope).

In practice the second stage suffers from the microscope imperfections and the diffracted beams acquire a phase shift χ with respect to the transmitted beam. This phase shift is position dependent and given by :

$$\chi(\bar{g}) = \frac{1}{2} \pi C_s \frac{\beta^4}{\lambda} + \pi \Delta f \frac{\beta^2}{\lambda}$$

with C_s : spherical aberration constant intrinsic to the instrument
Δf : the defocus value of the objective lens tunable with minimum steps of ~4 nm
β : the diffracted angle corresponding to $\bar{g}$

The final diffracted amplitude in the image plane therefore becomes :

$$\psi(\bar{r}) = \underset{g}{F^{-1}}\left(e^{-i\chi(\bar{g})} F(u,v)\right)$$

and the intensity is given by :

$$I(\bar{r}) = \left|\psi(\bar{r})\right|^2$$

In general this intensity expression is very complex and hardly shows any visual relationship with the projected crystal potential. However if the specimen is very thin and acts as a pure phase grating the total phase shift is given by :

$$\chi(x,y) = \sigma\ \phi(x,y)$$

where σ is constant and $\phi(x,y)$ represents the projected crystal potential. If the defocus value Δf is furthermore tuned such that $\sin\chi \approx 1$, the final intensity reduces to a very simple expression :

$$I(x,y) \approx 1 - 2\sigma\phi(x,y)$$

This assumption $\sin\chi = 1$ only holds for a very particular defocus value of the objective lens; the so called Scherzer defocus. At this particular defocus value the high resolution image will reflect the projected crystal potential; for other defocus values the image can be very complex and detailed interpretations should be based on a comparison between the experimental results and the calculated (computer simulated) images. In general, the more complex the structure, the more problems with an intuitive interpretation. A structure image calculation along the tetragonal axis of the relatively simple Au_4Mn is reproduced in fig. 9 at three different thickness values of 2.5, 5 and 7.5 nm for the same defocus value of -100 nm.

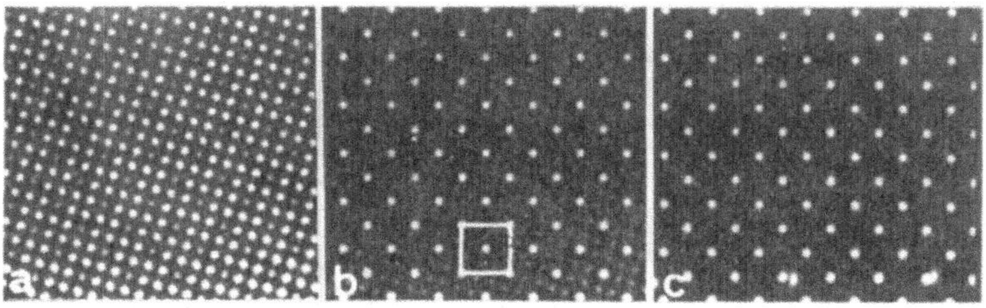

FIGURE 9. Simulated high resolution images of Au_4Mn along [001] for thicknesses of 2.5 nm (a); 5 nm (b) and 7.5 nm (c). The defocus value was -100 nm. The unit cell has been outlined; it is clear that the Mn configuration is imaged as brighter dots except for the smallest thicknesses (see fig.a).

Interesting and helpful is that the configuration of the Mn atoms appears as extra bright dots at most defocus values. It can even be shown that the bright dots not only exhibit the Mn-configuration but that they appear exactly at the position of the Mn-columns.
This simple code allows a straightforward interpretation of defects observed in $D1_a$ ordered alloys. Fig. 10 for example shows the presence of antiphase boundaries together with a twin boundary in Au_4Mn. Since bright dots are to be associated with Mn columns it is obvious that the translation defects are along (110) and (1$\bar{1}$0) while the displacement vector $\bar{R}$ of the APB's can be identified as $\bar{R}_1$ (see fig. 1). Also for the twin boundary having a common $\bar{A}_3$-axis the atomic structure can be observed down to the interface which in this case is not coherent.

The fact of imaging the minority atoms in alloys is not a unique case. Many alloys are based on a simple lattice (FCC, BCC or HCP) and upon ordering they acquire a column structure i.e. along a given orientation only atoms of the same chemical nature superimpose. When viewing along that direction in the electron microscope the minority atoms show up as bright dots; this is so for most thicknesses (except for the very thin areas where the basic lattice is also resolved). This is illustrated in fig. 11 for a wide range of alloys and compositions. This intuitive interpretation in function of the minority atoms is backed up by theoretical calculations using the real space description of electron microscopy [17]. It shows that the one to one correspondence between minority atom columns and bright dots holds far beyond the thin phase grating limit [18]. A further step, still assuming column type structures is to image different sublattices at different defocus values. In a complex material such as e.g. $Ba_2NaNb_5O_{15}$ one can separately image the Ba, Na and Nb configuration at different defocus values.[19] It has been shown that this is always possible whenever :

1) the structure is a column type structure along a given orientation
2) the average distance betweem the columns of a given sublattice is approximately equal
3) this average distance between columns is different for different sublattices.

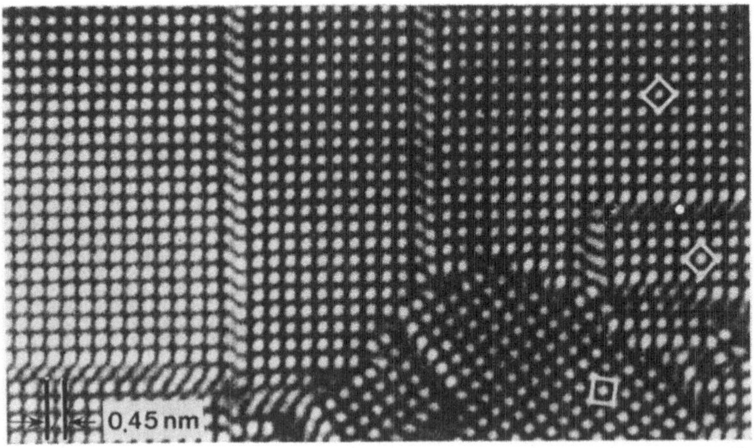

FIGURE 10. Defect structure in Au_4Mn; twin boundaries and antiphase boundaries can be easily recognized. Bright dots have the Mn-configuration.

For the example of $Ba_2NaNb_5O_{15}$ the Na-Na intercolumn distance is 0.88 nm, the Ba-Ba distance is ~0.55 nm while the Nb atoms are separated by ~0.27 nm.

The main reason for this particular behaviour is that every sublattice can roughly be considered as a two dimensional liquid having its diffraction intensity maxima along rings. Selective imaging of the different sublattices is obtained if one assumes that the two most intense rings of the diffraction pattern interfere in phase. This requirement pins down the defocus value. A good correspondence between experimental defocus values and calculated values is obtained [19].

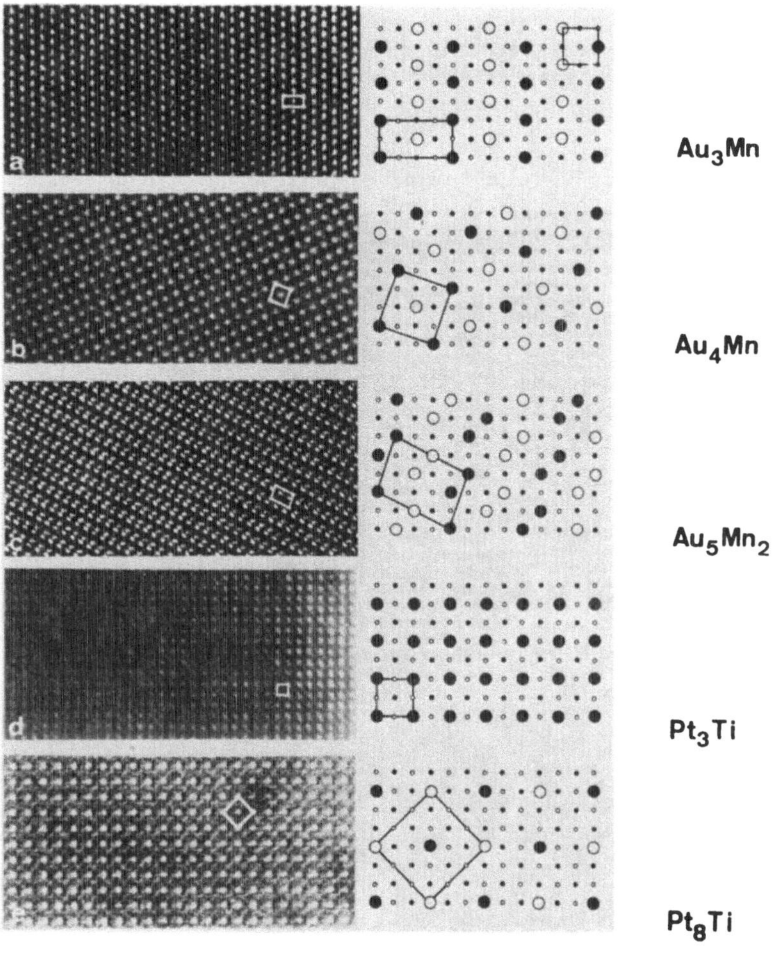

FIGURE 11. High resolution images of different fcc based superstructures where the minority atoms form a column structure along the viewing direction. The crystal structures projected along [001] are also shown. Notations are the same as in fig. 1.
The fcc unit cell, indicated in the top drawing, is approximately 0.4 nm wide.

5. HIGH RESOLUTION ELECTRON MICROSCOPY OF LONG PERIOD SUPERSTRUCTURES

Most planar defects we have discussed before may occur periodically; periodic twin boundaries are observed e.g. in NiMn [20] and periodic inversion boundaries are reported in γ-brass structures [12] or in the χ-phase [13]. When stacking faults are generated periodically they are termed polytypes (see e.g. [21]) and when antiphase boundaries occur periodically they are called long period superstructures (LPS).

LPS are invariably based on simple ordered structures such as $L1_2$ (Cu_3Au-type) or $L1_0$ (CuAu I-type) where periodic antiphase boundaries are introduced along one of the cube planes (see fig. 12a). Since they were first discovered in 1936 [22] they have attracted a lot of interest both from experimental and theoretical point of view. It is indeed intriguing how such simple ordered structures can produce such sometimes extremely long periodicities. Recently there is a renewed interest in these structures since the so called ANNNI-model in statistical physics provides new insights in the behaviour of the LPS (see the contribution of W. Selke) and at the

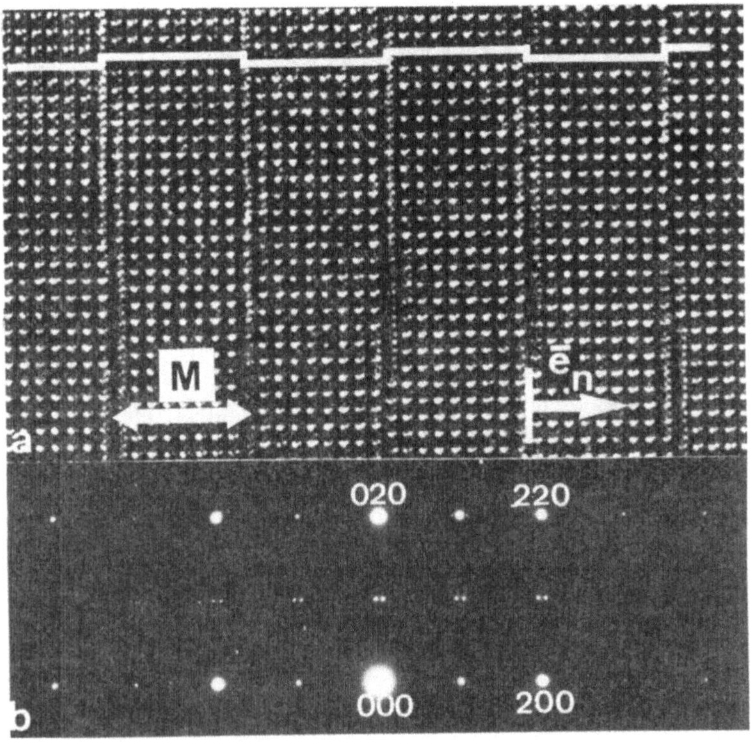

FIGURE 12. (a) High resolution image along [001] of a long period superstructure in Cu_3Pd. Bright dots indicate position of the Pd-atoms. The average distance M between successive antiphase boundaries is indicated.
(b) Corresponding diffraction pattern showing the characteristic spot splitting of certain superstructure reflections.

same time HREM observations allow detailed studies of the boundaries which were not possible before. D. Broddin in his contribution will mainly treat the link between the HREM observations in $Cu_{3\pm x}Pd$ and the theoretical models; in this part we will restrict ourself to general aspects of long period structures and the way to deduce information about them by E.M. techniques.

When one creates an interface modulated structure; i.e. a periodic sequence of interfaces, the electron diffraction pattern will be affected and reflect the extra periodicity. If one introduces periodic translation interfaces characterized by a constant displacement vector $\bar{R}$, the diffraction pattern will consist of arrays of "satellite reflections" associated with the spots of the basic structure. The direction of the arrays of satellites is parallel with the unit normal $\bar{e}_n$ on the interface. The spacing between satellite spots is determined by the average spacing M between interfaces. The diffraction vectors $\bar{g}$ of the superstructure are given in terms of the diffraction vectors $\bar{H}$ of the basic structure by the relation :

$$\bar{g} = \bar{H} + \frac{1}{M}(m-\bar{H}.\bar{R})\bar{e}_n$$

where m is an integer determining the order of the satellite.

This spot splitting is illustrated in fig. 12 where in the top part a high resolution image of Cu_3Pd along [001] is shown with the bright dots representing Pd-columns. The corresponding electron diffraction pattern of fig. b clearly shows intensity maxima around 100, 010 and 110. Some of these reflections are split others are unsplit, depending upon the dot product $\bar{H}.\bar{R}$. Two sections of reciprocal space are sufficient to determine $\bar{R}$ from the fractional shift $\bar{H}.\bar{R}$ of the satellite sequence with respect to the position of the basic spots [23]. The intensity of the satellite spots decreases with increasing order m i.e. with increasing distance from the basic reflections with which they are associated. Although the theory is made for a kinematical approximation, the position of spots remains unaltered under dynamic conditions.

Imaging of long period superstructures can essentially be performed in two different modes.

(a) in dark field by selecting in the objective aperture only the satellite reflections belonging to a single basic reflection (e.g. 110). This mode only reveals the antiphase boundaries but is perfectly suited e.g. for detecting irregularities. An example of one- and two-dimensional long period superstructures in Au_4Zn is reproduced in fig. 13a. This technique was first used about 30 years ago to image periodic antiphase boundaries in CuAuII [24].

(b) in bright field by selecting in the objective aperture a large number of reflections including or not the basic fcc reflections. Such an image of Au_4Zn is reproduced in fig. 14b. It reveals the atomic configuration up till the boundary (bright dots are again to be associated with the Zn-columns). The diffuse line observed along the translation interface can have its origin in two distinct phenomena.

1° a disorder of Zn and Au along the columns

2° an inclined antiphase boundary such that a given column is e.g. at the top filled by Zn and at the bottom by Au.

Image calculations have been performed in order to differentiate between both phenomena but the expected differences are so small that from the experimental images one can hardly decide for one of the possibilities. Moreover lattice relaxation, mostly associated with the antiphase boundaries will further blurr the image. All these items and much more is explicitly treated in recent publications of Au_4Zn [25], Al_3Ti [26] or Cu_3Pd [27].

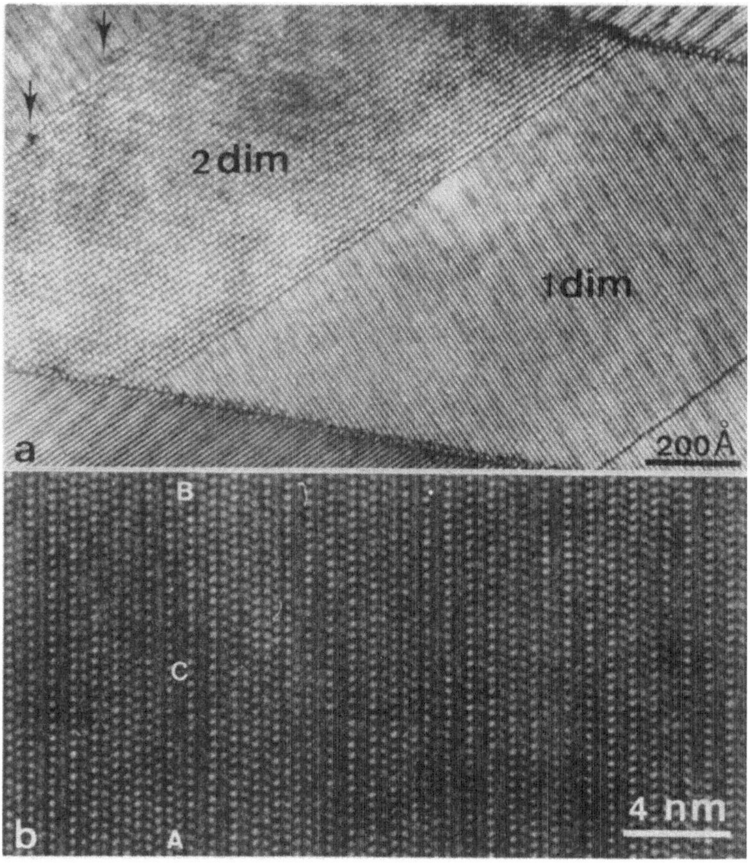

FIGURE 13. High resolution images along [001] of LPS in Au_4Zn.
(a) in dark field using only the satellites around 110 in the objective aperture.
(b) in bright field using a large number of diffraction beams, including the basic fcc reflections.

6. ELECTRON DIFFRACTION EFFECTS ASSOCIATED WITH ORDERING

When a perfect crystal is irradiated with X-rays, electrons or neutrons it will only diffract intensity in the Bragg reflections. Any deviation from its perfection will give rise to radiation out of the discrete Bragg positions. Substitutional disorder is one of such deviations and will be considered in more detail.

Diffraction through an imperfectly ordered crystal can be described using the kinematical or well the dynamical theory.

(a) in a kinematical theory it is assumed that the irradiating particle (X-ray, electron or neutron) will only be scattered once within the crystal;

(b) in a dynamical theory which is used when the interaction between irradiating particle and crystal is stronger, the particles can be scattered several times within the crystal.

The kinematical theory is found to give a good description of the diffraction processes for X-rays and neutrons since the interaction with the crystal itself is weak. High energy electrons interact much stronger with the crystal and the dynamical theory should be used to describe diffraction of electrons. However since multiple scattering has to be considered the mathematical description becomes more difficult and in a number of cases the kinematical theory is still used as a first approximation. When structural defects are included and more particularly when substitutional disorder has to be described the dynamical formulation becomes very complicated and the obtained expressions are of limited practical use [28, 29]. However as far as the geometrical aspects of the diffuse intensity distribution in reciprocal space is concerned one can show that the dynamical and the kinematical theory give the same results, assuming that the crystal potential can be considered as the sum of a periodic potential and a non periodic perturbation. The dynamical interaction will only give rise to a redistribution of the diffuse intensity over an idealised locus of diffuse intensity.

Let us consider a binary system of A and B atoms exhibiting substitutional disorder based upon a lattice with a primitive unit cell. A more general treatment valuable for non primitive unit cells and/or more than two kinds of atoms is given in [30, 31].

The occupation of site (j) ($\bar{r}_j$) can be described by the occupation operators σ_j^A (or σ_j^B) :

$$\sigma_j^A = \begin{cases} 1 \text{ when (j) is occupied by an A atom} \\ 0 \text{ when (j) is occupied by a B atom} \end{cases}$$

so that for all sites (j) :

$$\sigma_j^A + \sigma_j^B = m_A + m_B = 1$$

where m_A and m_B are the atom fractions of A and B atoms respectively.

The kinematical expression for the scattered amplitude can then be written as :

$$A(\bar{g}) = \sum_j f_j e^{2\pi i\bar{g}.\bar{r}_j}$$

or :

$$A(\bar{g}) = \sum_j (f_A\sigma_j^A + f_B\sigma_j^B)e^{2\pi i\bar{g}.\bar{r}_j}$$

where the sum extends over all lattice sites (j)

$\bar{g}$ is a vector in reciprocal space

f_A and f_B are the atomic scattering amplitudes of A and B atoms respectively.

Introducing the Flinn operators :

$$\bar{\sigma}_j = m_A - \sigma_j^A = \sigma_j^B - m_B \qquad (m_A > m_B)$$

the Bragg component and the diffuse scattering component of the scattered amplitude $A(\bar{g})$ can be separated :

$$A(\bar{g}) = A_B(\bar{g}) + A_D(\bar{g})$$

$$A_B(\bar{g}) = (m_Af_A + m_Bf_B) \sum_j e^{2\pi i\bar{g}.\bar{r}_j}$$

$$A_D(\bar{g}) = (f_B - f_A) \sum_j \bar{\sigma}_j e^{2\pi i\bar{g}.\bar{r}_j}$$

$A_B(\bar{g})$ contains all the information about the perfect lattice occupied by average atoms in the form of delta functions at the Bragg positions. $A_D(\bar{g})$ is the amplitude of the diffuse scattering containing information with respect to the ordering; its amplitude is zero at the Bragg positions, so that $A_D(\bar{g}).A_B(\bar{g}) \equiv 0$ for all $\bar{g}$.
Since $\langle\bar{\sigma}_j\rangle = 0$ (see definition of $\bar{\sigma}_j$) it follows that :

$A_D(\bar{g}_i) = 0$ $\bar{g}_i$ being the diffraction vectors of the Bragg reflections

The reduced amplitude :

$$a(\bar{g}) = \frac{A_D(\bar{g})}{f_B - f_A}$$

can be written as a Fourier sum :

$$a(\bar{g}) = \sum_j \bar{\sigma}_j\, e^{2\pi i\bar{g}.\bar{r}_j}$$

The reduced diffuse intensity is then obtained as :

$$I(\bar{g}) = a(\bar{g})\, a^*(\bar{g})$$

or :

$$I(\bar{g}) = \sum_{\ell\ell'} \bar{\sigma}_\ell \bar{\sigma}_{\ell'} e^{2\pi i\bar{g}.(\bar{r}_\ell - \bar{r}_{\ell'})} = N \sum_j \langle\bar{\sigma}_\ell \bar{\sigma}_{\ell+j}\rangle e^{2\pi i\bar{g}.\bar{r}_j}$$

N being the number of atoms in the crystal.
These averages of Flinn operators are related to the Warren-Cowley short range order parameters α_{oj}.

$$\alpha_{oj} = \frac{\langle\bar{\sigma}_\ell \bar{\sigma}_{\ell+j}\rangle}{\langle\sigma_\ell^2\rangle} = \frac{\langle\bar{\sigma}_\ell \bar{\sigma}_{\ell+j}\rangle}{m_A m_B}$$

Fourier inversion yields :

$$\alpha_{oj} = \frac{1}{m_A m_B N V^*} \int_{V^*} d\bar{g}\; I(\bar{g}) e^{-2\pi i\bar{g}.\bar{r}_j}$$

V^* being the volume of the reciprocal unit cell. This means that to determine the short range order (SRO) parameters α_{oj}, the diffuse intensity should be known over the complete reciprocal unit cell.

The short range order state is characterized in the diffraction pattern by a diffuse intensity, spread out over the whole of the reciprocal space; i.e. the diffuse intensity has a three dimensional distribution. In the final long range order state the intensity is concentrated in points forming the superstructure reflections (zero dimensional distribution).

Experiments teach us that in a large number of ordering systems an intermediate stage appears where the diffuse intensity is concentrated on a geometric locus, either a surface (a two dimensional distribution) or a curve (a one dimensional distribution). The ordering stage associated with a one or two dimensional distribution of the diffuse intensity is termed the "transition state". This state of order is clearly intermediate between SRO and LRO and can be considered as a prefiguration of the LRO state.

From the geometry of the diffuse intensity i.e. from the equation of the geometric locus, an ordering relation can be deduced and the clusters playing an important role in the ordering mechanism can be deduced.

Let $\chi(\bar{g}) = 0$

be the equation of diffuse intensity locus. Since the diffuse intensity should have the translation symmetry of the reciprocal lattice $\chi(\bar{g})$ can be written by means of its Fourier representation :

$$\chi(\bar{g}) \equiv \sum_k \omega_k \, e^{-2\pi i \bar{g}.\bar{r}_k} = 0$$

The requirement that the diffuse intensity should be located on this locus is expressed by the identity :

$$\chi(\bar{g}) \; a(\bar{g}) \equiv 0$$

This means that the diffuse scattering amplitude may only be different from zero along the geometric locus defined by $\chi(\bar{g}) = 0$.

Let us for simplicity reasons consider a binary ordering on a primitive unit cell (extrapolation to non-binary or non-primitive unit cells is straightforward and is described explicitely in [31]. Introducing expressions for $a(\bar{g})$ and $\chi(\bar{g})$ into the previous identity leads to :

$$\sum_j \sum_k \omega_k \bar{\sigma}_j \, e^{2\pi i (\bar{r}_j - \bar{r}_k).\bar{g}} \equiv 0$$

or :

$$\sum_j \left(\sum_k \omega_k \bar{\sigma}_{j+k}\right) e^{2\pi i \bar{g}.\bar{r}_j} \equiv 0$$

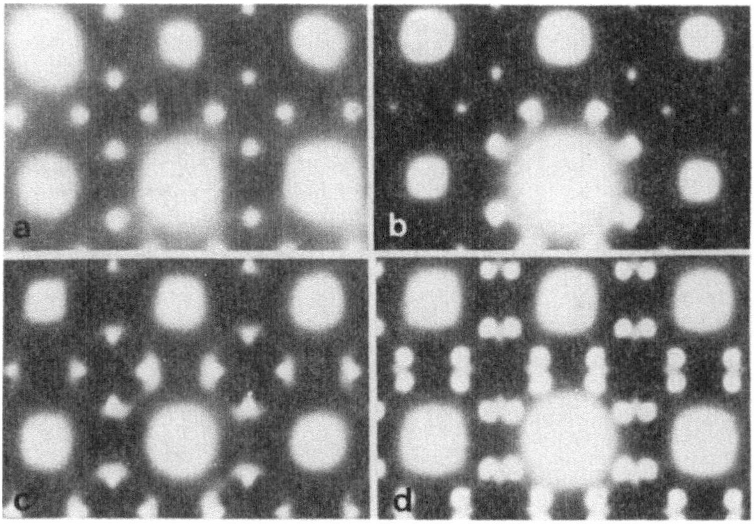

FIGURE 14. [001] diffraction sequence of ordering in Ni_4Mo upon irradiation at 300°C with 1MeV electrons.
(a) as quenched
(b)(c)(d) as function of continued irradiation for 15 sec, 2 minutes and 5 minutes.

which leads to a homogeneous linear relation between the Flinn operators :

$$\sum_k \omega_k \bar{\sigma}_{j+k} = 0 \quad \text{for all } j$$

This relation will be called the cluster relation; it defines a cluster with sites determined by the position vectors $\bar{r}_k$ ($k = 0, 1, \ldots, S_0-1$).

Ordering in $D1_a$-type alloys has been extensively studied since it is one of the few systems where the SRO reflections appear at a position different from the LRO-$D1_a$ positions. This perfectly allows to discern both ordering states. In all $D1_a$ type structures the final long range ordered structure is of course the same but the evolution from disorder or SRO toward LRO is different in most cases.

We will apply the previous formalism to the ordering of Ni_4Mo. Fig.14 shows an evolution sequence from the SRO state to the LRO state under electron irradiation. All diffraction patterns are recorded along one of the cube axes of the fcc solid solution. The typical SRO state with diffuse intensity maxima at 1 1/2 0 positions is obtained after quenching from the high temperature phase (fig. 14a). Upon irradiation with 1 MeV electrons at 300°C (or after annealing at more elevated temperatures) circular rings develop around the SRO spots (fig. 14b,c). Only later after further irradiation the long range order reflections at positions 1/5 2/5 0 develop (fig. 14d). The presence of this diffuse circle is typically a transition state between SRO and LRO.
The geometric locus of diffuse intensity is described by :

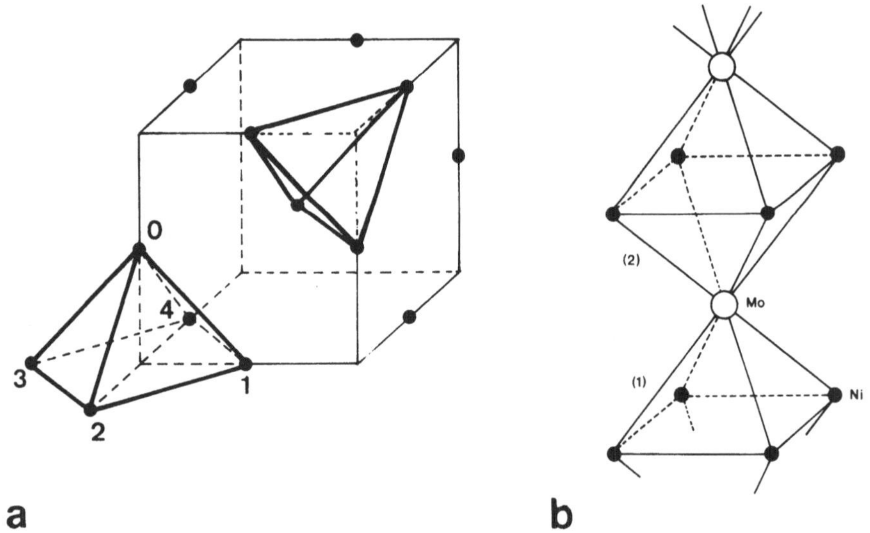

FIGURE 15. (a) Fcc lattice with indicated in heavy lines two important clusters in the ordering of Ni_4Mo : the tetrahedral cluster and the pyramidal cluster 0, 1, 2, 3, 4.
(b) The requirement that the pyramidal cluster should have 4 Ni at the base and one Mo at the top introduces Mo-Mo chains along the cube axes.

$$\begin{cases} \cos\pi h + \cos\pi k + \omega\cos\pi\ell = 0 \\ \sin \pi\ell = 0 \end{cases}$$

where ω determines the radius of the ring; for ω=1/2 it passes through the $D1_a$ reflections; for ω=1 it passes through the SRO spots. Following the general theory one finds the ordering relation (with respect to the FCC disordered phase).

$$\sigma_{(1/2\ 0\ 0)} + \sigma_{(-1/2\ 0\ 0)} + \sigma_{0\ 1/2\ 0} + \sigma_{0\ -1/2\ 0} + 2\omega\sigma_{0\ 0\ 1/2} = 0$$

where σ takes the values $-m_A$ and $+m_B$ depending on whether the site is occupied by a B or an A atom.
The ordering relation indicates that the important cluster is a five point pyramidal cluster (see fig. 15a) formed by a square base with corners (0,±1/2,0), (±1/2,0,0) and a top at (0,0,1/2). To optimally satisfy the cluster relation for m_A=4/5, m_B=1/5 and ω=1/2 a large number of clusters will have A atoms (Ni) at the base of the pyramid and a B atom (Mo) at the top.
This arrangement will lead to a rodlike structure along the cube axes. Indeed a first pyramid will have a nickel base and a Mo-atom at 0 0 1/2 but this atom is again the top of a second pyramid with, as a consequence, all Ni atoms as a base. This clustering will form Mo-Mo chains along [001], (fig. 15b); a result which has been confirmed by X-ray measurements [32] and which is important when considering high resolution microscopy.

7. HIGH RESOLUTION ELECTRON MICROSCOPY OF SHORT RANGE ORDERED ALLOYS AND THEIR EVOLUTION TOWARDS LONG RANGE ORDER

High resolution electron microscopy down to an atomic scale is till now only possible at room temperature, therefore all ordering states studied are "quenched in" situations. If one aims to study the SRO state one must be able to freese in this state at room temperature. For some alloys this is hardly possible, for others it is perfectly feasible. Au_4Cr (a member of the $D1_a$ series) requires annealing times of several hours close to the transition temperature to develop any long range order. To be on the save side we will consider this alloy as an example for obtaining SRO information through the HREM technique. Imaging is performed along the [001] section; a diffraction pattern similar to that of Ni_4Mo is reproduced in fig. 16. In order to image the weak ordering effects it is favourable not to use the

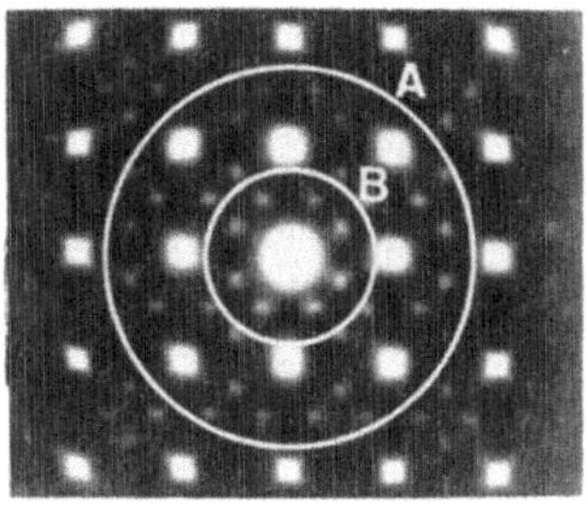

FIGURE 16. [001] diffraction pattern of Au_4Cr after quenching from the disordered state. SRO reflections are concentrated into 1 1/2 0 positions.

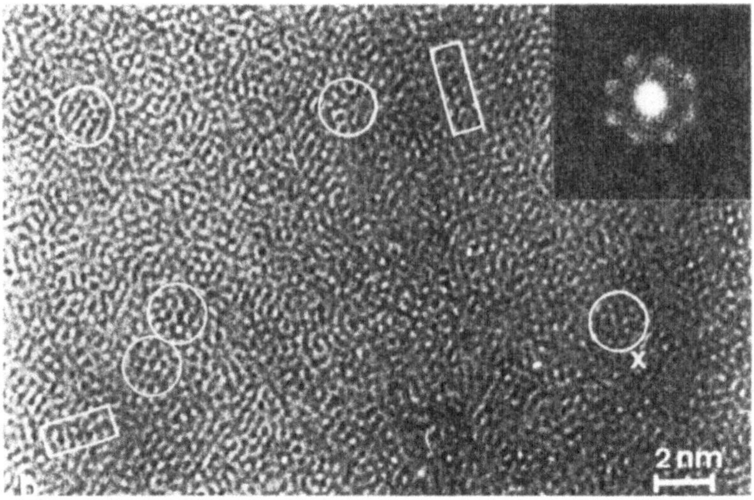

FIGURE 17. [001] High resolution image of Au_4Cr using the aperture indicated by B in fig. 16. Areas showing a lozenge type configuration of bright dots can be recognized in the encircled areas. The area marked X shows a square arrangement of bright dots with the same spacing and orientation as the Cr atoms in the $D1_a$ ordered phase.

larger objective aperture but rather a small one (indicated B in fig. 16) which excludes the Bragg reflections but contains the eight nearest SRO reflections inside its diameter. In the other case (aperture A in fig. 16) one would image a perfect fcc lattice but any possible ordering effects would be very weak. In the present case, using aperture B, an image of reduced resolution is obtained (see fig. 17). It clearly exhibits bright dots alligned in small clusters (encircled areas). In almost all clusters a triangular arrangement, characteristic of the minority atom configuration in a DO_{22} structure can be recognized. The borders of such clusters often show a square configuration as it occurs in the ordered $D1_a$ state. Occasionally (see circle X) a square arrangement as in the $D1_a$ ordering is also observed. It is very tentative to interpret these areas as $D1_a$ and DO_{22} ordered microdomains, however one should never forget that HR images only show a projected image and that in a SRO material ordering along the viewing direction is not obvious. In the previous paragraph however we have deduced from electron diffraction evidence that the alloys tend to form A-A chains along the cube directions; which is exactly the observation direction. Bright dots can therefore with a reasonable amount of credibility be considered as the reflection of a [001] Cr-chain. Chains which are formed along other cube directions will not be recorded in this image since for these there is no coherency along the electron beam. Overlap of clusters or incomplete [001] chains explain why the ordering effect is only relevant in thin foils and in restricted areas. Optical diffraction patterns obtained using the E.M. negative as a grating for the laser beam reflects the presence of the 1 1/2 0 SRO reflections (see inset of fig.17), indicating that the

information contained in these reflections is indeed recorded on the HR images. In a trial to simulate the SRO state we have constructed two dimensional models (assuming a perfect periodicity along the third dimension) based on $D1_a$, DO_{22} or $D1_a + DO_{22}$ clusters. The results are recorded in fig.18. A mixture of DO_{22} and $D1_a$ clearly gives the best results and produces maxima at different 1 1/2 0 positions.
It is not evident whether this state of order is the true SRO state for Au_4Cr (and Ni_4Mo) but it certainly reflects the tendency of ordering and indicates how this ordering will proceed.
In the initial stage clusters of DO_{22} and $D1_a$ with a large coherency along the cube directions are formed, later on the DO_{22} clusters gradually disappear and are replaced by $D1_a$ domains which then simply grow out to form the LRO state. More details on this subject can be found in [3].

Au_4Mn and Au_{4+x}, another member of the $D1_a$ series has also been thoroughly studied by X-ray and electron microscopy [33][34]. The short range order state is appreciably different from the one in Au_4Cr or Ni_4Mo; instead of having diffuse intensity maxima at 1 1/2 0 in reciprocal space, it exhibits diffuse streaks rather strictly bound between 1 1/4 0 and 1 3/4 0. A cube zone pattern is reproduced in fig. 19a.
High resolution electron microscopy images of the SRO state can be obtained in bright field with or without the fcc reflections [34]. Interpretation is far from straightforward (see e.g. the discussion in [34]) but it is in agreement with the local three dimensional Mn-arrangement in the SRO state as reconstructed from the SRO parameters by computer simulation [33]. It shows that Mn atoms tend to adopt chain like arrangements along three <100> directions. Upon further ordering the $D1_a$ structure will not develop

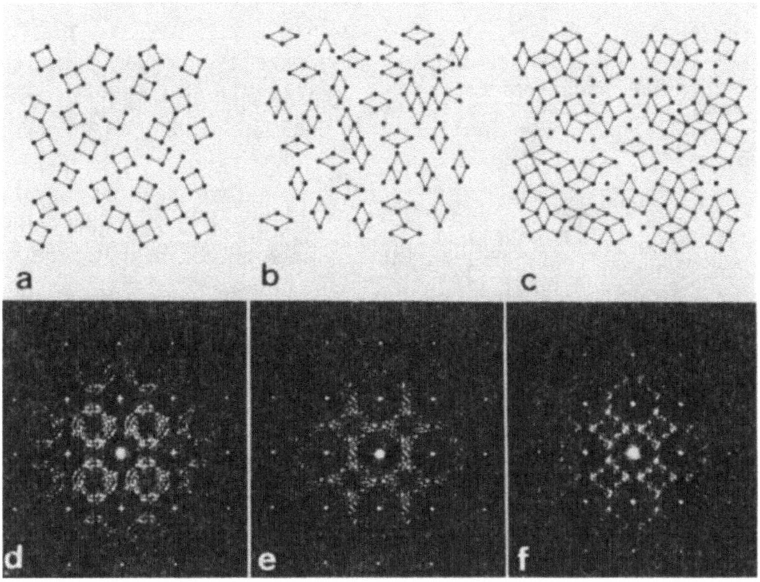

FIGURE 18. Two dimensional schematic representation of microclusters of $D1_a$ (a); DO_{22} (b) and $D1_a + DO_{22}$ (c) together with the optical diffraction of these models (d)(e)(f).

immediately but a transition phase consisting of islands of DO_{22} separated by $D1_a$ like boundaries will be formed [35]. The cube diffraction pattern over several of such variants is reproduced in fig.19b. Only later the long range order $D1_a$ phase will form out of this transition structure (fig.19c).

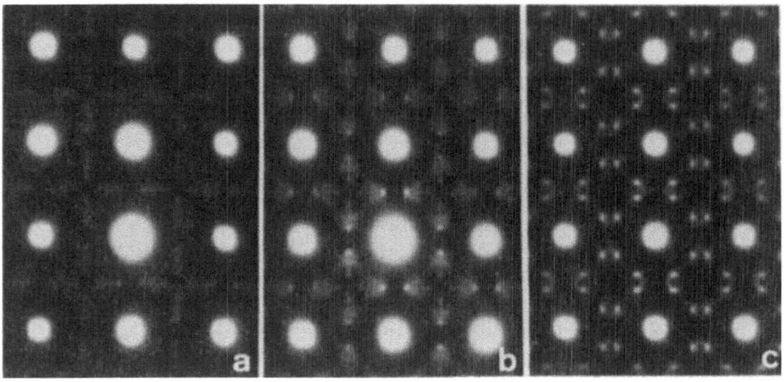

FIGURE 19. [001] electron diffraction patterns of Au_4Mn upon ordering under electron irradiation at 1 MeV.
(a) as quenched; (b) after 2 minutes at 473 K; (c) after 5 minutes at 473 K.

8. CONCLUSION

Electron microscopy has a great advantage over other techniques as it combines information from reciprocal space with information from direct space. Depending on the problem the most relevant information can be obtained by :

- electron diffraction
- convergent beam diffraction; this technique has not been discussed here but it is extremely useful for space group determination and nanoprobe diffraction information. More on convergent beam microscopy can be found e.g. in [36].
- conventional electron microscopy
- high resolution electron microscopy
- optical diffraction from the high resolution negative.

Side techniques associated with electron microscopy and benefitting from the electron-specimen interaction are :

- EDS (energy dispersive spectroscopy) i.e. an analysis of the produced X-rays allows to determine the local composition except for the lightest elements
- EELS (electron energy loss spectroscopy) an analysis of the energy loss suffered by the transmitted electrons allows to detect and sometimes to quantify the lighter elements.

Both techniques are treated explicitely in [37].

A study of ordered alloys, their SRO state and the evolution towards long range order is possible using a combination of these different techniques.

ACKNOWLEDGEMENT

Many ideas and figures of this collected work have been taken from original work together with S. Amelinckx, D. Van Dyck and J. Van Landuyt with whom the author forms a solid (state) group.

REFERENCES

1. P.B. Hirsch, A. Howie, R.B. Nicholson, D.W. Pashley and M.J. Whelan, Electron Microscopy of Thin Crystals, Butterworths 1965.
2. Diffraction and Imaging Techniques in Material Science, Eds. S. Amelinckx, R. Gevers and J. Van Landuyt, North Holland Publishing Co., Amsterdam (1978).
3. G. Van Tendeloo, S. Amelinckx and D. de Fontaine, Acta Cryst. B41, 281 (1985).
4. G. Van Tendeloo, S. Amelinckx, Acta Cryst. A30, 431 (1974).
5. R. Portier and D. Gratias, Journal de Physique 43, C4-17 (1982).
6. A. Finel, D. Gratias and R. Portier, in : L'ordre et le désordre dans les matériaux, Ecole d'Hiver (Aussois) 1984, Les Editions de Physique, Les Ulis Cedex (France).
7. J.W. Cahn and G. Kalonji in : "Solid-solid phase transformations", eds. H.I. Aaronson, D.E. Laughlin, R.F. Sekerka and C.M. Wayman, AIME publications (1982).
8. M.A. Melvin, Rev. Mod. Phys. 28, 18 (1956).
9. G. Van Tendeloo, S. Amelinckx, Phys. stat. sol. (a) 22, 621 (1974).
10. C. Boulesteix, Phys. stat. sol. (a) 86, 11 (1984).
11. R. Serneels, M. Snykers, P. Delavignette, R. Gevers, S. Amelinckx, Phys. stat. sol. (b) 58, 277 (1973).
12. A.J. Morton, Phys. stat. sol. (a) 44, 205 (1977). Acta Metal. 27, 863 (1979).
13. M. Snykers, P. Delavignette, S. Amelinckx, Phys. stat. sol. (a) 48, K1 (1971).
14. J.W. Edington, Practical Electron Microscopy in Materials Science nr. 5, Philips (1975).
15. J.C.H. Spence, "Experimental High Resolution Electron Microscopy", Clarendon Press Oxford (1981).
16. D. Van Dyck in [2], p. 355.
17. D. Van Dyck, J. Microscopy 119, 141 (1980).
18. D. Van Dyck, G. Van Tendeloo and S. Amelinckx, Ultramicroscopy 10, 263 (1982).
19. G. Van Tendeloo, D. Van Dyck, S. Amelinckx, Ultramicroscopy 19, 235 (1986).
20. I. Baele, G. Van Tendeloo and S. Amelinckx, Acta Metall. 35, 401 (1987).
21. "Crystal growth and characterization of polytype structures" ed. P. Krishna, Pergamon Press (1983).
22. C.H. Johansson and J.O. Linde, Ann. Phys. 25, 1 (1936).
23. J. Van Landuyt, R. De Ridder, R. Gevers and S. Amelinckx, Mat. Res. Bull. 5, 353 (1970).
24. A.B. Glossop and D. Pashley, Proc. Royal Soc. A250, 132 (1959).
25. D. Schryvers, G. Van Tendeloo, J. Van Landuyt, S. Amelinckx, Phys. stat. sol. (a) 75, 607 (1983).
26. A. Loiseau, G. Van Tendeloo, R. Portier and F. Ducastelle, J. Phys. (Paris) 46, 595 (1985).
27. D. Broddin, G. Van Tendeloo, J. Van Landuyt, S. Amelinckx, R. Portier, M. Guymont and A. Loiseau, Phil. Mag. A 54, 395 (1986).
28. J. Gjønnes, Acta Cryst. 20, 240 (1966).
29. J. Cowley, R. Murray, Acta cryst. A24, 329 (1968).

30. R. De Ridder, G. Van Tendeloo, S. Amelinckx, Acta Cryst. A32, 216 (1976).
31. R. De Ridder, G. Van Tendeloo, D. Van Dyck, S. Amelinckx, Phys. stat. sol. (a) 38, 663 (1976); Ibid. 43, 541 (1977).
32. B. Chakravarti, C.J. Sparks, E.A. Starke, R.O. Williams, J. Phys. Chem. Solids 35, 1317 (1974).
33. K. Ohshima, J. Harada, M. Matsui, K. Adachi, J. Magn. Mater. 54-57, 157 (1986).
34. N. Tanaka, J.M. Cowley and K. Ohshima, Acta Cryst. B43, 41 (1987).
35. G. Van Tendeloo, R. De Ridder, S. Amelinckx, Phys. stat. sol. (a) 49, 655 (1978).
36. J. W. Steeds in "Introduction to Analytical Electron Microscopy" Eds. J.J. Hren, J.I. Goldstain, D.C. Joy, Plenum Press, New York, 1979.
37. "Principles of Analytical Electron Microscopy", eds. D.C. Joy, A.D. Romig,Jr., J.I. Goldstein, Plenum Press, New York, 1986.

Quantitative statistical description of the long period antiphase boundary structures from their high resolution image

A. LOISEAU, J. PLANES, F. DUCASTELLE

Office National d'Etudes et de Recherches Aérospatiales,
B.P. 72, 92322 Chatillon Cedex, France

1 INTRODUCTION

Studied for years, the long period structures (LP) are a very intriguing order phenomenon occurring in numerous and very different metallic alloys such as noble metal alloys like Cu_3Au, Cu_3Pd, Ag_3Mg..., known for their long range interactions and such as $TiAl_3$ [1] or such as Pt_3V alloys, [2, 3] certainly characterized by short-range interactions [4, 5]. Regarding their structure, they can be described, basically, as one dimensional commensurate or incommensurate arrangements, along the direction of the long period, of $L1_2$ domains bounded by conservative antiphase boundaries (APB). From experimental electron microscopy studies, very different topologies emerge depending on the following factors : the value of the mean size of the domains, M, as compared to the $L1_2$ cell parameter, a_0, the nature, continuous or discontinuous, of the variation of M with concentration and temperature, and the degree of rigidity, or roughness, of the APB i.e. the degree of freedom in the APB positions around the mean position M. APB can be very straight as in Pt_3V, Al_3Cu [6] or jogging over one atomic plane as in $TiAl_3$ [1] or wavy and spread over a few atomic planes as in Cu_3Pd [7, 8].

LP structures are relevant to the theoretical framework of one-dimensional commensurate/incommensurate structures, on which extensive works have been done in recent years [see e.g. 9,10]. Although the nature and the range of the effective atomic interactions responsible for the stabilization of LP structures are not yet elucidated, their topological properties can be fruitfully apprehended within a microscopic thermodynamical model called the ANNNI model. This is a particular anisotropic Ising model which describes however general features which may apply to any system with competive interactions. A detailed review on this model and on its applications to the LP structures will be found in the contribution by Selke in this issue. As found in this model we have experimentally shown that, qualitatively , the APB distribution can in all cases be described by a smoothed square wave function : the discontinuities of the function define the mean positions of the APB and the roundings of its corners are related to the diffuseness in the APB positions i.e. physically to a local disorder confined to the APB and induced by the temperature [1, 11, 12].

In order to determine precisely the shape of this function and to compare quantitatively the different experimental situations, a statistical description of the HREM images of the structures is necessary. Such an analysis requires a numerical treatment of the images which is described in this paper. It is then applied to various situations.

2 STATISTICAL DESCRIPTION OF THE LP STRUCTURES

2.1 Definition and signification of the representation by a smoothed square wave function

In the low temperature regime LP structures are perfect and are characterized by an uniform distribution of APB, which are straight and without jogs. Their positions along the long period axis then completely describe the structure. A convenient representation is the perfect square wave function of period 2M, involved in Fujiwara's model and defined as follows :

$$f_M(x) = 1 \text{ if } 2kM + \alpha \le \xi < (2k+1)M + \alpha$$
$$= -1 \text{ if } (2k+1)M + \alpha \le \xi < (2k+2)M + \alpha$$

where $k = 0,1,2$ .... and α is an arbitrary phase factor.

This representation involves two lengths M and a_0. Discontinuities of the function univoquely define APB positions as shown in Fig1a. We recall that the structure is a stacking along the c axis of alternate pure A planes and AB planes. In the ordered AB planes, B atoms occupy one of the two atomic site types, labelled + and - (Fig

G. M. Stocks and A. Gonis (eds.), Alloy Phase Stability, 101–106.

1); the role of an APB is to change the nature of the site occupied by B atoms. If p(n) defines the occupation number of the sites of type + by the B atoms in the plane at position n, we then have :

$$p(n) = (1 + f_M(n))/2 \;; \quad p(n) = 1 \text{ if sites + are occupied by B atoms}$$
$$= 0 \text{ in the opposite case.}$$

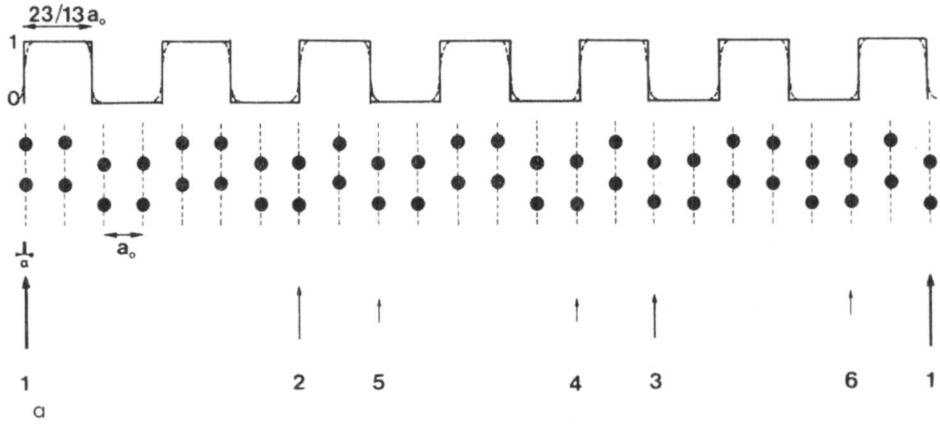

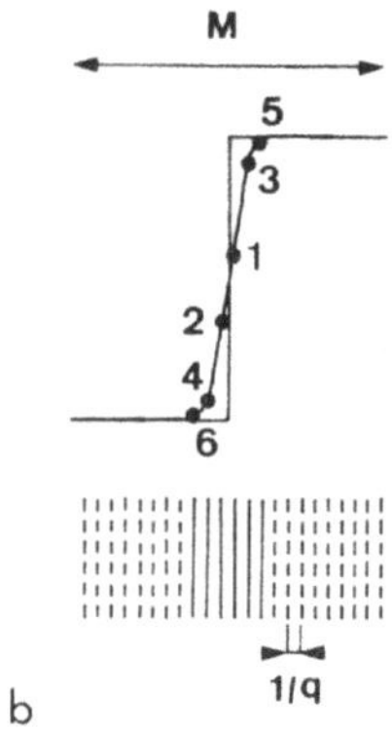

Fig 1 : a) One-dimensional representation p(n) for the LP structure M = 23/13 a_0 projected along its a(or b) axis. Streaks indicate positions of the mixed planes and black dots the B atoms located on atomic sites + or -. The solid curve corresponds to the low temperature case and the dashed line to the high temperature case : arrows indicate the more or less disordered planes classified with the numbers 1 to 6. b) Reduced representation p(n) modulo M = 23/13 a_0 : ordered planes are drawn in dashed lines and disordered planes in solid lines

In the high temperature regime, an entropic disorder arises : it is confined to the APB which are no more perfectly planar but can move by jogs around a mean position within a certain width. Provided that there is no correlation between the defects occurring on the APB, a one-dimensional description remains valid. On the basis of experimental images it has been shown that a convenient representation is a smoothed square wave function of period 2M, the discontinuities extending now on the interval 2 x_0, in which APB positions are free. As shown in Fig1a, this induces a local disorder in the atomic planes lying within the interval 2 x_0, and defects in the arrangements of domains: the occupation number is between 1 and 0 in this interval. This function involves three lengths : a_0, M and x_0, these two latter lengths being temperature dependent. One must keep in mind that this representation does not describe the nature of the fluctuations of the APB, which is a 3d-problem, but takes only into account their amplitude, the fluctuations being assumed to be randomly distributed.

The disordered part of the structure has striking topologic properties characterizing both the uniformity and the periodicity in the APB distribution which can be finely analyzed. Let us consider in Fig 1 the M = 23/13 structure in both low and high temperature regimes. Within a period p of the structure - here p = $23a_0$ - the atomic planes are all inequivalent and their relative disorder is univoquely defined by their distance to the symmetry centre of the APB profile. There is thus a precise hierarchy of the diffuse planes which is the arithmetical signature of the uniformity. The smoothing chosen in Fig 1 concerns thus six different planes, sketched by the arrows, which belong each to domains 2 bordering on a domain 1. As far as x_0 is smaller than a_0

- which is necessarily the case when M is between 1 and 2 - the partially disordered planes are never consecutive but at least separated by one ordered plane. Consequently, each APB is related at most to one diffuse plane. In order to display the hierarchy of the diffuse planes a very convenient way is then to plot the function modulo M (Fig 1b). In this reduced representation, the diffuse planes are now consecutive and naturally classified by their distance to the symmetry center of the profile. Defining α as the smallest distance (α is necessarily smaller than $1/2q$), these distances are α, $n/q \pm \alpha$ with $M = p/q$ and $n = 1,2..$, and the distances between consecutive planes is $1/q$. Thus, the larger q the larger the number of different diffuse planes.

2.2 Statistical analysis of a high resolution image

To determine the square wave function, the available data are the high resolution images, which are a 2d-projection of the structure along its a (or b) axis. For perfect LP structures, the projected atomic columns are pure A or pure B columns, the latter ones appearing as white dots for convenient defocus and thicknesses. At the location of APB jogs, atomic columns become somewhat disordered and appear as grey dots, sometimes slightly displaced, or completely diffuse, depending on their degree of disorder [1, 13](see ex in Fig 2).

The first step in the analysis is to determine from an image an 1d statistical description of the sequence of domains i.e. the 1d occupation numbers p(n). The numerical treatment of an image is as follows. We assume that bright dots correspond to pure B columns and diffuse or grey dots to completely disordered columns. One can thus define, in each plane n, an approximate occupation number $l(n,n_b)$ of the atomic column of type + at the position n_b along the b axis perpendicular to the c axis:

$$l(n,n_b) = \begin{cases} 1 & \text{if the column + is observed as a white dot} \\ 0 & \text{if the column + is observed as a dark dot} \\ 1/2 & \text{if the column + corresponds to a diffuse or grey dot} \end{cases}$$

Projecting now the plane n along the b axis, one obtains the occupation number p(n) of sites + by B atoms for this plane :

$$p(n) = \sum_{n=1}^{n=L} l(n,n_b) / L$$

where L is the total number of columns + of the plane n in the area over which the analysis is performed. Note that this analysis is only valid if the average estimated over the b axis is independant of that realized in the formation of the image in the microscope along the a axis. This condition is fulfilled if the thickness of the foil is large compared to the distance between jogs and small compared to L.

The second step consists, in determining through a last square analysis, the best continuous hull function p(x) interpolating between the values p(n) in the form of a periodic uniform function, i.e. a smoothed square wave function defined with three parameters : the periodicity 2M, the phase α and the broadening factor x_0, which is a measure of the APB width. In practice, the p(n) values are plotted modulo M for a set of M values. M is determined with a very high accuracy by minimizing the dispersion between the points. Once M is known, the best APB profile is found. If the APB distribution is well uniform, the p(n) values are well fitted by a smooth continuous function : the APB profile is explicitely fitted by the simple empirical form $f(x) = (1 + \tanh(x / x_0))/2$ that reproduces the asymptotical exponential variations, which are involved in classical soliton theories [9]. The better defined this function, the larger the number of p(n) values i.e. the number of inequivalent diffuse planes. In the limit of incommensurate M values, this number is infinite so that the APB profile is perfectly defined. Conversely, for commensurate $M = p/q$ values, the APB profile will be precisely determined for large q values only. In these cases, however, the phase α is inaccurately defined. On the contrary, for commensurate values with small q, the APB profile can not be determined but the phase α is known with high accuracy.

3 ANALYSIS OF LONG PERIOD STRUCTURES IN TI - AL ALLOYS

We now present applications of this method to the LP structures observed in Ti-Al alloys. We recall that, in this system, a large number of structures (about 20) has been observed by varying the temperature with M values between 1.33 and 1.77 in $Ti_{28}Al_{72}$ [1] and between 1.83 and 2 in $Ti_{33}Al_{67}$ (the structure M = 2 has been very recently observed, [14]). M can take very simple rational values such as 5/3, 7/4 or in between very complicated rational values such as 39/22. There is no order-disorder transition before melting but for temperatures higher than 1000°C, LP structures contain a small disorder in the form of jogging APB over one atomic plane. The complicated structures have been statistically analysed in order to determine the shape of the hull function as precisely as possible.

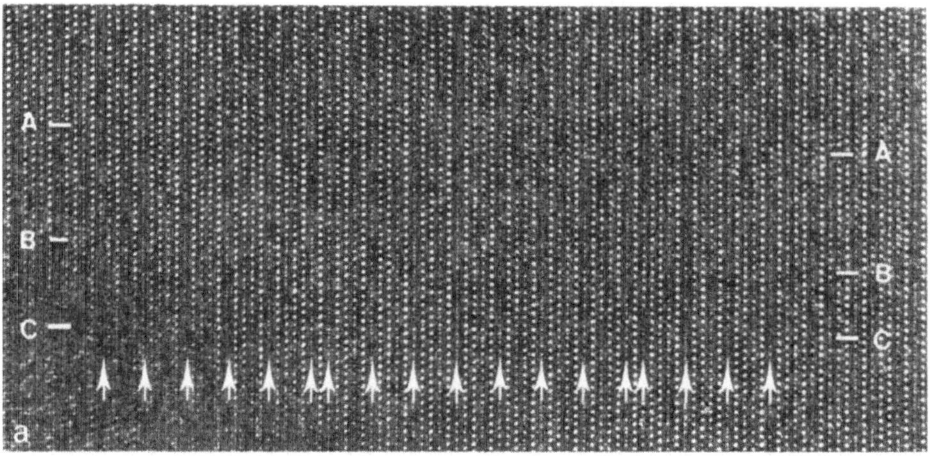

Fig 2 : High resolution image of a LP structure observed in $Ti_{28}Al_{72}$ annealed at 1150°C.

Fig 2 reproduces a high resolution image of a LP structure having a M value very close to 5/3 which shows well ordered planes with white dots and more or less diffuse planes . The statistical analysis has been performed on a wide area, 33nm x 17nm, of constant thickness and constant optical conditions. Results are given in Fig 3. Fig 3a is a plot p(n) and Fig 3b the plot p(n) modulo M. One can appreciate in Fig 3b the quality of the fit by the very weak dispersion of the points around the fitted curve. The highest dispersion lies in the extremities of the profile and concern experimental points which are less accurately estimated since they correspond to planes containing some weakly disordered columns assumed to be completely ordered in this analysis. M is determined with a very high accuracy (better than 5°/₀₀) and found to be very close to the rational number 27/16 = 1.683. The structure is thus periodic with a period $p = 27\ a_0$. The phase α is close to zero. 10 over the 27 atomic planes are partially disordered according to a hierarchy fully consistent with a uniform structure. The plot p(n) modulo p (Fig 3c) gives an other evidence for this hierarchy : planes, equivalent up to a translation λp, yield identical p(n), with a dispersion less than 5%. This effect is not visible by eye on the image and spectacularly emerges from the statistical analysis.

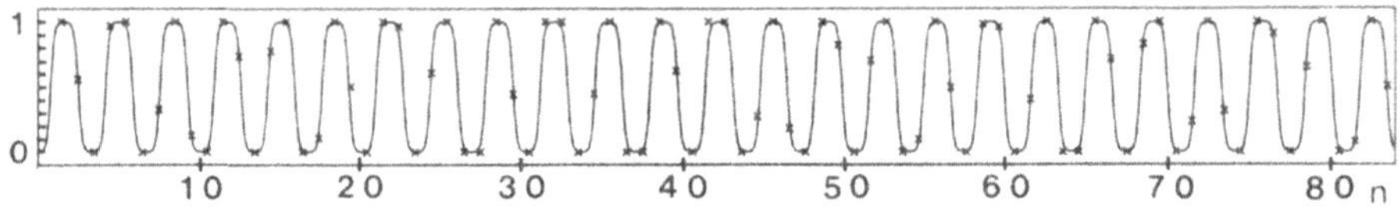

a

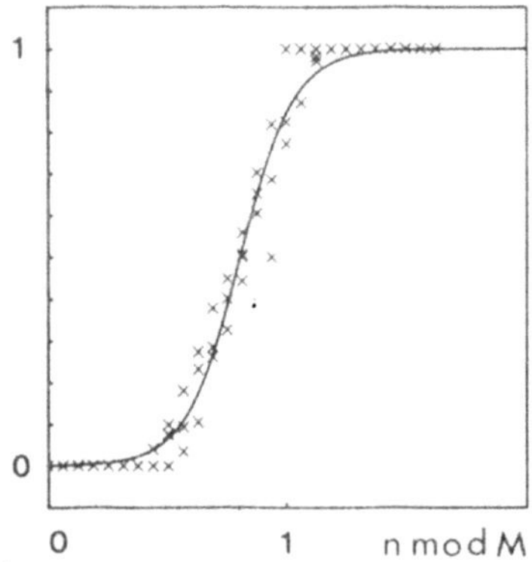

b

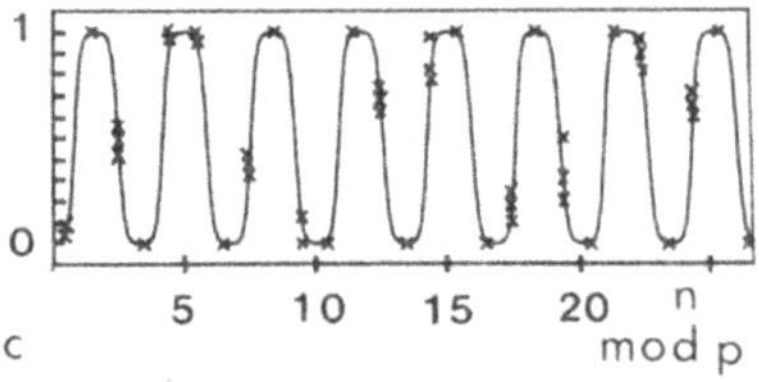

c

Fig 3 : Statistical analysis of the LPS of Fig 2 ; M = 27/16 = 1.683 , $x_0 \approx 0.2\ a_0$, $\alpha \approx 0.002$. The crosses are experimental occupation probabilities p(n) estimated numerically from the image and the solid line is the fitted curve.

a) extended representation. b) p(n) plotted modulo M.

c) p(n) plotted modulo the period $p = 27\ a_0$; Three successive periods are superimposed.

Having determined the hull function for a structure close to 5/3, we have extended it by continuity to the case M = 5/3 observed in very similar temperature conditions. Fig 4 shows the result of the statistical analysis performed on a high resolution image. In that case, the structure contains only one diffuse plane per period so that any hull function may a priori fit the experimental points provided the phase α is fixed. The hull function drawn in Fig 4 is defined with the APB profile tanh x/x_0 with $x_0 \approx 0.23$, value found for M = 27/16 : the phase is then univoquely determined and is very close to zero, $\alpha = -1.10^{-3}$. This result differs from that of the ANNNI model for which phases are maximum (1/2q). In such a case, the 2 planes bordering on the domain 1 in the sequence 221characterizing the M = 5/3 perfect structure should be equally diffuse. This is clearly not the case.

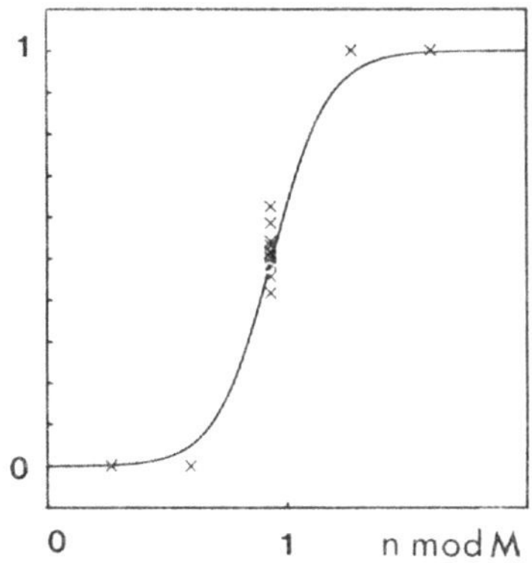

Fig 4 : Statistical analysis for the LP structure M = 5/3 ; The crosses are experimental occupation probabilities p(n) estimated numerically from the image and the solid line is the fitted curve., the APB profile being that found for the case M = 27/16.

4 CONCLUSION

These examples show the power of a statistical analysis of the HR images for determining with a good accuracy the hull function of the APB modulation and for testing its periodicity and its uniformity. It is thus confirmed that in Ti-Al alloys LP structures can remarkably be represented by commensurate smoothed step functions with an APB width $\approx 0.2\ a_0$ in the range 1150-1200°C. Similarly, we have shown [8,15] that this representation also nicely applies to incommensurate structures with large wavy APB observed in Cu-Pd with M $= 10.35\ a_0$ and $x_0 = 0.94\ a_0$. Topological differences between the two cases come from the relative values of the three lengths a_0,M, x_0. In both cases, $x_0 / M \approx 0.1$. However, in Cu-Pd, M / a_0 is large, i.e. larger than 2, so that diffuse planes can be consecutive ; in this case they reveal the shape and the width of each APB on the image. APB profiles are thus directly visible on the HR image. x_0 / a_0 is also much larger, so that the APB are less pinned to the lattice and the structures are incommensurate. This situation corresponds to a high temperature regime. On the other hand, the Ti-Al system is more typical of an intermediate temperature regime.

This work was supported in part by the Direction des Recherches et Etudes Techniques through Grant n° 86-34-001.

REFERENCES

1. Loiseau A, Van Tendeloo G, Portier R, Ducastelle F, J. Physique 46, 595 , 1985
2. Schryvers D, Amelinckx S, Acta Metall. 34, 43, 1986
3. Planès J, Loiseau A, Ducastelle F to be published in the Proceedings of the EMAG 87 Conference (Manchester 6-10 September 1987), 1987
4. Bieber A, Gautier F, Acta Metall. 34, 2291, 1986
5. Solal F, Caudron R, Ducastelle F, Finel A, Loiseau A, Phys. Rev. Lett. 58 n° 21, 2245, 1987
6. De Graef M, Broddin D, Van Humbeek J, Delay 1, Proc. XIth Int. Cong. on Electron Microscopy (Kyoto,1986) 845 and present issue
7. Broddin D, Van Tendeloo G, Van Landuyt, Amelinckx S, Portier R, Guymont M, Loiseau A, Phil.Mag. 54, 395, 1986
8. Broddin D, Van Tendeloo G, Van Landuyt J, Amelinckx S, Loiseau A to be published in Phil. Mag., 1987

9. Bak P, Rep. Prog. Phys. 45, 587, 1982
10. see e.g. Aubry S, J. Physique 44, 147, 1984
11. De Fontaine D, Kulik J, Acta Metall. 33, 145, 1985; Kulik J, PHD Thesis (Berkeley, USA, 1987)
12. Selke W, Fisher M E, Phys. Rev. B20, 257, 1979
13. Portier R, Gratias D, Guymont M, Stobbs W M, J. Micros. 119, 163, 1980
14. Loiseau A, Vannuffel C, submitted to Solid State Phys., 1988
15. Broddin D, Van Tendeloo G this issue

LONG-RANGE ORDER AND SHORT-RANGE ORDER IN Pd_3V: BREAKDOWN OF THE MEAN FIELD THEORY

F. SOLAL, R. CAUDRON, A. FINEL
Office National d'Etudes et de Recherches Aérospatiales
29 avenue de la Division Leclerc, B.P. 72, 92322 Châtillon Cedex (France)

1. INTRODUCTION

The electronic structure calculations of pair potential interactions have been widely developed in the case of transition metal alloys [1]. However, few experimental data exist in this case. Among the possible systems we choose the Pd-V one, which is not magnetic [2] and shows little size effects (small disparity between Pd and V atomic size). In this way, we hope to isolate chemical order effects and then, to perform suitable experiments to compare with theoretical calculations. We present results on Pd_3V which are quite complete. We performed diffuse scattering experiments of neutrons and X-rays which give rise to intensity maxima at positions (100). This indicates a local order different from that of the low temperature DO_{22} ordered phase, characterized by (1½0) diffraction spots. After comparing in detail both experiments, we show how this result is in contradiction with the usual mean field theory. However, the cluster variation method (octahedron-tetrahedron approximation) with first and second neighbor interactions can explain this effect. Introducing third and fourth neighbor interactions V_3, V_4 while keeping the tetrahedron-octahedron entropy expression, we propose a solution for the four first interactions in good overall agreement with experimental data.

2. DIFFUSE SCATTERING EXPERIMENTS ON Pd_3V

2.1. Neutron diffuse scattering experiments

Neutron diffuse scattering experiments have been performed on a Pd_3V single crystal on a two axis 64-detector spectrometer with time of flight analysis. The results for T = 840°C and 940°C are displayed on fig. 1a and b respectively.

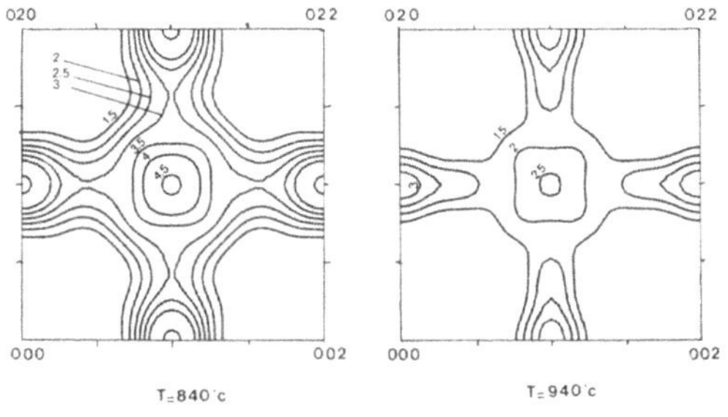

FIGURE 1. Short range order intensities for Pd_3V at T=840°C and T=940°C on the (100) plane obtained by in situ neutron diffuse scattering measurements

These figures show the short-range order intensities in the (100) plane after performing all the following corrections [3] (intensities are expressed in Laue units):

- the experimental intensities show little background due to the sample

G. M. Stocks and A. Gonis (eds.), Alloy Phase Stability, 107–111.

environment. We have evaluated and corrected it by measuring the intensity induced by the environment and by a good absorber of neutron (boron nitride) of the same shape as the sample;
- the elastic part of the signal has been separated;
- the Debye-Waller factor has been measured in situ on a powder sample with neutrons of wave-length λ = 1 Å; We found B = 1.66 Å^2 and B = 2.5 Å^2 for T = 840°C and T = 940°C respectively;
- the first order displacement term has been corrected by means of mean square fitting.

2.2. X-ray diffuse scattering experiments

X-ray diffuse scattering experiments have been performed on a four circle spectrometer on a Pd_3V single crystal quenched from T = 840°C. The measurements were made at liquid nitrogen temperature. The intensity was normalized via the intensities of Ni powder diffraction peaks. We also performed thermal diffuse scattering corrections [4] using the Pd elastic constants. We used the Sparks and Borie method [3] to eliminate the first order displacement term. The measurements were done in the whole necessary volume of the reciprocal space; fig. 2 shows the resulting intensities on the (100) plane.

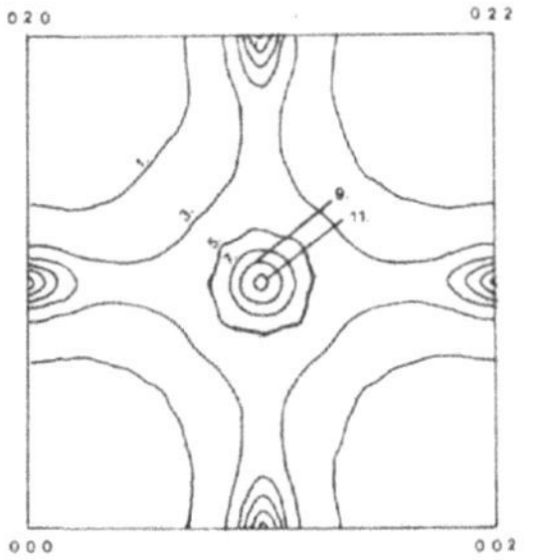

FIGURE 2. Short range order intensities for Pd_3V at T=840°C on the (100) plane obtained by X-ray diffuse scattering measurement on a quenched sample

Both kinds of measurements exhibit the same main features. The position of the intensity maxima is at (100) points. However, the maximum intensity as well as the contrast between the (100) and the (1½0) intensities are much higher for X-ray measurements than for neutron ones. To avoid any systematic error, we have checked this result as follows:
- we measured the neutron intensities at several temperatures down to T = 820°C (the order-disorder transition temperature of Pd_3V is 815°C). Even at this low temperature the neutron intensities are not as high as the X-ray ones;
- we measured the neutron intensity of the X-ray sample (quenched from T = 840°C) on the <110> line. The agreement between both techniques is quite good on the same sample. We conclude that the state of the quenched sample, even if obviously disordered, is not an equilibrium state.

Thus, in spite of these quantitative disagreements, there is still a remarkable qualitative results: the short-range order in Pd_3V is of (100) type instead of the expected (1½0) one.

3. DISCUSSION OF THE RESULTS ON Pd_3V

The mean field theory provides an analytical relation between the SRO diffuse scattering intensity and the Fourier transform of the pair interactions, the so-called Krivoglaz-Clapp-Moss relation.

To stabilize DO_{22} as a ground state, at least first and second neighbor interactions, V_1, V_2 are necessary with $V_1 > 0$ $V_1/2 > V_2 > 0$. According to the mean field theory, this regime of V_1, V_2 gives rise to maxima of diffuse intensity at (1½0) points, whereas they occur at (100) points when $V_2 < 0$, $L1_2$ being then the ordered phase (fig. 3). This is not what is experimentally seen in Pd_3V. Obviously we needed a better thermodynamical model; we then performed CVM calculations in the tetrahedron-octahedron approximation with the interactions V_1 and V_2. The resulting phase diagram is shown in fig. 4, compared to the equivalent mean field diagram fig. 3 [5].

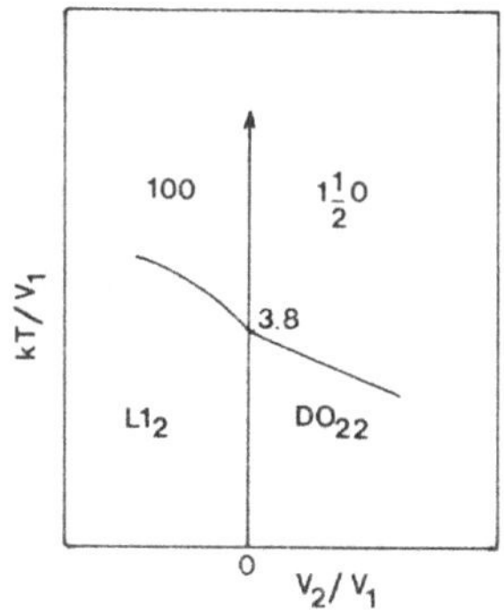

FIGURE 3. F.C.C. phase diagram obtained within Bragg-Williams mean field theory. In the disordered phase, the instability k vector (i.e. maximum of diffuse scattering) ⟨100⟩ or ⟨1½0⟩ is indicated

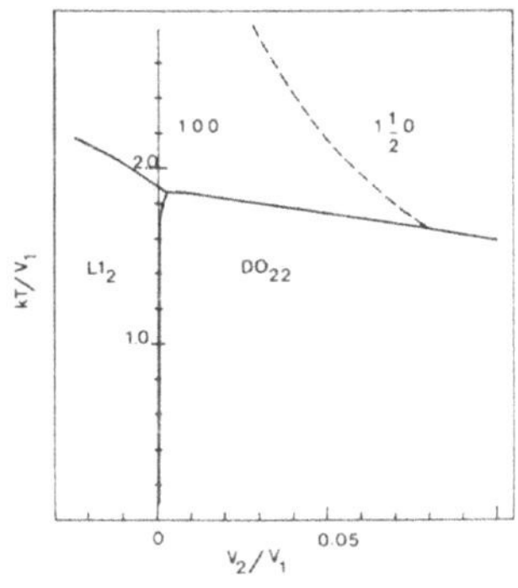

FIGURE 4. F.C.C. phase diagram obtained using CVM in the tetrahedron-octahedron approximation

We have proved experimentally that Pd_3V corresponds to the domain of V_1, V_2 where the ordered DO_{22} phase disorders towards a (100) type phase. In addition to the diffuse scattering intensities, we measured the variation of both order parameters η_1 and η_2 that describe the DO_{22} structure. η_1 corresponds to (100) diffraction spots and η_2 corresponds to (1½0) diffraction spots. We found the order-disorder transition being strongly of first order with a jump of order parameter of about .8 at the transition and a two-phase domain of about 10 degrees (fig. 5). This two-phase domain has also been observed by electron microscopy. These measurements are consistent with CVM calculations. This result is important as an order-disorder with such a change in local order should be of first order.

The experimental results on Pd_3V are qualitatively well explained using only first and second neighbour interactions. This is consistent with recent calculations showing that, in transition metal alloys, the ordering pair interactions should be short ranged [1]. In fact these calculations predict that interactions up to fourth neighbours are relevant. Qualitatively, it is evident that the phase diagram of fig. 4 would remain

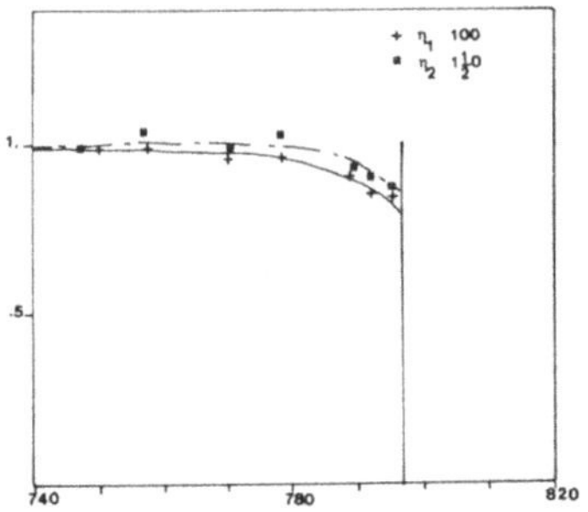

FIGURE 5. Both order parameters of Pd_3V versus temperature. This was obtained by meaning all equivalent (100) and ($1\frac{1}{2}0$) superstructures spots by in situ neutron diffraction

valid when changing V_2 into the antiphase energy $\xi = V_2 - 4V_3 + 4V_4$. In fact, it is the value of that energy that seems to play the main part in this effect even if this only parameter does not describe the whole phenomenon. When only $V_1 \neq 0$, all structures built on $L1_2$ by introducing conservative antiphase boundaries perpendicular to a ⟨100⟩ direction have the same ground state energy, DO_{22} being the simplest of these structures. All of them give rise to superstructure spots located on the [1ξ0] lines. In the disordered state, according to the mean field theory, when the interactions are short-ranged, the intensity is almost entirely concentrated on these lines and constant along them. Our calculations show that entropy effects lift this degeneracy in favour of ⟨100⟩ type ordering. When V_2 or ξ is small, this leads to a quasi-degeneracy of the system, Pd_3V gives an example of such degeneracy.

We then introduced third and fourth neighbours interactions in the CVM framework, while keeping the tetrahedron-octahedron entropy formula. We found not too bad an agreement with our neutron diffuse scattering data when $V_2 = -.08\ V_1$, $V_3 = .130\ V_1$, $V_4 = .170\ V_1$. Even if the agreement is not perfect, the improvement from fig. 6 to fig. 7a and 7b is very important and promising.

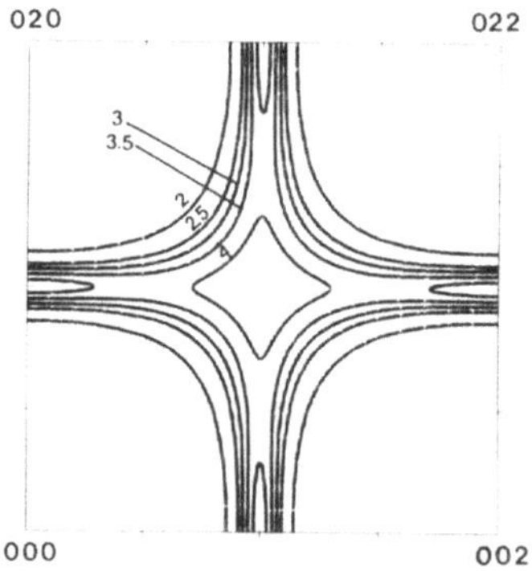

FIGURE 6. Calculated diffuse scattering intensities in the tetrahedron-octahedron CVM approximation (V_1 and V_2) $kT/V_1 = 2$ and $V_2/V_1 = .05$

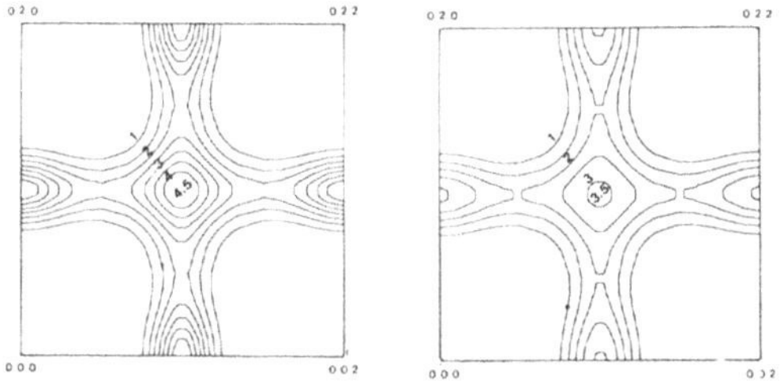

FIGURE 7. Calculated diffuse scattering intensities in the tetrahedron-octahedron CVM approximation with $V_2/V_1 = -.08$ $V_3/V_1 = .130$, $V_4/V_1 = .170$, $kT/V_1 = 1.835$ (fig. a) and $kT/V_1 = 2.$ (fig. b)

4. CONCLUSION

We have showed theoretically and experimentally that for A_3B alloys on the fcc lattice, there is a regime for which the order-disorder transition happens with a change in local order: the DO_{22} ordered structure, characterized by (1½0) concentration waves, disorders towards a (100) type disordered phase. This effect can only be explained with a thermodynamic model beyond the Bragg-Williams mean field theory, more precisely using a better approximation for the entropy. The CVM in the tetrahedron octahedron approximation provides a sufficient approximation to explain qualitatively this effect. Evidently the introduction of further interactions (V_3,V_4) in the model improves the understanding of the phenomenon. A better CVM approximation including the four first interactions on an equal footing is under study. We also aim to study other concentrations of Pd-V and similar systems, like Ni-V for instance.

ACKNOWLEDGEMENTS

This work was supplied in part by DRET (Direction des Recherches et Etudes Techniques) through the grant n° 86.34.011.

REFERENCES

1. see Ducastelle's lecture notes and references therein
2. Williams AR, Zeller R, Moruzzi VL & Gelatt(Jr) CD: J. Appl. Phys. **52**, 2067 (1981)
   Burmster WL & Sellmyer DJ: J. Appl. Phys. **53**, 2024 (1982)
   Mixmer DK, Auluck S, Sellmyer DJ, Jaswal SS & Arko AJ: Phys. Rev. B **31**, 3356 (1985)
   Xu Jian-Hua: Acta Phys. Sin. **34**, 1373 (1985)
3. de Novion CH: in **l'Ordre et le Désordre dans les Matériaux** edited by Reynaud F, Clement N and Couderc JJ (Les Editions de Physique, Paris 1984), p. 243
4. Walker CB & Chipman DR: Acta Cryst. **A28**, 572 (1972)
5. Solal F, Caudron R, Ducastelle F, Finel A & Loiseau A: Phys. Rev. Lett. **58**, 2245 (1987)

LONG PERIOD SUPERSTRUCTURES IN $Cu_{3+x}Pd$

D. BRODDIN, G. VAN TENDELOO and S. AMELINCKX*
University of Antwerp, RUCA, Groenenborgerlaan 171, B-2020 Antwerp, Belgium
*Also at S.C.K./C.E.N., B-2400, Belgium

ABSTRACT
Electron microscopy and electron diffraction results on the one-dimensional long period superstructures (1D-LPS) in $Cu_{3+x}Pd$ are presented. The observations strongly support the models (like the ANNNI model) in which the profile of the modulating function smoothens with rising temperature. Near the order-disorder transition temperature, the low PD content LPS have an incommensurate character.

1. INTRODUCTION

Recent results in the study of the simple magnetic ANNNI (Anisotropic Next Nearest Neighbour Ising) model (a review is given in [1]) reveal an extremely rich phasediagram with ordered, disordered and commensurate as well as incommensurate modulated structures. The possibility to use the model for the description of LPS in alloys [2,3] caused a renewed interest in the behaviour of long period modulated alloy systems such als Al_3Ti [3], Ag_3Mg [4] and Cu_3Pd [5,6].

In the context of the ANNNI model, Monte Carlo simulations by Selke and Fisher [7] indicate that square wave modulated LPS (like Ag_3Mg) are low temperature structures whereas smooth profile modulated LPS (like CuAuII,

	Ll_2	LPS	A_1
Schubert et al.	△	▲	⬔
Guymont and Gratias	○	◉	⊕
present study	▢	▣	⊞

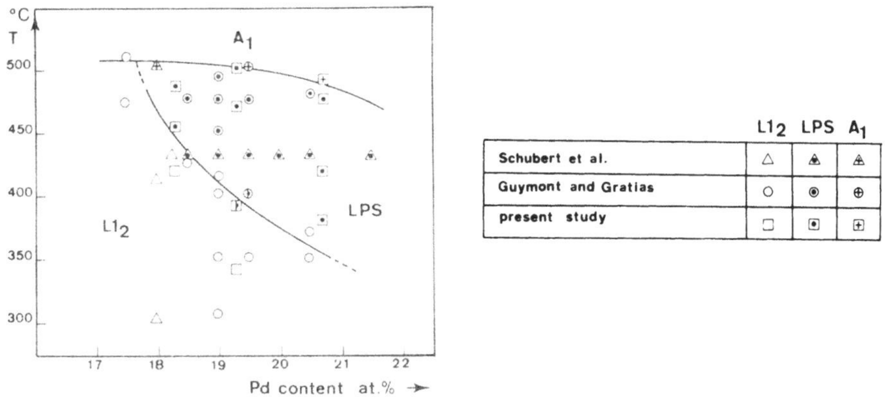

FIGURE 1. The Cu-Pd phasediagram for the composition range 18-21 at.Pd, combining our measurements with earlier work by Guymont and Gratias [9] and by Schubert, Kiefer, Wilkens and Haufler [10].

113

G. M. Stocks and A. Gonis (eds.), Alloy Phase Stability, 113–117.
© 1989 by Kluwer Academic Publishers.

Cu_3AuII) are high temperature structures. A well chosen alloy system might exhibit a transition from a square wave type modulation to a smooth profile type modulation with rising temperature.

We present some high resolution electron microscopy and electron diffraction results on $Cu_{3+x}Pd$ which confirm a significant smoothening of the modulation profile with rising temperature.

2. THE Cu-Pd SYSTEM

Three types of ordered structures can be found in the Cu-Pd system near the Cu_3Pd composition : the cubic $L1_2$ structure and one- and two-dimensional long period superstructures, which are derived from the latter by antiphase boundary (APB) modulation.

Previous results [5] show that for Pd-concentrations in the range 24-30 at.% Pd, commensurate 1D-LPS can be found at low temperature, whereas 2D-LPS are stable just below the order-disorder transition temperature (T_c). In this contribution, we consider in more detail the behaviour of the low Pd-content alloys in the range 18-22 at.% Pd. The corresponding phasediagram is shown in fig. 1. The alloys in this composition range have the $L1_2$ structure at low temperature, while the phase field of the 1D LPS extents up to the T_c. Near T_c, the modulation has a definite incommensurate character.

2.1. Incommensurate LPS near T_c

Fig. 2 shows a cube zone bright field high resolution image, corresponding to a Cu-19.3 at. %Pd alloy, annealed at 500°C (near T_c). In such an image [8, 5] bright dots correspond to minority (Pd) atom columns, which are shifted at the APB. It was shown in [5] that diffuse intensity at the APB is associated with the presence of mixed columns at the APB's. The mixed composition along the columns can be due to local disorder, to frequent APB-ledging or to a combination of these.

The boundary images in fig. 2 are diffuse and wavy and seem irregular and aperiodic to some extent. However, a statistical analysis, presented in [6], leeds to a very smooth 1D profile, with a nearly perfect incommensurate periodicity. The APB-width amounts to several atomic planes and its mean

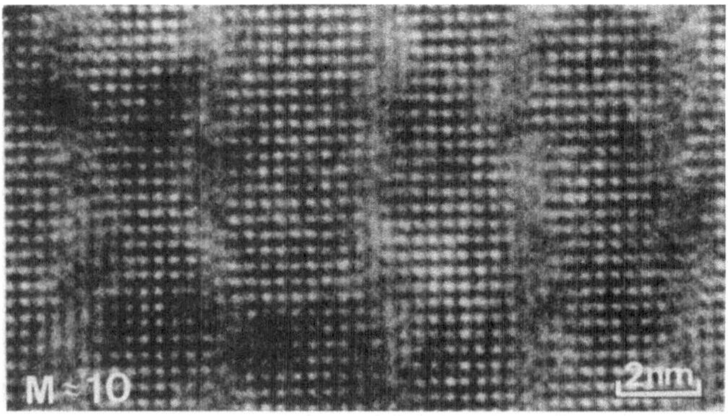

FIGURE 2. Cube zone high resolution image for the Cu-19.3 at. %Pd alloy annealed at 500°C. APB's have a wavy and diffuse character.

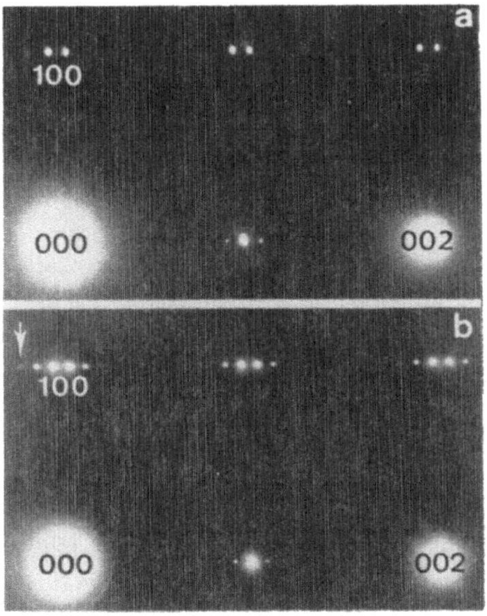

FIGURE 3. Comparison between cube zone electron diffraction patterns for the Cu-19.3 at. %Pd alloy, annealed
(a) near T_c (500°C) : second order satellites are extremely weak; the modulation profile is very smooth.
(b) near the $L1_2$-LPS transition (400°C) : third order satellites can be observed (arrow). The modulation profile tends to a more square wave type profile as the temperature is decreased.

position seems independent of the underlying lattice. The APB's are unpinned and the period was found to decrease in a continuous way with rising temperature.

2.2. The $L1_2$-LPS transition

It is clear from fig. 3 that higher harmonics of the modulation function profile (reflected by the intensities of higher order satellites) are more important at relatively low temperatures (400°C - near the $L1_2$-LPS transition temperature for the Cu-19.3 at. %Pd alloy) than near T_c (500°C). Therefore, the APB-width is much smaller at low temperature and the coupling of the APB to the lattice cannot be neglected under these conditions. The APB-configuration of fig. 4, which corresponds to the Cu-19.3 at. %Pd alloy annealed at 400°C for 4 days after a quench from the disordered state, can be considered metastable. The APB's are locked to the lattice and deviations from periodicity are important.

The dark field images in fig. 5 show that when an established LPS (as obtained after a treatment at higher temperatures) is annealed below 400°C, thin slabs of unmodulated $L1_2$ structure appear in the contact region between

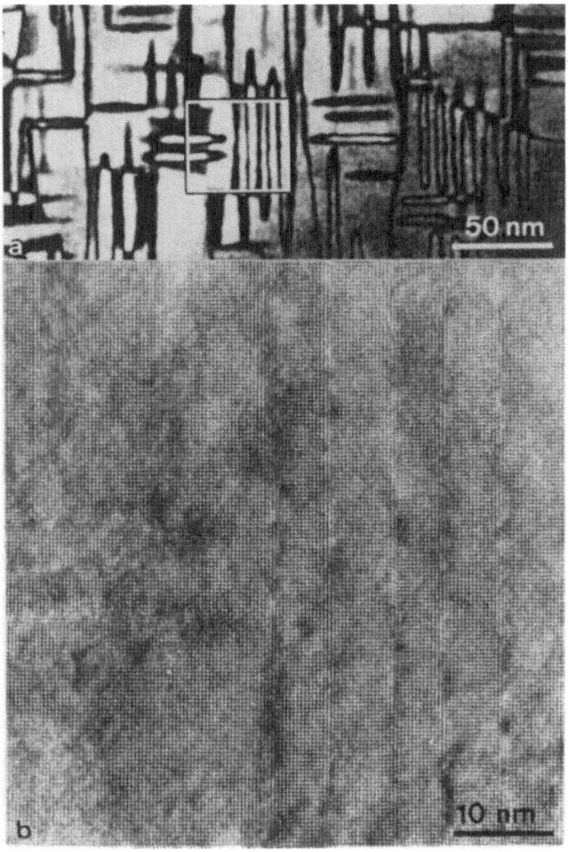

FIGURE 4. Dark field image using the pairs of satellites near (101) (fig. 4a) and bright field high resolution image (fig. 4b) of the same area (indicated square in a). This, probably metastable, configuration corresponds to the Cu-19.3 at. %Pd alloy, annealed at 400°C for 4 days.

the different orientation variants of the LPS. Electrical resistivity measurements performed for this alloy composition [11] indicate that the chosen annealing temperature still allows reasonable diffusion rates. The observed situation can be considered as indicative of the first order character of the $L1_2$-LPS transition (like the ferromagnetic-incommensurate transition within the ANNNI model [1]).

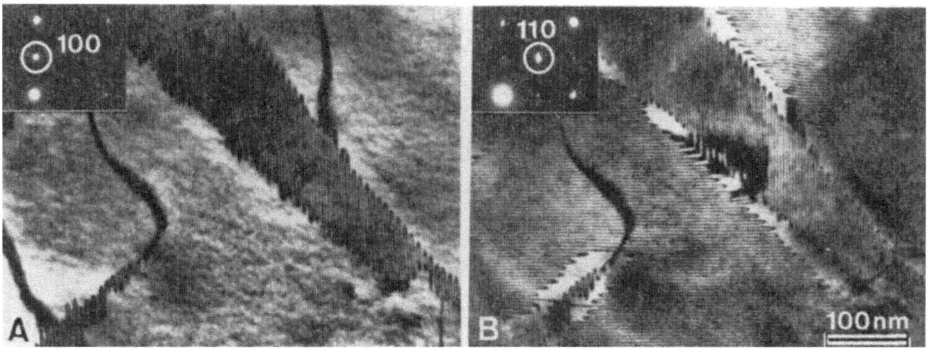

FIGURE 5. Dark field iamges for the Cu-19.3 at. %Pd treated at 400°C (4 days) after a previous annealing at higher temperature (establishing the LPS)
(a) 100 dark field image; one family of APB's (vertical fringes) is in contrast;
(b) 110 dark field image; the second family of APB's (horizontal fringes) is also in contrast.
The bright area in the contact region between the two orientation variants correspond to the $L1_0$ structure.

REFERENCES

1. J. Yeomans, 1986 Oxford University preprint.
2. D. de Fontaine and J. Kulik, 1985, Acta Metall. 33, 145-165.
3. A. Loiseau, G. Van Tendeloo, R. Portier and F. Ducastelle, 1985, J. Physique 46, 595-613.
4. J. Kulik, S. Takeda and D. de Fontaine, 1987, Acta Metall. 35, 1137-1147.
5. D. Broddin, G. Van Tendeloo, J. Van Landuyt, S. Amelinckx, R. Portier, M. Guymont and A. Loiseau, 1986, Phil. Mag. A54, 395-419.
6. D. Broddin, G. Van Tendeloo, J. Van Landuyt, S. Amelinckx and A. Loiseau, 1987, Phil. Mag. in the press.
7. W. Selke, this conference.
W. Selke and M. Fisher, 1979, Phys. Rev. B20, 257-265.
8. G. Van Tendeloo, this conference.
D. Van Dyck, G. Van Tendeloo and S. Amelinckx, 1982, Ultramicroscopy 10, 263.
9. M. Guymont and D. Gratias, 1976, Phys. stat. sol. (a) 36, 329.
10. K. Schubert, B. Kiefer, M. Wilkens, R. Haufler, 1955, Z. Metallk. 46, 692.
11. J. Planes, unpublished work at O.N.E.R.A., France.

LONG PERIOD SUPERLATTICE PHASES IN Cu-Al-Zn ALLOYS

M. DE GRAEF[1] & D. BRODDIN[2]

[1] Department of Metallurgy and Materials Engineering, De Croylaan 2, B-3030 Heverlee, Belgium
[2] Center for High Voltage Electron Microscopy, RUCA, Groenenborgerlaan 171, B-2000 Antwerpen, Belgium

1. INTRODUCTION

The phase diagram of the Cu-Al and Cu-Al-Zn systems shows a rich variety of different equilibrium phases and, as will be shown in this paper, several metastable phases. Among the stable phases occuring in the composition range used for this study are the face centered cubic α-phase (short range ordered), the cubic α_2-phase (fcc with long period anti-phase boundary lattice) and the γ_2-phase (complex cubic with 52 atoms per unit cell). (Fig. 1.) Upon a quench from the high temperature β-phase (body centered cubic) the alloy orders (DO_3-type of order) and transforms martensitically towards a close packed structure; depending on the electron to atom ratio this structure is 3R (α'-martensite), 9R (β'-martensite) or 2H (γ'-martensite). Since the martensitic transformation is by definition diffusionless the product phase inherits the DO_3-order from the parent β-phase, as well as all the defects. The industrial application of these alloys in 'two-way shape memory devices' is seriously obstructed by the fact that martensite is a metastable phase; when kept for some time at a temperature below the reverse transformation temperature the system will try to lower its free energy, thereby becoming more stable with respect to the high temperature phase. This so-called 'stabilisation' was first observed through an increase of the reverse transformation temperature (1) with time and a strong change of the damping capacity (2). According to the equilibrium phase diagram, martensite is expected to decompose into a mixture of α and γ_2-phases, depending on the exact composition. It has been reported that the martensitic phase slowly transforms into a so called X-phase (3), which is based on a long period superlattice in the basal planes, with the same original 18R stacking sequence. After prolonged tempering the 18R changes towards a 3R (fcc) stacking sequence with a slight modification of the superlattice period; in this way several LPS-phases have been found. This paper reports on some electron microscopic observations of phase tranformations from martensite towards these equilibrium phases.

2. EXPERIMENTAL PROCEDURE

A group of binary Cu-Al alloys with compositions between 19 and 25 at % Al and several ternary alloys with low Zn-content (upto 8 at %) were used in this study. All alloys were either betatised at temperatures around 1200 K and quenched to room temperature to obtain the martensitic state or slowly cooled from 1200 K to obtain the equilibrium phases. Subsequently they were annealed at different temperatures for different periods of time. All samples were then prepared for observation in a transmission electron microscope JEOL 200 CX (both in conventional and high resolution modes).

G. M. Stocks and A. Gonis (eds.), Alloy Phase Stability, 119–123.

3. OBSERVATIONS

During the quench from a high temperature the β-phase orders: first B2 ordering (CsCl-type structure) and at a lower temperature DO_3 : the martensitic phase inherits the order of this parent phase. The martensitic transformation in this and related systems can be formally decomposed into two steps: the $(110)_{bcc}$ planes become close packed and subsequently a stacking order is adopted which minimizes the transformation strains and also depends on the electron to atom ratio.

As can be seen in Fig. 2a all martensite plates contain DO_3 anti-phase domains; the anti-phase boundaries are sharply defined. However, within the martensite, this type of anti-phase boundary (combination of APB-plane orientation and displacement vector) is energetically unfavourable. After tempering for 15 minutes at 620 K these anti-phase boundaries start "meandering" (avoiding the original APB-orientation). These meanders (=spatial fluctuations, Fig. 2b) become hair pin-like on prolonged tempering and penetrate the anti-phase domains, as can be seen in Fig. 2c and d. The hairpins grow on the close packed basal planes and change the DO_3 order into an anti-phase boundary superlattice with average period depending on the composition; the symmetry of the martensitic phase being monoclinic, only one growth direction is possible within each martensite plate. The structure of this LPS-phase is depicted in Fig. 3.; full circles represent Al-atoms and open circles Cu and Zn. Every 3 unit cells an anti-phase boundary occurs with displacement vector $1/2[100]_{ort}$; this results in a monoclinic unit cell with anti-phase boundary planes parallel to the $(010)_{mon}$ plane.

By means of electron diffraction the LPS-period could be deduced from the distance between split reflections and was observed to vary from 3 to 5 (in units of b_{ort}); the alloys with the lower Al-content also showed the smaller period. The rather large differences of the growth kinetics in adjacent martensite plates could be caused by compositional inhomogeneities; it is also possible that the period never reaches a "stable" value, i.e. the system is continually changing.

High resolution electron microscopic observations revealed that the structure of the X-phase is consistent with the model put forward in literature. In this model the 18R stacking sequence of the original martensitic structure remains unchanged but the order in the basal planes is changed through a periodic anti-phase boundary modulation, the period of which depends on the composition. Fig. 4. shows a HR electron micrograph taken along the $(101)_{mon}$ orientation; the anti-phase boundaries are clearly visible and sharply defined. The inset shows an image simulation for a foil thickness of 21 nm at defocus -90 nm (Real Space method (4)).

Prolonged tempering results in a change of the stacking sequence from 18R to 3R through the growth of fcc lamellae in the 18R matrix (Fig. 5.). It is important to emphasize this observation because a previous model for the transformation of the X-phase to the α_2-phase assumed that the direction of the anti-phase boundary planes slowly changed from the $[102]_{ort}$ direction to the cubic [001] direction; this continuous change was never observed in our experiments. Instead the 3R grows in lamellae, so the APB-direction changes discontinuously at the interface plane. The electron diffraction patterns show a mixture of both patterns.

Starting from a polycrystalline high temperature β-phase it is also possible to obtain the equilibrium phases when the alloy is very slowly cooled down to room temperature; in this way the decomposition into α, α_2 and γ_2 phases has been studied. The α_2-phase is a LPS-phase with period

4/3 with respect to the DO_{22} unit cell (i.e. every 4 atomic planes contain 3 APB's). The decrease of the period with decreasing tempering temperature reported in (5) was confirmed by means of superlattice imaging.

Starting from a single-crystalline (or coarse grained) β-phase one can induce, by plastic deformation, another martensitic phase, α'-martensite, which is fcc stacked, but with the original "DO_3" order (no diffusion during the deformation). From this so-called "pink phase" (because of its pink colour) one can induce, for fixed compositions, several long period superstructures at different temperatures; in this way, a thourough study of the temperature dependence of the LPS-period can be carried out.

4. CONCLUSION

The Cu-Al(-Zn) system presents a real challenge for theoretical work, because of the rich variety of stable and metastable phases and phase transformations. Unfortunately, the abundance of experimental results is still contrasted by very little theoretical backup. In this paper it was attempted to give a brief overview of several phase transitions and phases that occur in a small region of the ternary phase diagram.

Martensitic Cu-Al(-Zn) alloys decompose into stable phases through a complex sequence of anti-phase boundary growth and dislocation migration; the X-phase nucleates at the DO_3 type anti-phase boundaries inherited from the high temperature parent phase. In (6) the period of this LPS-phase is related with the e/a ratio using the Sato & Toth theory; the result is satisfactory for e/a>1.45. One of the difficulties in the determination of the period is that the X-phase itself is not stable and through the migration of partial dislocations the stacking sequence changes in a lamellar way from 18R to 3R. The kinetics of this process depends on the electron to atom ratio as well as on the temperature; the smaller this ratio the faster the kinetics.

From the observations presented in this paper it is clear that the growth of LPS-phases in Cu-Al(-Zn) alloys is a complex process, for which the Sato & Toth model is not applicable without substantial modifications. The large number of LPS-phases found in the final state of the tempering process and the fact that the observed period also depends on the tempering temperature suggests that statistical models like the ANNNI model may provide a framework to study some aspects of the behaviour in this system.

REFERENCES.

1. J. Van Humbeeck, J. Janssen, N. Mwamba & L. Delaey, Scripta Met. 18, 893 (1984)
2. J. Van Humbeeck, A. Hulsbosch, L. Delaey & R. De Batist, Journ. de Phys., Supp 12, 46 (1985), C10-633
3. I. Lefever & L. Delaey, Acta Met. 20 (1972), 797
4. D. Van Dyck, J. Microsc. 119 (1980), 141
5. W. Gaudig & H. Warlimont, Acta Met. 26 (1978), 709
6. E. Bernard, These de doctorat (in French), Universite de Rouen, (1974)

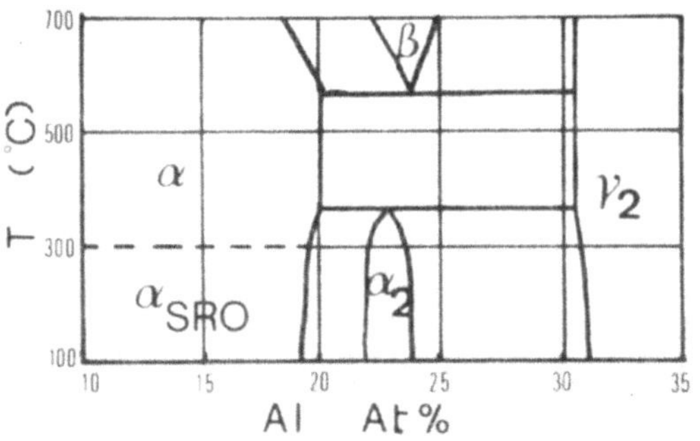

Fig. 1. Part of the binary Cu-Al phase diagram.

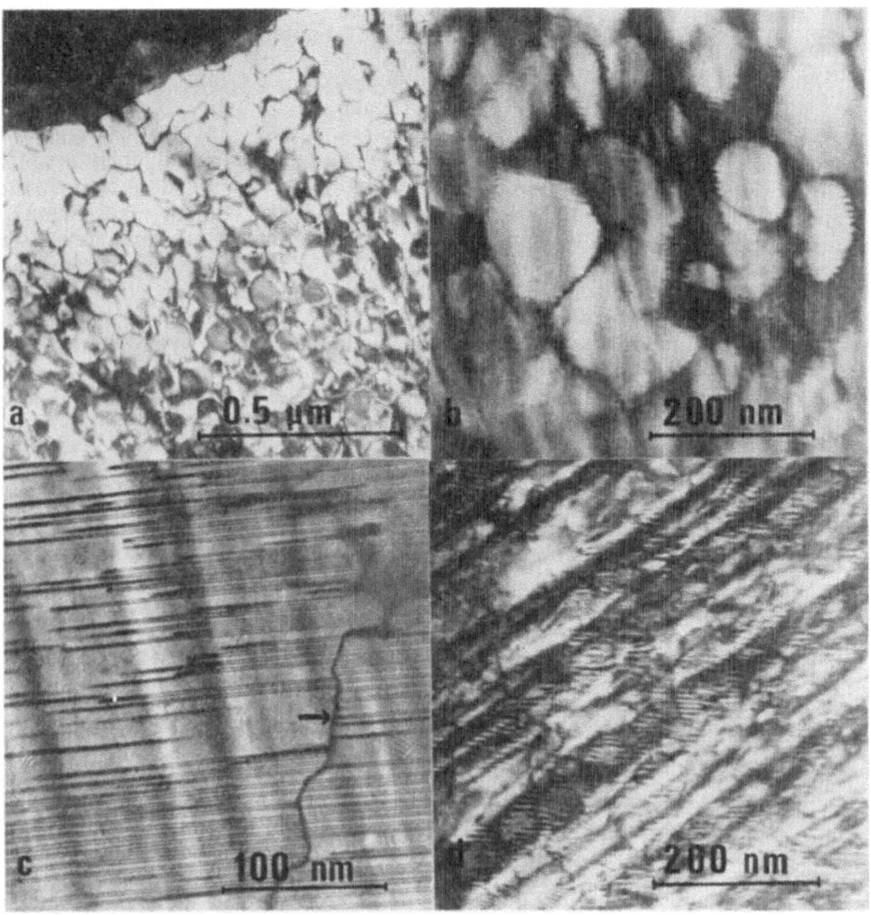

Fig. 2. (a) DO_3 anti phase domains in a martensite plate start to "meander" (b) after 15 min. at 620 K and finally grow in a hairpin-like way into the anti-phase domain (c–d)

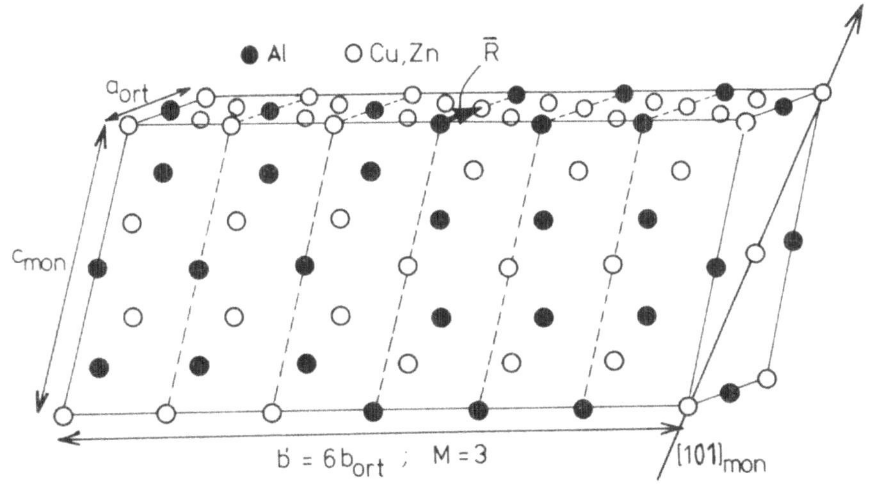

Fig. 3. Structure of the X-phase for a period of M=3; for clarity only the atoms on the border planes of the unit cell have been drawn. Dashed lines indicate the traces of the $(010)_{mon}$ planes.

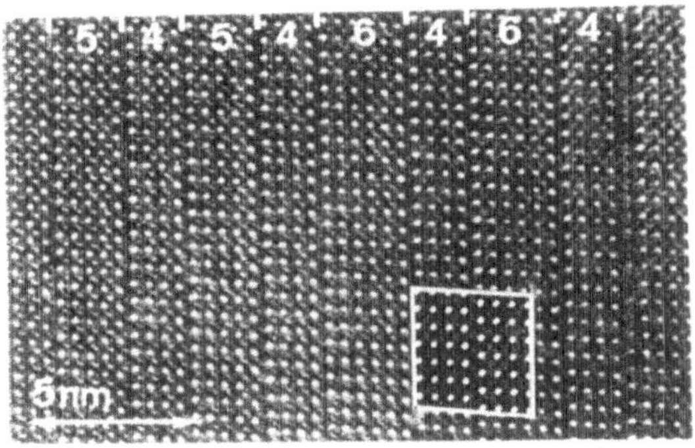

Fig. 4. High Resolution electron micrograph of the X-phase. Zone axis parallel to the direction indicated in Fig. 3.

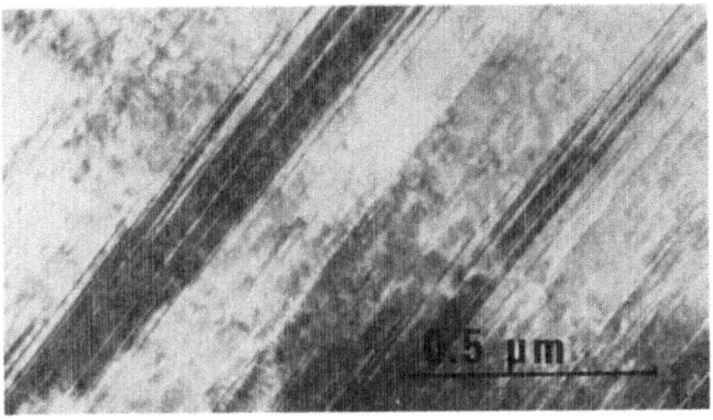

Fig. 5. Growth of fcc lamellae in the martensite 18R stacking sequence; dark regions correspond to fcc structure.

New antiphase superstructures in Nb-Ga binary system

M.Takeda[1,*], G.Van Tendeloo[1] and S.Amelinckx[1]

1) Univ.of Antwerp, RUCA, B-2020 Antwerp, BELGIUM
*) Toyohashi Univ.of Tech., 1-1 Hibarigaoka, Tenpakucho, Toyohashi 440, JAPAN

Abstract

The superstructures in $NbGa_{3-x}$ have been studied by electron diffraction and high resolution electron microscopy. Four long period superstructures, which correspond with Nb_5Ga_{13}, Nb_6Ga_{16}, Nb_7Ga_{19} and Nb_8Ga_{22}, have been found. The superstructures are closely related to the alloy compositions. The long period superstructure of Nb_5Ga_{13} comprises two types of antiphase boundaries alternately arranged, and has an unusually long c-axis (8.02nm). Other superstructures of Nb_6Ga_{16}, Nb_7Ga_{19} and Nb_8Ga_{22} involve one-type antiphase boundary, of which domain sizes only depend upon the compositions.

1. Introduction

Crucial works for the understanding of the long period superstructures (LPS) have been performed in recent years. From a theoretical point of view [1], a new development (the so-called ANNNI model) has been proposed, which enables one to describe several important features of order-disorder phenomena. In reality, however, there remain a number of experimental results on LPS, for which the theoretical interpretations are still lacking.

We have studied the superstructures of $NbGa_{3-x}$ by electron diffraction and high resolution electron microscopy. Several ordered structures have been reported in the phase diagram of the Nb-Ga alloy system [2]. However, few results are reported about the $NbGa_{3-x}$ region except for the study on Nb_5Ga_{13} by Schubert et al. [3]. We present here some new results on Nb-Ga superstructures which have never been observed before.

For the present study, bulk specimens with several different compositions have been made by alloying the two elements and subsequent tempering below the peritectic temperature of the phases (900-1100°C) for a few weeks. Finally specimens for electron microscopic investigation were obtained by crushing the bulk alloys. The high resolution observations have been made at 200kV with symmetrical illumination and with all electron waves included in the aperture with a size of $k=11.7\ nm^{-1}$.

G. M. Stocks and A. Gonis (eds.), Alloy Phase Stability, 125–130.

2.Ordered phases appearing in $NbGa_{3-x}$

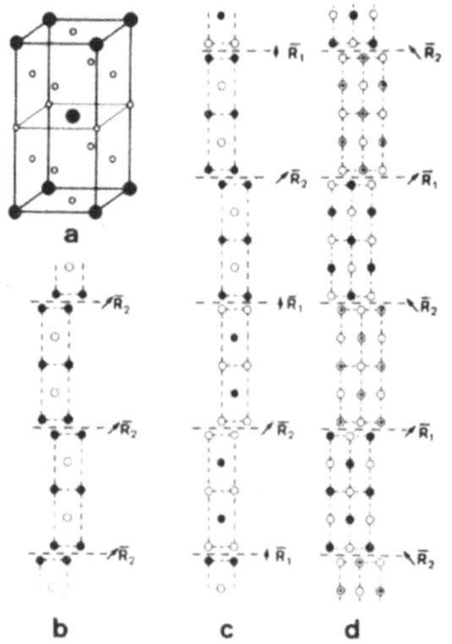

Fig.1 (a) Basic DO_{22}, (b) Schubert's model, (c) [100] and [110] view of the new model [5]

2-1.Model of Nb_5Ga_{13} proposed by Schubert et al.[3]

Schubert et al.[3] proposed their model for Nb_5Ga_{13} based upon X-ray diffraction analysis. The structure is built up by the periodic introduction of non-conservative antiphase domain boundaries in the fundamental DO_{22} structure of $NbGa_3$ as shown in Fig.1a,b. The domain boundaries are situated every 2.25 DO_{22} unit cells and the displacement vector is $a/4[201]DO_{22}$. Lattice parameters of the unit cell have been measured as a=b=0.378nm and c=4.01nm. Ten niobium and twenty-six gallium atoms, totally thirty-six atoms, are accomodated in the cell. The proposed structure has the space group Ammm.

2-2.Composition dependence of the LPS

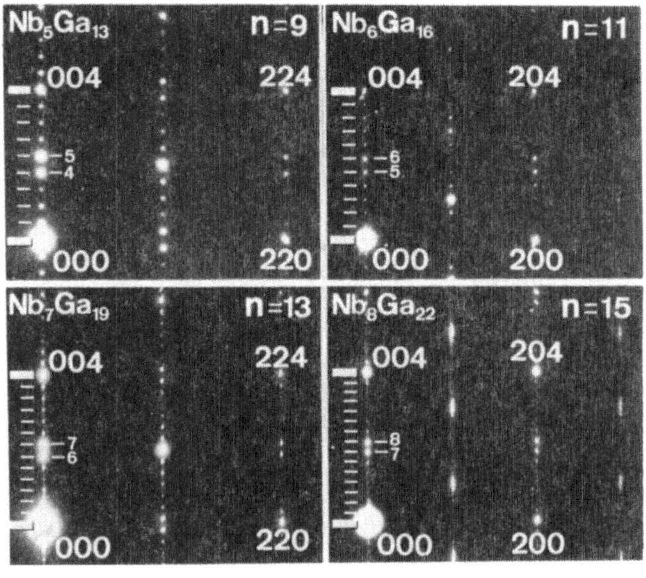

Fig.2 Electron diffraction patterns of four LPS in $NbGa_{3-x}$

Several LPS are formed in Nb-Ga system. Fig.2 shows the electron diffraction patterns obtained from different phases.

These patterns have been taken along the [100] or the [110] zone axis. All positions of strong diffraction spots are associated with the fundamental DO_{22} structure. Along the 001 direction these patterns show nine (in Fig.2a), eleven (in Fig.2b), thirteen (in Fig.2c) and fifteen (in Fig.2d) weak superreflections. This means that the LPS are based upon the DO_{22} structure of $NbGa_3$ and the long periodicities appear along the c axis of the fundamental structure. From the fractional shifts of the superstructure reflections in Fig.2b,c,d with respect to the positions of the basic DO_{22} reflections the displacement vector R=a/4[201] can be determined using the method proposed by Van Landuyt et al.[4].

The electron diffraction patterns shown in Fig.2 lead to one of the important results of this study that the long periodicity of the ordered structures in $NbGa_{3-x}$ is composition driven. Suppose that the structure of the antiphase domain boundary is the same for all boundaries (as Schubert proposed) we can determine the alloy compositions of the LPS with different boundary spacings. Let n be the average number of (004) planes of DO_{22} between two domain boundaries, then the niobium composition of the alloy phase is defined as x/(x+y), where x means the number of niobium atoms in the unit cell and y the number of gallium atoms in the cell. The values x and y are expressed as,

$x= [(n+1)/2], \quad y=n+[n/2]$ [ ]: Gauss's symbol

The alloy composition depends on the number n. The composition is exactly equal to $NbGa_3$ when n is an even integer. Therefore, only phases with odd n like

n=5,	Nb_3Ga_7	(Nb at%=0.300)	n=13,	Nb_7Ga_{19}	(Nb at%=0.269)
7,	Nb_4Ga_{10}	(0.286)	15,	Nb_8Ga_{22}	(0.267)
9,	Nb_5Ga_{13}	(0.278)		- - - - - - - - - - -	etc.
11,	Nb_6Ga_{16}	(0.273)			

will need structural modulations which accomodate the eccess niobium.

In Fig.2, the number of superreflections appearing between the origin and the (004) diffraction spot of the fundamental DO_{22} indicates the number of $(004)DO_{22}$ planes between two domain boundaries. As the above consideration predicts, only odd n numbers (n=9,11,13 and 15) of superstructures have been observed in Nb-Ga system thus far.

3.Structural analysis of $NbGa_{3-x}$ phases

3-1.Atomic structure of Nb_5Ga_{13}

Fig.3 shows the electron diffraction patterns obtained from the Nb_5Ga_{13} alloy. The patterns have been indexed to be [001], $[1\bar{1}0]$ and [100] zone axis diffraction patterns, respectively. Schubert proposed a structural model with space group Ammm; however, the extinctions do not agree with this model. As shown in Fig.3b,c,d the superreflections due to the long period structure are shifted over a half inter-spacing in two adjacent

001 rows. This means that the structure has a body-centered lattice instead of an end-centered lattice.

Fig.4a,b shows the high resolution images observed along the [100] and [110] zone axes. The images visualize the atomic structure in the crystal. The configuration of bright dots inside a domain corresponds with the DO_{22} structure. Fig.4a shows that two kinds of domain boundaries are alternately arranged. On the other hand, only one type of boundary is observed in Fig.4b. Based on the high resolution images from different orientations, we have presented a model different from Schubert's model, which enables us to interpret all experimental results (see Fig.2c,d) [5].

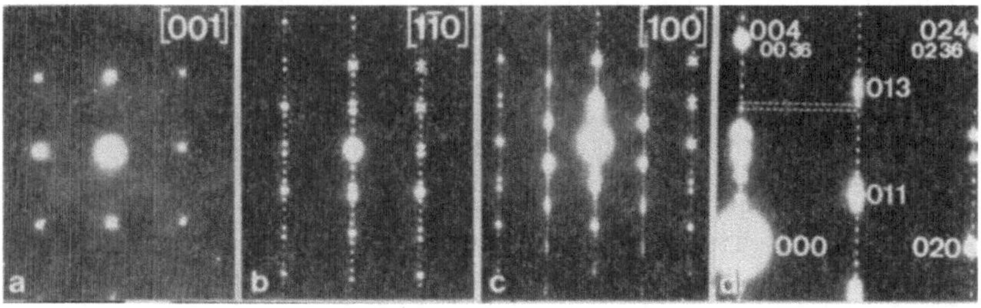

Fig.3 Electron diffraction patterns of Nb_5Ga_{13} (a) [001], (b) [1$\bar{1}$0], (c) [100] and (d) a portion of [100]

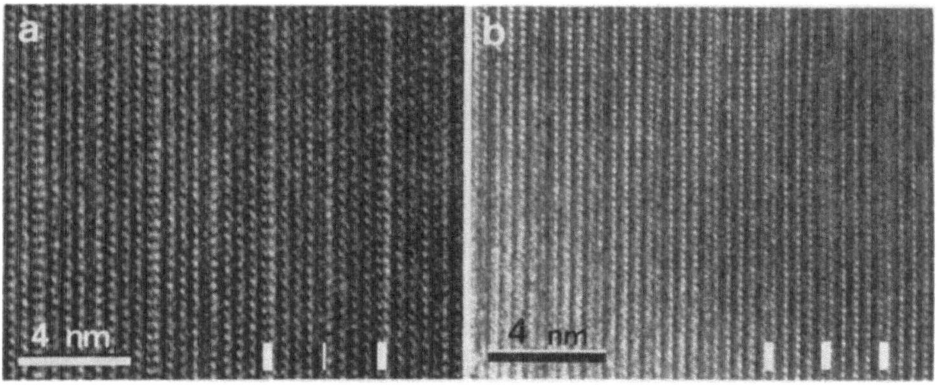

Fig.4 High resolution image of Nb_5Ga_{13} (a) [100] and (b) [110] zones

The structure has lattice parameters, a=b=0.378nm and c=8.02nm i.e., the model is twice as large as Schubert's model in the c-direction. The space group of the model is $I4_1$/amd. In this model the displacement vectors of the domain boundaries are both non-conservative, a/4[201] and a/4[210] of the DO_{22} cell. Because the two displacement vectors are symmetry-related and locally the atomic arrangement is similar in the two kinds of domain boundaries, the composition consideration in section 2-2 is still valid for this superstructure.

3-2.The LPS of Nb_6Ga_{16}, Nb_7Ga_{19} and Nb_8Ga_{22}

In a similar way, the LPS of other $NbGa_{3-x}$ alloys have been investigated. Figs.5-6 show the electron diffraction patterns and the high resolution images of the Nb_6Ga_{16} and Nb_7Ga_{19} crystals. In Fig.5, the electron diffraction pattern of Nb_6Ga_{16} shows the strong reflections due to the DO_{22} structure as is the case in Nb_5Ga_{13}. However, there is an important difference, i.e., no shift is observed between any two adjacent 001 rows. The patterns indicate that the crystal structure has an end-centered orthorhombic lattice with the parameters a=b=0.378nm, c=2.50nm.

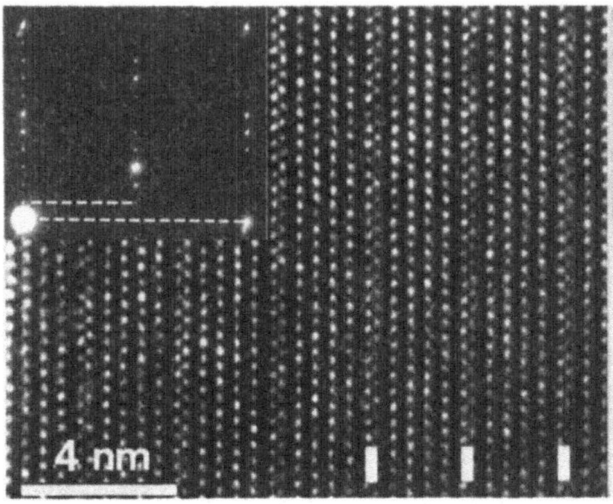

Fig.5 Electron diffraction pattern and high resolution image of Nb_6Ga_{16}

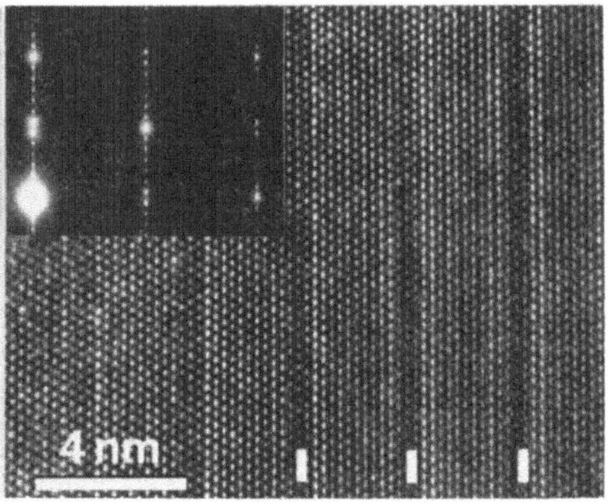

Fig.6 Electron diffraction pattern and high resolution image of Nb_7Ga_{19}

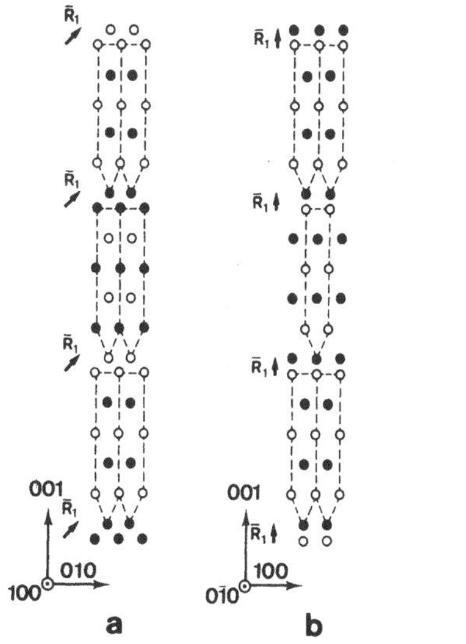

Fig.7 The model of LPS of Nb_6Ga_{16} (a) [100], (b) [010]

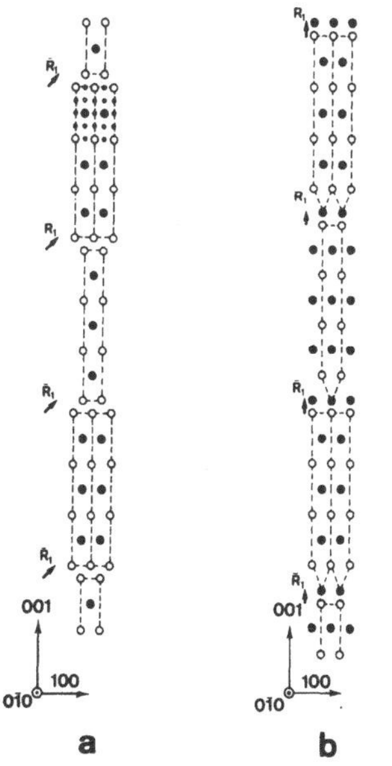

Fig.8 The models of LPS of (a) Nb_7Ga_{19} and (b) Nb_8Ga_{22}

The structural model of Nb_6Ga_{16} is presented in Fig.7. Fig.6 shows the electron diffraction pattern and the high resolution image of the LPS of Nb_7Ga_{19}. The atomic arrangements of Nb_7Ga_{19} and Nb_8Ga_{22} are the same at the domain boundaries as can be deduced from electron diffraction patterns. The structural models of Nb_7Ga_{19} and Nb_8Ga_{22} are shown in Fig.8a,b.
The lattice parameters have been measured as a=b=0.378nm, c=2.87nm in Nb_7Ga_{19}, and a=b=0.378nm, c=3.32nm in Nb_8Ga_{22}.
Comparing the models of Nb_5Ga_{13}, Nb_6Ga_{16}, Nb_7Ga_{19} and Nb_8Ga_{22} in Figs.1, 7 and 8, one can easily understand the relation between the superstructures. Unlike the Nb_5Ga_{13}, the Nb_6Ga_{16}, Nb_7Ga_{19} and Nb_8Ga_{22} have only one type of domain boundary with the same displacement vector. The space group Bmm2 (or Amm2) of the model for Nb_6Ga_{16} and Nb_8Ga_{22} is slightly different from Bmmm (or Ammm) for Nb_7Ga_{19} though the local boundary structures are the same in these three crystals. This is due to the difference in the domain sizes.

4.Conclusion

We have studied the atomic structures of $NbGa_{3-x}$ superstructures by means of electron diffraction pattern analysis and high resolution electron microscopy. The present work has revealed several interesting aspects of the superstructures. Four different LPS have been found in $NbGa_{3-x}$. Basically they have similar atomic structures with periodic non-conservative antiphase boundaries. The periodicities of the antiphase boundaries are sensitive to the composition; the four superstructures correspond with Nb_5Ga_{13}, Nb_6Ga_{16}, Nb_7Ga_{19} and Nb_8Ga_{22}, respectively. The Nb_5Ga_{13} superstructure is composed of two types of domain boundaries whereas the others Nb_6Ga_{16}, Nb_7Ga_{19} and Nb_8Ga_{22} have only one type of boundary. The alternate arrangement of two types of boundaries as realized in Nb_5Ga_{13} has never been observed in any LPS before.

References

1.M.E.Fisher and W.Selke, Phys.Rev.lett. 44 (1980) 1502.
D.de Fontaine and J.Kulik, Acta Met. 33 (1985) 145.
M.E.Fisher and A.M.Szpilka, Phys.Rev.B, 36 (1987) 644.
2.T.B.Massalski (ed.) "Binary Alloy Phase Diagrams vol.2" (A.S.M. 1986) 1147.
3.K.Schubert et al., Naturwiss. 50 (1963) 41.
4.J.Van Landuyt, R.De Ridder, R.Gevers and S.Amelinckx, Mat.Res.Bull.5 (1970) 353.
5.M.Takeda, G.Van Tendeloo and S.Amelinckx, Mat.Res.Bull.(1987) (in press), and Poc.EMAG'87 (Manchester, 1987) (in press)

ORDER-DISORDER TRANSITION IN CuPt-ALLOYS, INVESTIGATED BY RESISTIVITY MEASUREMENT

J. Banhart[1]), W. Pfeiler[2]) and J. Voitländer[1])

1) Institut für Physikalische Chemie der Universität München, Sophienstraße 11, D-8000 Munich, W. Germany
2) Institut für Festkörperphysik der Universität Wien, Strudlhofgasse 4, A-1090 Vienna, Austria

INTRODUCTION

It is evident that ordering effects are essentially dependent on temperature. The formation of long-range order (LRO) in most cases is a question of low enough temperature with respect to the critical temperature because of an increasing driving force towards ordering with increasing distance from the phase boundary. On the other hand the experimental observation of LRO-effects is linked to a high enough temperature to guarantee the atomic mobility for the rearrangement of the alloy atoms.

Effects of short-range order (SRO) frequently are investigated far above a phase boundary, the degree of order continously decreasing with increasing temperature. Because of rather small differences in tthe atomic arrangements the formation of these effects is much faster, though equally reflecting the atomic mobility (1).

The effect of SRO on the electronic structure of CuPt-alloys has been calculated by means of the KKR-CPA and the embedded cluster method (2) and it is hoped that the calculations can be extended to the electrical resistivity.

It was the aim of the present work to observe the formation of LRO and SRO in CuPt-alloys by measuring changes in the electrical resistivity during isochronal and isothermal annealing.

EXPERIMENTAL

Cu-alloys with 29,35,50 and 70 at% Pt were induction melted from 99.999 Cu and 99.99 Pt by Degussa.

Foils of about 0.2 mm thickness were rolled with several intermediate anneals in purified Argon atmosphere. Then serpentine shaped samples were cut out of the foils by spark erosion.

Before the isochronal annealing procedure the samples were homogenized just above the phase boundary and quenched into water of room temperature.

The resistivity measurements were made by usual potentiometric method in a bath of liquid nitrogen relative to a dummy specimen. The measuring accuracy was about $\pm$5x10E-5, but the accuracy of the results obviously is limited by the reproducibility of the annealing temperature of about $\pm$2°C. Therefore the accuracy for the alloys investigated is given by the resistivity change per temperature interval to: $\pm$5x10E-4 for Cu-29at%Pt and Cu-35at%Pt, $\pm$4x10E-3 for Cu-70at%Pt and $\pm$1x10E-2 for Cu-50at%Pt.

RESULTS AND DISCUSSION

Cu-alloy with 50 at% Pt

In fig. 1 the change of electrical resistivity of Cu-50at%Pt is plotted

G. M. Stocks and A. Gonis (eds.), Alloy Phase Stability, 131–135.

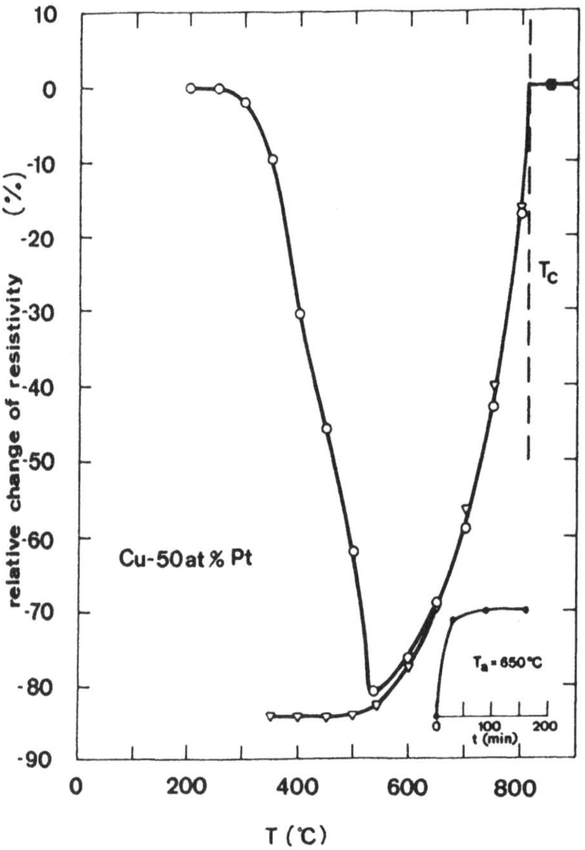

FIGURE 1. Isochronal annealing of Cu-50at%Pt. o rising ▽falling temperature; ■starting value. Insert: isothermal annealing at 650°C.

versus the annealing temperature. At about 250°C a drastic decrease of resistivity by 80% obviously indicates the formation of LRO. After a minimum at 540°C, where the equilibrium degree of SRO is reached within the isochronal time interval of about 20 min., the resistivity increases again, which must be connected with the beginning destruction of LRO. It is particularly interesting that this effect of disordering starts almost 300°C below the critical temperature at 827°C. Above the critical temperature no measurable change of resistivity was detected.

The triangles in fig.1 show the effect of isochronal annealing at falling temperatures (reversed isochrone): below 650°C the decreasing concentration of thermal vacancies prevents the ordering process to be completed.

The question whether the increasing values of resistivity between 540°C and the critical temperature are equilibrium values reflecting the dependence of the LRO parameter has been checked by an isothermal annealing experiment at 650°C: the change of electrical resistivity with annealing time (see inset of fig. 1) shows that a stable state is adjusted.

The results of isochronal annealing for this alloy are in good correspondence with an investigation by Torfs et al. (3).

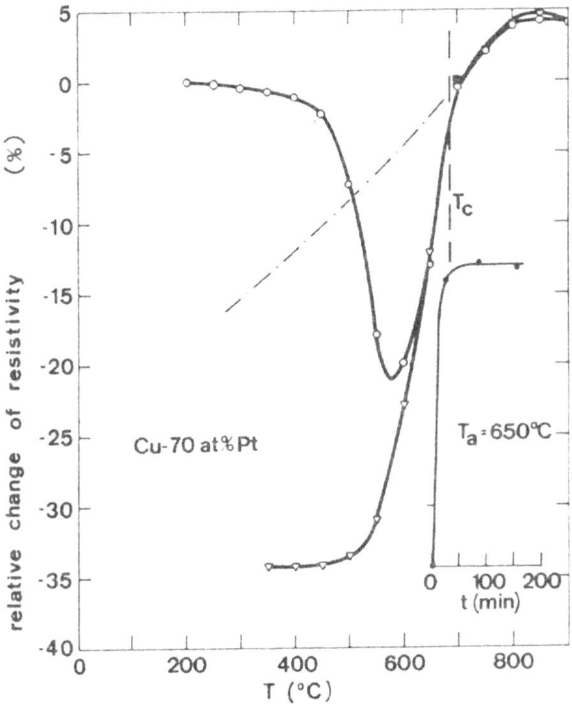

FIGURE 2. Isochronal annealing of Cu-70at%Pt. o rising ▽ falling temperature; ■ starting value. Insert: isothermal annealing at 650°C. Dash-dotted: hypothetical SRO-equilibrium line.

Cu-alloy with 70at%Pt

Fig.2 gives the change in electrical resistivity of Cu-70at%Pt as a function of the annealing temperature. There starts a slight decrease of resistivity already at 200°C, but the drastic drop of about 20%, which obviously indicates the formation of LRO, starts above 400°C. A minimum at 580°C ($=T_c$-100°C) is followed by a strong increase of resistivity.

As a difference to Cu-50at%Pt for this alloy the resistivity continues to increase above the phase boundary at about 685°C, though at a smaller rate. This can be interpreted by a decreasing degree of SRO with increasing temperature, if an increase of SRO means a decrease of the resistivity in this case. This could also account for the slight decrease of resistivity of the present sample at rather low temperatures.

A test by isothermal annealing at 650°C demonstrates the stability of the resistivity values of the increasing part of the isochronal curve (see inset of fig.2).

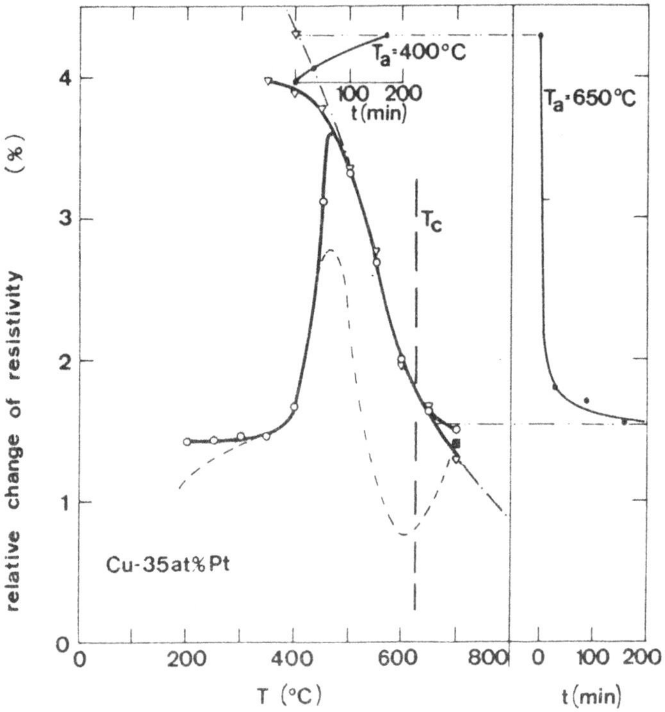

FIGURE 3. Isochronal annealing of Cu-35at%Pt. o rising ∇ falling temperature; ■ starting value. Inserts: isothermal annealing at 400°C and 650°C. Dashed line: previous run.

Cu-alloy with 35 and 29 at%Pt

Fig. 3 gives the change of electrical resistivity of Cu-35at%Pt versus annealing temperature during the isochronal annealing experiment. Here the resistivity first **increases** up to a maximum value at 460°C and then **decreases** in a reproducible way (isochrone and reversed isochrone). This can be interpreted by an increase of SRO below 460°C, which in this case increases the electrical resistivity. At the maximum the resistivity equilibrium line is met leading to a decrease of resistivity with decreasing degree of SRO. Note that the equilibrium line crosses the phase boundary at 625°C unchanged. An isothermal annealing experiment at 400°C and 650°C shows the stability of these SRO-equilibrium states (see inset of fig. 3). The absence of LRO is attributed to the short annealing times (20 min.) and the pretreatment of the sample.

That the thermal history is crucial for the observed result shows the dashed line in fig. 3, which represents another isochrone of the same sample. Here, after a short annealing at 820°C immediately following the initial cold-rolling for the sample preparation, an influence of LRO was observed above 450°C by a resistivity drop below the phase boundary. This interpretation was checked by an isothermal anneal at 500°C (not shown here) yielding a drastic decrease of resistivity by more than 15%.

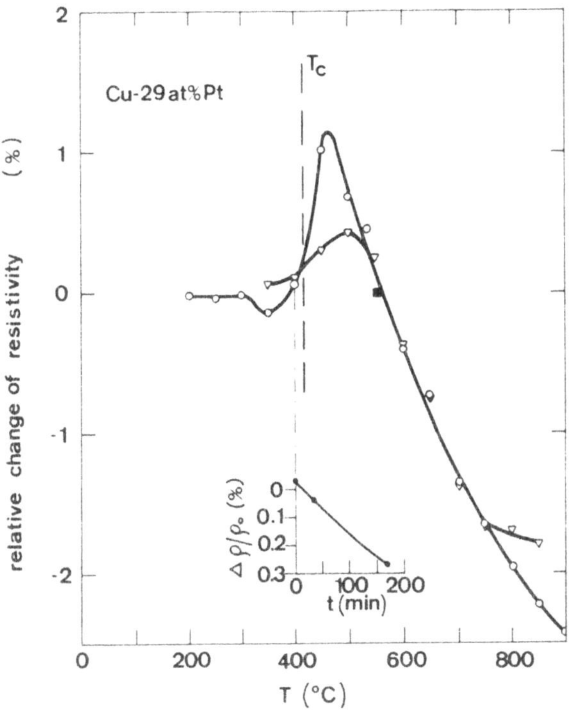

FIGURE 4. Isochronal annealing of Cu-29at%Pt. o rising ▽falling temperature; ■starting value. Insert: isothermal annealing at 650°C.

Fig. 4 shows that a very similar behaviour was observed for the Cu-29at%Pt alloy. After a first small decrease of resistivity between 300°C and 350°C the resistivity increases up to a maximum at 460°C just above the phase boundary at about 400°C. Then the resistivity decreases until quenching effects are observed above about 750°C. This can be interpreted as increasing and decreasing SRO, where an increase of SRO is accompanied by an increase of resistivity.

The reversed isochrone coincides with the isochrone at rising temperatures above 550°C (equilibrium line). At temperatures near the phase boundary a decrease of resistivity hints at an influence of LRO-formation. This was checked by an isothermal annealing at 400°: a steady decrease of resistivity confirms this presumption. Therefore the small decrease of resistivity at the beginning of isochronal annealing (350°C) also may be attributed to LRO-effects.

CONCLUSIONS

i) Cu-50,70 at%Pt: Both alloys show a prompt and strong formation of LRO. The destruction of order starts far below the critical temperature and equilibrium states are achieved for certain temperatures. Effects of SRO only were observed for the alloy with 70at%Pt leading to a decrease of resistivity with increasing SRO.

ii) The results for the Cu-35at%Pt and the Cu-29at%Pt alloys show a repeated change between SRO-behaviour and LRO-formation depending on the thermal treatment of the sample. In this case an increasing degree of SRO increases the electrical resistivity.

iii) SRO seems to form independently of LRO and not to be a pre-stage of LRO.

REFERENCES

1. Pfeiler W: to be published in Acta Met.
2. Banhart J et al: to be published in Sol. State Comm.
3. Torfs E: Phys. Stat. Sol. (a) 22, 45 (1974)

EFFECTIVE PAIR-INTERACTIONS IN BINARY ALLOYS

W. SCHWEIKA

Institut für Festkörperforschung der Kernforschungsanlage Jülich GmbH
Postfach 1913, D–5170 Jülich, Fed. Rep. of Germany

1. INTRODUCTION

A number of diffuse scattering studies either with x-rays or neutrons have been done in order to study the local or short-range order (SRO) in binary systems, and mostly in alloys. The SRO as an example of a generalized susceptibility can be related to the pair-interactions between the different atoms [1, 2] . The configurational energy may be described by (the lattice gas model or) the Ising model :

$$H = \frac{1}{4} \sum_{j,i} V(\vec{r_i} - \vec{r_j}) \sigma_i \sigma_j \tag{1}$$

where the site occupation is denoted by $\sigma_i = \pm 1$. The attractive idea of using such pair-interactions is that one expects that they are not only capable of describing the attendant measured SRO configuration but also, perhaps the whole coherent binary phase diagram. But the application of such pair-interactions for the calculation of phase diagrams is restricted, since in real alloys the interactions may not only be pairwise and may also depend on composition and temperature etc. A further problem that the pair-interactions could have been only determined within a mean-field approximation [1, 2], has meanwhile been solved by the Inverse Monte Carlo (IMC) method [3]. In the particular case of the $Ni_{.89}Cr_{.11}$ alloy a remarkable agreement between the numerically exact result of the IMC method and the mean-field solution has been found [4]. However in general, this agreement can not be expected. . But the observation that the range of interaction for a real alloy may extend further than only to nearest and next-nearest neighbor is really a new challenge for the standard tools of statistical mechanics, for the Monte Carlo (MC) and even more for the Cluster Variation method (CVM). Some IMC results for $CuAl$, $CuPt$, $NiCu$ and Cu_3Au have already been published [3]. An alternative method [5] to IMC has recently been applied to determine pair-interactions in Ni_3Fe. In this paper we give further results of the IMC method for $NiAl_{.9}$, $NiCu$ and Ni_3Fe. Especially for $NiCu$, the interaction model is discussed by MC results and give reliable information on the phase diagram. Then we will consider whether many-body interactions may be described by concentration dependent pair-interactions and whether it is possible to obtain sufficient information from the measured pair-correlation function to distinguish uniquely between interaction models.

2. DETERMINATION OF PAIR-INTERACTIONS

2.1. Inverse Monte Carlo method

The IMC method has been presented elsewhere [3, 4, 6] and only a brief description is given here. The first step is to simulate the experimental SRO parameters in a computer model, using the Gehlen Cohen procedure [7]. All parameters should be simulated

G. M. Stocks and A. Gonis (eds.), Alloy Phase Stability, 137–141.

within their range of confidence. This procedure also offers a first test as to whether the SRO parameters belong to a possible configuration at all. Assuming the SRO describes a configuration of thermal equilibrium, one can apply the principle of detailled balance to determine the unknown interactions from the SRO parameters. Thus one has to solve by numerical iteration techniques the system of nonlinear equations

$$\sum_{k}^{K} \Delta p_{kl} w_k = 0, \tag{2}$$

where Δp_{kl} is the change in the number of bonds with the interaction energy V_l and w_k is the attendant transition probability for a particular fluctuation k. Eq.2 only holds for the average over a large number of fluctuations. The transition probability is

$$w = \frac{e^{-\beta \Delta H}}{1 + e^{-\beta \Delta H}} \qquad \text{where} \qquad H = \sum_{l} p_l V_l. \tag{3}$$

In comparison to the direct MC method, the inverse is not so straightforward, since first one has to assume an interaction model. For transition metal alloys it is known from TB-CPA calculations [8] that pair-interactions should be rather dominant to cluster interactions. However it is important to prove by a further direct Monte Carlo simulation of the SRO whether the obtained results are self-consistent. Another point of interest are the limitations of the IMC method and these will be also discussed in the following examples. The accuracy of the IMC method depends on the size of the treated SRO model crystal. Usually a size is chosen between 10000 to 200000 (this work) atoms. This also fulfills another neccesary condition that the linear dimensions of the model crystal should be large compared to the range of the SRO.

2.2. Ni_3Fe

The SRO in the alloy Ni_3Fe has been measured very accurately at various temperatures by diffuse neutron scattering [9]. The interaction parameters have already been determined by another alternative method (Livet) [5]. It is a direct Monte Carlo method, where the interaction parameters are altered to model the measured SRO. For example we give only one result for the pair-interactions determined for the equilibrium SRO at 808K in fig.1a and confirm that the same results are obtained by both methods. Only the nearest and next-nearest neighbor interactions differ significantly from zero and favour the $L1_2$ ordering. From further MC simulations we can assure that the interactions are consistent with the measured SRO. However, the interaction parameters change significantly with temperature. One obvious reason for this behaviour is that the magnetism effects the ordering and thus the simple Ising model can not be appropriate and sufficient. In principle the additional magnetic interactions can be discovered as well, but this would require the measurement of the SRO of the magnetic moments by diffuse polarized neutron scattering.

2.3. $NiAl$

Various X-ray scattering data for the SRO in NiAl alloys have been published [10, 11, 12]. For X-ray scarttering the determination of the SRO is more difficult since the background of the thermal diffuse scattering can not be separated by experimental means in X-ray scattering. Especially for $NiAl_{.127}$ quenched from 1323K [10] it has not been been possible to simulate completely the SRO parameters in a model crystal. This indicates that the SRO do not describe a homogeneous state of thermal equilibrium.

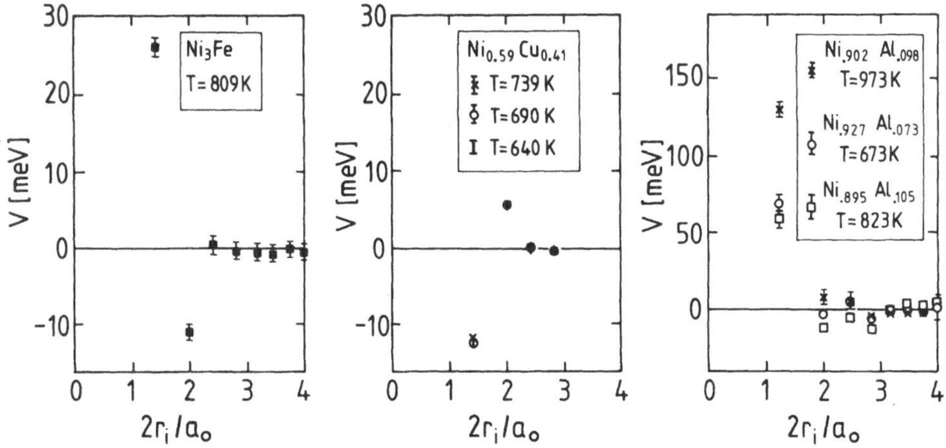

Fig. 1: IMC results on effective pair-interactions in Ni based alloys
a) Ni_3Fe *(9) b)* $Ni_{.59}Cu_{.41}$ *(13)*
c) $NiAl_{.098}$ *(11),* $NiAl_{.073}$ *(12),* $NiAl_{.105}$ *(12).*

The IMC results for the interaction parameters are shown in fig.1b. The deviations between the results should be taken as a measure of the reproducibility of the SRO measurements, rather than as real effects of the different temperatures or compositions. There is a remarkable large value of the nearest neighbor interaction. With further distance a sudden decrease in interaction is observed. A similar behaviour has been found for $CuAl$ alloys [3].

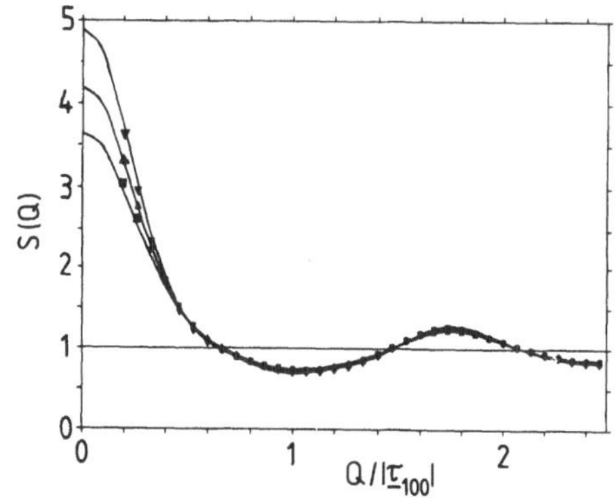

Fig. 2: S(Q) of $Ni_{.59}Cu_{.41}$ *polycrystals according to SRO data (13)*
■ $739K$, ▲ $690K$, ▼ $640K$,
– *MC simulation*
$V_1 = -12.2meV$ *and* $V_2 = 5.6meV$.

2.4. *NiCu*

The NiCu alloy shows the tendency for decomposition in the homogenoeous SRO states as observed by diffuse neutron scattering on polycrystalline samples [13]. Using isotopes, the average scattering length for neutrons was zero. Therefore the diffuse scattering was only due to SRO. In fig.1c the IMC results for the pair-interactions of the alloy $Ni_{.59}Cu_{.49}$ are presented. There is a remarkable agreement in the results obtained from SRO equilibrium states belonging to three different temperatures, thus emphasizing the good quality of the experimental data. Because the diffusion is slow in the low temperature region, the spinodal decomposition temperature is not known very well. By MC simulation the structure factors S(Q) due to the SRO at the three different temperatures have been consistently determined from only the two interaction parameters

$V_1 = -12.2meV$ and $V_2 = 5.6meV$. With MC simulation spinodal decomposition temperature $T = 495 \pm 5K$ is found, which is in very good agreement with the experimental observation of the decomposition in periodically layered structures of CuNi [14].

3. COMPARISON BETWEEN MANY-BODY INTERACTIONS AND EFFECTIVE CONCENTRATION DEPENDENT PAIR-INTERACTIONS

The (coherent) phase diagrams of binary alloys are almost never symmetric with respect to composition. Either many-body interactions or concentration dependent pair-interactions or may be even both would have to be taken into account to describe real alloys within models of statistical mechanics. Here the question arises whether we can distinguish between different interaction models and whether additional many-body interactions can be determined from the measured pair-correlation function (SRO). To answer these questions consider a simple many-body interaction : the tetrahedron interaction model used [15] to describe the CuAu phase diagram. The parameters are the bare nearest neighbor interaction $V^0 = 60meV$ and the additional contribution due to the occupation of a tetrahedron $\alpha = -0.08$ and $\beta = 0.01$. The phase diagram for this interaction model has been calculated by the Monte Carlo method in the grand canonical ensemble. The result shown in fig.3a compares favourably to the earlier calculated CVM phase diagram [15]. Note that the phase boundaries between $L1_0$ and $L1_2$ are determined with less precision than the transition temperatures of the stoichiometric phases. The tetrahedron many-body interactions can be approximated for high temperatures by composition dependent effective pair-interactions [16]:

$$V(c) = V^0(1 - \frac{3}{2}(\beta - \alpha) < \sigma > + \frac{3}{2}(\beta + \alpha) < \sigma >^2) \tag{4}$$

The Monte Carlo calculation of the attendant phase diagram reveal clear differences compared to the one for the tetrahedron interactions especially for the transition temperatures of the pure ordered phases. From this comparison it can be concluded that the considered parameterization is not fully consistent. This is not surprising since the eq. (4) should hold only in the high temperature limit.

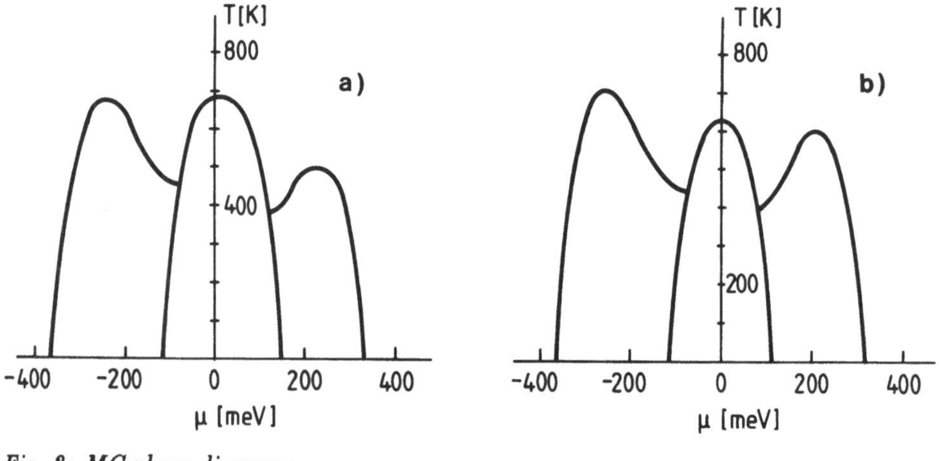

Fig. 3: MC phase diagrams
a) for the tetrahedron many-body interaction
b) for the concentration dependent pair-interaction.

Thus it is interesting to compare the results for the SRO for the two interaction models. For example the SRO for different temperatures have been considered, first 750K closely above the order disorder transition of $L1_0$ and second for the same composition at an intermediate but not very high temperature of 1000K. The SRO parameters as obtained from the MC simulation are presented in fig.4.

Fig. 4: MC simulation of the SRO above $L1_0$ tetrahedron many-body interaction
× 750*K* ■ 1000*K*
concentration dependent pair-interaction
+ 750*K* • 1000*K*.

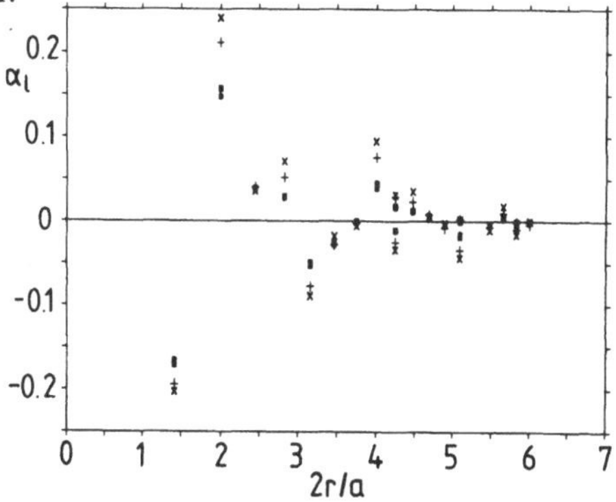

It is found that for the lower temperature, the amplitudes of the SRO parameters but not their sign differ for the two interaction models. At the higher temperature the agreement is already so close that it is not possible to distinguish between the interaction models. It is apparent that difficulties arise if one attempts to determine many-body interactions from a single set of measured SRO parameters. This parameterisation of many-body interactions may be improved if an intrinsic temperature dependence of the pair-interaction is taken into account. Since the effective pair-interactions themselves should be temperature dependent if real many-body interactions are present, we can conclude that the contribution due to many-body interactions is rather small in the above considered case of the NiCu alloy. If the IMC gives different results for the pair-interactions when different SRO configurations are analysed, this then offers the opportunity to determine assumed many-body interactions by using the different SRO configurations simultaneously in the IMC method.

REFERENCES

1. Krivoglaz M A *Theory of X-Ray and Thermal Neutron Scattering by Real Crystals* (Plenum Press, New York) 1969
2. Clapp P C and Moss S C Phys Rev 139A, 844 1966
3. Gerold V Kern J Acta Met 35 No2 393-400 1987
4. Schweika W and Haubold H-G to be published in Phys Rev B 1988
5. Livet F Preprint to be published 1987
6. Binder K Festkörperprobleme - Adv in Sol Stat Phys 26 Grosse P ed 133-168 1986
7. Gehlen P C and Cohen J B Phys Rev 139A 844 1965
8. Bieber A Gautier F J Phys Soc Jap 53 No6 2061-2074 1984
9. Lefebvre S Bley F Bessiere M Fayard M Roth M Cohen J Acta Cryst A36 1-7 1980
10. Epperson J E Fuernrohr P Acta Cryst A39 740-746 1983
11. Klaiber F Schoenfeld B Kostorz G Acta Cryst A43 525-533 1987
12. Chassagne F Thesis Univ Paris VI 1986
13. Wagner W Poerschke R Axmann A and Schwahn D Phys Rev B21 3087 1980
14. Jankowski A F Tsakalakos T this conference 1987
15. De Fontaine D and Kikuchi R NBS Publication SP 496,999 1978
16. Carlsson A Phys Rev B35 No10 4858-4864 1987

SECTION 3

THERMODYNAMICS AND STATISTICAL MECHANICS

COMPUTER BASED THERMOCHEMICAL MODELING OF MULTICOMPONENT PHASE DIAGRAMS

LARRY KAUFMAN

MANLABS, INC., 21 Erie St., Cambridge, Massachusetts 02139-4279, USA

1. RECENT PROGRESS IN CALCULATION OF PHASE DIAGRAMS

During the past twenty years, very substantial progress has been achieved in the development of methods and data bases for thermochemical calculation of phase diagrams (1). Our activities at ManLabs, Inc. (2,7) have been devoted to developing these methods and collaborating with many colleagues through the CALPHAD Groups in exchanging ideas and concepts concerning the application of such methods to technologically meaningful problems. Table 1 lists some of the systems which we have recently calculated.

Prior to the 1950's, efforts directed toward prediction of phase diagrams were largely confined to development of rules based on size effects, valence characteristics, and electron-atom ratios formulated by Hume-Rothery and his colleagues, which could serve to predict equilibrium phase diagrams in metal systems. Although the development of many of the current physical models is an outgrowth of this earlier work, we now have in place a well-developed methodology for coupling phase diagrams and thermochemical data by using models which seek to incorporate the most up-to-date results of "first principle" or "ab initio" calculation of phase stability (8-25). Moreover, these methods can readily deal with metastable phases and, as a consequence, can be readily employed to consider martensitic transformations, thin film and surface deposition effects, rapid solidifica-

G. M. Stocks and A. Gonis (eds.), Alloy Phase Stability, 145–175.

TABLE 1

List of Recently Calculated Binary and Ternary Systems

Si-C	Mn-C	B-C	W-B
Al-Si	Ti-N	Al-B	Fe-N
Ti-Si	Al-Sb	Si-B	Fe-Cr
Cr-Si	Al-Ga	Ti-B	Fe-Mo
Mn-Si	Sb-Ga	Cr-B	Cr-P
Fe-Si	Ti-Be	Mn-B	Mn-P
Co-Si	Zr-Be	Fe-B	Fe-P
Ni-Si	Hf-Be	Co-B	Co-P
Cu-Si	Fe-Be	Ni-B	Ni-P
Nb-Si	Co-Be	Cu-B	Cu-P
Mo-Si	Ni-Be	Nb-B	V-Ni
W-Si	Cu-Be	Mo-B	V-Cu
V-B	V-Si	V-Mn	V-Nb
V-C	V-Ti	V-Fe	V-Mo
V-Al	V-Cr	V-Co	V-W

Fe-Ti-Al	Co-Nb-W	Cr-Co-Mo
Fe-Ti-Mn	Co-Mo-W	Cr-Co-Ti
Cr-Ti-W	Al-Nb-Ti	Cr-Co-W
Ni-Co-Nb	Al-Mo-Ti	Cr-Nb-W
Ni-Co-Mo	Al-Ti-W	Fe-Mo-W
Ni-Co-W	Fe-Ni-Ti	Fe-Mo-Ti
Co-Nb-Mo	Cr-Ni-W	Fe-Ti-W
Ga-Sb-Al	Fe-Cr-Si	Ti-C-N
Fe-Cr-Mo	Ni-Cr-Mo	Fe-Nb-W
Fe-Cr-Nb	V-Al-Ti	Co-Cr-V
Fe-Ni-W	V-Ti-Cr	Al-Mn-V
Fe-Co-Nb	V-Cr-Fe	Al-Ni-V
Fe-Co-Ti	V-Cr-Ni	Mn-Si-V
Fe-Co-W	V-Co-Ni	W-Si-V
Fe-Nb-Mo	Fe-Ni-V	Fe-Ti-Nb
Ni-Si-Al	Fe-Co-V	Ni-Si-Cr
		Ni-Pt-Al

tion, neutron irradiation, and ion implantation perturbations of the equilibrium state. Moreover, since the method can be employed to describe the thermochemical properties in a detailed fashion, explicit definition of the activity of each of the participating species in the system of interest, as well as kinetic models of nucleation, growth and diffusion in multicomponent systems, can readily be developed.

The evolution of an extensive data base concerning stable, metastable, and unstable phases requires the description of metastable and unstable phases and their energetics. This requirement has fostered a closer interaction between physicists interested in "ab initio" and other model calculations (23-25, 8-22) and thermodynamicists concerned with developing methods for estimating data for condensed phases which are difficult to acquire experimentally (8-13, 21, 22). This interaction has generated additional activity, interest, and progress in this field. Thus theoreticians who can calculate the energetics of metastable and unstable structures now have audiences who are interested in the numerical values produced by such calculation and willing to provide feedback concerning their own perceptions concerning such numerical values.

The CALPHAD Group, formed in 1973, has focused on the calculation of phase diagrams based on thermochemical data by using models which attempt to describe the thermochemical properties of the phases of interest over the widest possible ranges of temperature, pressure, and composition.

Groups like CALPHAD, SGTE (Scientific Group Thermodata Europe), and the ASM-NBS Alloy Phase Data activity have fostered progress in this area through cooperative sharing of data, computer methods, and research techniques. Such interactions are quite usual at the annual CALPHAD meetings which have occurred during the past sixteen years. The ten-year subject index prepared for Volume X (1986) of the Journal illustrates the range of topics which has been studied. Of late, these annual conferences have been attended by more than one hundred researchers presenting 70-80 papers. Recently, a conference on USER APPLICATIONS of PHASE DIAGRAMS was held at the ASM Materials Conference in Orlando, Florida in October 1986. More than thirty papers were presented and published in a proceeding volume. To date analyses of all of the binary systems formed by combination of B, C, Al, Si, Ti, V, Cr, Mn, Fe, Co, Ni, Cu, Nb, Mo and W have been completed and published (2-11), or are being published. CALPHAD activities have stimulated substantial worldwide progress in multicomponent phase diagram calculations, teaching, solid state physics, and fused salt research, in addition to studies in metallic systems. Two recent examples of such work are the analyses of Fe-Ni-W (13) and Fe-Cr-Mn-Ni-C (14). These experimental and computational analyses of complex multicomponent systems are illustrative of the extent to which progress in this field has occurred.

Figures 1-5 and Tables 2-7 (11) display some of our recent results for vanadium alloy systems and the lattice stability of the sigma phase. In each case, the analytical description of the excess Gibbs energy of the solution and compound phases is shown along with the calculated and experimental binary phase diagram. In addition. the heat of formation of the fcc and bcc forms derived from the thermochemical description and the pair potential calculation (8,9) is displayed. Thus physical models and "ab initio" results are integrated into the process of characterization (15-25).

Figures 6-14 illustrate how the binary calculations are combined to compute ternary sections (11). For example, the Fe-V and Cr-V cases displayed in Figures 3 and 4 are combined with an earlier Fe-Cr analysis (3) to compute Fe-V-Cr. The

TABLE 2

LATTICE STABILITY VALUES FOR THE ELEMENTS

(Units of J/mol and J/mol K will be used throughout)

Element	(L=Liquid, S=Sigma, D-Diamond Cubic R=Rhombohedral)	Temperature Range (K)
V	$°G^{L}-°G^{bcc} = 21502-9.749T$	$300 < T$
	$°G^{bcc}-°G^{hcp} = -6276-3.347T$	$300 < T$
	$°G^{bcc}-°G^{fcc} = -8996-3.556T$	$300 < T$
	$°G^{S}-°G^{bcc} = 3347+3.682T$	$300 < T$
	$°G^{L}-°G^{D} = -30.00T$	$300 < T$
	$°G^{L}-°G^{R} = -21.84T$	$300 < T$
Cr	$°G^{S}-°G^{bcc} = 2531+3.682T$	$300 < T$
Mn	$°G^{S}-°G^{bcc} = -2008+1.674T$	$300 < T$
Fe	$°G^{S}-°G^{bcc} = 3347+3.682T$	$300 < T$
Co	$°G^{S}-°G^{fcc} = 10460+4.184T$	$1300 < T$
	$°G^{S}-°G^{fcc} = 15071+4.184T-0.6314X10^{-2}T^{2}+0.2804X10^{-5}T^{3}$	$300<T<1300$
Ni	$°G^{S}-°G^{fcc} = 10460+4.184T$	$300 < T$

TABLE 3

ANALYTICAL DESCRIPTION OF THE VANADIUM-ALUMINUM SYSTEM

Phase	${}^{E}H = H_M - x_V {}^{\circ}H_V - x_{Al} {}^{\circ}H_{Al}$	${}^{E}S = S - x_V {}^{\circ}S_V - x_{Al} {}^{\circ}S_{Al} + R\,[x_V \ln x_V + x_{Al} \ln x_{Al}]$	Composition Range	Comments
Liquid	$- x_V x_{Al} 48032$	$x_V x_{Al} 11.213$	$0 < x_{Al} < 1$	800<T<2300K °Refers To Liquid
bcc	$- x_V x_{Al} 149703$	$- x_V x_{Al} 28.535$	$0 < x_{Al} < 1$	300<T<2400K °Refers to bcc
fcc	$- x_V x_{Al} 149703$	$- x_V x_{Al} 28.535$	$0 < x_{Al} < 1$	300<T<2300K °Refers to fcc

Compound	$H = H - x_V^* {}^{\circ}H_V^{\theta} - x_{Al}^* {}^{\circ}H_{Al}^{\theta}$	$S = S - x_V^* {}^{\circ}S_V^{\theta} - x_{Al}^* {}^{\circ}S_{Al}^{\theta}$	Composition Range	Comments
$V_{0.390}Al_{0.610}$	-43857	-5.180	$x_{Al}^* = 0.610$	θ = bcc
$V_{0.250}Al_{0.750}$	-32857	-3.435	$x_{Al}^* = 0.750$	θ = bcc
$V_{0.150}Al_{0.850}$	-17326	+3.414	$x_{Al}^* = 0.850$	θ = fcc

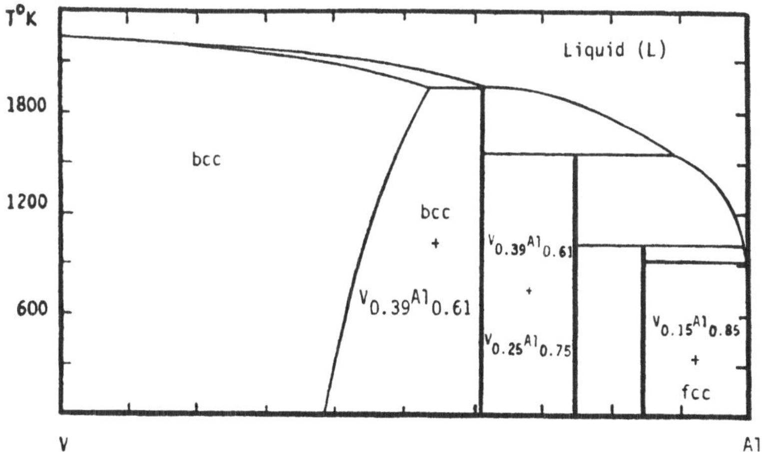

Figure 1a. Calculated V-Al Phase Diagram.

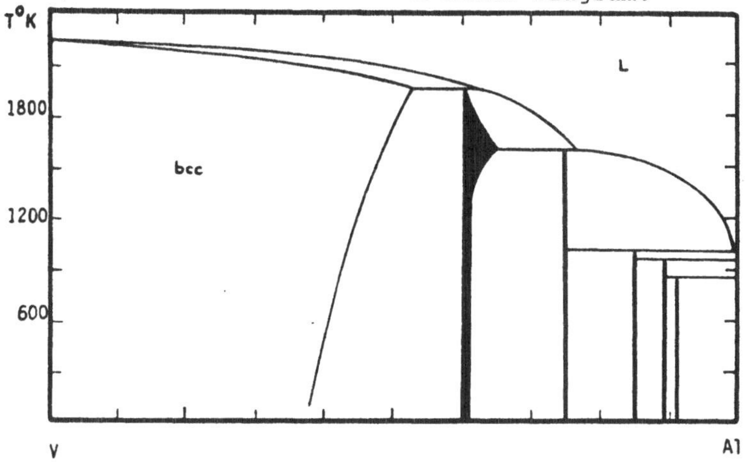

Figure 1b. Observed V-Al Phase Diagram.

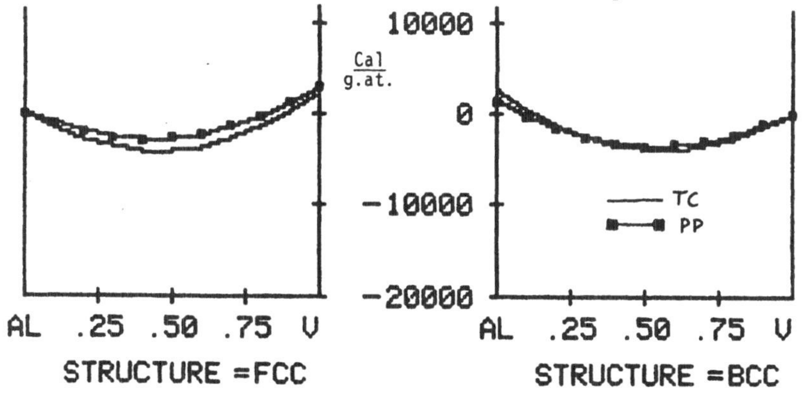

Figure 1c. Comparison of Heat of Formation of fcc and bcc phases Derived from Pair Potential (PP) and Thermochemical (TC) Analysis.

TABLE 4

ANALYTICAL DESCRIPTION OF THE TITANIUM-VANADIUM SYSTEM

Phase	${}^{E}H = H_M - x_{Ti}{}^{\circ}H_{Ti} - x_V{}^{\circ}H_V$	${}^{E}S = S - x_{Ti}{}^{\circ}S_{Ti} - x_V{}^{\circ}S_V + R[x_{Ti}\ln x_{Ti} + x_V\ln x_V]$	Composition Range	Comments
Liquid	$x_{Ti}x_V5422$	0	$0 < x_V < 1$	1500<T<2400K °Refers To Liquid
bcc	$x_{Ti}x_V11581$	0	$0 < x_V < 1$	300<T<2300K °Refers to bcc
hcp	$x_{Ti}x_V11995$	0	$0 < x_V < 1$	300<T<1500K °Refers to hcp

TABLE 5

ANALYTICAL DESCRIPTION OF THE CHROMIUM-VANADIUM SYSTEM

Phase	${}^{E}H = H_M - x_{Cr}{}^{\circ}H_{Cr} - x_V{}^{\circ}H_V$	${}^{E}S = S - x_{Cr}{}^{\circ}S_{Cr} - x_V{}^{\circ}S_V + R[x_{Cr}\ln x_{Cr} + x_V\ln x_V]$	Composition Range	Comments
Liquid	$- x_{Cr}^2x_V30125$	$x_{Cr}x_V[0.209x_{Cr} + 6.485x_V]$	$0 < x_V < 1$	1800<T<2400K °Refers To Liquid
bcc	$- x_{Cr}^2x_V20920$	$x_{Cr}x_V[0.209x_{Cr} + 6.485x_V]$	$0 < x_V < 1$	300<T<2300K °Refers to bcc
fcc	$- x_{Cr}^2x_V20920$	$x_{Cr}x_V[0.209x_{Cr} + 6.485x_V]$	$0 < x_V < 1$	300<T<2300K °Refers to fcc

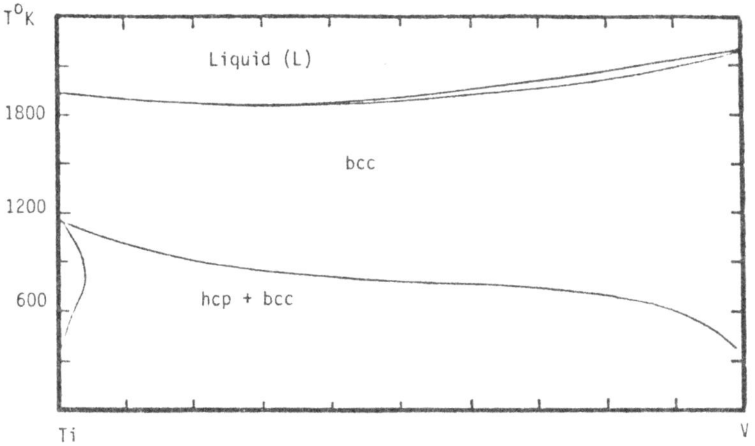

Figure 2a. Calculated Ti-V Phase Diagram.

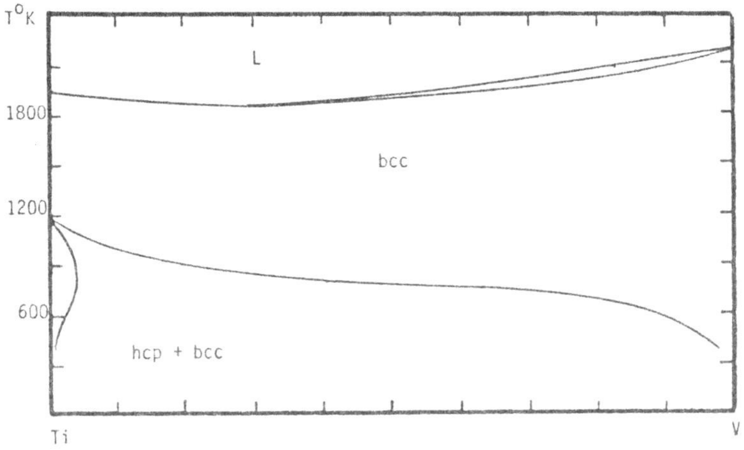

Figure 2b. Observed Ti-V Phase Diagram.

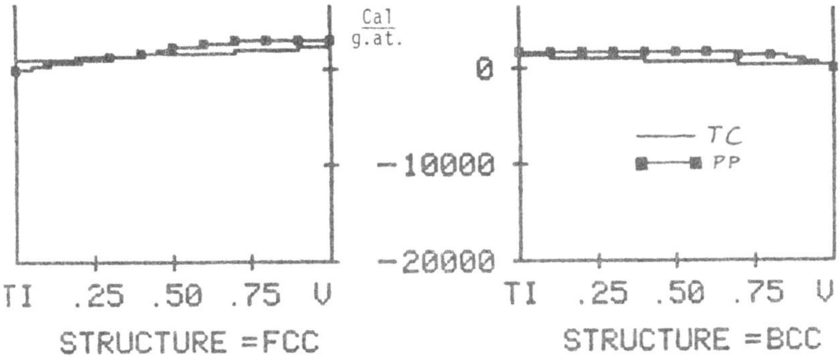

Figure 2c. Comparison of Heat of Formation of fcc and bcc phases Derived from Pair Potential (PP) and Thermochemical (TC) Analysis.

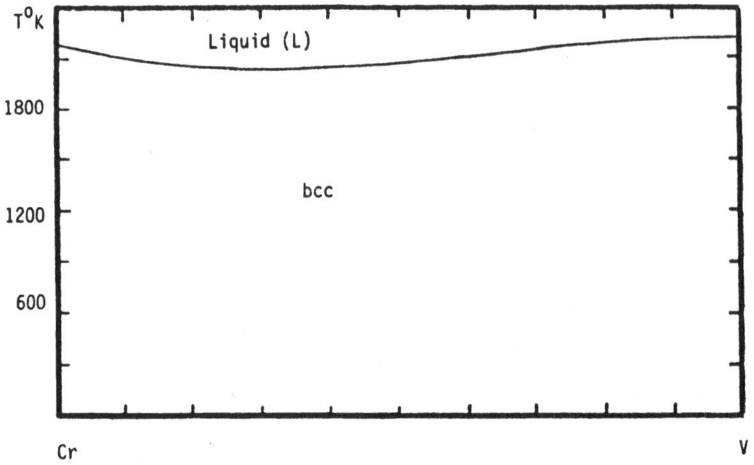

Figure 3a. Calculated Cr-V Phase Diagram .

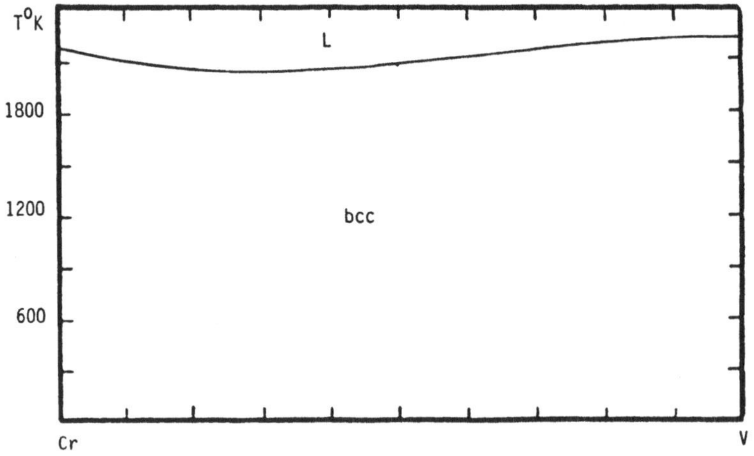

Figure 3b. Observed Cr-V Phase Diagram.

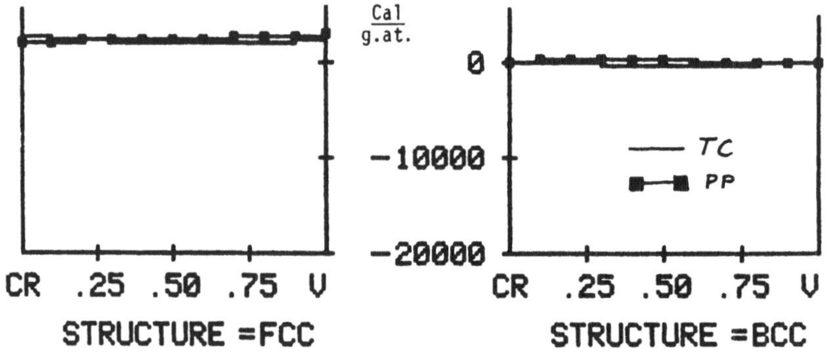

Figure 3c. Comparison of Heat of Formation of fcc and bcc phases Derived from Pair Potential (PP) and Thermochemical (TC) Analysis.

TABLE 6

ANALYTICAL DESCRIPTION OF THE IRON-VANADIUM SYSTEM

Phase	$^{E}H = H_M - x_{Fe}\,{}^{o}H_{Fe} - x_V\,{}^{o}H_V$	$^{E}S = S - x_{Fe}\,{}^{o}S_{Fe} - x_V\,{}^{o}S_V + R\,[x_{Fe}\ln x_{Fe} + x_V \ln x_V]$	Composition Range	Comments
Liquid	$x_{Fe}x_V[-17355x_{Fe} - 46476x_V]$	$- x_{Fe}x_V 3.933$	$0 < x_V < 1$	1600<T<2400K °Refers To Liquid
bcc	$x_{Fe}x_V[-15958x_{Fe} - 31999x_V]$	$x_{Fe}x_V[-7.029x_{Fe} - 2.008x_V]$	$0 < x_V < 1$	300<T<2300K °Refers to bcc
Sigma	$x_{Fe}x_V[-45175x_{Fe} - 49794x_V]$	$x_{Fe}x_V[-3.163x_{Fe} + 10.256x_V]$	$0 < x_V < 1$	300<T<1700K °Refers to Sigma
fcc	$- x_{Fe}x_V 17405$	$-x_{Fe}x_V 1.159$	$0 < x_V < 1$	600<T<1800K °Refers to fcc

TABLE 7

ANALYTICAL DESCRIPTION OF THE COBALT-VANADIUM SYSTEM

Phase	$^{E}H = H_M - x_{Co}\,^{o}H_{Co} - x_V\,^{o}H_V$	$^{E}S = S - x_{Co}\,^{o}S_{Co} - x_V\,^{o}S_V + R\,[x_{Co}\ln x_{Co} + x_V \ln x_V]$	Composition Range	Comments
Liquid	$x_{Co}x_V[-55229x_{Co} - 37823\,x_V]$	$x_{Co}x_V 4.519$	$0< x_V <1$	1000<T<2400K °Refers To Liquid
bcc	$x_{Co}x_V[-58994x_{Co} - 44350x_V]$	$- x_{Co}x_V 8.535$	$0< x_V <1$	300<T<2400K °Refers to bcc
hcp	$x_{Co}x_V[-45187x_{Co} - 51555x_V]$	$x_{Co}x_V[7.113x_{Co} - 5.439x_V]$	$0< x_V <1$	300<T<1000K °Refers to hcp
fcc	$x_{Co}x_V[-38911x_{Co} - 51555x_V]$	$x_{Co}x_V[19.665x_{Co} - 5.439x_V]$	$0< x_V <1$	300<T<2000K °Refers to fcc
Sigma	$x_{Co}x_V[-78241x_{Co} - 80751x_V]$	$x_{Co}x_V[10.837x_{Co} + 4.477x_V]$	$0< x_V <1$	300<T<2000K °Refers to Sigma

Compound	$H = H - x^{*}_{Co}\,^{o}H^{\theta}_{Co} - x^{*}_V\,^{o}H^{\theta}_V$	$S = S - x^{*}_{Co}\,^{o}S^{\theta}_{Co} - x^{*}_V\,^{o}S^{\theta}_V$	Composition Range	Comments
$C_{0.750}V_{0.250}$	-11581	4.456	$x^{*}_V = 0.250$	θ = fcc
$C_{0.250}V_{0.750}$	-10929	1.791	$x^{*}_V = 0.750$	θ = bcc

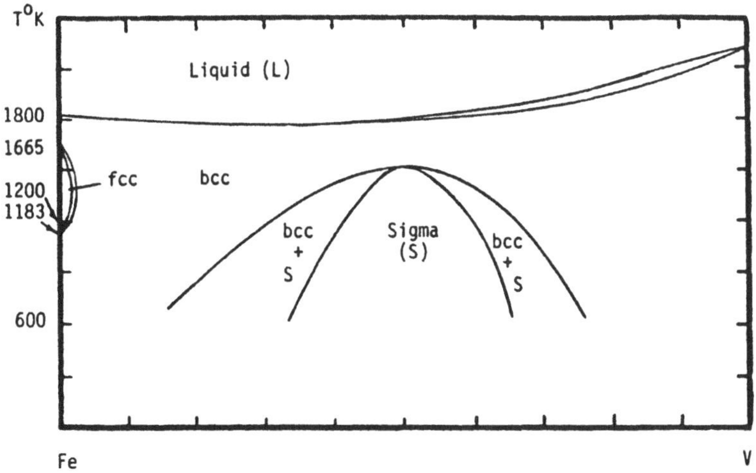

Figure 4a. Calculated Fe-V Phase Diagram.

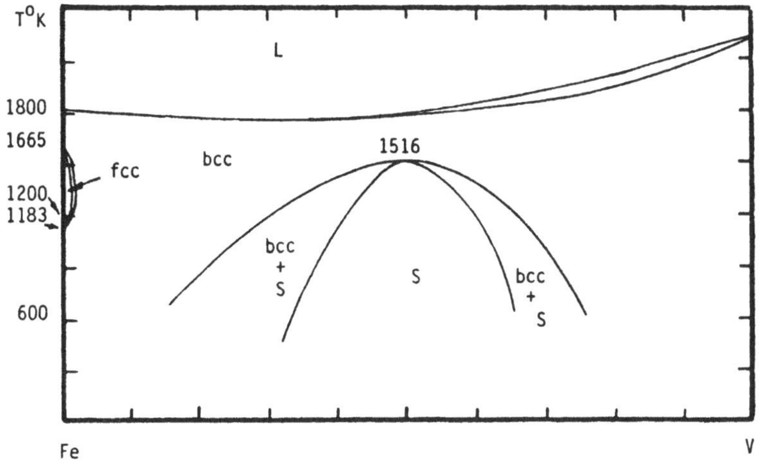

Figure 4b. Observed Fe-V Phase Diagram.

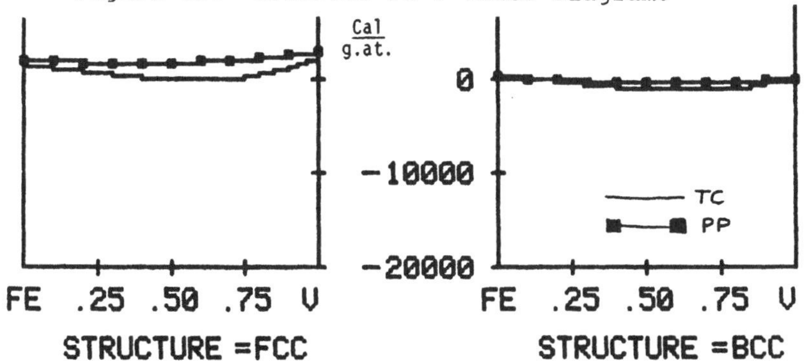

Figure 4c. Comparison of Heat of Formation of fcc and bcc phases Derived from Pair Potential (PP) and Thermochemical (TC) Analysis.

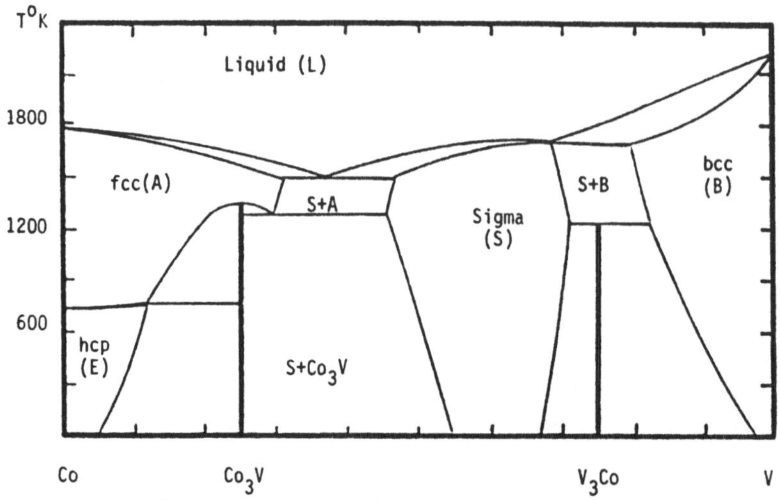

Figure 5a. Calculated Co-V Phase Diagram.

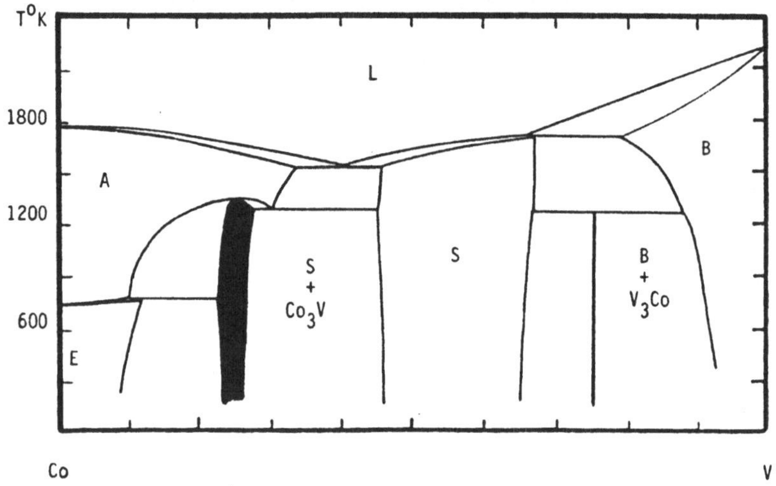

Figure 5b. Observed Co-V Phase Diagram.

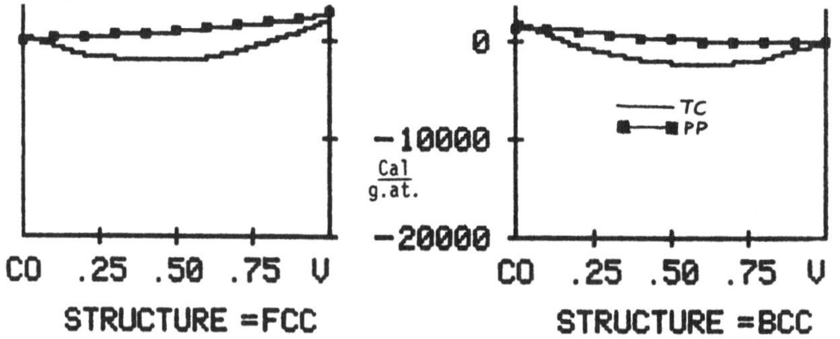

Figure 5c. Comparison of Heat of Formation of fcc and bcc phases Derived from Pair Potential (PP) and Thermochemical (TC) Analysis.

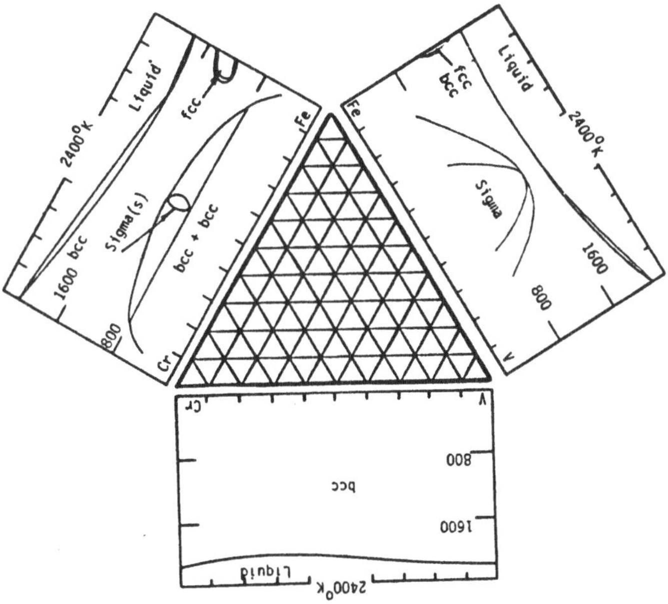

Figure 6. Calculated Isothermal Sections in the Fe-V-Cr System.

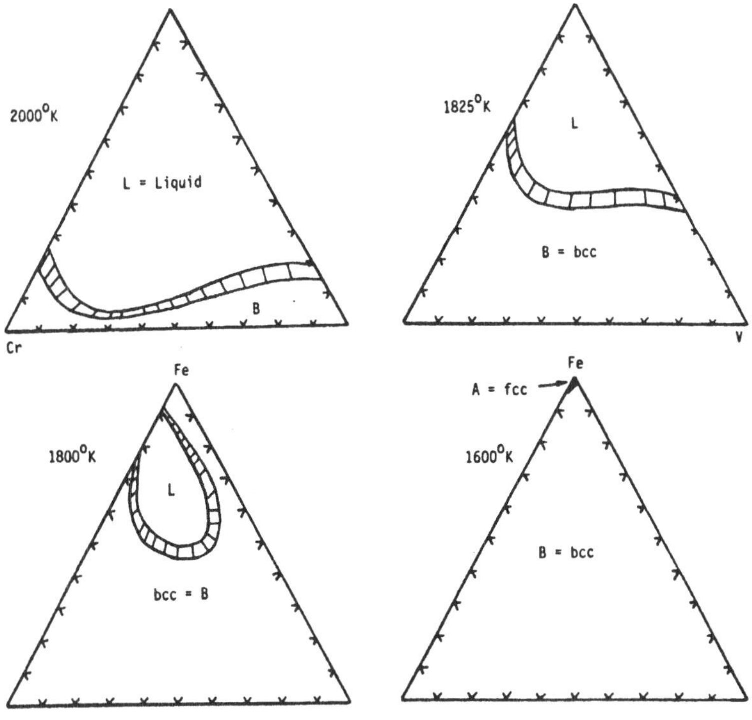

Figure 7. Calculated Isothermal Sections in the Fe-V-Cr System.

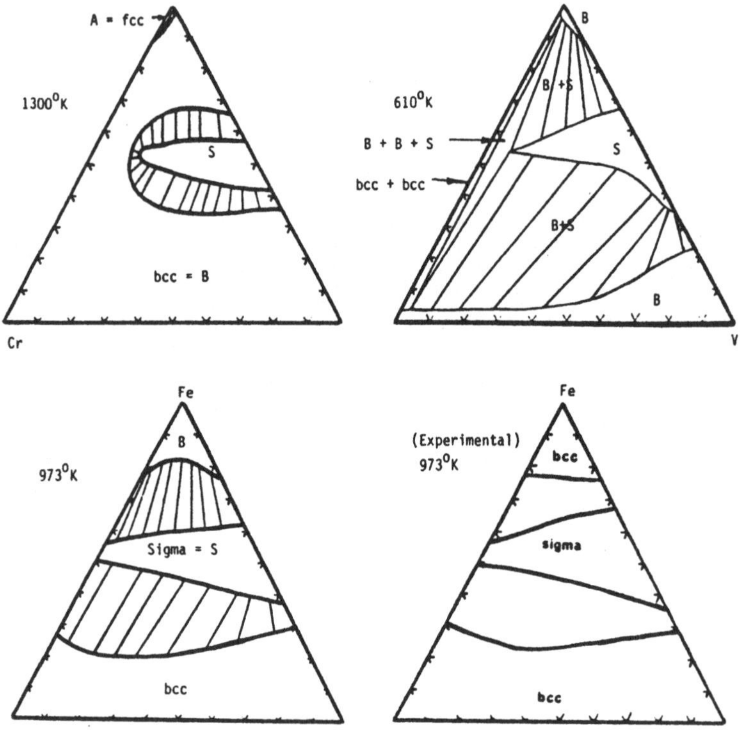

Figure 8. Calculated Isothermal Sections in the Fe-V-Cr System.

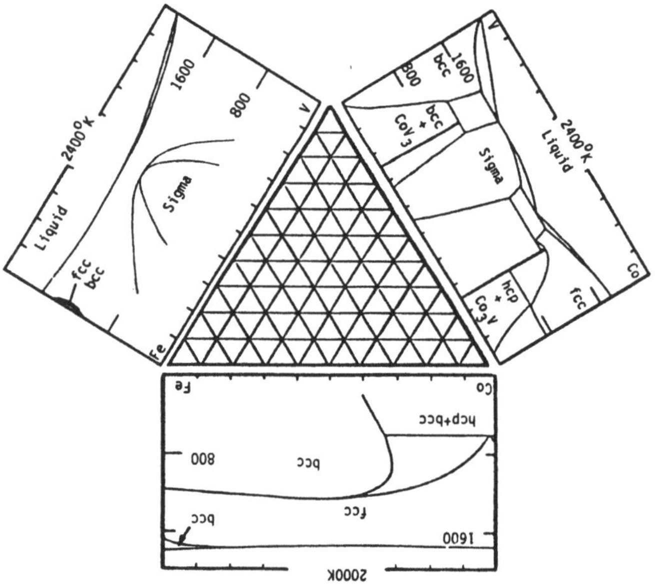

Figure 9. Calculated Isothermal Sections in the V-Co-Fe System.

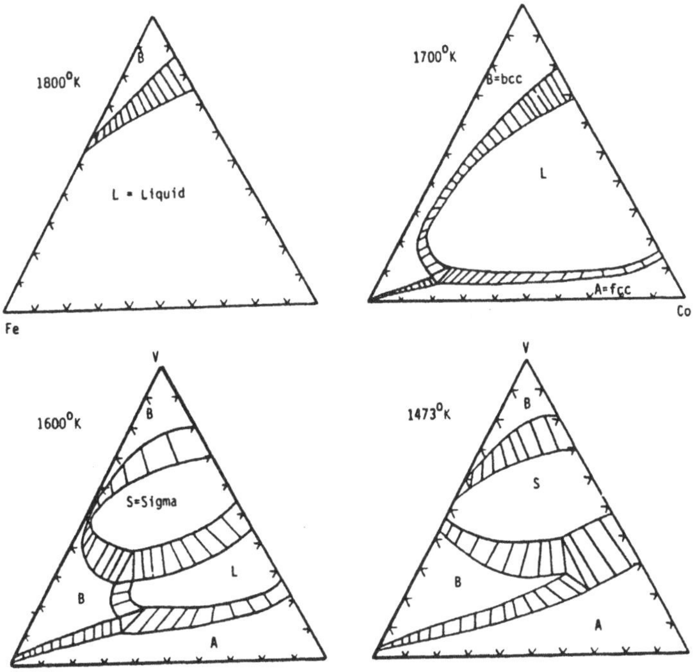

Figure 10. Calculated Isothermal Sections in the V-Co-Fe System.

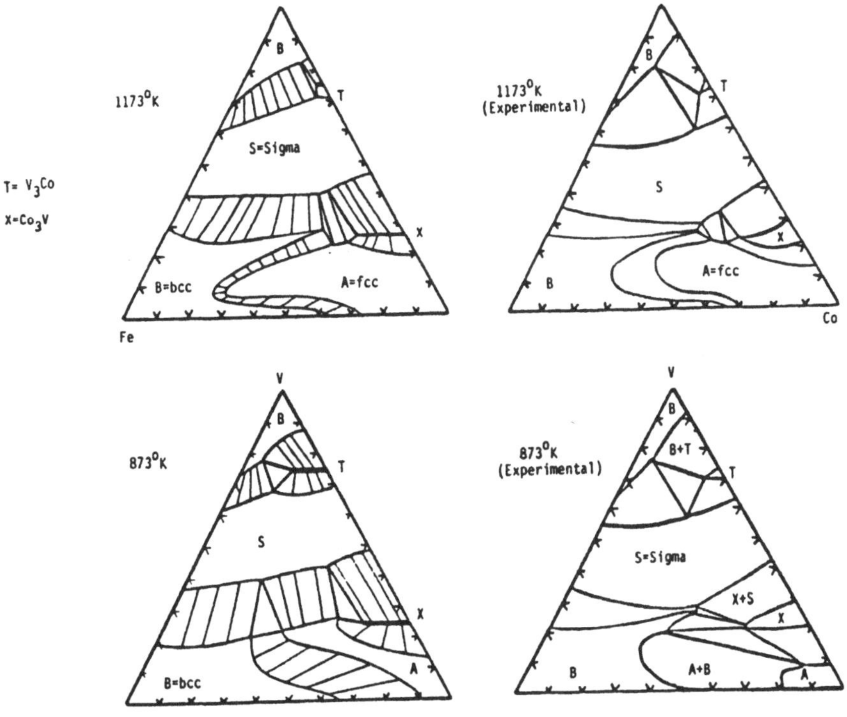

Figure 11 Calculated Isothermal Sections in the V-Co-Fe System.

lowest temperature in Figure 8 (i.e. 973K) shows a comparison of calculated and observed sections.

In Figures 9-11, calculated and experimental sections are shown at 1173K for V-Co-Fe, while the Ti-Al-V cases in Figures 12-14 display calculated and experimental sections at 900K. Comparable results obtain for all of the systems listed in Table 1. The addition of vanadium to the data base means that phase equilibria can be calculated on all of the binary and multi-component systems based on elements listed above.

2. CALCULATION OF THE LATTICE STABILITY OF METALS

Tables 8 and 9 and Figures 15 and 16 show the results of a recent study (12) of the lattice stability of a selected group of elements within our data base. In obtaining these values, recent thermochemical data, as well as the results of "physical model" and "ab initio" (16, 21-25) calculations, have been considered.

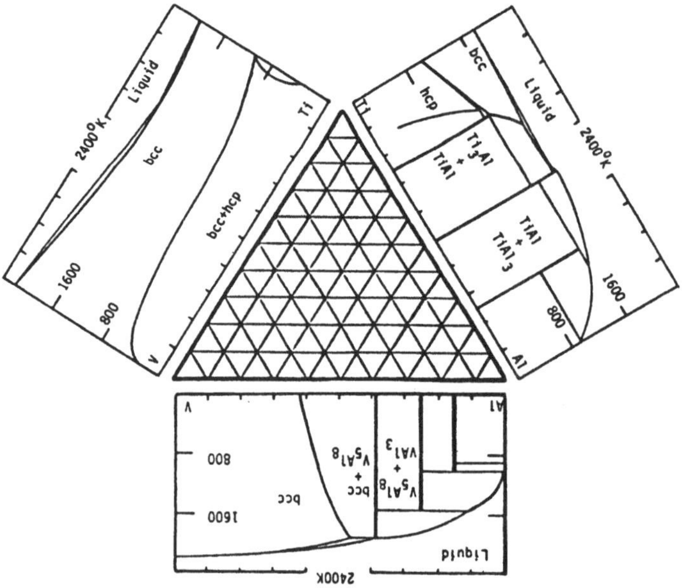

Figure 12. Calculated Isothermal Sections in the Ti-Al-V System.

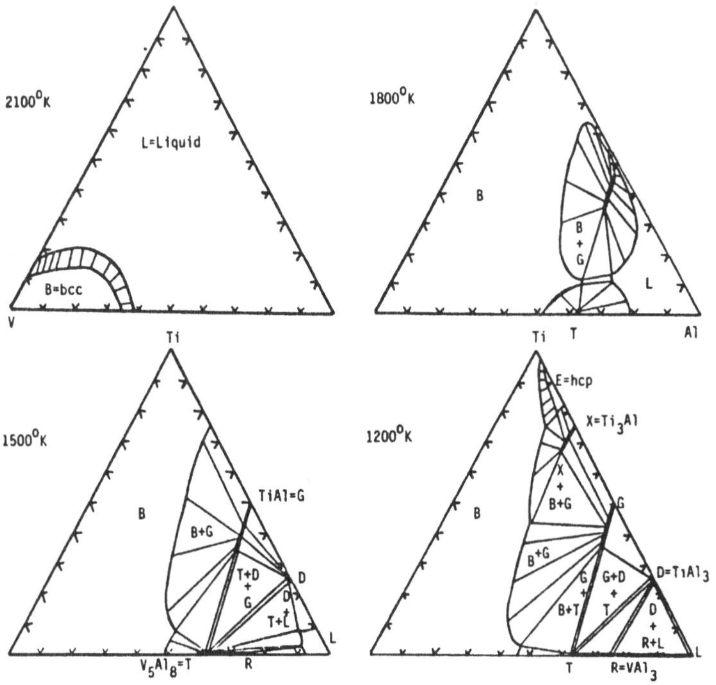

Figure 13. Calculated Isothermal Sections in the Ti-Al-V System.

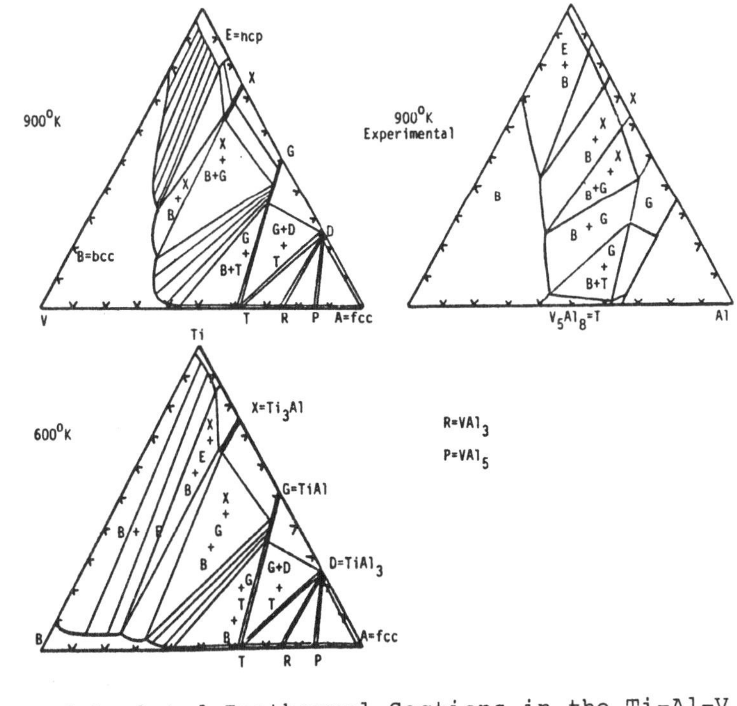

Figure 14. Calculated Isothermal Sections in the Ti-Al-V System.

TABLE 8

SUMMARY OF GIBBS ENERGY DIFFERENCES BETWEEN THE FCC, BCC AND HCP STRUCTURES IN THE ELEMENTS OF INTEREST

Element		ΔG(Joule/mole°K), T°K
Be	hcp → bcc	$4450 \pm 2000 - (2.91 \pm 1.29)T$
	hcp → fcc	$3500 \pm 1500 + (0.65 \pm 1.15)T$
Al	fcc → bcc	$10080 \pm 1500 - (4.18 \pm 1.0)T$
	fcc → hcp	$5480 \pm 1500 - (1.80 \pm 1.0)T$
Ti	hcp → bcc	$4260 \pm 90 - (3.69 \pm 0.08)T$
	hcp → fcc	$2100 \pm 1100 - (0.0 \pm 1.0)T$
V	bcc → fcc	$8950 \pm 3000 + (3.61 \pm 0.16)T$
	hcp → fcc	$400 \pm 2400 + (0.20 \pm 1.0)T$
Cr	bcc → fcc	$8900 \pm 1600 + (1.80 \pm 1.60)T$
	fcc → hcp	$1000 \pm 3000 - (0.6 \pm 1.0)T$

TABLE 9

SUMMARY OF GIBBS ENERGY DIFFERENCES BETWEEN THE FCC, BCC AND HCP STRUCTURES IN THE ELEMENTS OF INTEREST

Element		ΔG(Joule/mole°K), T°K
Mn	bcc → fcc	(See Figure 16) 0 ± 1000 at 0K -400 ± 500 at 800K Figure 16 ± 100 above 1000K
	hcp → fcc	$2500 \pm 1500 - (0.6 \pm 1.0)T$
Fe	bcc → fcc	(See Figure 15) 6200 ± 1000 at 0K Figure 31 ± 200 at 500K Figure 31 ± 10 at 1000K
	hcp → fcc	$2300 \pm 1600 - (5.75 \pm 4.0)T$
Co	fcc → bcc	$5450 \pm 1500 + (1.25 \pm 1.25)T$ $- 0.63137X10^{-2}T^2 + 2.8037X10^{-6}T^3$
	hcp → fcc	$457 \pm 3 - (0.636 \pm 0.01)T$
Ni	fcc → bcc	$4900 \pm 1000 - (0.76 \pm 0.80)T + 4.109X10^{-3}T^2$ $- 4.853X10^{-6}T^3 + 1.41X10^{-9}T^4$
	fcc → hcp	$2000 \pm 1000 + (1.30 \pm 1.00)T$
Cu	fcc → bcc	$5150 \pm 1030 - (2.30 \pm 1.05)T$
	fcc → hcp	$1600 \pm 1000 + (1.30 \pm 1.00)T$

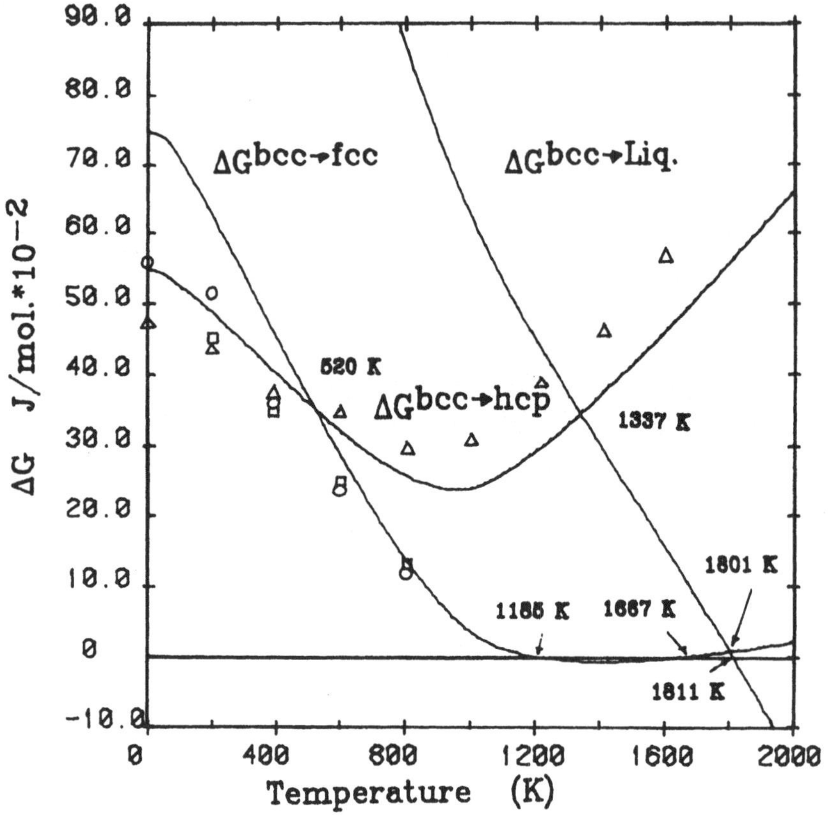

Figure 15. The Lattice Stability of Iron after Guillermet and Gustafson (41). Points are from (2) Δ bcc/hcp, ○ bcc/fcc and (26) □ bcc/fcc.

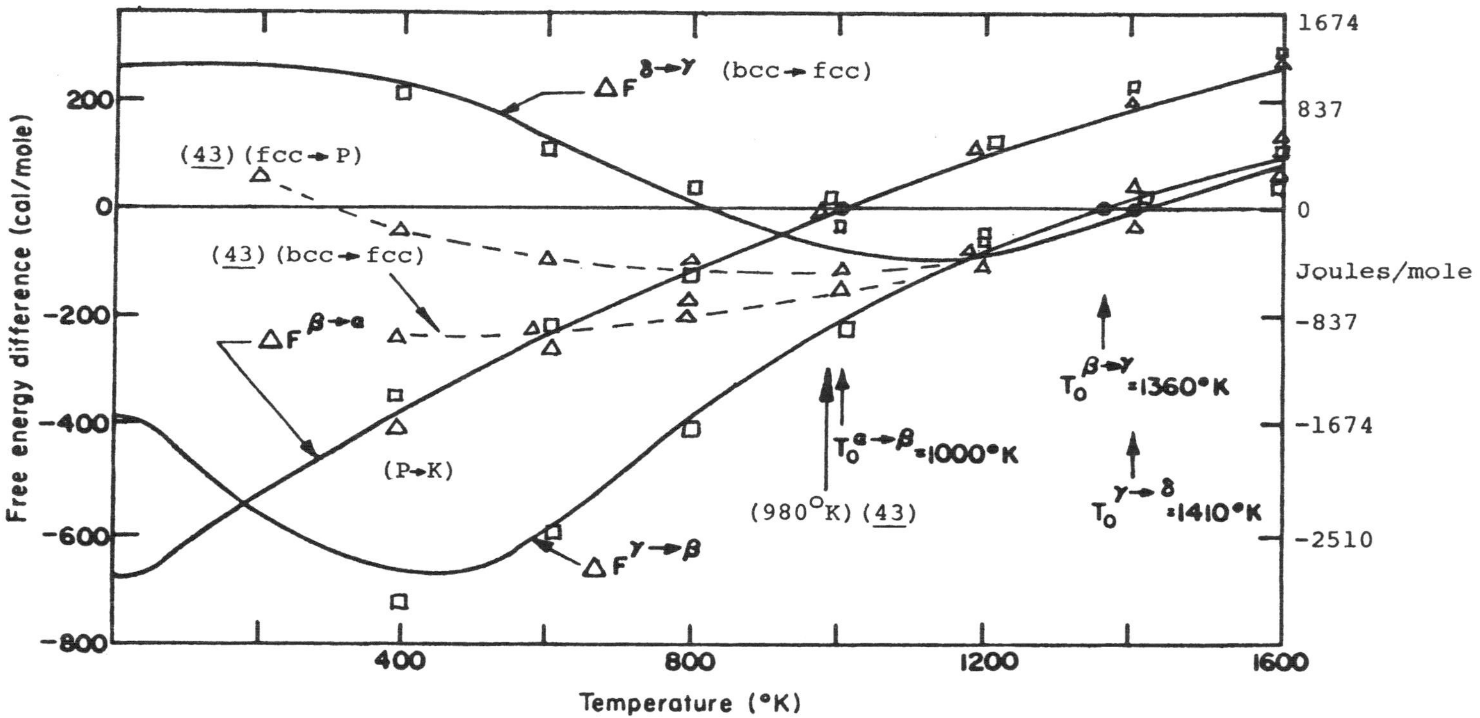

Figure 16. The Lattice Stability of Manganese after Tauer and Weiss (42) (Solid Curves and Points), Hillert (43) (triangles and dashed curves), and (6) Squares.

Recent studies of the relative stability of fcc and bcc iron and the efffects of magnetism on the stability published by Bendick and Pepperhoff (26) are in good general agreement with our earlier findings (1). A recent analysis of the Fe-Ru systems by Swartzendruber and Sundman (27) produced lattice stability results for the bcc, hcp, and fcc forms of iron and ruthenium as well as thermochemical descriptions of the liquid, bcc, hcp and fcc phases in the Fe-Ru system, in keeping with earlier findings (2). J.L. Murray (28) has reevaluated the thermochemistry of more than twenty binary systems by employing the lattice stability values derived previously (2) for Ti, Nb, Cr, Hf, Mo, W, Zr, Mn, Fe, Re, Al, Mg, Co, Cd, Li, Ir, Os, Ru, Pd, Pt, Rh and Zn with good results.

Application of thermochemical techniques for computing metastable extensions of solid-liquid equilibria relevant to the formation of metallic phases has been noted (29). Similar applications have been recently presented by Murray, Massalski, Miodownik and coworkers (30-33). Williams (34) has employed the database developed in the present program in order to evaluate the stability of the gamma prime and sigma phases in superalloys, while Fullman (35) has used the database in order to asses intergranular corrosion cracking in austenitic stainless steels. The measured heat of formation of copper-titanium alloys reported by Kleppa and Watanabe in 1982 (36) compares very well with the values calculated previously (6). Similar results obtain for the Hf-Be phase diagram calculated in 1979 and measured in 1980 (37).

A recent study of the Ni-Al-Mo system by D. Miracle, K. Lark, V. Strinivasan and H. Lipsitt (38) reveals a four phase equilibrium reaction between nickel-rich (fcc) and molybdenum-rich(bcc) pairs and Ni_3Al based and NiMo based compound pairs on heating and cooling through 1400K. The calculated Ni-Mo-Al ternary system (39) published in 1974 predicted fcc/bcc equilibrium at 1437K and Ni_3Al/NiMo equilibria at 1273K in agreement with the new findings of

Miracle et al but in opposition to the experimental data available in 1974!

These examples reinforce the previous experience with verification of the prediction of thermochemical and phase diagram results by subsequent experimental studies.

It is anticipated that further progress in this field will be made by coupling recent measurements of thermochemical data and improvements in the level of accuracy of first principle or "ab initio" calculations of the heat of formation of stable and unstable phases. Such efforts will provide more accurate data to enhance our capability of predicting phase equilibria in multicomponent systems over a wide range of conditions as well as a deeper understanding of the energetics of phase stability.

Tables 8 and 9 and Figures 15 and 16 provide the results of our recent study of the lattice stability of a series of elements including most of the first transition series. Comparison of these results with those derived from "ab initio" studies (21, 23, 25 and 40) shows relatively good agreement except for the case of the bcc/fcc or hcp difference in Cr and V. In these cases, phase diagrams/thermochemical studies yield substantially smaller values than do the "ab initio" results. In particular, Table 8 yields values of 8950 $\pm$ 3000 J/mol for the fcc/bcc enthalpy difference in vanadium and 8900 $\pm$ 1600 J/mol for the fcc/bcc difference in chromium. By comparison, the "ab initio" values are 26000 for vanadium and 39000 for chromium. Saunders and Miodownik (21) have attempted to close the gap between the thermochemical values and the "ab initio" results by choosing the maximum possible entropy differences between the bcc and fcc forms. However, for the first transition series, they find enthalpy differences of 15000 J/mol for V and 9192 J/mol for Cr, which do not agree with the "ab initio" results.

The original investigations of the lattice stability of the second and third transition series (2) was based on estimated heats of fusion for most of the elements of interest (21). The availability of measured heats of fusion permit a more accurate assessment to be made. Saunders and Miodownik have attempted such an analysis for Nb, Ta, Mo and W. In these cases extremely large bcc/fcc enthalpy differences result. Thus Scriver (25) gives 40KJ/mol for Mb and Ta and 50KJ/mol for Mo and W. In the case of W, Saunders and Miodownik (21) have suggested that an enthalpy difference of 35KJ/mol when coupled with an entropy difference of 4.2 J/mol^{o}K (fcc higher) could be employed to "accomodate" the "ab initio" and "thermochemical" values. This suggestion leads to a melting point of 1739K for fcc tungsten. The Saunders-Miodownik suggestion was based largely on a desire to accomodate the very large magnitudes suggested by the "ab initio" calculations (25, 40) and the "thermochemical" values without performing detailed phase diagram calculations. Figure 16 illustrates the consequences of taking such details into consideration for the case of tungsten. The four calculated phase diagrams shown in Figure 17 and labeled 17-1, 17-2, 17-3 , and 17-4 respectively are derived by using the lattice stability data suggested for bcc and fcc Pt by Saunders and Miodownik (21) along with the enthalpy of the liquid given by Miedema and co-workers (16). Combining these data with the heat and entropy of fusion for bcc tungsten fixes the excess Gibbs energy of the bcc W-Pt phase to conform with the experimental bcc + Liquid field shown in Panel 17-1. It is now possible to establish the properties of the fcc phase and the enthalpy difference between bcc and fcc tungsten as a function of the melting point of fcc tungsten. Figure 17-1 shows that the best results are obtained with a melting point of 3000K for fcc W corresponding to a fcc/bcc enthalpy difference of 22440 J/mol. In this case the calculated fcc/Liquid equilibrium coincides with the experimental values. The succeeding panels 17-2, 17-3 and 17-4 shown in Figure 17 disclose the result of keeping

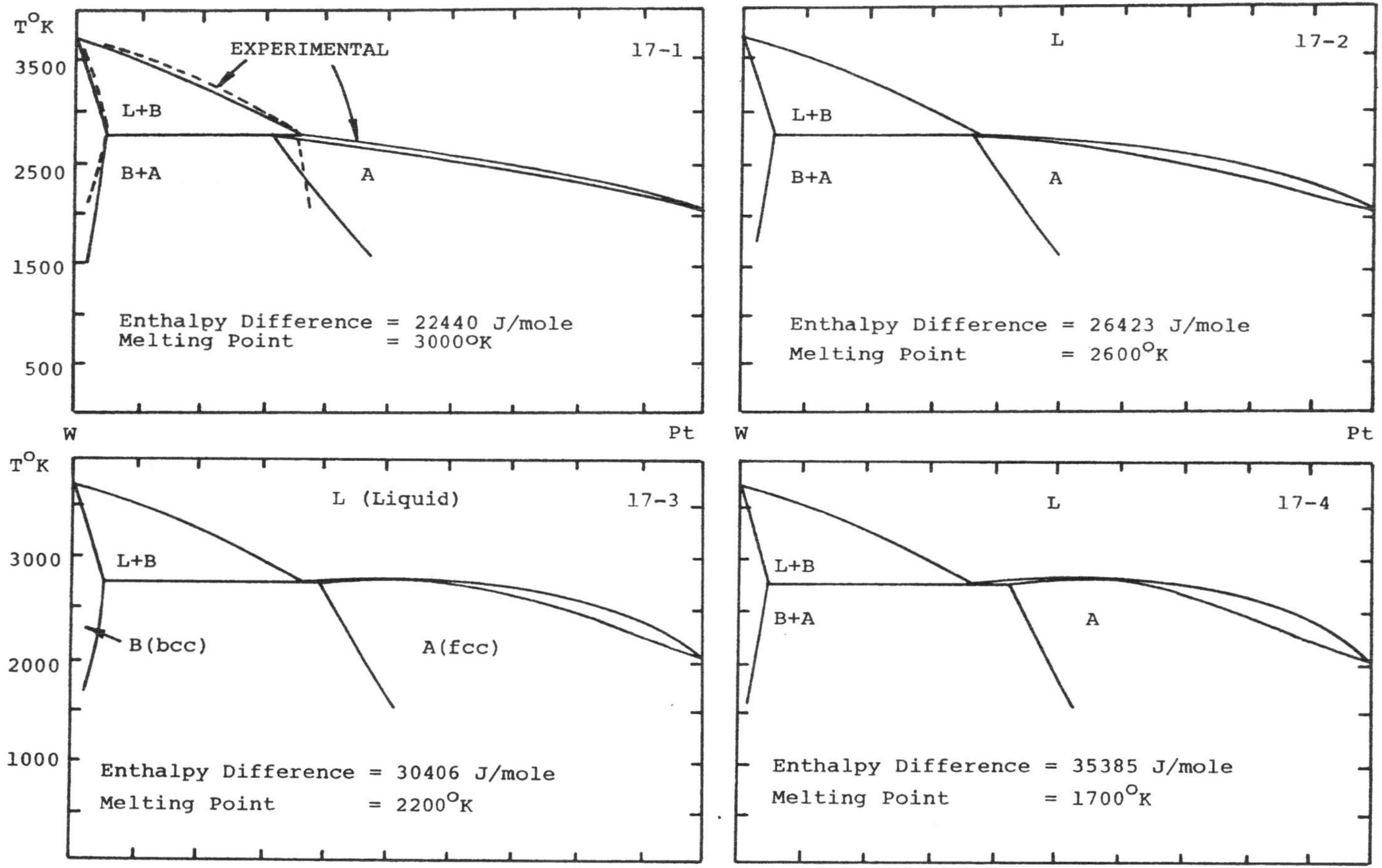

Figure 17. Calculated W-Pt Phase Diagrams with various melting points for fcc tungsten (and enthalpy differences between fcc and bcc tungsten) as compared with the experimental diagram shown in panel 17-1.

everything constant and varying the melting point of fcc tungsten (and the excess Gibbs energy of the fcc phase). It can now be seen that as the melting point is reduced to 2600K (in 17-2), 2200K (in 17-3), and finally to 1700K (in 17-4), the fcc/bcc enthalpy difference increases and the calculated phase diagram deviates farther and farther from the observed diagram. It is therefore apparent that, while the suggestion that the fcc/bcc enthalpy difference for W is near 35835 J/mol (corresponding to a melting point of 1700K for fcc W) may represent an accomodation with the "ab initio" results, such a finding cannot be reconciled with the experimental W-Pt phase diagram. Clearly, larger values of the fcc-bcc differences (i.e. 50,000 J/mol) leading to still lower melting point for fcc W would be even more difficult to accomodate.

It will therefore be necessary to reinvestigate the binary phase diagrams between second and third row transition elements along the lines previously considered (2) by employing the most recent measurements of the heats and entropies of fusion (21) and model calculations of the excess Gibbs energies of the liquid, fcc, bcc, and hcp phases (8, 16, 17, 19, and 20) along with analyses of the type presented above (i.e. Figure 16) for the W-Pt case in order to develop a consistent set of lattice stabilities for the second and third transitions series which parallels that presented for the first transition series in Tables 8 and 9 and Figures 15 and 16.

During the past five years, a great deal has been written about ARTIFICIAL INTELLIGENCE. In the experience of this writer, it is usually very difficult to obtain a meaningful definition of this field from authors (or speakers) discussing this topic. Nevertheless, it appears that the method developed here for predicting the behavior of higher order systems from elemental and binary data is a prime example of ARTIFICIAL INTELLIGENCE! Thus examination of V-Fe-Co ternary in Figure 11 shows that the current analytical description adequately reproduces the behavior measured at 873 and 1173K. At higher temperatures, the sections shown at 1800K, 1700K, 1600K and 1473K provide predictions of the melting behavior of the fcc, bcc and sigma phases in this ternary system. In the absence of any experimental data, these sections provide a reasonable guide to the behavior V-Co-Fe alloys. Moreover, the method and data base can be employed to deal with higher order systems containing these elements.

3. ACKNOWLEDGEMENT

This work was sponsored by the Metallurgy Program, Division of Materials Research, National Science Foundation under Grant DMR 87-11866.

4. REFERENCES

1. Hillert M: Computer Modeling of Phase Diagrams. Bennett LH(ed): The Metallurgical Society, Warrenton, Pennsylvania, 1986, p. 1.

2. Kaufman L and Bernstein H: Computer Calculation of Phase Diagrams. New York: Academic Press, 1970.

3. Kaufman L: CALPHAD 1, 7, 1977.

4. Kaufman L and Nesor H: CALPHAD 2, 55, 1978.

5. Kaufman L and Nesor H: CALPHAD 2, 81, 1978.

6. Kaufman L: CALPHAD 2, 117, 325, 1978.

7. Kaufman L: CALPHAD 3, 45, 1979.

8. Birnie D, Machlin ES, Kaufman L and Taylor K: CALPHAD 6, 93, 1982.

9. Kaufman L, Uhrenius B, Birnie D and Taylor K: CALPHAD 8, 25, 1984.

10. Kaufman L: Computer Modeling of Phase Diagrams. Bennett LH(ed): The Metallurgical Society, Warrenton, Pennsylvania, 1986.

11. Kaufman L: User Application of Phase Diagrams. ASM, Metals Park, Ohio, 1987.

12. Kaufman L: To be published in the Bulletin of Alloy Phase Diagrams, ASM, Metals Park, Ohio.

13. Guillermet AF and Ostland L: Met. Trans. 17A, 1809, 1986.

14. Kundrat D: Met. Trans. 17A, 1878, 1986.

15. Shao J and Machlin ES: CALPHAD 2, 279, 1978.

16. Miedema AR and Niessen AK: CALPHAD 7, 27 1983.

17. Niessen AK, de Boer FR, Boom R, de Chattel PF, Mattens WCM and Miedema AR: CALPHAD 7, 51, 1983.

18. Watson RE and Bennett LH: CALPHAD 9, 71, 1985.

19. Colinet C, Pasturel A and Hicter P: CALPHAD 9, 71, 1985.

20. Pasturel A, Colinet C and Hicter P: CALPHAD 9, 349, 1985.

21. Saunders N and Miodownik AP: Metastable Lattice Stabilities for the Elements. University of Surrey, July 1986 (presented at CALPHAD XV, London, England, July 1986).

22. Miodownik AP: Computer Modeling of Phase Diagrams. Bennett LH(ed): The Metallurgical Society, Warrenton, Pennsylvania, 1968, p. 253.

23. Pettifor D: Metallurgical Chemistry. Kubaschewski O(ed): London: HMSO, 191, 1972.

24. Pettifor D: CALPHAD 1, 305, 1977.

25. Scriver HL: Phys. Rev. Letters 49, 1768, 1982, Phys. Rev. B. 31, 1909, 1985.

26. Bendick W and Pepperhoff W: Acta Met. 30, 679, 1982 (Iron).

27. Swartzendruber LJ and Sundman B: Bulletin of Alloy Phase Diagrams 4, 155, 1983 (Iron-Ruthenium).

28. Murray JL: Bulletin of Alloy Phase Diagrams. (Ti-Ta), (Ti-Nb), (Ti-Cr), (Ti-Hf), (Ti-Mo), T(Ti-W), (Ti-Zr), (Ti-Mn), (Ti-Fe), (Ti-Re); 2, 55, 62, 174, 181, 185, 192, 197, 320, 334, and 462 respectively, 1981. (Al-Mg), (Ti-Co), (Al-Cd), (Al-Li), (Ti-Ir), (Ti-Os), (Ti-Ru), (Ti-Pd), (Ti-H), (Ti-Rb); 3, 60, 74, 205, 212, 216, 177, 1677, 321, 329 and 329 respectively, 1982. (Al-Zn), (Cu-Ti); 4, 55 and 81 respectively, 1983.

29. Kaufman L and Nesor H: Materials Science and Engineering 23, 119, 1976.

30. Murray JL: Alloy Phase Diagrams. Bennett LH, Massalski TB and Giessen BC(eds): Materials Research Society Symposia 19, 249, 1983.

31. Perepezko J: ibid., p. 223.

32. Massalski TB, Woychik CG and Murray JL: ibid. p. 241.

33. Saunders N and Miodownik AP: Berichte der Bunsen Gesellschaft f. Physicalishe Chemie 87, 830, 1983.

34. Williams RO: Met. Trans. 13A, 959, 1982.

35. Fullman RL: Acta Met 30, 1407, 1982.

36. Kleppa OJ and Wanatabe S: Met. Trans. 13B, 391, 1982.

37. Tanner L: Acta Met. 28, 1905, 1980.

38. Miracle D, Lark K, Strinivasan V and Lipsitt H: Met. Trans. 15A, 481, 1984.

39. Kaufman L and Nesor H: Met. Trans. 5, 1617, 1974.

40. Watson RE, Bennett LH, Davenport JW and Weinert M: Computer Modeling of Phase Diagrams. Bennett LH(ed): The Metallurgical Society, Warrenton, Pennsylvania, 1986, p. 223.

41. Guillermet AF and Gustafson P: An Assessment of the Thermodynamic Properties and P-T Phase Diagram of Iron TRITA-MAC-0229. Materials Center, Royal Institute of Technology, Stockholm, Sweden, August 1984.

42. Tauer KJ and Weiss RJ: J. Phys. Chem Solids 4, 135 (1958).

43. Hillert M: The Thermodynamic Properties of Manganese TRITA-MAC-D305. Materials Center, Royal Institute of Technology, Stockholm, Sweden, February 1984.

THE CLUSTER VARIATION METHOD AND THE CALCULATION OF ALLOY PHASE DIAGRAMS

D. de FONTAINE

University of California, Dept. of Materials Science and Mineral Engineering; and Lawrence Berkeley Laboratory, Berkeley, CA 94720

1. INTRODUCTION

I was asked by the Workshop organizers to write a tutorial on the Cluster Variation Method (CVM) in phase diagram calculations. Hence I shall present an elementary treatment of the essential features of the method rather than a comprehensive review.

As is well known, the CVM was originally proposed by Kikuchi[1] in 1951 as a hierarchy of approximation for the Ising model, in principle, in an arbitrary number of dimensions. The basic idea consists of handling local order as accurately as possible, then of expressing the configurational entropy in terms of small cluster probabilities. The corresponding free energy is then minimized with respect to independent configuration (cluster) variables. Higher cluster correlations are essentially treated by a superposition approximation. Hence, in the limit, the CVM is a classical theory. Much of the early work in the field was devoted by Kikuchi and collaborators to a systematic comparison of different levels of cluster approximations with known exact solutions such as that of Onsager's solution of the two-dimensional Ising model in zero field. In favorable cases, the CVM transition temperature turned out to be very close to the exact one. Unfortunately, the CVM being a classical theory, predicts critical exponents with classical values. That fact tended to discredit the CVM in the eyes of critical phenomena specialists.

Kikuchi's original approach to deriving CVM entropy formulas for various lattices and cluster approximations was an essentially geometrical one. Soon, however, other, more analytical (and simpler) methods appeared, those of Barker[2] and of Hijmans and de Boer[3]. Later, a very general and systematic method was developed by Sanchez[4-6]. This method, based on the idea of expressing the CVM free energy in terms of linearly independent correlation functions, is the most widely used today; it will be reviewed in this article.

A convenient method of minimizing the CVM free energy is the Natural Iteration Method, also developed by Kikuchi[7]. It was proposed just after Van Baal[8] had demonstrated the use of the CVM in the calculation of temperature/composition phase diagrams for binary alloys. Subsequently it was shown that Van Baal's ideas could be used to produce a rather faithful approximation to the crystalline equilibria in the Cu-Au phase diagram[9,10].

Because of its importance for alloy thermodynamics, development of the CVM is being pursued very actively in both theoretical and practical aspects. The most general and sophisticated formulation of the method is that developed initially independently by Sanchez and by Ducastelle and presented in a joint publication[11]. The present description is a simplified version of this approach. Recently, significant progress has been made in the derivation of suitable criteria for the selection of clusters[12,13]. Finel presented the essence of his method at the Workshop; a complete description thereof can be found in

G. M. Stocks and A. Gonis (eds.), Alloy Phase Stability, 177–203.

Finel's Doctoral Dissertation[13] which contains the most complete description of the CVM known to me.

The present article is divided into three main sections: the first one is a fairly detailed exposition of the CVM basics, the second one describes stability analysis, and the third one summarizes a recent application of the CVM to phase diagram calculations. A two-dimensional example will be worked out in some detail. The two-dimensional geometry is of course much simpler to visualize than the three-dimensional, also, the example chosen is relevant to the currently hot topic of vacancy ordering in the high-T_c superconductor $YBa_2Cu_3O_x$ ($6<x<7$).

2. THE CVM FORMALISM

Calculations of properties of pure elements and stoichiometric compounds has progressed considerably in recent years, but it is clear that alloys (mixtures of one or more elements) present additional difficulties. Pure crystals or compounds can be uniquely described by their unit cells, and completely disordered solid solutions can be described by specifying the average lattice and the average composition. For phase equilibrium calculations, however, it is essential to consider *states of partial order*: arbitrary degrees of long-range order (LRO), with short-range order (SRO) present as well.

Hence, the study of alloy thermodynamics must begin by an adequate description of the state of order (Sect. 2.1). Then, it is necessary to derive a statistical model, given that the statistical thermodynamics of three dimensional systems cannot be treated exactly. In particular, what is needed is an approximate but reliable expression for the configurational entropy (Sect. 2.2). Thirdly, the thermodynamics of the system must be completed by introducing physical parameters, the so-called effective cluster interactions which must be determined either empirically or by quantum mechanical calculations (Sect. 2.3).

2.1. State of Order

I shall now summarize, in simplified form, the elegant CVM treatment of Sanchez, Ducastelle and Gratias[11] (SDG). These authors show that the CVM actually provides a completely general and optimal way of describing partial order, the basic idea being that of representing local order by sampling configurational space by means of small clusters of crystal lattice sites. This is accomplished by expressing arbitrary functions of configuration in terms of a complete set of orthonormal functions.

In SDG, multicomponent systems were considered. Here, for simplicity, only binary systems will be treated. Each lattice site (p) can then be occupied by either an A or a B atom, with corresponding "spin" variable $\sigma_p=+1$ or -1. The complete crystal, of N sites, has instantaneous configuration fully specified by the vector $\sigma=\{\sigma_1,\sigma_2,\ldots\sigma_N\}$. The scalar product of two functions of configuration, $f(\sigma)$ and $g(\sigma)$, is defined as

$$< f, g > = \overset{\circ}{\rho}_N \mathrm{Tr}^{(N)} f \bullet g \qquad (1)$$

where the "trace" operation is defined as a sum over all configurations

$$\mathrm{Tr}^{(N)} = \sum_{\sigma_1=\pm 1} \sum_{\sigma_2=\pm 1} \cdots \sum_{\sigma_N=\pm 1} \qquad (2)$$

with normalization

$$\overset{\circ}{\rho}_N = 2^{-N} \quad .$$

The scalar product definition (1) allows the construction of a complete orthonormal set (CONS) of functions. For a single lattice point, the set of functions is simply

$$\Gamma(\sigma_p) = \{\Gamma_0, \Gamma_1\} = \{1, \sigma_p\} \tag{3}$$

such that

$$<\Gamma_i(\sigma_p), \Gamma_i(\sigma_p)> = \delta_{ij} \tag{4}$$

where the Kronecker delta is unity if i=j, zero otherwise. Now form the direct product, over all N lattice sites,

$$\Gamma(\sigma_1) \times \Gamma(\sigma_1) \times \ldots \times \Gamma(\sigma_N) = \Phi \quad . \tag{5}$$

By (5), each function of the set Φ, except $\Phi_0=1$, is itself a product

$$\Phi_\alpha = \sigma_{p_1}\sigma_{p_2}\cdots\sigma_{p_n}$$

over the cluster $\sigma=\{p_1, p_2, \ldots p_n\}$ of n points. From (1) and (2) we then have

$$<\Phi_\alpha, \Phi_\beta> = \delta_{\alpha\beta} \tag{6}$$

and the closure relation

$$\overset{\circ}{\rho}_N \sum_\beta \Phi_\beta(\sigma)\,\Phi_\beta(\sigma') = \delta_{\sigma,\sigma'} \quad . \tag{7}$$

Hence, by (4) and (7) the set Φ is a CONS. It is convenient to treat separately the configuration independent function $\Phi_0=1$. Then any function may be expanded as

$$g(\sigma) = g_0 + \sum_\alpha g_\alpha\,\Phi_\alpha(\sigma) \tag{8}$$

with

$$g_0 = <1, g> \quad , \quad g_\alpha = <\Phi_\alpha, g> \quad . \tag{9}$$

For example, the Ising Hamiltonian may be expanded in terms of *cluster functions* Φ_α as follows

$$H = \sum_\alpha V_\alpha \Phi_\alpha \tag{10}$$

where the V_α are *cluster interactions* defined by

$$V_\alpha = <\Phi_\alpha, H> \quad . \tag{11}$$

To evaluate averages, it is necessary to define a configuration density thus

$$\rho(\sigma) = Z^{-1}\, e^{-H/k_BT} \tag{12}$$

with partition function

$$Z = \mathrm{Tr}^{(N)}\, e^{-H/k_BT} \tag{13}$$

where k_BT has its usual meaning. The density can also be expanded in orthonormal functions

$$\rho(\sigma) = \overset{o}{\rho}_N [\, 1 + \sum_{\alpha} \Phi_\alpha(\sigma) \xi_\alpha] \tag{14}$$

where ξ_α is a *multiplet correlation* function for cluster α, defined as the average value of the product of σ variables over the cluster α

$$\xi_\alpha = < \Phi_\alpha , \rho > = \overset{o}{\rho}_N < \Phi_\alpha > \tag{15}$$

It is also useful to consider *reduced densities* obtained by performing the partial trace

$$\rho_\beta = \mathrm{Tr}^{(N-\beta)} \rho \quad , \tag{16}$$

i.e. by summing over all configurations except that, σ_β, of the n-point cluster β envisaged. The partial trace operating on Φ_α in (14) equals $(1/\overset{o}{\rho}_\beta)\Phi_\alpha$ if α is contained in β ($\beta \supset \alpha$) and zero otherwise, where $1/\overset{o}{\rho}_\beta = 2^n$. The reduced density can be regarded as the expectation value, in an ensemble of systems, of the cluster α having configuration σ_β. Thus, $\rho_p(\sigma_p)$ is simply the average concentration of A (σ_p=+1) or B (σ_p=-1) atoms, at site p, over the ensemble. If all lattice points p are equivalent, then ρ_p is the crystal average. If long-range order is present, distinct sub-lattices must be defined, and averages taken only over points of a given sublattice, p_i, say. By combining Eqs. (14) and (16) we then have, for cluster concentrations, or reduced densities, expressed in terms of correlations functions,

$$\rho_\beta(\sigma_\beta) = \overset{o}{\rho}_\beta \; [1 + \sum_{\alpha<\beta} \Phi_\alpha(\sigma_\beta)\, \xi_\alpha] \qquad , \tag{17}$$

where the summation is over all subclusters (α) of the cluster α. This important formula was first derived by Sanchez[4]. The notion of partial trace was introduced into the CVM by Morita[14]. In practice, it is convenient to group all ξ_α which are identical because of the symmetry of the crystal structure. Equations (14) or (17) then take on a slightly different form, with appropriate sums of Φ_α functions being regarded as elements of the so-called *configuration matrix* [see Eqs. (67) and (68), below]. The crystallography of the problem is thus introduced into the statistical formulation by means of this matrix.

2.2. Cluster Statistics

The energy and configurational entropy can be written as, respectively,

$$E[\rho] = \mathrm{Tr}^{(N)} \rho H \tag{18}$$

and

$$S[\rho] = -k_B^{(N)} \,\mathrm{Tr}^{(N)} \rho \ln \rho \tag{19}$$

The free energy is then given by

$$F[\rho] = \mathrm{Tr}^{(N)} \rho[H + k_B T \ln \rho] \tag{20}$$

E, S and F are to be regarded as functionals in $\rho(\sigma)$, and will take on their equilibrium (expectation) values if ρ corresponds to the correct equilibrium distribution of configurations. In traditional variational treatments, $F[\rho]$ is minimized with respect to ρ subject to the constraint Tr $\rho=1$. It was shown by SDG[11], however, that the equilibrium free energy could be obtained by a purely algebraic procedure, as will now be demonstrated. In a sense, the "Variation" has now been taken out of the variational treatment. It is, of course, impossible to consider the full density function $\rho(\sigma)$ over all configurations on all lattice sites. Hence, some method must be devised for handling only a small number of configurations. In the CVM, this is accomplished by considering configurations over a small number of small clusters, up to some maximum-size cluster(s). Usually, the larger the clusters retained, the better will be the approximation to the free energy.

In the energy expression, the approximation consists of neglecting interaction energies V_α in Eq. (10) for all clusters not contained in the maximal cluster. The same simple procedure will not do for the entropy expression, however: the ln ρ term cannot be written as a truncated sum of partial densities of the type (16), as no convergence is expected. Instead, following Morita[14], we define new functions $\tilde{\rho}_\beta$ by writing successively

$$\rho_1 = \tilde{\rho}_1 \quad , \quad \rho_{12} = \tilde{\rho}_1\,\tilde{\rho}_2\,\tilde{\rho}_{12} \quad , \quad \ldots \quad \rho_\alpha = \prod_{\beta<\alpha} \tilde{\rho}_\beta \quad , \quad \ldots$$

with, finally

$$\rho = \prod_{\alpha} \tilde{\rho}_\alpha \tag{21}$$

The CVM approximation consists in truncating the product (21), by assuming that the cumulant corrections $\tilde{\rho}_\gamma$ for γ not contained in the maximal cluster are equal to unity. Hence,

$$\ln \tilde{\rho} = \sum_{\beta}{}' \ln \tilde{\rho}_\beta \tag{22}$$

the accent on the summation denoting a truncated sum. It is now necessary to relate the ln $\tilde{\rho}_\beta$ ($=\Omega_\beta$, say) to ln ρ_α ($=\Lambda_\alpha$, say) which can be done by procedures already developed by Barker[2] and Hijmans and de Boer[3]. To this end, we write Eq. (22) as a sum over the Λ_α times some coefficients a_α, determined only by geometrical considerations:

$$\sum_{\beta}{}' \Omega_\beta = \sum_{\alpha}{}' a_\alpha \Lambda_\alpha = \sum{}' a_\alpha \sum_{\beta<\alpha} \Omega_\beta = \sum_{\beta}{}' \Big(\sum_{\alpha\supset\beta} a_\alpha \Big) \Omega_\beta \tag{23}$$

Therefore,

$$\sum_{\alpha\supset\beta} a_\alpha = 1 \quad , \tag{24}$$

there being a separate Eq. (24) for each subcluster α contained in the set of maximal clusters. Thus, the a_α can be determined uniquely by recursion.

We have therefore derived an important expression

$$\ln \rho = \sum_{\alpha}{}' a_\alpha \ln \rho_\alpha \tag{25}$$

whereby the logarithm of the density function is approximated by a weighted sum or reduced densities, i.e. cluster concentrations. By taking the logarithm of both sides of Eq. (12) we have

$$\ln \rho = -\ln Z - \frac{H}{k_B T}$$

or, using (25),

$$k_B T \sum_{\alpha}{}' a_\alpha \ln \rho_\alpha = (-k_B T' \ln Z)\,\Phi_o - \sum_{\beta} V_\beta \Phi_\beta \quad .$$

By properties of the CONS we therefore have, as in Eq. (9):

$$-k_B T \ln Z = Z \;<1, \sum_{\alpha}{}' a_\alpha \ln \rho_\alpha> \tag{26}$$

and

$$-V_\beta = Z <\Phi_\beta, \sum_{\alpha}{}' a_\alpha \ln \rho_\alpha> \quad . \tag{27}$$

the latter equation can be rewritten as

$$V_\beta = \sum_{\alpha \supset \beta}{}' a_\alpha \rho_\alpha^{\circ} \, \mathrm{Tr}^{(\alpha)} \, \Phi_\beta \ln \rho_\alpha \tag{28}$$

which is exactly the result which would have been obtained by direct minimization of the free energy with respect to the correlation functions ξ_β. In the Eqs. (28), the ρ_α must be replaced by their expressions in terms of the independent variables ξ_β. There results a set of non-linear equations in as many unknowns as there are maximal clusters and their distinct subclusters. Solving this set of equations by numerical techniques constitutes a major difficulty of the CVM, hence, in practice, clusters must be kept small and few in number.

By a well-known result of statistical mechanics, Eq. (26) gives directly the equilibrium free energy:

$$F = k_B T \sum_{\alpha} a_\alpha \, \mathrm{Tr}^{(\alpha)} \ln \rho_\alpha \quad . \tag{29}$$

The free energy functional itself is, from Eqs. (20), (10), (15) and (25),

$$F[\rho] = \sum_{\beta} V_\beta \, \xi_\beta + k_B T \sum_{\alpha} a_\alpha \mathrm{Tr}^{(\alpha)} \rho_\alpha \ln \rho_\alpha \tag{30}$$

which is the classical CVM expression for the free energy. Note that the two sums in Eq. (30) need not run over the same clusters; it is only required that the entropy sum contain the maximal cluster(s), and the energy sum include only clusters contained in the maximal one(s). Of course, some of the a_α subclusters may vanish, as explained by SDG.

2.3. Internal Energy

In order to calculate phase diagrams, it is not sufficient to treat Ising models alone: it is required to evaluate cohesive energies of various phases, in various states of order, referred to the same reference energy, for instance the energy of an infinitely dilute gas of pure A and pure B atoms, E_A^{∞} and E_B^{∞}, say. E_A^{α} and E_B^{α} are then the cohesive energies of pure A and B in the α phase and E_A^{β} and E_B^{β} are the corresponding cohesive energies in the β phase. Actually, the cohesive energy E_{dis} of random mixtures of A and B, in α or β as a function of concentration c of B atoms, will differ from the linear interpolation by amount ΔE_{dis} the energy of completely random mixing.

It is possible to use the concepts derived in Sects. 2.1 and 2.2 to obtain formal expressions for the physical energy parameters required. To that end, let $E(\sigma)$ denote the cohesive energy of a particular configuration (σ), on a given lattice ($\alpha,\beta,\ldots$). The expectation value of E, for distribution ρ, will be

$$<E> = \mathrm{Tr}^{(N)}\, \rho(\sigma)\, E(\sigma) \tag{31}$$

Inserting expression (14) for the density ρ into (31) then yields

$$<E> = E_0 + \sum_{\alpha} E_\alpha \xi_\alpha \tag{32}$$

with

$$E_0 = \overset{o}{\rho}_\alpha \mathrm{Tr}^{(N)} E(\sigma) \tag{33}$$

and

$$E_\alpha = <\Phi_\alpha, E> = \overset{o}{\rho}_N\ \mathrm{Tr}^{(N)} \Phi_\alpha(\sigma)\ E(\sigma) \tag{34}$$

Note that, in this *grand canonical averaging*, because of the Trace operation, the energy E_0 and the effective cluster interactions E_α are not only configuration independent but even concentration independent.

Let us rewrite (34) explicitly for the case of pair interactions E_r, where r denotes the spacing between lattice points p and q of the pair:

$$V_r \equiv E_{pq} = \frac{1}{2^2} \sum_{\sigma_p=\pm 1} \sum_{\sigma_q=\pm 1} \sigma_p\, \sigma_q\, \frac{1}{2^{N-2}} \mathrm{Tr}^{(N-2)} E(\sigma) \tag{35}$$

where the trace operation is carried out everywhere except at p and q. From Eq. (35) follow the definition of *Effective Pair Interactions* (EPI):

$$V_r \equiv \frac{1}{4}\left(V_{AA} - V_{AB} - V_{BA} + V_{BB}\right) \quad , \tag{36}$$

where V_{ij} (i,j=A,B) represents the energy of the rth (i,j) pair embedded in an artificial medium in which all configurations are equally represented.

Unfortunately, Eqs. (34) or (35) cannot be used to calculate the cluster interactions since the configuration energies $E(\sigma)$ are, of course, not known. Hence, a more direct method of computation for the E_α is required. It has been argued[15,16] that the proper way to calculate effective cluster interactions is by means of a perturbation expansion of a disordered medium of specified concentration c. It appears that the expansion is much more rapidly convergent if the states of (partial) order are considered as perturbations of the disordered states rather than as perturbations of the pure states or their linear interpolations. Quantum mechanical techniques suitable for performing such calculations are now available. For now, let us merely show how Eqs. (31) to (35) must be modified formally in order to obtain disordered state energies and concentration dependent cluster interactions.

In a sample containing a large number N of atoms, it is expected that, at equilibrium, the concentration c of the systems in a grand canonical ensemble will hardly ever depart significantly from the equilibrium concentration c°. In other words, the density function $\rho(\sigma)$ will be very sharply peaked about configurations having average concentration c°. Hence, in Eq. (31), it is practically equivalent to sum only over those configurations $\{\sigma°\}$ which all have concentration $\sigma°$:

$$<E> \cong \mathrm{Tr}^{(N)} \rho(\sigma°)\, E(\sigma°) \tag{37}$$

where the superscript (°) denotes *canonical averaging*, as it were. We now have

$$\langle E \rangle = \overset{o}{E}_o + \sum_{\alpha} \overset{o}{E}_{\alpha} \xi_{\alpha} \tag{38}$$

with

$$\overset{o}{E}_o = \overset{o}{\rho}_N \,\mathrm{Tr}^{(N)} E\,(\sigma°) \tag{39}$$

and

$$\overset{o}{E}_{\alpha} = \overset{o}{\rho}_N \,\mathrm{Tr}^{(N)} \Phi\,(\sigma°)\; E\,(\sigma°) \tag{40}$$

The total number of configurations having fixed number (N_A, N_B) of A and B atoms is

$$M = \frac{N!}{N_A! N_B!} \tag{41}$$

so that the energy of the completely random state of concentration $c° = N_B/(N_A+N_B)$ is, by Eq. (38),

$$E_{dis} = \frac{1}{M} \,\mathrm{Tr}^{(N)} E\,(\sigma°) = \overset{o}{E}_o + \sum_{\alpha} \overset{o}{E}_{\alpha}\; \xi^{R}_{\alpha} \tag{42}$$

where ξ^{R}_{α} denote multiplet correlation functions in the fully disordered state. It is now apparent that the disordered energy E_{dis} and the cluster interactions $\overset{o}{E}_{\alpha}$ are concentration dependent since, in Eqs. (40) and (42), the Trace operation samples different configurations at each concentration c.

The term $\overset{o}{E}_{\alpha}$ may be eliminated from Eq. (38) by means of Eq. (42):

$$<E> = E_{dis} + E_{ord} \tag{43}$$

where the disordered state energy E_{dis} is the energy of the completely disordered medium, in a given crystal structure, calculated in a single-site coherent potential approximation (CPA), for instance, and E_{ord} is given by

$$E_{ord} = \sum_{\alpha}{}' E^{o}_{\alpha}\,\delta\xi_{\alpha} \tag{44}$$

where

$$\delta\xi_{\alpha} = \xi_{\alpha} - (\xi_1)^{n} \quad , \tag{45}$$

since the correlation function, in the fully disordered state, for an α cluster of n points is practically equal to n^{th} power of the point correlation function $\xi_1=c_A-c_B=1-2c$. The accent on the summation in Eq. (44) indicates that, because of Eq. (45), the point clusters are not included in the sum. Eqs. (43) to (45) were given previously by Sigli and Sanchez[17]. The E_{α}, or V_{α} in the notation of previous Sections, must now be calculated. This can be accomplished by perturbing the single site CPA according to the so-called Generalized Perturbation Method (GPM)[15,16]. Alternately, the Embedded Cluster Method[18] may be used since, by Eq. (40), each cluster interaction V_{ij} in Eq. (36), rewritten for canonical averaging, as in Eq. (40), represents the energy of cluster α embedded in a medium of random configuration of concentration c.

The derivations given above may explain formally why pair interactions E_r, for large spacing r, tend to become small in magnitude: at large spacing in a random medium, V_{ij} is approximately given by the sum of point energies $V_i + V_j$, hence the linear combination $V_{AA} + V_{BB} - 2V_{AB}$ will tend to vanish.

To complete the calculation of the internal energy, E_{ord} must be evaluated, which requires, in addition to the E^{o}_{α}, knowledge of the equilibrium correlation functions ξ_{α}. These must be obtained by minimizing the free energy, at given temperature and concentration, by solving the system of algebraic equations (28). In summary, then, E^{o}_{dis} and E^{o}_{α} (or V_{α}) can all be calculated by Quantum Mechanical methods at absolute zero of temperature. The ξ_{α} are calculated by the CVM with temperature independent parameters. Hence, the procedure described here achieves a very convenient decoupling of the Quantum and Statistical Mechanical computations.

The determination of alloy phase equilibrium in the CVM framework will be deferred to Sect. 4. Before that, it is necessary to discuss the topic of configurational stability (Sect. 3).

3. STABILITY

3.1. Susceptibility

Stability theory is concerned with the response of a system to a small applied field, in the present case, a "configurational" field, created by appropriate chemical potential changes. Since such responses are best expressed in reciprocal, or k-space, formalism, it is advantageous to modify the cluster notation used up to now. For that purpose, we specify a cluster α by its type q and by the point p at which it is located: $\alpha \rightarrow q\,,p$. Equation (30) can now be expanded in powers of linearly independent cluster correlation functions $\xi_q(p)$:

$$F = F_o + F_1 + F_2 + \ldots \tag{46}$$

where each term groups like power of ξ. Stability analysis considers only small variation $\delta\xi$ so that the Taylor's expansion (46) may be terminated at the second-order term. Since

F_1 vanishes at equilibrium, a small variation about the unperturbed state may be expressed as the quadratic form

$$\delta F = F_2 = \frac{N}{2} \sum_{qq'} \sum_{pp'} f_{qq'}(p, p') \, \delta\xi_q(p) \, \delta\xi_{q'}(p')$$

where $F_{qq'}$ is the second derivative with respect to $\xi_q \, \xi_{q'}$, at the indicated points, of the free energy per unit cell. Because of translational symmetry in the disordered (unvaried) state, $f_{qq'}$ is a function of the distance between p and p′ only.

A first diagonalization is accomplished by Fourier transforming over space:

$$\delta F = \frac{N^2}{2} \sum_{qq'} F_{qq'}(\mathbf{k}) \, \delta X_q(\mathbf{k}) \, \delta X_{q'}(\mathbf{k}) \tag{47}$$

where $F_{qq'}$ and X_q are, respectively, the Fourier transforms of $f_{qq'}(p'-p)$ and $\xi_q(p)$. Expression (47) may be further diagonalized in cluster space:

$$\delta F = \frac{N^2}{2} \sum_{q} \Lambda_q(\mathbf{k}) \left|\delta Y_q(\mathbf{k})\right|^2 \tag{48}$$

where Λ_q denotes the eigenvalues of the Hermitian matrix F and Y_q are normal "cluster modes".

The system is unconditionally stable if δF is positive for all possible configurational variations i.e., if all eigenvalues Λ are positive, or equivalently if the matrix F is positive definite. Instability sets in for given **k** when the determinant of F vanishes

$$\mathrm{Det}\left[F(\mathbf{k})\right] = 0 \quad . \tag{49}$$

The matrix F is a function of temperature and concentration, so that, for each possible wave vector **k**, condition (49) represents a locus in phase diagram space. The locus lying at highest temperature will correspond to a particular wave vector $\mathbf{k}_0$, that of the *ordering wave*.

Above the highest stability limit, a generalized susceptibility $\chi_{qq'}(\mathbf{k})$ may be defined as the expectation value of the product of two cluster waves

$$\chi_{qq'}(\mathbf{k}) = \left\langle \chi_{q'}^{*}(\mathbf{k}) \, \chi_q(\mathbf{k}) \right\rangle = N \, k_B T \, G_{qq'}(\mathbf{k}) \tag{50}$$

where the star denotes a complex conjugate quantity and $G_{qq'}$ is the qq′ element of the matrix F^{-1}, the inverse of the second derivative matrix.

So-called short-range order intensity I_{SRO}, in appropriate units, is just proportional to the point-point susceptibility χ_q written as *the* susceptibility $\xi(\mathbf{k})$ for short:

$$I_{SRO}(k) = \chi(\mathbf{k}) = N \, k_B T \, G_{11}(\mathbf{k}) \tag{51}$$

In the Bragg-Williams (BW, mean field) approximation, valid at very high temperatures, Eq. (51) yields the well-known Krivoglaz-Clapp-Moss (KCM) formula with

$$G_{11}(\mathbf{k}) = 2V(\mathbf{k}) + k_BT/c(1-c) \tag{52}$$

in which $V(\mathbf{k})$ is the Fourier transform of the effective pair interactions defined previously. In the KCM approximation $G_{11}(\mathbf{k})$ is just the Fourier Transform of the second-order term in the Taylor's expansion of the free energy. By analogy, Eq. (52) for the general case, may also be written in KCM form

$$\chi(\mathbf{k}) = N\, k_BT / \Phi(\mathbf{k}) \tag{53}$$

with $\Phi = G_{11}^{-1}$ representing the Fourier transform of a generalized free energy second derivative. As will be explained in more detail in Sect. 3.2, $\Phi(\mathbf{k})$ must have symmetry dictated by that of the crystal's point symmetry group and the translational symmetry of the reciprocal lattice. Hence, for any crystallographic space group, $\Phi(\mathbf{k})$ can be written in symmetry adapted form, as a sum of trigonometric terms multiplied by parameters Φ_s calculated from the free energy second derivatives. As an example, for crystals of cubic symmetry class, the expansion is:

$$\Phi(\mathbf{k}) = \Phi_o + \sum_s \Phi_s\, \Phi^{(s)}(\mathbf{k}) \tag{54}$$

where $\Phi^{(s)}$ are "coordination shell functions"

$$\Phi^{(s)}(\mathbf{k}) = \frac{z_s}{6} \sum_{j=1}^{3} \cos(2\pi h_1 p_j^s)\, [\cos(2\pi h_2 p_{j+1}^s)]\, \cos(2\pi h_3 p_{j+2}^s) + \cos(2\pi h_3 p_{j+1}^s)\, [(\cos 2\pi h_2 p_{j+2}^s)] \tag{55}$$

where z_s is the coordination number for shell s, $p^s{}_j$ are integers and half-integers denoting the Cartesian coordinates of a point in the first octant of shell s, and h_i ($= k_i a_i^*/2\pi$) denote Cartesian coordinates in the first BZ, a_i^* being reciprocal lattice parameters. The subscripts on direct space coordinates must be taken as (j-1) modulo 3 + 1. By analogy with $V(\mathbf{k})$ itself, these parameters Φs may then be regarded as effective, temperature-dependent pair interactions. An interesting and unexpected consequence of this temperature dependence was recently discovered by Solal et al.[19]

As emphasized by Sanchez, in a remarkable paper[20], Eq. (53) for $\chi(\mathbf{k})$ is exact. Unfortunately, for all but the simplest one and two-dimensional Ising models, no exact expression is known for the free energy, hence the matrix $F(\mathbf{k})$, and consequently $G_{11}(\mathbf{k})$ can only be calculated for approximate models. One consequence is that the integration of $\chi(\mathbf{k})$ over the Brillouin zone will not be a conserved quantity. For the one-dimensional Ising chain with nearest-neighbor interaction, for which the CVM expression is exact, the integrated χ does remain rigorously constant with temperature. For two and three-dimensional cases, integrated χ, according to various CVM approximations, tends to increase without limit near the instability temperature T_0, although the temperature variation is less drastic than that for the KCM approximation. As shown elsewhere[21], SRO intensity maxima appear to be sharper in the CVM than in the KCM approximation..

3.2. Special Point Symmetry

As pointed out by Lifshitz[22] many years ago, symmetry-dictated extrema must be found, for any function, at points where k-space symmetry elements intersect at a point, provided that the function, say $V(\mathbf{k})$, possesses the symmetry of the crystal. At these special points (SP), the so called: "Lifshitz criterion" is satisfied. Later, Khachaturyan[23] showed how ordered structures could be constructed from SP ordering waves in fcc and bcc lattices. The role of SP waves in stability theory was then examined by the present author[9,10], again for the case of fcc and bcc crystals.

It was shown by Haas[24] then by Hornreich, Lubhan and Shtrichman[25] that second order transitions need not occur exclusively for ordering k-vectors located right *at* the SP: the free energy second derivative could present a saddle point at the lowest SP so that actual minima may be located at some distance away but still definitely associated with the SP. In fact, the term "Lifshitz point" was coined[25] to designate the point in phase diagram space at which the free energy second derivative ceases to have a minimum at the SP and just begins to develop the saddle. Thus, at the Lifshitz point, at least one of the eigenvalues of the matrix of spatial second derivatives of the free energy second derivative (with respect to ordering wave amplitudes) vanishes. "Beyond" the Lifshitz point, minima will develop symmetrically about the SP at a k-space distance $\mathbf{q}$, let us say, from $\mathbf{k}_0$. It follows that the actual ordering instability can be represented by the ordering wave $\mathbf{k}_0$ amplitude-modulated by a wave of wavelength $2\pi/q$. These ideas are described in detail by Takeda et al.[27] and applied to instabilities created by electron irradiation in CuPd samples. The notion of a metastable Lifshitz point is introduced.

From the foregoing, it follows that it is desirable to determine the SP for crystal structures of interest, and to determine the nature of the extrema (maximum., minimum, saddle point) at each SP, for various ranges of energy parameters. This determination was performed, in the BW framework, first for fcc and bcc lattices[24], then for general crystal structures[26]. In the general case, the recipe for finding the set of SP for any one of the 230 crystallographic space groups is the following one: introduce a center of inversion (if absent) into the symmetry point group G_x of the crystal, from which the point group G_p is derived. One then forms the direct product group

$$G_s = G_p \times \{g\} \tag{56}$$

where $\{g\}$ is the translational group of the reciprocal lattice. The resulting G_s must then be one of the 24 centered symmorphic groups, i.e. those containing a center of inversion but no glide planes or screw axes. The SP are found by locating the relevant G_s in the International Tables for Crystallography[27] and noting those Wyckoff positions with fixed coordinates. A complete list of all SP is given in Table 1. Since the translation groups are in k-space, the symmorphic groups are listed for reciprocal space. Actually, it was pointed out by F. Ducastelle, at the conference itself, that, in the case of multiple lattice systems (such as hcp), the scheme just described may produce extraneous "special" points. It is thus necessary to test whether gradients really do vanish at each special point found.

For each group, the origin 000 is always a SP and corresponds to an infinite wave instability. Such an instability would normally give rise to *spinodal decomposition*, as originally described by Cahn[28]. By extension, instabilities associated with other SP will give rise to *spinodal ordering*[9,10].

For a given crystal, which of the listed SP will actually be a *minimum*, and thus give rise to some spinodal-type reaction, will depend on the nature of the $V(\mathbf{k})$ in particular on the relative values of the effective pair interactions, V_1, V_2, V_3...which, in general, may depend on temperature, as mentioned above, as well as on concentration. If these variations are not too abrupt, it is expected that, in rather extensive regions of phase diagram (T,c) space, a given SP instability will predominate. Equilibrium superstructures found in those regions will then be associated with the corresponding SP instability, i.e. will be said to be members of the *SP family of superstructures*. Members of the family can be further subdivided into two classes, (a) those for which the structure factor spectrum consists exclusively of SP wave vectors (not necessarily *minimum* SP) and (b) those for which the spectrum contains k vectors located elsewhere in the BZ. The former, SP-structures, were discussed by Khachaturyan[23] for fcc and bcc parent lattices, but do not necessarily correspond to lowest energy states. The latter, non-SP-structures, were determined as true ground for restricted sets of V_n interactions for fcc, bcc and hcp parents[29,30,31].

In an actual phase diagram, phases other than ordered superstructures may, of course, be found at equilibrium. These compounds, not manifestly related to either fcc, bcc or hcp parents, may be regarded, in a sense, as "interlopers". Examples are A15, σ, Laves... phases.

In any case, the concept of SP families of ordered superstructures is an important one for classifying crystal structures in alloy systems; it is a particularly useful one when attempting to perform first principles calculations.

3.3. Example: Instabilities in the Perovskite Basal Plane

Thus far, the treatment has been rather formal. It is thus instructive to consider an example: that of ordering of filled and empty oxygen sites in the Perovskite basal plane, chosen for reasons mentioned in the Introduction.

The plane in question is shown in Fig. 1.[32].

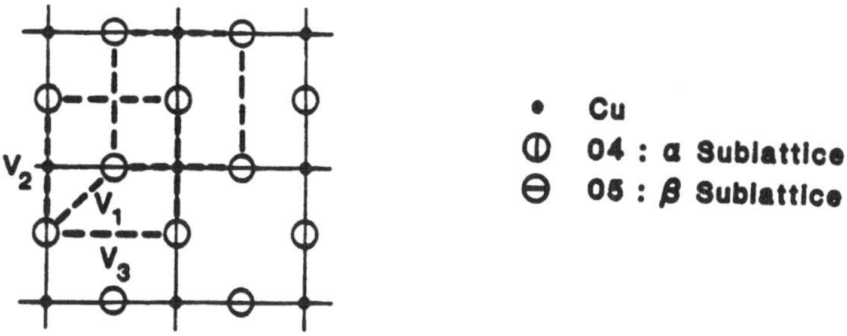

FIGURE 1. Oxygen sublattices on perovskite basal plane with indicated effective pair interactions. Large open circles denote oxygen sites, small filled circles denote Cu atoms.

The open circles represent two types of oxygen sites, occupying two interpenetrating sublattices, α and β, say. The small filled circles represent the captions, here referred to as Cu atoms because of the relevance of this model to the high-T_c superconductor $YBa_2Cu_3O_x$, with $6 \leq x \leq 7$. We now introduce three effective pair interactions V_r defined, following Eq. (36), as

$$V_r = \frac{1}{4}\left[V_r(O-O) + V_r(\square-\square) - V_r(O-\square)\right] \tag{57}$$

The strongest interaction is expected to be the nearest-neighbor one, V_1, which couples the two sublattices. It is also necessary to include two second-neighbor-interactions, V_2 which is mediated by the Cu ion, and V_3 which is not. Both V_2 and V_3 connect sites on the same sublattice as shown in Fig. 1. The signs and strengths of the effective pair interactions will, of course, depend on the electronic structure of the full three-dimensional crystal. We adopt the usual convention that $V_r>0$ favors "ordering" of the rth pair (unlike site occupation), and $V_r<0$ favors "clustering" (like-site occupation). We may then perform an ordering stability analysis of the 2D Ising problem by formally expanding the free energy to second order in the configuration variables as was done in Sect. 3.1. In a mean-field approximation (Bragg-Williams) the configuration entropy is site diagonal in the configuration variables so that all structural effects are determined by the pair interaction term written in its most general form as[26]

$$\Phi = \frac{1}{2} \sum_{nn'} \sum_{mm'} v\left(\mathbf{R}_m + \rho_n - \mathbf{R}_{m'} - \rho_{n'}\right) \sigma\left(\mathbf{R}_m + \rho_n\right) \times \sigma\left(\mathbf{R}_{m'} - \rho_{n'}\right) \quad , \tag{58}$$

where $\mathbf{R}_m$ designates a lattice vector and ρ_n a position inside the unit cell. The interaction parameters depend on the distance between lattice sites and the configuration variables σ denote site occupancy, i.e., +1 is filled, -1 is empty. The summations extend over all pairs of sites, compatible with the limited range of interactions v considered. The correspondence between these interactions and the V_1, V_2, and V_3 introduced above is established in Fig. 1.

The quadratic form (58) must now be diagonalized. The Fourier diagonalization in Sect. 3.1. was performed under the tacit assumption of translational invariance of all points p in the reference state. In the present case of interpenetrating lattices, this is no longer the case. Translational symmetry can be restored to the interactions, however, by converting v to a matrix of elements $v_{nn'}$[26], the order of the matrix being equal to the number of sublattices considered, here equal to two. Hence, the diagonalization proceeds very much as in Eqs. (47) and (48) except that now the indices (n,n′) denote sublattices rather than clusters (q,q′) as in Sect. 3.1.

The first step in the diagonalization thus gives

$$\Phi = \frac{1}{2} \sum_{mm'} \sum_{nn'} v_{nn'}\left(\mathbf{R}_m - \mathbf{R}_{m'} - \rho_{n'}\right) \sigma_n\left(\mathbf{R}_m\right) \sigma_{n'}\left(\mathbf{R}_{m'}\right) \quad . \tag{59}$$

The second step consists of a lattice Fourier transform over N sites of a suitably large region:

$$\Phi = \frac{N}{2} \sum_{\mathbf{k}} \sum_{nn'} V_{nn'}(\mathbf{k}) \, \sigma_n(\mathbf{k}) \, \sigma_{n'}(-\mathbf{k}) \tag{60}$$

in which $V_{nn'}(\mathbf{k})$ is the Fourier transform of the effective pair interactions, $v_{nn'}$, and $\sigma_n(\mathbf{k})$ is the amplitude of an "occupancy wave" on sublattice n. In the third step, the diagonalization is completed by defining "normal modes" $\Gamma(\mathbf{k})$

$$\sigma_n = \sum_{n'} U_{nn'} \Gamma_{n'} \tag{61}$$

where $\mathbf{U}$ is a unitary matrix diagonalizating $\mathbf{V}$. Let the eigenvalues of $\mathbf{V}$ be Λ_n. The fully diagonalized expression is thus

$$\Phi = \frac{N}{2} \sum_{\mathbf{k}} \sum_{n} \Lambda_n(\mathbf{k}) \left|\Gamma_n(\mathbf{k})\right|^2 \quad . \tag{62}$$

As explained in Sect. 3.2., symmetry-dictated extrema of $V_{nn'}(\mathbf{k})$, or of $\Lambda_n(\mathbf{k})$, must be located at the special points. In the present case, the group G_s, defined by Eq. (56), is the two-dimensional square p4m with fixed-coordinate Wyckoff positions at <00>,

<½0>, and <½½>. The short range of interactions postulated for the problem precludes the existence of minima away from the SP, hence the search for k-space minima will be limited to those three points.

The Fourier transforms of the interaction parameters are given by

$$V_{11}(\mathbf{k}) = 2V_2 \cos 2\pi h_1 + 2V_3 \cos 2\pi h_2$$

$$V_{22}(\mathbf{k}) = 2V_3 \cos 2\pi h_1 + 2V_2 \cos 2\pi h_2 \tag{63}$$

$$V_{12}(\mathbf{k}) = V_{21}(\mathbf{k}) = 2V_1\left[\cos \pi(h_1 + h_2) + \cos \pi(h_1 - h_2)\right] \quad .$$

In these equations, the V_r parameters are those defined in Fig. 1.

At the SP, the eigenvalues take on very simple forms:

$$\langle 0\,0\rangle: \quad \Lambda_{\pm}(0\,0) = 2(V_2 + V_3) \mp 4|V_1|$$

$$<\tfrac{1}{2}\,0>: \quad \lambda_{\pm}(\tfrac{1}{2}\,0) = \mp 2|V_2 - V_3| \tag{64}$$

$$<\tfrac{1}{2}\,\tfrac{1}{2}>: \quad \lambda_{\pm}(\tfrac{1}{2}\,\tfrac{1}{2}) = -2(V_2 + V_3) \ .$$

In the search for minimum eigenvalues, only the upper sign needs to be considered. Depending on the relative values of V_1, V_2, and V_3, one SP eigenvalue will be lower than the other two. Let us divide Eqs. (64) through by V_1 and define normalized interaction parameters

$$x = V_2/V_1 \quad \text{and} \quad y = V_3/V_1 \quad .$$

These parameters can then be used as coordinates in an "ordering instability map" which indicates the regions in interaction parameter space where a given SP wave will be most unstable. Boundaries between such regions are obtained by equating different SP eigenvalues. The resulting map is shown in Fig. 2 for the case $V_1>0$.

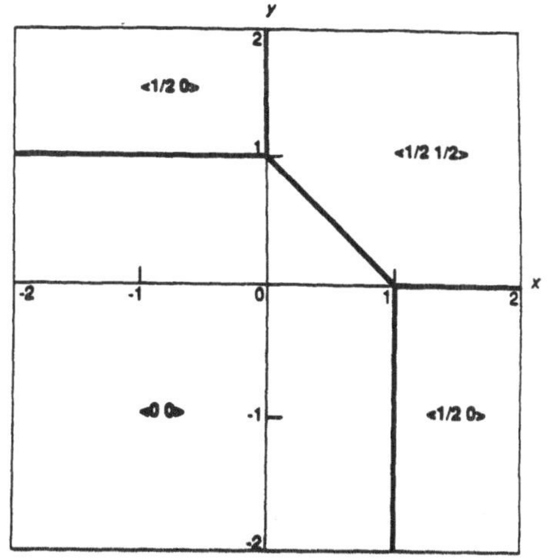

FIGURE 2. Ordering instability map for $V_1>0$. Coordinates are ratios $x=V_2/V_1$, $y=V_3/V_1$.

It is seen that the <0 0> instability is favored for "ordering" first-neighbor and "clustering" second-neighbor interactions. Conversely, the <1/2 1/2> instability is favored by large "ordering" second-neighbor interactions, and <1/2 0> is favored by V_2 and V_3 differing in sign.

When a given "ordering wave" (that with lowest eigenvalue) becomes unstable, the corresponding normal mode amplitude will increase, thereby modulating the sublattice site occupation. Since the normal mode Γ_+ (corresponding to Λ_+) will always have lowest energy, we have by Eq. (61), with $\Gamma_- = 0$,

$$\begin{aligned} \sigma_1(\mathbf{k}) &= u_{11}(\mathbf{k})\ \Gamma_+(\mathbf{k}) \\ \sigma_2(\mathbf{k}) &= u_{21}(\mathbf{k})\ \Gamma_+(\mathbf{k}) \end{aligned} \tag{65}$$

where u_{11} and u_{21} are the components of the eigenvector corresponding to $\Lambda_+(\mathbf{k})$. At the <0 0> SP, the eigenvectors are

$$\left[u_{11}, u_{21}\right] = \frac{1}{\sqrt{2}}\ [1, -\mathrm{Sgn}\,(V_1)],\ \ \mathrm{Sgn}\,(V_1) = \left|V_1\right|/V_1\ \ . \tag{66}$$

For the Brillouin zone center instability, infinite-wavelength modulations will be placed on the α and β sublattices; for $V_1>0$ the two waves will be out of phase, i.e., there will be maximum concentration of filled sites on one sublattice and minimum on the other. For $V_1<0$, the two waves will be in phase. In the former case, the resulting structure, for average concentration of filled sites, $c_0=1/2$ (x=7 in $YBa_2Cu_3O_x$), will produce the unit cell depicted in the lower left quadrant of Fig. 3a. This structure, with two-dimensional space group symbol p2mm, is characterized by O-Cu-O...chains along the b axis and produces the three-dimensional orthorhombic structure Pmmm of the superconducting phases.

For the zone boundary instabilities <1/2 0> and <1/2 1/2> intra-sublattice modulations are produced, leading to doubling and quadrupling of the original unit cell,

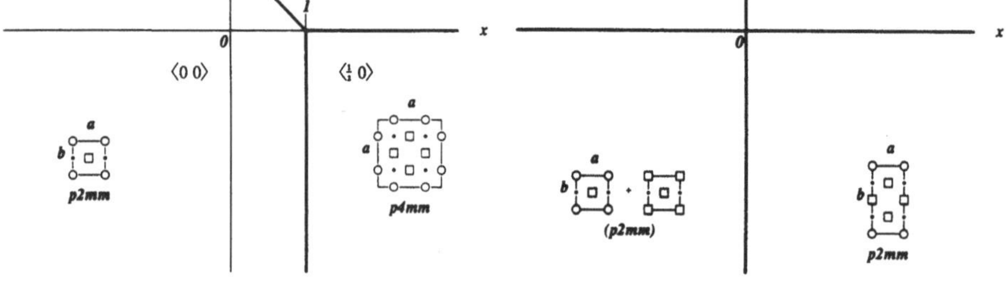

(a) (b)

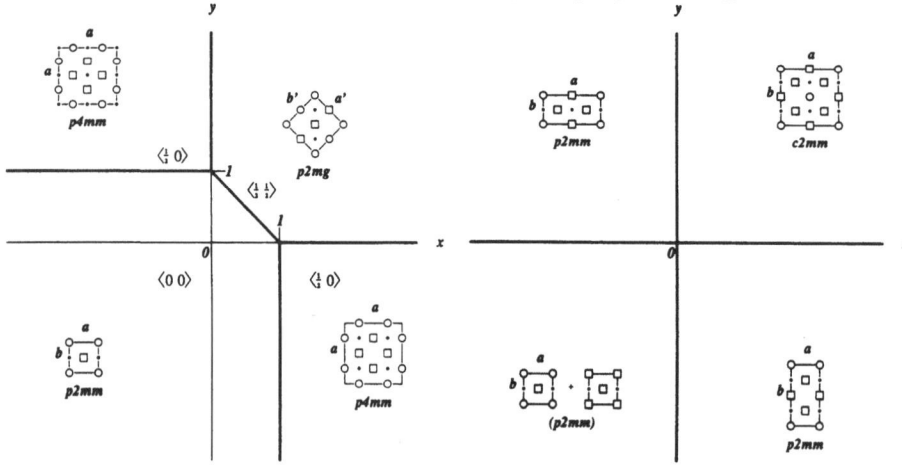

FIGURE 3. Grounds states as a function of parameter ratios $x=V_2/V_1$, $y=V_3/V_1$ at oxygen concentrations (a) $c_0=0.50$, (b) $c_0=0.25$.

respectively. For both cases, due to the vanishing of the off-diagonal element $V_{12}(\mathbf{k})$, the eigenvectors are [1,0] and [0,1]. Hence, by Eq. (9), we have for the <1/2 0> case, $\sigma_1(\mathbf{k})=\Gamma_+(\mathbf{k}), \sigma_2(\mathbf{k})=0$. The resulting structure may be interpreted as consisting of one sublattice modulated by a <1/2 0> wave, with the other sublattice containing a random distribution of filled and empty sites. Actual ground state structures corresponding to this SP instability will be described in the next Section. For the <1/2 1/2> case (for compositions near $c_0=1/2$), since $\Lambda_+=\Lambda_-$, both sublattices will be modulated by a <1/2 1/2 > wave, producing a structure with [1,1] rows populated alternately by filled and empty sites.

By a fortunate coincidence, the day before I presented, at the Crete meeting, an early version of the above stability analysis (later corrected by Dr. L. T. Wille), Dr. G. Van Tendeloo[33] presented electron diffraction and microscopy results which indicated the existence of a "cell doubling" ordering reaction evidenced by diffuse intensity at point <1/2 0> in the high-T_c superconductor. It was thus highly gratifying to see that such a simple analysis could predict actual structures in such complex materials.

4. PHASE DIAGRAMS

The exciting possibility now exists of deriving equilibrium phase diagrams virtually from first principles. The subject is now in its infancy but already some interesting results have been obtained. The CVM has proved to be a reliable tool for performing the required statistical thermodynamics, but requires as input, of course, the structure-independent energy E_0^o of Eq. (39) and the effective clusters interactions E_α^o, defined in Eq. (40), in particular, the effective pair interactions (EPI) V_r defined in Eq. (36). Several schemes have been proposed recently for calculating the EPI's by quantum mechanical means. A brief summary of these techniques has been given elsewhere[34].

The calculation of phase diagrams proceeds in two main steps: firstly, equilibrium crystal structures are determined at zero absolute temperature; secondly, equilibria between those structures are determined as a function of temperature. The first problem, one of energy minimization, can itself be subdivided into one of comparing total energies of dissimilar structures, such as fcc and bcc, or intermetallics $L1_2$ and A15, for example, at the same concentration, and one of determining ground states of order based on a given fixed lattice. It is this second sub-problem which is addressed in the next section (4.1), and was already alluded to at the end of Sect. 3.2.

The problem of thermodynamic equilibria at non-zero temperatures is to be solved by minimizing the relevant *free* energy. This is where the CVM comes in, as will be briefly described in Sect. 4.2.

4.1. Ground States of Order

Given a fixed lattice (or sublattice), two types of "atoms" (say), and a set of parameters V_r representing interactions between neighboring atoms, what will be the atomic arrangements (σ) minimizing the "ordering energy", that given by Eq. (10)? Such is the ground state problem. Thus, the problem is formulated simply, but, except in the simplest of cases, cannot be solved rigorously or completely. A very thorough description of the ground state problem has been given by Finel[13]. Here, only a brief summary will be given.

In Eq. (17), the partial trace ρ_β was expressed as a function of the correlations ξ_α. In this equation, β represents a cluster of n_β points and the indicated summation is over all subclusters α contained in β. In highly symmetric structures, several subclusters (α) may be crystallographically equivalent, in which case they will have the same correlation function ξ_α. All such clusters will be members of a class j, say, and denoted by α_j. It is then advantageous to group such clusters into a single term in the summation. Thus, the subcluster summation in Eq. (17) is now replaced by a sum over cluster classes j;

$$\rho_\beta(\sigma_\beta) = \overset{o}{\rho}_\beta[1 + \sum_j W_j(\sigma_\beta)\, \xi_j] \tag{67}$$

with

$$W_j(\sigma_\beta) = \sum_{\alpha_j} \Phi_{\alpha_j}(\sigma_\beta) \quad . \tag{68}$$

In the last equation, the cluster functions Φ, for given class j, differ by nature of the site occupancies σ_β. The latter cluster configurations can also be listed in some suitable order, and given an index k, say. The W_{jk} may then be considered as elements of a (generally rectangular) matrix, denoted here as *configuration matrix* [for historical reasons, this matrix is often called the "V-matrix," an unfortunate nomenclature since it has nothing to do with the V_r interaction parameters].

Equation (67) is the practical one used in CVM codes. It has its use in the ground state problem as well: since ρ_β represents a cluster probability, or concentration, for configuration σ_β, it must be a non-negative quantity. Hence, the following inequalities must hold

$$\sum_j W_j(\sigma_\beta)\; \xi_j \geq -1 \quad . \tag{69}$$

These inequalities, along with the conditions $-1 \leq \xi_j \leq +1$ define a convex region in multidimensional ξ-space (configuration space), the so called *configurational polyhedron*, which contains all realizable configurations σ on the given lattice. It can be shown[13] that the vertices of the polyhedron determine, in principle, all ordered ground states for the given *range* or cluster interactions, different ordered states being stable for different stoichiometries and different ratios of interaction parameters. Unfortunately, except for the case of short-range interactions, inequalities (69) often result in "non constructible crystal structures," which means that the set of inequalities is incomplete, the constraints "too loose." How many more inequalities must one seek to obtain correct ground states? That, unfortunately, appears to belong to a class of mathematically undecidable problems[35]. In simple cases, the problem can be solved in a straight-forward manner; an example, that leading to the ordering maps of Fig. 3, will be discussed in Sect. 4.3.

4.2. Phase Diagram Constructions

In elementary treatments of binary (A,B) phase equilibrium, the common tangent construction is used to derive (or at least to rationalize) phase diagrams. In the present case of free energies which are function of a great many independent variables, the common tangent construction is not appropriate. For this reason, Kikuchi[36] proposed a method of phase equilibrium determination based on minimizing what he called the *grand potential* defined by

$$\omega = f - \mu\xi \quad , \tag{70}$$

where ξ is actually a normalized sum of point correlation variables over the various sublattices and where μ is an appropriate chemical potential, shortly to be specified. In Eq. (70), f is the free energy (30) normalized to one lattice point. The grand potential or free energy minimization proceeds by solving the set of simultaneous non-linear algebraic equations (28) which, through Eq. (17), are implicit in the linearly independent correlation variables ξ. To make the meaning of Eq. (70) more explicit, consider a new function f_A defined as the free energy f in which $(2x_A - 1)$ [when x_A is the average concentration of A atoms] has been substituted for ξ_1, all of the other ξ variables having their equilibrium values. Another function f_B is defined similarly. The classical "intercept rule" may be then written in two alternate ways:

$$\mu_A = f - x_A \frac{df_A}{dx_A} , \tag{71a}$$

$$\mu_B = f - x_B \frac{df_B}{dx_B} , \tag{71b}$$

where also

$$\frac{df_A}{dx_A} = \mu_A - \mu_B = - \frac{df_B}{dx_B} \tag{72}$$

in which μ_A and μ_B are the classical potentials of A and B respectively. Making use of Eq. (72) and summing Eqs. (71a) and (71b), we obtain immediately Eq. (70), where ω now appears as a sum and μ as a difference of chemical potentials:

$$\omega = \frac{\mu_A + \mu_B}{2} \quad , \quad \mu = \frac{\mu_A + \mu_B}{2} . \tag{73}$$

Calculations of phase equilibrium then proceeds as follows: for fixed values of temperature T and chemical potential μ, the grand potential ω is minimized with respect to the independent ξ variables, the free energy f to be used being the CVM free energy appropriate for the ordered (or disordered) phase under consideration, i.e. the one for which the W-matrix reflects the proper sublattice structure. Next, the minimum values of ω are plotted as a function of μ for that phase and for other phases of interest. Points at which two ω vs. μ curves intersect determine phase equilibrium. The lowest intersections found are then used to construct the phase diagram: an intersection between curves for phases α and β, say, yields two values of ξ, hence, two values of equilibrium concentrations x^α and x^β, for example, at the chosen temperature. Two sets of all other correlation variables are also obtained, one set for each phase, thereby completely determining the state of order of each of the two phases in equilibrium. This procedure is repeated for all temperatures and for all ordered phases predicted by the ground state analysis. The calculated loci of phase boundary points, in T-x_B or T-μ space constitutes the required *coherent* (or *ordering*) equilibrium phase diagram.

The grand potential method is obviously equivalent to the traditional one since, at a point of intersection of ω curves, both the sum ω and the difference μ of chemical potentials are the same, hence the chemical potentials themselves are equal in the two phases, as required. It is also seen that the grand potential can be obtained from the free energy f by a Legendre transformation, μ which then appears as the intensive (field) variable conjugate to the extensive (density) variable ξ.

CVM phase equilibria with fixed values of the ratio V_2/V_1 of second to first neighbor pair interactions were explored systematically for the fcc[4,5,6,38] and bcc[37] lattices (for which the ground state problem had been solved exactly, for this range of pair interactions). In practice, EPI's are expected to vary with concentration, due to the influence of both electronic structure[39] and elastic effects[40,41]. Furthermore, pure components A and B may be stable in fcc, bcc, hcp or other crystal structures. The procedure for calculating a phase diagram should therefore include the following steps: select a lattice (fcc, say), then calculate free energy curves (actually, grand potential curves ω) for that lattice and its relevant ordered superstructures, based on ground state analysis. Repeat the calculation for other disordered state structures (bcc, hcp...) and all of their relevant superstructures. Finally, combine all grand potential curves, at various temperatures, and seek lowest curve intersections; plot intersections in temperature-concentration space. For completeness, a liquid free energy curve should also be included. This can be done by empirical means. This general procedure was followed in a recent calculation[42] of the Ti-Rh phase diagram; the competition between fcc and bcc lattices and their respective superstructures was well illustrated in this example. "Interloper" intermetallic phases (A15...) have not yet been incorporated into the computational scheme.

4.3. Example: Vacancy Ordering in $YBa_2Cu_3O_x$

In Sect. 3.3, a stability analysis of O- ❑ ordering in the basal Perovskite plane was presented. Such an analysis does not yield ordered ground states readily: One must instead proceed as was described in Sect. 4.1. In the simple case under consideration here, which is that modeled by Fig. 1, it actually suffices to perform a "brute force" calculation as follows: all possible configurations of filled and empty oxygen sites are tested on a 2x2 and a 2x4 lattice containing 8 and 16 oxygen sites, respectively[43]. Structures are then sought which minimized the Hamiltonian

$$\Phi = \frac{1}{2} \sum_{p} [V_1 \sum_{p_1} \sigma_p \sigma_{p_1} + V_2 \sum_{p_2} \sigma_p \sigma_{p_2} + V_3 \sum_{p_3} \sigma_p \sigma_{p_3}] \tag{74}$$

where $\sigma=\pm 1$, according to whether a site is filled or empty. In Eq. (74), p denotes any lattice site, and p_1, p_2, p_3 denote sites which are respectively nearest and next nearest (p_2, across a Cu atom; p_3 otherwise) neighbors to it.

With the range of interactions considered, only two stoichiometric compositions gave rise to ordered ground states: $c_0=0.50$ and $c_0=0.25$ (or 0.75). The resulting structures are indicated in the ordering map of Fig. 3a and 3b for these two concentrations, respectively. Minimum energy structures are designated by their two-dimensional space group symbols. For $c_0=0.5$, the ground state regions are separated by limits (heavy lines) which agree with those found in the stability analysis, Fig. 2. The structure of the orthorhombic superconducting phase is that found in the lower left region of Fig. 3a and clearly results from the <0 0> instability described earlier. In other regions, 2x2 and $\sqrt{2}$x$\sqrt{2}$ structures are predicted.

At stoichiometry $c_0=0.25$, the ordering map regions coincide with the four quadrants of the (x,y) plane [recall that $x=V_2/V_1$, $y=V_3/V_1$, $V_1>0$]. It is seen that phase separation is predicted in the (-,-) quadrant, cell quadrupling is predicted in the (+,+) quadrant, and two similar (yet distinct) cell doubling structures are predicted in the other quadrants. In particular, the structure which doubles the a direction is precisely the one observed by electron microscopy by Zandbergen et al.[33,44] and believed to be that identified by Cava et al.[45] as having a superconducting transition of about 60K.

The next step in the study of vacancy ordering is to formulate a CVM approximation of the free energy. For that, appropriate clusters must be chosen. That aspect of the problem was, until recently, something of a black art. Now, thanks to the work of Schlijper[12] and Finel[13], significant progress has been made.

At the Crete NATO Institute, Finel gave a special seminar on his method, the detailed exposition of which can be found in his Doctoral dissertation[13]. It is clear from the formulation of Sect. 2.2 that the CVM approximation is based upon the replacement of a density matrix ρ, for an infinite system (thermodynamic limit), by products of "reduced densities" ρ_α pertaining to small clusters. The resulting factorization is not expected to be exact (if it were, the three dimensional Ising model would be solved!) but at least it should be required to produce approximate density matrices such that they belong to a class defined by Tr $\rho=1$. Finel shows how one may use this criterion in selecting appropriate cluster approximations, the first step being that of reducing by unity the number of infinite dimensions of the system.

In the present case, consider again the Perovskite basal plane in the p2mm arrangement (orthorhombic phase) shown in Fig. 4a. A motiv (Fig. 4b) is selected and is translated over the actual structure (Fig. 4a) in such a way that the whole infinite plane is covered. It is seen that there are just four non-equivalent positions of the motiv on the two-dimensional structure: two large centered squares (full lines), one on the α, the other on the β sublattice, and two smaller squares, outlined by dashed lines. For the approximation chosen (motiv in 4b), these are the four basic clusters required in this CVM approximation. Of course, the motiv could be made larger, presumably ensuring higher accuracy, but at the cost of greatly increased numerical complications.

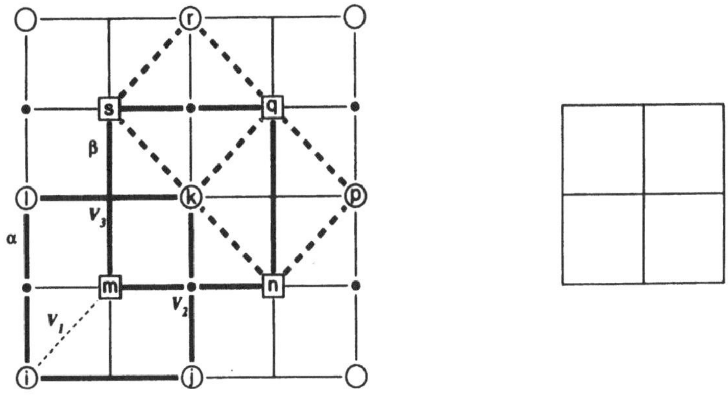

(a) *(b)*

FIGURE 4a. Perovskite basal plane in the orthorhombic phase. As in Fig. 1, but now open circles denote oxygen atoms, and squares denote empty sites. Sites are labeled in accordance to W-matrix notation of Table III. Fig. 4b. is CVM motiv.

In principle, all subclusters of the basic ones must be considered. In fact, only a certain number will appear in the entropy formula. To derive the latter, it is merely required to determine the coefficients a_α appearing in Eqs. (25) to (30). In practice, it is simpler to determine a completely equivalent set of integers, the "Kikuchi-Barker coefficients," obtained from the a_i (the index i now stands for a cluster *type*, or class, consisting of all α which are equivalent under the symmetry operations of the crystal) by the formula[11]

$$\gamma_i = m_i a_i \tag{75}$$

where m_i is the number of i-type cluster per lattice site. The following recursive method[2] is then used to find successive coefficients, starting from the one corresponding to the basic cluster(s) labeled I:

$$\gamma_I = -m_I$$

$$\gamma_i = -m_i - \sum_{j=i+1}^{I} m_i^j \gamma_j \quad . \tag{76}$$

To obtain the Kikuchi-Barker coefficients, it suffices to consider cluster types in the disordered state. In the various ordered states, cluster types split into subclasses, but the γ_i coefficients remain the same (except for a common multiplicative factor). Cluster types in the completely disordered state, with Cu atoms discarded, are listed in Table I

i	Cluster	m_i	j= 1	2	3	4	5	6	7	8	9	10	11	γ_i
0		1	1	2	2	2	3	3	3	4	4	4	5	0
1		2		1	0	0	2	2	0	4	3	0	4	–2
2,3		2			1	0	1	0	2	2	2	4	4	0
4		2				1	0	1	1	0	1	2	2	0
5		4					1	0	0	4	2	0	4	4
6		2						1	0	0	1	0	2	0
7		4							1	0	1	4	4	0
8		1								1	0	0	0	–1
9		4									1	0	4	0
10	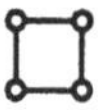	1										1	1	0
11		1											1	–1

m_i^j

Table I. Cluster type (in the disordered state), number of these per unit cell (m_i) and m_i^j coefficients required to derive Kikuchi-Barker coefficients γ_i according to Eq. (76).

along with the values of m_i and m_i^j. The γ_j, calculated by application of Eq. (76), are given in the last column. Many of these γ_j vanish, which could have been predicted from general rules[11,13]. The set of γ_i, leading to the required entropy formula, had been derived earlier by Kikuchi[46], Kulik[47] and Finel[13].

How clusters and subclusters split in the ordered orthorhombic (p2mm) phase is indicated, as an example, in Table II. For illustration, Table III gives the explicit form of

i	α	β	i	α	β
0	○	□			
1			6		
2			7		
3			8		
4			9		
5			10		
5′			11		

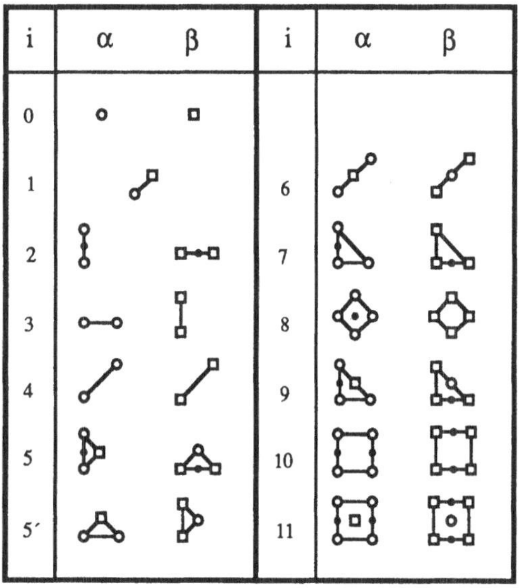

Table II. Cluster types required for the orthorhombic (p2mm) structure.

$$x_1(i,m) = \frac{1}{4}[1 + i\xi_0^{\alpha} + m\xi_0^{\beta} + im\xi_1]$$

$$x_5^{\alpha}(i,l,m) = \frac{1}{8}[1 + (i+l)\xi_0^{\alpha} + (m)\xi_0^{\beta} + (i\,m + l\,m)\xi_1$$
$$+ (i\,l)\xi_2^{\alpha} + (i\,l\,m)\xi_5^{\alpha}]$$

$$x_8^{\alpha}(k,r,s,q) = \frac{1}{16}[1 + (k+r)\xi_0^{\alpha} + (s+q)\xi_0^{\beta} + (k\,s + r\,s + r\,q + q\,k)\xi_1 + (k\,r)\xi_2^{\alpha}$$
$$(s\,q)\xi_2^{\beta} + (k\,r\,s + k\,r\,q)\xi_5^{\alpha} + (k\,s\,q + r\,s\,q)\xi_5^{\beta} + (k\,r\,s\,q)\xi_8^{\alpha}]$$

$$x_{11}^{\alpha}(i,j,k,l,m) = \frac{1}{32}[1 + (i+j+k+l)\xi_0^{\alpha} + (m)\xi_0^{\beta} + (i\,m+j\,m+k\,m+l\,m)\xi_1$$
$$+ (i\,l+j\,k)\xi_2^{\alpha} + (i\,j+k\,l)\xi_3^{\alpha} + (i\,k+j\,l)\xi_4^{\alpha} + (i\,l\,m+j\,k\,m)\xi_5^{\alpha}$$
$$+ (i\,j\,m+k\,l\,m)\xi_{5'}^{\alpha} + (i\,m\,k+j\,m\,l)\xi_6^{\alpha} + (i\,j\,l+i\,j\,k+l\,k\,i+l\,k\,j)\xi_7^{\alpha}$$
$$+ (i\,j\,l\,m+i\,j\,k\,m+l\,k\,i\,m+l\,k\,j\,m)\xi_9^{\alpha} + (i\,j\,k\,l)\xi_{10}^{\alpha} + (i\,j\,k\,l\,m)\xi_{11}^{\alpha}]$$

Table III. Cluster Probabilities x (or partial densities ρ) for orthorhombic phase expressed as linear functions of correlation variables ξ, according to Eq. (67).

Eq. (17) for the p2mm (superconducting) phase, more precisely, of the symmetry adapted from (67), the coefficients of the correlation functions ξ (corresponding to the clusters of Table II) yielding the elements of the configuration matrix W. In Table III, only one example each of cluster concentrations x_5, x_8 and x_{11} are given. Other cluster formulas (for x_5^{β}, $x_{5'}^{\alpha}$ ...), as required by Table II, can be derived similarly, with the help of Fig. 4a.

The resulting phase diagram[48], obtained from this cluster scheme according to the methods described above is shown in Fig. 5 for the particular choice $V_1>0$, $V_2/V_1 = V_3/V_1 = -0.5$. The values of these parameters were chosen so that comparisons could be made with diagrams obtained by Monte Carlo simulation[49] and renormalization group methods[50,51].

At low concentrations and high temperatures, the two-dimensional disordered phase 4mm (short symbol, long space group symbol is p4mm) of square symmetry is calculated to be the stable one. It corresponds to the three-dimensional tetragonal phase P4/mmm. A line of second-order transitions (heavy dashed line) separates the disordered phase region from that of the ordered phase mm. The upper ordering critical point at stoichiometry c_0=0.5 lies at a reduced temperature kT_0/V_1=4.03; whereas high-temperature series expansions[52] give the value kT_0/V_1=3.80.

The line of second-order transitions ends at a tricritical point (t) which has coordinates kT_t/V_1=1.41, c_0=0.19, and μ=3.94, where the field variable μ represents a difference of

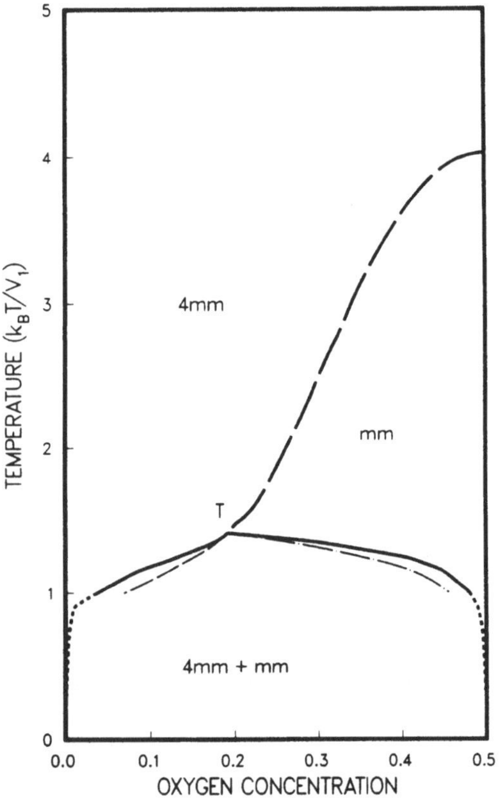

FIGURE 5. CVM phase diagram calculated for $V_1>0$, $V_2=V_3=-0.5V_1$. Tetragonal and orthorhombic phase regions are designated by symbols 4mm (full symbol p4mm) and mm (full symbol p2mm), respectively.

chemical potentials μ_O-$\mu_\square$. These values are to be compared to those obtained from renormalization group techniques by Rikvold et al.[51]: $kT_t/V_1=1.205\pm0.003$, $c_O\approx.30$, $\mu/V_1=3.965\pm0.001$ and by Claro and Kumar[50]: $kT_t/V_1=1.28$, $c_O\approx.31$, in reasonable agreement with Monte Carlo results[49]. An "interface method" calculation by Slotte[53] gives, for the tricritical temperature, $kT_t/V_1\cong-2.27V_2$ at $\mu/V_1=4$.

Detailed calculations of phase boundary lines in the immediate vicinity of the tricritical point indicate that properties derived by Allen and Cahn[54] on the basis of the Landau theory are well obeyed here: the disordered phase boundary joins the line of second-order transitions with no change in slope, unlike the case for the ordered phase boundary at point t. The fine dashed line is the metastable extension of the line of second-order transitions and represents an ordering spinodal[9], i.e. a line below which the disordered phase becomes marginally unstable to small-amplitude ordering fluctuations. The fine dot-dash curve is the locus of marginal instability for phase separation on the partially filled oxygen sublattice. In terms of the stability analysis presented above, the fine dashed curve is the stability limit for two <0 0> ordering waves operating, in phase opposition, on the two square sublattices of oxygen sites, and the dot-dash curve is the stability limit for a single <0 0> wave acting on the partially filled sublattice in the ordered phase. The two-phase region (4mm + mm) rapidly spreads out as the temperature is lowered so that, at absolute zero, the solubility of 0 in the disordered phase and ❑ in the ordered phase are nil.

These tangency rules and spinodals are mean-field features and are therefore not strictly valid from an equilibrium statistical mechanical standpoint. In practice, however, the spinodal concept is a very useful one as it provides simple interpretations of phenomena observed under the constraint of slow kinetics at low temperatures. In two dimensions, tricritical points are expected to have non-classical exponents[55], however, so that, in fact, the two-phase coexistence curve at t should be very flat, as calculated by renormalization group methods, and not pointed, as shown in Fig. 5.

This calculation was presented as a tutorial in the setting up and use of the CVM in an easy-to-visualize two-dimensional example. The case treated is also a very topical one, of course, since it pertains to the high-T_c superconductor $YBa_2Cu_3O_z$. Phase diagrams for those interesting cases, with $V_2\neq V_3$, have now been calculated[55], and show good agreement with experimental findings.

5. CONCLUSION

The CVM, though it is a classical approximation, and thus frowned upon by the critical phenomena specialists, has proved itself to be extremely useful, particularly in the area of phase diagram calculations. The method is by no means foolproof, the choice of correct clusters still being more in the nature of an art than a science. Numerical convergence for large-cluster approximations also can be a serious problem, particularly at low temperatures.

Still, the improvement over the standard mean field (Bragg-Williams) methods are so considerable that it is becoming feasible to derive certain classes of phase diagrams from first principles. For that, of course, it will be necessary to calculate physical parameters, such as the EPI (V_r), along with structural energies, by quantum mechanical methods. Much progress in this aspect of the problem is being realized currently, and I hope to be able to report on this important topic in the near future.

6. ACKNOWLEDGEMENTS

I am indebted to colleagues who have contributed very significantly to the work reported in this paper. Among many others, I should cite the following: Ryo Kikuchi, who taught me the CVM, Juan Sanchez who extended it, Alphonse Finel and Patrice Turchi, both from the research group of François Ducastelle in Paris, who collaborated on CVM and electronic band structure calculations, along with Marcel Sluiter. Also, Luc Wille corrected an earlier version of the perovskite stability analysis and, along with Arjun Berera, performed the $YBa_2Cu_3O_x$ phase diagram calculations. Early CVM work was funded, in part, by the U.S. Army Research Office (Durham). Current research on phase diagrams calculations is supported by a grant from the Director, Office of Energy Research, Materials Sciences Division, U.S. Department of Energy, under contract DE-AC03-76SF00098.

REFERENCES

1. R. Kikuchi, *Phys. Rev.* **81**, 988 (19851).
2. J. A. Barker, *Proc. Roy. Soc.*, A **216**, 45 (1953).
3. J. Hijmans and J. de Boer, *Physica* **21**, 471, 485, 499 (1955).
4. J. M. Sanchez and D. de Fontaine, *Phys. Rev.* B **17**, 2926 (1978).
5. J. M. Sanchez and D. de Fontaine, *Phys. Rev.* B **21**, 216 (1980).
6. J. M. Sanchez and D. de Fontaine, *Phys. Rev.* B **25**, 1759 (1982).
7. R. Kikuchi, *J. Chem. Phys.* **60**, 1071 (1974).
8. D. M. van Baal, *Physica (Utrecht)* **64**, 571 (1973).
9. J. M. Sanchez, D. Gratias and D. de Fontaine, *Acta Cryst.* A **38** (1982) 214.
10. D. de Fontaine, "Configurational Thermodynamics of Solid Solutions", in *Solid State Phys.*, H. Ehrenreich, F. Seitz and D. Turnbull, Eds., Vol. 34, pp. 73-294, Academic Press (1979).
11. J. M. Sanchez, F. Ducastelle and D. Gratias, *Physica (Utrecht)* **128A**, 334 (1984).
12. A. Schlijper, *J. Stat. Phys.* **35**, 285 (1984); **40**, 1 (1985).
13. A. Finel, *Thèse de Doctorat d'Etat*, Université Pierre et Marie Curie, Paris (1987), unpublished.
14. T. Morita, *J. Phys. Soc. Jpn.* **12**, 753, 1060 (1957); *J. Math. Phys.* **13**, 115 (1972).
15. F. Ducastelle, *J. Phys.* C **8**, 3297 (1975).
16. F. Ducastelle and F. Gautier, *J. Phys.* F **6**, 2039 (1976).
17. C. Sigli and J. M. Sanchez, *CALPHAD* **8**, 221 (1984).
18. A. Gonis, G. M. Stocks, W. H. Butler and H. Winter, *Phys. Rev.* B **29**, 555 (1984).
19. F. Solal, R. Caudron, F. Ducastelle, A. Finel and A. Loiseau, *Phys. Rev. Lett.* (in press).
20. J. M. Sanchez, *Physica (Utrecht)*, **111A**, 200 (1982).
21. T. Mohri, J. M. Sanchez and D. de Fontaine, *Acta Metall.* **33**, 1463 (1985).
22. E. M. Lifshitz, *J. Phys. USSR 7*, **61**, 251 (1942).
23. A. G. Khachaturyan, *Phys. Status Solidi* B **60**, 9 (1973).
24. C. Haas, *Phys. Rev.* A **140**, 863 (1965).
25. R. M. Hornreich, M. Lubhan and S. Shtrikman, *Phys. Rev. Lett.* **35**, 1678 (1975); *Phys. Rev.* B **19**, 3799 (1979).
26. J. M. Sanchez, D. Gratias and D. de Fontaine, *Acta Cryst.* A **38**, 214 (1982).
27. International Tables for Crystallography, Vol. A, Space Group Symmetry, T. Hahn, ed., Kluver Acad. Publisher (1987).
28. J. W. Cahn, *Acta Metall.* **9**, 795 (1961).
29. J. Kanamori, *Prog. Theor. Phys.* **35**, 16 (1986); J. Kanamori and M. Kaburagi, *J. Physique (Paris)* **38**, C7, 274 (1977).

30. S. M. Allen and J. W. Cahn, *Acta Metall.* **20**, 423 (1972); *Scripta Metall.* **7**, 1261 (1973).
31. T. Kudo and S. Katsura, *Progr. Theor. Phys.* **56**, 435 (1976).
32. D. de Fontaine, L. T. Wille and S. C. Moss, *Phys. Rev.* B **36**, 5709 (1987).
33. G. Van Tendeloo, H. W. Zandbergen and S. Amelinckx, *Sol. St. Comm.* **63**, 603 (1987).
34. D. de Fontaine, M. Sluiter and P. Turchi, *Proc. Phase Transformations Conference*, Cambridge, UK, 6-10 July, Institute of Metals (in press).
35. A. Schlijper, private communication.
36. R. Kikuchi and D. de Fontaine, NBS Special Publication No. 496, G. C. Carter, ed., p. 967 (1978).
37. T. Mohri, J. M. Sanchez and D. de Fontaine, *Acta Metall.* **33**, 1171 (1985).
38. M. Sluiter, P. Turchi, Fu Zezhong and D. de Fontaine, *Physica (Utrecht)*, (in press).
39. P. Turchi, M. Sluiter and D. de Fontaine, *Phys. Rev.* B **36**, 3161 (1978).
40. L. G. Ferreira, A. A. Mbaye and A. Zunger, *Phys. Rev.* B **35**, 6475 (1987).
41. K. Terakura, T. Oguchi, T. Mohri and K. Watanabe, *Phys. Rev.* B **35**, 2169 (1987).
42. M. Sluiter, P. Turchi, Fu Zezhong and D. de Fontaine, *Phys. Rev. Lett.* (in press).
43. L. T. Wille and D. de Fontaine, *Phys. Rev.* B **37** (1988).
44. H. W. Zandbergen, G. Van Tendeloo, T. Okabe and S. Amelinckx, *Physica Status Solidi* (a), (in press).
45. R. J. Cava, B. Batlogg, C. H. Chen, E. A. Rietman, S. M. Zahurak and D. Werger, *Phys. Rev.* B **36**, 5709 (1987).
46. R. Kikuchi (private communication).
47. J. Kulik (unpublished and at U.C. Berkeley).
48. A. Berera, L. T. Wille and D. de Fontaine, *J. Stat. Phys.* (in press).
49. K. Binder and D. P. Landau, *Phys. Rev.* B **21**, 1941 (1980).
50. F. Claro and V. Kumar, *Surf. Sci.* **119**, L371 (1982).
51. P. A. Rikvold, W. Kinzel, J. D. Gunton and K. Kashi, *Phys. Rev.* B **28**, 2686 (1986).
52. J. Oitmaa, *J. Phys. A: Math. Gen.* **14**, 1159 (1961).
53. P. A. Slotte, *J. Phys. C: Sol. St.* **16**, 2935 (1983).
54. S. M. Allen and J. W. Cahn, in "Alloy Phase Diagrams," L. M. Bennett, T. B. Massalski and B. C. Giessen, eds., *Materials Res. Soc. Proc.* **19**, 195 (1983).
55. J. W. Wheeler (private communication).

LONG PERIOD STRUCTURES IN ALLOYS-STATISTICAL MECHANICS OF THE ANNNI MODEL AND RELATED CONCEPTS

Walter Selke
Institut für Festkörperforschung der Kernforschungsanlage Jülich, Postfach 1913, D-5170 Jülich, Federal Republic of Germany

ABSTRACT
Results on the phase diagram of the axial next-nearest neighbour Ising (or ANNNI) model are reviewed and applications to strikingly similar findings on long-period superstructures in A_3B alloys (like Al_3Ti, Ag_3Mg, and Cu_3Pd) are discussed.

1. INTRODUCTION

Many materials are known to form complex spatially modulated superstructures including magnets, absorbed monolayers, reconstructed surfaces, ferroelectrics, as well as alloys. The actual structure may depend quite sensitively on external parameters such as pressure and temperature.

Usually, very careful measurements have to be perfomed to establish the (meta-)stability and details of these structures. Indeed, much of our current knowledge on fine features is due to recent advances in the experimental techniques such as high resolution electron microscopy in studying, for instance, alloys.

Of course, the aim of the theory is to identify and describe the typical aspects of superstructures and to explain their origin. Eventually, specific materials should be amenable to a microscopic description. - Two main mechanisms for the formation of spatially modulated structures can be distinguished. Firstly, superstructures may occur already at zero temperature, minimizing the energy. Several reasons are conceivable, for example, nesting of Fermi surfaces, short range competing interactions in systems where the local property or variable, S_α, (α refers to a site of the underlying lattice) subject to the modulation can assume a continuum of values, or long-range competing interaction in systems with a discrete variable S_α. Secondly, entropy effects may stabilize superstructures at non-zero temperature. In particular, this possibility exists in systems where S_α is discrete (like the occupation variable in alloys) and effectively short-range competing interactions are present.

In the following special emphasis will be put on one of the simplest statistical mechanics models where configurational entropy effects play a decisive role in stabilizing periodic

G. M. Stocks and A. Gonis (eds.), Alloy Phase Stability, 205–232.

superstructures: the axial next-nearest neighbour Ising (or ANNNI) model (1,2). The model will be introduced and its phase diagram will be presented in Section II. Variants of the model will be also mentioned. - As has been observed first by de Fontaine and Kulik (3) properties of the ANNNI model are strikingly similar to experimental findings on long-period superstructures in some binary AB and A_3B type alloys, as will be demonstrated in Sect. III. Of course, Fermi surface effects, as suggested by Sato and Toth (4) some time ago and supported by recent calculations on Fermi surfaces in random metallic alloys (5), are expected to be important in the formation of superstructures in alloys. This aspect of a theory on long--period superstructures in alloys will be also discussed in Sect. III. A brief summary will conclude this review.

2. STATISTICAL MECHANICS OF THE ANNNI MODEL

In ANNNI models, Ising spins $S_\alpha = \pm 1(+,-)$ are situated on a regular d-dimensional lattice formed of (d-1)-dimensional layers of coordination number $q_\perp$ normal to the z-axis. Within the layers each spin is coupled only by nearest neighbour (nn) ferromagnetic interactions, $J_0 > 0$. However, along the z-axis spins are coupled both by nn ferromagnetic interactions, $J_1 > 0$, and by competing next-nearest neighbour (nnn) antiferromagnetic interactions, $J_2 = -\kappa J_1 < 0$. The parameter κ controls the degree of competition. If both J_1 and J_2 are negative, the interactions are still competing, since different antiferromagnetic structures are favoured. In that case, the results follow from the ones obtained for $J_1 > 0$ by merely reversing the spins in each second layer. Explicitely the Hamiltonian may be written as

$$\mathcal{H} = -\sum_i (J_0 \sum_{(kl)} S_{ik}S_{il} - J_1 \sum_k S_{ik}S_{(i+1)k} - J_2 \sum_k S_{ik}S_{(i+2)k}) \qquad (1)$$

where the subscript i counts the layers, (kl) denotes a pair of nearest neighbour sites within a layer. - The three-dimensional version of the ANNNI model is depicted in Fig. 1.

At <u>zero</u> temperature T = 0, each individual layer is ferromagnetically ordered. For $\kappa < \frac{1}{2}$ the ground state is the ferromagnetic structure while for $\kappa > \frac{1}{2}$ it is a (2,2) antiphase (or <2>) structure with a layer pattern ... --++--++ ... . At $\kappa = \frac{1}{2}$ the competing interactions, J_1 and J_2, balance each other, and the ground state becomes infinitely degenerate in all periodic and non-periodic configurations where at least two "+" layers are followed by at least two "-" layers. The number of these configurations, D_N, is given for a system of N layers with periodic boundary conditions by $D_N = D_{N-1}+D_{N-2}$ (adding to (N-1) layers a layer with the same sign as the one of the last layer and adding to (N-2) layers two layers with the opposite sign) with $D_2 = D_3 = 2$, i.e., by the recursion relation of a Fibonacci sequence. In the thermodynamic limit, $N \rightarrow \infty$, the entropy per spin is then $S = N_1^{-1} k_B \ln[(\sqrt{5}+1)/2]$, where N_1 is the number of spins in a layer.

At non-zero temperatures, the dimensionality, d, of the ANNNI model plays a crucial role. In the one-dimensional case, the Ising chain with competing nn and nnn interactions, order is destroyed already at infinitesimally small temperatures, and the correlations between distant spins, $\langle S_o S_r \rangle$, decay exponentially.

From an exact transfer matrix calculation (substituting $S_\alpha S_{\alpha+1} = \tau_\alpha$ the ANNNI chain can be mapped onto an nn Ising chain in a field) (7,8) it follows that the exponential decay may be superimposed by an oscillatory modulation, $\langle S_o S_r \rangle \sim \exp(-r/\xi) \cos(qz)$, where $\xi = \xi(T,\kappa)$ is the correlation length and $q = q(T,\kappa)$ the wavenumber characterising the periodically modulated, albeit disordered spin pattern. For $\kappa < \frac{1}{2}$ and low T the wavenumber is equal to zero implying a purely exponential decay. For $\kappa > \frac{1}{2}$, q varies continuously with κ and T, i.e. the wavenumber is usually an irrational number and the wavelength is "incommensurate" to the lattice spacing of the chain. (The spacing along the z-axis is set equal to one in the following). The wavenumber of the <2> structure, $q = \pi/2$, is realized only at T = 0.

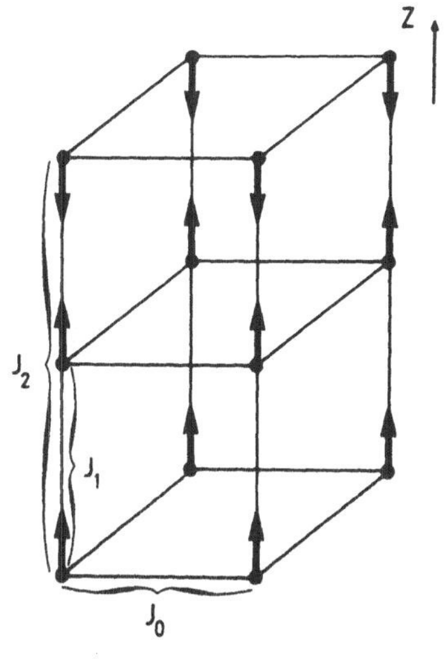

FIGURE 1. The three-dimensional ANNNI model with $q_\perp = 4$

The two-dimensional ANNNI model has not been solved exactly. Results of various approaches for the model on a rectangular lattice, $q_{\perp} = 2$, are summarized in the phase diagram depicted in Fig. 2 (9,10). For d = 2, the ferromagnetic and <2> ground states evolve into long range ordered phases at non-zero temperatures. At sufficiently high temperatures there is, of course, a disordered or paramagnetic phase with exponentially decaying correlations. A very interesting aspect of the phase diagram is the existence of a complex spatially modulated or "floating incommensurate" phase above the <2> phase extending down to the special highly degenerate point ($T = 0$, $\kappa = \frac{1}{2}$). The correlations are believed to decay algebraically with a superimposed modulation, $\langle S_o S_r \rangle \sim r^{-\eta} \cos(qz)$ with $\eta = \eta(T,\kappa)$ and $q = q(T,\kappa)$; i.e. the correlations decay more slowly than in the disordered phase, but there is no real long-range order in the modulated phase.

- Typical excitations driving the transition from the <2> to the floating phases and from the floating to the disordered phases have been depicted before (2,9,11), and the interested reader may look up these articles. Results on the 2d version of the ANNNI model and variants are of interest especially in connection with chemisorbed monolayers (11,12).

We now turn to the three-dimensional model. The gross features of the phase diagram in the (T,κ) plane are shown in Fig. 3, combining results of high (13) and low (1,14) temperature expansions as well as Monte Carlo simulations (15,16). As in two dimensions the ferromagnetic and <2> ground states give rise to stable ordered phases at $T > 0$. At high temperatures there is, of course, a disordered paramagnetic phase for all κ with exponentially decaying correlations. There, the wave-

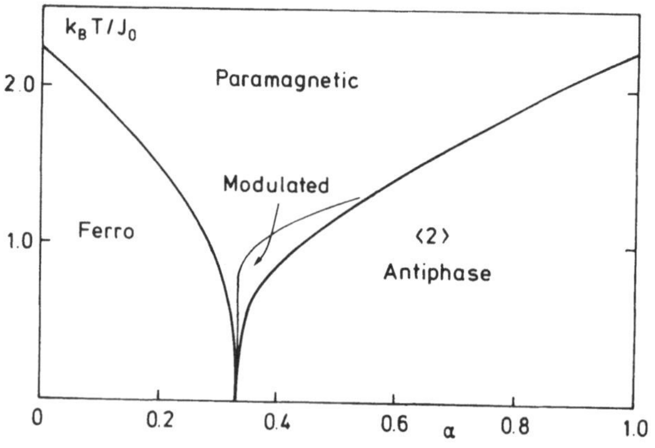

FIGURE 2. Phase diagram of the two-dimensional ANNNI model with $J_1 = (1-\alpha)J_o$ and $J_2 = -\alpha J_o$. Whether the modulated phase extends, possibly as a very narrow strip, up to $\alpha = 1$ is open to question. From (9).

vector-dependent susceptibility $S(\kappa,T,\underline{q}) \sim \sum_{\underline{r}} e^{i\underline{q}\underline{r}} \langle S_o S_{\underline{r}} \rangle$ exhibits maxima at non-zero $\underline{q}$, if $\kappa > 1/4$. Only one component of the full wavevector does not vanish, $\underline{q} = (0,0,q)$, because there is competition leading to modulated pattern only along the z-axis. The positions of these maxima vary continuously with κ and T, reflecting incommensurate correlations. As temperature falls, the maximal susceptibility eventually diverges on the critical line, $T_c(\kappa)$, to the ordered spatially modulated phase. The critical wavenumber, $q_c(\kappa)$, at which the susceptibility diverges on the critical line, increases continuously from its zero at the Lifshitz point, L, to $\pi/2$ as κ goes to infinity. In mean field theory one easily finds (2,6)

$$q_c = \cos^{-1}(1/4\,\kappa) \qquad (2)$$

implying a Lifshitz point at $\kappa_c = 0.25$, in good agreement with supposedly accurate estimates based on high temperature series expansions (13) and Monte Carlo studies (16) which locate the Lifshitz point at $\kappa_c \approx 0.27$. - In general, Lifshitz points are multicritical points at which modulated ordered, uniformly ordered and disordered phases meet (17). The ANNNI model exhibits a uniaxial Lifshitz point of discrete, Ising-like symmetry, which is characterised by a specific set of critical exponents. For example, the critical exponent, β, of the order parameter, the magnetization, is $\beta_L \approx 0.19 \pm 0.02$ (16) which may be compared to its value for the standard three-dimensional nearest neighbour Ising model, $\beta \approx 0.32$.

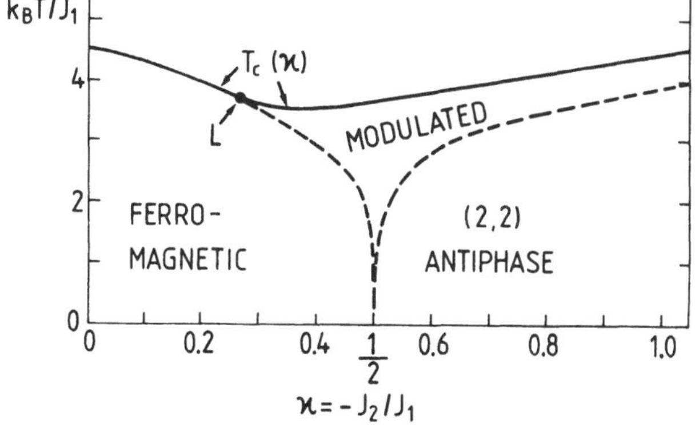

FIGURE 3. Main features of the three-dimensional ANNNI model on a tetragonal lattice with $J_o = J_1$. From (15).

For $T \lesssim T_c(\kappa)$ and $\kappa > \kappa_L$ the system enters some sort of ordered modulated phase with spatially varying modulation pattern, $M(z=i) \approx M_0 \cos q z + \ldots$ where $M(z=i)$ is the magnetization per layer

$$M(i) = \frac{1}{N_l} \sum_k \langle S_{ik} \rangle \tag{3}$$

with the sum running over all N_l spins in the i-th layer at position z. The characteristic wavenumber, $q(T,\kappa)$, exhibits an extremely complicated behaviour in the (κ,T) plane as a consequence of balancing the competing interactions, J_1 and J_2, and satisfying the discrete symmetry, $S_\alpha = \pm 1$, of the Ising model. - Note that even at very low temperatures where $M(z)$ is not of a sinusoidal form, the magnetization pattern are periodic and can be described by the wavenumber q. However, a more appropriate notation will be introduced below.

To give a first illustration of the modulated phase some typical findings of an early Monte Carlo (MC) study (15) are sketched. In Fig. 4 MC magnetization pattern, $M(z)$, at $\kappa = 0.6$, i.e. in the modulated phase on the <2> side of the phase diagram, are depicted. In that example, the lattice is composed of $N = 40$ layers. Because of full periodic boundary conditions the values of the wavenumber q tend to be "quantized" and changes in q occur by jumps of at least $\Delta q = 2\pi/N$. - The pattern of Fig. 4 are observed at fixed ratio κ and increasing temperature demonstrating clearly the breaking of the <2>-symmetry at a definite temperature. Above that temperature, $M(z)$ may be approximated as $M(z) = M_0\cos(qz) + M_1\cos(3qz)+\ldots$ with $q = n\ 2\pi/N$; $n = 9,8,7$ as T increases in the case displayed in Fig. 4. The content of the third harmonic, M_1, is quite large at the lower temperatures, but it decreases rapidly on approach to the transition line, $T_c(\kappa)$, to the disordered phase (15).

These and similar MC observations (15,18) do establish the extent of the ordered modulated phase on the <2> and ferromagnetic side. However, to obtain the precise equilibrium variation of $q(T,\kappa)$ one had to circumvent the quantization of q, possibly by performing very long runs on very large systems. Indeed, other methods are more suitable to uncover the spectacular complexity of the modulated phase. Nevertheless, simulations do provide useful information on a variety of static and kinetic (like the "squeezing process" (15), the local mechanism by which the wavenumber may be changed) phenomena, if interpreted properly.

The neighbourhood of the special degenerate ground state at $\kappa = \frac{1}{2}$ has been explored at low temperatures by a systematic low temperature series expansion valid for $q_\perp J_0 > J_1$ and $d > 2$ (1,14). The degeneracy is lifted at $T > 0$ by an infinite sequence of distinct commensurate phases springing directly from the "multiphase point", see Fig. 5. These phases may be denoted as $<2^{j-1}3>$ (or $<2^{j}1>$ in the case J_1, $J_2 < 0$), $j = 1,2,3,\ldots\infty$, corresponding to a pattern where (j - 1) pairs of lattice layers whose spins point predominantly two "+1" and two "-1" (or (j-1) "2-bands") are followed by three layers with spins pointing predominantly in the same direction (or one 3-band). For example, $<2^{2}3>$ corresponds to the repeating

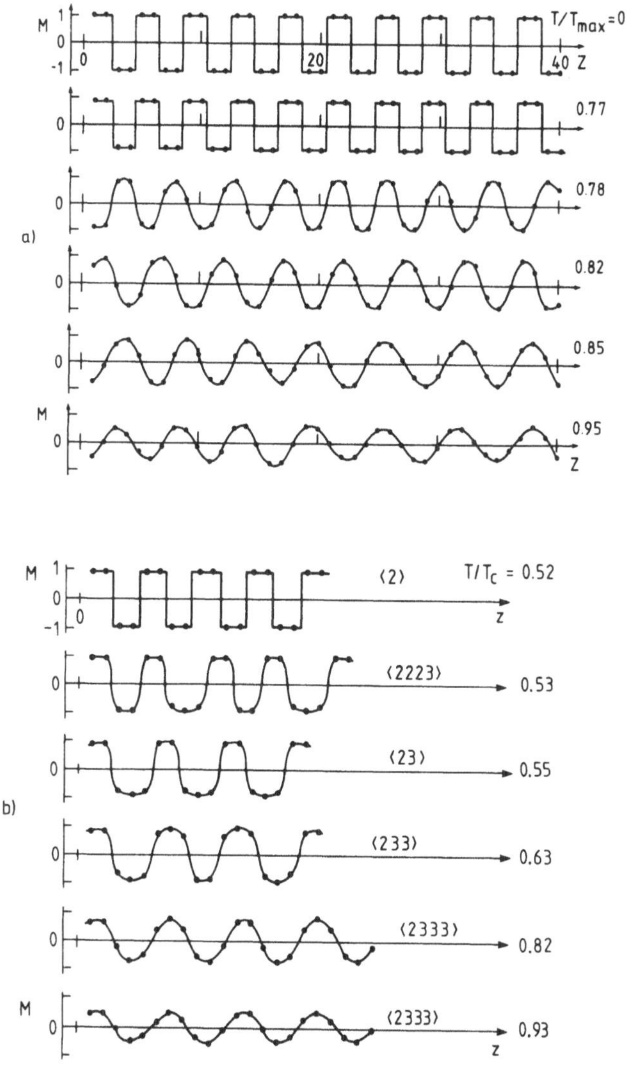

FIGURE 4. Magnetization pattern, M(z), at $J_2 = -0.6J_1$ and $J_o=J_1$ obtained from (a) Monte Carlo data for a system of size 6x6x40 (from (15)) and (b) mean field theory.

sequence ...++--+++--++---... and so on. One may also describe such arrangements in terms of equally spaced domain walls, the "3"s, separated by 2(j-1)+3 layers. - Each of these phases has a definite range of stability in the (T,κ) plane which vanishes exponentially on approach to the multiphase point, with the boundary lines, see Fig. 5, given by (1,14)

$$\kappa_{j+1}(T) - \kappa_j(T) \sim w^{q_\perp j} \tag{4}$$

where $w = \exp(-2J_0/k_BT)$, i.e. the phases become extremely narrow for increasing values of j. The wavenumber characterising the periodic structures takes on the discrete values

$$q_j = j\pi/(2j+1) \tag{5}$$

giving rise to a staircase-like variation of q as one goes from one phase to the other via a first order transition by changing the temperature or the competition ratio, κ.

In the following, the low temperature series expansion technique will be outlined (1,14). In that expansion about the multiphase point one considers all possible degenerate ground states at this point. To determine the stability in the (κ,T)-plane of a phase evolving from one of these ground states one calculates its free energy, F, associated with excitations due to consecutive (one, two, three, ..., n) spin flips. Of course, the stable phase is the one which minimizes the free energy. The various ground states may be characterized by "structural variables", l_k, denoting the number of times per

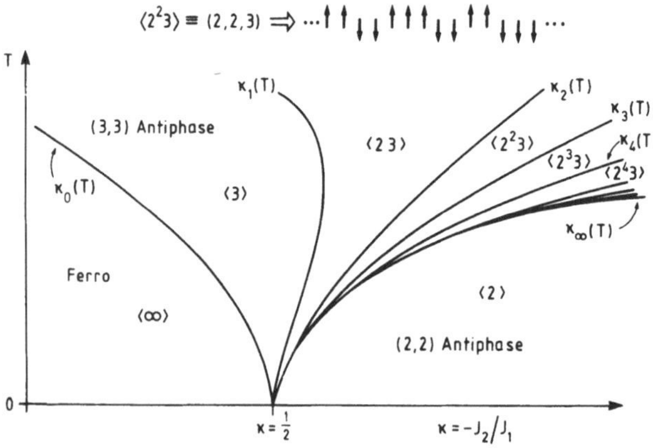

FIGURE 5. Sequence of the $<2^{j-1}3>$, j = 1,2,3,...∞, phases which spring directly from the multiphase point at T = 0 and $-J_2/J_1$ = 0.5. From (1).

spin a band of length k appears in the given state. For example, for the structure $<2^2 3>$, $l_2 = 2/7$, $l_3 = 1/7$ and $l_k = 0$, $k \neq 2,3$. The structural variables are related through the constraint

$$\sum_k k\, l_k = 1 \tag{6}$$

The ground state energy per spin, E_o, can be written as (14)

$$E_o\{l_k\} = -\frac{1}{2} q_\perp J_o - \frac{1}{2} J_1 - J_1 \delta [2l_2 + l_3 - \sum_{k\geq 5} (k-4) l_k] \tag{7}$$

where $\delta = \kappa - \frac{1}{2}$. Note that $l_1 = 0$, as has been mentioned before. The low temperature series expansion for the reduced free energy about a given state is then

$$f\{l_k\} = - \frac{\beta F\{l_k\}}{N} = -\beta\, E_o\, \{l_k\} - \sum_{m=1} \frac{\Delta Z_N}{N}(m, \{l_k\}) \tag{8}$$

where N is here the number of lattice sites and $\Delta Z_N(m,\{l_k\})$ is the contribution to the free energy from configurations obtained from the ground state by flipping m spins. - Defining the Boltzmann factors $w = \exp(-2K_o)$ and $x = \exp(-2K_1)$ (hence $\exp(-2K_2) = x^{-\frac{1}{2}-\delta}$) with $K_i = J_i/k_B T$, $i = 0,1,2$, and using equation (6) to eliminate l_3 the reduced free energy to first order, m = 1, reads

$$f\{l_k\} = \frac{1}{2} q_\perp K_o + \frac{1}{2} K_1 + K_1 \delta/3 + \frac{1}{3}(2 + x^{3+2\delta}) w^{q_\perp} + a_2(\delta) l_2 + \sum_{k\geq 4} a_k(\delta) k\, l_k \tag{9}$$

with the "structural coefficients"

$$a_2(\delta) = 4K_1\delta/3 - \frac{2}{3}(2 - 3x^{1+2\delta} + x^{3+2\delta}) w^{q_\perp} + 0(w^{2q_\perp - 2}) \tag{10}$$

and

$$k\, a_k(\delta) = -4K_1\, \delta(k-3)/3 - [2(k-3)/3 - (k-4)\, x^{1-2\delta} - 2x^2 + k\, x^{3+\delta}/3]\, w^{q_\perp} + 0(w^{2q_\perp - 2}), \quad k > 3 \tag{11}$$

The stable phases are found by maximising f with respect to the l_k. As there is a region in which both a_2 and a_k, $k > 3$, are negative the corresponding structural variables, l_2 and l_k, k > 3, have to be zero there implying the stability of the phase with $l_3 = \frac{1}{3}$, see eq. (6). Therefore the <3> phase is stable over a region of extent $O(w^{q_\perp})$ between the <2> and the ferromagnetic or <∞> phases, originating from the multiphase point, see Fig. 5.

By doing an analogous calculation up to second order (involving, in general, higher order structural variables and corresponding coefficients (14)) the boundary between the <3> and <∞> phases can be shown to be of first order at sufficiently low temperatures. Note, however, that the situation at the <2> : <3> boundary is completely different. Here, from equation (9) it is apparent that all phases which contain only 2- and 3-bands remain degenerate in first order of the expansion, independent of the way the bands are ordered. Hence, it is necessary to proceed to higher orders, m > 1, in the series expansion to ascertain whether some of those phases are stable ones springing directly from the multiphase point as the degeneracy is lifted. Indeed, the expansion up to second order, m = 2, proves the stability of the <23> phase intervening in between the <2> and <3> phases over a region $O(w^{2q_\perp})$, see equation (4), on approach to zero temperature. While one can show that the <23> : <3> boundary is of first order, the <23> : <2> boundary is a highly degenerate one, and one has to proceed to higher orders, m > 2, again. Indeed, in general order, m, of the expansion an inductive argument, based on the important structural coefficients, distinguishing between different axial orderings in lowest orders of w, demonstrates the stability of all $<2^{j-1}3>$ phases, j = 1,2,3,...∞, as summarized in Fig. 5. Only these phases (and all of them) originate directly from the multiphase point (14).

To study the range of stability of the $<2^{j-1}3>$ phases as temperature increases it is useful to picture them as determined by effective forces acting between the domain walls and to attempt to identify these forces. More precisely, let $\{n_1,n_2,...\}$ be the successive different separations between walls in a given state; for example, for $<2^23>$ one has $\{n_1 = 7\}$ or, for $<2322^23>$ one has $\{n_1 = 5, n_2 = 7\}$. Then the free energy may be expanded in the form (19)

$$F = F_0 + L^{-1} \sum_{i=1} [\sigma + W_2\{n_i\} + W_3\{n_i,n_{i+1}\} + ...] \tag{12}$$

where F_0 is a "background" term, L is the number of layers, σ the surface tension of a single, isolated domain wall and $W_n\{n_i,n_{i+1}...\}$ are n-wall interaction potentials. The sophisticated technical details in determining these potentials at low temperatures are quite similar to the ones of the low temperature series expansion (14) and will be also omitted. Taking into account terms up to W_3 (19) the stability of periodic phases with up to two different separations between walls has been investigated at low temperatures without the restriction that the phases have to spring directly from the multiphase point as had been assumed before (14). Based on the analysis of W_2 it is found that the basic $<2^{j-1}3>$ phases, arising from the multiphase point, are stable only up to a temperature $T_c(j)$ (if j is sufficiently large, $j \geq j_0$) which decreases monotonically for increasing values of j. This behaviour is due to the fact that on the boundary of the <2> phase W_2 has a unique negative absolute minimum at a finite separation of the walls, $l_c = 2j_c+3$, $l_c = l_c(T)$. Accordingly, there is a first order transition from the <2> phase to the $<2^{j_c}3>$ phase along the boundary line $\kappa_{<2>}(T)$, i.e. the wavenumber jumps from $q = \pi/2$ to $q = (j_c+1)\ \pi/(2j_c+3)$. However, $l_c(T)$ diverges as T approaches zero temperature ascertaining that all basic phases spring from the multiphase point. Note that this cut-off behavior affects only a tiny portion of the phase diagram, since the high order phases are very narrow, see equation (4), and j_0 is fairly large (equal to nine in mean field theory (20)).

Another interesting phenomenon occurs due to W_3: in addition to the basic $<2^{j-1}3>$ phases mixed phases of type $<2^{j-1}3\ 2^j3>$, j = 1,2,3,..., become stable over certain regions of the phase diagram. According to the low temperature series expansions it can be ruled out that these phases, for any finite value of j, extend down to the multiphase point. Indeed, for each value of j there exists a non-zero temperature, $T_b(j)$, below which that mixed phase is no longer stable in the (κ,T) plane. However, evaluation of W_3 shows that $T_b(j)$ approaches zero as j goes to infinity, i.e. the multiphase point is an accumulation point of the mixed $<2^{j-1}3\ 2^j3>$ phases. Note that this intriguing phenomenon affects also only a very small part of the phase diagram at low temperature, because there only mixed phases with large values of j are involved.

To elucidate the full complexity of the phase diagram, at least at low temperatures, n-wall interaction potentials, W_n, with $n \geq 4$ had to be calculated. So far, such presumably extremely tedious calculations have not been done. - However, a more convenient route to that aim may be to analyse mean field theory which has turned out to give a seemingly qualitatively correct description of the entire phase diagram of the three--dimensional ANNNI model. We shall present results of the mean field theory in the following (20,21).

The condition for an extremum of the free energy leads to the standard mean field equations for the magnetization of the i-th layer, M_i, (20,21)

$$M_i = \tanh[\beta(4J_o M_i + J_1(M_{i+1}+M_{i-1}) - \kappa J_1(M_{i+2}+M_{i-2}))] \quad (13)$$

To determine the stable magnetization pattern at a given point in the phase plane (κ,T) one has to solve the set of coupled equations (13) and find the one solution which minimizes the free energy, F, given by

$$F=\sum_i[-\frac{1}{2}(4J_o M_i^2+J_1 M_i M_{i\pm1}-\kappa J_1 M_i M_{i\pm2})+k_B T \int_o^{M_i} \tanh^{-1} m \, dm] \quad (14)$$

In general, one has to rely upon a numerical analysis for lattices of a finite number of layers with periodic boundary conditions, and great care is needed in choosing appropriate lattice sizes and in obtaining and interpreting the results (20,21). - Usually, solutions of equation (13) are generated iteratively by taking a low temperature configuration with $M_i^o \approx \pm 1$ as initial pattern. To ensure stability of the algorithm M_i^n may be calculated by using the already improved values of $M_1^n,\ldots M_{i-1}^n$, i.e.

$$M_i^n = \tanh[\beta(4J_o M_i^{(n-1)} + J_1 (M_{i+1}^{(n-1)} + M_{i-1}^n) - \kappa J_1 (M_{i+2}^{(n-1)} + M_{i-2}^n))] \quad (15)$$

Highest possible numerical precision may be needed in computing both $\{M_i\}$ and the corresponding free energy to resolve fine details of the phase diagram. - We shall review some of the findings; additional results can be found in Refs. 20 and 21.

The overall mean field phase diagram, assuming $J_1 = J_o$, is depicted in Fig. 6 showing a few of the main commensurate phases which fill up an appreciable portion of the modulated region. The remaining parts of the phase diagram are occupied by additional commensurate as well as incommensurate structures. - To discuss the systematics of the commensurate phases-as inferred from numerous calculations (20) - one may start with the basic $\langle 2^{j-1}3 \rangle$ phases which spring directly from the multiphase point, in agreement with the exact results (1,14,19).

Of course, numerically one can follow only the few dominant phases with small values of j down to very low temperatures: The extent of these phases is asymptotically given by equation (4). However, the prefactor in front of the exponential differs

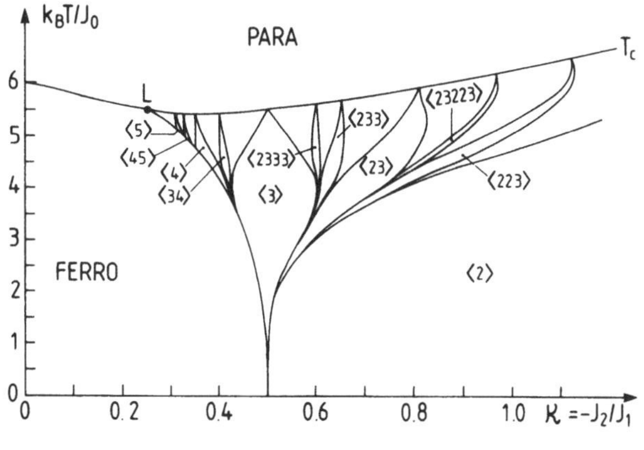

FIGURE 6. Mean field phase diagram of the ANNNI model showing some of the main commensurate phases. From (20).

in general (20), demonstrating that mean field theory is qualitatively, but not quantitatively correct at low temperatures, see also Ref. (19). More complicated mixed phases are generated at $T > 0$ by a "structure combination branching process" (20), see Fig. 7, in which two at low temperatures adjacent basic phases, $<2^{j-1}3>$ and $<2^j3>$ (or, symbolically, A and B) become separated beyond a definite (first order) branching point, at $T_b(j)$, by the formation of a new periodic phase which combines the two structures in the simplest way, i.e. $<2^{j-1}3\ 2^j3>$ (or AB), the mixed phase obtained from W_3 in the exact analysis (19). At a second order branching point, $T_b^{(2)}(j) > T_b(j)$, the boundary between the, say, A and AB phases (or AB and B -the corresponding branching point occurs at a somewhat different temperature, but also above $T_b(j)$; for brevity, in our short hand notation we do not distinguish between the different branching points at a given order) becomes unstable against the intervening new phase AAB (or ABB), and so on. In that way a cascade of mixed phases (corresponding to W_n-terms in the domain-wall analysis with increasing values of n) is generated:

$$
\begin{aligned}
&(1) \quad AB \\
&(2) \quad A^2B,\ AB^2 \\
&(3) \quad A^3B,\ A^2BAB,\ ABAB^2,\ AB^3 \qquad (16)\\
&(4) \quad A^4B,\ A^3BA^2B,\ (A^2B)^2AB,\ A^2B(AB)^2,\ (AB)^2AB^2,\\
&\quad AB(AB^2)^2,\ AB^2AB^3,\ AB^4
\end{aligned}
$$

up to the fourth order branching process. The new phases are more and more narrow, and accordingly very hard to detect in a numerical analysis even of high precision (for instance, using 128 bits per real number (20)).

The branching processes lead to interesting phenomena both at low and high temperatures. The first order branching points between the basic $<2^{j-1}3>$ and $<2^{j}3>$ phases approach zero as j goes to infinity, i.e. $T_b(j) \to 0$ as $j \to \infty$ (19,20). Similarly, mean field theory shows (20) that second order branching points, $T_b^{(2)}(j)$, between the $<2^{j-1}3>$ and $<2^{j-1}32^{j}3>$ phases (giving rise to $<(2^{j-1}3)^2\ 2^{j}3>$ phases obtainable only from W_4) occur at lower temperatures as j increases. One may speculate that these branching points may also accumulate at the multiphase point as j goes to infinity, i.e. $T_b^{(2)}(j) \to 0$ as $j \to \infty$, and so on for higher order branching processes, yielding possibly a very interesting singular character of that point. However, one should bear in mind that most of the phases involved in this phenomenon would fill up only a tiny portion of the phase diagram which is largely dominated by the basic phases with small values of j at low temperatures, see equation (4).

If the branching processes at finite, fixed value of j would continue to higher and higher order as temperature increases,

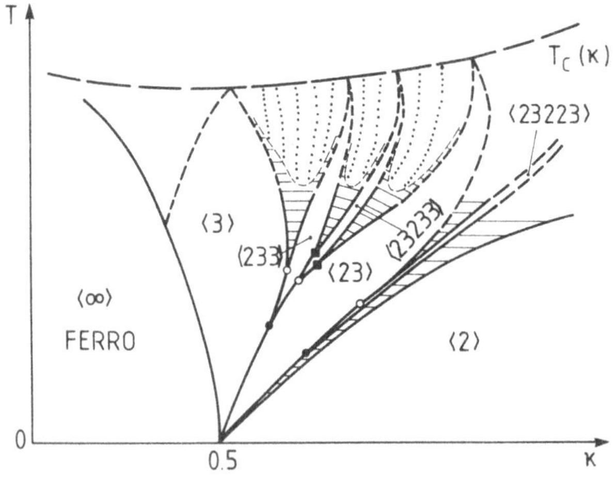

FIGURE 7. Sketch of the mean field phase diagram depicting branching processes of first (•), second (o) and third (■) order, pinned commensurate regions (shadowed) as well as incommensurate structures (dotted lines) occuring above the thin dashed line. Continuous transitions (dashed lines) and transitions of first order (solid lines) are distinguished.

starting from two basic phases, $\langle 2^{j-1}3\rangle$ and $\langle 2^j 3\rangle$, with branching points at $T_b^{(n)}(j)$ ($>T_b^{(n-1)}(j)$) and an accumulation point at $T_b^{(\infty)}(j)$, periodic structures would be generated with wavenumbers, $\bar{q}$, $\bar{q} = q/2\pi$, covering all rational numbers in between $j/2(2j+1)$ and $(j+1)/2(2j+3)$. For example, if $j = 1$, one obtains, up to third order branching processes, the following numbers in between $\frac{1}{6}$ and $\frac{1}{5}$:

$$(1.)\ \frac{3}{16},\qquad (2.)\ \frac{2}{11},\ \frac{5}{26},\qquad (3.)\ \frac{7}{36},\ \frac{4}{21},\ \frac{7}{38},\ \frac{5}{28}\ .$$

This simple example already demonstrates that care is needed in choosing reasonable lattices sizes to perform mean field calculations; if the number of layers is smaller than the denominator of these rational numbers one easily overlooks the corresponding phase (21). - Indeed, there is good numerical evidence for such sequences of branching processes of successively higher order, for example, along the path generating the $\langle 23^k\rangle$, $k = 2,3,\ldots$ phases where, within the numerical precision, the maximal value which can be resolved is $k = 13$ (20). Extrapolating to infinite order one finds $T_b^{(\infty)}(j=1)$ to be roughly at $\frac{1}{2}$ $T_c(\kappa = \frac{1}{2})$ (20). - Above $T_b^{(\infty)}$ devil's staircases (22) would then separate the basic phases.

The set of rational numbers generated by the branching mechanism may be also expressed in terms of appropriately devised continued fraction expansions (3).

Presumably, all basic and mixed commensurate phases are present close to the transition line to the disordered phase, $T_c(\kappa)$, vanishing on approach to T_c in form of a narrow tongue, see Figs. 6 and 7, as suggested by a Landau-like expansion of the mean field equations (21) and supported by renormalization group arguments (23). However, the region of stability of a given structure may not need to be a connected area in the (κ,T) plane; for instance, the basic $\langle 2^{j-1}3\rangle$ phases with $j \geq j_0$ (j_0 is equal to 9, see (20)) spring from the multiphase point, are cut off by the $\langle 2\rangle$ phase at some low temperature, but reappear at higher temperatures (20).

So far we discussed only commensurate phases, i.e. periodic phases with wavenumbers, $\bar{q}$, given by rational numbers, $\bar{q} = P/S$, P and S are integers, for which the magnetization pattern, M_i, repeats itself after a finite number of layers. If $\bar{q}$ is an irrational number such a repetition length does not exist, and the structure is an incommensurate one. Non-periodic or even chaotic structures seem to be at most metastable at $T > 0$ and will be not considered here.

Following the systematic low temperature approaches (1,14,19) only commensurate phases exist at low temperatures. Incommensurate magnetization pattern can be shifted without cost of energy (21,22) and therefore are expected to occur not below the ordering temperature of a single layer. On the other hand, $\bar{q}$ is known to vary continuously along the critical line, $T_c(\kappa)$, see equation (2), implying that the rational numbers, i.e. the commensurate structures, are of measure zero in each finite interval on that line. Indeed, it can be shown (21) that the total width of the commensurate phases vanishes as $(T_c - T)^{1/2}$ on approach to $T_c(\kappa)$. It is therefore of interest to deter-mine the borderline, $T_I(\kappa) < T_c(\kappa)$, down to which incommensu-rate structures persist, that is, below which they are of measure zero. Note that an incommensurate structure with a fixed value of q is stable only along a line in the (κ,T) plane - in contrast to the commensurate phases. In the range $T_I < T < T_c$ an "incomplete devil's staircase" (22) may exist, formed by commensurate structures generated by branching processes up to, possibly, indefinite order and incommensurate structures. To determine $T_I(\kappa)$ one may search for stable solutions of the mean field equations, equation (6), filling a closed one-dimensional orbit or Kolmogorov-Arnold-Moser (KAM) trajectory (21,22) in the (M_i,M_{i+1}) plane. This method is numerically difficult (21), and its accuracy is hard to estimate. Nevertheless, a similar analysis has been done successfully for a model closely related to the ANNNI model, the chiral clock model (24). Typically, $T_I(\kappa)$ has some sort of parabolic shape in between adjacent main commensurate phases. However, the line is broken by the various intervening high order commensurate phases and may have a fractal character. In the ANNNI model, T_I has been also obtained from equating the force pinning the domain walls to the lattice and the repulsion between the walls pushing them out of the commensurate positions (20,21). In particular, it has been found that incommensurate structures are of measure zero in the immediate neighbourhood of both the <2> and <3> phases. The domain of the pinned high order commensurate phases surrounding these simple commensurate phases vanishes only on approach to T_c (20), see Fig. 7.

Within mean field theory a coherent and consistent picture of the complex phase diagram can be given, see Figs. 6 and 7, although some details still need to be clarified. For example, at sufficiently high temperature (20) an expansion of the mean field equations, equations (13), around the main commensurate phases leading to a sine-Gordon equation ("soliton approximation") (21) may be useful in determining such details. Note, that this mapping onto the sine-Gordon equation does not hold, strictly speaking, in describing first order transitions, i.e. at low temperatures or, for instance, close to the <2> phase, as had been assumed in Ref. 21 and clarified in Ref. 20.

It should be emphasized again that mean field theory is expected to give the qualitatively correct topology of the phase diagram. (Improvements on the mean field theory such as the Kirkwood approximation seem to have only a minor effect on the phase diagram of the ANNNI model (33)). Indeed, agreement with results of accurate methods is found both at low temperatures (1,20) and close to T_c (21,23). However, it is conceivable that at intermediate temperatures some of the fine structures described above may be washed out due to thermal fluctuations underestimated in mean field theory. In particular, in mean field theory there is no roughening transition for an Ising model below T_c (25), above which walls (possibly also the domain walls) tend to be delocalised. That aspect deserves further attention.

To compare results on the ANNNI model with high resolution electron microscopy findings on alloys, see the following section, it is useful to study the magnetization pattern in some detail, especially at the boundary of adjacent bands of layers. These boundaries are quite sharp or straight at low temperatures becoming more wavy or diffuse as temperature increases, see Fig. 4 and references 2, 21 and 26 (note that in Fig. 4a the boundaries between differently oriented layers are fairly diffuse already slightly above the <2> phase, possibly because the pattern can shift freely along the z-axis due to the periodic boundary conditions used in the Monte Carlo simulation (15). Actually, the pattern corresponds to a high order mixed phase, $\langle 2^3 3\ 2^4 3 \rangle$, due to the quantization condition mentioned above, and may need a long equilibration time). Certainly, the boundary can be sharpened by strengthening the in-plane coupling, J_o. The changes in the sharpness with temperature are fairly gradual, and it is not clear whether they are affected in a non-analytic way by the onset of incommensurate structures at T_I, as has been proposed in Ref. 3. Obviously, a roughening transition is no prerequisite for the wavyness of the boundaries (3), see the results of the mean field theory in Fig. 4b. - At any rate, the boundaries are quite diffuse at temperatures where incommensurability is expected to dominate.

To determine the features of the ANNNI model which are stable against small perturbations preserving the discrete Ising symmetry (the weakening of that restriction has been also analysed (27,28)) one may study the effect of axial interactions of longer range, non-axial interactions as well as different lattice structures. Two modest steps in these directions have been done: (i) The low temperature approaches can be transcribed to the hexagonal closed packed and body centred rectangular lattices with interactions up to geometrically third nearest neighbours yielding essentially unchanged results (29); (ii) adding axial third nearest neighbour interactions, J_3, to the ANNNI model may lead to a stabilization of the <3> phase at zero temperature, but does not affect the appearance of long period commensurate phases, branching processes, incommensurate structures, and so on (30), see Fig. 8. Similar features are

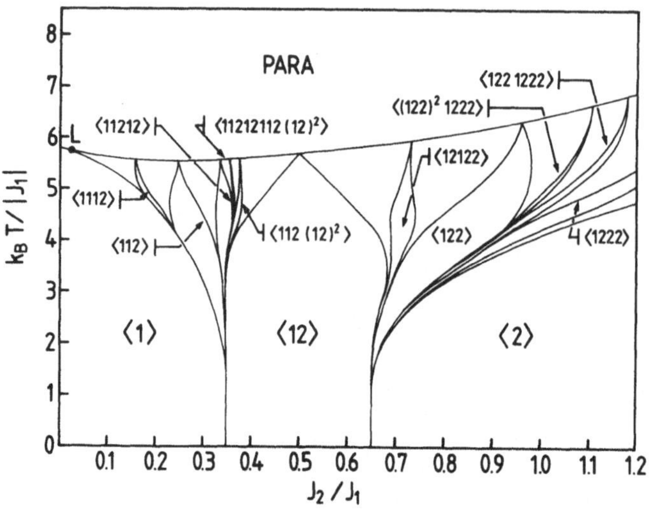

FIGURE 8. Mean field phase diagram of the axial Ising model with third-neighbour interactions for $J_0 = -J_1 > 0$ and $J_3/J_0 = 0.1$ showing some of the main commensurate phases. From (30).

also exhibited by the ANNNI model in a field (31), the chiral clock model (19,24) and the S = 1 variant of the ANNNI model (32). Hence the ANNNI model seems to display generic aspects of systems with discrete symmetry and effective short range competing interactions, and it is of interest to identify their relevance for real materials.

3. LONG PERIOD STRUCTURES IN ALLOYS

3.1. Experimental Results

The existence of long period structures in binary alloys is known for many years from diffraction data and, recently, from high resolution electron microscopy (3,33-41). Such structures usually occur in alloys of either AB or A_3B type (possibly off stoechiometry) with one-dimensional (1d) long period modulations along a single axis of the crystal or, in a few cases, with two-dimensional ones. Examples are, for the AB alloys CuAu (4,34), and, for the A_3B alloys Cu_3Pd (35,36), Cu_3Al (37), Ag_3Mg (38,39), Au_3Zn (40), and Al_3Ti (41).

In Fig. 9 the unit cells of the low temperature simply ordered phase of CuAu, CuAuI, and of its 1d superstructure, CuAuII, are shown. CuAuII is composed of two types of "antiphase domains" each of length Mb = 5b (b is the lattice constant in the direction of the modulation), separated by "antiphase domain boundaries", yielding a periodic $(5,\overline{5})$ (in the notation of Fujiwara

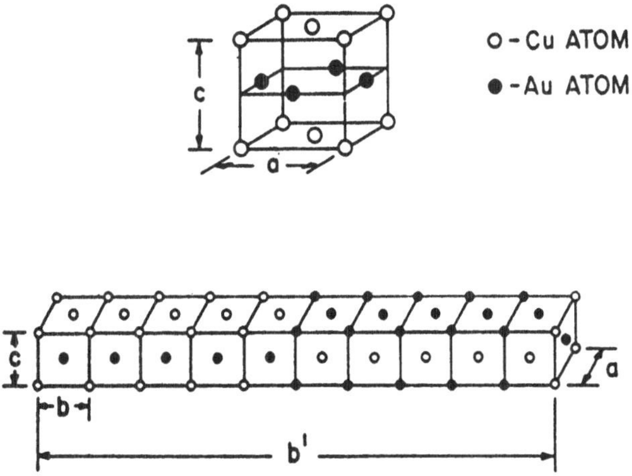

FIGURE 9. Unit cells of CuAuI (top) and CuAuII (bottom).

(42)) arrangement of period 10b for the superstructure. Identifying the unit cell of CuAuI with an "up" state of an Ising spin system and the one, building up the other type of domain, with a "down" state, CuAuII may be denoted as a <5> structure using the terminology of the previous section. The two unit cells may be transformed into each other by a translational shift perpendicular to the direction of the modulation.

In Fig. 10 two 1d superlattices for A_3B lattices based on the $L1_2$ structure are shown which may be denoted as <1> (or $(1,\bar{1})$ or DO_{22}) and <2> (or $(2,\bar{2})$ or DO_{23}) structures. As before there are two distinct unit cells forming the antiphase domains related to each other by a simple translation. Depending on the kind of translation vector the stoechiometry of the $A_{3-x}B_{1+x}$ alloys will remain unchanged ("conservative" antiphase domain boundary", as depicted in Fig. 10) or will be modified (non--conservative boundary, see (4) or (41)) by creating an antiphase domain boundary.

The average size of an antiphase domain is a dimensionless number (setting the lattice constant in the direction of the modulation equal to one), M. 2M is the mean period of the superstructure.

We will briefly describe typical features of phase diagrams of alloys with long-period superstructures, elaborating on specific systems. Of course, at sufficiently high temperatures the A- and B-sites of the lattice are occupied more or less randomly by the A- and B-atoms, and the alloy is disordered. On lowering the temperature a transition to the ordered phase takes place. Very interestingly, M is found to vary possibly both with temperature and composition, M = M(T,x). The variation may be continuous, for instance, as composition is changed in CuAuII (33), indicating incommensurate modulations. Lock-in phenomena to commensurate phases also do occur. At low temperatures there is often a tendency to order in a comparatively simple structure.

The, perhaps, most spectacular system is $Al_{3-x}Ti_{1+x}$. Results are depicted in Fig. 11 obtained from visual inspection of beautiful high resolution electron microscopy images (41) (other examples of such images will be presented at this workshop (43)). Each of the observed fifteen different commensurate superlattices is composed of a regular periodic array of anti-phase domains of length one or two (1-bands and 2-bands) based on the $L1_2$ structure, see Fig. 10.

Very long annealing times, up to more than a month, have been used to produce these equi-librium long-period superstructures. Insufficient care may easily result in quite irregular (or chaotic) arrays of anti-phase domains. From Fig. 11 it is readily seen that only a few comparatively simple commensurate structures lock in at low temperatures while at higher temperatures several quite complex configurations have been identified.

Another alloy which has been investigated carefully recently is $Cu_{3\pm x}Pd$ (35,36). The phase diagram, see Fig. 12, consists of a simply ordered ($L1_2$) phase (or <∞>, i.e. ferromagnetic phase in the analoguous spin system) at low concentrations of Pd, a 1d modulated phase at high concentrations and a narrow strip of

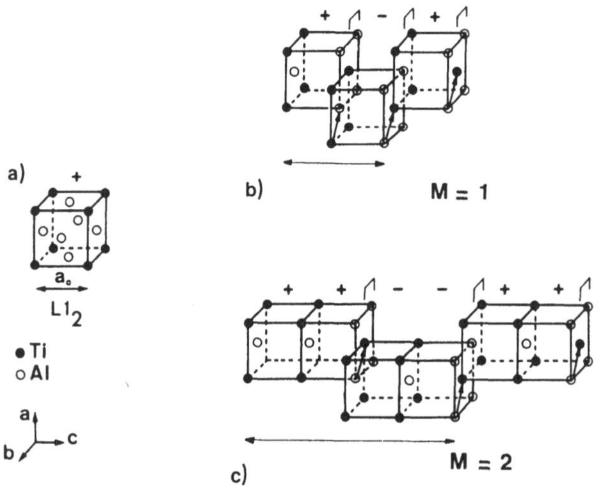

FIGURE 10. $L1_2$ unit cell (a) and <1> or DO_{22} (b) as well as <2> or DO_{23} superstructures of A_3B alloys on the fcc lattice. After (41).

2d modulated structures. The 1d modulated phase can be subdivided in two regions (35): (i) At fairly low concentrations and high temperatures the structures are seemingly incommensurate - with M varying continuously with temperature. The antiphase domain boundaries are diffuse or wavy, as in CuAuII. (ii) At higher concentrations commensurate long period superstructures lock in at simple values of M such as 9/2, 7/2, or 3 in the range of 23 to 30 at.% Pd over a wide range of temperatures. The antiphase boundaries are less diffuse and become even very sharp at the highest concentrations. - In addition to these equilibrium properties the effect of irridation due to high--energy electrons on modulated short and long range order has been studied in detail for $Cu_{3\pm x}Pd$ providing evidence for a "metastable Lifshitz point" (36).

Series of commensurate structures have been also observed in Ag_3Mg (38,39) and Au_3Zn (40) in close analogy to results of the ANNNI model. Both alloys exhibit the $\langle 2 \rangle$ (or DO_{23}) structure at low temperatures. In the case of Ag_3Mg, additional configurations of tpye $\langle 2^j 1 \rangle$ with j = 2,3,4,5,6, 7,8, and 12 as well as a few mixed structures like $\langle 2^2 1\ 2^3 1 \rangle$ (39) have been found. For Au_3Zn, the $\langle 2^3 3 \rangle$ and $\langle 3 \rangle$ configurations have been reported

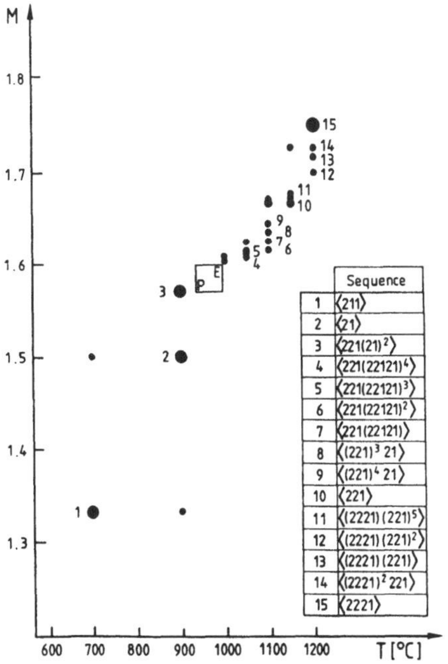

FIGURE 11. Period, M, as a function of the annealing temperature, T, in $Al_{3-x}Ti_{1+x}$. After (41).

(40). Detailed studies of the phase diagrams would be most desirable.

3.2. Theoretical Descriptions

The formation of long-period superstructures in alloys has been explained as being due to the "nesting" of the Fermi surface at its flat parts (4). In particular, several experimental results on the relation between the electron-atom ratio and the period, M, of the superstructure can be described convincingly. This classical theory of Sato and Toth has received additional back-up from first principle calculations of Gyorffy and Stocks (5) on Fermi surfaces in random metallic alloys. For example, the concentration dependence of M at low temperatures in the Cu_xPd_{1-x} system could be calculated in very good agreement with the experimental values.

Of course, a theory based solely on the energy of the electrons cannot explain the experimentally observed temperature dependence of, for instance, M. Therefore, Sato and Toth assumed an additional contribution to the free energy stemming from the antiphase domain boundaries (44). However, to my knowledge, so far no quantitative analysis has been performed including both contributions.

To describe order-disorder phenomena in alloys a commonly accepted procedure is to assume <u>effective</u> interactions between atoms; usually, only pair interactions of short range are considered (45,46). The Hamiltonian for a binary (A,B) alloy then reads

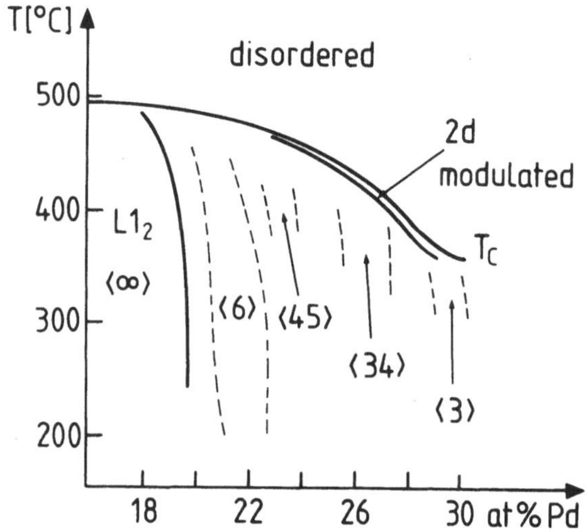

FIGURE 12. Tentative phase diagram of $Cu_{3\pm x}Pd$ based on experimental data of Broddin et al (45). The boundaries (solid and dashed lines) are approximate ones.

$$\mathcal{H} = \sum_{\alpha \neq \beta} [c_\alpha c_\beta v_{BB} (\underline{r_\alpha} - \underline{r_\beta}) + 2c_\alpha(1-c_\beta) v_{AB}(\underline{r_\alpha} - \underline{r_\beta}) + + (1-c_\alpha)(1-c_\beta) v_{AA}(\underline{r_\alpha} - \underline{r_\beta})] \quad (17)$$

where $c_\alpha = 1$, if site α of the lattice is occupied by a B-atom, $c_\alpha = 0$ otherwise. This Hamiltonian is isomorphic to one for an Ising model substituting the occupation variable, c_α, by a pseudo-spin, $S_\alpha = 1 - 2c_\alpha$. Using then standard statistical mechanical methods, like cluster-variation calculations or Monte Carlo simulations, phase diagrams can be mapped and compared to the experimental ones (47). - The effective pair interactions, v_{AA}, v_{AB}, and v_{BB}, may be obtained experimentally, for example, from neutron scattering data via an inverse Monte Carlo method (48). Quite often, the parameters may depend fairly strongly on temperature and concentration reflecting especially the relevance of multi-particle interactions. Following this line of analysis de Fontaine and Kulik (3) suggested to describe also long-period superstructures in alloys in the framework of Ising-type models with competing short range pair interactions. A prominent candidate to provide guidance in interpreting recent experiments has been the ANNNI model (3,35,36,40,43). Indeed, crucial features of the model and experimental findings discussed above show striking similarities: (i) Tendency to form comparatively simple structures at low temperatures and more complicated ones as temperature increases (entropy driven mechanism) (35,40,43). (ii) Lock-in phenomena to structures with regularly placed domain walls describable by the basic $<2^j3>$ or $<2^j1>$ phases or mixed phases which originate from a structure combination branching process (35,39,41). (iii) Strong temperature dependence of the period, M, with quite large jumps in M(T) at low temperatures and a devil's staircase-like behaviour at higher temperatures (41). (iv) Occurrence of sharp as well as wavy or diffuse antiphase domain boundary depending, in the ANNNI model, on temperature and/or strength of the in-plane coupling, J_o (35,36,39). Therefore it is, indeed, tempting to describe alloy systems by ANNNI-type models. Results of such attempts have been already reported in the literature. For example, it has been shown how an ANNNI-type model might be applied to an A_3B ordered structure on an fcc lattice (3). More concretely, Ag_3Mg has been analysed in that spirit and values of the pair interaction parameters have been estimated (39). Because the various structures observed in Ag_3Mg correspond to ones found in the straight ANNNI model (J_1, $J_2 < 0$), effective interactions reaching up to next nearest layers (which should not be confused with geometrically next nearest neighbours) seem to be sufficient in describing that alloy. (Note that elastic forces may also yield an effective ANNNI-Hamiltonian (33)). Similarly, effective interactions of only comparatively short range need to be included to mimic $Al_{3-x}Ti_{1+x}$ - possibly ones acting up to third nearest layers to stabilize also the <112> structure at low temperatures (30), see Figs. 8 and 11. - On the other hand, in Cu_3Pd domains consisting of 3, 4, and 5 unit cells have been observed (corresponding to <3>, <34> and <45> structures) already at low temperatures. Therefore one may have to consider effective interactions of somewhat longer range (35,36). The variation of M with composition is expected to reflect the composition dependence of the effective interactions (35,36,4,5). - Likewise, the <5> phase of the CuAu-alloy may be explained by invoking effective interactions up to axial fifth neighbors (3). However, it is also conceivable that the superstructure, CuAuII, is formed in a way similar to the one stabilizing the <5> phase on the ferromagnetic side (corresponding to CuAuI) of the phase diagram of the ANNNI model, see Fig. 6. Note that in CuAu the transition to the disordered high temperature phase seems to be of first order in contrast to the one in the model suggesting the importance of additional, but not necessarily longer range interactions.

To substantiate this analysis of long-period superstructures in alloys in the framework of ANNNI-type models further work has to be done. In particular, it would be of great interest to determine the effective pair interaction parameters experimentally using, for instance, the inverse Monte Carlo method. Very preliminary results for Cu_3Pd have been already obtained (49) suggesting fairly long range interactions (of course, it might be legitimate to truncate such interactions appropriately, assuming strongly damped ones) - in agreement with the above considerations. Furthermore, it might be feasible to estimate the effective couplings, J_1, J_2, ..., from computations of the electron energy, see (5), as has been done for SiC-polytypes confirming an ANNNI model-like mechanism for the formation of long-period structures in that case (50); see also the recent analysis on Ag_3Mg (39). One might even speculate whether one could attribute effective interactions to unit cells instead of to individual atoms to simplify the description.

Once realistic effective interaction parameters are available for specific alloy systems detailed quantitative comparison between theory and experiment should be possible. At any rate, the present state of the art may be considered as a very encouraging and promising one.

4. DISCUSSION AND CONCLUSION

In this contribution our present knowledge on the ANNNI model has been reviewed and possible applications to long-period superstructures in alloys have been discussed.

Despite its simplicity the 3d model has not been analysed completely and exactly (exact solutions have been achieved only for related decorated or mock models (51) as well as models on the Bethe lattice (52). Note that even the nearest neighbour Ising model ist still not amenable to a full analysis of, for example, its critical properties). Nevertheless, a coherent and consistent picture of the entire phase diagram can be given in mean field theory in qualitatively correct agreement with the true phase diagram as far as it is known. The model describes the thermally or entropy driven formation of long-period superstructures due to competing short range forces. While at low temperatures only distinct commensurate phases appear, either springing directly from a multiphase point or generated by a structure combination branching process, implying a staircase--like variation of the characteristic wavenumber, also incommensurate structures occur at higher temperatures with a continuous change in the wavenumber characterising the perodicity.

Recently very careful and beautiful experiments on A_3B-alloys (like Ag_3Mg, Al_3Ti, and Cu_3Pd) have been performed revealing features strikingly similar to the ones exhibited by the ANNNI model, in particular, the tendency towards simple structures at low temperatures, lock-in phenomena of phases with regularly spaced domains, strong temperature dependence of the period of the superstructure and occurrence of sharp as well as diffuse antiphase domain boundaries. Indeed, following the commonly accepted approach to describe order-disorder properties of alloys in the framework of Ising-type models with effective short range pair interactions ANNNI-like models have provided guidance in designing and interpreting these recent experiments (it may be worth mentioning that crucial features of the ANNNI model have been also observed in a variety of other real systems including magnets (2,53), ferroelectrics (20,27,54) and polytypes (2,50)).

The next step is to determine the effective interaction parameters for specific alloy systems either experimentally using, for instance, the inverse Monte Carlo method or theoretically from, e.g., computations of the electron energy (in that way one would combine the powerful theory of Sato and Toth (4) and modern concepts of statistical mechanics). Knowing the values of these parameters, detailed quantitative comparison between calculations on ANNNI-type models (with possibly various ranges of interaction) and experimental results is feasible. First work in that direction has already been done (39).

I thank my colleagues for cooperation and discussions on this topic, in particular Michael E. Fisher, Julia M. Yeomans, Phillip M. Duxbury and Didier de Fontaine.

REFERENCES

1. Fisher, M.E. and W. Selke, Phys. Rev. Lett. 44 (1980) 1502.
2. For previous reviews, see Bak, P., Rep, Prog. Phys. 45 (1982) 587; Fisher, M.E. and D.A. Huse, Melting, Localization, and Chaos. Kalia, R.K., Vashishta, P. (eds.) p. 259. New York: Elsevier 1982; Selke W., Modulated Structure Materials. Tsakalakos, T. (ed.) p. 23. Dordrecht, Boston, Lancaster: Martinus Nijhoff 1984; Yeomanns, J., preprint 1987.
3. de Fontaine, D. and J. Kulik, Acta Metall. 33 (1985) 145.

4. Sato, H. and R.S. Toth, Alloying Behavior in Concentrated Solid Solutions. Massalski, T.B. (ed.) p. 295. New York: Gordon and Breach 1965.
5. Gyorffy, B.L. and G.M. Stocks, Phys. Rev. Lett. 50 (1983) 374.
6. Elliott, R.J., Phys. Rev. 124 (1961) 346.
7. Stephenson, J., Can. J. Phys. 48 (1970) 1724.
8. Hornreich, R.M., R. Liebmann, H.G. Schuster and W. Selke, Z. Physik B 35 (1979) 91.
9. Selke W., Z. Physik B 43 (1981) 335; Selke, W. and M.E. Fisher, Z. Physik B 40 (1980) 71; Barber, M.N. and W. Selke, J. Phys. A 15 (1982) L 617.
10. Recent references include, Oitmaa, J., J. Phys. A 18 (1985) 365; Finel, A. and D. de Fontaine, J. Stat. Phys. 43 (1986) 645; Duxbury, P.M., J.M. Yeomans, and P.D. Beale, J. Phys. A 17 (1984) L 179; Saqi, M.A.S. and D.S. McKenzie, J. Phys. A 20 (1987) 471; Rujan, P., W. Selke, and G. Uimin, Z. Physik B 65 (1986) 235.
11. Selke, W., K. Binder and W. Kinzel, Surface Sci. 125 (1983) 74; Rujan, P., W. Selke and G. Uimin, Z. Physik B 53 (1983) 221.
12. Van Hove, M.A., W.H. Weinberg and C.M. Chan, Low-Energy Electron Diffraction. Heidelberg, Berlin, Tokyo, New York: Springer 1986; Zhadanov, V.P. and K.I. Zamarayev, Usp. Fiz. Nau. 149 (1986) 635.
13. Redner, S. and H.E. Stanley, Phys. Rev. B 16 (1977) 4901; Oitmaa, J., J. Phys. A 18 (1985) 365.
14. Fisher, M.E. and W. Selke, Philosophical Trans. R. Soc. 302 (1981) 1.
15. Selke, W. and M.E. Fisher, Phys. Rev. B 20 (1979) 257; J. Magn. Magn. Mat. 15-18 (1980) 403.
16. Selke, W., Z. Physik B 29 (1978) 133; Kaski, K. and W. Selke, Phys. Rev. B 31 (1985) 3128.
17. Hornreich, R.M., M. Luban and S. Shtrikman, Phys. Rev. Lett. 35 (1975) 1678.
18. Rasmussen, E.B. and S.J. Knak Jensen, Phys. Rev. B 24 (1981) 2744; Kawasaki, T., Prog. Theor. Phys. 71 (1984) 246.
19. Szpilka, A.M. and M.E. Fisher, Phys. Rev. Lett. 57 (1986) 1044; Fisher M.E. and A.M. Szpilka, Phys. Rev. B (1987); for a related domain wall analysis using a harmonic approximation to the mean field equations at low temperatures keeping correctly only terms of lowest order, W_2, see Villain, J. and M. Gordon, J. Phys. C 13 (1980) 3117.
20. Selke, W. und P.M. Duxbury, Z. Physik B 57 (1984) 49; Duxbury, P.M. and W. Selke, J. Phys. A 16 (1983) L 741.
21. Bak, P. and J. von Boehm, Phys. Rev. B 21 (1980) 5297; Jensen, M.H. and P. Bak, Phys. Rev. B 27 (1983) 6853.
22. Aubry, S., Soliton and Condensed Matter Physics. Bishop, A.R., Schneider, T. (eds.) p. 264 Berlin, Heidelberg, New York: Springer 1978; similar ground state analyses on related models include Chou, W., and R.B. Griffiths, Phys. Rev. B 34 (1986) 6219; Marchand, M., K. Hood and A. Caille, Phys. Rev. Lett. 58 (1987) 1660; Yokoi, C.S.O., L. Tang, and W. Chou, Preprint (1987).

23. Aharony, A. and P. Bak, Phys. Rev. B 23 (1981) 4770.
24. Siegert, M. and H.U. Everts, Z. Physik B 60 (1985) 265; Z. Physik B 66 (1987) 227 and references therein. For other Potts models with competing interactions, see Peschel, I. and T.T. Truong, J. Stat. Phys.45 (1986) 233.
25. For a review, see Weeks, J.D., Ordering in Strongly Fluctuating Condensed Matter Systems. Riste, T. (ed.) p. 293. New York: Plenum Press 1980.
26. Tentrup, T. and R. Siems, J. Phys. C 19 (1986) 3443. Domain wall properties of the ANNNI model are also considered in Widom, B., K.A. Dawson and M.D. Lipkin, Physica A 140 (1986) 26.
27. For a review, see Janssen, T., Microscopic Theories of Incommensurate Crystal Phases. Blinc, R., Levanyuk, A.P. (eds.) p. 67. New York: Elsevier 1986.
28. Benkert, C., V. Heine and E.H. Simmons, Europhysics Lett. 3 (1987) 833.
29. Selke W., J. Phys. C 14 (1981) L 17.
30. Selke W., M. Barreto and J. Yeomans, J. Phys. C 18 (1985) L 393; Barreto, M. and J. Yeomans, Physica A 134 (1985) 84; see also Randa, J., Phys. Rev. B 32 (1985) 413.
31. Yokoi, C.S.O., M.D. Coutinho-Filho and S.R. Salinas, Phys. Rev. B 24 (1981) 4047; Smith, J. and J.M. Yeomans, J. Phys. C16 (1983) 5305; Pokrovsky, V.L. and G.V. Uimin, Sov. Phys. Zh.E.T.F. 82 (1982) 1640; Szpilka, A.M., J. Phys. C 18 (1985) 569.
32. Jensen, P.J., K.A. Penson and K.H. Bennemann, Phys. Rev. B 35 (1987) 7306.
33. De Simone, T. and R.M. Stratt, Phys. Rev. B 32 (1985) 1537; De Simone, T., R.M. Stratt and J. Tobochnik, Phys. Rev. B 32 (1985) 1549.
34. Guymont, M., R. Portier and D. Gratias, Acta Cryst. A 36 (1980) 792.
35. Broddin, D., G. Van Tendeloo, J. Van Landuyt, S. Amelinckx, R. Portier, M. Guymont and A. Loiseau, Phil. Mag. A 54 (1986) 395.
36. Takeda, S., J. Kulik and D. de Fontaine, Preprint 1986; de Fontaine, D., A. Finel, S. Takeda and J. Kulik, Noble Metal Alloys. Massalski, T.B., Pearson, W.B., Benett, L.H. and Change, Y.A. (eds.) p. 49 (1986).
37. De Graef, M., D. Broddin, J. Van Humbeeck and L. Delaey, Proc. XIth Int. Congr. on Electron Microscopy, Kyoto (1986); De Graef, M. and D. Broddin, private communication.
38. Portier, R., D. Gratias, M. Guymont and W.M. Stobbs, Acta Cryst. A 36 (1980) 190. Fujino, Y., H. Sato, M. Hirabayashi, E. Aoyagi and Y. Koyama, Phys. Rev. Lett. 58 (1987) 1012.
39. Kulik, J., S. Takeda and D. de Fontaine, Acta Metall. 35 (1987) 1137.
40. Van Tendeloo, G. and S. Amelinckx, Phys. Stat. Sol. A 43 (1977) 553.
41. Loiseau, A., G. Van Tendeloo, R. Portier and F. Ducastelle, J. de Physique 46 (1985) 595; Loiseau, A., Ph.D. thesis, O.N.E.R.A. (1985).
42. Fujiwara, K., J. Phys. Soc. Japan 12 (1957) 7.
43. Van Tendeloo, G., these proceedings.

44. For example, Cahn, J.W. and R. Kikuchi, J.Phys. Chem. Solids 20 (1961) 94; Kikuchi, R., and Cahn, J.W., J. Phys. Chem. Solids 23 (1962) 137; Inden, G., S. Bruns and H. Ackermann, Phil. Mag. A 53 (1986) 87.
45. Bieber, A. and F. Gautier, Z. Physik B 57 (1984) 335; Acta Metall. 34 (1986) 2291.
46. Finel, A. and F. Ducastelle, Phase Transformations in Solids. Tsakalakos, T., (ed.) p. 293. New York: North Holland 1984; Sanchez, J.M. and D. de Fontaine, Structure and Bonding, Vol. II, p. 117, New York: Academic Press 1981; Ducastelle, F., Rech. Aerosp. 4 (1986) 11.
47. Binder, K., Festkörperprobleme XXVI (1986) 133.
48. Gerold, V. and J. Kern, Acta Metall. 35 (1987) 393; Maysenhölder, W., Phys. Stat. Sol. B 139 (1987) 399.
49. Schweika, W. and H.G. Haubold, Preprint (1987); Schweika, W., Ph.D. thesis, RWTH Aachen (1985)
50. Cheng, C., R.J. Needs, V. Heine and N. Churcher, Europhysics Lett. 3 (1987) 475.
51. Huse, D.A., J.M. Yeomans and M.E. Fisher, Phys. Rev. B 23 (1981) 180; Selke, W. and F.Y. Wu, J. Phys. A 20 (1987) 703.
52. Yokoi, C.S.O., M.J. de Oliveira, and S.R. Salinas, Phys. Rev. Lett. 54 (1985) 163; Vannimenus, J., Z. Physik B 43 (1981) 141; da Silva, C.R. and S. Coutinho, Phys. Rev. B 34 (1986) 7975; Mariz, A.M., C. Tsallis, and E.L. Albuquerque, J. Stat. Phys. 40 (1985) 5777; Inawashiro, S., C.J. Thompson and G. Honda, J. Stat. Phys. 33 (1983) 419.
53. Rossat-Mignod, J., P. Burlet, J. Villain, H. Bartholin, W. Tscheng-Si and D. Florence, Phys. Rev. B16 (1977) 440; Fischer P., B. Lebech, G. Meier, B.D. Rainford and O. Vogt, J. Phys. C 11 (1978) 345; Hälg, B. and A. Furrer, Phys. Rev. B 34 (1986) 6258.
54. Müller, K.A., W. Berlinger, J.Y. Buzare and J.C. Fayet, Phys. Rev. B 21 (1980) 1763; Currat, R., Multicritical Phenomena. Pynn, R., Skjeltrop, A. (eds.) p. 177. New York: Plenum Press.

MECHANISMS FOR THE DECAY OF UNSTABLE AND METASTABLE PHASES: SPINODAL DECOMPOSITION, NUCLEATION AND LATE-STAGE COARSENING

Kurt Binder

Institut für Physik, Johannes Gutenberg-Universität
D-6500 Mainz, Postfach 3980, W-Germany

ABSTRACT

The basic concepts on the kinetics of phase separation in alloys are introduced, and the current status of the theory is briefly reviewed. Particular emphasis is given to questions such as the conditions under which the linearized theory of spinodal decomposition is valid, the significance of spinodal curves, the possible description of coarsening in terms of power laws and structure-factor scaling, and non-equilibrium percolation phenomena.

I. INTRODUCTION

We consider a solid binary (AB) mixture whose temperature (T)-concentration (c) phase diagram has a miscibility gap (Fig. 1): The coexistence curve (binodal) described by two branches $c^{(1)}_{coex}(T)$, $c^{(2)}_{coex}(T)$ which merge at the critical point ($T = T_c$, $c = c_{crit}$), separates the one phase region from the two phase coexistence region. In thermal equilibrium, the system is macroscopically homogeneous in the one phase region, but macroscopically inhomogeneous in the two phase region: there a fraction of $[c - c^{(1)}_{coex}(T)]/[c^{(2)}_{coex}(T) - c^{(1)}_{coex}(T)]$ of the volume is in domains of composition $c^{(2)}_{coex}$ and the remaining fraction $[c^{(2)}_{coex}(T) - c]/[c^{(2)}_{coex}(T) - c^{(1)}_{coex}(T)]$ of the volume has the composition $c^{(1)}_{coex}$ ("lever rule"). We now focus interest on quenching processes where the system is cooled from an initial state at temperature T_0 in the one phase region to a final state at temperature T in the two phase region: then some microscopic concentration inhomogeneities will become magnified as the time t after the quench proceeds and thus the system gradually evolves towards the phase-separated equilibrium state.

Following ideas of van der Waals and Cahn and Hilliard[1-4], it is common to divide the two phase region into a metastable region, in between the coexistence curve and the so-called "spinodal curve" $c_{sp}(T)$ [i.e., for $c^{(1)}_{coex} < c < c^{(1)}_{sp}$ and for $c^{(2)}_{sp}(T) < c < c^{(2)}_{coex}(T)$ the macroscopically homogeneous one phase state is "metastable"], and an unstable region, inside the spinodal [for $c^{(1)}_{sp}(T) < c < c^{(2)}_{sp}(T)$]. The branches of the spinodal should merge with the coexistence curve at the critical point.

G. M. Stocks and A. Gonis (eds.), Alloy Phase Stability, 233–262.

It now is widely believed that this distinction has its counterpart in the early stage dynamics of phase separation: the decay of metastable states is assumed to start by the formation of localized large-amplitude "heterophase" fluctuations ("droplet formation", "nucleation") while unstable states are assumed to decay by the growth of weak delocalized long-wavelength fluctuations ("spinodal decomposition").

As will be emphasized below, this sharp distinction between different kinds of concentration fluctuations is based on a mean field-type theory, and cannot really be maintained: rather the transition from nucleation to spinodal decomposition is gradual: in this sense, the spinodal is smeared out over a nonzero concentration interval δc [and moreover the center of this interval typically is somewhat shifted away from the mean-field prediction for the spinodal towards the coexistence curve, as shown qualitatively in Fig. 1].

A related view on this subject is that the signature of spinodal decomposition is the formation of an interconnected ("percolating") structure, while the signature of nucleation is the formation of isolated droplets: hence the spinodal curve is interpreted as a kind of "percolation transition".

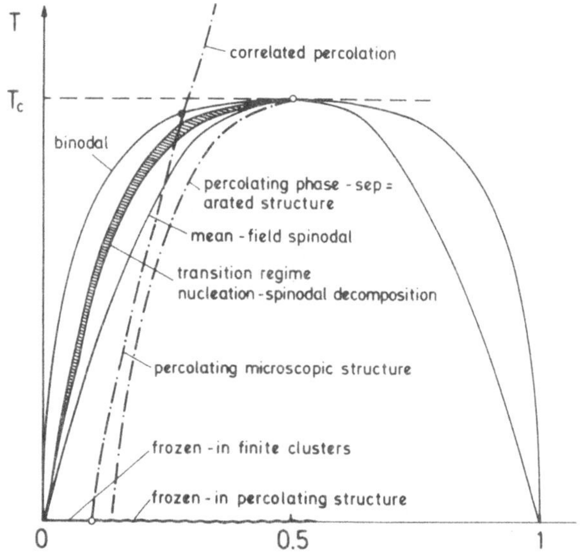

FIGURE 1. Schematic phase diagram of a binary mixture, assuming symmetry around $c_{crit} = 0.5$ for simplicity. Full curves indicate the coexistence curve (binodal) and (mean-field) spinodal, the shaded area denotes the regime where a gradual transition from nucleation to spinodal decomposition occurs. Dash-dotted curves describe various percolation transitions, as described in the text. From Hayward et al.[5]

Again we shall have to point out that also this view is partially misleading: although percolation phenomena in phase separation indeed can be identified[5-7], there are in fact several percolation lines in the phase diagram (Fig. 1) and not just a single one. Moreover, at low temperatures there is a regime where the growth indeed proceeds by well-separated droplets, but there is no nucleation free energy barrier which needs to be overcome to form these droplets: thus this process is not described as usual nucleation.

What is the problem which has led to all this confusion? Well, when one talks about fluctuations and the formation of structure, one needs to specify with which spatial resolution one is looking at the system! On the atomic scale, the concentration is rapidly varying from lattice site to lattice site ($c_i = 1$ if site i is taken by a B-atom while $c_i = 0$ if it is taken by an A-atom), and hence "weak delocalized long wavelength fluctuations" are not readily identified on that scale. To identify them, a coarse-graining must be possible such that the $\{c_i's\}$ are averaged over a d-dimensional coarse-graining volume L^d, where the length L is much larger than the lattice spacing a and much smaller than the scale set by the structures dominating the phase separation process [e.g., by[3] the critical wavelength λ_c]. The questions then arise: are fluctuations on this intermediate scale weak or strong? Well separated or interconnected? Etc.

The question of which spatial scale one shall choose to study this problem crucially depends on both temperature and the time stage of the process. Near the critical point, the scale for L is set by the correlation length ξ which diverges at T_c, $c = c_{crit}$, and hence near T_c a continuum theory approach makes sense. At low temperatures, however, both ξ and λ_c are not very different from the lattice spacing and the continuum theory can at best hold qualitatively. Percolation, in turn, can be studied on every length scale: on the atomic scale, one may define a "cluster" of B-atoms by the condition that every B-atom must have at least one nearest neighbor B-atom also belonging to the same cluster; at the percolation threshold an infinite cluster, spanning from one boundary of a macroscopic system to its opposite edge, for the first time appears. In the phase diagram, this line marking the onset of a percolating "microscopic" (= atomic) structure starts out at about $c \cong 0.16$ (at the simple cubic lattice[5]) and hits the binodal at about $c \cong 0.22$ ($T \approx 0.96\ T_c$)[8]: even in the macroscopically homogeneous one phase region, this line continues (the so-called "correlated percolation" in the Ising model[9]) und ends for $T \longrightarrow \infty$ at the random site percolation concentration c_p (≈ 0.31 for the simple cubic lattice[10]). On the other extreme, we may ask whether the macroscopic structure formed for $t \longrightarrow \infty$ from phase separation consists of well separated "islands" (of composition $c^{(2)}_{coex}$ in a "sea" of composition $c^{(1)}_{coex}$) or forms an interconnected "continent". Experimental techniques which involve large characteristic lengths, such as light microscopy, will yield information on the latter question only {via a difference in refractive index proportional to $c^{(2)}_{coex} - c^{(1)}_{coex}$}. With the scanning tunneling microscope, on the other hand, individual atoms can be seen. So the questions which one may like to ask can also be linked to the experimental tools one wishes to apply.

In Sec. II we now recall the simple predictions of the Cahn-Hilliard linearized theory of spinodal decomposition and discuss their limitations (including effects due to finite cooling rate, coupling of the concentration fluctuations to other slow variables, etc.). Sec. III discusses nucleation, including the behavior close to the spinodal curve and critically examining the significance of the latter. Sec. IV is devoted to the later stages of the growth of the phase separated structures, summarizing recent ideas about coarsening and paying attention to the percolation effects mentioned above. Sec. V contains our conclusions.

II. SPINODAL DECOMPOSITION

As mentioned above, we obtain a continuum description by coarse-graining the concentration variables $c_i(t)$, defining $c(\vec{x},t) \equiv L^{-d} \sum_{i \in L^d} c_i(t)$, $\vec{x}$ being the center of gravity of the cell L^d. Since in the total volume V the average concentration $\bar{c} = V^{-1} \int d\vec{x}\, c(\vec{x},t)$ is conserved, the concentration field $c(\vec{x},t)$ satisfies a continuity equation, $\partial c(\vec{x},t)/\partial t + \nabla \cdot \vec{j}(\vec{x},t) = 0$. This equation is exact but one needs the concentration current density $\vec{j}(\vec{x},t)$; the standard assumption expresses $\vec{j}$ in terms of a mobility M(T) and the local chemical potential difference $\mu(\vec{x},t)$ as $\vec{j}(\vec{x},t) = -M(T)\ \nabla\mu(\vec{x},t)$. Since in equilibrium μ is given by a partial derivative of the free energy F(c,T) as $\mu = (\partial F/\partial c)_T$, we generalize this relation to an inhomogeneous situation far from equilibrium, defining a free energy functional

$$\frac{1}{k_B T} \Delta\mathcal{F} \equiv \int d^d x \left\{ f[c(\vec{x})]/k_B T + \frac{1}{2} r^2 [\nabla c(\vec{x})]^2 \right\}, \tag{1}$$

where f is the coarse-grained free energy density, and r turns out to be related to the interaction range of the Hamiltonian coupling the $\{c_i\}$ $\Big[$e.g., if $\mathcal{H} = \frac{1}{2} \sum_{i \neq j} \Big\{ \varphi_{BB}(\vec{r}_i - \vec{r}_j)\, c_i c_j + \varphi_{AB}(\vec{r}_i - \vec{r}_j) \big[c_i(1-c_j) + c_j(1-c_i) \big] + \varphi_{AA}(1-c_i)(1-c_j) \Big\}$, r is defined as $r^2 = \sum_j (\vec{r}_i - \vec{r}_j)^2 (\varphi_{AA} + \varphi_{BB} - 2\varphi_{AB}) / \{2d \sum_j (\varphi_{AA} + \varphi_{BB} - 2\varphi_{AB})\}\Big]$. Notice that in reality <u>f will depend on L</u> and is hard to obtain explicitly in practice; therefore, often f is identified with the mean field result for the free energy of a mixture

$$\frac{1}{k_B T} f(c) = c \ln c + (1-c) \ln (1-c) + 2 \frac{T_c^{MF}}{T} c(1-c),$$

$$k_B T_c^{MF} \equiv \sum_j (\varphi_{AA} + \varphi_{BB} - 2\varphi_{AB})/4. \tag{2a}$$

or its Landau expansion (f_0, A,B being suitable coefficients)

$$f(c) = f_0 + A(c-c_{crit})^2 + B(c-c_{crit})^4 + \ldots,\ A \quad \alpha \quad \frac{T}{T_c^{MF}} - 1 < 0,\ B > 0. \tag{2b}$$

From Eq.(1) $\mu(\vec{x})$ follows as a functional derivative,

$$\mu(\vec{x}) \equiv \delta(\Delta\mathcal{F})/\delta c(\vec{x}) = (\partial f/\partial c)_T - r^2\, k_B T\, \nabla^2 c(\vec{x}); \tag{3}$$

using this in the continuity equation yields the Cahn-Hilliard equation[1]

$$\partial c(\vec{x},t)/\partial t = M(T)\nabla^2 \left[\left[\partial f\left[c(\vec{x},t)\right]/\partial c\right]_T - r^2 k_B T \nabla^2\, c(\vec{x},t)\right]. \tag{4}$$

We now assume that in the initial stages of unmixing the fluctuation $\delta c(\vec{x},t) \equiv c(\vec{x},t) - \bar{c}$ is everywhere small in the system (since L can not be made arbitrarily large, this assumption typically is not true, as we shall see below, but it is nevertheless instructive to study it!). Then Eq.(4) becomes

$$\partial\, \delta c(\vec{x},t)/\partial t = M(T)\nabla^2 \left[(\partial^2 f(c)/\partial c^2)\Big|_{T,\bar{c}} - r^2 k_B T\, \nabla^2\right] \delta c(\vec{x},t), \tag{5}$$

and introducing fourier transforms $\delta c_{\vec{k}}(t) \equiv \int d^d x \exp(i\vec{k}\cdot\vec{x})\ \delta c(\vec{x},t)$ Eq.(5) is solved by

$$\delta c_{\vec{k}}(t) = \delta c_{\vec{k}}(0)\exp[R(\vec{k})t],\ R(\vec{k}) \equiv -M(T)k^2[(\partial^2 f/\partial c^2)_{T,\bar{c}} + r^2 k_B T\, k^2]. \tag{6}$$

The equal-time structure factor $S(\vec{k},t)$ at time t after the quench then shows exponential growth for $0 < k < k_c$, $k_c \equiv 2\Pi/\lambda_c = [-(\partial^2 f/\partial c^2)_{T,\bar{c}}/(r^2 k_B T)]^{1/2}$,

$$S(\vec{k},t) \equiv \langle\, \delta c_{-\vec{k}}(t)\delta c_{\vec{k}}(t)\, \rangle_T = S_{T_0}(\vec{k}) \exp[2R(\vec{k})t], \tag{7}$$

with $S_{T_0}(\vec{k}) \equiv \langle\, \delta c_{-\vec{k}}\ \delta c_{\vec{k}}\, \rangle_{T_0} = \langle\, \delta c_{-\vec{k}}(0)\ \delta c_{\vec{k}}(0)\, \rangle_T$ being the structure factor in equilibrium at temperature T_0, since we consider an infinitely rapid quench from T_0 to T at t = 0. Note that $2R(\vec{k})/k^2$ plotted vs. k^2 ["Cahn plot"] should be linear $2R(k)/k^2 = -D_0(1-k^2/k_c^2)$, with a negative diffusion constant $D_0 \equiv 2M(T)(\partial^2 f/\partial c^2)_{T,\bar{c}}$ ["uphill diffusion"]. In addition, at $k = k_c$ the structure factor is time-independent $S(\vec{k}_c, t) = S(\vec{k}, 0)$, so that curves $S(\vec{k}, t)$ recorded at different times t should intersect in a common point.

In reality, however, such a behavior is hardly ever seen: typically the Cahn plot is distinctly curved, and there is no common intersection point. There are several possible reasons why the simple linearized Cahn[1] theory of spinodal decomposition, Eqs.(1)-(7), is invalid:

(i) Fluctuations in the final state at the temperature T must be included[11].

(ii) The concentration field $c(\vec{x}, t)$ is coupled to another slowly relaxing variable[12].

(iii) There is already appreciable relaxation of the structure factor occurring during the quench from T_o to T, if the cooling rate $g = -dT/dt$ is finite[13,14].

(iv) Nonlinear effects are important already during the early stages of the quench[2-4,15-18].

In the following, we shall consider all these points briefly. Fluctuations in the final state are usually represented by a random force $\eta_T(\vec{x}, t)$, so Eq.(4) is replaced by[11]

$$\frac{\partial c(\vec{x},t)}{\partial t} = M(T)\nabla^2[(\partial f(c(\vec{x},t))/\partial c)_T - r^2 k_B T \nabla^2 c(\vec{x},t)] + \eta_T(\vec{x},t); \quad (8)$$

one assumes $\eta_T(\vec{x}, t)$ to be delta-correlated gaussian noise, and the mean-square amplitude $\langle \eta^2 \rangle_T$ is then linked to M by a fluctuation-dissipation relation

$$\langle \eta(\vec{x},t)\eta(\vec{x}',t')\rangle_T = \langle \eta^2\rangle_T \nabla^2 \delta(\vec{x}-\vec{x}')\delta(t-t'), \quad \langle \eta^2\rangle_T = 2k_B T\, M(T). \quad (9)$$

In the framework of the same linearization approximation that led from Eq.(4) to Eq.(5), one then obtains

$$\frac{d}{dt} S(\vec{k},t) = -2M(T)k^2\left\{\left[(\partial^2 f/\partial c^2)_T \Big|_{\bar{c}} + r^2 k_B T k^2\right] S(\vec{k},t) - k_B T\right\}. \quad (10)$$

It is seen that an effective diffusion constant for the "uphill diffusion" defined as

$$D_{eff}(\vec{k},t) \equiv (1/k^2)\, d[\ln S(\vec{k},t)]/dt \quad (11)$$

now yields

$$D_{eff}(\vec{k},t) = -2M(T)\ (\partial^2 f/\partial c^2)_T \Big|_{\bar{c}}\ (1-k^2/k_c^2) + 2M(T)k_BT[S(\vec{k},t)]^{-1}. \tag{12}$$

Since Eq.(10) is integrated as $\{S_T(\vec{k}) = k_BT/[(\partial^2 f/\partial c^2)_T \big|_{\bar{c}} + r^2T\ k^2]$ = "virtual structure factor"}

$$S(\vec{k},t) = S_{T_0}(\vec{k})\exp\ [2\ R(\vec{k})t] + S_T(\vec{k})\ \{1 - \exp\ [2\ R(\vec{k})t\}, \tag{13}$$

Eq.(12) would reduce to the Cahn result formally for $t \longrightarrow \infty$, but of course in this limit the theory is never applicable, due to nonlinear effects. At $t = 0$, on the other hand, Eq.(12) leads to a linear relation between $D_{eff}(\vec{k},0)$ and k^2 again, because $[S(\vec{k},0)]^{-1} = [(\partial^2 f/\partial c^2)_{T_0} \big|_{\bar{c}} + r^2T_0k^2]/k_BT_0$ is linear in k^2 also; the point where $D_{eff}(\vec{k},0)$ changes sign is not k_c, however, but shifted to a larger value. These points are illustrated in Fig. 2, where we compare the scaled structure factor $\tilde{S}(q,\tau)$ and scaled effective diffusion constant $\tilde{D}_{eff}(q,\tau)$ following from Eqs.(12,13) to corresponding results obtained[14] in the framework of the Langer-Baron-Miller[15,16](LBM)-approximation, which accounts for nonlinear effects approximately $\{q = k/k_c,\ \tau = 2M(T_f)r^2T_fk_c^4\ t,\ \tilde{S} = r^2k_c^2\ S\}$. It is seen that already on the level of the CHC approximation there is initially some shift in the position where $\tilde{S}(q,\tau)$ has its maximum. The main distinction between the CHC and LBM approximations during the early stages is the lack of a common intersection point in the LBM approximation, and also the growth of $\tilde{S}(q,\tau)$ with τ is generally slower (note the difference in ordinate scale!). Qualitatively, however, the behavior is rather similar, and so is the behavior of $\tilde{D}_{eff}$, which (apart from $\tau \longrightarrow 0$) is distinctly curved in both cases.

Now we consider briefly the <u>coupling of the concentration to a slowly relaxing variable</u>. According to Eq.(6), the maximum amplification rate $R_m = R(k_m)$ occurs for $k_m = k_c/\sqrt{2}$. Suppose now the concentration couples to a non-conserved variable a(t), whose fluctuations in the absence of any coupling would decay exponentially, proportional to $\exp(-\gamma t)$. The decay of concentration fluctuations will be affected if (i) the rates R_m,γ are of the same order, and (ii) if the coupling between the variables a(t), $c(\vec{x},t)$ is sufficiently strong. Measuring the strength of this coupling in terms of a parameter $1-D_\infty/D_0$ (see Ref. 12 for a definition of D_∞), a simple model calculation[12] on the Cahn-Hilliard level yields the mode spectrum in Fig. 3. It is seen that the coupling between the modes has a strongest effect near their intersection point where the degeneracy is lifted and mode mixing occurs, similar to other coupled-mode problems in condensed matter

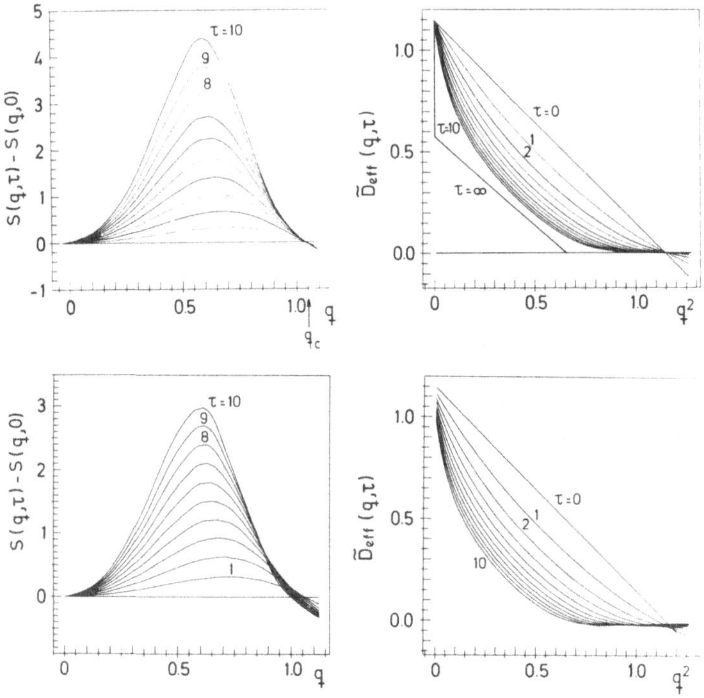

FIGURE 2. Scaled structure funtion (left) and diffusion coefficient $\tilde{D}_{eff}$ (right) plotted vs. q or q^2, respectively, for an instantaneous quench from infinite temperature to $T_f/T_c = 4/9$. Upper part is the Cahn-Hilliard-Cook (CHC) approximation, lower part the LBM approximation. 10 times τ are shown; $\tilde{D}_{eff}(q,\tau) \equiv q^2 d[\ln \tilde{S}(q,\tau)]/d\tau$. Note that in the CHC approximation we use the renormalized value μ_- (Fig. 7b) of $(\partial^2 f/\partial c^2)|_{T,\bar{c}}$ instead of $\partial^2 f/\partial c^2)|_{T,\bar{c}}$ itself, $\mu_-(T,\tau = 0) = 0.65$ in our units. From Carmesin et al.[14]

physics. Plotting now the unmixing mode" $R(k) = \Gamma^-(k)$ in the form of a Cahn plot, $R(k)/|D_0|k^2$, one again encounters pronounced curvature (Fig. 4). Such a coupling to slow variables might occur if one studies spinodal decomposition in glasses (e.g.[19]) or in fluid polymer mixtures near their glass transition (e.g.[20]). Of course, one then expects the slow variables to relax with a broad spectrum of rates rather than with a single rate γ; in addition, fluctuations in the final state need to be included in this case also: a theory thus combining the effects shown in Figs. 2 and 4 is just being developed[21].

Now we consider effects due to finite quench rates. If the mobility follows an Arrhenius law $\{M(T) = M(T_f) \exp(-E_{act}/k_BT)/\exp(-E_{act}/k_BT_f)$ with $E_{act} \gg k_BT_f\}$, most of the relaxation may in fact already occur in the one phase region even before the spinodal region is crossed[14]. But even if in the region between T_o and T the temperature dependence of $M(T)$ can be neglected (this is reasonable for fluid mixtures; note that then as well as in polymer mixtures one has a lower consolute critical point, the coexistence curve in the T-c plane being inverted in comparison with Fig. 1), one has drastic effects. As an example, Fig. 5 considers the same situation as in Fig. 2 but now the quench is performed in several steps:

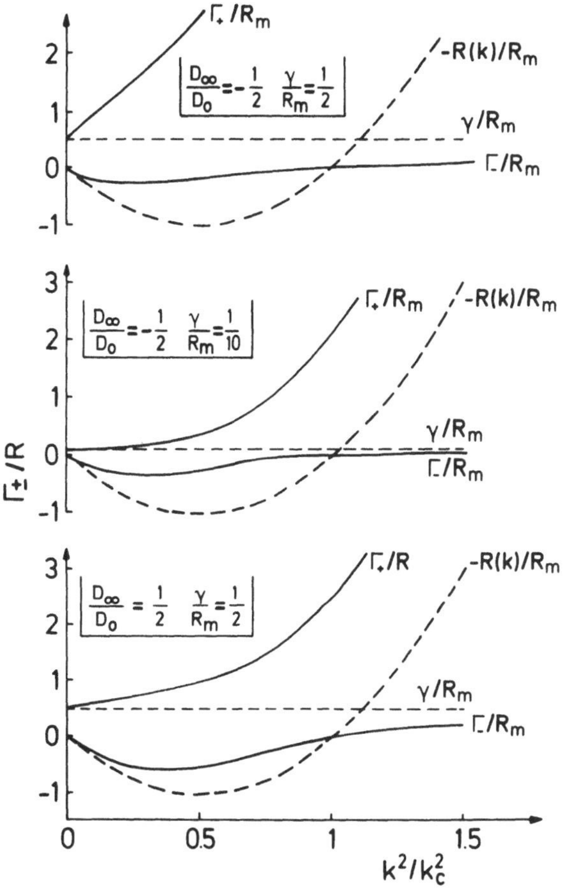

FIGURE 3. Mode spectrum $\{\Gamma_+(k), \Gamma_-(k)\}$ of an unmixing system {which in the absence of any coupling would relax with the rate $R(\vec{k})$, broken curve} coupled to a slow variable {which in the absence of any coupling would relax with the rate γ, broken straight line} plotted vs. k^2/k_c^2, for three parameter choices. All rates are normalized by R_m. Full curves represent the coupled modes. From Binder et al.[12]

first the system is cooled instantaneously from infinite temperature to $T_1/T_c = 0.75185$, at $\tau = 1$ to $T_2/T_c = 0.67667$, at $\tau = 2$ to $T_3/T_c = 0.60148$, at $\tau = 3$ to $T_4/T_c = 0.5263$, and at $\tau = 4$ to $T/T_c = 4/9$, at $\tau = 3$ to $T_4/T_c = 0.5263$, and at $\tau = 4$ to $T/T_c = 4/9$, where the system is kept. It is seen that now also in the CHC approximation there is no longer a unique intersection point in the $S(q,\tau)$ curves; apart from $\tau = 0$, $\tilde{D}_{eff}(q,\tau)$ depends now rather weakly on time for both models, but is again strongly curved. Note that in the LBM approximation one can no longer see any shift of the maximum position of $\tilde{S}(q,\tau)$ for the times shown: this happens because first (at the intermediate temperature steps) smaller q get more amplified and later on (at the final temperature) amplification occurs at larger q values than initially, and this behavior just happens to offset the coarsening tendency. This example shows that one has to be very careful to draw any conclusions on the validity of the CHC approximation from experimental data: the latter always will be affected to some extent by fluctuations in the final state, finite quench rate effects, and nonlinearities - these effects come in altogether, and are hard to disentangle in practice.

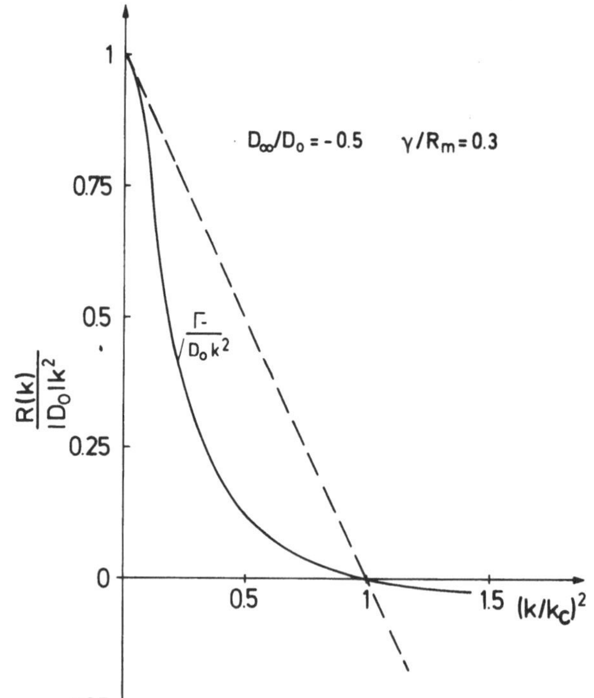

FIGURE 4. Cahn plot, $\Gamma_-(k)/|D_0k^2|$ plotted vs. k^2/k_c^2, for the mode-coupling model of Fig. 3, and a typical choice of parameters. Broken straight line indicates normal straight line in the absence of any coupling to the slow variable, in the Cahn[1] approximation. From Binder et al[12].

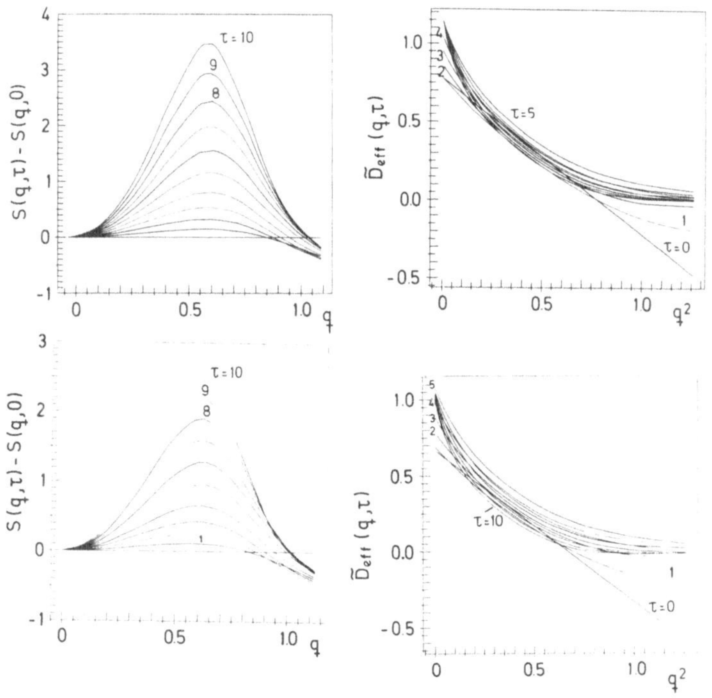

FIGURE 5. Scaled structure function (left) and diffusion coefficient $\tilde{D}_{eff}$ (right plotted vs. q or q^2, respectively, for the stepwise quenching procedure described in the text. Upper part is the CHC approximation, lower part the LBM approximation. 10 times τ are shown. From Carmesin et al.[14].

We conclude this section by a discussion of nonlinear effects. In 1975, computer simulations of spinodal decomposition in the nearest neighbor Ising model[17] as well as the LBM theory[15,16] indicated that nonlinear effects are very important already immediately after the quench, and the linearized CHC theory[1,11] is invalid even during the earliest stages of unmixing. Since all these treatments[15-17] use parameters appropriate to the Ising model, it would also be interesting to have simulations directly of the model Eq.(1), which is the starting point of the theory. Such a direct simulation of a lattice version of the Cahn-Hilliard model [Eq.(1) together with (2b)] has recently been given[18], see Fig. 6. There a square lattice was treated, where the concentration variable at each site i is in the interval $-\infty < c_i < +\infty$, the Hamiltonian being

$$\mathcal{H}/k_BT = \sum_i [A(c_i-c_{crit})^2 + B(c_i-c_{crit})^4] + \sum_{\langle ij\rangle} \frac{1}{2} C(c_i-c_j)^2. \quad (14)$$

Eq.(14) reduces to Eq.(1) if one expands $c_j = c_i + (\vec{x}_j-\vec{x}_i)\cdot\nabla c_i$ and identifies $\frac{C}{4} \sum_{j(\neq i)} [(\vec{x}_j-\vec{x}_i)\cdot\nabla c_i]^2 \equiv \frac{1}{2} r^2(\nabla c_i)^2$ and replaces $\sum_i$ by an integral $\int d\vec{x}$. Thus the lattice in Eq.(14) has nothing to do with the original lattice of the solid (if one considers crystals) rather one thinks of dividing the system up into a lattice of coarse-graining cells, with $L \sim r$ (assuming that r is much larger than interatomic spacings). After rescaling the model putting $\tilde{c}_i = (c_i-c_{crit})/[-(A+dC)/(2B)]^{1/2}$, $\tilde{\alpha} = (A + dC)/(2B)$ [d= dimensionality, and we assume A is sufficiently negative such that A+dC < 0] one sees that two parameters $\tilde{\alpha}$, $\tilde{\beta}$ remain, which are related to $1-T/T_c$ and the interaction range r. The parameters chosen in Fig. 6 still correspond to a fairly short range r, and hence it is not too surprising that one finds nonlinear effects very important during the earliest stages of the quench already.

It turns out, however, that the linearized CHC theory becomes valid[22] if one would consider systems with a sufficiently large range of interaction, r. This is seen by applying a Ginzburg criterion[22]: the linearized theory will hold for initial times if

$$\langle[\delta c(\vec{x},t)]^2\rangle_{T,L} \ll [\bar{c} - c_{sp}(T)]^2 \tag{15}$$

where $c_{sp}(T)$ is the concentration of the spinodal curve. Eq.(15) simply expresses the fact that the root mean square amplitude of the growing concentration fluctuations must be small in comparison to the concentration difference over which f(c) in Eqs.(1,2) has appreciable nonlinearities. Now remembering that $c(\vec{x},t)$ is obtained by coarse-graining of a lattice model, we obtain

$$\langle[\delta c(\vec{x},0)]^2\rangle_{T,L} = L^{-2d} \sum_{i,j\in L^d} [\langle c_i c_j\rangle - \bar{c}^2] \propto L^{-2d} \sum_{i,j\in L^d} |\vec{x}_i-\vec{x}_j|^{-(d-2)} r^{-2} \propto L^{-d} \int_0^{L} x^{d-1} dx\, x^{-d+2}\, r^{-2} \propto L^{2-d}\, r^{-2} \tag{16}$$

using for $\langle c_i c_j\rangle - \bar{c}^2$ the mean field result near criticality. Now the maximum choice for the coarse-graining length L is the critical wavelength λ_c, which near the spinodal behaves as $\lambda_c \propto r(1-T/T_c)^{-1/2}\left[\frac{\bar{c} - c_{sp}(T)}{c_{crit}-c_{sp}(T)}\right]^{-1/2}$. Assuming further $\langle[\delta c(\vec{x},t)]^2\rangle_{T,L} \approx \langle[\delta c(\vec{x},0)]^2\rangle_{T,L} \exp(2R_m t)$, we obtain from Eqs.(15),(16) the condition for the validity of Cahn's theory[1] as[22]

$$\exp(2R_m t) \ll r^d(1-T/T_c)^{(4-d)/2} \left[\frac{\bar{c} - c_{sp}(T)}{c_{crit} - c_{sp}(T)}\right]^{(6-d)/2} \tag{17}$$

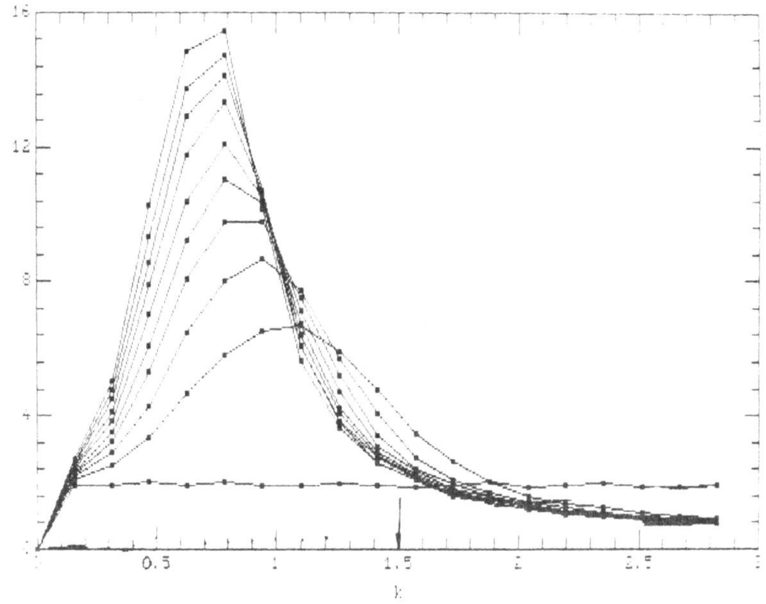

FIGURE 6. Time evolution of the normalized structure factor $\tilde{S}(k,t)$ plotted vs. k for the model with $\tilde{\alpha} = 0.175$ at $\tilde{c}_i = 0$ quenched instantaneously from $\tilde{\beta} = 0$ to $\tilde{\beta} = 0.3$. Data are for a 40 x 40 lattice, with periodic boundary conditions. Note that the possible $\vec{k}$-values are $k_x = (2\Pi/40)n_x$, $k_y=(2\Pi/40)n_y$, with n_x, n_y integers. In this figure, the structure factors with the same (k) are averaged together, and curves are drawn through the discrete points at which $\tilde{S}(k,t)$ only is defined to indicate the expected behavior for $N \rightarrow \infty$. The curves shown are for t = 0–100 Monte Carlo steps/site and are averaged over 100 individual runs. Arrows show estimate for the wavevector at which R(k) is maximal, as obtained from Eq.(6). Note that in Fig. 6 $\tilde{S}(k,t) = |2B/(A + dC)|S(k,t)$. From Milchev at al.[18]

Thus we see that off the spinodal and sufficiently below T_c for sufficiently large r the rhs of Eq.(17) is much larger than unity, and then the linear theory[1] holds initially. The range of times where Eq.(17) holds increases only slowly with increasing r, namely $t \lesssim (d/2R_m) \ln r$. This validity of Cahn's linearized theory for long-range systems has been beautifully verified by Monte Carlo simulations[23]. Since the long wavelength problems of polymer mixtures[24] can be mapped[22] onto long range Ising models, i.e. in Eq.(17) one has to replace r^d by $a^d\, N^{(d-2)/2}$ where a is the size of a polymer segment and N is the number of segments of a polymer chain, one can understand why the validity of the linearized theory has been established recently for polymer mixtures[25]. For most other systems (as well as for polymer mixtures at later times) nonlinear effects must be accounted for. A systematic treatment (to first order in r^{-1}) is only available for $\bar{c} = c_{crit}$ and is rather complicated[26]. A qualitative account of nonlinear effects is achieved by the LBM approximation[15,16] which replaces Eq.(10) by

$$\frac{d}{dt} S(\vec{k},t) = -2M(T)k^2\{[(\partial^2 f/\partial c^2)_T \Big|_{\bar{c}} + a(t) + r^2 Tk^2]S(\vec{k},t) - k_B T\}, \tag{18}$$

where the correction term $a(t)$ depends on $S(\vec{k},t)$ in a nonlinear fashion.

The strength of the nonlinear effects in this approach is controlled by a parameter f_o which near T_c is expressed as

$$f_o \quad \alpha \quad \hat{\xi}^d \, \hat{B}^2 \, \hat{C}^{-1} \, (1-T/T_c)^{\gamma+2\beta-d\nu}, \tag{19}$$

where critical amplitudes[45] and exponents[45] enter {order parameter $(c^{(1)}_{coex}-c_{crit})/c_{crit} = \hat{B} \quad (1-T/T_c)^{\beta}$, critical scattering intensity $\chi_{coex} = \hat{C}(1-T/T_c)^{-\gamma}$, and the correlation length $\xi_{coex} = \hat{\xi} \; (1-T/T_c)^{-\nu}$}. In the mean field critical region $\hat{\xi} \;\; \alpha \;\; r$, $\gamma = 1$, $\beta = \nu = \frac{1}{2}$ and hence $f_o \;\; \alpha \;\; r^d(1-T/T_c)^{(4-d)/2} \gg 1$, consistent with Eq.(17). In the non-mean-field critical region {where $r^d(1-T/T_c)^{(4-d)/2} \lesssim 1$, which is reached very close to T_c}, on the other hand, the hyperscaling relation[45] $d\nu = \gamma + 2\beta$ eliminates the temperature dependence of f_o: in fact, there f_o becomes a universal constant ($f_o \approx 9.45$)[14,16]. Fig. 7a shows the time evolution of the structure factor[14] for two choices of f_o. It is seen that the linear theory is indeed valid for $f_o \approx 9.45$ from the start, but holds initially for large f_o, consistent with experiment[25] and simulation[23]. Note that in Figs. 2,5 also the Ising model value ($f_o = 9.45$) was chosen, which leads to a very pronounced nonlinear behavior.

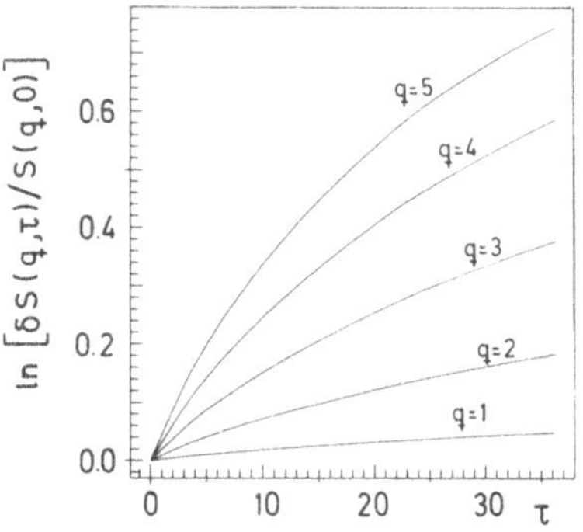

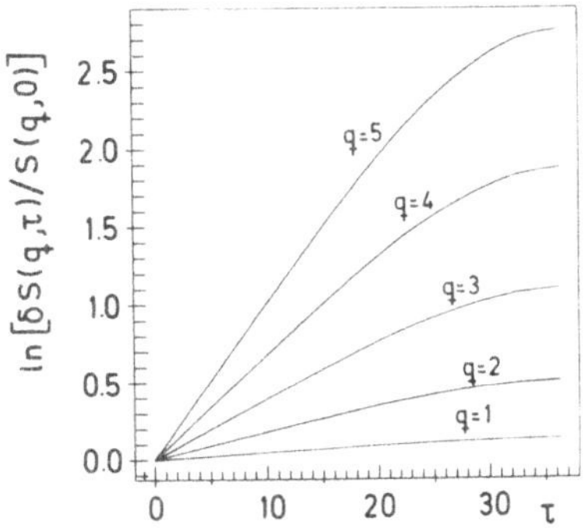

FIGURE 7a). Plot of $\ln \{[\tilde{S}(q,\tau) - \tilde{S}_T(q)]/\tilde{S}_{T_0}(q)$ vs. τ for five different values of a, for a quench from infinite temperature at c_{crit}, using $f_0 = 9.45$ (upper part) and $f_0 = 9450$ (lower part). From Carmesin et al.[14].

III. BRIEF COMMENTS ABOUT NUCLEATION AND SPINODALS

The decay of metastable state by the growth of a droplet means that the system moves from a local minimum in phase space (the metastable state without the droplet) over a saddle point (the metastable state plus one critical droplet) towards the stable minimum. As in the calculation of the free energy from Eq.(1) [Z = partition function]

$$F = -k_BT \ln Z = - k_BT \ln \int d\{c(\vec{x})\}\exp[-\Delta\mathcal{F}\{c(\vec{x})\}/k_BT] \tag{20}$$

we consider a functional integral but restrict the phase space to states near the metastable minimum of the coarse-grained free energy density f[c] and to nonuniform solutions $c(\vec{x})$ tending to $\bar{c}$ at large distances from the origin. Just as in the mean field theory of stable or metastable states where one replaces the actual distribution $\exp[-\Delta\mathcal{F}\{c(\vec{x})\}/k_BT]$ by a delta

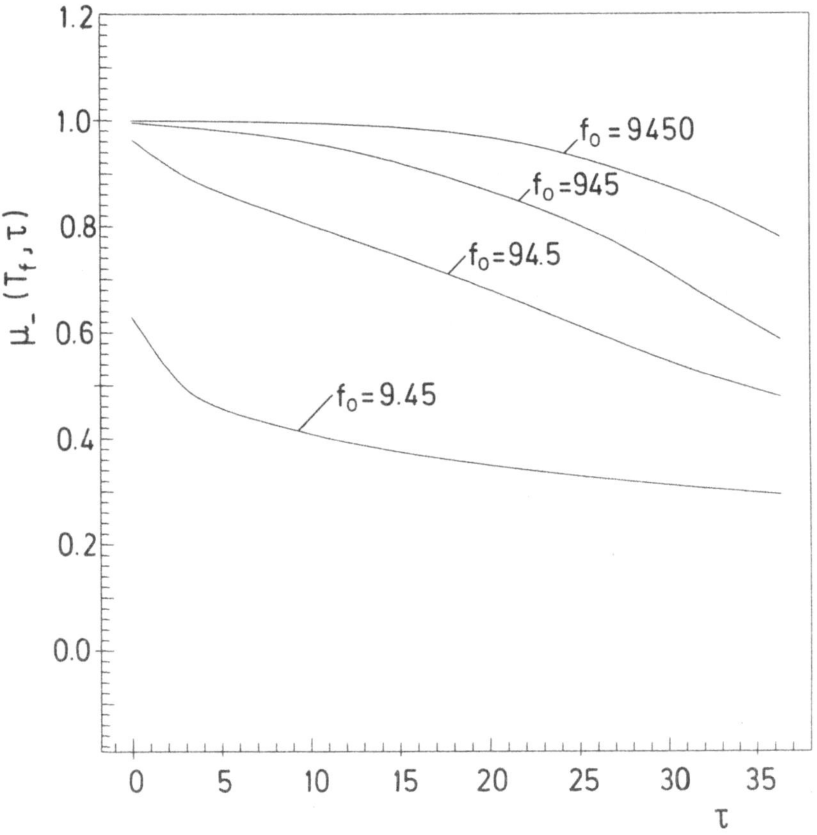

FIGURE 7b). Rescaled effective second derivative $\mu_-(T,\tau) \equiv 1 + a(t)/[(\partial^2 f/\partial c^2)_T |_{\bar{c}}]$ plotted versus scaled time τ for four choices of f_o. Only when $\mu_-(T,\tau) \approx 1$ the CHC approximation is accurate. In Figs. 2,5 the CHC approximation is applied with $\mu_-(T,0) \approx 0.65$. From Carmesin et al.[14].

function $\delta(c-\bar{c})$ where $\bar{c}$ minimizes $\Delta\mathcal{F}$, the mean field theory of nucleation[1] amounts to replacing the functional integral by a saddle point solution of $\delta\Delta\mathcal{F}\{c(\rho)\}/\delta c(\rho) = 0$, where $c(\rho)$ is a spherically symmetric concentration profile of the critical droplet. In the mean-field critical region, i.e. for $r^d(1-T/T_c)^{(4-d)/2} \gg 1$, this yields a free energy barrier ΔF^* to form one critical droplet given by[1,22,27]

$$\Delta F^*/k_BT_c \quad \alpha \quad r^d(1-T/T_c)^{(4-d)/2}\{[\bar{c}-c^{(1)}_{coex}]/[c^{(2)}_{coex}-c^{(1)}_{coex}]\}^{-(d-1)}, \quad \bar{c} \longrightarrow c^{(1)}_{coex}, \tag{21a}$$

$$\Delta F^*/k_BT_c \quad \alpha \quad r^d(1-T/T_c)^{(4-d)/2}\{[c_{sp}(T)-\bar{c}]/[c^{(2)}_{coex}-c^{(1)}_{coex}]\}^{(6-d)/2},$$
$$\bar{c} \longrightarrow c_{sp}(T). \tag{21b}$$

Thus for $r^d(1-T/T_c)^{(4-d)/2} \gg 1$ the barrier $\Delta F^* \gg k_BT_c$, even if one comes close to the spinodal. The ultimate limit of metastability is reached for $\Delta F^*/k_BT_c \approx 1$; i.e., the width of the region over which the spinodal is smeared out is given by

$$\delta c_{sp}/[c^{(2)}_{coex}-c^{(1)}_{coex}] \quad \alpha \quad [r^d(1-T/T_c)^{(4-d)/2}]^{-2/(6-d)} \underset{(d=3)}{\alpha}$$
$$[r^3(1-T/T_c)^{1/2}]^{-2/3}. \tag{22}$$

Eq.(22) is also obtained on the unstable side if we put $\bar{c}-c_{sp}(T) = \delta c_{sp}$ in Eq.(17) and treat this inequality rather as an equality for $t = 0$.

Again we have to ask under which conditions the above mean-field approximation is valid {evaluating the functional integral, we take only the one "path" $c(\rho)$ over the saddle point which minimizes $\Delta\mathcal{F}$ and disregard all other possible paths, thus neglecting again fluctuations}. This question also can be answered by a Ginzburg criterion[22]: the mean square amplitude of the coarse-grained order parameter along the radial droplet concentration profile must be smaller than the square of the order parameter difference between $\rho = 0$ and $\rho = \infty$ as described by the profile itself,

$$\langle[\delta c(\vec{x})]^2\rangle_{T,L} \ll [c(\rho \longrightarrow \infty) - c(\rho=0)]^2 \tag{23}$$

For $\bar{c} \to c^{(1)}_{coex}$ we may put $L \approx \xi_{coex}$, $c(\rho=0) \approx c^{(2)}_{coex}$, $c(\rho \to \infty) = \bar{c} \approx c^{(1)}_{coex}$ and then Eq.(23) yields the standard Ginzburg criterion; for $\bar{c} \longrightarrow c_{sp}(T)$, on the other hand, $L \approx \xi \;\; \alpha \;\; r(1-T/T_c)^{-1/2}\{[c_{sp}(T)-\bar{c}] / [c^{(2)}_{coex}-c^{(1)}_{coex}]\}^{-1/2}$, and $c(\rho = 0) - c(\rho \longrightarrow \infty) \;\; \alpha \;\; c_{sp}(T)-\bar{c}$, and then Eq.(22) yields

$$1 \ll r^d(1-T/T_c)^{(4-d)/2}\{[c_{sp}(T)-\bar{c}]/[c^{(2)}_{coex}-c^{(1)}_{coex}]\}^{(6-d)/2}, \bar{c} \to c_{sp}(T). \tag{24}$$

Comparison of Eqs.(21b),(24) shows that as long as $\Delta F^* \gg k_BT$ the mean-field treatment of nucleation is in fact self-consistent. Comparison of Eqs.(24),(17) shows further that the validity criteria are essentially the same on both sides of the spinodal curve. It is also gratifying that computer simulations of metastable states in Ising models with long-range interactions[28] confirm this description.

Fig. 8 now summarizes the various regions that can be identified in the phase diagram of a mixture with large but finite range of interactions near its critical point.

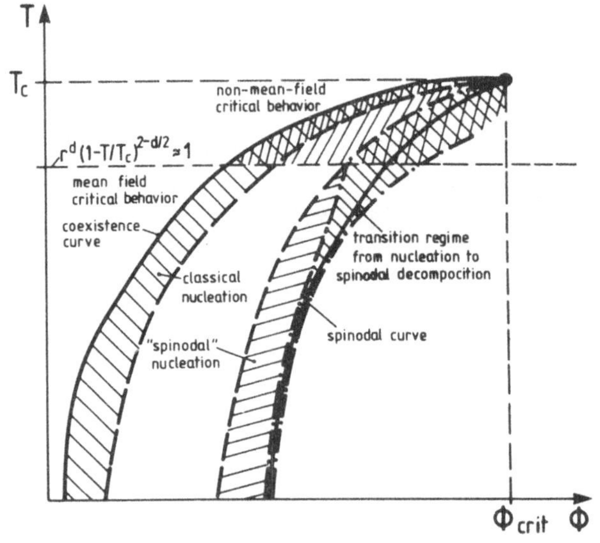

FIGURE 8. Various regimes in the plane temperature T-volume fraction ϕ of a binary mixture near T_c (only volume fractions $\phi < \phi_{crit}$ are shown). The regime inside the two dash-dotted lines around the spinodal line is the regime where a gradual transition from nucleation to spinodal decomposition occurs. The regime between the coexistence curve and the left of the two broken curves is described by classical nucleation theory (compact droplets), the regime between the right broken curve and the left dash-dotted curve is described by "spinodal nucleation" (ramified droplets[27]); it exists only in the regime of mean-field critical behavior just as the regime to the right of the right dash-dotted curve, where the linearized theory of spinodal decomposition holds during the early stages. From Binder[22].

While the mean-field theory of nucleation, as sketched above, is no longer valid if one enters the nonclassical critical region, $r^d(1-T/T_c)^{(4-d)/2} \lesssim 1$, near the coexistence curve one can apply the "classical" Becker-Döring-type theory of nucleation[2-4,29] at all temperatures: there the formation free energy $\Delta F(R)$ of a spherical droplet of radius R is written in terms of bulk and surface free energies,

$$\Delta F(R) = -V_d R^d [\bar{c} - c^{(1)}_{coex}][c^{(2)}_{coex} - c^{(1)}_{coex}]/\chi_{coex} + S_d R^{d-1} f_{int}, \qquad (25)$$

V_d and S_d being volume and surface area of a d-dimensional unit sphere, respectively, and f_{int} is the surface tension for a flat interface between coexisting phases ("capillarity approximation"[29]). Extremizing $\Delta F(R)$ yields the critical cluster radius $R^*\{\partial \Delta F(R)/\partial R \,\big|_{R^*} = 0\}$ and the associate nucleation free energy barrier $\Delta F^* \equiv \Delta F(R^*)$, namely

$$\Delta F^* = [S_d/d]^d [d-1)/V_d]^{d-1} \frac{f_{int}^d \, \chi_{coex}^{d-1}}{[c_{coex}^{(2)} - c_{coex}^{(1)}]^{2(d-1)}} \left[\frac{\bar{c} - c_{coex}^{(1)}}{c_{coex}^{(2)} - c_{coex}^{(1)}}\right]^{-(d-1)} \tag{26}$$

Noting that in the mean-field critical regime one has $f_{int} \propto r(1-T/T_c)^{3/2}$, Eq.(26) is found to coincide with Eq.(21a). In the non-mean field critical region, however, one has instead[30] $f_{int} = \hat{f}(1-T/T_c)^{(d-1)\nu}$, and Eq.(26) yields

$$\frac{\Delta F^*}{k_B T_c} = \left[\frac{S_d}{d}\right]^d \left[\frac{d-1}{V_d}\right]^{d-1} \frac{\hat{f}^d \, \hat{C}^{d-1}}{\hat{B}^{2(d-1)}} \left[\frac{\bar{c} - c_{coex}^{(1)}}{c_{coex}^{(2)} - c_{coex}^{(1)}}\right]^{-(d-1)}, \tag{27}$$

where the amplitude combination $\hat{f}^d \, \hat{C}^{d-1} / \hat{B}^{2(d-1)}$ is another universal constant of order unity[31,32], similar to the constant f_o encountered in Eq.(19). This result means that the scale for $\Delta F^*/k_B T_c$ is no longer given by a large number {as it happens in the mean-field critical regime, where the scale is set by the factor $r^d(1-T/T_c)^{(4-d)/2} \gg 1$, see Eq.(21)}, but the scale is now of order unity: $\Delta F^* \gg k_B T_c$ only for $\bar{c} \to c_{coex}^{(1)}$; further inside in the metastable region we have $\Delta F^* \approx k_B T_c$ and hence the regime where a gradual transition from nucleation to spinodal decomposition occurs is very broad (Fig. 8).

The classical theory of nucleation, Eqs.(25),(26), obviously neglects fluctuation in droplet shape. For $R^*/\xi_{coex} \to \infty$ the dominating fluctuations are due to capillary waves and give rise to[33] a logarithmic correction to Eq.(27). For the experimentally interesting region $10 \lesssim \Delta F^*/k_B T \lesssim 60$, where R^* is only a few times larger than ξ_{coex}, a reliable analytical theory yielding the corrections to Eq.(27) does not yet exist, and various attempts to answer this problem by Monte Carlo simulations[34,35] contradict each other: while Ref. 34 predicts that ΔF^* is larger than expected from Eq.(27), Ref. 35 predicts that ΔF^* is smaller than Eq.(27) says. Clearly more work is required to resolve this unsatisfactory situation.

IV. LATE STAGES OF PHASE SEPARATION

While there are many regimes with different behavior during the initial stages of phase separation (Fig. 8), less distinctions occur during the late stages: there the system is locally unmixed ($c \cong c^{(1)}_{coex}$ or $c \cong c^{(2)}_{coex}$, respectively, irrespective of whether it started to unmix by nucleation or by spinodal decomposition. the typical linear dimensions $l(t)$ of unmixed regions grow with time according to a power law[36], where x is some exponent which will be discussed below

$$l(t) \quad \alpha \quad t^x, \tag{28}$$

and the structure factor in this coarsening regime is believed to satisfy a scaling law[37-43]

$$S(\vec{k},t) = [l(t)]^d \, \tilde{S}_{\vec{e}} \{kl(t)\}, \; t \longrightarrow \infty, \; kl(t) \text{ finite.} \tag{29}$$

Here $\tilde{S}_e$ is a scaling function which also[18,44] depends on the direction of $\vec{k}$, $\vec{e} \equiv \vec{k}/|\vec{k}|$. This scaling behavior is somewhat reminiscent of the critical point scaling[45]. In fact, if we consider a quench to $T = T_c$ we expect, instead of Eq.(29)[46]

$$S(k,t) = [l'(t)]^{2-\eta} \, \tilde{S}' \{kl'(t)\}, \; l'(t) \quad \alpha \quad t^{1/z}, \tag{30}$$

where the exponent z describing the growth of the time-dependent correlation length $l'(t)$ is the dynamic critical exponent[47] while the exponent η reflects the fractal[48] structure of critical correlations[45] $\left\{\text{note also } \tilde{S}'(\chi \longrightarrow \infty) \quad \alpha \quad \chi^{-(2-\eta)} \text{ and hence } S(k,\infty) = S_{T_c}(k) \quad \alpha \quad k^{-2+\eta}\right\}$.

Alternative scaling descriptions of phase separation consider the number $n_s(t)$ of clusters of size s at time t after the quench[36,39]

$$n_s(t) = t^{-dx} \, \tilde{n} \, (st^{-dx}), \tag{31}$$

where $\tilde{n}$ is another scaling function and it is assumed that x is the same exponent as in Eq.(28). So far, analytical derivations of Eqs.(28),(29) and (31) are suffering from many short-comings; one believes that Eqs.(28),(29) and (31) are valid because they agree with simulations[41,49,50] and each other: while Ref. 34 predicts that ΔF^* is larger than expected from Eq.(27), Ref. 35 predicts that ΔF^* is smaller than Eq.(27) says. Clearly more work is required to resolve this unsatisfactory situation.

V. LATE STAGES OF PHASE SEPARATION

While there are many regimes with different behavior during the initial stages of phase separation (Fig. 8), less distinctions occur during the late stages: there the system is locally unmixed ($c \cong c^{(1)}_{coex}$ or $c \cong c^{(2)}_{coex}$, respectively, irrespective of whether it started to unmix by nucleation or by spinodal decomposition. the typical linear dimensions $l(t)$ of unmixed regions grow with time according to a power law[36], where x is some exponent which will be discussed below

$$l(t) \quad \alpha \quad t^x, \tag{28}$$

and the structure factor in this coarsening regime is believed to satisfy a scaling law[37–43]

$$S(\vec{k},t) = [l(t)]^d \, \tilde{S}_{\vec{e}} \, \{kl(t)\}, \; t \longrightarrow \infty, \; kl(t) \text{ finite.} \tag{29}$$

Here $\tilde{S}_e$ is a scaling function which also[18,44] depends on the direction of $\vec{k}$, $\vec{e} \equiv \vec{k}/|\vec{k}|$. This scaling behavior is somewhat reminiscent of the critical point scaling[45]. In fact, if we consider a quench to $T = T_c$ we expect, instead of Eq.(29)[46]

$$S(k,t) = [l'(t)]^{2-\eta} \, \tilde{S}' \, \{kl'(t)\}, \; l'(t) \quad \alpha \quad t^{1/z}, \tag{30}$$

where the exponent z describing the growth of the time-dependent correlation length $l'(t)$ is the dynamic critical exponent[47] while the exponent η reflects the fractal[48] structure of critical correlations[45] $\left\{\text{note also } \tilde{S}'(\zeta \longrightarrow \infty) \quad \alpha \quad \zeta^{-(2-\eta)} \text{ and hence } S(k,\infty) = S_{T_c}(k) \quad \alpha \quad k^{-2+\eta}\right\}$.

Alternative scaling descriptions of phase separation consider the number $n_s(t)$ of clusters of size s at time t after the quench[36,39]

$$n_s(t) = t^{-dx} \, \tilde{n} \, (st^{-dx}), \tag{31}$$

where $\tilde{n}$ is another scaling function and it is assumed that x is the same exponent as in Eq.(28). So far, analytical derivations of Eqs.(28),(29) and (31) are suffering from many short-comings; one believes that Eqs.(28),(29) and (31) are valid because they agree with simulations[41,49,50] and experiments[51-59,44]. This "agreement" though is not fully convincing, however, since often it is not clear that the asymptotic region where this scaling behavior holds has actually been reached (and some model calculations[60,61] do in fact indicate very long lived transients). While in fluid mixtures for large enough values of the volume fraction of the minority phase $\Psi \equiv [\bar{c} - c^{(1)}_{coex}]/[c^{(2)}_{coex} - c^{(1)}_{coex}]$ it is agreed that ultimately $x = 1$ [due to hydrodynamic mechanisms], in solids most workers assume that the Lifshitz-Slyozov[36] result $x = 1/3$ is true, though the original derivation holds in the limit $\Psi \longrightarrow 0$ only. In fact, both some simulations[17,18,49,50] and experiments[44,56] often yield significantly smaller values of the exponent x and it is not entirely clear whether this is a transient intermediate regime (as suggested in[39,56]) or whether there is possibly a regime with a different asymptotic behavior. Mazenko et al.[62] propose even that the asymptotic law instead of Eq.(28) is $l(t) \;\alpha\; \ln t$ at low temperatures. We feel, however, that the phenomenological renormalization group arguments used to "derive" this logarithmic law are inconclusive, as well as the simulations of Ref. 62, because the latter use by far too short times and too small sizes. Furthermore this logarithmic law clearly is incompatible with experimental data such as[56] and recent simulations[63].

Here we shall not describe the theoretical work dealing with the theory of coarsening in any detail, but only present a simplified derivation[39] of the Lifshitz-Slyozov[36] result, $x = 1/3$. From classical nucleation theory[29], one can justify the following "continuity equation in cluster size space{s}" for the cluster size distribution $n_s(t)$,

$$\frac{\partial}{\partial t} n_s(t) + \frac{\partial}{\partial t}\left\{n_s(t)\, R_s\, [\Delta\mu(t) - Cs^{-1/d}]\right\} = 0, \qquad (32)$$

where R_s is a cluster reaction rate, $\Delta\mu(t)$ a time-dependent chemical potential difference $\{\mu(t) \longrightarrow 0$ if one moves towards phase coexistence$\}$ in units of k_BT, and the constant C is related to the surface free energy $\Big\{$Eq.(32) is based on assuming that the free energy of a cluster would be $F_s/k_BT = -\Delta\mu(t)s + \frac{c}{1-1/d}s^{1-1/d}\Big\}$. Also R_s is estimated implying that the growing clusters are essentially compact spherical domains, with radius $\rho_s \;\alpha\; s^{1/d}$: the current of B-atoms impinging on the surface and of a B-cluster per unit area is $j = D_T[\partial c(\rho)/\partial\rho \,\big|_{\rho=\rho_s}] \;\alpha\; \rho_s^{-1}$ $\Big\{D_T$ is a diffusion constant$\Big\}$. The total current, and hence R_s, is found by multiplying j with the cluster surface area ($\alpha\; \rho_s^{d-1}$). Thus $R_s \;\alpha\; \rho_s^{d-2}$, or $R_s = \hat{R}\, s^{1-2/d}$, with $\hat{R}$ an amplitude factor. Assuming now Eq.(31) which is consistent with the conservation law of the total concentration, $d\bar{c}/dt = \frac{d}{dt} \int_0^\infty ds\ s\ n_s(t) = 0$, one finds from Eq.(32), integrating by parts and using once more the conservation law, that

$$\Delta\mu(t) = C \int_0^\infty ds\ n_s(t) R_s\ s^{-1/d} / \int_0^\infty ds\ n_s(t)\ R_s = C'\ t^{-x}, \qquad (33)$$

where C' is another constant. Using Eq.(33) in Eq.(32) and rewriting it as differential equation for $\tilde{n}(\chi)$ we find

$$-2dx\ \tilde{n} - dx \frac{d\tilde{n}}{d\chi} = ct^{1-3x} \frac{d}{d\chi} [\tilde{n}\ \chi^{1-2/d} (\chi^{-1/d} - C'/C]. \qquad (34)$$

Thus Eq.(31) solves Eq.(32) only if $x = \frac{1}{3}$. The typical linear dimension $l(t)$ follows from

$$[l(t)]^d \equiv \int_0^\infty dss^2\ n_s(t) / \int_0^\infty ds\ sn_s(t) \quad \alpha \quad t^{dx} = t^{d/3}. \qquad (35)$$

One can go further and obtain from Eq.(34) both the scaling function explicitly, as well as the proportionality constant in Eq.(35)[36]. Neither the scaling function $\tilde{n}(\chi)$ nor the rate Ω in the law $l(t) = (\Omega t)^{1/3}$ are in agreement with most experimental results, however. This discrepancy is attributed to the fact that the Lifshitz-Slyozov theory[36] treats the cluster dynamics on a mean-field level only; the theory should become correct for $\Psi \longrightarrow 0$ only. A lot of work has been done[64-70] to generalize the Lifshitz-Slyozov theory to finite but small volume fraction (it turns out that correction terms are of order $\sqrt{\Psi}$ already). Fig. 9 shows that the various theories proposed are not in mutal agreement with each other.

We here also note that the range of volume fractions Ψ where these theories are applicable is widely overestimated. See e.g. Fig. 9: all theories invoke the growth of well-separated compact spherical droplets. This picture breaks down, however, when one comes close to the percolation line in the phase diagram (Fig. 1, Fig. 10). Fig. 10 shows the line $c_p^{(corr)}(T,t)$ where in the nearest-neighbor simple cubic Ising model of a binary alloy at times t = 120 Monte Carlo steps per particle (or later) a percolating infinite cluster appears. This percolation line refers to the atomic scale (a "cluster" of B-atoms is defined geometrically by the condition that every B-atom in the cluster must have at least one nearest-neighbor B-atom belonging to the same cluster). For $\bar{c} \geq c_p^{corr}(T,t)$ the system may start out with only finite clusters, but as time proceeds these clusters aggregate together and an infinite cluster forms (in a finite system the size of the largest cluster S_{max} is strictly limited by

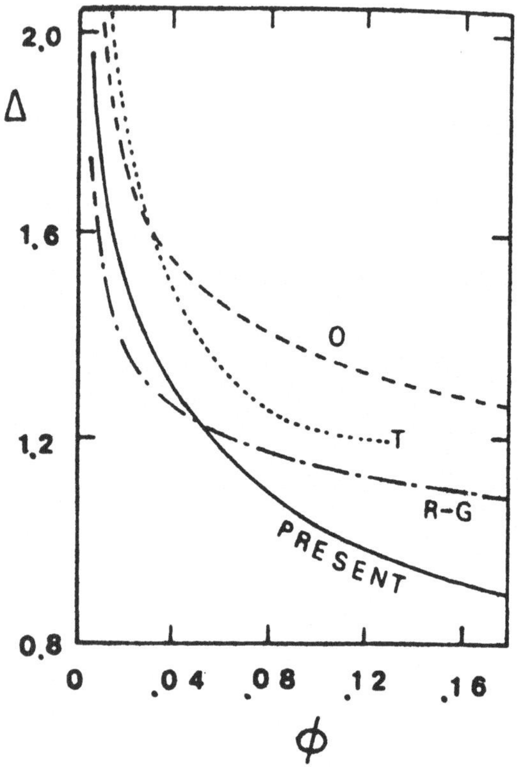

FIGURE 9. Dependence of the halfwidth of the structure function on the volume fraction φ of the minority phase, according to several recent theories: Ohta (O)[65], Tomita (T)[71], Rikvold and Gunton (RG)[72], and Furukawa (present)[70]. From Furukawa[70].

the system size L^d, $s_{max} \lesssim \bar{c}\, L^d$, of course), see Fig. 11. From the corresponding "susceptibility" $\chi \equiv \sum_s' s^2 n_s(t)$ [the largest cluster being excluded from the sum], see Fig. 12, and the percolation probability $P_\infty \equiv s_{max}/L^d$ one can infer the critical exponents associated with this percolation transition. While at T = 0 the analysis is consistent with exponents appropriate for standard random percolation[10] ($\chi \;\alpha\; |\bar{c} - c_p^{(corr)}(T,t)|^{-\gamma_p}$, $P_\infty \;\alpha\; (c_p^{(corr)}(T,t) - \bar{c})^{\beta_p}$, with $\gamma_p \cong 1.40$, $\beta_p \cong 0.45$), at nonzero T the data seem to imply somewhat different exponents. But in any case, the "fractal dimensionality" $d_f = d - \beta_p/\nu_p$ [$\nu_p = (\gamma_p + 2\beta_p)/d$] differs appreciably from d, which implies that at $\bar{c} = c_p^{(corr)}(T,t)$ the clusters are not compact, as anticipated in the

Lifshitz-Slyozov-type theories[36,39,64-69]. We believe that such percolation phenomena are in fact responsible for observations of fractal clusters in other phase separation models[70], too. At this point, it is not known whether this percolation transition at $c_p^{(corr)}(T,t)$ has any consequences for the behavior of the structure factor $S(\vec{k},t)$. Another interesting problem is the behavior in two space dimensions. Grest and Srolovitz[74] have studied cluster properties at a phase-separating triangular lattice for c = 0.5, which is a rather special case as the system stays at its percolation threshold for all times.

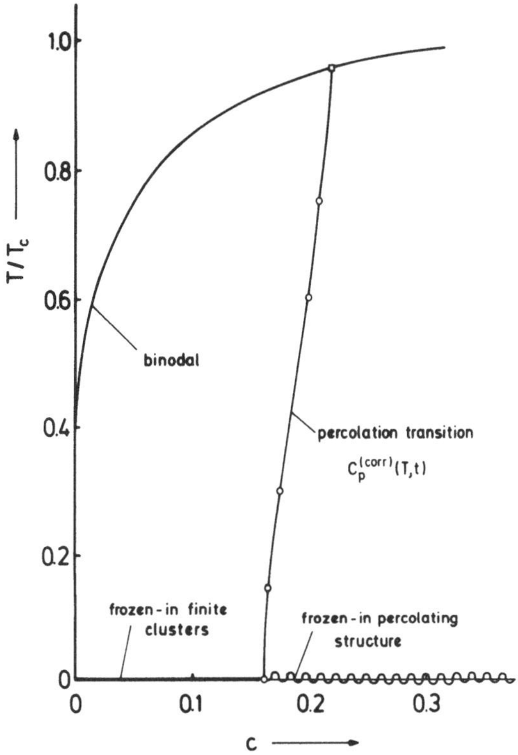

FIGURE 10. Low-concentration part of the phase diagram of the nearest neighbor simple cubic Ising model of a binary alloy, showing the location of the percolation transition at t = 120 MCS/particle. From Hayward et al.[5].

VI. CONCLUSIONS

These lectures have emphasized the Cahn-Hilliard mean-field theory of spinodal decomposition and nucleation, and the conditions for which this mean-field theory is valid have been investigated in detail. It has been shown that the theory in fact is practically useful for the understanding of very early stages of phase separation in polymer mixtures, since the latter can be mapped to systems with a long range interaction. No such long range character of the forces normally is expected in solid alloys or in fluid small molecule mixtures, on the other hand, and in these systems therefore the concept of the spinodal line has little use; spinodal decomposition is an intrinsically nonlinear phenomenon there, one has virtually no time range where the linear theory holds, and the transition from nucleation to spinodal decomposition proceeds completely gradual, no dramatic changes can be detected in the structure function describing the small angle scattering from these systems. These conclusions were first obtained from qualitative theories, and are now amply confirmed by many experiments. Unfortunately, no quantitatively reliable theory is at hand which could be fitted to the data. This lack of a theory applies also to the later stages of unmixing in polymer mixtures.

symbol	c
•	0.156
×	0.163
+	0.170
△	0.177
✳	0.184
○	0.191

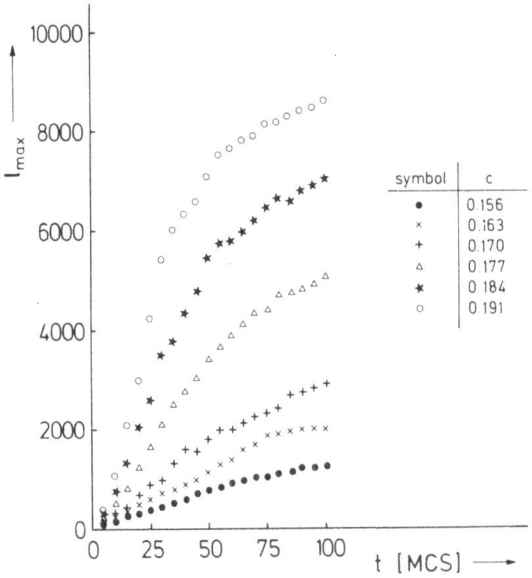

FIGURE 11. Evolution of the size of the largest cluster (s_{max}) as a function of time after a quench from infinite to zero temperature, for various concentrations, in a finite lattice (L = 40). From Hayward et al.[5].

In the very late stages of phase separation, theory predicts a power-law growth of the characteristic linear dimension $l(t)$ describing the coarsening structure, and the structure factor is found to satisfy a scaling hypothesis. While the characteristic exponent x probably has for all temperatures $T > 0$ and all volume fractions ψ the universal value x = 1/3 already predicted by the Lifshitz-Slyozov theory, there exist still many open questions: Are there well-defined intermediate time regimes with different exponents? What are the leading correction terms to the Lifshitz-Slyozov law? What are the precise consequences of the percolation transition which separates the two regimes of structures with different morphology (well-separated droplets versus interconnected structures)? How

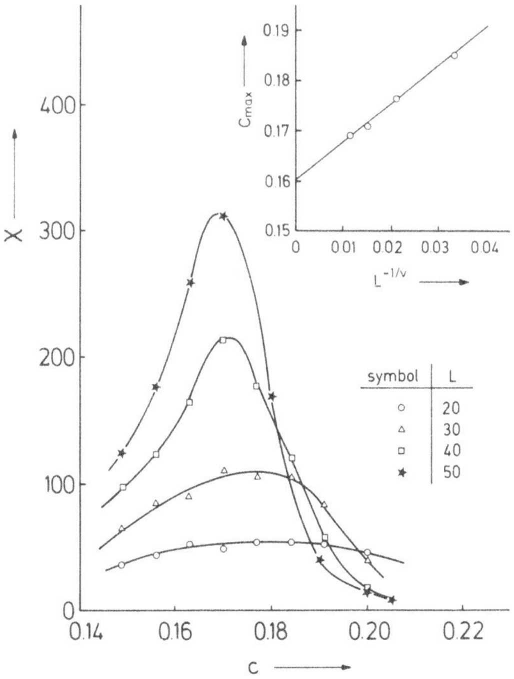

FIGURE 12. Percolation susceptibilities χ plotted versus concentration c at T = 0 in the sc lattice for several lattice sizes. The inset shows a plot of the position c_{max} of the maximum of χ versus L^{-1/ν_p} where the exponent ν_p has the value[10] of random percolation, $\nu_p \approx 0.88$. From Hayward et al.[5].

can one reliably predict the various scaling functions $\tilde{n}$ and $\tilde{S}$, their dependence on temperature and volume fraction, etc.? Although computer simulations have given valuable hints to some of these questions, the lack of a complete analytical theory answering these questions hampers the analysis of experiments.

ACKNOWLEDGEMENTS: The author has profited from a fruitful collaboration with D.W. Heermann, A. Milchev, S. Hayward, H.-O. Carmesin, H.L. Frisch and J. Jäckle. This research is supported in part by the Deutsche Forschungsgemeinschaft, Sonderforschungsbereich 41.

REFERENCES

1. J. W. Cahn and J. E. Hilliard, J. Chem. Phys. 28, 258 (1958); 31, 688 (1959); J. W. Cahn, Acta Metall. 9, 795 (1961)
2. For recent reviews, see Refs. 3,4 and J. D. Gunton, M. San Miguel, and P. S. Sahni, in Phase Transitions and Critical Phenomena, Vol 8, C. Domb and J. L. Lebowitz, eds., Academic Press, New York (1983) p. 267.

3. K. Binder, in Condensed Matter Research Using Neutrons, S. W. Lovesey and R. Scherm, eds., Plenum, New York (1984) p. 1; K. Binder and D. W. Heermann, in Scaling Phenomena in Disordered Systems, R. Pynn and A. Skjeltorp, eds., Plenum, New York 1985.
4. K. Binder, Rep. Progr. Phys. (1987, in press)
5. S. Hayward, D. W. Heermann and K. Binder, to be published.
6. K. Binder, Solid State Comm. 34, 191 (1980)
7. D. W. Heermann, Z. Physik B 55, 309 (1984)
8. H. Müller-Krumbhaar, Phys. Lett. 50, 27 (1974); D. W. Heermann and D. Stauffer, Z. Physik B 44, 339 (1981)
9. A. Coniglio and W. Klein, J. Phys. A 13, 2775 (1980)
10. D. Stauffer, An Introduction to Percolation Theory, Taylor and Francis, London 1985.
11. H. E. Cook, Acta Metall 18, 297 (1970)
12. K. Binder, H. L. Frisch and J. Jäckle, J. Chem. Phys. 85, 1505 (1986)
13. E. L. Houston, J. W. Cahn and J. E. Hilliard, Acta Metall 14, 1685 (1966)
14. H.-O. Carmesin, D. W. Heermann and K. Binder, Z. Phys. B 65, 89 (1986)
15. J. S. Langer, M. Baron and H. D. Miller, Phys. Rev. A 11, 1417 (1975)
16. C. Billotet and K. Binder, Z. Phys. B 32, 195 (1979); K. Binder, C. Billotet and P. Mirold, Z. Phys. B 30, 183 (1978)
17. J. Marro, A. B. Bortz, M. H. Kalos and J. L. Lebowitz, Phys. Rev. B 12, 2000 (1975)
18. A. Milchev, D. W. Heermann and K. Binder, Acta metall.,in press.
19. A. F. Craievich and J. P. Olivieri, J. Appl. Cryst. 14, 444 (1981);R. J. Acuna and A. F. Craievich, J. Non-Cryst. Solids 34, 13 (1979); H. Yokota, J. Phys. Soc. Jpn. 45, 29 (1978)
20. H. Meier and G. R. Strobl, preprint; G. R. Strobl, priv. communication
21. J. Jäckle, priv. communication.
22. K. Binder, Phys. Rev. A 29, 341 (1984)
23. D. W. Heermann, Phys. Rev. Lett. 52, 1126 (1984); Z. Phys. B 61, 311 (1985)
24. K. Binder, J. Chem. Phys. 79, 6387 (1983)
25. T. Izumitani and T. Hashimoto, J. Chem. Phys. 83, 3694 (1985); N. Okada and C. C. Han, J. Chem. Phys. 85, 5317 (1986); H. Meier and G. R. Strobl, preprint; T. Hashimoto, M. Itakura and N. Shimitzu, J. Chem. Phys. 85, 6673 (1986)
26. M. Grant, M. San Miguel, S. Vinals and J. D. Gunton, Phys. Rev. B 31, 302 (1985)
27. W. Klein and C. Unger, Phys. Rev. B 28, 445 (1983); C. Unger and W. Klein, Phys. Rev. B 29, 2698 (1984)
28. D. W. Heermann, W. Klein and D. Stauffer, Phys. Rev. Lett. 49, 1262 (1982)
29. A. C. Zettlemoyer (ed.) Nucleation, M. Dekker, New York 1979; K. Binder and D. Stauffer, Adv. Phys. 25, 343 (1976)
30. B. Widom, in Phase Transitions and Critical Phenomena, Vol.2 (C. Domb and M. S. Green, eds.) Academic, New York 1972; D. Jasnow, Rep. Progr. Phys. 47 1059 (1984)
31. D. Stauffer, M. Ferer and M. Wortis, Phys. Rev. Lett. 29, 245 (1972)
32. K. Binder, Phys. Rev. A 25, 1699 (1972)
33. J. S. Langer, Ann. Phys. (N.Y.) 41, 108 (1967); N. J. Günther, D. A. Nicole and D. J. Wallace, J. Phys. A 13, 1755 (1980)
34. D. Stauffer, A. Coniglio and D. W. Heermann, Phys. Rev. Lett. 49, 1299 (1982)

35. H. Furukawa and K. Binder, Phys. Rev. A 26, 556 (1982)
36. I. M. Lifshitz and V. V. Slyozov, J. Phys. Chem. Solids 19, 35 (1961)
37. This type of scaling was first suggested in Refs. 38,39 and 16. Recent reviews can be found in Refs. 3 and 40, which contain more complete references than given here.
38. K. Binder and D. Stauffer, Phys. Rev. Lett. 33, 1006 (1974)
39. K. Binder, Phys. Rev. B 15, 4425 (1977)
40. H. Furukawa, Adv. Phys. 34, 703 (1985)
41. J. Marro, J. L. Lebowitz and M. H. Kalos, Phys. Rev. Lett. 43, 282 (1979)
42. H. Furukawa, Phys. Rev. Lett. 43, 136 (1979); Phys. Rev. A 23, 1535 (1981); Progr. Theor. Phys. 59, 1072 (1978)
43. A. Milchev, K. Binder and D. W. Heermann, Z. Physik B 63, 521 (1986)
44. J. R. Simon, P. Guyot and A. Ghilarducci de Silva, Phil. Mag. A 49, 151 (1984)
45. H. E. Stanley, An Introduction to Phase Transitions and Critical Phenomena, Oxford University Press, Oxford 1971
46. A. Sadiq and K. Binder, J. Stat. Phys. 35, 617 (1984)
47. P. C. Hohenberg and B. I. Halperin, Rev. Mod. Phys. 49, 435 (1977)
48. B. B. Mandelbrot, The Fractal Geometry of Nature, Freeman, San Francisco (1982)
49. J. L. Lebowitz, J. Marro and M. H. Kalos, Acta met. 30, 297 (1982); P. Fratz, J. L. Lebowitz, J. Marro and M. H. Kalos, Acta met. 31, 1849 (1983); D. A. Huse, Phys. Rev. (1987, in press)
50. F. F. Abraham, S. W. Koch and R. C. Desai, Phys. Rev. Lett. 49, 923 (1982); S. W. Koch, R. C. Desai and F. F. Abraham, Phys. Rev. A 2, 2152 (1983); S. W. Koch and R. Liebmann, J. Statist. Phys. 33, 31 (1983)
51. Y. C. Chou and W. I. Goldburg, Phys. Rev. A 23, 858 (1981); C. M. Knobler and N. C. Wang, J. Phys. Chem. 85, 1972 (1981)
52. A. Craievich and J. M. Sanchez, Phys. Rev. Lett. 47, 1308 (1981)
53. D. N. Sinha and J. K. Hoffer, Physica 107 B, 155 (1981)
54. O. Blaschko, G. Ernst, P. Fratzl, M. Bernole and P. Auger, Acta met. 30, 547 (1982)
55. S. Katano and M. Iizumi, J. Phys. Soc. Japan 51, 347 (1982); Physica 120 B, 302 (1983)
56. S. Katano and M. Iizumi, Phys. Rev. Lett. 52, 835 (1984)
57. M. Hennion, P. Guyot and D. Ronzaud, Acta met. 30, 599 (1982)
58. G. Kostorz, Physica 120 B, 387 (1987); S. Komura, K. Osamura, H. Fujii and T. Takeda, Physica 120 B, 397 (1983)
59. S. Komura, K. Osamura, H. Fujii and T. Takeda, Phys. Rev. B 30, 2944 (1984)
60. J. S. Langer and A. J. Schwartz, Phys. Rev. A 21, 948 (1980)
61. E. Siggia, Phys. Rev. A 20, 595 (1979)
62. G. F. Mazenko, O. T. Valls and F. C. Zhang, Phys. Rev. B 31, 4453 (1985); G. F. Mazenko and O. T. Valls, Phys. Rev. B 33, 1823 (1986)
63. J. Amar, R. D. Mountain and F. Sullivan, to be published
64. K. Tokuyama and K. Kawasaki, Physica 123 A, 386 (1984); K. Kawasaki and T. Ohta, Physica 118 A, 175 (1983); M. Tokuyama, Y. Enomoto and K. Kawasaki, preprints
65. T. Ohta, Ann. Phys. 158, 31 (1984); Progr. Theor. Phys. 71, 1409 (1984); H. Tomita, Progr. Theor. Phys. 71, 1405 (1984); Progr. Theor. Phys. 72, 656 (1984)
66. P. W. Voorhees and M. E. Glicksman, Acta met. 32, 2001, 2013 (1984); P. W. Voorhees, J. Stat. Phys. 38, 231 (1985)
67. J. A. Marqusee and J. Ross, J. Chem. Phys. 80, 536 (1984)

68. C. W. J. Beenakker, preprint; C. W. J. Beenakker and J. Ross, preprint; see also C. W. J. Beenakker and J. Ross, J. Chem. Phys. 83, 4710 (1985); M. P. Marder, preprint
69. H. Furukawa, Phys. Rev. A 29, 2160 (1984); A 30, 1052 (1984); Physica A 123, 497 (1984); Progr. Theor. Phys. 73, 586 (1985)
70. H. Furukawa, Prog. Theor. Phys. 74, 174 (1985)
71. H. Tomita, Prog. Theor. Phys. 71, 1405 (1984)
72. P. A. Rikvold and J. D. Gunton, Phys. Rev. Lett. 49, 226 (1982)
73. M. Schöbinger, S. W. Koch and F. F. Abraham, J. Statist. Phys. 42, 1071 (1986); S. W. Koch, preprint.
74. G. S. Grest and D. J. Srolovitz, Phys. Rev. B 30, 5150 (1984)

MONTE CARLO CALCULATIONS OF PHASE DIAGRAMS OF MAGNETIC ALLOYS ON THE BODY-CENTERED CUBIC LATTICE

B. Dünweg and K. Binder

Institut für Physik, Johannes Gutenberg-Universität Mainz
D-6500 Mainz, Postfach 3980, Federal Republic of Germany

ABSTRACT

We treat a model for a binary (AB) alloy, where species A is magnetic (Ising spin $\sigma_i = \pm 1$) while species B is not, and repulsive interactions are assumed between first and second neighbors of the same kind, in addition to a nearest-neighbor ferromagnetic exchange interaction. Both the mean-field approximation, the cluster variation (CV) method in the tetrahedron approximation and the Monte Carlo (MC) method are applied; comparing the phase diagrams obtained by the various approximations their accuracy is tested. It is shown that the CV method is in rather close agreement with the MC method for the present problem.

The calculation of phase diagrams for simple models of alloys is still a challenge in statistical mechanics[1–3]. In the present contribution, we briefly describe[4] a study of the following model Hamiltonian expressed in terms of occupation variables $\{c_i\}$ [$c_i = 1$ if lattice site i is taken by an A-atom (which is a magnetic species, carrying an Ising spin $\sigma_i = \pm 1$), while $c_i = 0$ if it is taken by a (nonmagnetic) B-atom]

$$\mathcal{H} = \sum_{\langle i,j\rangle_{nn}} \left\{c_i\, c_j\, (v_{nn}^{AA} - J\sigma_i\sigma_j) + \left[(1-c_i)c_j + c_i(1-c_j)\right] v_{nn}^{AB} + (1-c_i)(1-c_j)v_{nn}^{BB}\right\} + \sum_{\langle i,j\rangle_{nnn}} \left\{c_i c_j\, v_{nnn}^{AA} + \left[(1-c_i)c_j + c_i(1-c_j)\right] v_{nnn}^{AB} + (1-c_i)(1-c_j)v_{nnn}^{BB}\right\} \tag{1}$$

Here the first sum runs once over all pairs of nearest neighbors, the second over next nearest neighbors. Strictly pairwise interactions are assumed throughout: v_{nn}^{AA}, v_{nn}^{AB}, v_{nn}^{BB} are crystallographic interactions between nearest neighbor AA, AB and BB pairs, respectively, and likewise v_{nnn} for next-nearest neighbor pairs, while J is the magnetic exchange interaction. For the description of order-disorder phenomena, only the reduced interactions W_{nn}, W_{nnn} and J matter

$$W_{nn} = (v_{nn}^{AA} + v_{nn}^{BB} - 2\, v_{nn}^{AB})/4, \quad W_{nnn} = (v_{nnn}^{AA} + v_{nnn}^{BB} - 2\, v_{nnn}^{AB})/4. \tag{2}$$

G. M. Stocks and A. Gonis (eds.), Alloy Phase Stability, 263–268.

We study only one particular choice, $W_{nnn}/W_{nn} = 0.5$ and $J/W_{nn} = 0.7$, where $W_{nn} > 0$, $W_{nnn} > 0$, $J > 0$. Fig. 1 illustrates the ordered structures which are stable for these interactions.

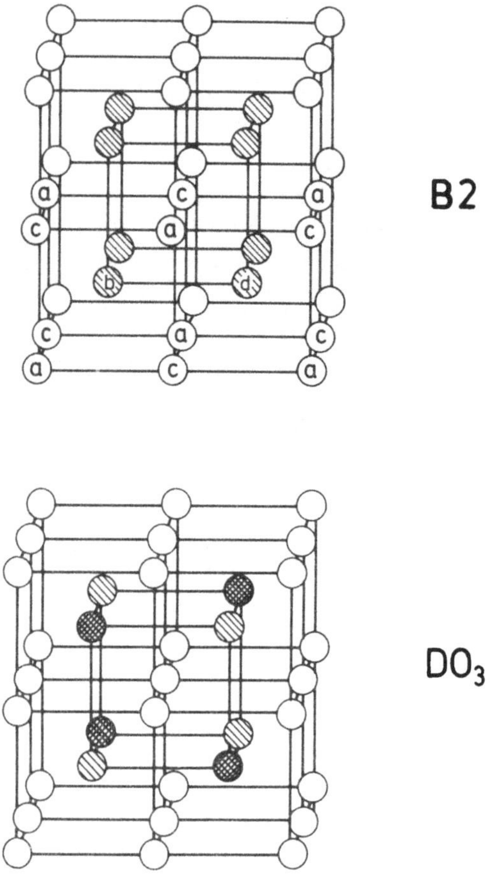

FIGURE 1. Body-centered cubic lattice showing the B2 structure (upper part) and DO_3 structure (lower part). Upper part shows assignment of four sublattices a,b,c,d. In the A2 structure, the average concentrations of A and B atoms are identical at all four sublattices, while in the B2 structure the concentrations at the b and d sublattices are the same, but differ from the concentrations at the a and c sublattices, which again are the same (example: stoichiometric FeAl). In the DO_3 structure, the concentrations at the a and c sublattices still are the same, while the concentration at sublattice b differs from the concentration at sublattice d (example: stoichiometric Fe_3Al).

Previous related work has studied similar models on magnetic alloys on the bcc lattice with mean-field approximations[5,6] (intending to fit[6] the real Fe-Al phase diagram) and on the fcc lattice with the Kikuchi[7] cluster variation (CV) method[8]. The present paper is the first, however, which applies three methods to the same model, namely the mean field approximation (MFA), the CV method in the tetrahedron approximation, and the Monte Carlo (MC) method[3,9]. The tetrahedron is put onto the lattice such that the atoms at each corner belong to different sublattices. Since each lattice site can be in 3 states (A↑,A↓,B), there are up to $3^4 = 81$ variational parameters, which need to be handled in the minimization procedure of the free energy in the CV method. Details of these calculations can be found in Ref. 4; here we only wish to compare the final results and to discuss the merits of the various methods.

Figs. 2-4 show the phase diagram, plotted in the plane of the variables temperature and concentration 1-<c> of the nonmagnetic atoms. It is seen that the MFA overestimates the ordering tendency distinctly (note different ordinate scales), and also some topological features of the phase diagram go wrong in MFA: it does not yield a direct transition A2 para - DO_3 para, but predicts always an intermediate B2 para phase. On the other hand, the CV phase diagram (Fig. 3) and the MC phase diagram (Fig. 4) are in rather close agreement with each other: their topology is precisely the same, and the transition temperatures predicted by the CV method exceed those of the MC method (which should be exact, apart from statistical errors[9]) never by more than 5% relative error. This good agreement is rather remarkable for a phase diagram exhibiting 6 distinct phases, transition lines of both second and first order, and several multicritical points.

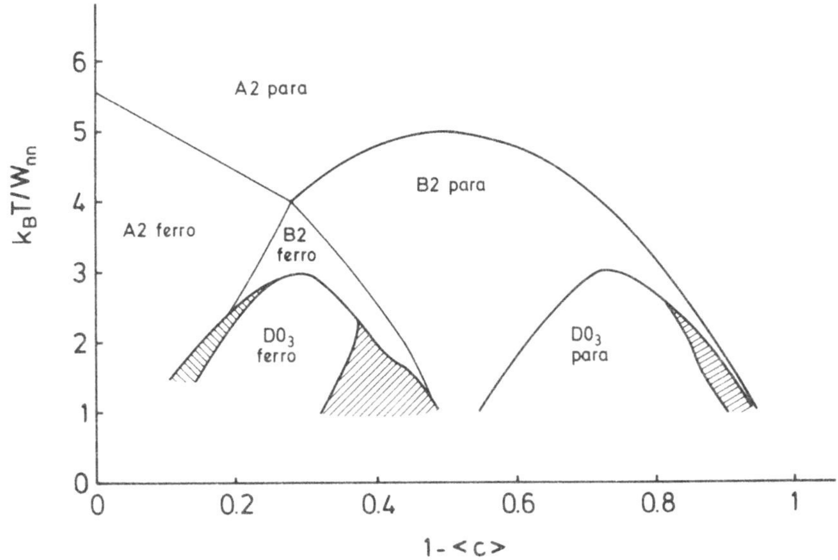

FIGURE 2. Phase diagram as predicted by the MFA. Nature of the various pure phases is indicated in the figure; the shaded regions denote two-phase coexistence regions between the two adjoining pure phases.

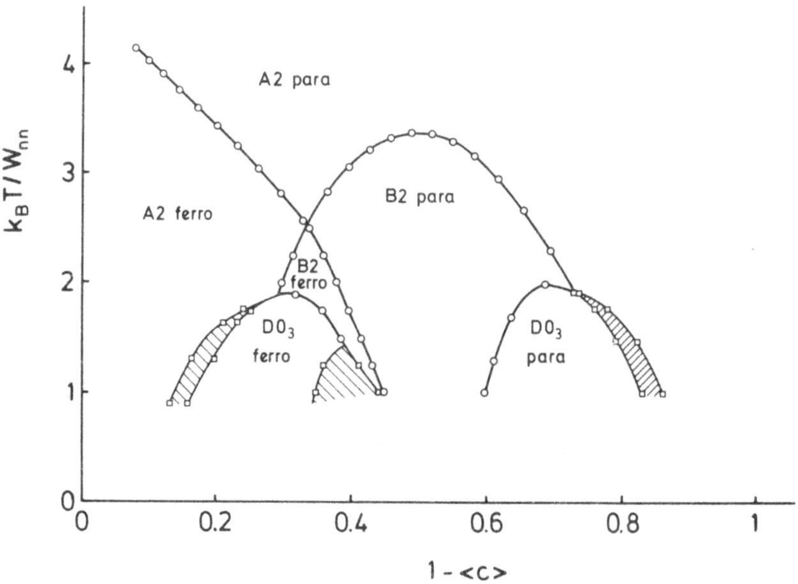

FIGURE 3. Phase diagram as predicted by the CV method in tetrahedron approximation.

For the accurate location of second order phase boundaries, we have found it convenient in the Monte Carlo study to use the cumulant intersection method[10]. The cumulant U_F for ferromagnetic order is defined in terms of the second and fourth moments of the magnetization M as $U_F = 1 - \langle M^4\rangle/(3\langle M^2\rangle^2)$, the cumulant U_1 for the B2 phase in terms of the second and fourth moments of its order parameter ϕ_1 as $U_1 = 1 - \langle\phi_1^4\rangle/(3\langle\phi_1^2\rangle^2)$, etc. Finite size scaling theory implies that for finite systems of different linear dimensions L the various cumulants should intersect at the respective critical points, and this is in fact borne out nicely by the data (Fig. 5).

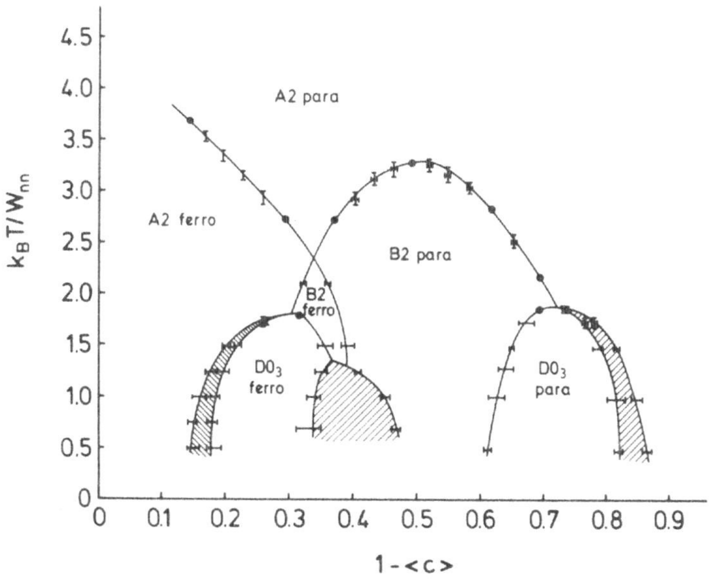

FIGURE 4. Phase diagram as predicted by the MC method. Note that the actual calculation is done in the grandcanonical ensemble; therefore, data on transition lines obtained by temperature variation at fixed chemical potential difference now get a horizontal error bar in addition to the vertical one, due to the statistical error of <c>. Data are shown as dots if error is too small for being resolved, and curves are drawn as guides to the eye only.

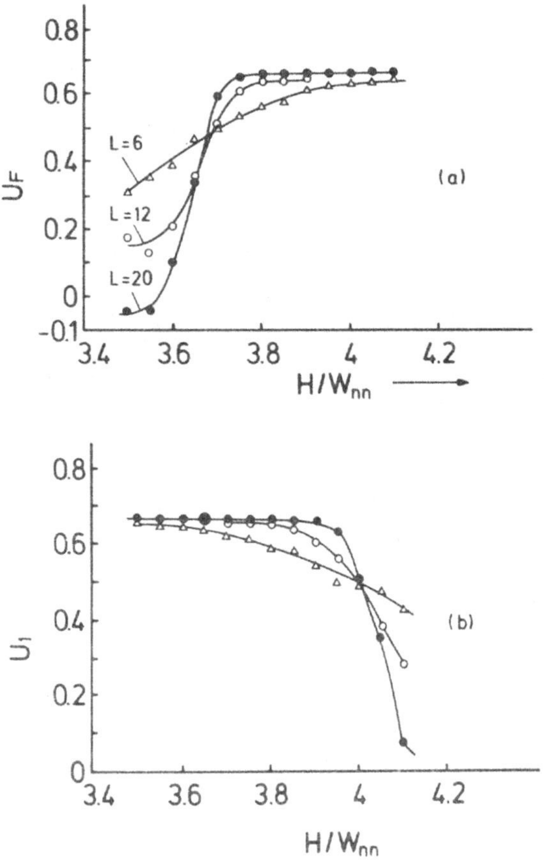

FIGURE 5. Cumulant intersection plot for the ferromagnetic order (a) and the B2 order (b) for $k_BT/W_{nn} = 2.1$, varying the chemical potential difference H/W_{nn}. Data for L = 6,12 are based on runs with 20000 MCS/site, while only 15000 MCS/site were recorded for L = 20.

REFERENCES

1. D. de Fontaine, in Solid State Physics (ed. by H. Ehrenreich, F. Seitz, and D. Turnbull) Vol. 34, p. 73 (Academic Press, New York 1979)
2. T. Mohri, J. M. Sanchez, and D. de Fontaine, Acta Metall. 33, 1171 (1985)
3. K. Binder, in Festkörperprobleme (Advances in Solid State Physics) Vol 26, p. 133 (P. Grosse, ed.) Vieweg, Braunschweig (1986)
4. For a more detailed account, see B. Dünweg and K. Binder, Phys. Rev. B (1987, in press)
5. G. Schlatte, G. Inden, and W. Pitsch, Z. Metallkde 65, 94 (1974)
6. S. V. Semenovskaya, phys. stat. sol. (b) 64, 291 (1974)
7. R. Kikuchi, Phys. Rev. 81, 998 (1951)
8. C. E. Dahmani, M. C. Cadeville, J. M. Sanchez, and J. L. Moran-Lopez: Phys. Rev. Lett. 55, 1208 (1985)
9. K. Binder (ed) Monte Carlo methods in statistical physics (Springer, Berlin 1979)
10. K. Binder, Z. Phys. B 43, 119 (1981)

HIERARCHY OF CLUSTER VARIATIONAL METHODS ON 3-DIMENSIONAL LATTICES AND APPLICATION TO THE STUDY OF FCC PHASE DIAGRAMS

A. FINEL

Office National d'Etudes et de Recherches Aérospatiales
ONERA, BP 72, 92322 Chatillon Cedex, France

1. INTRODUCTION

Closed form approximations for statistical systems, such as mean field theories or cluster variational technics, are very useful to calculate phase diagrams. The most popular and succesful closed form technic to date is Kikuchi's Cluster Variation Method (CVM) [1]. The variety of problems that have been analysed with the CVM shows the importance of that technic (for a recent review, see [2]).

Thus far, most of the CVM studies have been done with relatively small clusters (up to six points) and with short ranged interactions (first and second neighbors). When compared to Monte Carlo simulations, the CVM results are generally very accurate, except of course near critical points. However, in some cases, as on the FCC lattice with antiferromagnetic interactions between first neighbors, there is still some discrepancy between the CVM phase diagram and Monte Carlo simulations [3,4,5]. The only way to improve the CVM results is obviously to use bigger clusters. Moreover, to increase the size of the basic cluster is necessary if we want to introduce more distant interactions. Hence, one could greatly benefit from an optimized procedure for selecting the basic clusters. Obviously, this needs to clarify the precise nature of the approximations involved in the cluster methods.

This paper proceeds as follows. In Section 2, we develop two sequences of 3 dimensional cluster approximations, which converge towards the exact free energy with the size of the basic cluster. In Section 3, we present a numerical study of these hierarchies on the FCC lattice. In Section 4, we use one of these approximations to calculate the phase diagram of the FCC lattice with antiferromagnetic interactions between first neighbors. Finally, Section 5 contains some final remarks.

2. HIERARCHY of CVM APPROXIMATIONS FOR 3 DIMENSIONAL LATTICES

One of the main interest of the CVM is to provide a general formalism that allows to define a level of approximation for each selected basic cluster. It is then attractive to define a sequence of approximations as a function of the size of the basic clusters. However, numerical studies show that large clusters do not necessarely lead to more accurate results than smaller ones and that, more surprisingly, some CVM approximations lead to unphysical results , even if lower order approximations are rather precise [5]. This leads to the important question of the convergence of the CVM when the size of the basic cluster increases indefinitely.

G. M. Stocks and A. Gonis (eds.), Alloy Phase Stability, 269–279.

In contrast to the huge amount of applications, little effort has been spent on that point. It concerns the following questions: Which criteria should be used to judge the quality of a CVM approximation ? Which sequence of clusters and, more generally, which entropy approximation should be chosen in order to improve systematically the results and to converge towards the exact solution ?

These questions have already been studied and solutions proposed for 2 dimensional lattices by Kikuchi and Brush [6] and, from a more fundamental point of view, by Schlijper [7]. We present here very briefly a general solution for 3 dimensional lattices [5].

2.1 A simplified exact variational principle

As we are dealing with variational technics, a good starting point to build a hierarchy is the exact variational principle, which says the following.

Let ρ be the density matrix of any state of a system. Then $\rho(C)$ is the probabily that the system presents the configuration C. The free energy per site associated to the state ρ is

$$f(\rho) = e(\rho) - T.s(\rho) \qquad (1.a)$$

where $e(\rho)$ and $s(\rho)$ are respectively the internal energy and entropy per site :

$$e(\rho) = \frac{1}{N} \sum_C \rho(C)\, H(C) \qquad (1.b)$$

$$s(\rho) = -\frac{k}{N} \sum_C \rho(C)\ \mathrm{Ln}\, \rho(C) \qquad (1.c)$$

H stands for the Hamiltonian of the system and N for the number of sites; the sums run over all the configurations C of the system. According to the variational principle, the exact free energy, f^{ex}, is the minimum of the functional $f(\rho)$ over all the states ρ of the system; in other words :

$$f^{ex} = \min \{\, e(\rho) - T.s(\rho)\ ;\ \rho \in S \,\} \qquad (2.a)$$

where S represents the set of all the " admissible " states, i.e. such that :

$$\rho \in S\ :\ \sum_C \rho(C) = 1 \qquad (2.b)$$

A state ρ which minimizes (2.a) is called an equilibrium state. As we are interested by the thermodynamic limit of an infinite system ($N \to \infty$), ρ is a function of an infinite number of correlation functions and we do not know how to perform the minimization.

However, the variational principle (2) can be simplified : when the interaction range is finite, the minimization over the infinite system can be replaced exactly by a minimization over a cluster which is finite along one direction.

In order to be more explicit, we continue with a practical case. Consider a simple cubic lattice with first neighbor interactions and in the disordered phase (the generalisation to more complicated lattices, ordered phases and interactions is straightforward). We note P a plane infinite along directions X and Y and D the infinite cluster that consists in two consecutive planes along Z :

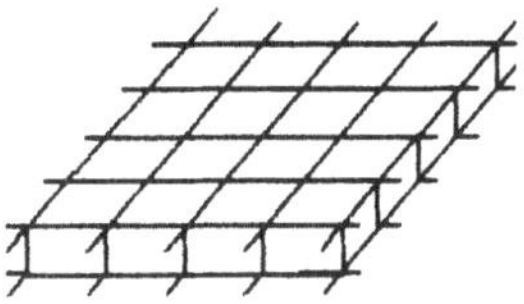

D

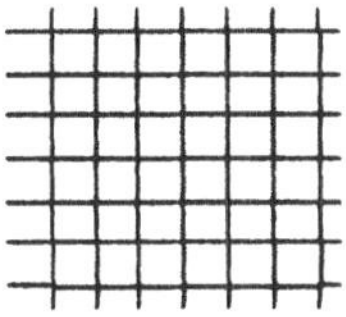

P

It is easy to show, with transfer matrix methods, that there exists an exact equilibrium state ρ^{ex} which factorizes along Z as follows[1] [5]:

$$\rho^{ex} = \frac{\prod_D \rho_D^{ex}}{\prod_P \rho_P^{ex}} \tag{3.a}$$

where the products run respectively over all the clusters D and P along Z, and ρ_D and ρ_P are the density matrices of D and P obtained through partial sommation over ρ^{ex} :

$$\rho_D^{ex} = \mathrm{Tr}_{\infty - D}\ \rho^{ex} \qquad\qquad \rho_P^{ex} = \mathrm{Tr}_{\infty - P}\ \rho^{ex} \tag{3.b}$$

The notation ($\infty-\alpha$) means a sum over all the configurations of the exterior of cluster α. Let then S^* be the set of all the admissible states ρ which factorize in the way given by (3) :

$$S^* = \{ \rho \in S\ ,\ \rho = \frac{\prod_D \rho_D}{\prod_P \rho_P} \} \tag{4}$$

Eq. (3) shows that S^* contains an exact equilibrium state. Thus, the exact free energy can be obtained by a minimization reduced to the subset S^* :

$$f^{ex} = \min\ \{ e(\rho) - T.s(\rho)\ ;\ \rho \in S^* \} \tag{5}$$

The interesting point of introducing the subset S^* is that the previous formulation of the exact variational principle can still be simplified. First, for a state $\rho \in S^*$, the entropy $s(\rho)$ (eq.1c) has a particular form :

$$\rho \in S^* :\ s(\rho) = 2\, s_D(\rho_D) - s_P(\rho_D)$$

where s_D and s_P are the entropies per site of clusters D and P [2]. Secondly, there is an obvious

[1] This result is intimetaly related to the fact that the cluster D, infinite along X and Y, but finite along Z, separates the whole lattice into two regions that do not interact.

[2] We recall that we consider here the case of a disordered phase of an infinite lattice : the entropies per site s_D (resp. s_P) of the subsets D (resp. P) are then all equal and functions of the same reduced density matrix ρ_D.

one-to-one correspondance between the subset S^* and S_D, the set of all the states ρ_D defined directly on the infinite cluster D. Hence, the exact variational principle (5) can be rewritten as follows :

$$f^{ex} = \min \{ e(\rho_D) - T.s^*(\rho_D) ; \rho_D \in S_D \} \tag{6.a}$$

with

$$s^*(\rho_D) = 2\, s_D(\rho_D) - s_P(\rho_D) \tag{6.b}$$

where the asterisk means that $s^*(\rho_D)$ is not the entropy per site of cluster D, but the entropy per site of the infinite system in the state ρ related to ρ_D by (4). In other words, the exact free energy can be obtained by a minimization of a functional where the entropy $s(\rho)$ has been replaced by a functional $s^*(\rho_D)$ and the minimization set S by S_D, the set of all the states defined on the subcluster D, finite along one direction and infinite along the others.

This result is important : it shows that we can reduce, exactly, the dimension of our problem by one unity. However, the cluster D remains infinite along two directions, which still prevents us from minimizing. It should then be interesting to continue the processus of factorization along these directions. Unfortunately, and that is not a surprise, the state ρ_D which minimizes the r.h.s. of (6a) cannot be factorized on density matrices of subclusters of D.

Nevertheless, the last form of the exact variational principle, and the way we followed to established it, give us a nice way for introducing approximations. The important point is that we need to approximate the exact variational principle along two directions only.

2.2 Hierarchy of approximations and convergence towards the exact solution

As suggested by the previous simplification, we continue to reduce the minimisation set S_D as follows. Instead of minimizing over all the states ρ_D of D, we try to restrict the minimization to a subset of states which can be factorized, in some way, on a finite cluster α included in D. In other words, we try to replace the minimization other the states of an infinite cluster by a minimization over the states of a finite cluster α. But, and that is the key point, we must choose a finite cluster α in D in such a way that any density matrix ρ_α, defined directly on α, can be related, through a processus of factorization, to an "admissible " state $\tilde{\rho}_D$ of D, i.e. a state that belongs to S_D. If such a cluster α can be found, we note S_α the set of the states defined on α, and $\tilde{S}_D(\alpha)$ the subset of S_D defined through the processus of extension we look for; in short :

$$\rho_\alpha \in S_\alpha \xrightarrow[\text{factorization}]{\text{processus of}} \tilde{\rho}_D \in \tilde{S}_D(\alpha) \tag{7}$$

Hence, a minimisation restricted to $\tilde{S}_D(\alpha)$ will still concern states of the infinite cluster D, but through the states of a finite cluster. Thus, in principle, we should be able to perform this restricted minimization and obtain a higher bound of the exact free energy.

We would like to point out that the finite cluster α we look for, needs only to be included in D, which is finite along Z. Hence, whatever its sizes along X and Y are, the height of α along Z can be fixed to one, i.e. to the interaction range along Z. This fact is a direct consequence of the formulation (6) of the exact variational principle.

There is a very simple cluster α for which a relation of type (7) exists, namely the cluster D_{nm} that consists in n cubes along direction X and m cubes along direction Y:

D_{nm}

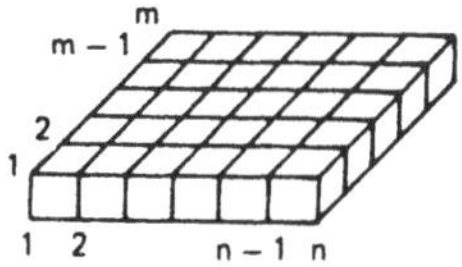

This cluster contains n*m unit cubes. It is possible to show that, for any density matrix $\rho_{D_{nm}}$, defined directly on D_{nm}, there exists, through a simple "processus of extension"[3], a state $\tilde{\rho}_D$ on the cluster D which verifies relation (7)[5] (more details will be published elsewhere). We recall that, as n and m increase, we keep the size of D_{nm} along Z fixed and equal to one. For given n and m, let then S_{nm} be the set of all the states of the cluster D_{nm}, and $\tilde{S}_D(nm)$ the set of all the extensions $\tilde{\rho}$ obtained from the elements of S_{nm}. By construction, $\tilde{S}_D(nm)$ is a subset of S_D. Hence, a minimization over $\tilde{S}_D(nm)$ instead of S_D leads to the inequality:

$$f^{ex} < \min \{ e(\tilde{\rho}_D) - T.s^*(\tilde{\rho}_D) ; \tilde{\rho}_D \in \tilde{S}_D(nm) \} \tag{8}$$

where the states $\tilde{\rho}_D$ depend only on the states $\rho_{D_{nm}}$ of the cluster D_{nm}. In other words, even if they concern the infinite system, the states $\tilde{\rho}_D$ depend only on a <u>finite</u> number of correlation functions, those included in the finite cluster D_{nm}. Hence, at first sight, we should be able to perform the minimization that occurs in the r.h.s. of (8). Unfortunately, the entropy functionnal $s^*(\tilde{\rho}_D)$ is still to much complicated [5]; in particular, it cannot be written as a linear combination of entropies of finite clusters included in D_{nm}. More precisely, $s^*(\tilde{\rho}_D)$ depends on $\rho_{D_{nm}}$ through entropies of infinite clusters included in D.

We need then an extra approximation : we must replace the entropy $s^*(\tilde{\rho}_D)$by a simpler functional which depends only and explicitly on clusters included in D_{nm}. However, this choice must preserve the inequality (8) in order to keep the comparison with the exact free energy. In other words, we must choose an approximate entropy expression $s^{app}(\rho_{D_{nm}})$ in such a way that:

$$s^{app}(\rho_{D_{nm}}) < s^*(\tilde{\rho}_D) \tag{9}$$

where the state $\tilde{\rho}_D$ is the extension of $\rho_{D_{nm}}$ to the infinite cluster D. Moreover, this approximation should lead to an approximate free energy which converges monotically towards f^{ex} when n and m

[3] This "processus of extension" consists in two consecutives factorizations along X and Y.

tend to infinity.

We have found only one approximate entropy which satisfies these two criteria, namely :

$$T_{nm}(\rho_{D_{nm}}) = S_{D_{nm}} - S_{D_{n-1,m}} - S_{D_{n,m-1}} + S_{D_{n-1,m-1}} - S_{P_{nm}} + S_{P_{nm}/1} \qquad (10.a)$$

where D_{kl} represents a cluster of sizes (in unit cubes) k and l along X and Y, P_{kl} a plane of sizes (in unit squares) k and l (P_{kl} can be viewed as the intersection of two consecutive clusters D_{kl} along direction Z), and $P_{kl/1}$ is a cluster P_{kl} from which we extract one corner (see Fig.1); S_α represents the total entropy of cluster α.

Fig.1

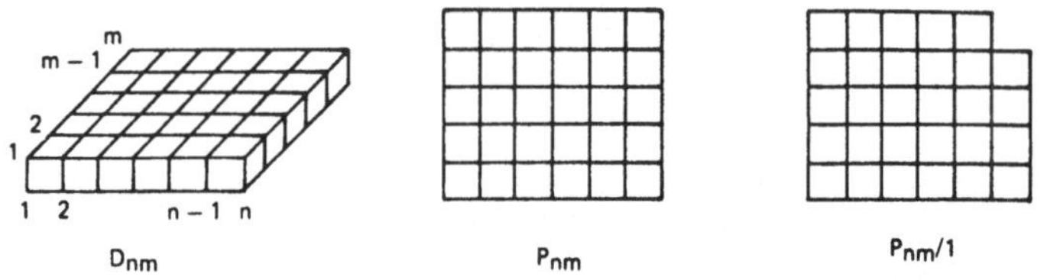

We note f^T_{nm} the approximate free energy obtained through the restricted variationnal principle associated with this approximation :

$$f^T_{nm} = \min \{ e(\rho_{D_{nm}}) - T \, . \, T_{nm}(\rho_{D_{nm}}) \; ; \; \rho_{D_{nm}} \in S_{D_{nm}} \} \qquad (10.b)$$

where the minimization is over the set $S_{D_{nm}}$ of all the states defined on cluster D_{nm}. This approximate free energy verifies [5] :

$$\text{a)} \quad f^{ex} < f^T_{nm}$$

$$\text{b)} \quad f^T_{n'm'} < f^T_{nm} \quad \text{if} \quad n' \geq n \; , \; m' \geq m$$

$$\text{c)} \quad \lim_{n,m \to \infty} f^T_{nm} = f^{ex} \qquad (11)$$

In other words, f^T_{nm} decreases and converges monotically towards f^{ex}. We will refer to this approximation as the T-hierarchy. We recall that, as n and m increase, the size of the basic cluster D_{nm} along Z is kept fixed : the previous hierarchy consists in increasing the size of D_{nm} along two directions only. This choice is intimetaly related to the fact that the cluster D, infinite along X and Y but finite along Z, separates the whole lattice into two regions that do not interact. This is the reason why this approximation is exact, from a formal point of view, for the basic cluster D.

We have defined the approximate entropy T_{nm} by using a criterion on the free energy : we wanted that the approximate free energy converges monotically towards the exact one.

Another way is to choose directly the usual CVM approximations [1] associated with the same sequence of basic cluster, i.e. :

$$f^{C}_{nm} = \min\{ e(\rho_{D_{nm}}) - T\,.\,C_{nm}(\rho_{D_{nm}})\ ;\ \rho_{D_{nm}} \in S_{D_{nm}} \} \tag{12.a}$$

with the approximate entropy :

$$C_{nm}(\rho_{D_{nm}}) = S_{D_{nm}} - S_{D_{n-1,m}} - S_{D_{n,m-1}} + S_{D_{n-1,m-1}} - S_{P_{nm}} + S_{P_{n-1,m}} + S_{P_{n,m-1}} - S_{P_{n-1,m-1}} \tag{12.b}$$

We will refer to these approximations as the C-hierarchy. We can easily show that property (11.c) is still verified by the C-hierarchy :

$$\lim_{n,m\to\infty} f^{C}_{nm} = f^{ex}$$

Unfortunately, we could not proved properties (11.a) and (11.b). However, we can show that, for any basic cluster D_{nm}, we have :

$$f^{C}_{nm} < f^{T}_{nm}$$

Hence the C-hierarchy converges more rapidely than the T-hierarchy, but perhaps not monotically.

3. NUMERICAL STUDIES OF HIERARCHIES C AND T ON THE FCC LATTICE.

The C- and T- hierarchies have been introduced above on the simple cubic (SC) lattice. They can be easily transposed to the FCC lattice as follows. A SC lattice can be subdivided into two FCC sublattices. Suppose that one of these sublattices is empty : the initial lattice reduces then to a unique FCC lattice with half the number of sites. Any entropy expression defined on the SC lattice can then be directly transposed to the FCC lattice with the following rule : a cluster of the SC lattice becomes a cluster of the FCC lattice by extracting from the initial cluster the sites of the empty sublattice. As example, a simple cube becomes a tetrahedron, a simple square becomes a first neighbor pair.......
More generally, approximations based on the clusters D_{22} (quadruple cube), D_{21} (double cube), D_{11} (simple cube) becomes approximations based on the quadruple tetrahedron (QT), the double tetrahedron (DT) and the tetrahedron (T).

We have studied on the FCC lattice, the approximations T and C based on the previous clusters (QT,DT and T). The corresponding entropy expressions are given in Fig.2. In order to limit the number of correlation functions, we have considered only the disordered phase (which is defined by 82 correlation functions in the approximations based on the QT). We present in Fig.3 the different free energies in hierarchies C and T, with antiferromagnetic interactions between first neighbors. We have also reported on Fig.3.b, lower and higher bounds of the exact free energy, obtained by high temperature expansion [19]. These bounds defined a width smaller than 0.08% ; the free enrgy given by the QT in the C-hierarchy lies just inbetween the lower and higher bounds !

Finally, we have determined the critical temperature T_C with ferromagnetic interactions between first neighbors. For the QT in the C-hierarchy, we obtained :

$$kT_C/12J = 0.82659\ ,$$

which is to be compared to the "exact one" obtained by high temperature expansion [8] : $kT_C/12J = 0.81638$.

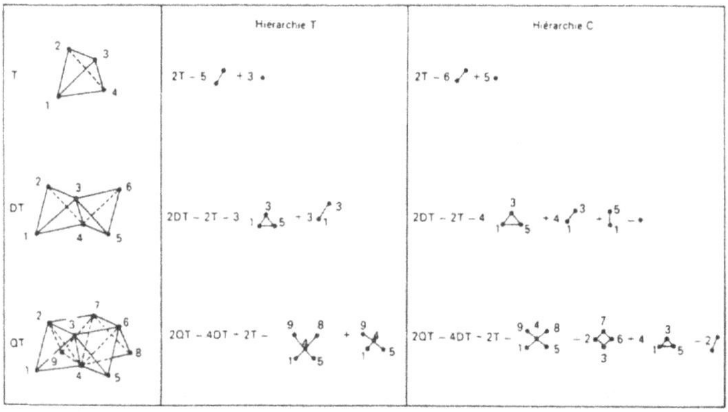

Fig. 2 : *Entropy expressions for the tetrahedron (T), double tetrahedron (DT) and quadruple tetrahedron (QT) in the T and C hierarchies, for the disordered phase of the FCC lattice; the notation "aα" represents the quantity a.S_α, where S_α is the total entropy of cluster α.*

These results show that the C-hierarchy converges very rapidely toward the exact free energy (the QT approximation is only the third step of the hierarchy). We recall that the C- and T-hierarchies, as presented here, are valid with first neighbor interactions only. More precisely, an approximation based on a cluster D_{nm} is valid when the interactions are precisely limited to the size of D_{nm}. As example, on the FCC lattice, an approximation based on the QT can be used if the interactions are limited to clusters included in the QT, i.e. all the 1st neighbor pairs, but only the 2nd and 4th neighbors which lie in the X-Y planes, the 3rd neighbors of type $(\pm 2, \pm\frac{1}{2}, \pm\frac{1}{2})$ and $(\pm\frac{1}{2}, \pm 2, \pm\frac{1}{2})$, etc.

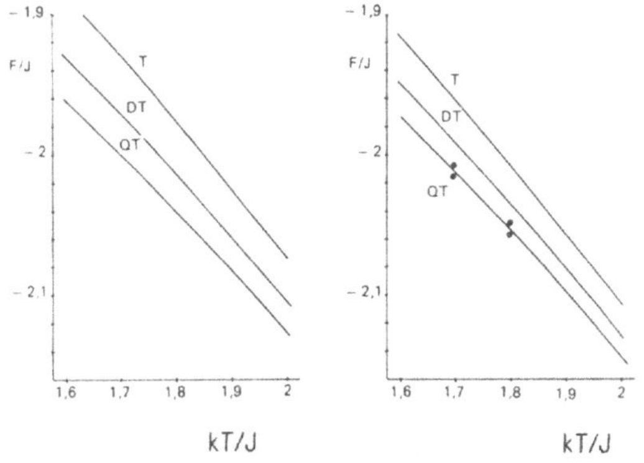

a) hierarchy T   b) hierarchy C

Fig. 3 : *Free energy per site F of the FCC lattice with antiferromagnetic interactions J between first neighbors, as a function of temperature; we present the results for the approximations based on the clusters T,DT and QT in the hierarchies T (a) and C (b). The points in (b) represent higher and lower bounds of the exact free energy, determined by high temperature series* [19].

4. THE FCC ISING MODEL WITH ANTIFERROMAGNETIC FIRST NEIGHBOR INTERACTIONS.

The Ising model on the FCC lattice has been intensively studied, in part because it models many binary alloys. When the interactions are limited to the first neighbors, the Hamiltonian reads :

$$H = J \sum_{nm} \sigma_n \sigma_m - h \sum_n \sigma_n$$

where the first summation extends over nearest neighbor pairs and $\sigma_n = \pm 1$. The phase diagram of this model has been investigated by such theoritical tools as the mean field theory [9], the CVM [3,10], Monte Carlo simulations [4,11-16], high and low temperature expansions [17-19] ....(for a recent review, see [20]). Most of those studies were motivated by the fact that the precise topology of the phase diagram was controversial. In particular, the most important discrepancy concerned the location of the triple point, where the ordered phases $L1_0$ and $L1_2$ and the disordered one meet. Early Monte Carlo and CVM studies suggested that the triple point lies at zero temperature at h=4J [10,11,12], whereas more recent investigations agree to locate that point at finite temperature (kT/J~1) [3,4,13-16].

The origin of the problem is that for h=4J, the ground state is so strongly degenerated that the residual entropy per site is finite. In such a case, the usual low temperature expansions cannot be used and nothing exact, even qualititavely, can be said about the equilibrium states at finite temperature around h=4J.

We have undertaken an improved CVM study of the whole phase diagram. We have used simultaneously two different approximations : the tetrahedron-octehedron (TO) approximation for the ordered phases, and the quadruple tetrahedron (QT) one for the disordered phase. The use of this "mixed" CVM is justified as follows.

As explained above, a CVM approximation consists in using a free energy functional where the exact entropy has been replaced by an approximate expression that depends only on the density matrices of a finite cluster α. For a given phase, the CVM free energy is then obtained by a minimization over the density matrices ρ_α compatible with the symmetry of that phase (more precisely, we minimize over a set of independant correlation functions whose number depends on the symmetry). As we use an approximate entropy, the larger the "exact" entropy is, the less precise the CVM free energy should be. Consider now a first order transition between an ordered phase and a disordered one. The entropy, which increases with temperature, presents a finite jump at the transition. Thus, for a given CVM approximation and near the transition, the error on the free energy is larger in the disordered phase than in the ordered one. Hence, with the size of the basic clusters, a sequence of approximations converges more rapidely in the ordered phase than in the disordered one (note that this argument is valid only for first order transitions). As a result, if we want to achieve the same degree of accuracy in both domains, we must use a higher order approximation in the disordered phase. That is what we call a "mixed" CVM.

The phase diagram of Fig.4 has been obtained with the TO approximation for $L1_0$ and $L1_2$ and the QT one for the disordered phase (the order-disorder transition are always first order, although very weakly near the triple point). We refer to this mixed CVM as the TO-QT approximation. As shown in Fig. 4, the TO-QT results are in very good agreement with the Monte Carlo simulations, even near the triple point (in the "pure" TO approximation, that point is off by 50% [3]). For the transition temperature at h=0, we obtain $kT_t/J = 1.745$, whereas the "exact" value, obtained by high and low temperature expansions, is $kT_t/J = 1.746 \pm 0.005$ [19] (the TO approximation leads to $kT_t/J = 1.810$ [3]).

These results show that a mixed CVM can lead to very precise phase diagrams, without too much computational effort, as only the disordered phase needs a high order approximation.

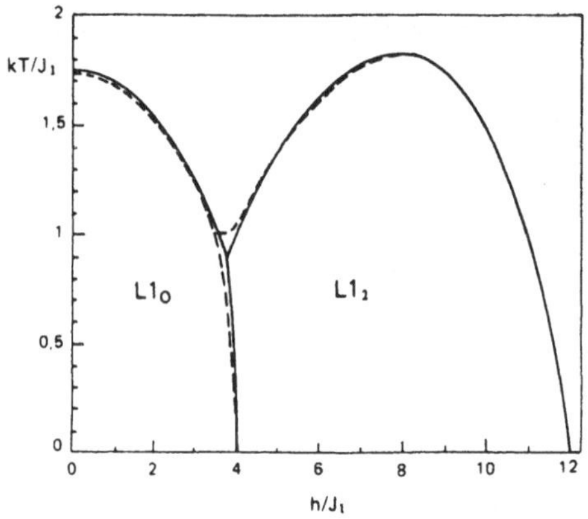

Fig. 4 : *Phase diagram of the FCC lattice with antiferromagnetic interactions J between first neighbors : the CVM result (TO-QT method) are represented in full lines, the Monte Carlo one in broken lines.*

5. FINAL REMARKS

We have presented very briefly a derivation of two series of cluster approximations, the C and T hierarchies, that have interesting properties. As noted above, these hierarchies are valid with first neighbor interactions only. They can be easily transposed when the interaction range is larger. Consider as example the simple cubic lattice with interactions up to the second neighbors along the three cubic directions. The exact reformulation (6) of the variational principle is still valid if we replace the infinite cluster D by the cluster D^2, which consists in three consecutive infinite planes along Z, and the infinite cluster P by D. As a result, the approximate entropies T_{nm} and C_{nm} can be directly transposed as follows : a cluster D_{nm} is replaced by a cluster D^2_{nm}, of size two along Z (in unit cube), and the cluster P_{nm} by D_{nm}. In other words, the C and T hierarchies are now based on the clusters D^2_{nm} (note that the first step of a hierarchy is now the cluster D^2_{22}).

We have seen above that any approximation defined on the simple cubic lattice can be easily transposed to the FCC lattice : we just need to remove one of FCC-sublattice of the original lattice. Through this operation, a cluster D^2_{22} becomes alternatively a face-centered cube, that contains 14 points, and a 13-point cluster formed by one site surrounded by its twelve first neighbors. Thus, any approximation based on D^2_{22} becomes an approximation based on these two clusters, referred to as the 13-14-point approximation (see also [21]).

Preliminary results in the disordered phase show that the 13-14-point approximation is very precise. Moreover, it can be used with interactions up to the fourth neighbors. Unfortunately, the number of correlation functions increases dramaticaly for an ordered phase (there are already 742 correlation functions in the disordered phase !). However, this difficulty can be by-passed by using a mixed CVM . a study of a phase diagram with interactions up to the fourth neighbors could be done with the 13-14-point approximation for the disordered phase and the TO approximation for the ordered phases (even if they are not included in the basic clusters, interactions between third and fourth neighbors can be taken into account in the TO approximation).

This work was supported in part by DRET through grant n° 86.34.011

REFERENCES

1. KIKUCHI R., Phys. Rev., 81 ,988 (1951)
2. MOHRI T., SANCHEZ J.M. and DE FONTAINE D., Acta Metall., 33 , 1171 (1985)
3. FINEL A. and DUCASTELLE F., Europhys. Lett., 1 ,135 (1986)
4. DIEP H.T., GHAZALI A., BERGE B. AND LALLEMAND P., Europhys. Lett., 2 , 603 (1986)
5. FINEL A., Thèse de Doctorat d'Etat, Université Paris VI, 1987
6. KIKUCHI R. and BRUSH S.G., J. Chem. Phys., 47 ,195 (1967)
7. SCHLIJPER A.G., J. Stat. Phys., 35 , 285 (1984)
   J. Stat. Phys., 40 , 1 (1985)
8. MACKENZIE N.D., J. Phys., A8 , 102 (1975)
9. SHOCKLEY W. , J. Chem. Phys. , 6 , 130 (1938)
   GAHN U. , Z. Metallkunde, 64 , 268 (1973)
   BUTH J. and INDEN G., Acta Met., 30 , 213 (1982)
10. SANCHEZ J. M., DE FONTAINE D. and TEITLER W. , Phys. Rev. B , 26 , 1465 (1982)
11. BINDER K. , Z. Phys. B , 45 , 61 (1981)
12. BINDER K., LEBOWITZ J. L., PHANI M. K. and KALOS M. H. , Acta Met., 29 , 1655 (1981)
    BINDER K., Phys. Rev. Lett. , 45 , 811 (1980)
13. LEBOWITZ J. L., PHANI M. K. and STYER D. F., J. Stat. Phys., 38 ,413 (1985)
14. STYER D. F., PHANI M. K. and LEBOWITZ J. L., Phys. Rev. B, 34 , 3361 (1986)
15. ACKERMANN H., CRUSIUS S. and INDEN G., Acta Met. , 34 , 2311 (1986)
16. GAHN U., J. Phys. Chem. Solids , 47 , 1153 (1986)
17. SLAWNY J., J. Stat. Phys., 20 , 711 (1979)
18. MACKENZIE N. D. and YOUNG A. P., J. Phys. C , 14 , 3927 (1981)
19. STYER D. F., Phys. Rev. B, 32 , 393 (1985)
20. KIKUCHI R., Prog. Th. Phys. Japan (in press)
21. SANCHEZ J. M. and DE FONTAINE D., Phys. Rev. B, 17 , 2926 (1978)

A CRITERION FOR DETERMINING THE TRICRITICAL POINT

Ryoichi Kikuchi
Max Planck Institut fur Eisenforschung
Dusseldorf, Fed. Rep. Germany
and Department of Materials Science and Engineering, FB-10
Advanced Materials Technology Program, Washington Technology Center
University of Washington, Seattle, WA 98195

INTRODUCTION

In phase diagram calculations, we sometimes come across a situation in which a second-order transition part and first-order transition part sit next to each other. The dividing point between the two parts is called a tricritical point, which we write as TRCP. Examples are in a bcc disorder-B2 phase boundary [1], and in Fe-Al systems [2]. In this note we derive a criterion to determine the TCRP.

THE "INSTABILITY" HESSIAN DETERMINANT

In calculating phase diagrams, a second-order boundary can be determined accurately using the "instability" Hessian determinant which vanishes at the boundary. The elements of this determinant are second derivatives of the free energy F with respect to the long-range order variables (l.r.o.) ξ_i, and are calculated in the disordered state.

Figure 1. Schematic plots of ξ_i against μ_i for a second-order transition (a) and a first order (b).

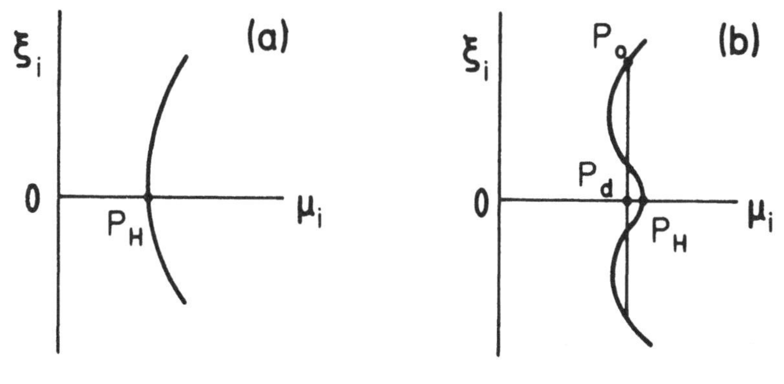

G. M. Stocks and A. Gonis (eds.), Alloy Phase Stability, 281–284.

When this determinant vanishes, the disordered state becomes unstable and a stable ordered phase sets in. When we schematically plot the l.r.o. ξ_i for the B2 phase against the temperature on the chemical potential, the graph looks like Fig. 1(a). We may call a point at which the "instability" Hessian determinant vanishes an H-point for short, and indicate it as P_H.

When the transition is first-order, approximate theories like the cluster variation method (CVM) [2, 3] lead to the shape of ξ_i near the transition point as in Fig. 1(b). In equilibrium, P_d and P_o coexist. They are different from P_H (on the $\xi_i = 0$ axis), where the "instability" Hessian determinant vanishes.

THE FREE ENERGY FUNCTIONS NEAR P_H

The Helmholtz free energy F is plotted against the composition ρ schematically near the transition point in Fig. 2(a) and (b). The corresponding curves of the grand potential Ω are in Fig. 3. The two free energy functions are defined as $F = E - TS$ and $\Omega = F - \Sigma_i \mu_i N_i$, where μ_i and N_i are the chemical potential and the number of the ith species, respectively. (In the following, we write $\mu = \mu_2 = -\mu_1$.) The two functions are connected by the Lengendre transform: $\Omega = F - 2\mu\rho$. The important relations are $\partial F / \partial \rho = 2\mu$ and $\partial \Omega / \partial \mu = -2\rho$.

Figure 2. Schematic plots of F vs. ρ for a second order (a) and a first order (b).

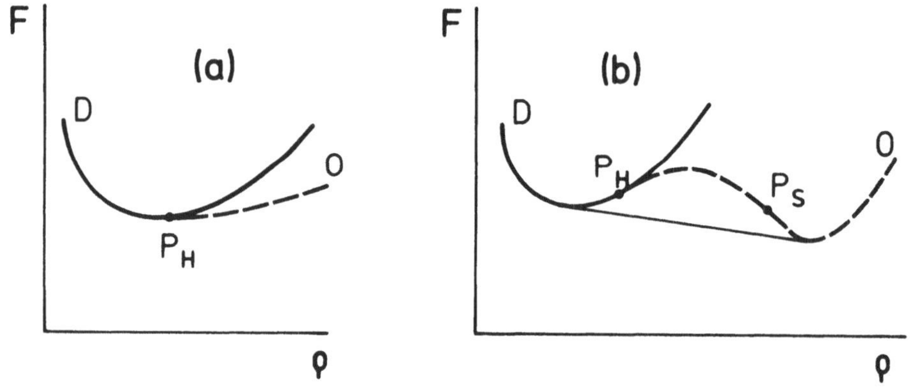

Figure 3. Schematic plots of Ω vs. μ for a second order (a) and a first order (b).

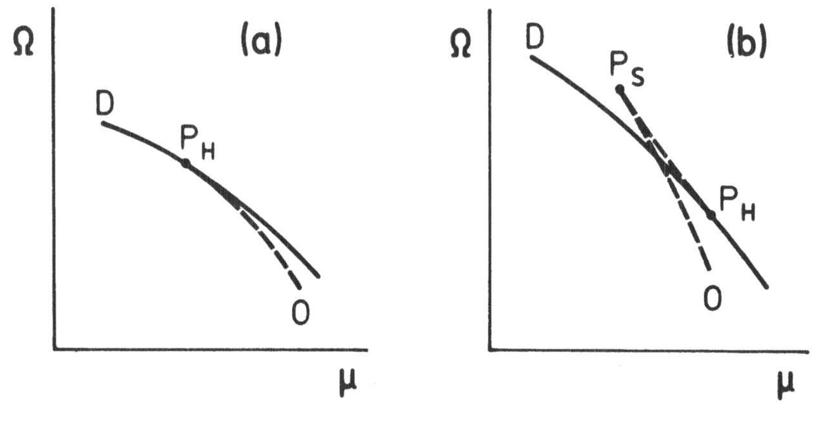

CRITERION OF THE TCRP [3]

One difference between the O curves in (a) and (b) is the existence of the spinodal P_s in (b). As the (b) curve moves into (a), the spinodal moves to the H-point and disappears. Therefore, we can characterize the TRCP as the point at which the spinodal comes to the H-point. This characterization was pointed out by Allen and Cahn [4].

In our criterion, however, we cannot use this spinodal for the following reason. The spinodal is inside the ordered phase and its position is to be calculated using the variables in the ordered state. Calculation of an ordered state becomes increasingly harder as we come closer to TRCP. It is not practical to find an H-point which coincides with the spinodal.

Table 1. Sign of the second derivatives of the O (ordered state) curves of Figs. 2 and 3 at P_H

At P_H	(a) Second Order	(b) First Order
Fig. 2 $\partial^2 F / \partial \rho^2 = \partial \mu / \partial \rho$	+	–
Fig. 3 $-\partial^2 \Omega / \partial \mu^2 = \partial \rho / \partial \mu$	+	–

Table 1 suggests that we can use the sign change of $\partial\rho / \partial\mu$ of the O curve at H as the criterion of the TRCP. Since this quantity is to be evaluated at H, the values of the variables in the expression are those of the disordered state and can be calculated without the convergence difficulty.

The cluster variation method (CVM) [4, 5] can calculate $\partial\rho / \partial\mu$. From Fig. 2(b) we can derive that when it changes sign, it goes through infinity, or $\partial\mu / \partial\rho = 0$. In the formulation, we express $\partial\rho / \partial\mu$ as a ratio and use the denominator $= 0$ as the criterion for the TRCP. The formulation of $d\rho / d\mu$ is similar as that of the specific heat on the ordered-phase side to the critical point of the Ising model [5]. The derivation is done in Ref. [2]. The resulting criterion can be expressed as vanishing of an $(m+1) \times (m+1)$ Hessian determinant whose elements are $\partial^2\Omega / \partial\eta_i\partial\eta_j$ where η_i's are defined as follows: $\eta_1, \eta_2, ..., \eta_m$ are the short-range order variables. The last variable η_{m+1} is ξ^2 where ξ is one of the long-range order variables. The TRCP in Ref. [1] is calculated using this criterion and fits nicely with the rest of the calculations.

It is a pleasure of the author to acknowledge the U.S. Senior Scientist Award from the Alexander von Humboldt Foundation, which made the author's stay at MPI in Dusseldorf possible, where the main part of this work was performed.

REFERENCES

1. G. Inden and R. Kikuchi, CALPHAD XV (7-11 July 1986) Fulmer Grange, UK.

2. S.M. Allen and J.W. Cahn, ***Mat. Res. Soc. Symp. Proc.***, Vol. 19. (Elsevier Sci. Publ., Amsterdam 1983), p. 195.

3. R. Kikuchi and D. de Fontain, *Scripta Met.* **10,** 995 (1976).

4. R. Kikuchi, *Physica A* **142A,** 321 (1987).

5. R. Kikuchi, *Phys. Rev.* **81,** 988 (1951).

THE COORDINATION CLUSTER THEORY FOR METALLIC SOLUTIONS

Milton Blander and Marie-Louise Saboungi
Chemical Technology Division
Argonne National Laboratory
Argonne, Illinois 60439-4837

1. INTRODUCTION

The coordination cluster theory (CCT) is a statistical mechanical theory for the thermodynamic properties of a dilute solute in a binary or higher order solvent.[1–4] Although the CCT has been applied to liquids, it can be modified for solids. The theory permits one to describe the thermodynamic properties of a solute in a binary solvent in terms of the solute properties in the pure solvent components, the properties of the binary mixture and one unknown parameter. In principle, the properties of solutes in multicomponent solvents can be calculated from those in the subsidiary binary solvents. Modern quantum mechanical methods for the calculation of the energetics of metal solutions could be used to place the theory on a firm fundamental basis by testing the approximations in the theory and determining the unknown parameters.

2. THE THEORY

The dissolution of a solute S in a binary solvent $A - B$ leads to the formation of coordination clusters $S(A_{Z-i}, B_i)$, where S is coordinated by Z solvent atoms of which i are B atoms and the remainder are A atoms. If one considers the equilibria for the formation of these coordination clusters,

$$S + (Z - i)A + iB \rightleftharpoons S(A_{Z-i}, B_i). \tag{1}$$

then one can write a set of $(Z + 1)$ equilibrium constants, K_i, given by

$$K_i = \frac{a_i}{a_S a_A^{Z-i} a_B^i}, \quad i = 0, \cdots, Z. \tag{2}$$

where the a's are activities of the species indicated. In dilute enough solutions of S, all the coordination clusters will obey Henry's law. If standard states are defined so that the activity coefficients of the clusters are unity at infinite dilution, then the activity of the ith coordination cluster species, a_i, is equal to its atomic fraction, X_i. It follows that

$$\frac{1}{\gamma_S} = \sum_{i=0}^{Z} K_i a_A^{Z-i} a_B^i = \sum_{i=0}^{Z} K_i (X_A \gamma_A)^{Z-i} (X_B \gamma_B)^i . \tag{3}$$

where the γ's are activity coefficients of the species in solution. The calculation is thus reduced to evaluating the equilibrium constants, K_i, or the standard Gibbs free energy of formation of the ith species at infinite dilution, $\Delta G_i^o (= -RT ln K_i)$. This evaluation

G. M. Stocks and A. Gonis (eds.), Alloy Phase Stability, 285–289.

can be performed by relating the equilibrium constants to identifiable energetic and statistical factors for the formation of the ith species and the interaction of that species with the solvent. A statistical mechanical calculation[1] leads to the relation:

$$-lnK_i = -ln\frac{Z!}{(Z-i)!i!} + \frac{(Z-i)}{Z}ln\gamma_{S(A)} + \frac{i}{Z}ln\gamma_{S(B)} + \frac{g_i^E}{RT} + (1-t)\left[(Z-i)\,ln\gamma_A + iln\gamma_B\right]. \quad (4)$$

The first term on the right hand side is a statistical factor. The second and third terms represent simple additivity of the pair interactions between S and the solvent atoms, and the fourth term, g_i^E/RT, represents the extent to which the bonding is nonadditive, that is, the extent to which the linear relationship given by the second and third terms does not hold. When $i = 0$ and $i = Z$, g_i^E is zero, so that this fourth term is analogous to an excess Gibbs free energy of mixing of A and B in the coordination shell. Thus, the first four terms on the right-hand side represent the standard free energy of formation of the ith species from the end-member species, $S(A_Z)$ and $S(B_Z)$, in the end-member solvents A and B. The last term on the right-hand side represents the difference between the interactions of the A or B atoms in the ith species with the mixed solvent and the interactions of these same atoms in the end-member species with the respective end-member solvents (pure A or pure B). The factor $(1-t)$ represents the relative extent to which the interaction with the solvent is the same as the interaction of a "free" A or B atom that has no solute atoms as nearest neighbors. The statistical mechanical calculation[1] leads to an expression for the activity coefficient of a solute, γ_S, in the binary solvent $A-B$

$$\frac{1}{\gamma_S} = \sum_{i=0}^{Z}\frac{Z!}{(Z-i)!i!}\left(\frac{X_A\gamma_A^t}{\gamma_{S(A)}^{1/Z}}\right)^{Z-i}\left(\frac{X_B\gamma_B^t}{\gamma_{S(B)}^{1/Z}}\right)^{i} exp\left(\frac{-g_i^E}{RT}\right) \quad (5)$$

where X_A, γ_A, and X_B, γ_B are the mole fractions and activity coefficients of components A and B, respectively; $\gamma_{S(A)}$ and $\gamma_{S(B)}$ are the activity coefficients of component S at high dilution in the pure solvents A and B; and t is a geometric factor. The value of t is 1/Z for a substitutional alloy and can range from 1/Z$'$ to 1/2 for an interstitial alloy, with Z$'$ being defined as the coordination number of the solvent atoms. It has been shown that calculations based on Eq.(5) are relatively insensitive to a reasonable choice of values for the parameters Z and t.[2] We have used a value of t = 1/3 for interstitial alloys, and t = 1/Z for substitutional alloys, with Z ranging from 6 to 12.[2]

Since g_i^E is an excess free energy of mixing of solvent atoms in the coordination shell, it can be expressed as a power series involving the concentrations of the atoms, ((Z-i)/Z) and (i/Z), in the coordination shell of the α cluster. Up to second-order terms it is given by the expression

$$g_i^E = \sum_{J<J'}\frac{(Z-i)ih}{2} + \cdots \quad (6)$$

where h is an unknown parameter. Equation (6) implies that the mixture of atoms in the coordination shell forms a regular solution. We have shown that Eq. (6) can

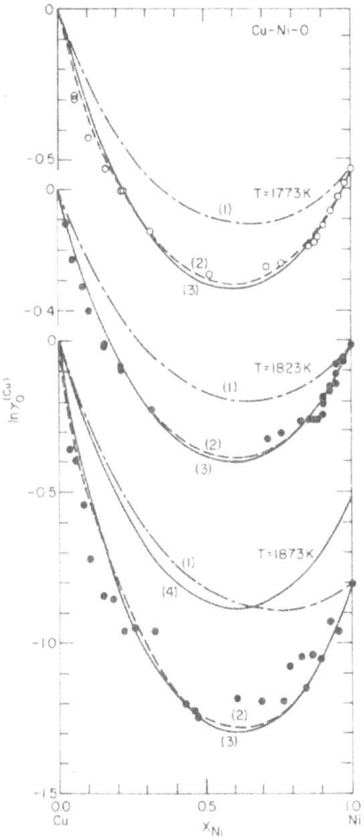

Fig. 1.Relative activity coefficients of oxygen in the Cu-Ni-O system. Curves (1), (2), and (3) refer to calculations from equations in References 5-7, in Reference 8, and from the CCT, respectively.

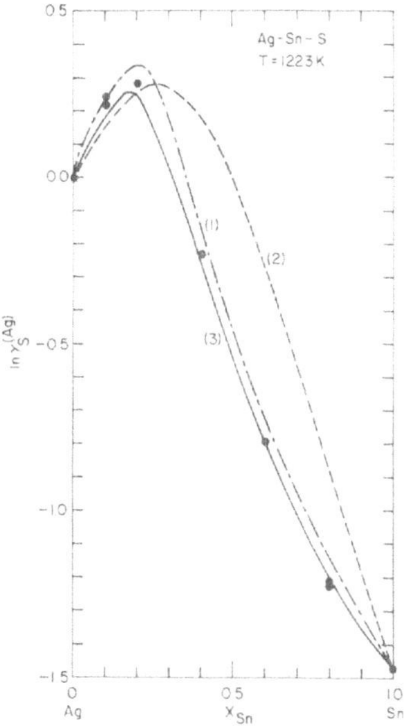

Fig. 2.Relative activity coefficients of sulfur in the Ag-Sn-S system. Curves (1), (2), and (3) refer to calculations from equations in References 5-7, in Reference 8, and from the CCT, respectively.

be used to accurately represent the solubility of oxygen or sulfur in a large number of binary alloys.[2]

3. COMPARISONS WITH EXPERIMENT

A comparison of all reliable data on a number of solutions of oxygen and sulfur in binary liquid alloys indicates that all the data in each system could be very well represented by the use of a single unknown parameter. Figure 1 illustrates this for the Cu-Ni-O system. It also illustrates the fact observed here and in the other cases where measurements were made at more than one temperature, that the same param-

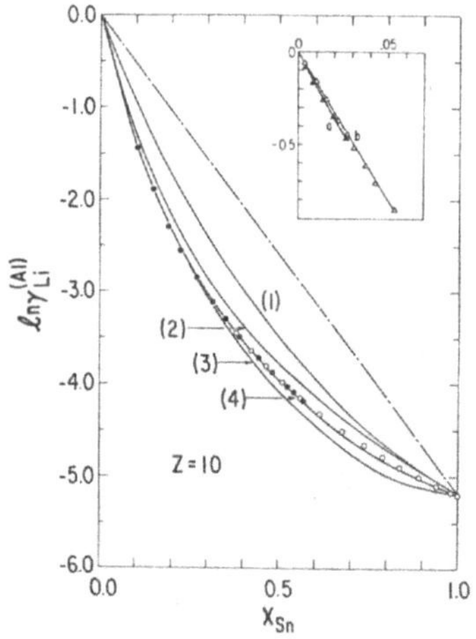

Fig. 3. Relative activity coefficients of Li in Al-Sn alloys. Curves (1), (2), (3), and (4) refer to calculations from equations in References 5-7, modified from References 5-7, in Reference 8, and from the CCT, respectively.

eter correctly predicts the temperature dependence of the activity coefficients. This means that the configurational entropy of solution predicted by the theory adequately describes the entropies in these melts. Figure 1 illustrates the fact that the theory in References 5-7 does not describe the data well and Figure 2 similarly illustrates this fact for Reference 8. Thus, the CCT is the only theory which provides a simple accurate description of a large number of measurements on alloy solutions. With the very accurate measurements on dilute solutions of Li in Al-Sn,[9] it can be seen in Figure 3 that the CCT provides the only good representation of the data.

4. THE POSSIBLE ROLE OF QUANTUM MECHANICAL THEORIES

Modern quantum mechanical calculations of the energetics of alloys present the possibility for testing the approximations and enhancing the utility of the CCT. There are two aspects of the energetics which can be investigated. One aspect is related to the relative energetics for the equilibria between the different coordination clusters in a solution

$$S(A_{Z-i+1}B_{i-1}) + B \rightleftarrows S(A_{Z-i}B_i) + A \tag{7}$$

where A and B are atoms in the bulk solvent which are not near an S atom. The theory and its ability to represent data well implies that, to a good approximation, the energies for reaction (7) change in increments of h as a function of i so that

$$\Delta E_i \sim \Delta E_0 - ih \tag{8}$$

This first order approximation for the non-additivity of "pair bond" energies requires a theoretical examination. In addition to this, the absolute values of the energetic parameters could, in principle, be predicted *a priori* by quantum mechanical methods. This would require calculations of the energetics of solutions of Z+1 clusters with central S atoms and 2(Z'+1) clusters with central A or B atoms. The initial implementation of such calculations could be made on a solid or liquid solution which is simple enough to readily perform calculations with current capabilities. There are several other requirements for initial work. The solution could be substitutional or interstitial and the expected energetics should be unique enough to provide an adequate test of the CCT. In addition, it should be possible to make precise measurements on the chosen alloy system. The coupling of quantum mechanical calculations with the theory could greatly enhance the usefulness of theories such as the CCT and provide guidance for improving their ability to describe real systems.

5. REFERENCES

1. Blander M., Saboungi M.-L., and Cerisier P., Metall. Trans., Vol. 10B, pp. 613-622, 1979.
2. Saboungi M.-L., Cerisier P., and Blander M., Metall. Trans., Vol. 13B, pp. 429-37, 1982.
3. Blander M. and Saboungi M.-L., Chemical Metallurgy-A Tribute to Carl Wagner, Gokcen N.(ed), Trans. Met. Soc., AIME, Warrendale, PA, pp. 223-31, 1981.
4. Saboungi M.-L., Caveny D., Bloom I., and Blander M., Metall. Trans. Vol. 18A, pp. 1779-1783, 1987.
5. Alcock C. B. and Richardson F. D. , Acta Met., Vol. 6, p. 385, 1958.
6. Alcock C. B. and Richardson F. D. , Acta Met., Vol. 8, p. 882, 1960.
7. Jacob K. T. and Alcock C. B., Acta Met., Vol. 20, p. 221, 1972.
8. Wagner C., Acta Met., Vol. 21, p. 1297, 1973.
9. Saboungi M.-L. and Blander M., J. Electrochem. Soc., Vol 124(1), pp. 6-13, 1977.

ACKNOWLEDGMENT

This work was supported by the Division of Materials Science, Office of Basic Energy Sciences, U.S. Department of Energy under contract W-31-109-ENG-38.

SECTION 4

ELECTRONIC THEORIES OF PHASE STABILITY, SEMI-PHENOMENOLOGICAL AND MODEL HAMILTONIANS

ELECTRONIC STRUCTURE, EFFECTIVE PAIR INTERACTIONS AND ORDER IN ALLOYS

F. DUCASTELLE

Office National d'Etudes et de Recherches Aérospatiales
ONERA, BP 72 92322, Châtillon Cedex, France

1. INTRODUCTION

Due to the presence of interatomic interactions, any alloy should eventually order or segregate at low temperature. This is observed in many systems. When the ordering interactions are strong, the alloy may remain ordered up to the melting point (Ni_3Al for instance). When they are very weak, the solid solution may be the only observable phase since at low temperature atomic diffusion is no longer efficient (Cu-Ni). In the intermediate regime we have the most interesting situation for our present purpose, i.e. the occurrence of one or several ordered phases at low temperature which disorder at a critical temperature before melting (Cu-Au, Pd-V, Ni-Fe,...). In the simplest situation all phases have a single simple underlying lattice (fcc, bcc,...), but more frequently several structures are involved and some compounds display structures different from those of the pure metals (Laves phases, σ phases, A15 phases,...).

It is of great fundamental interest to understand what stabilizes a particular structure at given concentration and temperature. This has also practical implications, in metallurgy in particular. Quite generally, it is obvious that it is very important and useful to understand and predict the phase diagrams of alloys.

Up to recent years most theories in this field were based on phenomenological models. The stability of several crystalline structures has been interpreted using geometrical arguments, by looking for the most efficient way of filling space with hard spheres of different radii. Size effects can account for the different stabilities of solid solutions in different systems. Chemical or electronic parameters have also been used. The celebrated Hume-Rothery rules for instance relate the stability of several structures to definite values of the electronic ratio (number of valence electrons per atom). Finally various electronegativity scales have been introduced to classify the structures [1]. More details can be found in Pettifor's lectures.

As far as phase diagrams are concerned, most approaches have been based on phenomenological thermodynamic models using regular solution models or various improvments on them.

There are now reasonable hopes to be able to attack these problems from first principle calculations. This confidence is based on the success of such calculations for pure metals. In most cases the one-electron local density functional approximation has been shown to be quite reasonable and crystalline structures, atomic volumes at equilibrium,..., have been calculated to a good accuracy without any adjustable parameter [2]. Although the task is much more difficult, similar calculations have already been performed in the case of alloys [3], and more and more of them will be available in the future, (see the other contributions to this volume). However there is still room for more phenomenological microscopic models. It is obvious that simple and well chosen models are always useful to understand the results of a full numerical calculation. Furthermore the latter calculations are still expensive and time consuming and cannot easily be coupled with subsequent models, in particular with thermodynamic models.

We thus have two extreme types of microscopic models to calculate for example a phase diagram. The ideal first principle method would be a theory in which the energy of any (ordered or disordered) atomic configuration would be calculated from the electronic

G. M. Stocks and A. Gonis (eds.), Alloy Phase Stability, 293–327.

structure and then inserted into the partition function to obtain the free energy. In practice some approximations should be introduced since it is not possible to explicitly calculate the energy of all possible atomic configurations and also to calculate the partition function of an infinite system, but in principle in this approach there is no need to assume a particular model for the interatomic interactions. An extreme phenomenological approach is to derive such interatomic interactions from simplified electronic models, and to insert them into appropriate thermodynamic theories. For example a method which has recently been used [4] consists in calculating the energy of different types of ordered structures at different concentrations, and in deducing from them the value of a limited, arbitrarily chosen set of interactions, for example between first neighbours. The energy of a given configuration can then be splitted into two parts: one part including these interactions which has just the form of an Ising model in the case of a binary alloy A_cB_{1-c} (with the analogy spin up → A, spin down → B) and a part independent of the configuration but which depends on the concentration and which is assumed to include in a phenomenological way several other contributions (elastic energy,...).In this way it is possible to reproduce reasonably well several phase diagrams [5] ; see also [6].

The disadvantage of the latter method is that the range and the nature (pair, triplet,...) of the interactions are assumed more than proved, so that these interactions are to a large extent phenomenological parameters which probably cannot be used to describe many different physical properties. More precisely such methods can be useful if one is only interested in gross properties of phase diagrams or of ordering phenomena which actually depend on the most important interactions. However they cannot be reliable to account for more subtle effects, for example those related to the competition between various types of interactions, which are theoretically predicted and experimentally observed (long period structures in Cu-Pd, Ti-Al,...; for a discussion see [7][8] and the contributions by Selke , Van Tendeloo and Loiseau et al. in this volume).

In the following we shall review the main features of a theory proposed during the last decade, whose main purpose is to determine whether effective ordering interatomic interactions can be derived from electronic structure calculations. In many cases this is actually possible, and the nature and range of the interactions can be related to the topology of the underlying lattice and to a few electronic parameters. The theory explains well the ordering trends of many binary transition alloys or substoichiometric transition carbides and nitrides. Its application to various situations is described in several articles so that we shall not enter into all the details. Our main purpose here is to insist on the principles of the method, on its range of applicability and on its limitations.

Basically the main idea is that even if the total energy of an alloy cannot be expanded in terms of interatomic interactions, the part of this energy depending on the atomic configuration can be written in this way if one takes as a reference the energy of the fully disordered state. We then obtain effective interactions and an effective Ising model for the configurational part of the energy. The trouble is that this is not exactly an Ising model because the interactions are concentration dependent. This raises several questions which will be discussed later. The interest of the method is to split the calculation of the total free energy into two parts. First one treats the electronic degrees of freedom in the disordered state, generally within the coherent potential approximation (CPA). Then the atomic degrees of freedom are inserted into an effective Ising model. One then takes full advantage of all the achievments of statistical physics applied to this model. When the intermediate step (derivation of an effective Ising model with a few relevant interactions) can be firmly justified, the method does not contain adjustable parameters if the electronic structure is itself calculated from first principles. We then have a theory which can be at the same time accurate and physically transparent.

In practice the calculations are fairly easy only if some restrictive conditions are imposed. The method applies simply only to ordering processes on a fixed underlying lattice, but it could probably be extended in the future to deal with structural effects. It is also very difficult to treat alloys when size effects are important. Then the atomic positions deviate from those of the rigid lattice, and both the electronic calculations and the thermodynamic treatment become very complicated. On the other hand some contributions (mean elastic energy, phonons, magnetism, spin-orbit coupling,...) can be included using natural extensions of the basic theory.

Most of the practical electronic calculations have been performed up to now within the tight-binding approximation, which has many technical and pedagogical advantages, and we shall use this formalism here, but there is no difficulty (in principle if not in practice!) to use more sophisticated techniques. Such calculations are currently under progress [9].

We then begin (§2) by presenting the tight-binding and CPA formalisms and show how a generalized perturbation method (GPM) can be introduced to define effective pair interactions. This approach is compared with a similar but more restrictive method which directly defines a mean field approximation both for the electrons (CPA) and the atoms (mean field Bragg-Williams approximation). In §3 we describe the main characteristics of these interactions and in §4 we discuss the influence of additional contributions (self-consistency, magnetism,...) neglected in the simplest scheme. Finally in §5 we show how these interactions can be used to account for the observed ground states and phase diagrams of alloys.

2. COHERENT POTENTIAL APPROXIMATION AND GENERALIZED PERTURBATION METHOD

2.1. Tight-binding model

The simplest tight-binding hamiltonian H for a pure metal writes

$$H = \sum_n |n> \varepsilon_n < n| + \sum_{m \neq n} |n> \beta_{nm} < m| \tag{2.1}$$

where $|n\rangle$ is an atomic state centred on site n, $\varepsilon_n = \varepsilon$ the corresponding atomic level and β_{nm} the so-called hopping or transfer integral. In the following we shall have principally in mind transition alloys which are well described within this approximation. In this case there are in fact five orbitals per site and per spin direction so that $|n\rangle$ should be replaced by $|n, \lambda\rangle$ and β_{nm} by $\beta_{nm}^{\lambda\mu}$, $\lambda, \mu = 1,...,5$; the hopping integrals being themselves linear combinations of three independent integrals for each distance m-n, $dd\sigma < 0$, $dd\pi > 0$, $dd\delta < 0$, the latter being negligible in general. Furthermore, by comparison with first principle calculations, one can verify that the crude approximation $dd\sigma \approx -2\, dd\pi$, $dd\delta \approx 0$ is reasonable, so that we are left with two parameters per distance m-n. In practice, only a few β_{nm} involving short distances m-n are taken into account; first neighbour distances are generally sufficient in the case of compact fcc or hcp structures whereas first and second neighbour distances are required for the bcc structure. More generally, in compact metallic systems one must take into account the first coordination shell. If a law of variation with distance of the integrals β_{nm} is given (generally of type $|m-n|^{-\alpha}$ or $\exp -q|m-n|$) we finally have canonical electronic structures for each crystalline structure which depend on this law and on two parameters fixing the origin of energies and the energy scale respectively [10]. Similar canonical electronic structures are defined in [11].

In the case of alloys, the parameters ε_n and β_{nm} depend on the atomic configuration and in the simplest scheme they are assumed to only depend on the occupation of sites n, and n and m respectively. In the case of a binary alloy A_cB_{1-c} we then write

$$\varepsilon_n = \sum_{i=A,B} \varepsilon^i p_n^i$$
$$\beta_{nm} = \sum_{i,j=A,B} \beta^{ij} p_n^i p_m^j \qquad (2.2)$$

where the p_n^i are the occupation numbers taking values 1 or 0 depending on the presence or not of an atom of type i at site n. Since $p_n^A + p_n^B = 1$ we shall define p_n as

$$p_n = p_n^A = 1 - p_n^B \qquad (2.3)$$

Taking as the origin of energy the mean atomic level $\bar{\varepsilon} = c\,\varepsilon^A + (1-c)\,\varepsilon^B$ we get

$$\varepsilon_n = (p_n - c)\,\delta$$
$$\text{with} \qquad \delta = \varepsilon^A - \varepsilon^B \qquad (2.4)$$

δ is the so-called diagonal disorder parameter. Along a transition series $\bar{\varepsilon}$ is approximatively a linear decreasing function of the mean number of valence electrons $\bar{N}_e$ and δ obviously plays the part of an electronegativity difference. The so-called off-diagonal disorder is related to the differences between the hopping integrals β^{AA}, β^{BB} and β^{AB}. A very convenient and reasonable approximation is to set $(\beta^{AB})^2 \approx \beta^{AA}.\,\beta^{BB}$. The widths W^A, W^B, of the d bands of the pure metals, whose dependence on the crystalline structure can be neglected here, are proportional to β^{AA} and β^{BB}. The off-diagonal disorder in this scheme is then characterized by $W^A/\,W^B$, or $(W^A - W^B)/\bar{W}$ where $\bar{W}$ is the mean bandwidth. The bandwidth of transition metals decreases along a transition series and increases from the 3d to the 5d series with extreme values of the order of 4 to 8 eV. The energy level on the other hand decreases about one eV per element (see Fig.1). We conclude that in many alloys built from elements with very

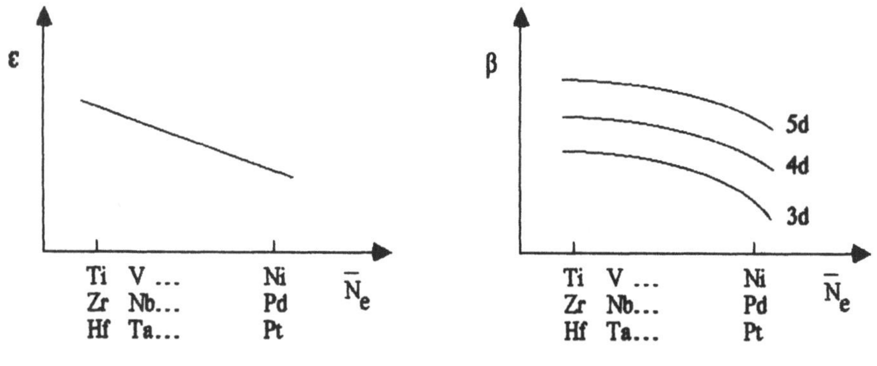

FIGURE 1. Typical variations of ε and β along the transition series.

different electron numbers N_e^A , N_e^B , the parameter δ is of the order of the mean bandwidth $\overline{W}$ whereas the bandwidth fluctuations $W^A - W^B$ are smaller. Thus, in most interesting cases off-diagonal disorder can be neglected, which will be done in the following except in §4.1 . In fact there is no basic difficulty in including it, but the formalism is simpler if we forget it. Using canonical units, the electronic structure of a given configuration of an alloy then only depends on $\delta/\overline{W}$. If this parameter is weak, standard perturbation theory applies, and to lowest order the density of states is that of a pure metal of bandwidth $\overline{W}$ centred on $\bar{\varepsilon}$. When it is large, a gap should open. There is indeed a theorem telling us that the electronic states of an alloy are necessarily within the bounds provided by the spectra of the pure metals A and B . From very general arguments, it is then easy to guess the qualitative shape of the density of states as a function of the degree of order and of the concentration [12] (Fig.2). Non perturbative methods are required to treat this high disorder regime. The coherent potential

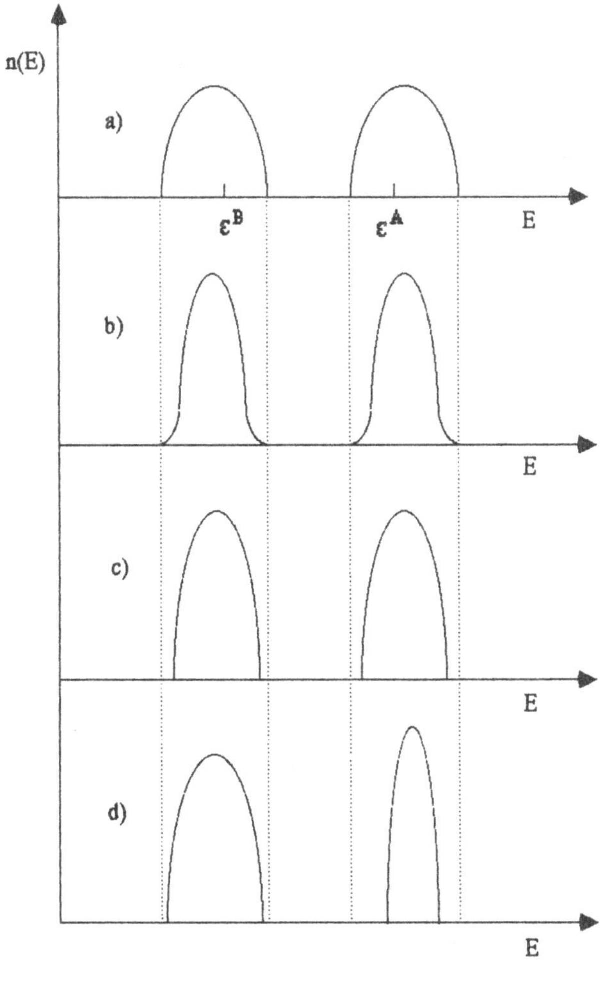

FIGURE 2. Schematic densities of states : a) for a segregated alloy, b) for a disordered alloy, c) for an ordered alloy. In all cases, c = 1/2 . The density of states of an ordered alloy when c < 1/2 is shown in d).

approximation (CPA) that we shall now briefly describe is undoubtedly one of the most simple and efficient methods of this type.

2.2. CPA

Suppose then that we want to calculate the density of states of a completely disordered alloy. By this, we mean that there are no interatomic correlations and that each site is occupied by an A or B atom with a probability c or (1-c). Using the ensemble average ‹ ... › over all possible atomic configurations for a given concentration, we have

$$\begin{aligned} \langle p_n \rangle &= c \\ \langle p_n p_m \rangle &= c^2 \quad \text{if } n \neq m \\ &= c \quad \text{if } n = m \end{aligned} \tag{2.5}$$

To calculate the average density of states $\langle n(E) \rangle$, we start from the formula

$$\langle n(E) \rangle = \mathrm{Tr}\ \delta(E-H) = \lim_{\varepsilon \to 0} -\frac{\mathrm{Im}}{\pi} \mathrm{Tr}\ G(E+i\varepsilon) \tag{2.6}$$

where Tr denotes the trace over the N_a atomic states $|n\rangle$, Im the imaginary part, and G the so-called resolvent or Green function

$$G(z) = (z-H)^{-1} \tag{2.7}$$

where z is a complex number. Let us split H into two parts

$$\begin{aligned} H &= H^0 + V \\ H^0 &= \sum_{m \neq n} |n\rangle \beta_{nm} \langle m| \\ V &= \sum_n |n\rangle \varepsilon_n \langle n| \end{aligned} \tag{2.8}$$

and define the bare resolvent $G^0 = (z - H^0)^{-1}$. We write the Dyson equation for the average resolvent $\langle G \rangle$

$$\begin{aligned} \langle G \rangle &= G^0 + G^0 \langle V G \rangle \\ &= G^0 + G^0 \langle V \rangle G^0 + G^0 \langle V G^0 V \rangle G^0 + \ldots \end{aligned} \tag{2.9}$$

An important point is that $\langle V G^0 V \rangle$ does not reduce to a product of averages. Using (2.8) we have

$$\langle V G^0 V \rangle = \sum_{n,m} |n\rangle \langle \varepsilon_n G^0_{nm} \varepsilon_m \rangle \langle m| \tag{2.10}$$

where G^0_{nm} is the matrix element of G^0. According to (2.5) the value of the average is different when n=m or n≠m; as a result we have

$$\langle V G^0 V \rangle - \langle V \rangle G^0 \langle V \rangle = \sum_n |n\rangle c(1-c)\delta^2 G^0_{nn} \langle n| \tag{2.11}$$

It is then useful to define an effective potential $\Sigma(z)$ through

$$\langle G \rangle = G^0 + G^0 \Sigma \langle G \rangle = (z - H^0 - \Sigma(z))^{-1} \tag{2.12}$$

$\Sigma(z)$ is a complex operator which can be expanded in successive powers of δ. Identifying (2.9) and (2.12), and using (2.11), we get

$$\Sigma(z) = \sum_n |n> c(1-c)\delta^2 F^0(z) < n| + \ldots \tag{2.13}$$

where $F^0(z) = G^0_{nn}$ is the diagonal matrix element of G^0. Using the Bloch basis $|k> = N_a^{-1/2} \sum_n e^{ik.n} |n>$ which diagonalizes $\langle G \rangle$, we then get, to this order

$$< G >_k = < k | < G > | k > = [z - c(1-c)\delta^2 F^0(z) - \varepsilon_k]^{-1} \tag{2.14}$$

where ε_k is the Bloch eigenvalue of H^0. This indicates that the spectral function $-(\mathrm{Im}/\pi) \langle G \rangle_k$ has changed from a δ-function towards a lorentzian of width $c(1-c)\delta^2 n^0(E)/\pi$, where $n^0(E)$ is the density of states of the bare hamiltonian H^0. The electronic states are no longer Bloch states. The scattering by the potentials ε_n introduces a lifetime $h/c(1-c)\cdot\delta^2 n^0(E)$.

To lowest order $\Sigma(z)$ is a diagonal operator. In fact this holds up to order δ^3. It is obviously impossible to sum up the full perturbation series (2.13), but one may try to sum the most significant terms, and in particular to sum a class of diagonal terms. This can be done using diagrammatic techniques, but fortunately more simple mean field arguments can be used [13]. Assume then that we look for the "best" single-site approximation

$$\Sigma(z) \approx \sum_n |n> \sigma(z) < n| \tag{2.15}$$

Then, if we consider $H^0 + \Sigma(z)$ as an effective hamiltonian characterizing the disordered alloy, $\sigma(z)$ plays the part of an effective correction to the mean atomic level $\bar{\varepsilon}$. Assume now that the occupation of site o is fixed. A new effective hamiltonian corresponding to this situation can be defined from the conditional average $\langle G \rangle_0^i$ of G (i.e. we average over the occupation of all other sites, o being occupied by an i atom). We also have to define a new effective potential. The coherent potential approximation consists in assuming that this potential simply derives from that of the uniform medium by replacing $\sigma(z)$ by ε^i at site o : the fixed atom is assumed to be embedded in the uniform medium of the disordered state. This gives an approximate relationship between $\langle G \rangle_0^i$ and $\langle G \rangle$

$$< G >_0^i = < G > + < G > | o > t^i < o | < G >$$

with

$$t^i = (\varepsilon^i - \sigma)[1 - (\varepsilon^i - \sigma)F]^{-1} \tag{2.16}$$

where t^i is the t-matrix describing the scattering of electrons by the potential $\varepsilon^i - \sigma$, difference between the potential at site o and that of the uniform medium, and F the diagonal matrix element of $\langle G \rangle$. Forcing the average of $\langle G \rangle_0^i$ to be equal to $\langle G \rangle$ we obtain

$$< t > = c t^A + (1-c) t^B = 0 \tag{2.17}$$

which is a self-consistent equation for σ since F is obtained from F^0 by replacing z by $z - \sigma(z)$, (see eq.(2.12)).

The properties of the CPA are described in many places [13][14], and we just mention here a few of them. The CPA properly interpolates between the limits $c \to 0$, $(1-c) \to 0$ and reproduces these limits to first order. For example, when $c \to 0$

$$\sigma(z) \to c\,\delta\,(1 - \delta F^0)^{-1} = c\,t^0 \tag{2.18}$$

The CPA is also exact up to δ^3, and reproduces the eight first moments of the exact average density of states.

The effect of the imaginary part of $\sigma(z)$ is to smooth out all fine structures of the density of states of the pure metals and to introduce times of life for the Bloch states. Because of the dependence on the energy ($z \to E + i\varepsilon$) all states are not equivalently affected. States close to band edges or in the subband related to impurity atoms are the most sensitive to disorder, and this is also in these regions that the CPA is the less accurate. This is easy to understand. States in minority bands tend to be localized on clusters of minority atoms which are necessarily poorly described in a single site mean field approximation.

To summarize, the CPA effective potential $\sigma(z)$ is a complex quantity that depends on $\delta/\overline{W}$, c, and the energy of the states, its dependence on the precise shape of the bare density of states being rather weak.

2.3. CPA and long-range order

Long-range order (LRO) in alloys is characterized by deviations of the local concentrations $\langle p_n \rangle = c_n$ from the mean concentration c. If short-range order effects are neglected, this is the only modification to introduce in equations (2.5). In practice, when studying definite ordered structures, c_n takes only a few different values on the sites of the new unit cell, and the deviations $\delta c_n = c_n - c$ are the long-range order parameters of these structures. It is clear that the whole CPA formalism applies as well in this case provided the averages correspond to definite values of the LRO parameters. The average medium is now non-uniform and has the symmetry of the ordered state

$$\Sigma(z) = \sum_n |n\rangle \sigma_n \langle n| \tag{2.19}$$

where σ_n depends on n.

The CPA equation (2.17) becomes a set of equations for all non-equivalent n provided that n labels are introduced everywhere: $c \to c_n$, $\sigma \to \sigma_n$, $F \to F_n$, $t \to t_n$. Notice however that now $F_n(z)$ is not so simply related to $F^0(z)$ since $\Sigma(z)$ is no longer proportional to the identity operator.

Thus, in the presence of LRO the extension of the CPA is straightforward even if the equations are more difficult to solve [15][16]. On the other hand the CPA, being a single site theory, cannot handle short-range order effects. Deeper modifications should be introduced to do that, but we shall see later how one can get round this difficulty.

2.4. Energy

It is now well admitted that the total energy of a pure transition metal is principally related to its band energy U_b

$$U_b = \int^{E_F} dE\, E\, n(E) \tag{2.20}$$

where E_F is the Fermi level. All other contributions (ion-ion, electron-electron interactions) nearly cancel each other or do not modify the qualitative trends induced by the bare band term. Charge transfer effects will be discussed in §4.2 . It is often more useful to handle the grand potential Ω_b

$$\Omega_b = U_b - E_F \overline{N}_e = -\int^{E_F} dE\, N(E)$$
$$N(E) = \int^{E} dE'\, n(E') \tag{2.21}$$

The integrated density of states N(E) can be related to the Green function through

$$N(E) = \frac{\text{Im}}{\pi} \text{ Tr Log } G(E+i\varepsilon) \tag{2.22}$$

What is the corresponding expression for the average of N(E) within the CPA ? Let us first compare G with its average $\overline{G}$ calculated within the CPA. We have the Dyson equation

$$G = \bar{G} + \bar{G}(V - \Sigma) G \tag{2.23}$$

Inserting this into (2.22) and using the identity Tr Log A B = Tr Log A + Tr Log B , valid for any pair of operators, we get

$$\text{Tr Log } G = \text{Tr Log } \bar{G} - \text{Tr Log } [1 - (V-\Sigma)\bar{G}] \tag{2.24}$$

Let us now split $\overline{G}$ into its diagonal $\hat{F}$ and off-diagonal $\overline{G}_{off}$ parts; we obtain

$$\text{Tr Log } [1 - (V-\Sigma)\bar{G}] = \text{Tr Log}[1 - (V-\Sigma)\hat{F}] + \text{Tr Log}[1 - \hat{t}\,\bar{G}_{off}] \tag{2.25}$$

where $\hat{t}$ is the diagonal operator

$$\hat{t} = \sum_n |n> t_n < n| \qquad ; \qquad t_n = \sum_i p_n^i\, t_n^i$$
$$t_n^i = t^i = (\varepsilon^i - \sigma)[1 - (\varepsilon^i - \sigma) F]^{-1} \tag{2.26}$$

Since $1 - (V-\Sigma)\hat{F}$ is also a diagonal operator, (2.24) can be rewritten

$$\text{Tr Log } G = \text{Tr Log } \overline{G} - \sum_n \text{Log}[1 - (\varepsilon_n - \sigma) F] - \text{Tr Log } [1 - \hat{t}\,\overline{G}_{off}] \tag{2.27}$$

Let us expand the third term in the r.h.s.

$$-\text{Tr Log}[1 - \hat{t}\,\overline{G}_{off}] = \left\{\frac{1}{2} \text{Tr } \hat{t}\,\overline{G}_{off}\,\hat{t}\,\overline{G}_{off} + \dots\right\}$$
$$= \frac{1}{2} \sum_{n \neq m} t_n \bar{G}_{nm} t_m \bar{G}_{mn} + \dots \tag{2.28}$$

Because $n \neq m$ the average of $t_n t_m$ is equal to $\langle t_n \rangle \langle t_m \rangle$ and therefore vanishes (remember the CPA equation (2.17)). In the spirit of the CPA we neglect all other terms, even those which do not strictly vanish and therefore obtain

$$\langle \text{Tr Log } G \rangle \approx \text{Tr Log } \bar{G} - \sum_n \langle \text{Log}[1 - (\varepsilon_n - \sigma) F] \rangle \tag{2.29}$$

The r.h.s. of (2.29) is formally a function of z and σ if $\overline{G}$ and therefore F are themselves considered as such functions : $\overline{G} = G^0(z-\Sigma)$. It is straightforward to verify using again the CPA equation that the partial derivative with respect to σ vanishes (this is a very important variational property of the CPA) and we get

$$d/dz < \mathrm{Tr\,Log}\, G > \; = \; -\mathrm{Tr} <G> \; \approx \; \partial/\partial z\,(\mathrm{Tr\,Log}\,\bar{G}\,) \; = \; -\mathrm{Tr}\,\bar{G} \tag{2.30}$$

In other words, the r.h.s. of (2.29) is exactly the integral of the CPA quantity $-\mathrm{Tr}\,\bar{G}$. Thus, within the CPA

$$< \Omega_b >_{CPA} \equiv \bar{\Omega}_b = -\frac{\mathrm{Im}}{\pi} \int^{E_F} dE \left\{ \mathrm{Tr\,Log}\,\bar{G} - \sum_n < \mathrm{Log}[1-(\varepsilon_n - \sigma)\,F] > \right\} \tag{2.31}$$

2.5. <u>General mean field theory at finite temperature</u>

In the presence of LRO, eq.(2.31) is generalized as said in §2.3, by introducing n-dependent quantities σ_n, F_n, ..., and gives us a closed form for the band energy [16]. At finite temperature we have the formula

$$\begin{aligned} \bar{\Omega}_b &= U_b - \nu \bar{N}_e - T S_e \\ &= -\frac{\mathrm{Im}}{\pi} \int dE\, f(E) \left\{ \mathrm{Tr\,Log}\,\bar{G} - \sum_n < \mathrm{Log}[1-(\varepsilon_n - \sigma_n)\,F_n] > \right\} \end{aligned} \tag{2.32}$$

where ν is the chemical potential (different from the Fermi level at $T \neq 0$) and S_e the electronic entropy. In order to obtain the full grand-potential including the atomic degrees of freedom, we must calculate the configurational entropy S_c. To be consistent with the single-site CPA we must use a single-site approximation for the entropy, i.e.

$$-S_c / k_B = \sum_{n,i} c_n^i \,\mathrm{Log}\, c_n^i = \sum_n c_n \,\mathrm{Log}\, c_n + \sum_n (1-c_n)\,\mathrm{Log}\,(1-c_n) \tag{2.33}$$

where k_B is the Boltzman constant. Finally, introducing the atomic chemical potentials μ^i we get

$$\begin{aligned} \Omega^*(T, \nu, \mu^i) &= U_b - \nu \bar{N}_e - T S_e - T S_c - \sum_i \mu^i N_a^i \\ &= -\frac{\mathrm{Im}}{\pi} \int dE\, f(E) \left\{ \mathrm{Tr\,Log}\,\bar{G} - \sum_{n,i} c_n^i \,\mathrm{Log}\,[1-(\varepsilon^i - \sigma_n)\,F_n] \right\} \\ &\quad + k_B T \sum_{n,i} c_n^i \,\mathrm{Log}\, c_n^i - \sum_{n,i} \mu^i c_n^i \end{aligned} \tag{2.34}$$

Ω^* is a functional of the LRO parameters $\delta c_n^i = c_n^i - c^i$, and can also be considered as a functional of σ_n if they are not assumed to take their CPA values. This functional is denoted $\Omega^*(\{\delta c_n^i\}, \{\sigma_n\}, T, \nu, \mu^i)$. The actual potential $\Omega(T, \nu, \mu^i)$ is obtained by minimizing Ω^* with respect to the σ_n and δc_n^i. As pointed out above, the variation of Ω^* with respect to σ_n gives the CPA equation. Thus

$$\partial \Omega^* / \partial \sigma_n = 0 \qquad \text{implies} \qquad \sum_i c_n^i t_n^i = 0 \tag{2.35}$$

To calculate the derivative with respect to c_n^i, we rewrite Ω^* as

$$\Omega^* = -\frac{\mathrm{Im}}{\pi}\int dE\, f(E)\, \mathrm{Tr}\, \mathrm{Log}\, \bar{G} + k_B T \sum_{n,i} c_n^i \,\mathrm{Log}\, c_n^i - \sum_{n,i} \mu^i_{n,\mathrm{eff}}\, c_n^i \tag{2.36}$$

$$\mu^i_{n,\mathrm{eff}} = \mu^i + E_n^i = \mu^i - \frac{\mathrm{Im}}{\pi}\int dE\, f(E)\, \mathrm{Log}\,[1-(\varepsilon^i - \sigma_n) F_n]$$

Varying (2.36) with respect to c_n^i yields

$$c_n^i = e^{\beta\mu^i_{n,\mathrm{eff}}} / \sum_j e^{\beta\mu^j_{n,\mathrm{eff}}} \quad ; \quad \beta = 1/k_B T \tag{2.37}$$

In the case of a binary alloy, defining $\mu = \mu^A - \mu^B$, we have

$$c_n = (1 + \exp - \beta\mu_{n,\mathrm{eff}})^{-1}$$

with

$$\mu_{n,\mathrm{eff}} = \mu - \frac{\mathrm{Im}}{\pi}\int dE\, f(E)\, \mathrm{Log}\{[1-(\varepsilon^A - \sigma_n) F_n] / [1-(\varepsilon^B - \sigma_n) F_n]\} \tag{2.38}$$

and (2.38) exactly takes the form of the familiar mean-field Bragg-Williams equations where $\mu_{n,\mathrm{eff}}$ is the mean field at site n, sum of the external field and of the average contribution of the medium.

Thus, we have a global consistent scheme where both the electronic and atomic degrees of freedom are treated on an equal footing, and which may be called the CPA-Bragg-Williams or CPA-mean-field approximation. Another equivalent formulation is given in [17].

Let us now study the case of weak LRO parameters and expand Ω as a function of the concentration fluctuations δc_n^i ($\delta c_n^A = -\delta c_n^B \equiv \delta c_n$). Choosing the canonical representation $\Omega(T, v, c^i)$ and then dropping the term involving the atomic chemical potentials, we have to vary $\bar{\Omega}_b$ and S_c

$$-T\, \delta S_c = (k_B T/2) \sum_{n,i} (\delta c_n^i)^2 / c^i + \ldots \tag{2.39}$$

To obtain $\delta\bar{\Omega}_b$, we first notice that to first order

$$\delta\bar{\Omega}_b = -\sum_{n,i} E_n^i\, \delta c_n^i \tag{2.40}$$

It remains to vary E_n^i which is a function of the δc_n^i through σ_n and F_n. The result is

$$\delta E_n^i = \frac{\mathrm{Im}}{\pi}\int dE\, f(E)\, \{ t_n^i\, \delta F_n - F_n(1 + t_n^i F_n)\, \delta\sigma_n \} \tag{2.41}$$

Using now the CPA equations it is easy to calculate δF_n and $\delta\sigma_n$ in Fourier space

$$\delta\sigma_n = \sum_q \delta\sigma_q \, e^{i\,q.n}$$
$$\delta c^i_n = \sum_q \delta c^i_q \, e^{i\,q.n} \tag{2.42}$$

Using then the CPA equation $\langle t_n \rangle = 0$ and the identity

$$\delta F_n = \sum_m \overline{G}^2_{nm} \, \delta\sigma_m \tag{2.43}$$

we get

$$\delta\sigma_q = \left\{ \sum_i t^i \, \delta c^i_q \right\} / \{ 1 - \langle t^2 \rangle \, A'_q \} \tag{2.44}$$

with

$$A'_q = \sum_{R\neq 0} \overline{G}(R)^2 \, e^{i\,q.R} \quad ; \quad \overline{G}(R) = \overline{G}_{OR} \tag{2.45}$$

Inserting this into (2.41), we finally obtain

$$\delta\overline{\Omega}_b = \frac{1}{2} N_a \sum_q e(q) \, | \delta c_q |^2 + \ldots \tag{2.46}$$

with

$$e(q) = -\frac{\mathrm{Im}}{\pi} \int dE\, f(E) \, \frac{A'_q (\Delta t)^2}{1 - \langle t^2 \rangle A'_q}$$
$$\Delta t = t^A - t^B \quad ; \quad \langle t^2 \rangle = c(1-c)\,(\Delta t)^2 \tag{2.47}$$

Thus

$$\delta\overline{\Omega}_b(T, v, c, \{\delta c_n\}) = \frac{1}{2} N_a \sum_q \left\{ e(q) + \frac{k_B T}{c(1-c)} \right\} | \delta c_q |^2 + \ldots \tag{2.48}$$

Notice that in this Landau-like expansion, the coefficient in front of $|\delta c_q|^2$ is simply $\chi^{-1}(q)$ where $\chi(q)$ is the susceptibility that measures the concentration fluctuation induced by a variation of chemical potential

$$\chi(q) = \partial c_q / \partial \mu_q = \left\{ e(q) + \frac{k_B T}{c(1-c)} \right\}^{-1} \tag{2.49}$$

$\chi(q)$ is itself related to the Fourier transform of the Warren-Cowley parameter $\alpha(q)$; this is the mean field Krivoglaz-Clapp-Moss formula

$$\alpha(q) = \frac{k_B T}{c(1-c)} \, \chi(q)$$
with
$$c(1-c)\, \alpha_{nm} = \langle (p_n - c)(p_m - c) \rangle \tag{2.50}$$

Let us compare the above formalism with the standard mean field formalism applied to the Ising model. For our alloy problem the Ising model may be written , up to a constant

$$H(\{p_n\}) = \frac{1}{2} \sum_{n,m} p_n p_m V_{nm} - \mu \sum_n p_n \tag{2.51}$$

where $H(\{p_n\})$ is the (internal) energy of a given configuration, and the V_{nm} are pair interactions ($= V_{nm}^{AA} + V_{nm}^{BB} - 2V_{nm}^{AB}$). The mean field equations are similar to (2.38) where now $\mu_{n,eff}$ is given by

$$\mu_{n,eff} = \mu - \sum_m V_{nm} c_m \tag{2.52}$$

Similarly $\chi(q)$ is given by (2.49) if $e(q)$ is replaced by $V(q)$, the Fourier transform of V_{nm} . It is then tempting to identify $e(q)$ with the Fourier transform of effective pair interactions e_{nm} . In fact, the identification is complete if in (2.38) we expand $\mu_{n,eff}$ to first order in δc_n (which corresponds to the second order expansion of $\delta\overline{\Omega}_b$).

To summarize, the CPA-mean-field scheme reduces to the usual Bragg-Williams equations with effective pair interactions e_{nm} given by (2.47) if the second order expansion (2.46) is valid. It has been checked that in fact this expansion is in general numerically valid [16][18], and we shall see below how this can be understood.

At this point it is worth discussing the importance of defining effective interactions since after all, the CPA-mean-field equations can be solved without further approximations. The difficulty lies in the configurational entropy which in a single-site treatment is necessarily of the form (2.33) . Numerous calculations have shown that this approximation very badly describes many ordering phase diagrams, in particular on the fcc lattice. A natural extension of the Bragg-Williams theory is the cluster variation method (CVM) of Kikuchi which basically is a mean field theory for clusters instead of single sites [56]. Thus a global satisfactory theory would be a theory where the electronic structure would be described by a similar cluster extension of the CPA. Unfortunately this is extremely difficult, and such a theory with all the nice features of the CPA does not exist at the moment. It is also clear that in any case such an ideal theory would imply heavy numerical calculations, certainly beyond the present possibilities.

Thus, instead of defining more complicated self-consistent approximations, we shall try to define a generalized perturbation expansion for the energy Ω_b of a given configuration, which then can be inserted into appropriate thermodynamical treatments.

2.6. Generalized perturbation expansion

We again consider the electronic grand-potential and compare its value $\Omega_b(\{p_n\})$ for a given configuration with that provided by the CPA *without* LRO, $\overline{\Omega}_b$. In the case of lattices with a single atom per unit cell, the second term in the r.h.s. of (2.27) is independent of the configuration, and from (2.21) to (2.28) we have

$$\begin{aligned} \Omega_b(\{p_n\}) &= \overline{\Omega}_b + \frac{\mathrm{Im}}{\pi}\int dE\, f(E)\, \mathrm{Tr\,Log}[1 - \hat{t}\,\bar{G}_{off}] \\ &= \overline{\Omega}_b - \frac{1}{2}\sum_{\substack{n\neq m \\ i,j}} \frac{\mathrm{Im}}{\pi}\int dE\, f(E)\, t_n^i\, \bar{G}_{nm}\, t_m^j\, \bar{G}_{mn} + \ldots \end{aligned} \tag{2.53}$$

Since $t_n^i = p_n^i\, t^i$, we obtain

$$\Omega_b = \overline{\Omega}_b + \frac{1}{2}\sum_{\substack{n\neq m \\ i,j}} p_n^i p_m^j V_{nm}^{ij} + \ldots$$

$$V_{nm}^{ij} = -\frac{\mathrm{Im}}{\pi}\int dE\, f(E)\, t^i t^j\, \overline{G}_{nm}^2 \tag{2.54}$$

Using the occupation number p_n we then obtain effective pair interactions V_{nm}

$$V_{nm} = -\frac{\mathrm{Im}}{\pi}\int dE\, f(E)\, (\Delta t)^2\, \overline{G}_{nm}^2 \tag{2.55}$$

Notice that because of the relation

$$\sum_i c^i V_{nm}^{ij} = 0$$

all V_{nm}^{ij} can be related to V_{nm}. The expansion to higher order also generates multiatom interactions, and corrections to V_{nm}. We now argue that the expansion (2.54) is rapidly convergent in most cases. There is no mathematical proof for that, but several pieces of evidence. A first test is to compare the expansion with the exact value of Ω_b when this is possible. Thus one may calculate Ω_b for a perfectly ordered state. This can be done either by using Bloch's theorem and the full band structure, or more simply in the tight-binding scheme, by using the recursion method [20] (Fig.3). In most cases the pair expansion (2.53)

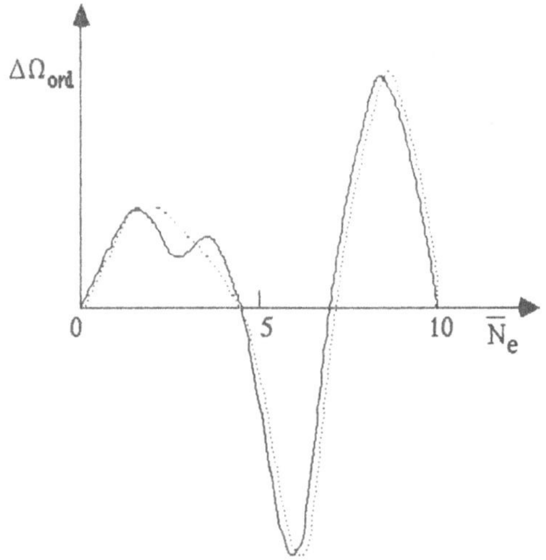

FIGURE 3. Ordering energy $\Delta\Omega_{ord} = \Omega(\text{order}) - \overline{\Omega}$ as a function of the band filling, for $c = 1/4$, $\delta/\overline{W} = 0.45$. Solid line: calculation with effective pair interactions. Dotted line: "exact calculation"; after [20].

has been found to reproduce Ω_b with a general accuracy of a few percents. How is this possible? To see that, let us consider the standard expansion of Ω_b as a function of

$\delta = \varepsilon^A - \varepsilon^B$. Then we get a formula similar to (2.53) where $\overline{G}$ is replaced by G^0 and t^i by $\varepsilon^i/(1 - \varepsilon^i F^0)$. To lowest order in δ (i.e. to second order) we then obtain

$$\Omega_b(\{p_n\}) \underset{\delta \to 0}{\approx} \Omega_b^0 + \frac{1}{2}\sum_{n,m}(p_n - c)(p_m - c)\, V_{nm}^0$$

$$V_{nm}^0 = \delta^2 \left(-\frac{\mathrm{Im}}{\pi}\right) \int dE\, f(E)\, (G_{nm}^0)^2 \tag{2.56}$$

From a simple dimensional analysis, V_{nm}^0 is of the order of $\delta^2/\overline{W}$, and this standard expansion converges in the weak disorder regime $\delta/\overline{W} \ll 1$. It obviously breaks down if $\delta/\overline{W} \approx 1$. Now, in (2.55), δ is replaced by Δt and G_{nm}^0 by $\overline{G}_{nm}$ and the product $|\Delta t^2\, \overline{G}_{nm}^2|$ does remain bounded when δ increases. In fact, it can strictly be proved that [14]

$$g = \left|\sum_{n \neq 0} \langle t^2 \rangle\, \overline{G}_{on}^2\right| \leq 1$$

Thus, in some respect, $\hat{t}\, \overline{G}_{off}$ is the small parameter of our expansion which insures its convergence. More precisely it can be verified numerically that $g \ll 1$ for most energies, so that the integrals over energies involving $\hat{t}\, \overline{G}_{off}$ are actually small [18]. Unfortunately it is difficult to go beyond these statements and one can just say that in most cases the generalized perturbation expansion (2.53) should be rapidly convergent. It will be clear in the following where it could fail to converge, in which case better expansions should be used.

Thus by changing the reference, (CPA medium for the disordered state instead of the energy Ω_b^0 of the mean crystal) we have improved the convergence of the perturbation expansion. The interactions generated by this procedure depend on the CPA medium and in particular are concentration dependent. This dependence is a way of including in some average manner the strong multiatom interactions of the usual perturbation method. It is clear that with other references other effective interactions would be defined. Thus the effective interactions are not completely intrinsic. What can be said is that they allow to calculate Ω_b with a good accuracy. Strictly speaking Ω_b does not take the form of a genuine Ising hamiltonian because of this concentration dependence.

Let us now compare the interactions V_{nm} with those, e_{nm}, obtained in the previous section. If g is small, the above arguments show that the expansion in powers of δc_q is convergent, *not because* δc_q *is small, but because the coefficients in front of them are small.* Furthermore $\delta\sigma_q$ in (2.44) can be simplified, which leads to

$$\delta\sigma_n \approx \sum_i t^i\, \delta c_n^i$$

Similarly in $e(q)$, the denominator can be replaced by unity, so that $e(q)$ precisely reduces to $V(q)$, the Fourier transform of our pair interactions V_{nm}. This shows that the mean field treatment of the previous section includes more contributions than the simple interactions V_{nm}. It is easy in particular to see that e_{nm} is a partial sum of the general perturbation (GP) expansion [19]. But notice that there is no guarantee that this amounts to summing the most important contributions of the GP development. Actually in some cases the GP interactions V_{nm} have been shown to lead to more accurate results than the renormalized interactions e_{nm} [18].

2.7. LCAO and KKR

Before discussing the characteristics of the pair and multiatom interactions, let us give some indications on the generalization of the previous formulation to more elaborated electronic structure techniques such as the KKR-CPA method, widely developed during the last years. In fact from a formal point of view, the formalism within the KKR method is exactly similar to ours, but this requires some translation, in particular in the intermediate calculations. If we refer to the review articles by Gyorffy and Stocks and by Stocks and Winter [21], the CPA equation is written in terms of quantities denoted t^A, t^B, $m = t^{-1}$, τ^c, $g(k)$,... In the KKR formalism we have site indices as in the tight-binding formalism, but the degeneracy labels λ, μ,... of the latter are replaced by orbitals labels L, L',...Also, all quantities are referred to the bare medium G^0 which then is the propagator of free particles. The t matrices are then (formally) related to ours through

$$t^{A(B)} = \varepsilon^{A(B)} (1 - \varepsilon^{A(B)} F^0)^{-1} \quad ; \quad t^c = \sigma (1 - \sigma F^0)^{-1}$$

$$\tau^c = \Sigma (1 - G^0 \Sigma)^{-1} = (\Sigma^{-1} - G^0)^{-1} = (t^{c^{-1}} - G^0_{off})^{-1} \tag{2.57}$$

$$g(k) = \langle k | G^0_{off} | k \rangle \quad \text{etc...}$$

With this glossary (see also [22]) almost all formulae obtained above have also been derived in the KKR-CPA language, the integral formula (2.31) in particular. The mean field effective pair interaction $e(q)$ has also been obtained, being then called $S^2(q)$ [23].

3. EFFECTIVE PAIR AND CLUSTER INTERACTIONS

3.1. Introduction

The generalized perturbation method generates effective pair (and cluster) interactions whose characteristics will now be described. We first recall that they are concentration dependent through $\sigma(z)$, the CPA effective potential. In the presence of diagonal disorder only, they then depend on $\delta/\overline{W}$. When this parameter is weak, standard perturbation theory applies and the pair interactions are of the order of $\delta^2/\overline{W}$. These pair interactions display familiar long-range Friedel oscillations of amplitude varying as $|m-n|^{-3}$, which is related to the asymptotic form of G^0_{nm}

$$|G^0_{nm}| \underset{|m-n|\to\infty}{\approx} |m-n|^{-1}$$

In this regime the interactions are weak and long-ranged. When $\delta/\overline{W}$ is large, G^0_{nm} is replaced by $\overline{G}_{nm}$ which is exponentially damped because of mean free path effects (i.e. because $\mathrm{Im}\,\sigma \neq 0$). In this regime the interactions are strong and short-ranged. Thus a few interaction parameters should be sufficient for most transition alloys in which ordering effects are important. The interactions also depend on the Fermi level and therefore on the mean filling $\overline{N}_e$ of the band. For a given alloy $\overline{N}_e$ is a definite function of the concentration, but in many cases it is useful to consider it as an independent parameter. Strictly speaking one must distinguish between the development at constant number of electrons and the development made before at constant chemical potential ν (= E_F at zero

calculated by these methods. Except in simple cases [52] they are still not comparable with real phase diagrams because of various assumptions (fixed lattice, ...) but they do indicate that the simple method described here can produce diagrams of various shapes (see de Fontaine's lectures and [53]). For example, in Fig.12 we reproduce schematically a phase diagram on the fcc lattice when only first neighbour interactions V_1 are taken into account. The three familiar phases A_3B, AB_3 ($L1_2$) and AB ($L1_0$) are present, but we notice the strong asymmetry induced by the concentration dependence of V_1 and also a surprising fact: the appearence of a two phase regime involving solid solutions at high temperature.

At first sight this phenomenon seems rather strange since the interactions favour ordering. Actually it cannot occur using the Bragg-Williams approximation since then the only possible instability is given by the q vector that minimizes V(q) (or e(q), see eq.(2.49)). With first neighbour interactions these q vectors are found to be of type (1x0). There is no segregation (q=0) instability. When short range order effects are included as in the CVM, there is a corresponding contribution in the internal energy which may strongly depend on c if $V_1(c)$ does. As a result the second derivative of the free energy with respect to c can vanish at some temperature, indicating a second order transition. Below this temperature, we have a coexistence of two solid solutions with different (strong and weak) short range order. Thus, we may have a coexistence of ordering and segregation processes in a given diagram even if the explicit interactions favour ordering. In this case ordered phases should of course appear at lower temperature (see Fig.12). Similar effects

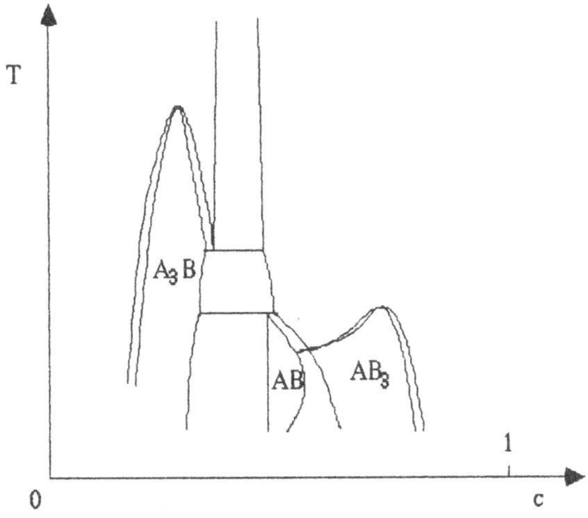

FIGURE 12. Schematic prototype phase diagram on a fcc lattice with concentration dependent first neighbour interactions; after [53].

could also appear if we add in the Ising hamiltonian concentration dependent terms (elastic effects,...,see [41]). This effect should not be confused with that which obviously occurs if the interactions V_1 change of sign with the concentration; such an effect has been predicted and observed in FeCr [54].

Phase diagrams where ordering and phase separation coexist do exist, for example in Pd-H [55] or in pseudobinary semiconductor alloys [41], but from our discussion a detailed analysis of the short range order as a function of the concentration and of the temperature would be very interesting.

5.3. Conclusion

This review has concentrated more on the principles than on practical applications which are described in detail elsewhere. It is hoped that by reading the corresponding articles the reader will be convinced that we now have a scheme which allows to understand many aspects of ordering phenomena in alloys.

Most of the material presented here is the result of multiple collaborations with A Bieber, F Gautier, J P Landesman, G Tréglia and P Turchi who are gratefully acknowledged. The author is also indebted to J Douin for fruitful discussions and to A Loiseau, F Solal and G Tréglia for careful readings of the manuscript.

REFERENCES

1. Pettifor D G : J.Phys.C: Solid State Phys. 19, 285 and 315, 1986.
2. Moruzzi V L, Janak J F and Williams A R : Calculated Electronic Properties of Metals. New York: Pergamon, 1978.
   Skriver H L : Phys.Rev. B31, 1909, 1985.
   Davenport J W, Watson R E and Weinert M : Phys.Rev. B32, 4883, 1985.
   Min B I, Oguchi T, Jansen H J F and Freeman A J : Phys.Rev. B34, 654, 1986.
3. Williams A R, Gelatt C D and Moruzzi V L : Phys.Rev.Lett. 44, 429, 1980.
   and : Phys.Rev. B25, 6509, 1982.
   Gelatt C D, Williams A R and Moruzzi V L : Phys.Rev. B27, 2005, 1983.
   Pettifor D G : Physical Metallurgy, Ch.3, R W Cahn ed. Amsterdam: North Holland, 1983.
   Johnson D D, Nicholson D M, Pinski F J, Gyorffy B L and Stocks G M : Phys.Rev.Lett. 56, 2088, 1986.
   Xu J H, Oguchi T and Freeman A J : Phys.Rev. B35, 6940, 1987.
4. Conolly J W D and Williams A R : Phys.Rev. B27, 5169, 1983.
5. Mohri T, Terakura K, Oguchi T and Watanabe K : Phys.Rev. B35, 2169, 1986.
6. Sigli C and Sanchez J M : Acta Metall. 33, 1097, 1985 and 34, 1021, 1986.
7. Loiseau A, Van Tendeloo G, Portier R and Ducastelle F : J.Physique 46, 595, 1985.
   Broddin D, Van Tendeloo G, Van Landuyt J, Amelinckx S, Portier R, Guymont M and Loiseau A : Phil.Mag. A54, 395, 1986.
8. De Fontaine D and Kulik J : Acta Metall. 33, 145, 1985.
9. Stocks G M, Nicholson D M, Pinski F J, Butler W H, Sterne P, Temmerman W M, Gyorffy B L, Johnson D D, Gonis A, Zhang X G and Turchi P E A : to be published in MRS Proceedings, Boston Dec 1986.
10. Ducastelle F and Cyrot-Lackmann : J.Phys.Chem.Solids 31, 1295, 1970 and 32, 285, 1971.
11. Andersen O K and Jepsen O : Phys.Rev.Lett. 53, 2571, 1984.
12. Ducastelle F : Mat. Res. Soc. Symp. Proc. Phase Transition in Solids, Tsakalakos T ed. New York, Amsterdam, Oxford : North Holland, 21, 375, 1984.
13. Velicky B, Kirkpatrick S and Ehrenreich H : Phys.Rev. 175, 747, 1968.
    Elliott R J, Krumhansl J A and Leath P L : Rev.Mod.Phys. 46, 465, 1974.
14. Ducastelle F : J.Phys. C : Solid State Phys. 7, 1795, 1974 and 8, 3297, 1975,
    Solid State Transformations in Metals and Alloys. Les Ulis : Les Editions de Physique, 51, 1980,
    Electronic Structure of Crystal Defects and of Disordered Systems, Gautier F, Gerl M, and Guyot P ed. Les Ulis : Les Editions de Physique, 235, 1981.
15. Brouers F, Giner J and Van der Rest J : J.Phys.F : Metal Phys. 4, 214, 1984.
16. Gautier F, Ducastelle F and Giner J : Phil.Mag. 31, 1373, 1975.
17. Ducastelle F and Tréglia G : J.Phys.F : Metal Phys. 10, 2137, 1980.
    Ducastelle F : Theory of Alloy Formation, Bennett L H ed. AIME, 194, 1980.
18. Tréglia G, Ducastelle F and Gautier F : J.Phys.F : Metal Phys. 8, 1437, 1978.
19. Ducastelle F and Gautier F : J.Phys.F : Metal Phys. 6, 2039, 1976.
20. Bieber A, Gautier F, Tréglia G and Ducastelle F : Solid State Commun. 39, 149, 1981.
21. Gyorffy B L and Stocks G M : Electrons in Disordered Metals and at Metallic Surfaces, Phariseau P ed. NATO ASI Series B, Vol.42. Plenum Press, 42, 1979
    Stocks G M and Winter H : The Electronic Structure of Complex Systems, Phariseau P and Temmerman W M ed. NATO Series B, Vol.113. Plenum Press, 463, 1984.
22. Butler W H : Phys.Rev. B31, 3260, 1985.

23. Gyorffy B L and Stocks G M : Phys.Rev.Lett. 50, 374,1983.
24. Ducastelle F : J.Physique C7, 79, 1975.
25. Hodges C H : J.Phys.F : Metal Phys. 9, L89, 1979.
26. Ducastelle F : Thèse de Doctorat d'Etat, Orsay, 1972.
27. Turchi P : Thèse de Doctorat d'Etat, Paris, 1984.
28. Hirai K and Kanamori J : J.Phys.Soc.Japan 50, 2265, 1981.
29. Bieber A and Gautier F : Z.Phys. B57, 335, 1984 and J.Phys.Soc.Japan 53, 2061, 1984.
30. Turchi P and Ducastelle F : The Recursion Method and its Applications, Pettifor D G and Weaire D L ed. Springer Series in Solid State Sciences Vol.58, Springer, 104, 1985.
31. Sluiter M, Turchi P and De Fontaine D : Prediction of ordering in binary transition alloys, preprint 1987.
32. Gautier F : High Temperature Alloys, Theory and Design. Stiegler J O ed. AIME, 163, 1984.
33. Bieber A and Gautier F : Segregation and order in binary substitutional alloys II Ground state and phase stability diagrams, preprint 1987.
34. Turchi P, Tréglia G and Ducastelle F : J.Phys.F : Metal Phys. 13, 2543, 1983.
35. Bieber A and Gautier F : Physica 107B, 71, 1981.
36. Jesser R, Bieber A and Kuentzler R : J.Physique 42, 1157, 1981 and 44, 631, 1983.
37. Bieber A and Gautier F : Phase stability and local order in magnetic transition metal alloys, preprint 1987.
38. Ducastelle F : J.Physique 31, 1055, 1970.
39. Moraitis G J and Gautier F : J.Phys.F : Metal Phys. 8, 1421, 1977.
40. Tréglia G and Ducastelle F : J.Phys.F : Metal Phys. 17, 1935, 1987.
41. Martins J L and Zunger A : Phys.Rev.Lett. 56, 1400, 1986.
Mbaye A A, Ferreira L G and Zunger A : Phys.Rev.Lett. 58, 49, 1987.
42. Bieber A, Ducastelle F, Gautier F, Tréglia G and Turchi P : Solid State Commun. 45, 585, 1983.
43. Bieber A and Gautier F : Acta Metall. 34, 2291, 1986.
44. Bieber A and Gautier F : Solid State Commun. 38, 1219, 1981.
45. Kanamori J and Kakehashi Y : J.Physique 38, C7-274, 1977.
46. Finel A and Ducastelle F : to be published.
47. Finel A and Ducastelle F : Mat.Res.Soc.Symp.Proc., Phase Transformations in Solids, Tsakalakos T ed. New York, Amsterdam, Oxford: North Holland, 21, 293, 1984.
48. Landesman J P, Tréglia G, Turchi P and Ducastelle F : J.Physique 46, 1001, 1983.
49. De Novion C H and Landesman J P : Pure and Applied Chem. 57, 1391, 1985.
50. Finel A and Ducastelle F : Europhys. Lett. 1, 135, 1986.
Finel A : Thèse de Doctorat d'Etat, Paris, 1987.
Ducastelle F : Rech.Aérosp. 4, 11, 1986.
51. Gautier F, Van der Rest J and Brouers F : J.Phys.F : Metal Phys. 5, 1884, 1975.
Van der Rest J, Gautier F and Brouers F : J.Phys.F : Metal Phys. 5, 2283, 1975.
Pettifor D G : Phys.Rev. Lett. 42, 849,1979
Watson R E, Bennett L H and Goodman D A : Acta Metall. 31, 1285, 1983
Colinet C, Pasturel A and Hicter P : Physica 128B, 5, 1985, Calphad 9, 71 and 349, 1985.
52. Hawkins R J, Robbins M O and Sanchez J M : Phys.Rev. B33, 4782, 1986
Sigli C, Kosugi M and Sanchez J M : Phys.Rev.Lett. 57, 253, 1986.
53. Turchi P, Sluiter M and De Fontaine D : Phys.Rev. B36, 3161, 1987.
54. Mirebeau I, Hennion M and Parette G : Phys.Rev.Lett. 53, 687, 1984.
55. Blaschko O : J.Less Common Met. 100, 307, 1984.
56. Sanchez J M, Ducastelle F and Gratias D : Physica 128A, 334, 1984 and references therein.

QUANTUM MECHANICS IN ALLOY DESIGN

D. G. PETTIFOR

Department of Mathematics, Imperial College, London SW7 2BZ.

1. INTRODUCTION

The purpose of this NATO Advanced Study Institute on Alloy Phase Stability is to clarify and enhance our understanding of the mechanisms responsible for phase formation and stability in metallic alloys. In these lectures I will address the particular problem of phase stability at absolute zero, discussing the microscopic origin of the heat of formation and structural stability of binary and ternary alloys. These lectures are given in parallel to C. T. Liu's on "Phase Stability and design of ordered intermetallic alloys". The overall title of my three lectures "Quantum Mechanics in Alloy Design" reflects C. T. Liu's influence on and encouragement of quantum theorists to become directly involved in materials design.

How can quantum theory help alloy designers in their search for new materials? At the outset alloy designers are faced with an immediate problem: that of a multitude of choice. For example, suppose they were interested in developing a new cubic Cu_3Au-type alloy with possible good mechanical properties at high temperatures (1) or tetragonal $BaCd_{11}$, $ThMn_{12}$, or $NaZn_{13}$ iron-based alloys for possible use as permanent magnets (2). Then, even for binary systems, there would in principle be about 10,000 possible candidates for each stoichiometry A_mB_n, since A and B can be chosen from the one-hundred elements in the periodic table. And, of course, real alloys consist of ternary, quaternary ... additions so that there are in principle millions of possibilities.

Quantum mechanics is the only theory which can predict the structure and stability of phases, since it describes correctly the bonding behaviour of the valence electrons. In these lectures I will show that quantum theory may be used in two different but complementary ways for guiding alloy designers in their search for new materials. The first approach orders the known structural data base (3) with the aid of two-dimensional structure maps, thereby allowing the alloy designer to interpolate more reliably. The second approach solves the equations of quantum mechanics, thereby predicting properties and providing reliable concepts for understanding structure and stability.

The use of phenomenological structure maps in alloy design will be discussed in the next section, whereas the use of microscopic quantum mechanics in understanding cohesion and structure will be presented in sections 3 and 4. Section 3 will provide a quantum mechanical critique of the Miedema rules (4) of alloy formation. Section 4 will discuss the origin of the structural stability of binary transition metal-metalloid AB compounds, illustrating the importance of atomic size, atomic energy level mismatch, and average number of valence electrons per atom respectively. In section 5 we conclude.

G. M. Stocks and A. Gonis (eds.), Alloy Phase Stability, 329–350.

2. PHENOMENOLOGICAL STRUCTURE MAPS

The crystal structures of tens of thousands of binary, ternary, and quaternary phases have been determined since the advent of X-ray crystallography in 1910. To date more than two thousand different structure types have been identified (3). Following the pioneering work of Mooser and Pearson (5), numerous authors (6-13) have attempted to order this very large empirical data base with the aid of two- or three-dimensional structure maps. These maps were constructed by choosing co-odinates based on those physical factors which were felt to be important in controlling structural stability. For example, Villars (13) has recently plotted three-dimensional maps (ΔX, ΔR, e/a) for binary compounds with the AB, AB_2, AB_3 and A_3B_5 stoichiometries. The three co-ordinates reflect the importance of the electronegativity difference ΔX, the atomic size difference ΔR, and the average number of valence electrons per atom e/a in the determination of structural stability (14, 15). However, these three-dimensional plots suffered several drawbacks. Firstly, the fifth most common AB structure type, hP4(NiAs), was unable to be separated and was, therefore, omitted from the AB structure map (13). Secondly, the structures of the pure elements themselves can not be ordered within these plots because the elements lie along the e/a axis corresponding to (ΔX = 0, ΔR = 0), so that tetrahedrally co-ordinated Si and close-packed Ti, for example, would both be characterized by the same point with e/a = 4. These failings are related to the neglect of the angular dependence of the valence orbitals, since whether the electrons have s, p, or d-like quantum character is not reflected in the choice of classical co-ordinates (ΔX, ΔR, e/a) (16).

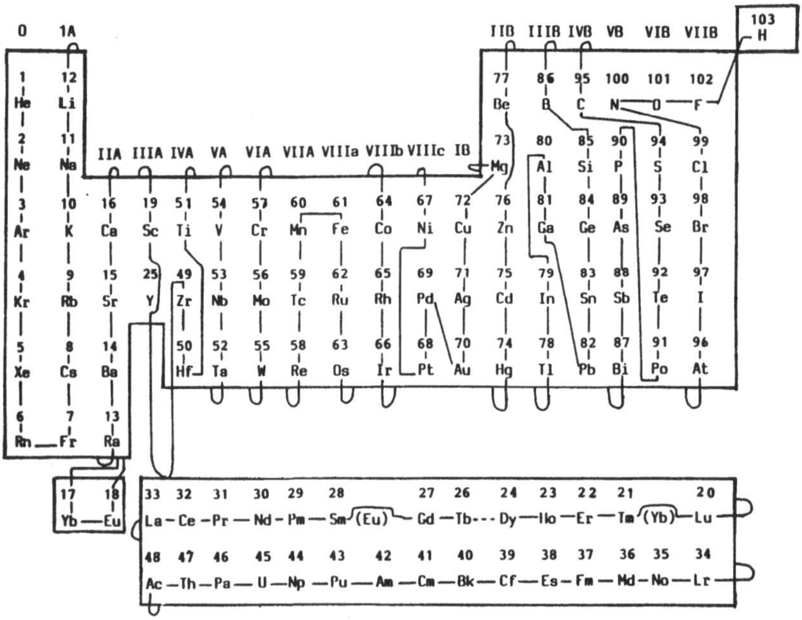

TABLE 1. The string running through this modified Periodic Table puts all the elements in sequential order, given by the Mendeleev number. Note that the group IIA elements beryllium and magnesium have been grouped with group IIB and that the divalent rare earths have been separated from the trivalent.

An alternative approach to the construction of structure maps has been proposed by the author (16, 17, 18). Rather than trying to find a set of microscopic co-ordinates which will produce a separation of the structure types within some n-dimensional space, he looked instead for a single phenomenological co-ordinate which will lead to good structural separation of the empirical data on binary systems within two-dimensions (19). This was achieved by running a one-dimensional string through the two-dimensional periodic table as shown in table 1 (16). Pulling the ends of the string apart orders all the elements along a one-dimensional axis, their sequential order being termed the Mendeleev number $\mathcal{M}$. This simple procedure is found to provide excellent structural separation of all binary compounds with a given stoichiometry A_mB_n within a single two-dimensional plot $(\mathcal{M}_A, \mathcal{M}_B)$. The following stoichiometries have been plotted and discussed (18): AB, AB_2, AB_3, AB_4, AB_5, AB_6, AB_{11}, AB_{12}, AB_{13}, A_2B_5, A_2B_{17}, A_3B_4, A_3B_5, A_3B_7, A_4B_5 and A_6B_{23}. The Mendeleev number also successfully demarcates the structural domains of molecules displaying, for example, the region where AB_2 trimers are either bent or linear (20).

Figs. 1 and 2 show the AB and AB_3 structure maps for the pure binaries which were plotted using the data base of Villars and Calvert (3) for the intermetallic phases. The bare patches correspond to regions with positive heats of formation where compounds do not form (4). We see that excellent structural separation has been achieved between the 52 AB and 52 AB_3 structure-types which have more than one representative compound each. In particular, the AB plot demarcates between the very similar structure-types hP4 (NiAs), cP8 (FeSi), and oP8 (MnP); oC8 (CrB) and oP8 (FeB); and cP2 (CsCl), tP4 (CuTi), hP18 (AuCd), and tP4 (CuAu) respectively. The AB_3 plot demarcates between the very similar 12-fold co-ordinated close-packed structure-types cP4 (Cu_3Au), hP8 (Ni_3Sn), tI8 (Al_3Ti) and oP8 (Cu_3Ti). The former two are based on the stacking of triangular-type ordered AB_3 close-packed layers, whereas the latter two are based on rectangular-type ordered AB_3 close-packed layers (1).

The structure maps for the binary compounds suggest (21) that it should be possible to move from one structural domain to another by suitable addition of a third or more alloying element. For example, it is well known (3) that Al_3Ti changes from its tetragonal structure to the cubic Cu_3Au-type structure on alloying with the 3d elements iron, nickel, copper or zinc, which is not unexpected since the Cu_3Au domain is adjacent to the Al_3Ti domain in fig. 2 If we assume that the ternary and quaternary additions C and D go preferentially to the A and B sites respectively, then we may regard the alloy $(A_x C_{1-x})(B_y D_{1-y})_3$ as the pseudo-binary $\bar{A}\,\bar{B}_3$ characterized by the average Mendeleev numbers

$$\overline{\mathcal{M}}_A = x\mathcal{M}_A + (1-x)\mathcal{M}_C \quad (1)$$

and

$$\overline{\mathcal{M}}_B = y\mathcal{M}_B + (1-y)\mathcal{M}_D \quad (2)$$

Figures 3 and 4 show that simple scheme orders the pseudo-binaries within the same structural domains as for the pure AB and AB_3 binaries in figs. 1 and 2 respectively (21).

These structure maps may, therefore, be used to help focus the search for new alloys with a given crystal structure. For example, we see that the cubic Cu_3Au-type alloys, which often have desireable high temperature mechanical properties (1), are located in five well-defined domains in figs.

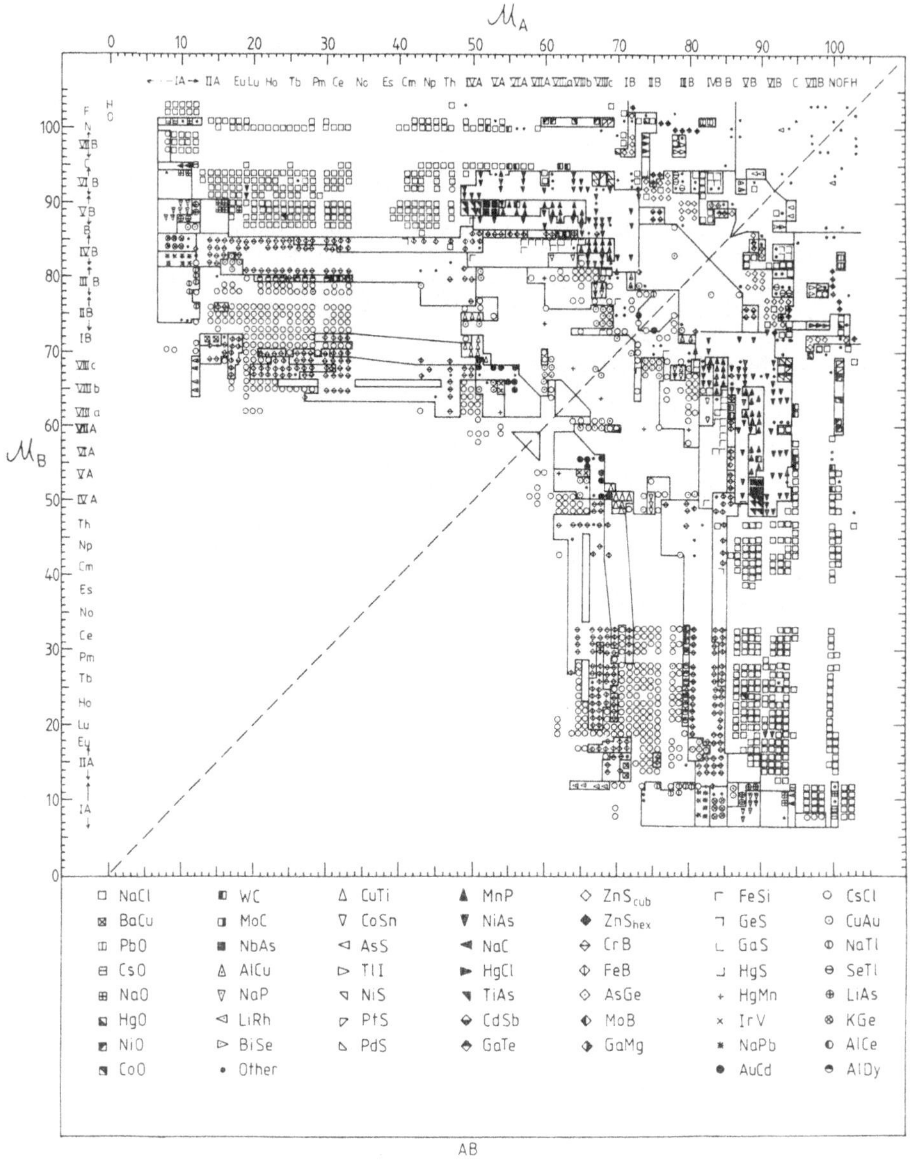

FIGURE 1. The AB structure map.

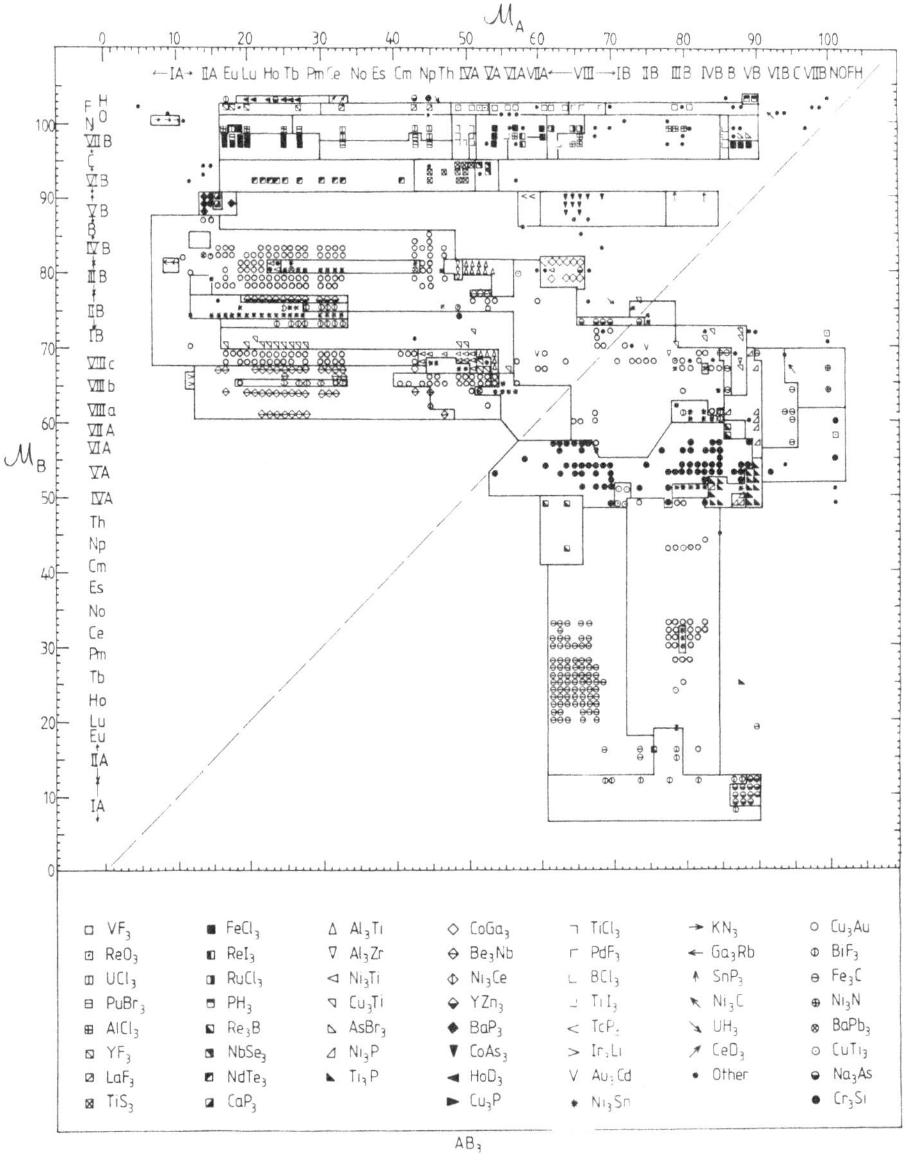

FIGURE 2. The AB_3 structure map.

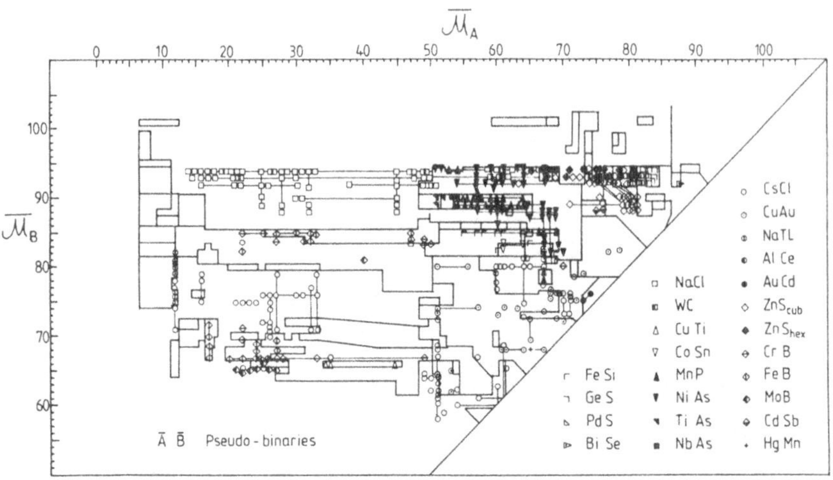

FIGURE 3. The structure map for AB pseudo-binaries.

2 and 4, thereby reducing the number of possible alloy combinations to be considered. Similarly, the AB_{11}, AB_{12}, and AB_{13} alloys demarcate clearly the domains where the tetragonal $BaCd_{11}$, $ThMn_{12}$, and $NaZn_{13}$ structure-types are stable, thereby focussing the search for new iron-based permanent magnets with these structures (2). However, these phenomenological maps cannot guarantee that another phase will not precipitate out in preference to the pseudo-binary structure-type chosen nor can they predict the existence of unknown structure-types (such as, for example, tetragonal $Nd_2Fe_{14}B$ discovered recently by chance). For this, microscopic quantum mechanical calculations are required.

3. A QUANTUM MECHANICAL CRITIQUE OF THE MIEDEMA RULES FOR ALLOY FORMATION.

A fundamental question in the formation of alloys is whether the alloy is more or less stable than the elemental systems separately i.e. whether the heat of formation is negative or positive. It is clear from figs. 1 and 2 that about half the binary combinations have negative heats of formation, corresponding to the observation of known compounds, whereas the other half have positive heats of formation corresponding to the absence of compounds. The Miedema rules (4) give a simple prescription for predicting not only the sign of the heat of formation ΔH, but also its magnitude.

The Miedema scheme considered atoms in the metallic state as being characterized by two co-ordinates, namely the work function ϕ^* and the cube root of the electronic charge density at the metallic Wigner-Seitz sphere radius $\rho_m^{1/3}$, respectively. An AB alloy would then be formed by cutting out the Wigner-Seitz cells from the pure metals A and B and packing them together to create the binary system as shown in fig. 5. By treating the atoms as macroscopic pieces of metal, they argued that there were two contributions to the heat of formation. The first was attractive and arose from the flow charge from one atom to another due to the difference in the work function $\Delta\phi^*$. The second was repulsive and arose from the removal of the discontinuity in the charge density $\Delta\rho_m^{1/3}$ across the interface between neighbouring A and B Wigner-Seitz cells.

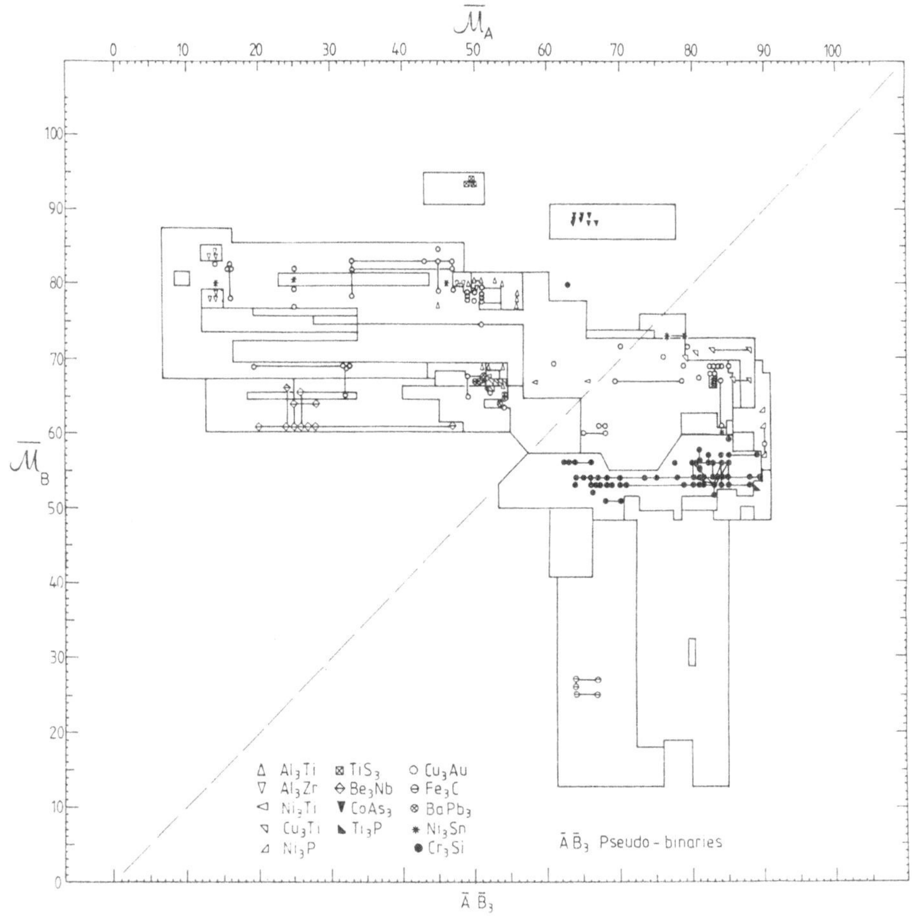

FIGURE 4. The structure map for $\bar{A}\bar{B}_3$ pseudo-binaries.

FIGURE 5. The "macroscopic atom" model for alloy formation.

Thus, the heat of formation ΔH of a disordered binary alloy is written within the Miedema scheme as

$$\Delta H = 2\, C_A C_B\, g(C_A,C_B)\, \lambda_m^{-1} \Delta\hat{H} \qquad (3)$$

where

$$g(C_A,C_B) = V_A^{2/3}\, V_B^{2/3}/(C_A V_A^{2/3} + C_B V_B^{2/3}) \qquad (4)$$

$$\lambda_m = (\rho_m^A)^{-1/3} + (\rho_m^B)^{-1/3} \qquad (5)$$

and

$$\Delta\hat{H} = -P(\Delta\phi^*)^2 + Q(\Delta\rho_m^{1/3})^2 . \qquad (6)$$

The atomic concentration and volume of the A and B constituents are represented by C_A, C_B and V_A, V_B respectively. The pre-factor $g(C_A,C_B)$ gives the deviation in ΔH from regular solution behaviour due to the difference in the sizes of the atoms V_A and V_B. The heat of formation within the Miedema model is proportional to the surface area shared by dissimilar cells, which accounts for the factor $V_A^{2/3}\, V_B^{2/3}$ in eq. (4). The pre-factor λ_m^{-1} is argued to reflect the influence of the electronic screening length on the width of the dipole layer at the A-B interface.

It follows from eqs. (3) and (6) that the sign of the heat of formation is fully determined by the ratio of $\Delta\phi^*$ to $\Delta\rho_m^{1/3}$. In particular,

$$\Delta H \lesseqgtr 0 \quad \text{for } \Delta\phi^*/\Delta\rho_m^{1/3} \gtreqless (Q/P)^{1/2}. \qquad (7)$$

We see from fig. 6 that the Miedema scheme reproduces the sign of ΔH for all binary alloys of two transition metals and of transition metals alloyed with noble, alkali, and alkaline earth metals. Each binary system is characterized by some point $(\Delta\rho_m^{1/3}, \Delta\phi^*)$ on the figure. If the binary system contains stable intermetallic compounds it is represented by a minus sign, whereas if no compounds exist and there is only a limited range of solid solubility then it is represented by a plus sign. By a judicious choice of the parameters ϕ^* and ρ_m the plusses and minuses in fig. 6 have been separated nearly perfectly by the straight line ΔH = 0 corresponding to $Q/P = 9.4\ V^2/(\text{d.u.})^{2/3}$.

The success of the Miedema scheme represented by fig. 6 has led to a belief that the underlying physical concepts upon which it is based are correct. In particular, the attractive term is assumed to be ionic in character. This supposition is central to the model of Alonso and Girifalco (23) which was proposed in order to account for the attractive and repulsive terms in the Miedema scheme. They argued that the electronic ground state of a binary alloy can be reached in a two-step process. First the Wigner-Seitz cells of the two dissimilar metals are "prepared" by either expanding or contracting them until they have identical charge density at their

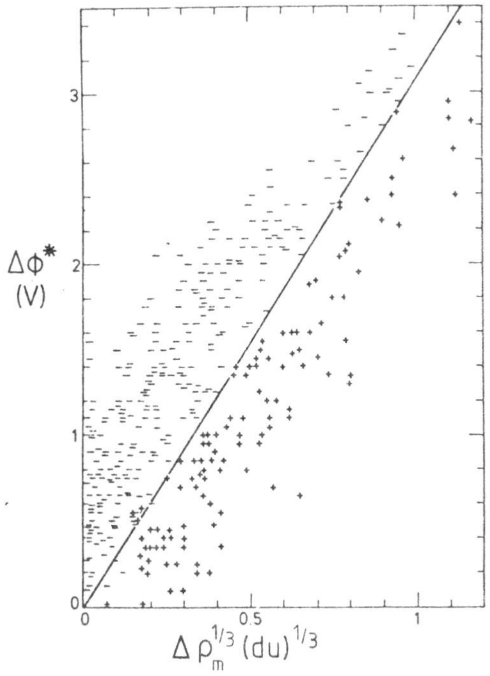

FIGURE 6. Demonstration that the Miedema scheme reproduces the sign of ΔH for solid binary alloys consisting of a transition metal and a transition, noble, alkali, or alkaline earth metal.

Wigner-Seitz boundaries, as illustrated schematically in fig. 7. This step leads to a positive contribution to ΔH which is dependent on the bulk moduli of the elemental metals and can be identified with the second term in eq. (6).

The second step comprises the relaxation of the electronic charge of the prepared atoms into the ground state of the binary alloy. The ionic model assumes that this redistribution of the charge is accompanied by a change in energy which is given by

$$\Delta H_{ionic} = -\frac{1}{4} q(\mu_A - \mu_B) \tag{8}$$

where μ_A and μ_B are the chemical potentials associated with the prepared A and B Wigner-Seitz cells. qe is the amount of charge which flows from one atomic cell to another in order to equilibrate the chemical potentials, as shown in fig. 7. Assuming that the charge transfer qe is proportional to $(\mu_A - \mu_B)$, the ionic contribution is negative and proportional to $(\mu_A - \mu_B)^2$. Thus, the attractive contribution to the Miedema heat of formation in eq. (6) is identified with the ionic energy eq. (8), which results from the flow of charge from one atom to another on alloy formation.

Unfortunately, however, the ionic model is at variance with first principles quantum mechanical calculations for metallic systems (24, 25). These calculations, which are based on the local-density-functional approximation (26), allow a detailed breakdown of the various contributions

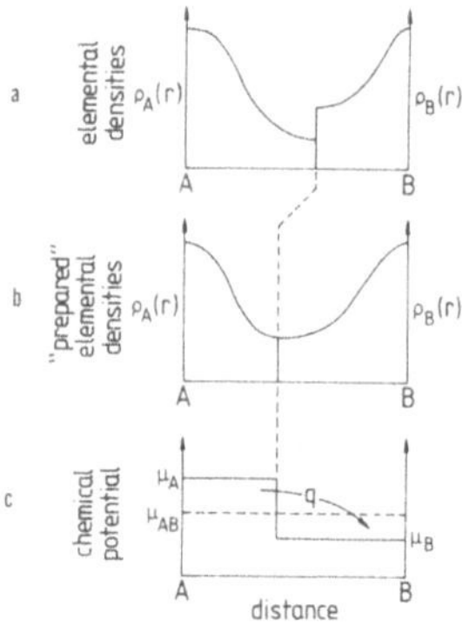

FIGURE 7. Schematic representation of the two steps constituting the ionic model of alloy formation. In the first step from (a) to (b) the constituent possessing the smaller (larger) Wigner-Seitz electron density is compressed (expanded) until the boundary-density mismatch is eliminated. In the second step, which is represented by (c), the discontinuity in the chemical potential between the elemental Wigner-Seitz cells is removed by the transfer of q electrons to the cell possessing the lower chemical potential (after Williams et al. (24)).

to the metallic bond, both for the elemental metals and for binary systems. In particular it is found that the energy change accompanying the second step in fig. 7 is not well represented by the ionic expression eq. (8), since this leads to errors as large as or larger than the heat of formation itself (24). The ionic model completely fails to describe the quantum mechanical metallic bond.

The nature of the metallic bond in simple metals and transition metals has been described in detail elsewhere (27), so it will suffice here to note three essential differences between the ionic and quantum mechanical bond models. Firstly, the ionic model is based on the rigid-band approximation, whereas quantum mechanics predicts a common-band in the binary alloy. This is illustrated schematically in fig. 8. The ionic rigid-band model assumes that the alloy density of states is a superposition of the elemental densities of states. The binding energy is lowered as charge flows from the one band to the other in order to equilibrate the chemical potentials as shown in fig. 7. Harrison (28) has demonstrated that this leads to values for the cohesive energy of the alkali halides that are consistent with the predictions of the classical ionic model of Born. The common-band model, on the other hand, assumes that the local density of states on a given atom changes when the atom is taken from its elemental environment to that of the alloy. The band in the alloy is formed by the quantum-mechanical overlap of neighbouring atomic wave functions so that the local densities of states have a common band width. This is illustrated in fig. 8 within the rectangular band approximation for transition metals in which the total AB alloy density of states is assumed to be constant (29-31) like that in the elemental metals (32).

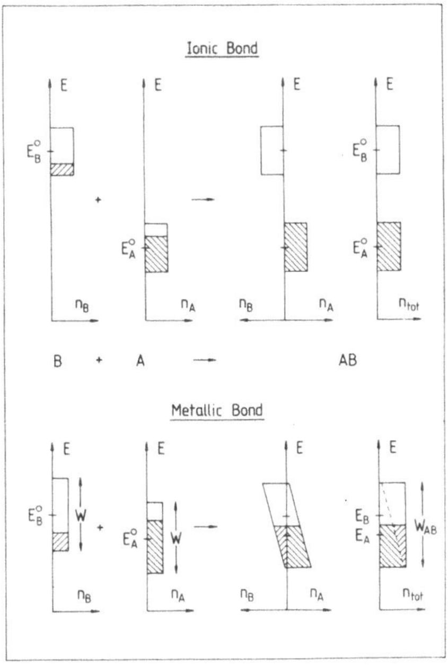

FIGURE 8. Schematic representation of the ionic rigid-band and metallic common-band models of bonding in binary AB systems. E_A^0 and E_B^0 give the free-atom energy levels, whereas the positions of E_A and E_B in the metallic bond reflect the small shift which takes place on alloy formation in order to maintain local charge neutrality.

The second essential difference between the ionic and simple quantum mechanical models of the metallic bond is that the latter assume local charge neutrality. It is well-known that the charge density of simple metal alloys is given by the super-position of neutral pseudo-atoms, since each individual ion is perfectly screened by the metallic response function of the alloy (34). The constraint of local charge neutrality may be imposed on the rectangular band for transition metal alloys shown in fig. 8 by a small adjustment of the atomic level difference ΔE from the free atom value ΔE^0.

The third essential difference between the ionic and quantum mechanical models of ΔH is that the attractive term in the former is due to charge flowing from one atom onto another, whereas in the latter it is due to the change in the quantum mechanical bonding between atoms. Whereas the ionic contribution is given by eq. (8), the quantum mechanical contribution reflects the change in bond order on alloy formation in fig. 8 and may be written (27) for the transition metal 4d series (in eV/atom) as

$$\Delta H^{bond}_{order} = -\frac{1}{600}\left[3\,\overline{N}_d(10-\overline{N}_d) - 50\right](\Delta N_d)^2 \qquad (9)$$

where $\overline{N}_d$ and ΔN_d are the average and difference in number of valence d electrons on the A and B sites respectively. As shown by the dotted curve in fig. 9, the bond order contribution for a given ΔN_d is most attractive for $\overline{N}_d = 5$. Thus, the most stable transition metal alloys would comprise elements from groups at opposite ends of the transition metal series such as Y and Pd, as was originally suggested by Brewer (35). These groups have very

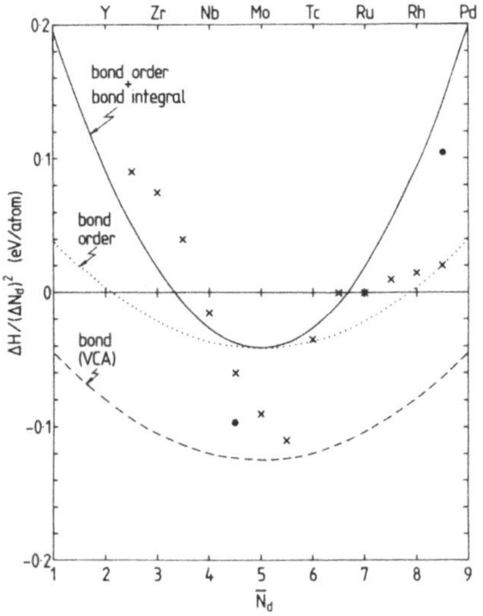

FIGURE 9. The contributions to the normalized heats of formation $\Delta H/(\Delta N_d)^2$ for the case of 4d transition metal alloys with either ΔN_d = 1 or 2. (×) and (●) represents the Miedema and experimental values, respectively, for disordered alloys.

few bonding electrons, since they have nearly empty or full d shells. Mixing these elements together would result in a dramatic increase in the bond order, because the electrons would be shared in the bonding states of the alloy corresponding to a half-full band. The behaviour of the bond order contribution cannot be reproduced by the ionic rigid-band model. The former contribution changes sign and is repulsive for nearly empty or full d bands. The latter is always attractive. The difference accounts for the oscillation in the error of the ionic model which the first principles calculations of Williams et al. (24) observed across the 4d transition metal series.

We see, therefore, that quantum mechanics provides no justification for the physical basis of the "Macroscopic atom" model of Miedema (4). The negative contribution to the heats of formation in metallic alloys is not driven by the difference in the elemental work functions or chemical potentials as the rigid-band ionic model assumes. Although there is a redistribution of charge on the formation of the alloy, the constituent atoms may be taken to be perfectly screened by the metallic environment and, hence, to remain charge neutral. The heat of formation arises from the creation of a common band within the alloy.

The Miedema parameters ϕ^* and $\rho_m^{1/3}$ are phenomenological co-ordinates which have been adjusted to reproduce the separation of experimental points shown in fig. 6. They should not be identified with the physical co-ordinates, the work function and cube root of the charge density respectively, even though they display similar trends across the periodic table as illustrated in fig. 10.

The heat of formation is sensitive to even very small changes in the values of ϕ^* and $\rho_m^{1/3}$. Consider, for example, the alloys of iron with s and d elements. ΔH will be positive or negative depending on whether $\Delta\phi^*$ is less than or greater than $(Q/P)^{1/2}\,\Delta\rho_m^{1/3}$. This is illustrated in fig. 11

temperature). This should introduce corrective terms in the GP development, but, to lowest order, expanding the grand-potential at constant ν is equivalent to expanding the electronic free energy at constant $\overline{N}_e$.

3.2. Moments

By definition, the moments μ_n of the (normalized) density of states n(E) are given by

$$\mu_n = \int dE\, E^n\, n(E) \tag{3.1}$$

Within the tight-binding formalism they can also be written [10] [24]

$$\mu_n = \frac{1}{N_a} \mathrm{Tr}\, H^n \tag{3.2}$$

Let us calculate the first moments, the concentration c being fixed. Obviously $\mu_0 = 1$ Then

$$\begin{aligned} \mu_1 &= \frac{1}{N_a}\mathrm{Tr}\,(H^0 + V) = \frac{1}{N_a}\mathrm{Tr}\,V = \bar{\varepsilon} = 0 \\ \mu_2 &= \frac{1}{N_a}\{\mathrm{Tr}\,(H^0)^2 + \mathrm{Tr}\,V^2\} = \mu_2^0 + c(1-c)\,\delta^2 \end{aligned} \tag{3.3}$$

where μ_2^0 is the moment of the bare density of states $n^0(E)$. Similarly

$$\mu_3 = \frac{1}{N_a}\mathrm{Tr}\,(H^0)^3 + \frac{1}{N_a}\mathrm{Tr}\,V^3 = \mu_3^0 + c(1-c)(1-2c)\,\delta^3 \tag{3.4}$$

Thus, up to μ_3, all moments are *independent of the configuration* [12] [20] [25]. This is easily seen in the path representation of the moments which are associated with closed circuits of n steps on the lattice, each step being associated with a matrix element of H. We have intersite jumps, using $H^0_{nm} = \beta_{nm}$ (independent of the configuration) and on-site jumps $V_{nn} = \varepsilon_n = (p_n - c)\,\delta$ (which depend on the configuration). If all on-site jumps concern a single site, the sum over sites implied by the trace yields

$$\sum_n p_n = N_a c$$

independent of the configuration. The same type of argument shows that the only contributions which are configuration dependent in μ_4 comes from $\mathrm{Tr}\, H^0 V H^0 V$. More precisely

$$\mu_4 = C + \frac{2}{N_a}\mathrm{Tr}\, H^0 V H^0 V \tag{3.5}$$

where C is independent of the configuration. Calculating the trace yields $\Delta\mu_4 = \mu_4 - C$

$$\Delta\mu_4 = \frac{2}{N_a}\sum_{n,m}(p_n - c)(p_m - c)\,\beta_{nm}^2\,\delta^2 \tag{3.6}$$

$\Delta\mu_4$ is also related to the moment m_n of the configurational part of the band energy. To lowest order in perturbation theory this part is that of Ω_b, the grand-potential. We thus define m_n as

$$m_n = \int dE_F\, E_F^n\, \Omega_b(E_F) \tag{3.7}$$

where $\Omega_b(E_F)$ is considered as a function of the Fermi energy (at zero temperature). If the moments $\Delta\mu_0$ and $\Delta\mu_1$ vanish, which is the case here, Δm_n is related to $\Delta\mu_n$ through [10]

$$\Delta m_n = -\, N_a \frac{1}{(n+1)(n+2)}\, \Delta\mu_{n+2} \tag{3.8}$$

We thus conclude that $\Delta m_0 = \Delta m_1 = 0$. The first non vanishing moment of $\Delta\Omega_b = \Omega_b - \overline{\Omega}_b$ is Δm_2. This implies that $\Delta\Omega_b$ considered as a function of E_F and therefore as a function of $\overline{N}_e$ has at least two zeros [10]. To lowest order in perturbation $\Delta\Omega_b$ is given by a sum of pair interactions. We then deduce that the two first moments of V_{nm} considered as a function of E_F vanish, and that its second moment M_2 is given by

$$M_2 = -\frac{\delta^2}{3}\, \beta_{nm}^2 \tag{3.9}$$

If the orbital degeneracy is taken into account, β_{nm}^2 should be replaced by $\frac{1}{5}\sum_{\lambda,\mu}(\beta_{nm}^{\lambda\mu})^2$.

In practice, we have seen that β_{nm} only differs from zero if m is on the first coordination shell of n. Thus, for the corresponding interactions, the simplest shape of V_{nm} as a function of E_F (or of $\overline{N}_e$) is that shown in Fig.4, with just two zeros. Since

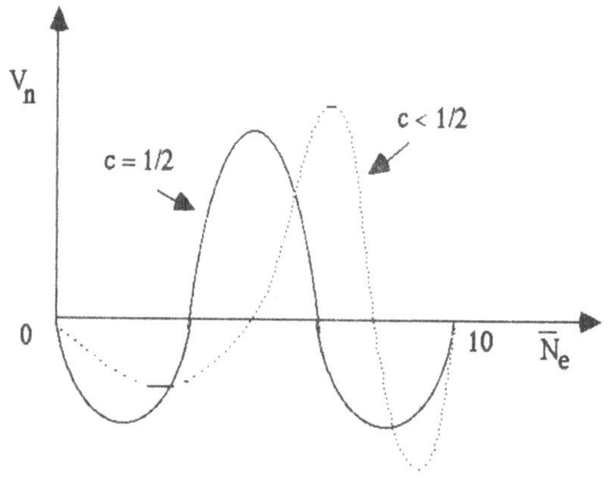

FIGURE 4. Schematic variations of the near-neighbour pair interactions V_n as a function of the band filling $\overline{N}_e$.

$M_2 < 0$, V_{nm} should be positive for intermediate fillings of the band and negative for small or large fillings. With our sign convention for V_{nm} , this implies that there is a tendency to ordering in the first case, and to phase separation in the second one.

It is possible to be a little bit more precise, and to guess how these interactions vary with the concentration. Calculating Δm_5 and therefore M_3 , the first contribution depending on c is that related to Tr $H^0\, V\, H^0\, V^2$ proportional to $(1\text{-}2c)\ \delta^3\ \beta_{nm}^2$. This indicates an asymmetry of $V_{nm}(\overline{N}_e)$ when $c \neq 1/2$ (Fig.4).

3.3. Hierarchy between the interactions

For simplicity, consider the fcc lattice. The above arguments show that the first neighbour interaction V_1 should have at least two zeros. Keeping only β integrals between first neighbours, the second moment M_2 of all other interactions vanishes so that these interactions are expected to be much smaller than V_1. It is easy to realize that the second, third and fourth neighbours can be reached with two first neighbour jumps. The first non vanishing moment is then M_4 and is related to paths $(H^0)^2 V (H^0)^2 V$. Consider for example the self-retracing paths shown in Fig.5a. With each step R is associated a 5×5

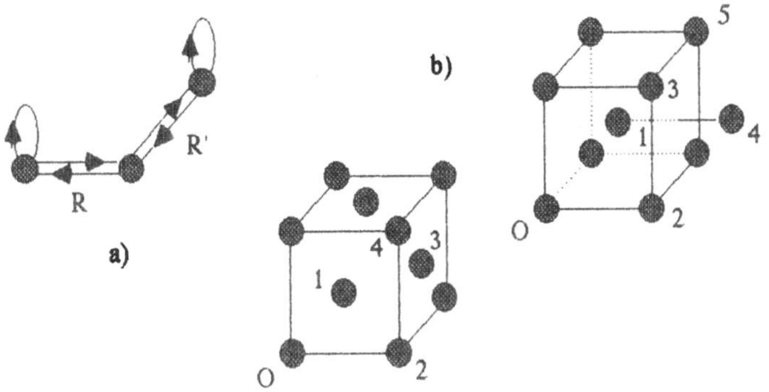

FIGURE 5. a) A path contributing to μ_6 and therefore to M_4
b) Position of near neighbours of the atom located at the origin on the fcc and bcc lattices

matrix β (R). The contribution of this path to M_4 is proportional to $\delta^2 \,\mathrm{Tr}\, \beta^2(R)\, \beta^2(R')$ where the trace is over the degeneracy indices. R' is obtained from R through a rotation, so that

$$\beta(R') = O^+ \beta(R)\, O \qquad (3.10)$$

where O is the corresponding unitary rotation operator in the space spanned by the d states. Thus, the above contribution is proportional to $\mathrm{Tr}\, O^+\beta^2(R)\, O\, \beta^2(R)$. Now, from the identity $\mathrm{Tr}\, A A^+ > 0$ valid for any operator A, and replacing A by $O\beta - \beta O$ we deduce that $\mathrm{Tr}\, O^+ \beta^2 O \beta^2 \leq \mathrm{Tr}\, \beta^4$, the latter contribution being that of a straight retracing path. In fact, using realistic values for $dd\sigma$, $dd\pi$ and $dd\delta$, it is found that the contribution of the path rapidly decreases when the angle between R and R' deviates from zero [26-29]. This effect, directly related to the anisotropy of the d states, is important since it shows for example that the paths reaching fourth neighbours yield larger contributions than those reaching third neighbours (see Fig.5b). In the case of second neighbours we also have self-avoiding "square" paths which can be shown to contribute less than self-retracing ones. We have in fact two opposite effects: the path contributions are larger for fourth neighbours than for second or third ones, but there are more paths visiting second than third and fourth ones. As a result all M_2 moments are of the same order of magnitude, those of the third neighbours being smaller than the others, so that we expect that $|V_2| \approx |V_4| \geq |V_3|$. To summarize, second, third and fourth neighbour interactions should be much smaller than $|V_1|$ and should exhibit more oscillations as a function of $\overline{N}_e$ (four zeros at least). Similar arguments apply to neighbours reached in three steps, etc..., and finally the pair interactions should on the average rapidly decrease with the distance, but all interactions corresponding to a given number of first neighbour steps should be of the same order of magnitude. This shows in particular that in the case of transition alloys, analysis based on first and second neighbour interactions are not reasonable. One should consider either V_1 or V_1 and (V_2, V_3, V_4). Similar arguments applied to the bcc lattice show that $|V_1| \geq |V_2| \gg |V_3|, |V_4|, |V_5| \gg ...$

Multiatom (cluster) interactions can also be studied. This is done in detail elsewhere [28][29], but from the previous analysis it is clear that "linear" cluster interactions should be important. Thus, in contradistinction with naive intuition, the important cluster contributions are generally *not* those of compact clusters.

To conclude this section, one can say that the main characteristics of the pair interactions are to a large extent universal and should not depend too much on the details of the electronic structure . In fact it is possible to calculate approximate pair interactions from the knowledge of the first moments, using the continuous fraction formalism [30](see also [31]), and they do look like those obtained using the full formula.

3.4. Some examples

The calculated first neighbour interactions always vary as a function of $\overline{N}_e$ as predicted by the above arguments (see Fig.4); the predicted hierarchy between the interactions is also well observed. As a function of δ ($\overline{N}_e$ and c being fixed) they first increase as δ^2 (since δ is equivalent to an electronegativity difference, this is a familiar law) and saturate at a constant value for high values of δ (except for a half-filled band,

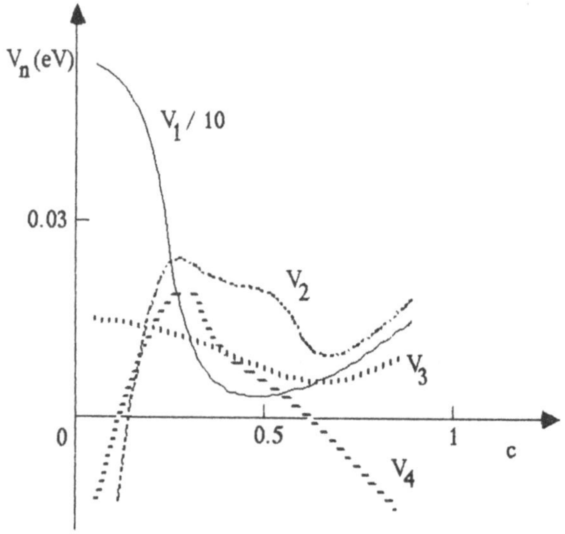

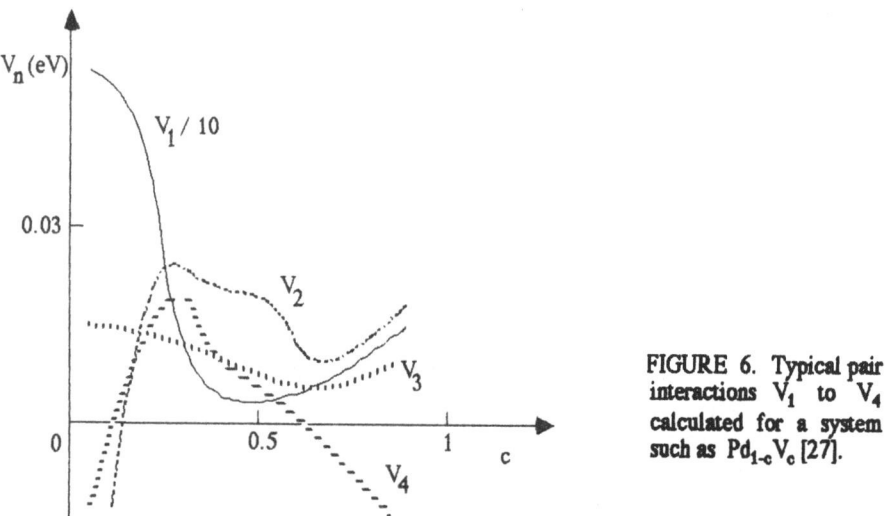

FIGURE 6. Typical pair interactions V_1 to V_4 calculated for a system such as $Pd_{1-c}V_c$ [27].

$\overline{N}_e = 5$, in which case they eventually vanish [18]). In between, their sign may change !

For a particular alloy $\overline{N}_e$ and c are related. There are then two origins for the concentration dependence of the interactions for a given system. The first one is related to the variation of $\overline{N}_e$ which exists even in the genuine perturbation limit $\delta/\overline{W} \ll 1$. The second one is related to the concentration dependence of the CPA effective medium. As a result many different behaviours can be obtained. Numerous examples can be found for instance in [20][27][29][32][33]. One of them is schematically reproduced in Fig.6 .

3.5. Validity of the GPM

We now have more elements to discuss the validity of the GP expansion. When the parameters $\delta/\overline{W}$, c ,... of the alloy are such that $|V_1|$ takes an extremum value, the latter should be the predominent interaction. Problems arise if the value of $\overline{N}_e$ is close to a zero of $|V_1|$; $V_{2,3,4}$ are then competitive. Since these interactions display more zeros, some of them can be very weak too. Therefore, other pair and multiatom interactions should be considered. In many cases it is possible to identify a set of important interactions, but it is easy to understand that problems arise when the first terms of the GP expansion nearly vanish. In particular it is almost hopeless to classify with this method pair interactions of very long range or interactions associated with large clusters since they display a lot of oscillations as a function of $\overline{N}_e$.

The method is very powerful for classifying the evolution trends when one parameter is varied (c, $\delta/\overline{W}$; this will be illustrated in §5) but requires a very high accuracy when $\overline{N}_e$ is close to a zero of V_1 . It is quite conceivable that the GP expansion does not converge at all in some cases, which means either that any analysis based on a finite number of interatomic interactions should break down, or that a better state of reference should be used.

3.6. Segregation limit

If the GP expansion is valid for any configuration, we can for example calculate the energy of a fictitious state made up of very large A and B blocks with linear size L . When $L \to \infty$ the interface interactions can be neglected and the energy of the system is given by

$$\Omega_b = c\,\Omega_b\,(p_n = 1) + (1-c)\,\Omega_b\,(p_n = 0) = \overline{\Omega}_b + \frac{c(1-c)}{2}\sum_{n,m} V_{nm} \tag{3.11}$$

The density of states of such a system is the average of the pure A and pure B densities of states. On the other hand, the energy should also be equal to the average band energy $\Omega_b = c\,\Omega_b^A + (1-c)\,\Omega_b^B$. The trouble is that when these energies are calculated separately, the Fermi levels of the pure systems are different and there is apparently no reason why the average energy calculated for separate systems should be equal to the energy of the average system defined above, with a single Fermi level. The answer is that in our fictitious system there should be charge transfers and potential barriers at each interface which adjust in such a way that the relative position of the Fermi level with respect to the atomic levels is exactly that of the pure metal in each pure block. Thus, both energies should be equal in a complete self-consistent scheme, but they are not necessarily identical in the non self-consistent method developed up to now. In fact the test has been made and the difference is not very important but not negligible [27][34]. This shows that accurate calculations should be self-consistent (see §4.2). Thus, finally, the GP expansion could in principle be written

$$\Omega_b(\{p_n\}) = c\,\Omega_b^A + (1-c)\,\Omega_b^B - \frac{c(1-c)}{2}\sum_{n,m} V_{nm} + \frac{1}{2}\sum_{n,m} p_n^i\, p_m^j\, V_{nm}^{ij} \tag{3.12}$$

but remember that the pair interactions are calculated for the disordered system, not for the above average system.

3.7. Dependence on the crystalline structure

The moment analysis of §3.1 shows that the moment M_2 of the first neighbour interactions does not depend on the geometry of the lattice. This is because $\mathrm{Tr}\ \beta^2(R) = (dd\sigma)^2 + 2\,(dd\pi)^2 + 2\,(dd\delta)^2$ does not depend on the orientation of R. Thus, these interactions should be similar in all crystalline structures. When they are much larger than the other ones, which is the general situation, this shows that the tendency to order or to segregate is practically independent of the crystalline structure, in the case of compact metallic structures at least. The situation is somewhat different for further interactions which are more sensitive to the geometry of the lattice. In fact, if the interactions calculated on various lattices are compared (assuming a constant atomic volume and a specific law for the dependence on R of the β integrals), their evolution as a function of

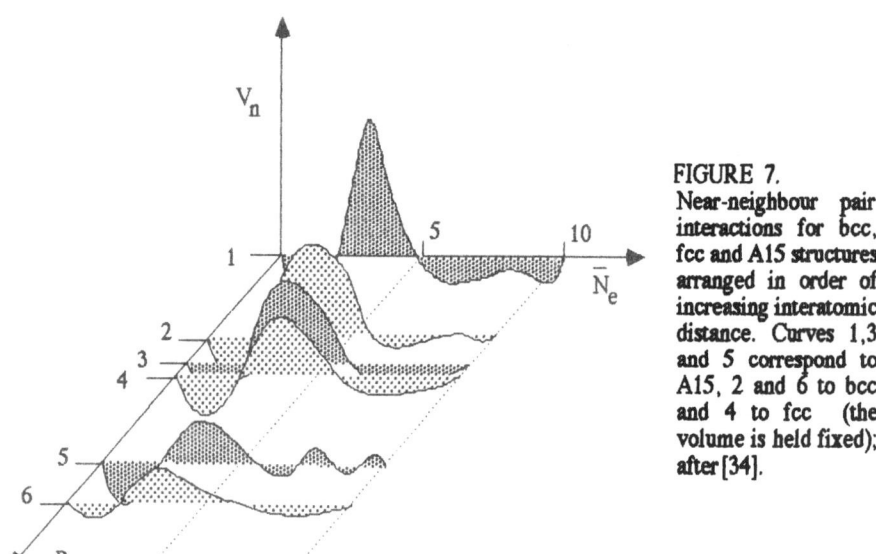

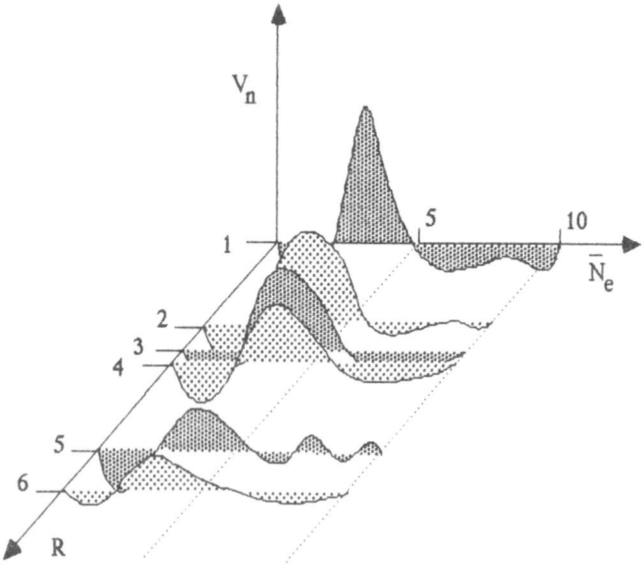

FIGURE 7. Near-neighbour pair interactions for bcc, fcc and A15 structures arranged in order of increasing interatomic distance. Curves 1,3 and 5 correspond to A15, 2 and 6 to bcc and 4 to fcc (the volume is held fixed); after [34].

the distance is relatively smooth (Fig.7), which suggests that it should also be possible to treat structural effects by perturbation theory. Some progresses have recently been made in this direction [27][30].

4. ADDITIONAL EFFECTS

4.1. Off-diagonal disorder

A more accurate theory should take into account differences in bandwidth, which means that H^0 becomes itself a function of the configuration. As a consequence the second moment becomes configuration dependent

$$\mu_2 = c(1-c)\,\delta^2 + \frac{1}{N_a}\,\mathrm{Tr}\,(H^0)^2 = c(1-c)\,\delta^2 + \frac{1}{N_a}\sum_{\substack{n,m\\ i,j}} (\beta^{ij}_{nm})^2\, p^i_n\, p^j_m \qquad (4.1)$$

which induces a pair contribution proportional to $(\beta^{AA})^2 + (\beta^{BB})^2 - 2(\beta^{AB})^2$. It is very difficult within the tight-binding method to get accurate values for β^{AB} but a reasonable guess is $(\beta^{AB})^2 = \beta^{AA}\,\beta^{BB}$, in which case this contribution becomes $(\beta^{AA} - \beta^{BB})^2$. The moment M_0 of the corresponding pair interaction then becomes negative. The simplest function with a negative M_0 is negative everywhere. Thus, if the above approximation for β^{AB} is valid, the difference in bandwidth induces a tendency to phase separation.

Another indirect effect of the off-diagonal disorder is the following. If the bandwidths W^A and W^B are different, the mean bandwidth $\overline{W}$ varies with the concentration. This modifies in turn the dimensionless parameter $\delta/\overline{W}$ and therefore the pair interactions.

In fact, although some preliminary calculations have been performed [31][35], a complete study of the role of off-diagonal disorder remains to be done.

4.2. Self-consistency

In an alloy the electronic density in a A or B cell should deviate from that of the pure metals. There are charge transfers correlated with potential variations which should modify the matrix elements of the hamiltonian. In particular, the atomic energy levels should be renormalized. In a simple Hartree-Fock scheme (see [18] for the details) we define the renormalized level ε_n^{i*} of an atom i at site n through

$$\varepsilon_n^{i*} = \varepsilon^{ion,i} + \varepsilon_n^{el,i}$$

with

$$\varepsilon_n^{ion,i} = \varepsilon^{ion,0,i} + \sum_{\substack{m\neq n\\ j}} U_{nm}^{el\text{-}ion}\, p_m^j Z^j \tag{4.2}$$

$$\varepsilon_n^{el,i} = U_0\,\rho_n^i + \sum_{\substack{m\neq n\\ j}} U_{nm}^{el\text{-}el}\, p_m^j \rho_m^j$$

where U_0 and U_{nm} are the intra- and interatomic Coulomb integrals which, for simplicity, are assumed to be independent of the configuration (in the following we also assume that $U_{nm}^{el-el} \approx -U_{nm}^{el-ion} \equiv U_{nm}$) ; ρ_m^j and Z^j are the electronic and ionic charges, and $\varepsilon^{ion,0,i}$ is the energy level of an ion of type i (atom without d electrons). The electronic charges ρ_n^i (positive by definition) are given by

$$\rho_n^i = -\frac{\mathrm{Im}}{\pi}\int^{E_F} dE\; G_{nn}^i(E) \tag{4.3}$$

where G_{nn}^i is the Green function for a configuration with an atom i at site n ($p_n^i = 1$).

Within the CPA we replace p_m^j in the r.h.s. of (4.2) by its average c^j, and since the average charge transfer should vanish, we get

$$\varepsilon_n^{i*} \equiv \varepsilon^{i*} = \varepsilon^{ion,0,i} + U_0\,\rho_i = \varepsilon^i - U_0\,Q^i \tag{4.4a}$$

with

$$\rho^{i} = -\frac{\mathrm{Im}}{\pi}\int^{E_F} dE \frac{F}{1-(\varepsilon^{i*}-\sigma)F} \quad ; \quad Q^{i} = Z^{i} - \rho^{i} \tag{4.4b}$$

The parameter $\delta = \varepsilon^A - \varepsilon^B$ is now renormalized into δ^*

$$\delta^* = \delta - U_0(Q^A - Q^B) \tag{4.5}$$

The new parameter δ^* and the charge transfers Q^i can be calculated using the coupled equations (4.4). In practice the electronic charge goes towards the element with the highest ionic charge. Assume then that $Z^A < Z^B$ so that $\varepsilon^A > \varepsilon^B$. Then Q^A increases and ε^A decreases. Similarly ε^B increases so that the effective δ^* is smaller than δ. Let us now compare ε^i_n for a given configuration with its CPA estimate

$$\delta\varepsilon^{i}_{n} = \varepsilon^{i*}_{n} - \varepsilon^{i}_{CPA} = U_0\,\delta\rho^{i}_{n} - \sum_{\substack{m\neq n\\ j}} U_{nm}\, p^{j}_{m}\, Q^{j}_{m}$$

with

$$Q^{j}_{m} = Z^{j} - \rho^{j}_{m} = Q^{j} - \delta\rho^{j}_{m} \tag{4.6}$$

where Q^j is the CPA charge transfer. The corresponding charge variation $\delta\rho^i_n$ is obtained from (4.3)

$$\delta\rho^{i}_{n} = -\chi^{i}_{0}\,\delta\varepsilon^{i}_{n} \quad ; \quad \chi^{i}_{0} = \frac{\mathrm{Im}}{\pi}\int dE\, f(E)\, F^2 / [1-(\varepsilon^{i*}-\sigma)F]^2 \tag{4.7}$$

Eliminating the $\delta\varepsilon^i_n$ in the system (4.6)-(4.7) yields

$$\delta\rho^{i}_{n} = \sum_{\substack{m\neq n\\ j}} \chi^{i}\, U_{nm}\, p^{j}_{m}\,(Q^{j} - \delta\rho^{j}_{m}) \quad ; \quad \chi^{i} = \chi^{i}_{0} / [1 + U_0\chi^{i}_{0}] \tag{4.8}$$

Using vector and matrix notations with respect to both set of indices n,m and i,j , we get

$$\delta\rho = (1+\chi U)^{-1}\chi U Q$$

with

$$(\delta\rho)^{i}_{n} = p^{i}_{n}\,\delta\rho^{i}_{n} \quad ; \quad (Q)^{i}_{n} = p^{i}_{n}\, Q^{i} \tag{4.9}$$

$$(\chi)^{ij}_{nm} = \delta_{nm}\,\delta^{ij}\, p^{i}_{n}\,\chi^{i} \quad ; \quad (U)^{ij}_{nm} = U_{nm}$$

We now calculate the self-consistent energy Ω , given by

$$\Omega = \Omega_b - \frac{1}{2}\sum_{n,i} p^{i}_{n}\,\varepsilon^{el,i}_{n}\,\rho^{i}_{n} + \frac{1}{2}\sum_{\substack{n\neq m\\ i,j}} p^{i}_{n}\, p^{j}_{m}\, U^{ion\text{-}ion}_{nm}\, Z^{i} Z^{j} \tag{4.10}$$

$$\Omega_b = -\frac{\mathrm{Im}}{\pi}\int dE\, f(E)\,\mathrm{Tr\,Log}\,\overline{G} + \sum_{n,i} p^{i}_{n}\frac{\mathrm{Im}}{\pi}\int dE\, f(E)\,\mathrm{Log}\,[1-(\varepsilon^{i}_{n}-\sigma)F] + \frac{\mathrm{Im}}{\pi}\int dE\, f(E)\,\mathrm{Tr\,Log}\,[1-\hat{t}\,\overline{G}_{off}]$$

Expanding all terms to lowest order in $\overline{G}_{nm}$, $n \neq m$ (notice that the second term in the r.h.s. of the second equation has to be expanded too) many terms cancel each other, and, setting $U_{nm}^{ion-ion} = U_{nm}$, the final result is

$$\Omega \approx \overline{\Omega} + \frac{1}{2}\sum_{\substack{n,m\\i,j}} p_n^i p_m^j V_{nm}^{ij} + \frac{1}{2}\sum_{\substack{n,m\\i,j}} p_n^i p_m^j Q^i U_{nm} (Q^j - \delta\rho_m^j)$$

$$\text{with} \qquad \overline{\Omega} = \overline{\Omega}_b - \frac{N_a}{2} U_0 \sum_i c^i (\rho^i)^2 \tag{4.11}$$

Note that the medium of reference is that of the self-consistent CPA. The pair interactions V_{nm}^{ij} are therefore calculated as before with δ^* replacing δ . The last term in (4.11) has the form of an Ewald term which, as usual in perturbation theory, involves a coupling between the charges Q^i of the unperturbed state and the screened charges $Q^j - \delta\rho_m^j$. Using (4.9) this can also be written

$$Q^i U_{nm} (Q^j - \delta\rho_m^j) = Q^i Y_{nm} Q^j \tag{4.12a}$$

where in matrix notations Y is given by

$$Y = (1 + U\chi)^{-1} U \tag{4.12b}$$

Y_{nm} is a screened Coulomb interaction which *depends on the configuration*. This is clear from the definition of χ in (4.9). Thus, in principle, the corrective Ewald term is not a pair contribution. This difficulty is related to the range of the Coulomb interaction. It is not allowed to treat U_{nm} as a small parameter as we did for $\overline{G}_{nm}$, and therefore to replace Y_{nm} by U_{nm} . However, if charge transfers are weak enough, it is certainly reasonable to replace p_n^i by its average c^i in the definition of χ . Then Y_{nm} becomes independent of the configuration and can be calculated in Fourier space. If $U(q)$ is the (discrete) Fourier transform of U_{nm} (which only depends on $m-n$, $m \neq n$) $Y(q)$ is given by

$$Y(q) = U(q)\,[1 + \overline{\chi}\,U(q)\,]^{-1} \quad ; \quad \overline{\chi} = \sum_i c^i \chi^i \tag{4.13}$$

and we have the renormalized effective pair interactions

$$V_{nm}^{*ij} = V_{nm}^{ij} + Q^i Y_{nm} Q^j$$
$$V_{nm}^{*} = V_{nm} + Y_{nm} (Q^A - Q^B)^2 \tag{4.14}$$

Charge transfer effects thus induce two effects. First δ is replaced by δ^* , which in most cases decreases $|V_{nm}^{ij}|$. Then, we have to add a positive Ewald contribution. As a consequence the main positive interactions and therefore the ordering energies are not strongly modified. The other interactions may be more affected.

Exactly as we did in §2.5 it is also possible to define a mean field theory, now including charge transfer effects. The calculations are given in detail in [18]. There are three contributions to Ω , one of which being related to products of U_{nm} and $\overline{G}_{nm}$ and which has been found to be negligible as expected and the two others corresponding exactly to the band and Ewald terms derived above.

To summarize, self-consistent effects are not expected to introduce important modifications provided that charge transfers are weak enough. This is always the case if the intraatomic Coulomb interaction U_0 is strong enough. Within the tight-binding formalism the Coulomb intregrals U_0 and U_{nm} should be considered as phenomenological parameters, which is a strong limitation. For example, screening effects by s-p electrons are not taken into account. More sophisticated electronic calculations are required, but it is important to realize that any consistent scheme should either neglect charge transfers or include them both in the calculation of the medium of reference *and* in that of the ordering part of the energy. Otherwise some compensations as those described above may be missed.

4.3. Magnetism

Magnetic effects can strongly modify the ordering tendencies. This is easy to understand on the following example. Consider a ferromagnetic alloy at zero temperature. In a Hartree-Fock-Stoner (or spin polarized local density formalism) we then have two different band structures, one for each spin direction, and two different band fillings $\overline{N}_e^{\uparrow}$ and $\overline{N}_e^{\downarrow}$. There are also two diagonal disorder parameters $\delta^{\uparrow}$ and $\delta^{\downarrow}$. The total band energy is obtained by summing the energies of both bands. Assume for example that we have a strong ferromagnet with a filled up spin band, which is the case of Ni_3Fe. Then $\overline{N}_e^{\uparrow} = 5$ and $\overline{N}_e^{\downarrow} = \overline{N}_e - 5 \approx 3.5$. A filled band does not contribute to the energy.

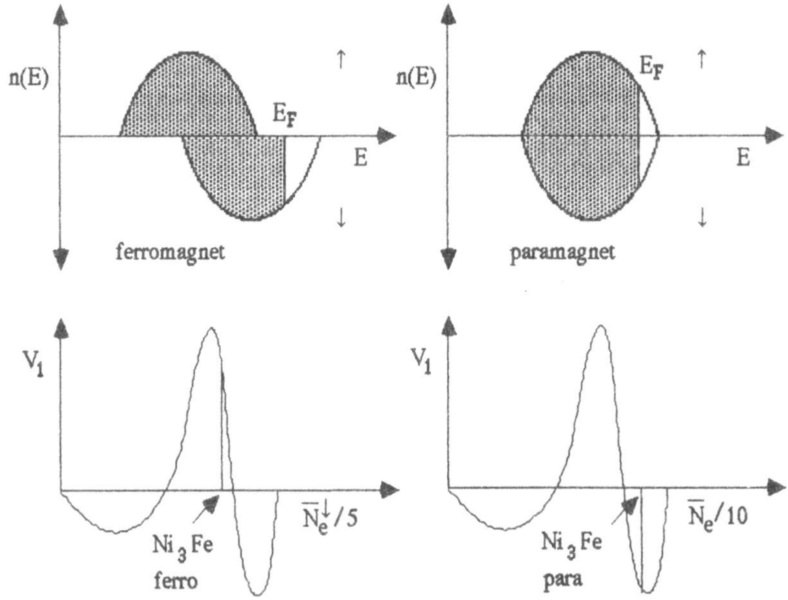

FIGURE 8. Schematic densities of states and first neighbour interactions for ferromagnetic and paramagnetic Ni_3Fe.

Therefore the sign of the first neighbour interactions is determined by the position of the Fermi level within the down spin band. Using normalized densities of states the filling of this band is $3.5/5 = 0.7$ and for the composition Ni_3Fe the corresponding interaction V_1

is positive. On the other hand, assuming a paramagnetic state, one finds $\overline{N}_e^{\uparrow} = \overline{N}_e^{\downarrow} = 8.5/2$, which corresponds to a normalized filling of 0.85 for each band. In that case V_1 is negative (tendency to phase separation). Thus, in this case, ordering is driven by ferromagnetism (Fig.8).

Conversely the magnetic properties should be influenced by ordering effects if the magnetic moments are not very large. For example Pt_3V is ferromagnetic in the ordered phase and paramagnetic in the disordered one [36]. In the limit where each atom carries a definite magnetic moment, it is possible to extend the GP theory to account for these effects even in the fully disordered phase (disordered and paramagnetic) [32][37]. Assume then a definite configuration of atoms (variables $\{p_n\}$) and of moments, these moments being characterized by a direction $\hat{h}_n$ or in simplified models (Ising model) by the sign $\tau_n = \pm 1$ of its projection along a fixed axis. Instead of using moments, we can handle the local exchange fields $h_n^i = h^i \hat{h}_n$. On each site this introduces a coupling $-h_n^i \cdot S_n$ where the $(S_n)_{x,y,z}$ are the 2×2 Pauli matrices defined in the space spanned by $| n \uparrow \rangle$ and $| n \downarrow \rangle$. This term should be added to the atomic energy levels, and finally on each site we have four levels $\varepsilon^i \pm h^i$. The CPA formalism is easily generalized using everywhere 2×2 matrices on each site, and a GP expansion can be defined from the fully disordered state. This generates effective pair and multiplet interactions. In the case of pair interactions we then get for the configurational part of the energy

$$\frac{1}{2}\sum_{\substack{n,m\\i,j}} p_n^i p_m^j V_{nm}^{ij} + \frac{1}{2}\sum_{\substack{n,m\\i,j}} p_n^i p_m^j \hat{h}_n \cdot \hat{h}_m W_{nm}^{ij} \tag{4.15}$$

where V_{nm}^{ij} is a "chemical" interaction as before and W_{nm}^{ij} a magnetic interaction between moments with directions $\hat{h}_n$ and $\hat{h}_m$ on sites occupied by i and j atoms. This allows to define an effective Ising-Heisenberg model that can be inserted into the appropriate thermodynamic theory. Usually this is replaced by a simpler double Ising model (Ashkin-Teller model for example, see [37]).

Note however that both quantities V and W depend on the chemical and on the magnetic disorder. In particular, the chemical interaction V_{nm} depends on the magnitude h^i of the fields, and we expect strong effects if the fields are of the order of δ, the bare chemical parameter. This is the case for Ni_3Fe and here again magnetic effects change the sign of V_1.

The fully disordered state is the appropriate state of reference at high temperature, and the theory then allows to discuss the interplay between chemical and magnetic orderings, but just as the usual GPM can account for phase separation, the method should give correct energies in the low temperature magnetic ordered state.

This simple theory applies strictly speaking only if local moments can be defined and if these moments do not strongly depend on the local environment. For a more extensive discussion, see [37]; see also Staunton's lectures.

4.4. <u>Elastic effects</u>

When the A and B atoms have very different "sizes" we expect significant elastic contributions which have been neglected up to now [33]. A first class of effects is related to the atomic volume differences. In a fully self-consistent scheme, the total energy of the disordered state can be calculated as a function of the volume, which allows to determine the equilibrium volume as a function of the concentration. This yields a new concentration dependent term in the energy (but independent of the configuration) which is generally positive and therefore favours segregation. Within a simple tight-binding model, such calculations are difficult and one has to postulate some form for the repulsive interatomic interactions [38]. There are other elastic contributions. There is again an homogeneous contribution related to the dependence on the configuration of the equilibrium volume. In principle, this can be determined from the volume dependence of the pair interactions (actually, both δ and $\overline{W}$ should vary with the volume). Finally there is a relaxation contribution, the positions of the atoms deviating from the perfect lattice positions. This may be analysed as follows [33]. An atom i at a given site embedded in the average medium induces a force on this medium and therefore on another atom j at another lattice site. The resulting interatomic coupling can be calculated within the harmonic approximation from the knowledge of the phonon spectrum of the average medium and of the local forces. In fact the formalism is very similar to that used in the GP expansion, the electronic Green functions being replaced by phonon Green functions. We then obtain elastic pair interactions, but it is very difficult to obtain them from a fully self-consistent calculation.

To summarize, homogeneous elastic effects can be taken into account through a concentration dependent term, and this can induce strong effects, but much remains to be done to calculate relaxation energies.

4.5. Phonons

Even if we neglect the variation with temperature of the previous effects, it is not possible in principle to neglect the phonon contribution to the free energy of an alloy. These contributions are only important at high temperature (compared to a Debye or an Einstein frequency). The corresponding free energy F writes

$$F = k_B T \sum_{\alpha} \mathrm{Log}\,[1 - \exp(-\beta \hbar\omega_{\alpha})] \approx k_B T \sum_{\alpha} \mathrm{Log}\,(\hbar\omega_{\alpha} / k_B T) \qquad (4.16)$$

where the ω_α are the eigenfrequencies of the system. To lowest order with respect to a development in $1/T$, the main (non constant) contribution then comes from the vibrational entropy. The eigenfrequencies are themselves obtained by diagonalizing the so-called dynamic matrix D. We write

$$\sum_{\alpha} \mathrm{Log}\,\omega_{\alpha} = \frac{1}{2}\mathrm{Tr\,Log\,D} = \frac{1}{2}\{\mathrm{Tr\,Log\,D^0} - \mathrm{Tr\,Log\,M}\} \qquad (4.17)$$

where D^0 and M are matrices defined from

$$D^0_{nn} = \sum_{m \neq n} \gamma_{nm} \;\; ; \;\; D^0_{nm} = -\gamma_{nm} \text{ when } m \neq n \;\; ; \;\; M_{nm} = m_n \delta_{nm} \qquad (4.18)$$

the γ_{nm} being the force constants and m_n the mass at site n.

Since M is a diagonal matrix Tr Log M does not depend on the configuration for a given concentration: *the mass disorder does not contribute to the excess entropy*. Only the force constant disorder does, so that in many cases the dependence on the configuration of the phonon entropy should be weak. Numerical estimates can be found in [39].

4.6. Spin-orbit coupling

Spin-orbit effects are important for transition elements of the 5d series. Within a tight-binding formalism, this coupling is local and adds a new contribution to the random potential V. The corresponding operator $\xi\, L\,.\,S$ is no longer proportional to the identity operator and in principle full calculations involving both the orbital and the spin degeneracy are required. However simple moment arguments give an idea of the role of these effects. By repeating the calculation of $\mathrm{Tr}\, H^0 V H^0 V$ (see §3.1) using for V the spin-orbit matrix, it is found that $\Delta\mu_4$ is proportional to ξ^2 and *negative*, which means that chemical and spin-orbit effects (characterized by δ and ξ) play in opposite directions, as far as first neighbour interactions are concerned at least. This might explain why Pt-Ni alloys order whereas practically all effects considered before predict that they should segregate [40].

4.7. Summary

Some other effects could still be invoked. For example the role of s-p electrons or of electronic correlations should be studied. It is clearly impossible to include systematically all contributions. One has first to decide on physical grounds what are the main ones. For example, for most of the transition alloys it is sufficient to include chemical and magnetic contributions. In some particular cases (Zr-Ni, Au-Cu,...) elastic effects are certainly important. They also play a predominant role in other systems (semiconductor compounds or pseudo-binary alloys [41]). Thus, it is very difficult at the moment to make accurate predictions on a given system, even when using sophisticated electronic calculations, but we shall see that the simplest scheme already accounts quite well for many ordering trends.

5. GROUND STATES AND PHASE DIAGRAMS

5.1. Ground states

The observed ordered structures of transition alloys can be classified according to so-called phenomenological structural maps, using the variables $\overline{N}_e$ and $\Delta N_e = N_e^A - N_e^B$. It turns out that for a fixed concentration the observed structures occur in well defined domains of the corresponding plane. For example at the concentration $c = 1/4$ most alloys on the fcc lattice display either the $L1_2$ or the DO_{22} structure. On the hcp structure we have the analogous structures, DO_{19} and DO_a respectively. Finally the DO_{24} structure is the equivalent of $L1_2$ on a double hcp structure. Looking at the structural map, one sees two domains for the $L1_2$ or related structures, but the upper one corresponds to magnetic

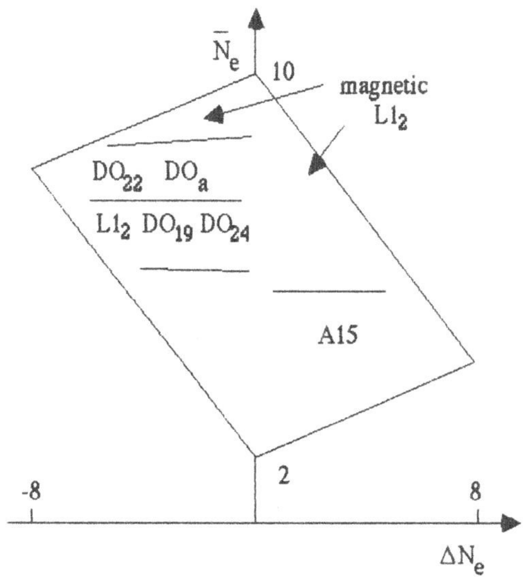

FIGURE 9. Schematic phenomenological structural map for AB_3 compounds.

alloys, the DO_{22} domain being in between. On the other hand the fcc and hcp structures occur principally for alloys AB_3 when $N_e^A < N_e^B$ whereas the corresponding ordered structures A_3B are principally of type A15 (Fig.9).

According to our analysis, the pair interactions V_1 to V_4 are relevant on the fcc and hcp structures. Since $|V_1| \gg |V_{2,3,4}|$ in general, the sign of V_1 first determines a domain where ordered states are to be expected. Then the difference between $L1_2$ and DO_{22} is proportional to $\zeta = V_2 - 4V_3 + 4V_4$. As a function of $\overline{N}_e$, ζ has at least two zeros and the result of this calculation (also compared to an exact calculation) is shown in Fig.10. For the

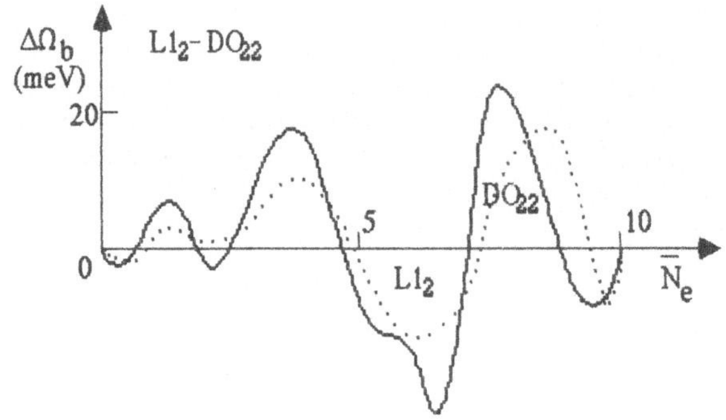

FIGURE 10. Relative stability $\Delta\Omega_b = \Omega_b(L1_2) - \Omega_b(DO_{22})$ of $L1_2$ and DO_{22} as a function of the band filling for $\delta/\overline{W} = 0.45$; after [42] (in this reference the energy scales of Fig.3 and 4 have to be multiplied by a factor of ten).

fillings of the band for which V_1 is positive and for which the compact fcc and hcp structures are stable a qualitative agreement is found, i.e. DO_{22} is more stable than $L1_2$ for large values of $\overline{N}_e$ [42][43]. The relative stability of the A15 structure has also been calculated, and this gives the very satisfactory theoretical map shown in Fig.11 [34] [43][44].

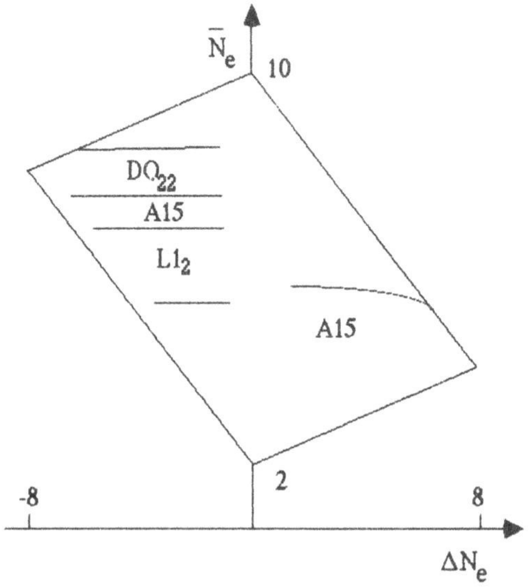

FIGURE 11. Schematic theoretical structural map after [34]. The stability of A15 when $\Delta N_e < 0$ is weak and is not observed experimentally.

In these calculations, the simplest scheme has been used (no charge transfers, no elastic term,etc...). When magnetism is introduced, the $L1_2$ structure is indeed found for large values of $\overline{N}_e$ [32][37]. Many other calculations at other concentrations have been performed recently and they always reproduce quite well the trends indicated by the structural maps [43][31]. Let us briefly describe the origin of this success.

First one has obvious structural effects. The fcc and hcp structures always occur when $\overline{N}_e$ is large, whereas the bcc and the more complex structures A15, Laves phases, σ phases,... appear for intermediate values (we recall that when $\overline{N}_e$ is small or large, ordering cannot occur without magnetism or spin-orbit coupling). This effect is clearly inherited from the stability of the pure transition metals, which is consistent with arguments given in §3.6 indicating that the ordering energies should be fairly independent of the nature of the lattice. Looking in more detail, there are in some cases clear indications of a coupling between ordering and structural effects. For example, ordering effects favour bcc over fcc when $c=1/2$. In the bcc based B2 structure we have eight B first neighbours and six A second neighbours around A (first coordination shell), to be compared with the corresponding numbers, eight and four in the $L1_0$ structure. Because the second neighbours in the bcc structure are rather far away, one can say that in some average way there are more AB pairs in B2 than in $L1_0$. Similar effects have been studied at the A_2B and AB_2 compositions. It is striking to observe that the $MoSi_2$ structure is very differently distorted in both cases: $c/a \approx 2.5$ for A_2B, $c/a \approx 3.4$ for AB_2. This again has been explained in part [27].

The only lattices on which there appears a large variety of ordered structures are the fcc,(hcp) and bcc lattices. In the case of the fcc lattice we have interactions up to at least fourth neighbours. All possible structures and their stability as a function of the value of the interactions $V_{2,3,4}$ have been classified by Kanamori and Kakehashi, assuming V_1 to be large enough to have the largest possible number of first neighbour AB pairs [45], but 40 structures are already found. Most observed structures belong to these 40 structures whereas a simple analysis with V_1 and V_2 only, misses many observed structures (for example Ni_4Mo). Furthermore when $V_2 > V_1/2 > 0$ in the latter analysis, several structures at compositions A_5B, A_2B and AB (CuPt structure) are predicted which are not observed in transition alloys. All these facts are in complete agreement with our predicted hierarchy between the interactions. For example the CuPt structure is never observed except in CuPt, but in that case s-p electrons, sp-d coupling and elastic effects should play an important part. A detailed study of the stability of different structures is given in [33][43].

A similar analysis has been made on the bcc lattice. Here the hierarchy is $|V_1| \approx |V_2| \gg |V_{3,4,5}|$. The possible structures have been investigated in [47]. Here again the structural maps are well described provided magnetism is included when necessary [43]. All observed structures do correspond to the above hierarchy except the B11 structure (TiCu,...) but it is in fact found that for these alloys we are precisely close to a zero of V_1 and V_2.

A detailed analysis of the ordered structures that appear in substoichiometric carbides and nitrides has also been made [48]. Here the basic structure is the NaCl structure with a metallic fcc sublattice and another fcc metalloid (C or N) sublattice. Off stoichiometry vacancies appear on the latter sublattice and we are reduced to an ordering problem of atoms and vacancies on a fcc lattice [48][49]. Defining the appropriate pair interactions, simple moment arguments then yield the hierarchy $V_2 > V_1 \gg \dots$, which precisely implies that structures such as A_5B, CuPt forbidden in transition alloys should be stable here. This is actually what is observed. It is highly satisfactory that such a simple scheme can account for the behaviour of very different systems.

Let us terminate this section by discussing the problems that may arise because of the concentration dependence of the medium of reference and of the interactions. In a genuine Ising model with interactions of finite range, the ground state energy can easily be shown to be a piecewise linear convex function of the concentration $E(c)$, the coefficients of the equation of each linear part being linear combinations of the interactions. If the interactions depend on the concentration, the curve $E(c)$ is distorted, and the common tangent construction is required to obtain the lowest energy at a given concentration [33]. What does this mean if the points of contact correspond to concentrations for which there is no ordered structure in the ground state analysis of the Ising model? We must go back to the GP expansion. Its convergence is basically related to the fact that the dependence on the concentration of the interactions compensates for the multiatom and/or long range interactions that would appear in a standard perturbation treatment. At zero temperature, we must in fact have $p_n = 0$ or 1 (whereas at finite temperature we deal with averages). So, when the previous situation happens, this means that an exact ground state analysis would involve many interactions, if they can be defined at all! Thus, when the $V_{nm}(c)$ strongly depend on the concentration, one should consider the possibility of having stable structures which differ from those of a real Ising model.

5.2. Phase diagrams

Except in very particular situations (one-dimensional systems or two-dimensional systems with first neighbour interactions) the Ising model cannot be solved exactly, i.e. it is not possible to obtain the exact free energy. Various approximate techniques are available. Basically there are two extreme methods: mean field methods which provide analytical formulae or numerical methods (Monte Carlo simulations) which should be considered as numerical experiments. The latter methods are very useful and will develop with the increasing possibilities of the computers, but the mean field methods will always keep specific advantages: simplicity, physical insight,...

The simplest mean field theory applied to the alloy problem is called the Bragg-Williams approximation, and is known to be quantitatively inaccurate. It can even be qualitatively inaccurate, in particular when applied to the fcc lattice. Then, generalizations of the method should be used, and it now seems that the most efficient method is the cluster variation method (CVM) of Kikuchi [56]. The problem is to approximate the configurational entropy. The usual mean field theory being a single-site approximation completely neglects short range order effects so that the high temperature disordered phase is assumed to be completely disordered. The CVM treats the occupancy correlations within small clusters (tetrahedra or tetrahedra and octahedra in the recent calculations for fcc) and are able to deal with the most important frustration effects on a fcc lattice [50]. It is still difficult to define a clear hierarchy of CVM approximations that would converge to the exact result, and then, to list which clusters (beyond tetrahedra or octahedra) play the most important part. Anyway there is no reason that these clusters correspond to the most important clusters involved in the GP expansion.

In other words, the problem of approximating at best the energy of a configuration and that of approximating at best the configurational entropy are not directly related. For example, effective pair interactions can be sufficient whereas clusters should be dealt with for the entropy. A fully consistent theory would be evidently preferable, but this seems a formidable task beyond the single site approximation (see §2.5) and the previous arguments show that mixed approximations are certainly quite valuable.

Thus, one can calculate phase diagrams using an effective Ising hamiltonian and the CVM approximation (previous calculations of enthalpies of formation of disordered alloys can be found in [51]). Because the effective interactions depend on the concentration, very different types of phase diagrams can be obtained. In particular there is no longer the forced symmetry $c \leftrightarrow (1-c)$. Some prototype phase diagrams have been recently

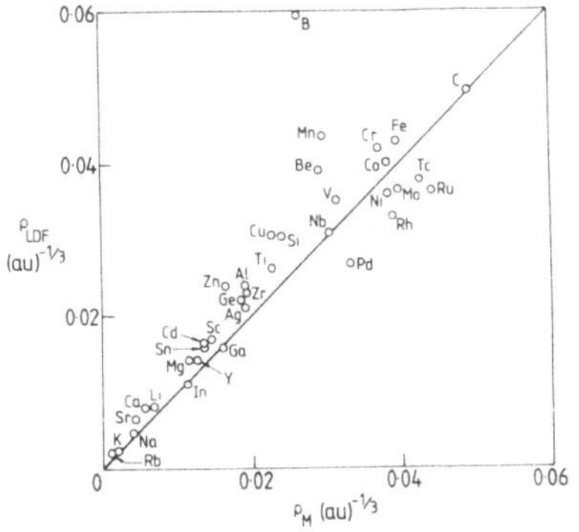

FIGURE 10. The similarity between the theoretical LDF values of the Wigner-Seitz electron density and the Miedema coordinate ρ_M in atomic units.

where the two lines with slopes $\pm(Q/P)^{1/2}$ separate the regions of positive and negative heats of formation (4). As a first approximation we see that ϕ^* and $\rho_m^{1/3}$ vary linearly across the figure with the slope $(Q/P)^{1/2}$. There-fore as Hodges (36) has stressed, it is the deviation from this linearity which is responsible for the observed heats of formation. Since the deviations in fig. 11 are only about 10% of the absolute values, the identification of the Miedema parameters ϕ^* and $\rho_m^{1/3}$ with the work function and charge density respectively is misleading. For example, the sign of the FePd, FeRh, and FeCa heats of formation would be changed by taking the theoretical values of the charge density (24). Similarly, the sign of the FeNb, FeTa, and FeZr heats of formation would be changed by taking the experimental values of the work function (37).

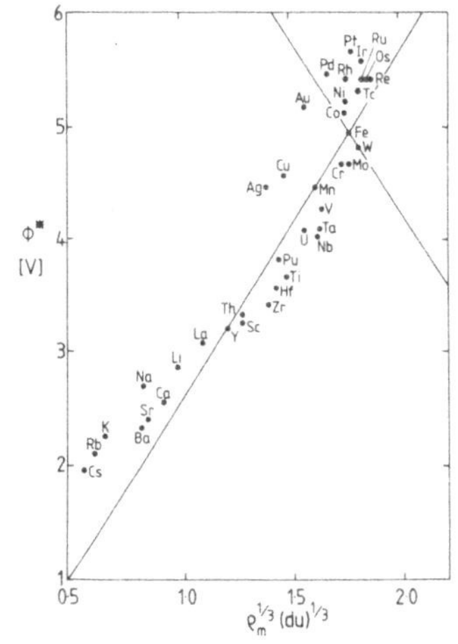

FIGURE 11. The graphical construction which determines whether the heats of formation of iron alloys with s and d elements are either positive of negative. The straight lines have slopes $\pm\sqrt{Q/P}$, respectively, and separate the regions of positive and negative heats of formation.

However, even though the physical concepts underpinning the Miedema model are not supported by quantum mechanics, the importance of the scheme rests in its compatibility with known experimental data and its ease of applicability to a multitude of problems in alloy cohesion (4).

4. THE QUANTUM PREDICTION OF STRUCTURE.

There are many problems in alloy design such as, for example, predicting solubility limits, site preferences, and grain boundary cohesion or embrittlement which require an understanding of the microscopic origin of structural stability. In this section, therefore we will explore the origin of the structural domains observed across the AB structure map in fig. 1. Although first principles local density functional theory would probably predict the correct crystal structure for any particular compound, the calculated total energies by themselves would not provide a physical explanation for the trends observed across the map. The energy must be broken down in order to relate with one's physical intuition regarding the importance of factors such as relative atomic size, electronegativity difference, average electron per atom ratio or the angular momentum character of the valence electrons.

The equilibrium energy difference theorem (18) allows direct contact to be made between the microscopic calculations and our physical intuition. Suppose the binding energy has been written as the sum of two terms, namely

$$U = U^{rep} + U^{bond} \tag{10}$$

where the labels rep and bond imply that a division has been made so that around the equilibrium volume the first contribution is repulsive and the second attractive. For example, in the ionic limit the bond contribution would be the Madelung electrostatic energy whereas in the covalent or metallic limit for Tight-Binding systems it would be the quantum mechanical bonding energy

$$U_{qm}^{bond} = \Sigma_i \int^{\varepsilon_F} (\varepsilon-\varepsilon_i)\; n_i(\varepsilon)\; d\varepsilon \tag{11}$$

where $n_i(\varepsilon)$ is the local density of states associated with the orbital on atom i with energy ε_i (assuming for simplicity of notation one orbital per site) and ε_F is the Fermi energy. Then, it can be shown that the difference in energy between two equilibrium structures is given to first order in $\Delta U/U$ by the difference in the bond energy alone provided the lattices have been fixed to show the same repulsive energy i.e.

$$\Delta U^{(1)} = \left[\Delta U^{bond}\right]_{\Delta Urep=0} \tag{12}$$

The importance of this theorem is that it allows the structural energy difference to be interpreted within a two-stage process. In the first step the volumes of the different structures are prepared to guarantee the same repulsive energy. This stage depends only on the nature of the repulsive interaction and reflects the atomic sizes of the constituents (38). It generalizes the usual classical procedure of packing together hard spheres until they touch. In the second step the bond energies are compared at these prepared volumes in order to see which stucture is the most stable. This corresponds in the ionic limit to the customary practice of comparing the electrostatic Madelung energies (see, for example, fig. 3.08 of (39)).

Away from the extreme ionic limit this stage depends on the nature of the quantum mechanical bonding between the atoms and is controlled by the angular momentum character of the bonding orbitals, the atomic energy level mismatch, and the band filling or average electron per atom ratio.

This theorem has been used to examine the microscopic origin of the structural domains of the pd-bonded AB compounds which are shown in the upper panel of fig. 12 (38). The total binding energy, eq. (10) is represented by

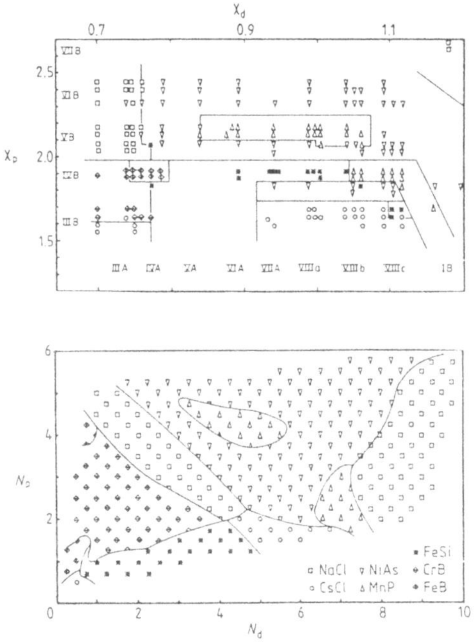

FIGURE 12. Upper panel: The structure map (χ_p, χ_d) for 169 pd-bonded AB compounds, where χ_p and χ_d are the values of the chemical scale for the A and B elements respectively. The earlier χ-scale orders the elements in a similar fashion to the present Mendeleev number. The 2p elements are not included in the present figure.
Lower panel: The theoretical structure map (N_p, N_d), where N_p and N_d are the number of p and d valence electrons respectively on the CsCl lattice.

the Tight-Binding Bond (TBB) model in which the repulsive energy is given by the sum of central pair potentials and the bond energy by eq. (11) under the assumption that local charge neutrality is maintained. This TBB model can be justified within density functional theory by approximating the exact ground state charge density which enters the Schrodinger equation by the sum of overlapping atomic charge densities (40, 41, 42). The error involved is second order in the difference between the overlapping free atom and the exact ground state charge densities. The errors are less than 15% for the equilibrium separation and vibrational frequency of diatomic molecules such as C_2, N_2, and Cu_2 (40). By identifying the second contribution in eq. (10) with the bond energy rather than the band energy which enters the density functional scheme, the first contribution is just the sum of the Coulomb, exchange and correlation interactions between individual pairs of atoms (42).

We have taken the simplest possible TBB model for the pd-bonding systems (38). The valence s electrons are neglected, so that we need only consider the bonding between the valence p and d states on the metalloid and transition element sites respectively. Their hopping or bond integrals are given by the canonical TB parameters that result from the Atomic Sphere Approximation of Andersen (43). They are given explicitly (44) by

$$dd(\sigma,\pi,\delta) = (-6,4,-1)\,(r_d/R)^5$$
$$pp(\sigma,\pi) = (2,-1)\,(r_p/R)^3 \qquad (13)$$
$$pd(\sigma,\pi) = (-3,3^{1/2})\,(r_p^3 r_d^5)^{1/2}/R^4$$

where R is the internuclear separation.

The repulsive pair potentials were chosen to fall off with distance as the square of the corresponding bond integrals, i.e. $\phi_{dd} = c_{dd}^2/R^{10}$, $\phi_{pp} = c_p^2/R^6$ and $\phi_{pd} = (\phi_{pp}\phi_{dd})^{1/2}$.

The repulsive energy per formula unit may then be written

$$U^{rep} = (c_p c_d/V_{AB}^{8/3})\,(\alpha_{pd} + 1/2\,\mathcal{R}^{-1}\,\alpha_{dd} + 1/2\,\mathcal{R}\,\alpha_{pp}) \qquad (14)$$

where

$$\alpha_{\ell\ell'} = \mathcal{N}^{-1}\sum(V_{AB}^{1/3}/R)^{2(\ell+\ell'+1)} \qquad (15)$$

with $\alpha_{dd} = \alpha_{22}$, $\alpha_{pp} = \alpha_{11}$ and $\alpha_{pd} = \alpha_{12}$ and the sum in (15) extending over the relevant dd, pp or pd interactions on the lattice. V_{AB} is the volume per formula unit. The coefficients $\alpha_{\ell\ell'}$ depend on structure and not on volume. The <u>relative size factor</u> $\mathcal{R}$ is defined by

$$\mathcal{R} = (c_p/c_d)\,V_{AB}^{2/3} \qquad (16)$$

where the volume dependence enters as the result of the different distance behaviour of the p and d potentials ϕ_{pp} and ϕ_{dd} respectively. It is a measure of the relative size of the p and d atoms as can be seen as follows. Consider a pair of transition metal atoms a distance of $2R_d = V_{AB}^{1/3}$ apart. Then a pair of metalloid atoms will show the same repulsive energy when they are separated by the distance $2R_p$ such that

$$(R_p/R_d)^3 = \mathcal{R}. \qquad (17)$$

Thus, $\mathcal{R}$ generalizes the concept of the relative size of hard spheres to the case where the mutual interaction varies smoothly rather than discontinuously.

The relative preparation volumes, which are required by the equilibrium energy difference theorem, may be obtained from (14) by setting $\Delta U^{rep} = 0$. The resultant first-order change in volume ΔV is given by

$$\frac{\Delta V}{V} = \frac{6\,\Delta\alpha_{pd} + 3\,\mathcal{R}^{-1}\,\Delta\alpha_{dd} + 3\mathcal{R}\,\Delta\alpha_{pp}}{16\,\alpha_{pd} + 10\mathcal{R}^{-1}\,\alpha_{dd} + 6\mathcal{R}\,\alpha_{pp}} \qquad (18)$$

where the $\Delta\alpha_{\ell\ell'}$ are the corresponding changes in the repulsive coefficients. Thus, the fractional change in volume is a function only of the relative size factor $\mathcal{R}$ so that a universal curve may be plotted for each structure with a given set of internal co-ordinates. Fig. 13 shows these curves with the CsCl lattice as reference. As expected the NaCl lattice has

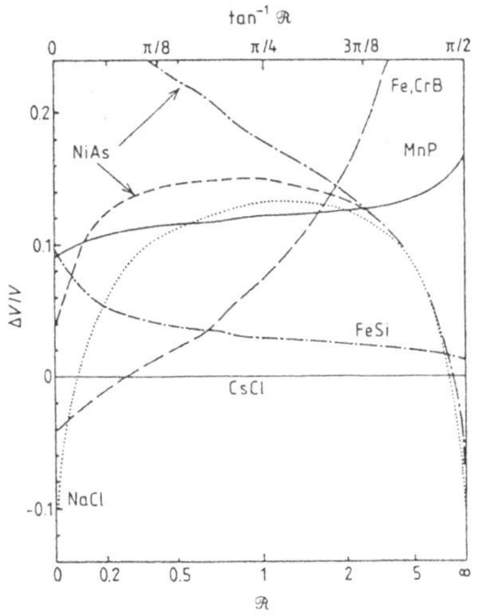

FIGURE 13: The fractional change in prepared volume ΔV/V with respect to the CsCl lattice versus the relative size factor $\mathcal{R}$. The upper and lower NiAs curves correspond to c/a = 1.39 and $(8/3)^{1/2}$ respectively.

the smallest volume at either end of the $\mathcal{R}$-scale, because as c_p or c_d tends to zero the repulsion is dominated by one or other of the close-packed fcc sublattices. On the other hand, in the middle of the scale where the nearest-neighbour (NN) pd repulsion dominates, the volume of the NaCl lattice with six NNs is about 13% larger than the CsCl with eight NNs. The packing of hard spheres would have predicted the much larger volume difference of 30%.

The structural stability of the pd-bonded AB compounds may now be studied by comparing the bond energy of the different lattices at the volumes determined by the relative size factor $\mathcal{R}$. Since we are interested in the global features of fig. 12 rather than predicting the structure of a particular compound, we have chosen for simplicity that value of $\mathcal{R}$ implicit in the assumption that the volumes are prepared so that there is no change in the second moment μ_2 of the TB density of states from one lattice to another i.e. $\Delta\mu_2 = 0$. This corresponds to the choice of $\mathcal{R} \simeq 0.8$ when hopping integrals appropriate to the 4d and 5p elements are taken. The fact that, apart from the borides and NiAs, the fractional volume changes are not too sensitive to the particular choice of $\mathcal{R}$ for $0.3 < \mathcal{R} < 3$ is responsible for the success of the theory in treating the global trends in fig. 12 with a single choice of $\mathcal{R}$ (38).

Fig. 14 shows the relative stabilities of the seven different lattices as a function of the band filling N and the p and d atomic energy level separation $\varepsilon_{pd} = \varepsilon_p - \varepsilon_d$ which were obtained from the computed TB density of states. We see at once the importance of both the average number of valence electrons per atom and the atomic energy level mismatch difference ε_{pd} on structural stability. Moreover, the shape of these curves would be very different for valence orbitals with different angular momentum character. The resultant structural stability predicted by this TBB model is displayed explicitly beneath each set of curves in fig. 14. Together the points constitute a theoretical ε_{pd} versus N structure map which may be compared with Burdett's (12) phenomenological (ε_{pd},N) structure map. However, because the ε_{pd} that enter the TB calculations are those appropriate to the binary compound and not isolated free atoms, we have chosen the rotated frame of axes (N_p,N_d), where N_p and N_d are the number of p and d valence electrons on the CsCl lattice for a given choice of (ε_{pd},N). This allows direct comparison with the phenomenological structure map in the upper panel of fig. 12.

Thus, the simple TBB model can account qualitatively for the observed domains of NaCl, CsCl, NiAs, MnP and boride stability amongst the transition-metal-metalloid AB compounds. Just as for the heat of formation, we see that the classical electrostatic Madelung energy plays no role in structural stability as the atoms may be taken to be perfectly screened and, therefore, charge neutral. The structural stability is determined by the quantum mechanical bond energy, eq. (11), which may be expressed directly in terms of the contributions from individual bonds by writing

$$U^{bond}_{qm} = \mathcal{N}^{-1} \sum_{\substack{i,j \\ i\neq j}} h_{ij} \left[-1/\pi \; \mathrm{Im} \int^{\varepsilon_F} G_{ji}(\varepsilon)\, d\varepsilon\right] \qquad (19)$$

where h_{ij} is the bond integral between neighbouring atoms i and j, G_{ij} is the inter-site Green function, and the expression inside the large brackets is the bond order (27). Eq. (19) is the starting point for calculating interatomic forces which may be used in atomistic relaxation packages for simulating defects in metals and alloys (42, 45, 46).

5. CONCLUSION

Quantum mechanics can help guide the alloy designer in two different but complementary ways. Firstly, the <u>phenomenological</u> structure maps can focus the search for new alloys with a given structure type. Moreover the Mendeleev number may also be used to display certain properties which are not very sensitive to the principal quantum number or period. For example, fig. 15 shows the weldability (47) of certain alloy pairs, where those which form good welds are separated from those which do not. Secondly, <u>microscopic</u> quantum mechanics is now starting to predict differences in energy between stable and metastable phases which are of direct interest to the materials scientist (16). In addition, quantum theory is providing concepts for understanding bonding in metals and alloys which will slowly replace the inappropriate classical concepts of ionicity and electronegativity difference. For the first time in the history of alloys, a meaningful dialogue has started between quantum theorists, materials scientists, and engineers, as illustrated schematically in fig. 16. Whether this cooperative venture will result in "New Alloys from the quantum engineer" - as the journalists entitled ref. 16 - remains to be seen.

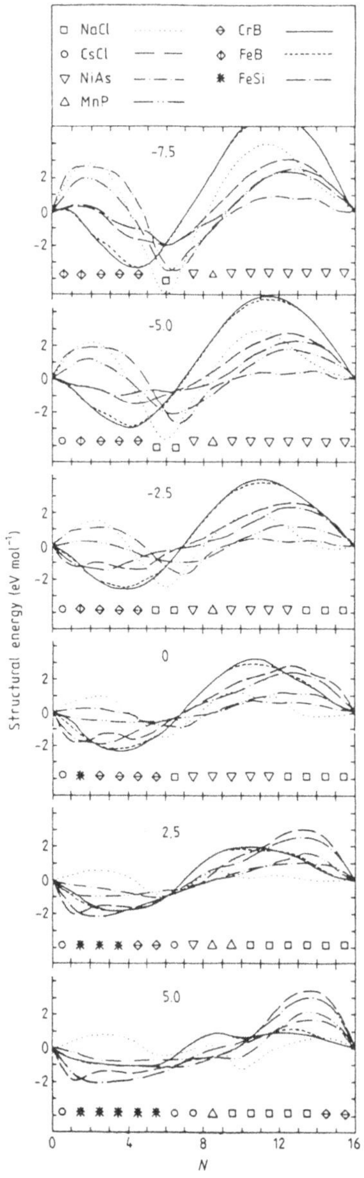

FIGURE 14. The structural energy curves as a function of band filling N and atomic energy level separation ε_{pd}. The value of ε_{pd} in eV is given at the top of each panel. The predicted structural trend for a given choice of ε_{pd} is shown by the symbols along the bottom of the corresponding panel.

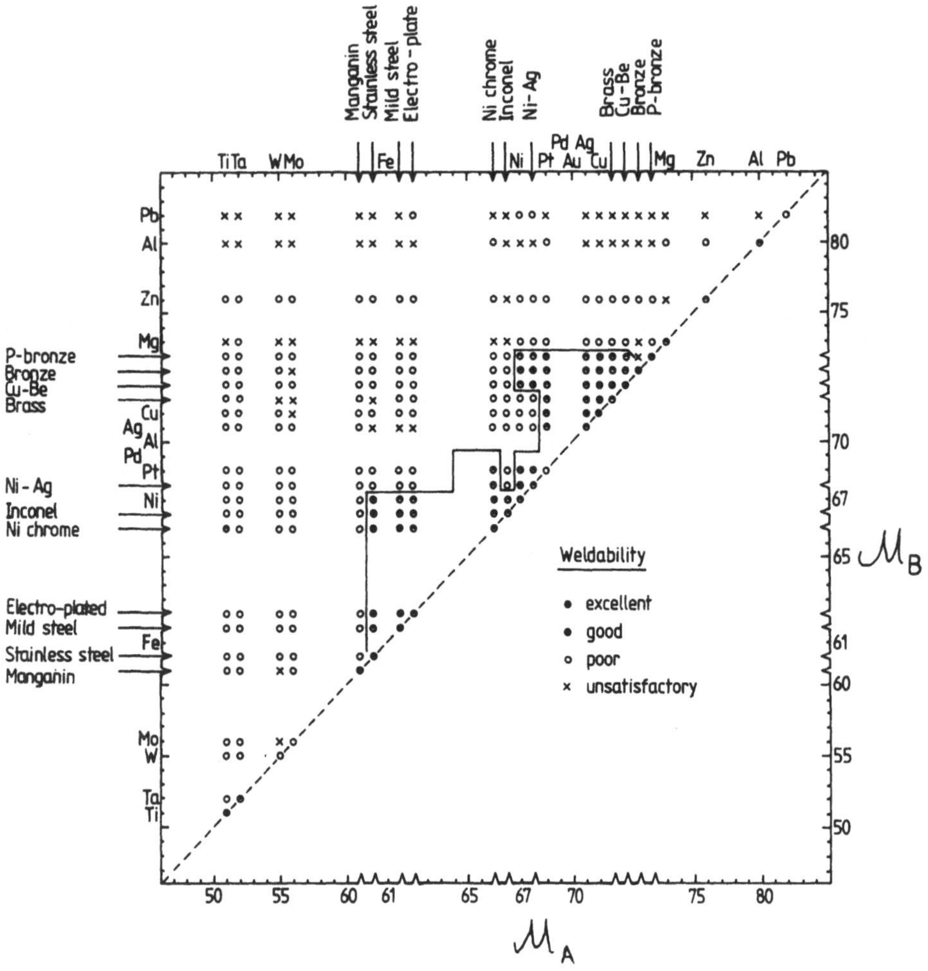

FIGURE 15: The weldability plot (data from Amor C (47)).

ACKNOWLEDGEMENTS

The author's research on the application of structure maps to alloy design was sponsored by the U.S. Department of Energy, Energy Conversion and Utilization Technologies (ECUT) Materials Program, under the subcontract No. 19X - 55992V through the Martin Marietta Energy Systems, Inc. The subcontract was technically monitored by the Oak Ridge National Laboratory.

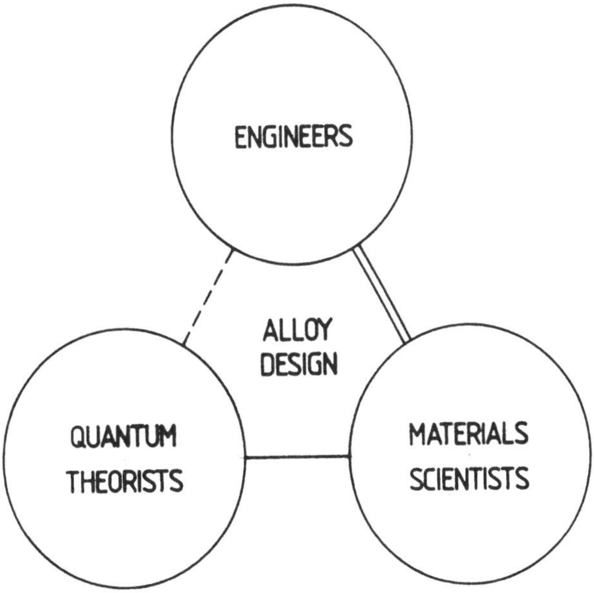

FIGURE 16: The alloy design triangle.

REFERENCES.

1. Liu CT: High Temperature Alloys: Theory and Design. Ed. Stiegler JO. AIME: Pennsylvania, 289 (1984).
2. Pettifor DG: Physica, May (1988).
3. Villars P and Calvert LD: Pearson's Handbook of Crystallographic Data for Intermetallic Phases, Vols. 1-3: ASM, Ohio, 1 (1985).
4. Miedema AR, de Chatel PF, and de Boer FR: Physica 100, 1 (1980).
5. Mooser E and Pearson WB: Acta Cryst. 12, 1015 (1959).
6. Phillips JC and van Vechten JA: Phys. Rev. Lett. 22, 705 (1969).
7. St. John J and Bloch AN: Phys. Rev. Lett. 33, 1095 (1974).
8. Watson RE and Bennett LH: Phys. Rev. B18, 6439 (1978).
9. Machlin ES and Loh B: Phys. Rev. Lett. 45, 1642 (1980).
10. Bieber A and Gautier F: Sol. State Commun. 38, 1219 (1981).
11. Zunger A: Phys. Rev. B22, 5839 (1980).
12. Burdett JK: J. Sol. State Chem. 45, 399 (1984).
13. Villars P: J. Less-Common Metals 92, 215 (1983); 99, 33 (1984); 102, 199 (1985).
14. Pearson WB: The Crystal Chemistry and Physics of Metals and Alloys. Wiley: New York (1972).
15. Pettifor DG: First Suppl. to the Pergamon Encyclopaedia of Materials Science and Engineering. Ed. Cahn RW. Pergamon: Oxford. Structure maps (1988).
16. Pettifor DG: New Scientist 110, No. 1510, 48 (1986).

17. Pettifor DG: Physical Metallurgy. Eds. Cahn RW and Haasen P. North-Holland: Amsterdam, Ch. 3, 73 (1983).
18. Pettifor DG: J. Phys. C19, 285 (1986).
19. Pettifor DG and Podloucky R: Phys. Rev. Lett. 55, 261 (1985).
20. Pettifor DG: Microclusters, Springer Series in Materials Science 4, 37 (1987).
21. Pettifor DG: Mat. Science and Tech. (to be published).
22. see, for example, Chelikowsky JR: Phys. Rev. B25, 6506 (1982).
23. Alonso JA and Girifalco LA: J. Phys. F12, 2455 (1978).
J. Phys. Chem. Solids 38, 869 (1977).
24. Williams AR, Gelatt CD, and Moruzzi VL: Phys. Rev. Lett. 44, 429 (1980);
Phys. Rev. B25, 6509 (1982); and unpublished.
25. Gelatt CD, Williams AR, and Moruzzi VL: Phys. Rev. B27, 2005 (1983).
26. Hohenberg P and Kohn W: Phys. Rev. 136, B846 (1964);
Kohn W and Sham LJ: Phys. Rev. 140, A1133 (1965).
27. Pettifor DG: Solid State Phys. 40, 43 (1987).
28. Harrison WA: Phys. Rev. B31, 2121 (1985).
29. Pettifor DG: Phys. Rev. Lett. 42, 846 (1979);
J. Magn. Magn. Mater. 15-18, 847 (1980).
30. Varma CM: Sol. St. Commun. 31, 295 (1979).
31. Watson RE and Bennett LH: Phys. Rev. Lett. 43, 1130 (1979).
32. Friedel J: The Physics of Metals. Ed. Ziman JM. Cambridge Univ. Press: London, 494 (1969).
33. Cyrot M and Cyrot-Lackmann F: J. Phys. F6, 2257 (1976).
34. Ziman JM: Adv. Phys. 13, 89 (1964).
35. Brewer L: Science 161, 115 (1968).
36. Hodges CH: J. Phys. F7, 1687 (1977).
37. Michaelson HB: J. Appl. Phys. 48, 4729 (1977).
38. Pettifor DG and Podloucky R: Phys. Rev. Lett. 53, 1080 (1984).
J. Phys. C19, 315 (1986).
39. Evans RC: An Introduction to Crystal Chemistry. 2nd ed. Cambridge Univ. Press: Cambridge, (1966).
40. Harris J: Phys. Rev. B31, 1770 (1985).
41. Foulkes WMC: Ph.D. thesis, Cambridge (1987).
42. Sutton AP, Finnis MW, Pettifor DG, and Ohta Y: J. Phys. C, Dec. (1987).
43. Andersen, OK: Phys. Rev. B21, 3060 (1975).
44. Pettifor DG: J. Phys. F7, 613 (1977);
Andersen OK, Klose W, and Nohl M: Phys. Rev. B17, 1209 (1978).
45. Ohta Y, Finnis MW, Pettifor DG and Sutton AP: J. Phys. F17, L273 (1987).
46. M. Saqi and D. G. Pettifor: Phil. Mag. Lett., Dec. (1987).
47. Data supplied by Amor C: The Design Council, 28, Haymarket, London SW1, England (1986).

TIGHT-BINDING HAMILTONIANS

D.A. Papaconstantopoulos

Condensed Matter Physics Branch
Naval Research Laboratory, Washington, DC 20375-5000

1. INTRODUCTION

The use of tight-binding (TB) Hamiltonians as a starting point of many theories is common place in condensed matter physics and materials science. The construction of such Hamiltonians is done with two different philosophies in mind. One is based on atomic level values and the idea of choosing a very small number of hopping integrals usually involving only d orbitals.[1-3] The other approach is in the spirit of the Slater-Koster (SK) interpolation[4,5] method which consists of a fit to the results of a band structure calculation and can easily incorporate hybridization of the d with the s-p orbitals. The practitioners of the first approach wish to retain simplicity in their model and physical significance of the values of their TB parameters. They achieve this at the expense of using a TB Hamiltonian that produces energy bands and densities of states which are of limited accuracy. On the other hand the second approach with the possible deficiency that occasionally produces opposite signs of certain small parameters, can be made to agree, especially for transition metals, almost exactly to the band structure derived from the first principles methods.[5]

2. REPRESENTATIVE RESULTS

In Fig. 1 we demonstrate[6] for Nb the very substantial differences found between the energy bands based on the Harrison solid state table and a 6x6 Hamiltonian (s+d orbitals) and those obtained from a SK type fit that corresponds to a 9x9 Hamiltonian (s+p+d orbitals). One can see that along the directions ΓH, ΓN, and ΓP the energy bands have entirely different shapes. In particular the position of the Fermi level, E_F, to be above the state $\Gamma_{25'}$ for the Harrison Hamiltonian (HH) and below $\Gamma_{25'}$ in our calculation is a striking difference that produces the wrong Fermi surface in the HH calculation. In addition, the non-inclusion of p-states in HH places the higher energy bands at too high energies which would prevent any serious comparison with experiments probing the excited states.

Reference 5 contains tabulations of SK parameters of 53 elements in the periodic table. These parameters were derived by fitting to first principles band structure calculations. Four different sets of SK parameters are given involving two and three center integrals as well as orthogonal and non-orthogonal basis. In most cases the energy bands are fitted very well with an rms deviation of the order of 1 mRy. The Fermi level values of the densities of states (DOS), its angular momentum decomposition, the Fermi velocity and plasmon energy together with energy band and DOS figures are presented for each element.

These TB Hamiltonians (TBH) can be used in a variety of applications concerning disordered materials, surfaces, interfaces, defects, phonon spectra, etc. We have applied the TB coherent potential approximation

G. M. Stocks and A. Gonis (eds.), Alloy Phase Stability, 351–356.

(CPA) formalism in many disordered systems. Our general conclusion[7] is that for substitutional alloys between neighboring elements in the periodic table the TB-CPA results are in excellent agreement with those of the KKR-CPA theory.[8] This is demonstrated in Fig. 2 where a comparison of the DOS between TB-CPA and KKR-CPA is given for the disordered alloy $Pd_{0.5}Ag_{0.5}$. Indeed the two calculations agree exceptionally well in all the details of the density-of-states spectra. We have found that the TB-CPA (within the diagonal disorder formulation) doesn't work as well when elements from different rows of the periodic table are involved. For example in the Cu-Pd system we have found substantial differences from the KKR-CPA results. It appears that in such alloys significant charge transfer occurs and therefore charge self-consistency and off-diagonal disorder are important. Indeed we have included in the formalism[9] the effects of off-diagonal disorder and results showed a substantial improvement.

Another interesting application of the TB-CPA is in hydrogenerated amorphous Si.[10] In this case the TB-CPA is particularly convenient because of the ease that one can incorporate a variety of atomic configurations that involve the replacement of Si atoms by either vacancies or vacancies whose dangling bonds may be decorated by 1-4 hydrogen atoms. This kind of approach despite its underlying crystallinity gives remarkable agreement to the experimental results of the amorphous state including XPS and absorption coefficient data.

We have also applied TBH to the study of interfaces.[11] Specifically, we have studied the Fe-Ge (110) interface which required 7 layers of Fe with 4 atoms each and 11 layers of Ge with 2 atoms each for a total of 50 atoms in the unit cell. Such a calculation is beyond the capabilities of present supercomputers if it were to be carried out in a self-consistent first principles way. Our results clearly identify the interface states and are consistent with the reactive nature of this interface seen in experimental studies.

Another use of TBH is that described in this volume by de Fontaine[12] where the generalized perturbation method and the cluster variation method are used together with the CPA. This approach brings together quantum and statistical mechanics to evaluate phase diagrams for a variety of alloys.

Finally TBH become a very valuable basis for the study of the new high temperature superconductors. We have already[13] constructed a TBH for the La_2CuO_4 compound. In this work we have fit first principles LAPW band structure results by the SK method using an orthogonal 31-orbital basis that includes La-d, Cu-s,p,d and O-p orbitals. This fit requires 44 two-center parameters obtained by fitting to the LAPW results at 71 k-points in the irreducible tetragonal Brillouin zone. The rms deviation from the LAPW eigenvalues is 14 mRy for the lower 17 bands. In order to obtain an accurate DOS near the Fermi level, E_F, we used an increased weight around E_F. The resulting DOS is shown in Fig. 3 where it is important to note that E_F falls just above a van Hove singularity that has been associated with the enhancement of the superconducting transition temperature upon doping with Ba or Sr. Calculations that account for oxygen vacancies and rare earth substitutions using the CPA are now in progress.

Other applications of the use of accurate TBH are in the evaluation of the electron-phonon interaction and the first principles determination of phonon dispersion curves.[14] Also, recently Fry et al.[15] have employed the two-center orthogonal SK parameters of Ref. 5 to compute the many-body-enhanced magnetic susceptibility for paramagnetic bcc Mn.

Acknowledgement
This work was partially supported by the Office of Naval Research.

References

1. Harrison WA: Electronic Structure and the Properties of Solids. W.H. Freeman and Co., San Francisco, 1980.
2. Ducastelle F: This volume.
3. Pettifor DG: J. Phys. C5, 97 1972.
4. Slater JC and Koster GF: Phys. Rev. 94, 1498 1954.
5. Papaconstantopoulos DA: Handbook of the Band Structure of Elemental Solids. Plenum, New York, 1986.
6. Shore JD and Papaconstantopoulos DA: Phys. Rev. B35, 1122 1987.
7. Laufer PM and Papaconstantopoulos DA: Phys. Rev. B35, 9019 1987.
8. Winter H and Stocks GM, Phys. Rev. B27, 882 1983.
9. Papaconstantopoulos DA, Gonis A and Laufer PM: to be published.
10. Papaconstantopoulos DA and Economou EN: Phys. Rev. B24, 7233 1981.
11. Pickett WE and Papaconstantopoulos DA: Phys. Rev. B34, 8372 1986.
12. de Fontaine D: This volume.
13. Papaconstantopoulos DA, DeWeert MJ and Pickett WE: Journal of Materials Research 1988.
14. Varma CM and Weber W, Phys. Rev. Lett. 39, 1094 1977; Fry JL, Flecther G, Pattnaic PC, and Papaconstantopoulos DA, Physica 135B, 473 1985.
15. Fry JL, Zhao YZ, Brener NE, Fuster G, and Callaway J, Phys. Rev. 36 868 1987.

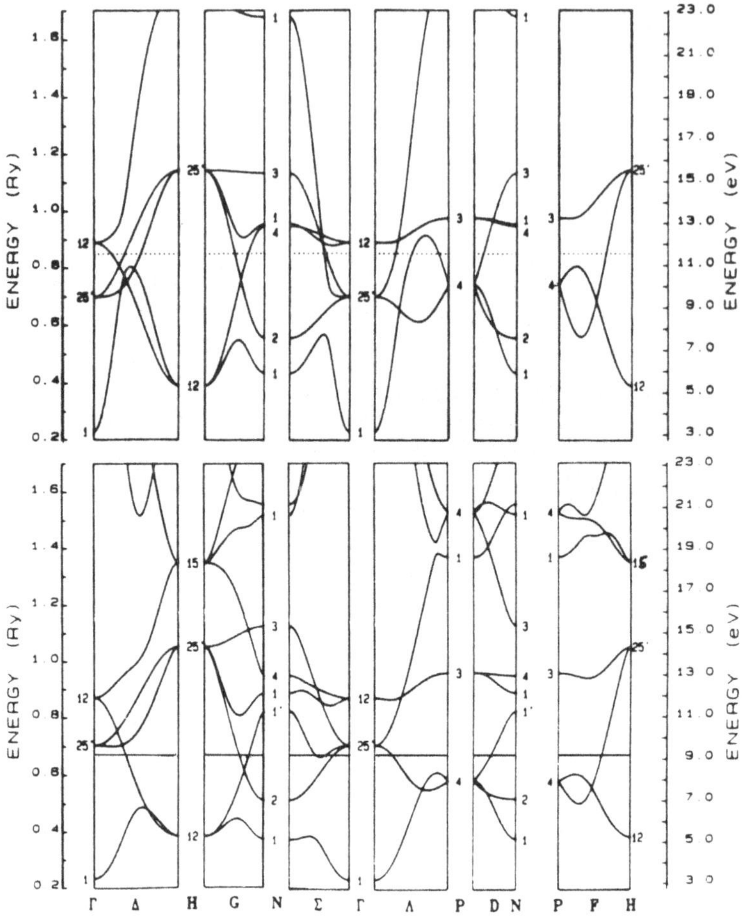

FIGURE 1. The top panel shows the energy bands of Nb calculated using Harrison's SK parameters and 6x6 Hamiltonian. The bottom panel shows the energy bands of Nb calculated using the Sk parameters of Ref. 5.

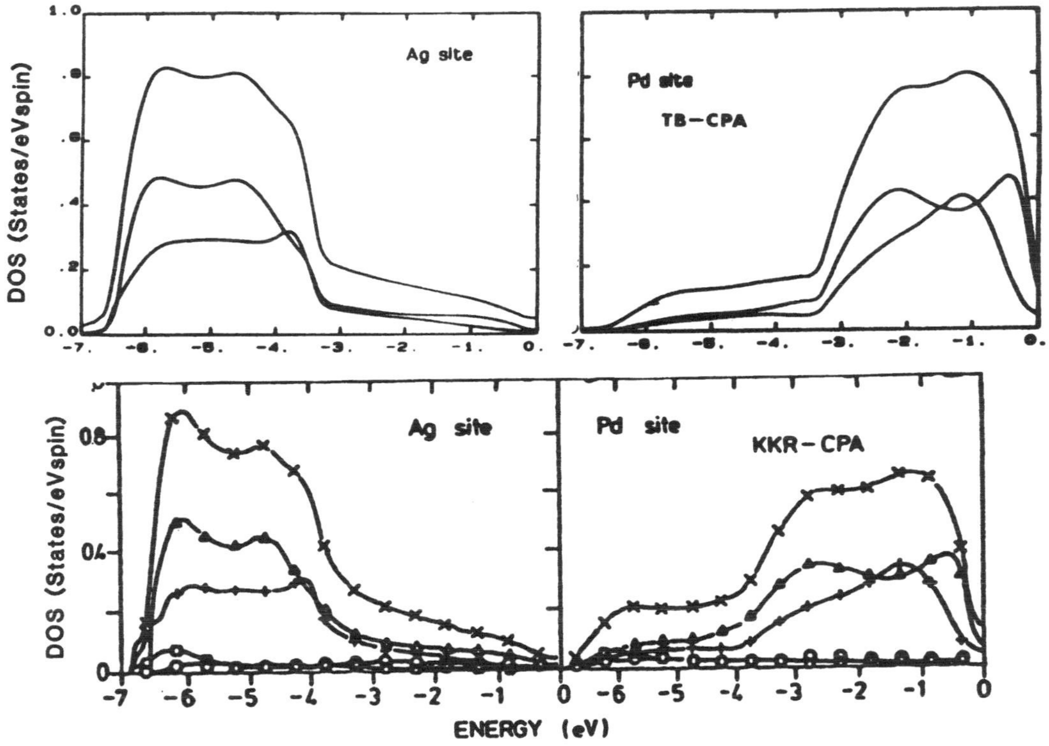

FIGURE 2. Comparison of TB-CPA and KKR-CPA densities of states for $Pd_{0.5}Ag_{0.5}$. Each panel shows the total, t_{2g} and e_g DOS per site.

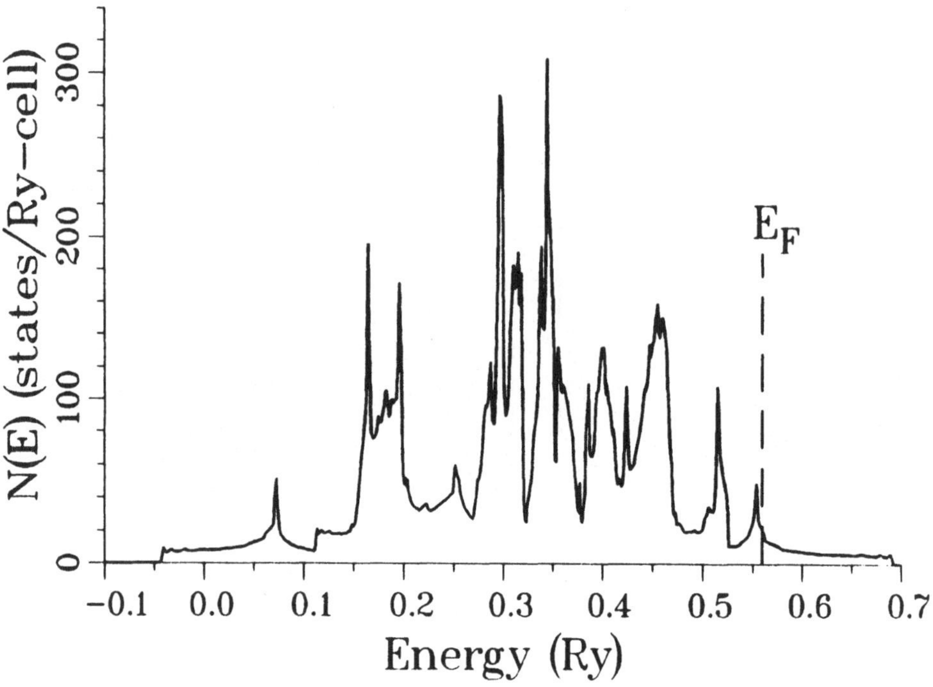

FIGURE 3. DOS of La_2CuO_4 resulting from a TB fit to the LAPW band structure.

CLUSTER BETHE LATTICE APPROACH TO CHEMICALLY DISORDERED ALLOYS WITH SHORT RANGE ORDER

A. N. ANDRIOTIS

Theoretical and Physical Chemistry Institute, National Hellenic Research Foundation, 48 Vassileos Constantinou Av., 116 35 Athens, Greece

J. E. LOWTHER

Physics Dept., University of the Witwatersrand, Johannesburg 2001, South Africa

1. INTRODUCTION

One simple way to incorporate short range order effects within the Hamiltonian of a binary alloy A_xB_{1-x} can be achieved by parameterizing the Coherent Potential Approximation (CPA). According to this parameterization[1], a correlation parameter $P_{A/B}$ is introduced to specify the probability a lattice site to be occupied by an atom of type A when a nearest lattice site is occupied by an atom of type B. The correlation parameter $P_{A/B}$ appears as an independent parameter of the total energy, E_{tot}, of the system and therefore the energetically favored state of the A_xB_{1-x} alloy is associated with the minimum of the energy function $E_{tot} = E_{tot}(P_{A/B})$. This theory has been used[2-3] within the Bethe lattice approximation to binary alloys which exhibit only diagonal disorder.

In order to allow for off-diagonal disorder too, in this work, the $P_{A/B}$ parameterization scheme is incorporated within a self-consistent scheme of the Cluster Bethe Lattice Approach (CBLA)[4-9] to binary systems.

Two types of short range order effects are studied. Namely, short range correlation and charge transfer effects. The present theory is applied to the alloy $Cs_{0.50}Au_{0.50}$ and it is shown that charge transfer effects are responsible for the metal insulator transition to occur in this system. As the formalism of the CBLA, we follow, has been extensively discussed elsewhere,[8-9], we will proceed by describing the parameterization of CBLA and present our results.

2. INCORPORATION OF SHORT RANGE ORDER

Given the parameter $P_{A/B}$ and the concentration x of the A species we can obtain the related parameters

$$P^{(1)}(B;B) = P_{B/B} = 1-P_{A/B} \qquad (1a)$$

G. M. Stocks and A. Gonis (eds.), Alloy Phase Stability, 357–362.

$$P^{(1)}(A;A) = P_{A/A} = 1- \frac{1-x}{x} P_{A/B} \qquad (1b)$$

$$P^{(1)}(B;A) = P_{B/A} = 1-P_{A/A} \qquad (1c)$$

For completeness, we define

$$P^{(1)}(A;B) = P_{A/B} \qquad (1d)$$

The above parameters specify $P^{(k)}(i;j)$, the probability of the kth nearest neighbour site (with respect to a given central (zeroth) lattice site) to be an atom of species i (=A or B) when the central site is occupied by an atom of species j (=A or B). In particular

$$P^{(k)}(i;j) = \sum_{l = A,B} P^{(k-1)}(i;l)P^{(1)}(l;j) \qquad (2)$$

$$k \geq 2, \; i,j = A,B$$

Our generalization is based on the observation that $P^{(k)}(i;j)$ tends to 0.5 for $k \geq 5$ for most values of the pairs (x, $P_{A/B}$). Based on this observation, we have generalized the parameter k (which indicates the number of atoms in a cluster which are of the same species as the atom occupying the central site of the cluster[8-9]) to:

$$k^{(i)} = Z_1P^{(1)}(i;i) + Z_2P^{(2)}(i;i), \; i = A,B \qquad (3)$$

where Z_1 and Z_2 are the number of the first and second nearest neighbors respectively. Similarly we generalized the parameter λ (see references 8-9 for more details) to:

$$\lambda^{(i)} = 1 - 2P^{(3)}(i;i) \; , \; i = A,B \qquad (4)$$

Thus the whole formalism of the CBLA[8-9] is generalized only through the parameters $k^{(i)}$ and $\lambda^{(i)}$ which allow us to incorporate short range order effects in a consistent way.

3. CHARGE TRANSFER EFFECTS

The present model approximation allows one to include the charge transfer effects. This can be achieved by integrating the site-electron dos $n_i(E)$ of the species i and subtracting from it the valence charge of the species i, i.e.

$$\Delta n_i = \int_{-\infty}^{E_F} n_i(E)dE - n_i^{(0)} \; , \; i = A, B \qquad (5)$$

where $n_A^{(0)} = n_B^{(0)} = 1.0$ in the present case of the Cs-Au alloy. It becomes obvious that a self-consistent solution becomes necessary as the site energy terms depend on the charge transfers Δn_i, i=A,B which in turn depend on both the inter- and intra-site interactions. One basic assumption in our calculations is that the cluster under consideration is electrically neutral. This assumption imposes a relation between $\Delta n_A^{(i)}$ and $\Delta n_B^{(i)}$, i = A or B denoting the species of the zeroth site of the cluster.

Due to the charge transfer and the intra-site and inter-site interactions, the site energies U_i, i = A,B, take the form (including up to 2nd n.n.)

$$U_A = U_A^{(o)} + U_o \Delta n_A^{(A)} + 2\{U_{AA}^{(1)} Z_1 \; P^{(1)}(A;A) + U_{AA}^{(2)} Z_2 \; P^{(2)}(A;A)\} \Delta n_A^{(A)} + 2\,\{U^{(1)}{}_{AB} Z_1 \; P^{(1)}(B;A) + U_{AB}^{(2)} Z_2 \; P^{(2)}(B;A)\}\, \Delta n^{(A)}{}_B \tag{6}$$

and similarly for U_B.

In equation (6) U_o is the intra-site Coulomb integral and $U^{(k)}{}_{ij}$ is the Coulomb potential between point charges at sites i and j the jth site being the kth nearest neighbour to the ith one. In our calculations we have taken $U^{(2)}{}_{ij} = U_{ij}^{(1)}/\sqrt{2}$ i, j = A or B and $U^{(1)}{}_{ij} = 1/_d$ where d is the average distance between nearest neighbors. On the other hand, the total energy, E_{tot}, of the system, takes the form[7,10]

$$E_{tot}(P_{A/B};x) = E_{el} + E_{inter} + E_{intra} \tag{7}$$

where

$$E_{intra} = -U_o\{x_A[n^{(o)}{}_A + \Delta n^{(A)}{}_A]^2 + x_B[n^{(o)}{}_B + \Delta n^{(B)}{}_B]^2\} \tag{8}$$

$$E_{inter} = -2n_A x_A Z_1\{U^{(1)}{}_{AA} P_{A/A} \Delta n^{(A)}{}_A + U^{(1)}{}_{AB} P_{B/A} \Delta n^{(A)}{}_B\} \tag{9}$$

$$-2n_B x_B Z_1\{U^{(1)}{}_{BA} P_{A/B} \Delta n^{(B)}{}_A + U^{(1)}{}_{BB} P_{B/B} \Delta n^{(B)}{}_B\} \tag{10}$$

$$E_{el} = \int_{\infty}^{E_F} n(E)\,E\,dE$$

In the above, E_F denotes the Fermi energy of the system,[2,8,9] x_A = x and $x_B = 1 - x_A$. Equations (5)-(10) dictate the self-consistent iteration scheme to be followed. We start with the assumption that $\Delta n^{(i)}{}_A = \Delta n^{(i)}{}_B = 0$, i = A,B. New $\Delta n^{(i)}$ or $\Delta n_B^{(i)}$ are calculated at the end of each interaction using equation (5). These are used as the input for the next iteration. The procedure is repeated till self-consistency is achieved for $\Delta n^{(i)}{}_A$ or $\Delta n^{(i)}{}_B$, i = A, B.

4. RESULTS AND DISCUSSION

The theory presented in the previous sections has been applied to the Cs_xAu_{1-x} alloy for x = 0.5. In table I we give the matrix elements $U_i^{(o)}$, V_{ij}, i = A, B as well as the values U_o and d used in the present calculations.

TABLE I

$U_{Au}^{(0)} = - U_{Cs}^{(0)} = - 0.099$ Ry	$V_{CsAu} = V_{AuCs} = - 0.0519$ Ry
$U_o = 0.09$ Ry	$V_{AuAu} = - 0.0192$ Ry
$d = 3.13$ Å	$V_{CsCs} = - 0.0476$ Ry

The results of the present work are in qualitative agreement with our previous ones[3] obtained within the generalized CPA. This becomes apparent in fig. 1 where we have plot the function $E_{tot}(P_{A/B})$ obtained within the non-self-consistent scheme of CBLA (i.e. $\Delta n_A^{(1)} = \Delta n_B^{(1)} = 0.0$, i = A,B above) and within the generalized CPA.

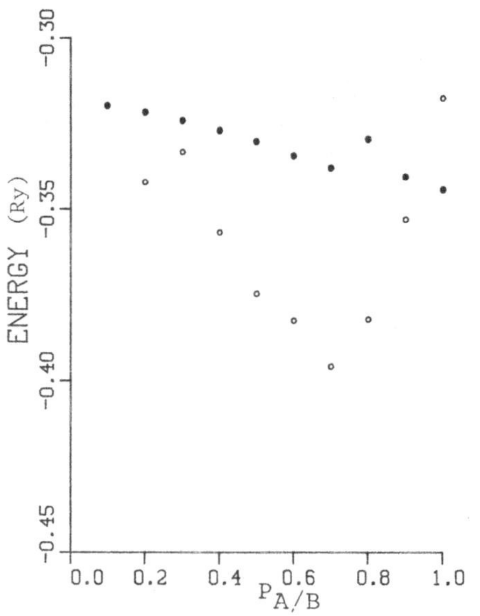

Fig. 1. Results for the total energy obtained within the generalized CPA (filled circles) and the non-selfconsistent generalized CBLA (open circles).

In figures 2a, 2b and 3a, 3b we present the computed electron DOS within the non-self-consistent and self-consistent schemes of the generalized CBLA discussed above. The most interesting feature emanating from these results is the implied different reason for a metal-insulator (MI) transition to occur in the $Cs_{0.5}Au_{0.5}$ alloy. In particular, in the non-self-consistent CBLA, the MI transition seems to result from the establishment of a local binary ABABA . . . order in the system. This order and the physical quantities of the system

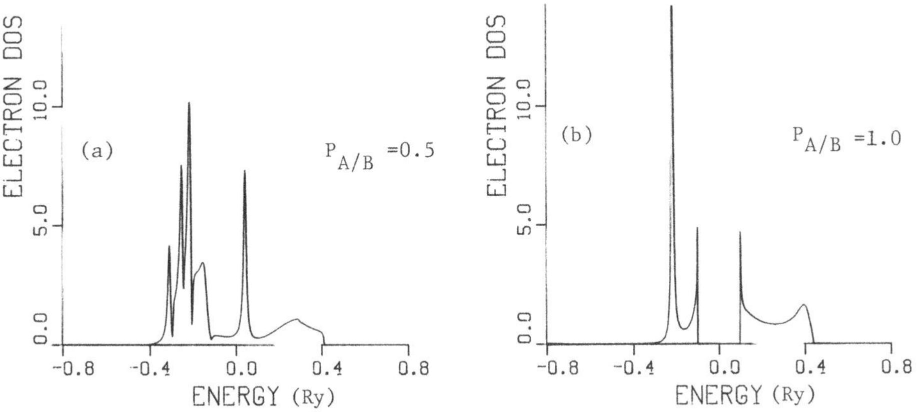

Figs. 2a, 2b. Electron DOS for $Cs_{0.5}Au_{0.5}$ alloy obtained within the non-self-consistent CBLA.

($U_i^{(0)}$ and V_{ij}, i, j = A, B) imply two well separated bands in the electron DOS. On the other hand, the charge transfer effects, allowed within the self-consistent scheme of the CBLA, make the site diagonal elements of the Hamiltonian to increase in magnitude as $P_{A/B}$ increases. As a result, the centers of the valence and the conduction bands move more and

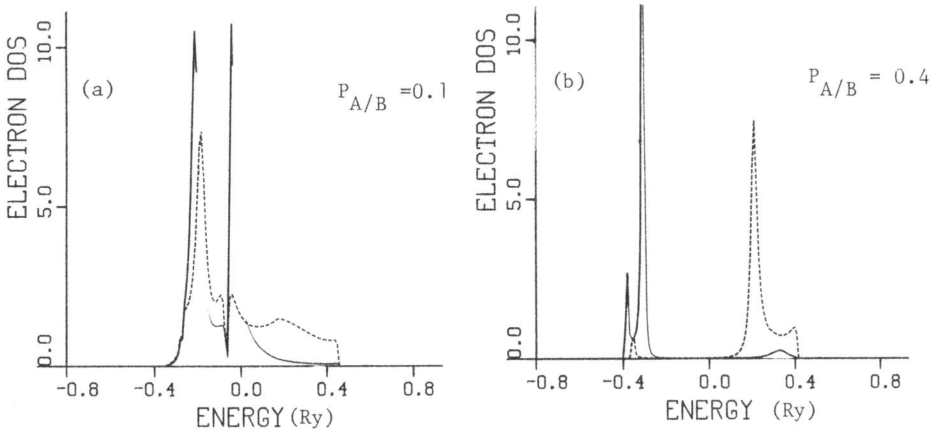

Figs. 3a, 3b. Electron DOS for $Cs_{0.5}Au_{0.5}$ alloy obtained within the self-consistent CBLA.

more apart as $P_{A/B}$ increases and allow the development of a large gap in the electron DOS even at complete randomness. This seems to be the physical reason for the MI transition in $Cs_{0.5}Au_{0.5}$ alloy and this conclusion is supported from the results of fig. 1.

REFERENCES

1. White C T and Economou E N : Phys. Rev. B15, 3742 (1977)
2. Andriotis A N, Poulopoulos P N and Economou E N : Solid St. Commun. 39, 1175 (1981)
3. Andriotis A N and Lowther J E : J. of Phys. F, 16, 1189 (1986)
4. Falikov L M and Yndurain F : Phys. Rev. B12, 5664 (1975)
5. Ioannopoulos J D nd Cohen M L : Solid State Physics, 31, 71 (1976)
6. Kittler R C and Falikov L M : J. Phys. C, 9, 4259 (1977)
7. Lowther J E : Physics Letters 81A, 470 (1981)
8. Lowther J E : J. Phys. F, 12, 895 (1982)
9. Andriotis A N : J. Phys. F, 17, 75 (1987)
10. Ten Bosch A, Moran-Lopez J L and Bennemann K H : J. Phys. C, 11, 2959 (1978)

SECTION 5

ELECTRONIC THEORIES OF PHASE STABILITY:

FIRST PRINCIPLES THEORY

STRUCTURAL STABILITY OF INTERMETALLIC COMPOUNDS: A COMPUTATIONAL METALLURGICAL APPROACH

A.J. FREEMAN
Department of Physics and Astronomy, Northwestern University, Evanston, Illinois 60208

1. INTRODUCTION

The driving force behind the research reported here is the recognition that technological advances depend strongly on a thorough understanding of the thermodynamic, mechanical, and electronic properties of materials. These, in turn, depend on our predictive ability regarding these properties. Understanding the structure and stability of the phases of compounds and alloys is an area of vital importance to materials science and technology. Such an understanding is starting to emerge from our first steps into carrying out all-electron quantum mechanical investigations on materials of aerospace interest. Briefly put, our emphasis has been on obtaining an understanding of the effects of alloying on bonding, crystal structure and phase stability of structural materials and to use this information to help design new alloy systems - in close collaboration with experimental efforts at a number of laboratories. In this paper, a brief indication is given of the progress made in a number of illustrative areas.

2. THEORETICAL/COMPUTATIONAL APPROACH: SELF-CONSISTENT LOCAL DENSITY FUNCTIONAL THEORY

The fundamental quantities of density functional theory are two observables, the total electron density (charge density) and the total energy of the system (due to the kinetic energy of electrons and the potential energy due to electrostatic interactions). An important theorem of density functional theory [1] states that given the correct ground state density, n(r), this density rigorously determines all electronic properties of the system, and in particular its total energy. Furthermore, the total energy of a system can be expressed as a functional of the density n'(r) and this functional, E[n'(r)], has a minimum when n'(r) becomes equal to the ground state density.

The total energy as a functional of the density can be written in the form

$$E[n] = T[n] + U[n] + E_{xc}[n] \qquad (1)$$

where T[n] is the kinetic energy of a system of non-interacting particles of density n; U[n] is the classical electrosatic energy due to the Coulomb interactions of a negative charge distribution of density n with itself, the interation of the charge density n with the positive nuclei, and the Coulomb interactions between the nuclei, and $E_{xc}[n]$ includes all many-body contributions to the total energy, in particular the exchange and correlation contributions.

To obtain single-particle Schrodinger equations, we decompose the total density into single particle densities which are expressed by

G. M. Stocks and A. Gonis (eds.), Alloy Phase Stability, 365–375.

single particle wave functions as

$$n(r) = \sum_i |\Psi_i(r)|^2 \qquad (2)$$

where the sum goes over all occupied states. Using this decomposition into single particle wave functions, the condition for E[n] to have its minimum leads to the so-called Kohn-Sham equations of the form

$$[-1/2\Delta + V_{eff}(r)]\Psi_i = \varepsilon_i \Psi_i \qquad (3)$$

where the effective potential is written as a sum of the Coulomb potential and the exchange-correlation potential:

$$V_{eff}(r) = V_C(r) + V_{xc}(r) \qquad (4)$$

The Coulomb potential is related to the charge density via Poisson's equation

$$-\Delta V_C(r) = 4\pi e^2\, n(r) \qquad (5)$$

and the exchange-correlation potential is given by

$$V_{xc}(r) = \delta E_{xc}(r)/\delta n(r) \qquad (6)$$

So far, the expressions given are formally rigorous. Clearly, the last term in equation (1) requires approximation. From a great number of calculations it became obvious that the following approximation, called the local density approximation (LDA), works surprisingly well: Using the known many-body energy, $\varepsilon(n)$, of an electron in a homogeneous, interacting electron gas of density n, this exact result can be used to approximate the inhomogeneous case by setting

$$E_{xc}[n] \simeq \int n(r) \quad [n(r)]dr \qquad (7)$$

In a sense, this approximation assumes that the leading terms in the exchange and correlation effects between electrons are of a fairly short range nature. Practical applications of the method as discussed below provide evidence that this may be indeed a realistic assumption. Using the LDA, the effective potential in the Kohn-Sham equations (3) can be evaluated explicitly via equations (4)-(6), provided the charge density is known. In turn, the charge density is defined by the solutions of the Kohn-Sham equations (3) via equation (2). Thus, the Kohn-Sham equations can be solved self-consistently: an initial charge density is constructed, e.g., by using overlapping atomic charge densities. From this ("input") charge density, the effective one-particle potential is calculated by solving Poisson's equation (5) and by evaluating the exchange-correlation potential (6) which is given by an analytic function of the density, n.

In the next step, which is numerically the most demanding, the differential equations (3) are solved. One possible way using a variational expansion of the single particle wave functions, Ψ_i, in augmented plane waves is described below. Alternatively, an expansion in atomic orbitals can be employed for localized systems. After the eigenvalues and eigenfunctions are found, a new ("output") charge density can be constructed using Fermi-Dirac statistics in equation (2). This closes one iterative step. The output charge density is fed back into the evaluation of the effective-potential energy operator and the procedure is repeated until the output charge density equals the input charge density to within a given tolerance. To construct the input density for a given iteration, the input and output charge densities from previous iterations are combined to ensure damping and an efficient extrapolation. Depending on the system, between 10 and 50 iterations are necessary (magnetic systems such as Ni are more difficult to converge than systems with a low density of states at the Fermi level).

Once the self-consistent charge density is evaluated for a given choice of nuclear coordinates, the corresponding total energy can be found using equations (1)-(7). Next, the nuclei can be displaced and the procedure can be repeated. In this way, the energy hypersurface of the system can be established point by point. In principle, derivatives of the total energy with respect to displacements of atomic nuclei could be evaluated. However, this has not yet been fully implemented in the approaches discussed here.

3. ILLUSTRATIVE RESULTS

3.1. Structural phase stability of Ni_3Al and Ni_3V.

To begin with, the advent of supercomputers has had the added benefit of permiting new methodologies and computational approaches to be implemented. In particular, it has led to a demonstration of the high precision and reliability of our total energy calculations. As a first example of our recent work, we describe briefly extensive studies of the structural phase stability and properties of Ni_3Al with and without V additions. These studies may be considered as illustrating our basic approach to understanding the aluminides.

3.1.1. Structural stability and structural properties of Ni_3Al.

The nickel-rich aluminide, Ni_3Al, has well-known attractive properties for structural applications at elevated temperatures [2]. The flow stress increases with increasing temperature to a maximum value, which occurs in the temperature region between 600-800°C; Ni_3Al is the most important strengthening (γ') phase of the Ni-base superalloys; further, the density of Ni_3Al is significantly lower than that of the Ni-base superalloys. However, it has been widely recognized that unlike its single crystal form, polycrystalline Ni_3Al is extremely brittle; hence, to be useful as an engineering material, one has to overcome its severe embrittlement and to improve its ductility.

Since embrittlement in the ordered intermetallic compounds is considered to be mainly caused by an insufficient number of slip systems or poor grain-boundary cohesion, considerable effort has been made to increase the grain-boundary cohesion or to change the crystal structure from low symmetry to high symmetry (more slip systems, and hopefully, greater ductility). The results obtained showed that the brittleness associated with ordered intermetallic compounds (such as Ni_3Al) can be solved using physical metallurgical methods: for example, ternary additions (such as boron) into Ni_3Al greatly promotes its ductility [3]. However, since the crystal structure may be changed due to ternary additions, the central task is to keep the aluminide in a high symmetry (cubic) structure. In other words, understanding crystal stability may have important significance in order to develop materials with sufficient ductility. Over the last decade, the electronic structure and the magnetic properties of Ni_3Al in its cubic $L1_2$ structure have been extensively studied both experimentally [4] and theoretically [5-7]; still, a rather poor understanding of the structural stability of Ni_3Al based on the electronic structure remains [5,7,8].

In our studies of the structural stability of Ni_3Al, the total energy of the three different crystal structures - cubic $L1_2$, tetragonal DO_{22} and hexagonal DO_{19} - were calculated by means of the all-electron semi-relativistic linear muffin-tin orbital (LMTO) method [9] based on the local density functional approach. We find that of these three different structures, the (weakly ferromagnetic) $L1_2$ structure is the most stable in Ni_3Al (cf. figure 1). It is characterized by a small density of states at the Fermi level (4.42 states per eV - formula unit)), and shows a weak ferromagnetic instability with a moment magnetic of about 0.15 μ_B per atom. Further, the exchange energy gained by inducing the magnetic moment in $L1_2$ is much smaller than the band energy gain due to the structural transition into DO_{22} (or DO_{19}). The higher order coupling between the Ni-d and Al-p states may play an important role in accounting for the structural stability of Ni_3Al. The calculated lattice constant (3.55Å), the bulk modulus (2.1 Mbar), and the heat of formation (44.4 kcal/mole) are in good agreement with experiment.

In order to make use of our understanding of the phase stability in a practical way, we have studied the density of states for the different structures. We find that there is a correlation between the stability and the density of states at the Fermi energy in that the energetically favorable atomic arrangement of Ni_3Al leads to a low density of states at the Fermi energy. This appears to be a universal feature in all the aluminides we have studied and, as we shall see later (cf. figures 2 and 3), permits us to make predictions about the possible stabilization of the cubic ($L1_2$) phase by means of ternary additions in other aluminide systems.

3.1.2. Site Preference and Solid Solution Strengthening: $Ni_3(Al,V)$.

Early on [8], we focused on solid solution strengthening as one of the technologically important features of Ni_3Al: Ni_3Al can accept into solution substantial alloying additions without losing its long range order. Clearly, knowledge of the preferential occupancy by ternary additions in Ni_3Al is fundamental for understanding strengthening affects in Ni_3Al and for the development of superalloys for practical use [11]. We therefore set out [8] to understand the effects of ternary additions in Ni_3Al focussing, in particular, on the effect of V on the stability of the $L1_2$ structure of Ni_3Al. (This was combined with separate studies of [10] Ni_3Al and [12] Ni_3V. For these pure systems [10,12], we found that the lattice constants and bulk modulus were in good agreement with experiment and other calculations). In these all-electron total-energy local density studies, we compared the total energies of two different structures [$L1_2$-like and DO_{22}-like of $Ni_3(Al_{0.5}V_{0.5})$]. We found that ternary additions of V in the $L1_2$-like structure (i.e., V substitution on Al) are more stable than for V additions in the DO_{22}-like structure (i.e. V substitution on Ni sites). Thus, these model calculations indicate that the Li_2-like Ni_3Al can dissolve up to 50% V replacing Al in an ordered way and still retain the $L1_2$-like structure with a lower energy than in the DO_{22}-like structure. In addition, we found a slight hardening effect caused by the V addition in Ni_3Al. Further, this hardness increment agrees with the experimental measurement on a Ni - 12.3% Al - 2.2% V ternary alloy studied by Westbrook and by Dimiduk [11] and earlier by Decker and Mihalisin [13] (in 1969).

3.1.3. Role of Density of States in Stabilizing Intermetallics.

We have also carried out a careful analysis of the total and the

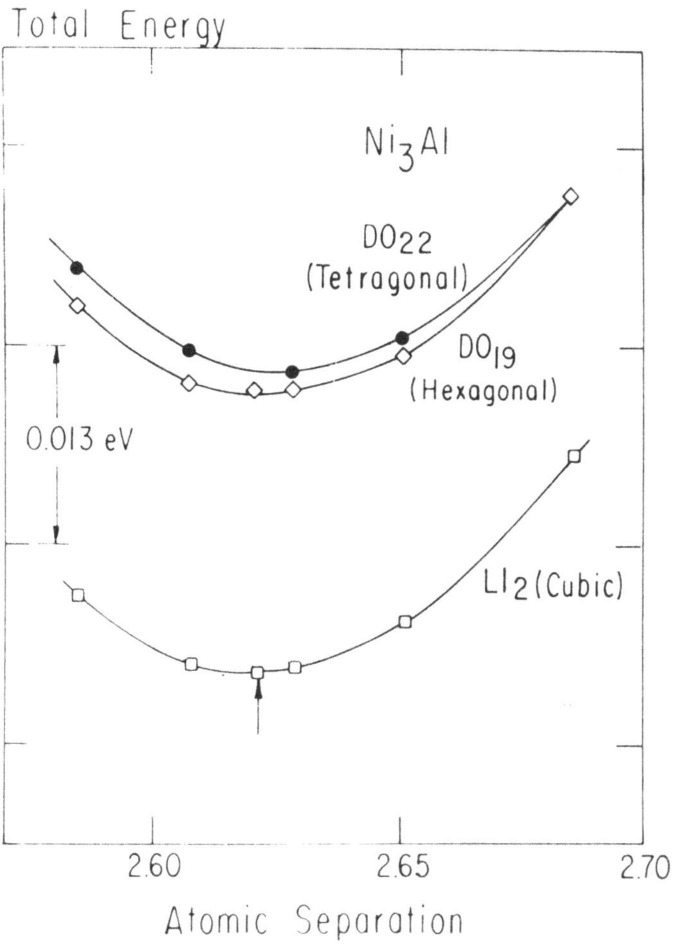

FIGURE 1. The total energy as a function of the Wigner-Seitz radius and its structure dependence for Ni_3Al open and closed circles, triangles indicate $L1_2$, DO_{22}, and DO_{19} structures, respectively.

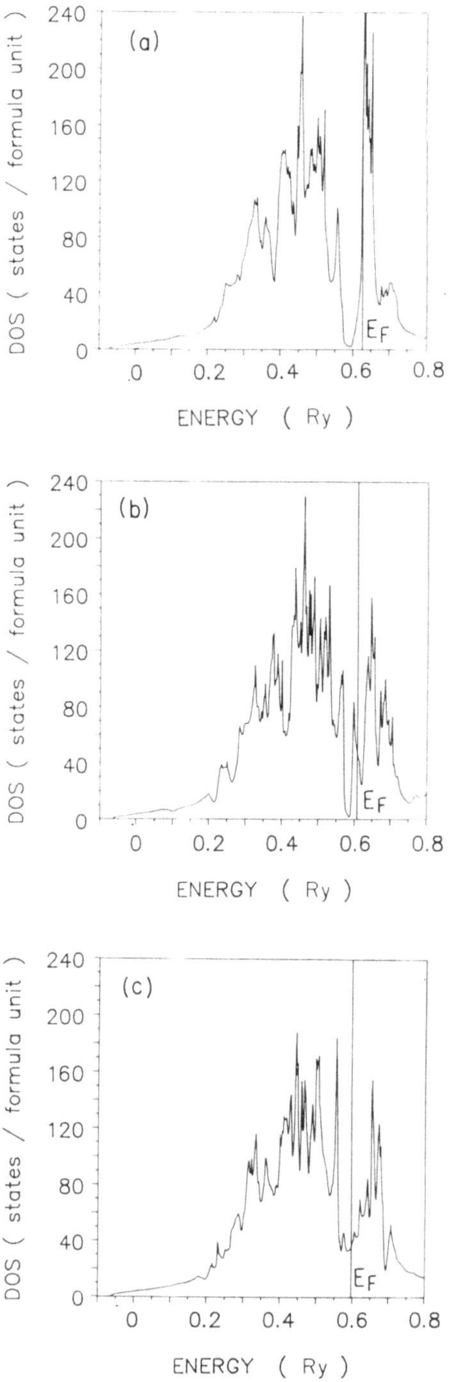

FIGURE 2. The total density of states for Ni_3V in the three different structures (a) $L1_2$, (b) DO_{22}, (c) DO_{19}.

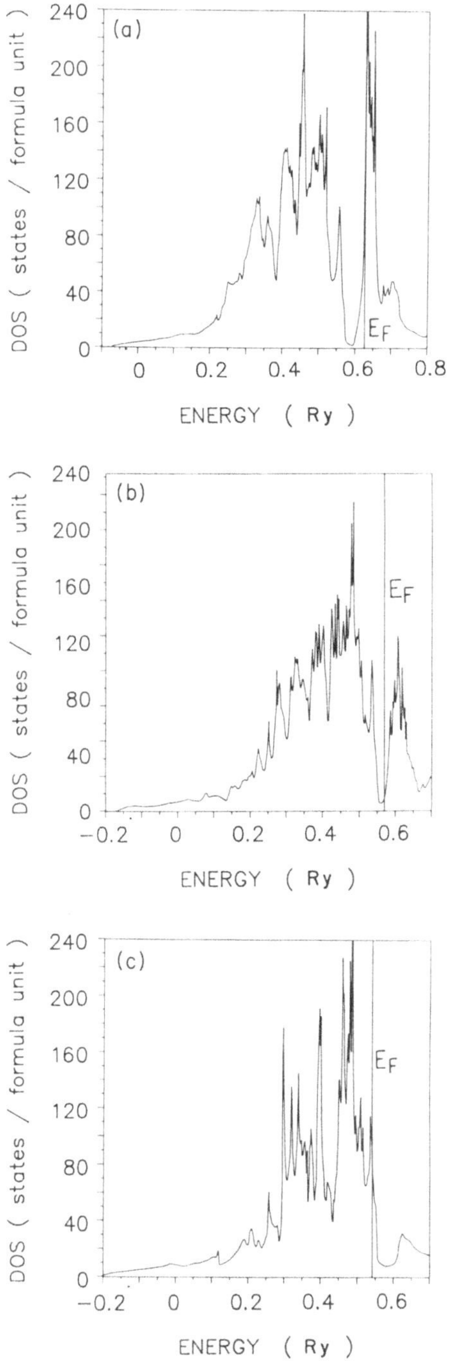

FIGURE 3. Comparison of the DOS for the $L1_2$ structure of Ni_3Al, $Ni_3(Al,V)$ and Ni_3V.

Ni-d and V-d partial density of states for $Ni_3(Al,V)$ in the $L1_2$- and DO_{22}-like structures [12]. We found that the d-d hybridization between V and the host (Ni) appears in both the $L1_2$-like and DO_{22}-like structures; however, the d-d interactions exhibit different features for the $L1_2$-like and DO_{22}-like structures of $Ni_3(Al,V)$. The strong hybridization between Ni-d and V-d in the $Ni_3(Al,V)$ $L1_2$-like structure has a special prominent feature: a well separated bonding and antibonding region. This turns out to also be a noticeable common feature for the $L1_2$ structures of Ni_3Al, Ni_3V and $Ni_3(Al,V)$ - cf. figures 2 and 3. A deep valley separates the bonding and antibonding parts of the DOS. Obviously, the stability of the compound depends on the position of the Fermi level. It is thus expected, and found from our total energy studies, that the cubic Ni_3V phase, in which the Fermi level is located in the antibonding region, is an unstable compound and Ni_3Al will be a stable compound in the $L1_2$ structure, because the Fermi level lies in the bonding region of the density of states. For $Ni_3(Al,V)$ in the $L1_2$-like structure there are enough valence electrons due to the V addition to bring the Fermi level to a position just below the antibonding d-d hybrid states. Thus, the valence electrons fill all the bonding states and leave all the antibonding states empty; therefore, the strongest bonding effect occurs in $Ni_3(Al,V)$ in the $L1_2$-like structure. For this reason, it is also expected, from the rigid band sense, that Ti, Zr, Hf, Nb, Ta, Si and Ge will stabilize the $L1_2$ structure in Ni_3Al in much the same way as V, and possibly preferentially occupy Al sites.

3.2. Stabilization of the $L1_2$ phase of $ZrAl_3$.

As a further effort to understand the stabilization of high temperature structural materials in highly ductile phases, we have investigated the electronic structure and the cohesive properties of $ZrAl_3$ in the metastable $L1_2$ structure using the total energy local density linear muffin-tin orbital approach [10]. ($ZrAl_3$ was suggested to us by M. Fine of Nortwestern based on his extensive work on this material). For this metastable structure, the calculated lattice constant (4.073 Å) is in excellent agreement with experiment (4.0731 ± 0.0008 Å) and the bulk modulus (1.0 Mbar) is consistent with the observed Young's modulus (196 GPa). The calculated heat of formation (39.0 kcal/mole) for cubic $ZrAl_3$ is consistent with the experimental value (34.0 kcal/mole) for the similar compound $TiAl_3$.

Encouraged by these results, we set out to determine how to stabilize the cubic ($L1_2$) phase of $ZrAl_3$. Following the above described procedure we first determined the density of states (DOS) at the Fermi energy, $N(E_F)$. We found that $N(E_F)$ was high because E_F occurs in the antibonding peak in the DOS. This suggested that ternary additions, like Li, would lower E_F into the minimum in the DOS and hence stabilize the $L1_2$ phase. This prediction was studied with detailed calculations designed to study site preference for Li substitution. In these calculations, a supercell approach is employed for the $L1_2$-like structure of (i) $ZrLiAl_6$ (i.e., Zr substitution) and (ii) Zr_2LiAl_5 (Al substitution). The results of our total energy studies show clearly a preference for Al substitution and a lower $N(E_F)$ which coincides with the prediction of $L1_2$ stabilization. Thus, these supercell calculations support the simple rigid band concept and thus may be more generally applied to the aluminides.

3.3. Structural phase stability of Titanium Aluminides .

Most recently, we have initiated an in-depth study of the structural phase stability and electronic properties of several TiAl intermetallics (Ti_3Al, TiAl, and $TiAl_3$). To begin with, we studied the properties of pure Ti metal and have demonstrated that the total energy of the hcp phase is lower than that of the bcc and fcc structures. An examination of the density of states at the Fermi energy for the three structures shows, once again, that the $N(E_F)$ is lowest for the hcp structure.

For the three TiAl compounds studied using our all-electron total energy approach, we find calculated equilibrium lattice constants in quite good agreement with experiment; the heat of formation and the bulk modulus also agree fairly well with the experimental results. In all cases for the stable phase of the compounds, the Fermi energy is on the bonding side of the DOS curve and a lower DOS value appears again at the Fermi energy - in agreement with our earlier studies cited above. This means that we may now have a good method for predicting which ternary additions would stabilize the $L1_2$ phase in $TiAl_3$ and in Ti_3Al. Some of these insights and predictions are now being fed to our experimental collaborators.

3.4. Combined statistical mechanical and electronic structure approach to first principles determination of alloy phase diagrams.

Stimulated by the complexity of alloy phase diagrams of the intermetallics, notably Ti-Al, we have undertaken a first step towards the calculation of the phase diagrams of binary alloys entirely from first principles [15]. This has been a long-sought and important goal (which started with the work of Kikuchi [16]). In our approach, we start from a local density total energy supercell formulation. After constructing the grand partition function which permits the determination of the entropy from our calculated total electronic energies, we are, in principle, able to obtain all thermodynamic quantities. While this approach faces the same problem as obtaining the entropy in cluster variational methods [17], since one is presently restricted to relatively small unit cells, there are a number of important advantages: (1) solid solution and ordered phases can be calculated with the same numerical method and precision; (2) local environment effects such as charge transfer and chemical bonding are accurately described; (3) any crystal structure for which the total energy has been calculated can be easily included in the grand partition function; (4) the model is self-contained and only needs the total energy; and (5) no fitting and no breaking into pairwise interactions is necessary.

The result is a combined statistical mechanical and electronic structure approach which shows some promise. It should be emphasized that the grand partition function is constructed from volume dependent internal energies obtained from local density total energy supercell calculations. The illustrative results of first calculations for the Al-Li system show: (i) structural properties versus concentration in very good agreement with experiment and (ii) features on the Al rich side of the phase diagram of the fcc solid solution which are important for alloy formation. Since we are limited to small supercells at present, our method should only be applied to regions in experimentally known phase diagrams where complicated structures do not occur and where concentration waves with long wave lengths are not important. Within these restrictions it appears that our combined statistical, mechanical and electronic structure method shows promise for studying alloy phase diagrams entirely from first principles.

3.5. Interfacial properties of intermetallics: (001) APB Energy of Ni_3Al.

We have started some first principles studies of the mechanical properties of the intermetallic compounds [18,19]. Because of their importance in understanding the high temperature of Ni_3Al, we have calculated from first principles the energy of formation of an antiphase boundary (APB). Two different approaches were used: (i) a thin film method [18] and (ii) a supercell method [19]. In the supercell approach, this required 16 atoms per supercell and in the thin film approach we set up a comparison of the total energy of the ordered single phase Ni_3Al with that for a 13 layer film representing the APB interface. Surprisingly, both methods gave the same result, namely a $E = 140 \pm$ erg/cm^2 - in very good agreement with the experimental result of Veyssiere et al. We are presently also analyzing these results in terms of our ability to compare charge densities so as to better understand the interface. The next step, i.e., undertaking a similar calculation of the (111) APB in Ni_3Al, is now in progress.

ACKNOWLEDGEMENTS

This work was supported by the Air Force Office of Scientific Research (grant no. 85-0358). I am grateful to my collaborators for carrying out the work cited in this lecture.

REFERENCES

1) Kohn, W. and L.J. Sham, Phys. Rev., 1965, 140 , A1133
2) Westbrook, J.H., ed. Intermetallic Compounds (Wiley, New York, 1967)
3) Aoki, K. and O. Izumi, Nippon Kinzok Takkaishi, 1979, 43 , 1190; Liu, C.T. and C.C. Koch, in Proceedings of a Public Workshop on Trends in Critical Materials Requirements for Steels of the Future: Conservation and Substitution Technology for Chromium (NBSIR-83-2679-2, Washington, DC, June 1983)
4) Bernhoeft, N.R., I. Cole, G.G. Lonzarich, and G.L. Squires, J. Appl. Phys., 1982, 53 , 8207; Sigfusson, T.I., N.R. Bernhoeft, and G.G. Lonzarich, J. Appl. Phys., 1982, 53 , 8207; Sigfusson, T.I., N.R. Bernhoeft, and G.G. Lonzarich, J. Phys., 1984, F14 , 2141 and references therein.
5) Hackenbracht, D. and J. Kubler, J. Phys., 1980, F10 , 427
6) Buiting, J.J.M., J. Kubler, and F.M. Mueller, J. Phys., 1983, F13 , L179
7) Min, B.I., T. Oguchi, H.J.F. Jansen, and A.J. Freeman, J. Mag. Magn. Matls., 1986, 54-57 , 1091
8) Xu, J.-H., T. Oguchi, and A.J. Freeman, Phys. Rev. (in press) about the substitution of V in Ni_3Al and the structural stability of $Ni_3(Al,V)$.
9) Andersen, O.K., Phys. Rev., 1975, B12 , 3060
10) Xu, J-H., A.J. Freeman and T. Oguchi, (to be published)
11) Westbrook, J.H., Ordered Alloys, Physical Metallurgy and Structural Applications , Claitors, Baton Rouge (1970), p. 1; Dimiduk, D.M., Solid Solution Strengthening of Ordered Ni_3Al, (unpublished).
12) Xu, J.-H., T. Oguchi, and A.J. Freeman, Phys. Rev., 1987, B35 , 6940
13) Decker, R.F. and J.R. Mihalisin, Trans. ASM, 1969, 62 , 481
14) Hong, T., T.J. Watson-Yang, A.J. Freeman and T. Oguchi, Bull. Am. Phys. Soc., 1987, 32 , No. 3, p. 413
15) Podloucky, R., H.J.F. Jansen, X.Q. Guo and A.J. Freeman, Phys. Rev. B., (to appear)
16) Kikuchi, R., Phys. Rev., 1951, 81 , 988
17) See, eg., D. de Fontaine in "Alloy Phase Diagrams", eds. L.H. Bennett, T.B. Massalski, and B.C. Giessen (North-Holland, 1983),

p. 149
18) Freeman, A.J., C.L. Fu, and J.I. Lee, Bull. Am. Phys. Soc., 1987, 32 , No. 3, p. 772.
19) Xu, J.-H., A.J. Freeman, and T. Oguchi, (to be published)
20) Veyssiere, P., Phil. Mag. 50 , 189 (1984).

ELECTRONIC STRUCTURE AND MAGNETIC PROPERTIES OF IMPURITIES IN METALS

P.H. Dederichs, H. Akai+, S. Blügel, N. Stefanou and R. Zeller

Institut für Festkörperforschung, Kernforschungsanlage Jülich, D-5170 Jülich, Fed. Rep. Germany

+ permanent address: Nara Medical University, Kashihara Nara 634, Japan.

1. INTRODUCTION

In this lecture we will consider the electronic structure of single impurities in metals. The calculations are relevant for dilute alloys, where interaction effects between the impurities are negligible. Then all physical properties of the alloy scale with the impurity concentration, so that it is in principle sufficient to study the electronic structure of a single impurity in an otherwise ideal host.

A large amount of experimental studies have been devoted to dilute alloys, probably more than to concentrated systems. This is partly due to the fact that dilute alloys are convenient model systems representing the simplest kind of disorder. Partly it is due to the fact that many more dilute alloys can be formed than disordered concentrated ones. There always exists a small, but finite solubility for most impurities and even if the solubility is too small one may introduce the impurities by ion bombardment. For instance practically all elements of the periodic table have been studied as impurities in iron using NMR or Mößbauer methods [1]. Thus the experimental information about dilute alloys is especially large.

In the last 10-15 years there has been an enormous progress in electronic structure calculations, which was initiated by the development of density functional theory. For ideal crystals and ordered compounds the introduction of linearized band structure methods was another important step since they efficiently reduce the computing time. For impurities and in general for disordered alloys Green's function methods play a key role. In an ideal manner they take into account that the host crystal is periodic and the impurity represents a localized perturbation in an otherwise ideal crystal. In the simplest case it is sufficient to change only the potential at the impurity site and to leave the neighboring host potentials unperturbed. This so called single site approximation is also the basis of the Kohn-Korringa-Rostoker coherent potential approximation (KKR-CPA) for concentrated alloys. However for a single impurity it is also possible to explicitly calculate the perturbations on the neighboring site, e.g. the Friedel oscillations of the charge density around the impurity. In this way one can also check the reliability of KKR-CPA calculations.

G. M. Stocks and A. Gonis (eds.), Alloy Phase Stability, 377–420.

In the lecture we will mostly be concerned with transition metal impurities in noble metals (Cu, Ag) and in transition metals (Fe, Ni and Pd). Of special interest will be the magnetic behavior of 3d-impurities, i.e., the local moments and the host polarization.

The organization of the paper is as follows. In sect. 2 we outline theoretical models used in impurity calculations and especially introduce the KKR-Green's function method which is based on a muffin-tin approximation of the atomic potentials. In sect. 3 we discuss the Anderson model for magnetic impurities and compare it with realistic calculations for impurities in Cu, Ag, Al and Pd. Sect. 4 deals with impurities in the elemental ferromagnets as well as sp-impurities and discusses the role of two different screening mechanisms in these alloys. A large amount of experimental work is connected with the measurements of hyperfine fields and isomer shifts. We will report about calculations of such properties in sect. 5. Especially we will discuss the relation between these nuclear signals and the valence properties and show what one can learn about the electronic structure from hyperfine field and isomer shift measurements.

2. THEORETICAL METHODS

2.1. Models for impurity calculations in metals

A large variety of methods has been used to study impurities in metals. In the following we will shortly mention the most important ones and discuss their advantages and disadvantages. Then we treat in detail Green's function methods, especially the KKR version.

For simple metals most calculations are based on the jellium model. For instance to simulate a vacancy one cuts a spherical hole in the otherwise uniform background of positive charge and then calculates the rearrangement of the electronic charge using density functional theory. Such calculations are relevant only for simple metals since the effect of the lattice potential is totally neglected.

Within the (2nd order) pseudopotential theory one considers the impurity as a weak perturbation. The energy is normally calculated up to second order in the pseudopotential. Homovalent impurities in simple metals are an example where this approach works well. Heterovalent impurities or vacancies are stronger perturbations and the validity of pseudopotential theory is questionable. A better approach would then be to treat the impurity in a jellium calculation and introduce the host pseudopotentials as a perturbation or to use the spherical solid model [2] where the crystal potential is spherically averaged around the impurity site. Clearly all these models are restricted to simple metals.

In cluster calculations one considers only a finite number of atoms, e.g. the impurity and one or two shells of neighboring atoms. Mostly one considers an arrangement of atoms surrounded by a repulsive spherical well potential so that the energy eigenvalues are discrete. The most serious problem in these calculations is the finite size of the cluster. Since in small clusters practically all atoms are at the surface it is very difficult to distinguish between impurity induced and surface induced features. In general clusters with one or two shells of host atoms are not large enough to give reliable results for the impurity. An interesting variant of cluster calculations is obtained if the outer potential is lowered to the muffin-tin zero thus immersing the cluster into a free electron sea. In this way one can take profit of some of the advantages of the jellium model. Especially Ries and Winter [3] and Winter et al. [4] have been able to perform selfconsistent calculations for very large cluster containing up to 100 atoms.

In supercell calculations a certain defect structure is repeated periodically so that a periodic crystal with a rather large unit cell is obtained, which can be handled by standard band structure codes. In this way Gupta and Siegel [5] and Chakraborty and Siegel [6] have calculated vacancies and divacancies in Al using a unit cell of 27 atoms. Still in these calculation spurious interaction effects between adjacent vacancies cannot be excluded, which might influence the value of the calculated vacancy formation energy.

These problems are avoided by the Green's function method where one really treats a single defect in an infinite and otherwise periodic crystal. The Green's function $G(\underline{r},\underline{r}',E)$ of the system is defined as the solution of the Kohn-Sham equation with a source at position $\underline{r}'$:

$$(-\partial_{\underline{r}}^2 + V_{eff}(\underline{r}) - E)\, G(\underline{r},\underline{r}',E) = -\delta(\underline{r}-\underline{r}') \qquad (1)$$

where the effective potential consists of the nuclear part, the Hartree potential and the exchange correlation potential V_{XC}.

$$V_{eff}(\underline{r}) = \sum_n \frac{-Z^n}{|\underline{r}-\underline{R}^n|} + \int d\underline{r}'' \frac{n(\underline{r}'')}{|\underline{r}-\underline{r}''|} + V_{XC}(\underline{r}) \qquad (2)$$

Here Z^n is the nuclear charge and $n(\underline{r})$ the electronic charge-density. For impurity calculations it is very convenient and essential to split the potential $V_{eff}(\underline{r})$ up into the ideal crystal potential $\overset{o}{V}(\underline{r})$ and a perturbation $\Delta V(\underline{r})$ which is spatially restricted to the vicinity of the impurity. Then the introduction of the ideal crystal Green's function $\overset{o}{G}(\underline{r},\underline{r}';E)$ allows one to transform the differential equation into an integral one, the Dyson equation:

$$G(\underline{r},\underline{r}';E) = \overset{o}{G}(\underline{r},\underline{r}';E) + \int d\underline{r}''\overset{o}{G}(\underline{r},\underline{r}'';E)\, \Delta V(\underline{r}'')\, G(\underline{r}'',\underline{r}';E) \qquad (3)$$

For the solution of this equation the most essential assumption is that the perturbation $\Delta V(\underline{r})$ is well localized, so that e.g. only the perturbation in the impurity cell and in the neighboring cells has to be taken into account. This is indeed a very good approximation and we will discuss this point later on. The main advantage compared to cluster and supercell calculations is that due to the introduction of the host Green's function the embedding of the cluster of perturbed potentials into the pure host crystal is described correctly.

Once the Dyson equation is solved, the charge density $n(\underline{r})$ can be easily obtained from the Green's function by integrating over all occupied states. This can be seen by introducing the spectral representation of $G(\underline{r},\underline{r}';E)$.

$$G(\underline{r},\underline{r}';E) = \sum_{\alpha} \frac{\psi_\alpha(\underline{r})\ \psi_\alpha^*(\underline{r}')}{E+i\epsilon - E_\alpha} \qquad \epsilon \to +0 \tag{4}$$

Here ψ_α are the eigenfunctions and E_α the eigenvalues of the Kohn-Sham equation. The charge density is then given by

$$n(\underline{r}) = 2\sum_{\substack{\alpha \\ E_\alpha < E_F}} |\psi_\alpha(\underline{r})|^2 = -\frac{2}{\pi}\int^{E_F} dE\ \mathrm{Im}\ G(\underline{r},\underline{r};E) \tag{5}$$

Similarly we can obtain the local density of states $n_{loc}(E)$ in a given cell V.

$$n_{loc}(E) = -\frac{2}{\pi}\int_V d\underline{r}\ \mathrm{Im}\ G(\underline{r},\underline{r};E) \tag{6}$$

The central problem of the Green's function method is the solution of the Dyson equation (3). Different versions exist depending on the basis set used for the expansion of the wave-functions. The Green's function method is attributed to Koster and Slater [7] who used Wannier functions of the host for the expansion. However the construction of the Wannier functions proved to be very tedious and difficult hence severely limiting the success of the method. A hybridized tight-binding nearly-free electron expansion has been proposed by Riedinger [8] and applied by him and others [9] in impurity calculations. In the last years extensive calculation for defects in semiconductors have been performed by several groups [10] using a LCAO-basis. Another alternative are Gaussian orbitals [11].

A Green's function method based on the KKR-band structure method has been proposed by Dupree [12], Beeby [13] and Morgan [14]. The theory of this socalled KKR-Green's function has been worked out in more details by Holzwarth [15], Lehmann [16], Harris [17] amd Hamasaki [18]. This formalism is based on a muffin-tin approximation and therefore well suited for defects in close-packed structures, e.g. in simple or transition metals. Extensive calculations with this method have been performed by Terakura et al [19], Katayama-Yoshida et al [20], Akai et al [21] and by our group [22-23]. Some of the results obtained with this method will be reviewed in the following. The generalization to concentrated alloys is the KKR-CPA as introduced by Shiba [24] and others [25]. A variant based on the LMTO-band structure method was developed by two groups in Straßburg [26] and Stuttgart [27].

All these expansions are aimed at transforming the integral equation (3) into an algebraic Dyson equation. Since in actual calculations the solution of this equation takes most of the computing time, an optimal and atom specific basis set is very important since it minimizes the rank of the matrices and consequently the computing time needed.

2.2. The KKR-Green's function method

For an infinite array of muffin-tin potentials we expand the wavefunction and the Green's function in each cell into eigenfunctions of the local muffin-tin potential V(r). For each energy E and each angular momentum ℓ there exist two linear independent eigenfunctions, a regular solution $R_\ell(r,E)$ varying as r^ℓ at the origin and a non-regular solution $H_\ell(r,E)$ diverging as $1/r^{\ell+1}$. Outside the muffin-tin sphere, $r \geq R$, the regular wavefunction is given by

$$R_\ell(r,E) = j_\ell(\sqrt{E}r) - \sqrt{E}\; t_\ell(E)\; h_\ell(\sqrt{E}r) \qquad r \geq R \tag{7}$$

with

$$t_\ell(E) = -\frac{1}{\sqrt{E}}\, e^{i\delta_\ell(E)} \sin\delta_\ell(E) = \int_0^\infty r'^2 dr'\; j_\ell(\sqrt{E}r')\; V(r') R_\ell(r',E)$$

Here $j_\ell(x)$ and $h_\ell(x)$ are the spherical Bessel and Hankel functions, the analogones to R_ℓ and H_ℓ in the potential free case. Thus $R_\ell(r,E)$ is the solution of the Lippman-Schwinger equation for a single potential V(r) and an incident spherical wave $j_\ell(\sqrt{E}r)$. The non-regular solution $H_\ell(r,E)$ coincides for $r \geq R$ with the Hankel-function h_ℓ.

$$H_\ell(r,E) = h_\ell(\sqrt{E}r) \qquad r \geq R \tag{8}$$

The Green's function $G_s(\underline{r},\underline{r}';E)$ for the single potential V(r) in free space can be constructed analogously to the spherical wave expansion of the free space Green's function $g(\underline{r},\underline{r}';E)$.

$$g(\underline{r},\underline{r}';E)=-\frac{1}{4\pi}\frac{e^{i\sqrt{E}|\underline{r}-\underline{r}'|}}{|\underline{r}-\underline{r}'|}=-\sqrt{E}\sum_L j_\ell(\sqrt{E}r_<)h_\ell(\sqrt{E}r_>)Y_L(\underline{r})Y_L(\underline{r}') \tag{9}$$

Here $r_<$ and $r_>$ denote the smaller and the larger values of the pair r and r'. $L=(\ell,m)$ is an abbreviation for the two angular momentum indices and $Y_L(\underline{r})$ are spherical harmonics. It is straightforward but lengthy to show that $G_s(\underline{r},\underline{r}';E)$ is given by

$$G_s(\underline{r},\underline{r}';E) = -\sqrt{E}\sum_L R_\ell(r_<,E)\ H_\ell(r_>,E)\ Y_L(\underline{r})\ Y_L(\underline{r}') \tag{10}$$

For the general case of many non-overlapping muffin-tin potentials V_n centered at the position $\underline{R}^n$ we introduce cell-centered coordinates ($\underline{r}\to\underline{r}+\underline{R}^n$). In each cell we then make a double expansion of the Green's function $G(\underline{r}+\underline{R}^n,\ \underline{r}'+\underline{R}^{n'};E)$ into radial eigenfunctions of the local potentials $V_n(r)$ and $V_{n'}(r)$. For n=n' we split-off the solution $G_s^{(n)}$ for the single muffin-tin potential V_n in free space.

$$G(\underline{r}+\underline{R}^n,\ \underline{r}'+\underline{R}^{n'};E) = \delta_{n,n'}\ G_s^{(n)}(\underline{r},\underline{r}';E) + \\ + \sum_{L,L'} R_\ell^n(r,E)\ Y_L(\underline{r})\ G_{LL'}^{nn'}(E)\ R_{L'}^{n'}(r',E)\ Y_{L'}(\underline{r}') \tag{11}$$

The single potential Green's function $G_s^{(n)}$ takes care of the source term $\delta(\underline{r}-\underline{r}')$ in eq. (1). Therefore the remaining term satisfies the source-free Schroedinger equation so that in the double expansion only the regular solutions are needed. Clearly eq. (11) satisfies the Schroedinger equation (1) in each cell separately. The socalled "structural Green's function coefficients" $G_{LL'}^{nn'}(E)$ have to be chosen such, that the solutions for the different cells match properly at the cell boundaries.

The above representation is valid for an arbitrary ensemble of muffin-tin potentials, especially both for the ideal as well as the defect crystal. By introducing both representations into the Dyson equation (3) one obtains after some lengthy algebra (see e.g. the appendix of ref. [22]) an algebraic Dyson equation relating the structural coefficients $G_{LL'}^{nn'}(E)$ for the crystal with defect to the coefficients

$\overset{o}{G}{}^{nn'}_{LL'}(E)$ of the ideal crystal.

$$G^{nn'}_{LL'}(E) = \overset{o}{G}{}^{nn'}_{LL'}(E) + \sum_{n''L''} \overset{o}{G}{}^{nn''}_{LL''}(E)\, \Delta t^{n''}_{\ell''}(E)\, G^{n''n'}_{L''L'}(E) \tag{12}$$

with $\quad \Delta t^{n}_{\ell}(E) = t^{n}_{\ell}(E) - \overset{o}{t}_{\ell}(E)$

Here the difference Δt^{n}_{ℓ} between the t-matrices of the perturbed and ideal crystal enters as perturbation. The size of this system of algebraic equations is determined by the number of non-zero elements Δt^{n}_{ℓ}. For instance, if for an impurity in an fcc crystal we take potential perturbations on the impurity site and on the nearest neighbor sites into account and if moreover we consider s-, p- and d-electrons, we have to invert matrices of size 117×117.

The major problem of the KKR-Green's function method is the determination of the structural coefficients $\overset{o}{G}{}^{nn'}_{LL'}$ of the ideal crystal, which are directly related to the band structure. Inserting the KKR-ansatz for the Bloch wavefunction

$$\psi_{\underline{k}\nu}(\underline{r}+\underline{R}^{n}) = e^{i\underline{k}\underline{R}^{n}} \sum_{L} i^{\ell}\, \Phi_{L}(\underline{k}\nu)\, \overset{o}{R}_{\ell}(r,E)\, Y_{L}(\underline{r}) \tag{13}$$

into the spectral representation (4) of the Green's function the imaginary part of $\overset{o}{G}{}^{(n-n')}_{L\,L'}$ can be directly obtained by a Brillouin zone integration.

$$\mathrm{Im}\left\{e^{i(\overset{o}{\delta}_{\ell}+\overset{o}{\delta}_{\ell'})}\, \overset{o}{G}{}^{(n-n')}_{L\,L'}(E)\right\} = \sqrt{E}\,\theta(E)\,\delta_{nn'}\delta_{LL'}$$
$$-\pi\sum_{\nu}\int_{BZ} d\underline{k}\;\delta(E-E_{\underline{k}\nu})\,\Phi_{L}(\underline{k}\nu)\,\Phi^{*}_{L'}(\underline{k}\nu)\, i^{\ell-\ell'}\, e^{i\underline{k}(\underline{R}^{n}-\underline{R}^{n'})} \tag{14}$$

Here we have set $R_{\ell}(r,E) = e^{i\delta_{\ell}(E)}\,\tilde{R}_{\ell}(r,E)$ where $\tilde{R}_{\ell}(r,E)$ is real. The real part is obtained from the imaginary one by a Kramers-Kronig integration.

$$\mathrm{Re}\left\{\overset{o}{R}_{\ell}(r,E)\, \overset{o}{G}{}^{(n-n')}_{L\,L'}(E)\, \overset{o}{R}_{\ell'}(r',E)\right\} =$$

$$= \frac{1}{\pi}\int_{-\infty}^{+\infty} dE' \; P\left(\frac{1}{E-E'}\right) \; \mathrm{Im}\left\{ \overset{o}{R}_{\ell}(r,E') \; \overset{o}{G}{}^{(n-n')}_{L\,L'}(E') \; \overset{o}{R}_{\ell'}(r',E') \right\} \qquad (15)$$

This equation holds for arbitrary r and r'. In actual calculations the E'-integration in (15) requires the knowledge of the band structure up to infinite energies. Since this is impossible one has to approximate the high energy behavior. For diagonal coefficients n = n' we introduce a cut-off energy E_c. This means that we replace the Green's function for $E \geq E_c$ by the single site Green's function $G_s^{(n)}$ in eq. (11). For non-diagonal elements $n \neq n'$ we replace the Green's function for $E \geq E_c$ by the Green's function for two muffin-tins at $\underline{R}^n$ and $\underline{R}^{n'}$. For details about this procedure and for an alternative construction of the Green's function we refer to [22,23].

In metals the impurity nuclear charge is totally screened by the valence electrons. Therefore the Friedel sum rule $\Delta Z = \Delta N(E_F)$ must be satisfied, where ΔZ is the change of the nuclear charge and $\Delta N(E_F)$ is the change of the integrated density of states.

$$\Delta N(E_F) = \frac{-2}{\pi} \int_{-\infty}^{+\infty} d\underline{r} \; \mathrm{Im}\left\{ G(\underline{r},\underline{r};E) - \overset{o}{G}(\underline{r},\underline{r};E) \right\} \qquad (16)$$

The integral can be calculated analytically and leads to the following result [16]

$$\Delta N(E) = \frac{2}{\pi} \sum_{n,L} \left(\delta_\ell^n(E) - \overset{o}{\delta}_\ell(E)\right) - \frac{2}{\pi} \,\mathrm{Im}\, \mathrm{Tr}_{nL} \ln\left(\underline{\underline{1}} - \underline{\underline{\overset{o}{G}}}(E)\Delta \underline{\underline{t}}(E)\right) \quad (17)$$

Due to approximations involved, the Friedel sum rule cannot be satisfied exactly in actual calculations. The deviation from neutrality is then an indication of the quality of the calculation.

In the case of spin polarization we have different potentials $V_+(\underline{r})$ and $V_-(\underline{r})$ for each spin direction. Consequently two different Dyson equations for the Green's functions $G_\pm(\underline{r},\underline{r}';E)$ have to be solved. The spin densities $n_\pm(\underline{r})$ are given by

$$n_\pm(\underline{r}) = -\frac{1}{\pi} \int^{E_F} dE \; \mathrm{Im}\, G_\pm(\underline{r},\underline{r};E) \qquad (18)$$

The charge and magnetization densities follow as the sum and the difference of these quantities.

$$n(\underline{r}) = n_+(\underline{r}) + n_-(\underline{r}) \quad , \quad m(\underline{r}) = n_+(\underline{r}) - n_-(\underline{r}) \qquad (19)$$

2.3. Complex energies etc.

The large matrices occuring in the solution of the Dyson eq. (12) ask for special methods and tricks. Note that due to strong structure in the density of states of transition metals reliable charge densities (5) can only be obtained if Dyson's equation and the radial Schroedinger equation are solved for at least 1000 energy points. Moreover in order to achieve selfconsistency for the potential, typically several hundreds of iterations are needed.

Group theory allows one to exploit the symmetry of the defect and leads to a block diagonalization of the matrices. For instance, for cubic symmetry the size of the submatrices is typically a factor 10 smaller than the original matrix. Since the computing time varies with the third power of the size of the matrices, about a factor of 100 can be gained by operating with the ten smaller submatrices instead of the original matrix. The number of iterations can also be reduced by using special iteration techniques [28,29].

Of special importance is the use of complex energies [30,31] in order to perform the energy integration in eq. (5). For complex energies the Green's function G(z) is analytical in the whole complex plane with the exception of the real axis where it may have poles or branch cuts. This can be seen from the spectral representation

$$G(z) = \sum_{\alpha} \frac{\psi_\alpha(\underline{r})\psi^*_\alpha(r')}{z-E_\alpha} \qquad (20)$$

Note that on the real axis Re G(z) is continuous, whereas ImG(z) changes sign (Im G(E+iε) = -Im G(E-iε)). Therefore since Im G(E+iε) is negative on the upper side of the branch cut along the real energy axis (see Fig.1), the charge density $n(\underline{r})$ (5) can be written as a line integral up to E_F along the upper side

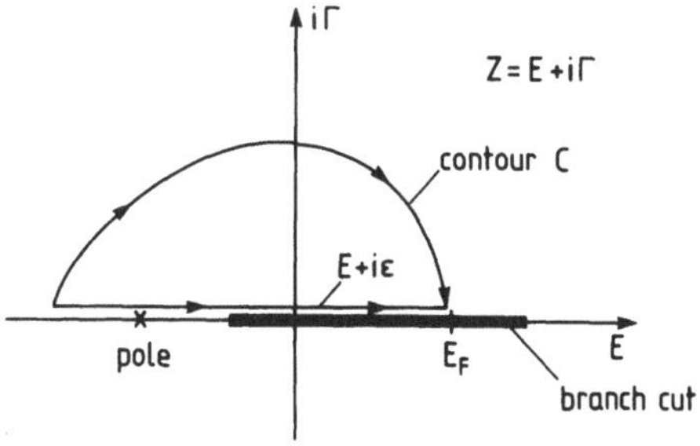

Fig.1 Real axis integration along the path E+iε and a complex integration contour C, both ending at E_F+iε.

$$n(\underline{r}) = -\frac{2}{\pi} \operatorname{Im} \int^{E_F} dE\, G(\underline{r},\underline{r};E+i\epsilon) \qquad (21)$$

Now the integral along the branch cut can be arbitrarily deformed to a contour C in the complex plane, since G(z) is analytical in this region. The only requirement is that the initial and final points are the same, i.e. that the path ends at $E_F+i\epsilon$.

$$n(\underline{r}) = -\frac{2}{\pi} \operatorname{Im} \int dz\, G(\underline{r},\underline{r};z) \tag{22}$$

The advantages gained by this contour arise from the analyticity of G(z). While on the real axis G(z) is non-analytical and strongly structured, it is quite smooth for complex energies so that the contour integral can be performed with few complex energy points. Typically 20-40 complex energies yield an accuracy which is superior to more than 1000 energy points on the real axis. Along the path C the energy point density should be higher close to the Fermi energy since there the Green's function is again strongly structured.

3. MAGNETIC IMPURITIES

3.1. Virtual bound states of non-magnetic impurities

The d-level of transition metal atoms develops into a resonance when inserted into a simple metal. For the local density of states (LDOS) one expects a Lorentzian:

$$n_{loc}(E) = \frac{5}{\pi} \frac{\Gamma}{(E-E_d)^2+\Gamma^2} \tag{23}$$

centered at E_d. The half width Γ of this "virtual bound state" can be calculated by the golden rule.

$$\Gamma = \frac{2\pi}{\hbar^2} |V_{sd}|^2 n_{host}(E_d) \tag{24}$$

Here $n_{host}(E)$ is the host density of states (DOS) and V_{sd} the sd-matrixelement containing the information about the overlap between the localized d-wavefunctions of the host.

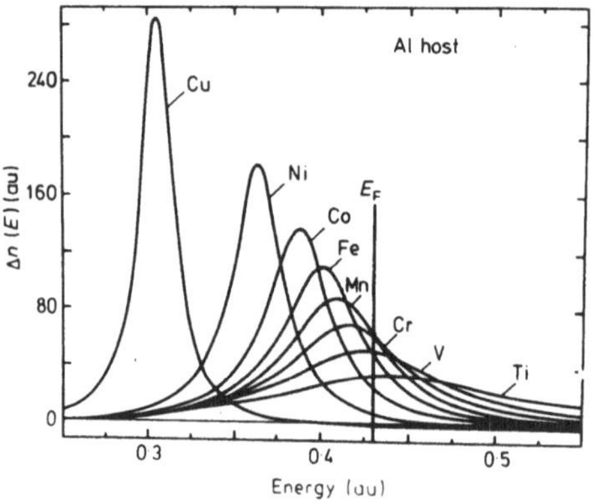

Fig.2a LDOS of 3d-impurities in Al as calculated by Nieminen and Puska [32] in the jellium model.

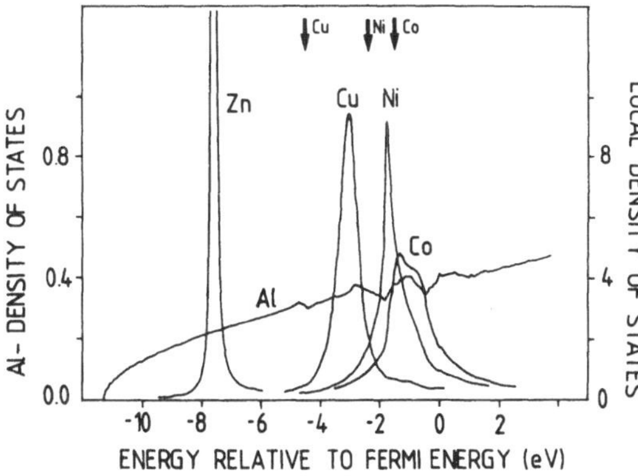

Fig.2b LDOS for some 3d-impurities in Al according to Green's function calculations [33]. The arrows indicate the positions of the virtual bound states as obtained by XPS [34]. For comparison the DOS of pure Al is also shown.

As an illustration Fig. 2a shows the LDOS of 3d-impurities in Al according to a jellium calculation of Nieminen and Puska [32]. It is seen that the resonance becomes narrower when the 3d-level decreases, which is expected since both the host DOS and the sd overlap decrease. For comparison Fig. 2b shows the results of a Green's function calculation for some 3d-impurities in Al [33]. Whereas the 3d-levels and the halfwidths are similar in both calculations, the LDOS of Ni and Co impurities show appreciable deviations from a Lorentzian behavior. These are due to van Hove singularities in the host DOS and can be understood as follows. The wave scattered at the impurity is intermediately backscattered ("Bragg reflected") from the host lattice planes. Depending on the coherence or incoherence of the backscattered wave with the incident one, sharp spikes or depressions can occur in the LDOS at the van Hove singularities. This shows that even for the simple metal Al band structure effects are important. The arrows indicate the positions of the virtual bound states as obtained by XPS-measurements of Steiner et al [34]. The calculated levels are consistently too high, a typical error of density functional theory which is e.g. also observed for the energy bands of the noble metals.

Fig. 3 shows the LDOS of Ni and Co impurities in Cu [35]. One finds a virtual bound state between the Cu-d-band edge and the Fermi energy. However in addition, appreciable intensity is also found within the range of the d-band. This is even more pronounced if we consider 4d-impurities in Cu. Fig. 4 shows the LDOS of a Mo impurity in Cu together with the result of a jellium calculation. Instead of a single peak in the jellium model the intensity is splitted into an upper virtual bound state like peak and into intensity within the d-band of Cu. For 4d-impurities the overlap between the impurity 4d-wave function and the host 3d-wave functions is very important so that by hybridization pronounced bonding and antibonding states are formed, with the bonding states energetically localized in the d-band range and the antibonding states forming the "virtual bound state". Thus we see that the concept of the virtual bound state in a jellium begins to break down due to hybridization effects.

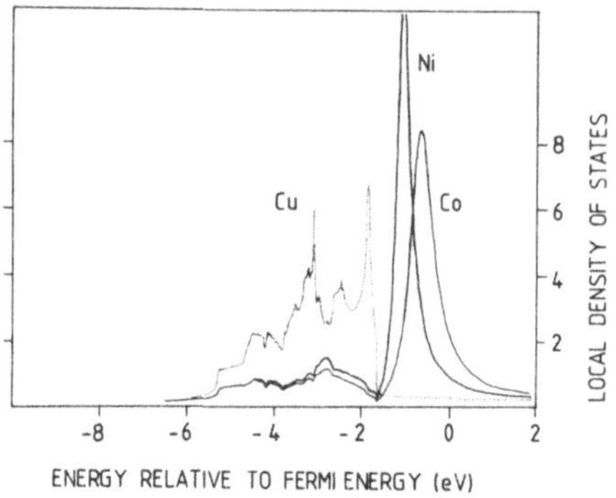

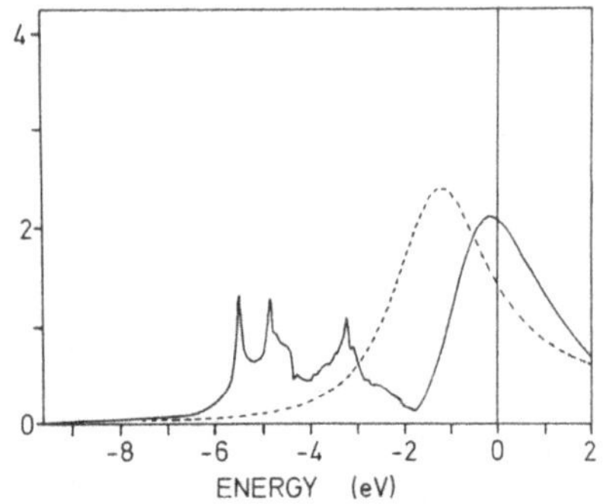

Fig.3 LDOS in the impurity Wigner Seitz sphere for Ni and Co impurities in Cu. The dotted line represents the DOS of pure Cu.

Fig.4 LDOS of Mo impurities in Cu. The dotted line refers to a jellium calculation.

3.2. The Anderson model

The 3d-electrons can gain exchange energy by spin alignment. This leads to a local moment of the impurity, if the gain of exchange energy overcomes the loss of kinetic energy. As was pointed out by Anderson [36] the s-d interaction and the host DOS are decisive for the formation of a local moment. Here we will give a somewhat simpler derivation which is however equivalent to the Anderson model in the Hartree-Fock approximation [36].

If the impurity becomes magnetic, the majority electrons gain energy, so that the resonance level E_d^+ decreases to lower energies

$$E_d^+ = E_d^o - IM/2 + U\Delta N \qquad (25a)$$

Here M is the local moment and I the exchange integral. U is the Coulomb integral and $\Delta N = N-N_o$ the deviation of the impurity charge from the paramagnetic N_o. Contrary the resonance level of the minority electrons is shifted to higher energies

$$E_d^- = E_d^o + IM/2 + U\Delta N \qquad (25b)$$

Assuming that the form $n_{loc}(E)$ of the paramagnetic LDOS of the impurity (Fig. 5a) is unchanged by the shifts of the resonance, we obtain the spin-splitted LDOS as shown in Fig. 5b. Since all states up to E_F are occupied the impurity charge N and the local moment M are to be determined by the following equations.

$$N = \int^{E_F} dE \left\{ n_{loc} \; (E - U\Delta N + IM/2) + n_{loc} \; (E - U\Delta N - IM/2) \right\} \tag{26}$$

$$M = \int^{E_F} dE \left\{ n_{loc} \; (E - U\Delta N + IM/2) - n_{loc} \; (E - U\Delta N - IM/2) \right\} \tag{27}$$

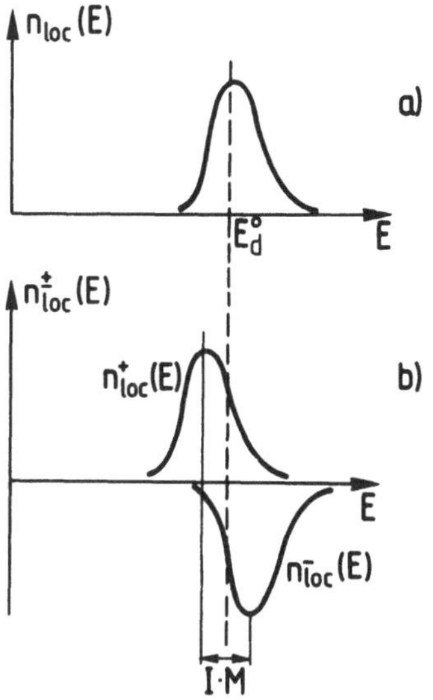

Fig.5a Schematic representation of LDOS of a non-magnetic 3d-impurity.

Fig.5b LDOS of a magnetic 3d-impurity according to the Anderson-model.

The first equation (26) allows to determine N as a function of M. Knowing N(M) the second equation (27) determines then the correct value of the local moment. It has the form M = F(M). The graphical solution of this equation is illustrated in

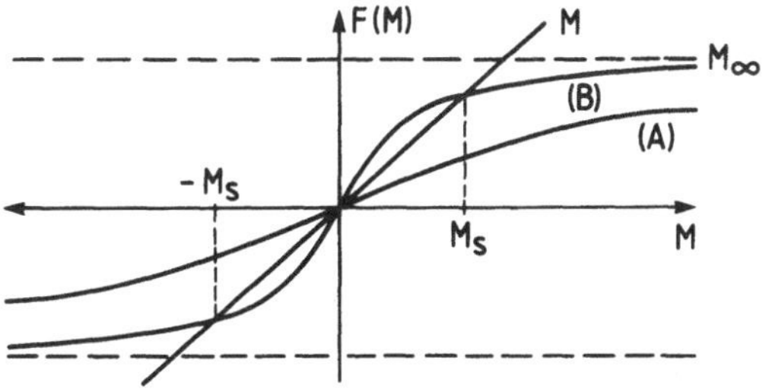

Fig.6 Graphical solution of the selfconsistency condition M = F(M) for the local moment (Anderson model). In case A only the trivial solution M = 0 exists. In case B in addition two degenerate solutions $\pm M_s$ are found.

Fig.6. F(M) is a monotonically increasing function of M, vanishes for M = 0 and approaches a saturation value M_s for large M. The intersection of F(M) with the straight line M gives the selfconsistent value M_{loc} for the local moment. For symmetry reasons also $-M_{loc}$ is an allowed solution. A sufficient condition for a non-zero moment is

$$F'(o) = I\, n_{loc}(E_F) > 1 \qquad (28)$$

This is very similar to the Stoner condition for elemental ferromagnets, with the difference that $n_{loc}(E)$ is the paramagnetic LDOS of the impurity, not the DOS of the host. Thus favorable conditions for the occurrence of a local moment are a large exchange integral I and a high LDOS at E_F. This directly explains why in metals in general only 3d-impurities are magnetic but not 4d- and 5d-ones. For these the exchange integral I is somewhat smaller and the LDOS at E_F are normally appreciably smaller due to the larger spatial extent of the 4d- and 5d-wave functions. For instance, in the jellium model the s-d matrixelement V_{sd} of eq. (24) is for 4d-impurities appreciably larger than for 3d-ones, leading e.g. to the rather broad virtual bound state of Mo in Fig. 4. Within the 3d-series the condition for magnetism is most favorable, when the peak of the LDOS is close to the Fermi energy, since the exchange integral I does not vary strongly within the 3d-series. Therefore impurities in the middle of the series, i.e. Cr, Mn and Fe, show the strongest tendency for magnetism and the largest local moments.

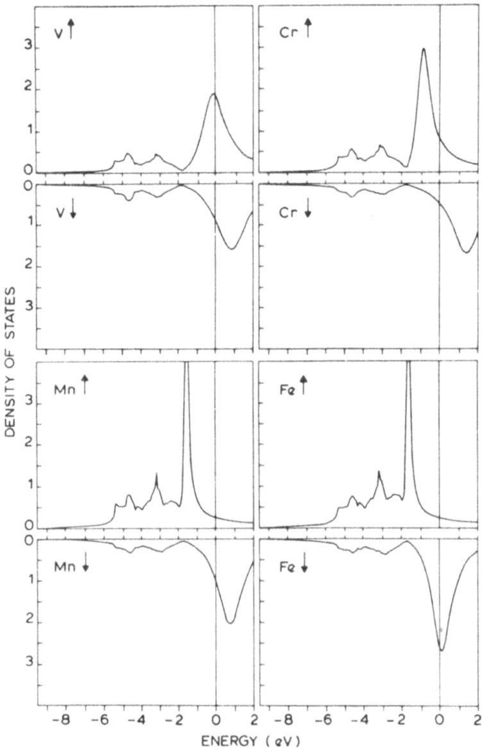

Fig.7 LDOS for both spin directions of V, Cr, Mn and Fe impurities in Cu.

3.3. Magnetic impurities in Cu and Ag

Fig. 7 shows the result of KKR-Green's function calculations [37] for the LDOS of V, Cr, Mn and Fe impurities in Cu. Contrary to the Anderson-model band structure effects are fully included and no adjustable parameters are involved. For V and Cr the host band structure plays a minor role. We obtain the spin-splitted virtual bound states as expected by the Anderson model and for both spin directions some additional intensity in the region of the d-band of Cu. For Mn and Fe impurities the majority virtual bound state is pushed to the upper edge of the d-band, becomes very narrow and a large amount of intensity is spread over the whole d-band. For the Ag-host the calculated virtual bound states are similar, except that the d-band is lower and contains locally much less intensity. Also the peaks are in general narrower.

Experimental information about the spin-splitted virtual bound states have been obtained only recently. Reehal and Andrews [37] and Andrews and Brown [38] found clear evidence for the majority virtual bound state of Fe and Cr in Au. For $Ag_{0.8}$ $Mn_{0.2}$ [37] they identified a majority peak at -2.8 eV below E_F. From XPS and BIS measurements van der Marel et al [39,40] find the Mn majority peak in Ag at -3.1 eV and the minority peak at +2.1 eV. The resulting exchange splitting of 5.2 eV is much larger than the calculated one of 3.0 eV [22] Jordan et al [41] find a slightly smaller exchange splitting (4.7 eV) than van der Marel et al., however the basic discrepancy with the calculations remains.

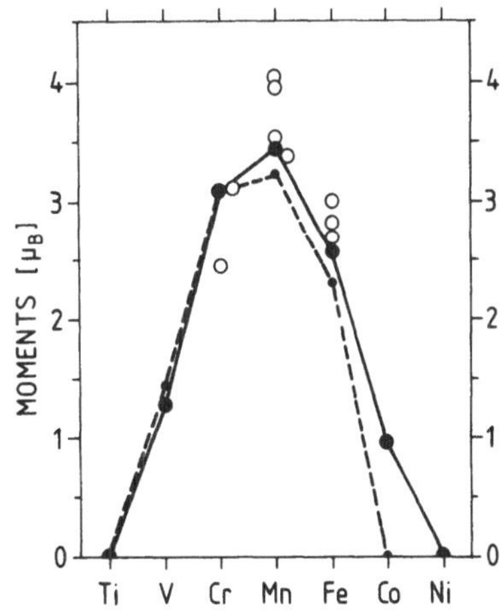

Fig.8 Local moments of 3d impurities in Cu: theoretical values (connected by full lines) single site approximation (connected by dashed lines) experimental results o

Fig. 8 summarizes the results for the calculated local moments of 3d-impurities in Cu [23]. The open circles are experimental results from neutron scattering and susceptibility measurements. In view of the scattering in the experimental data the agreement is satisfactory. The dashed line is the result of a single site calculation [22] whereas the full line refers to the results including the potential perturbation of the neighboring host atoms. In view of its simplicity the agreement of the single site calculation with the more elaborate and accurate "cluster" calculation is surprising. With the exception of Co the error for the local moments is less than 10 %. According to the single site approximation a Co impurity is just on the verge of becoming magnetic, so that the large disagreement is not too surprising.

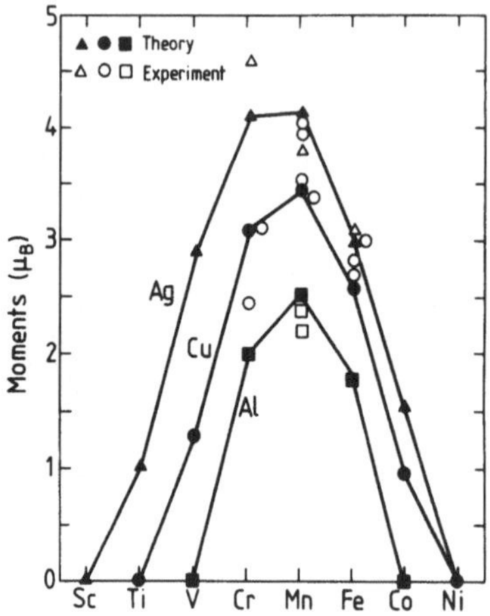

Fig.9 Theoretical results for the local moments in Ag, Cu and Al. The open symbols refer to experimental values.

Fig. 9 compares the theoretical results for the local moments of 3d-impurities in Ag, Cu and Al. The open symbols represent experimental values. It is seen that the tendency for magnetism is largest in Ag, intermediate in Cu and smallest in Al. The larger moments in Ag compared to Cu are due to the about 10 % larger lattice constant of Ag, which reduces the s-d interaction as well as the host density of states. The rather small moments in Al are a consequence of the larger conduction electron density.

3.4. Giant moments in Pd

Alloys of palladium with low concentrations of transition metals from the middle of the 3d series are typical examples of giant moment ferromagnetic systems where the moment per impurity can be as large as 10 μ_B. Since the local moment on the impurity site behaves normal the giant moment is usually attributed to a large, rather extended (≃ 10 A) polarization cloud on the surrounding Pd atoms. Despite the large amount of experimental work on giant moments in Pd (see, e.g. Nieuwenhuys [42]), the experimental information is not particularly clear. For Fe impurities values for the moments between 4.5 μ_B and 12.9 μ_B are reported [41], it is not known if Cr and V impurities are magnetic, and disagreement exists whether the magnetism of Ni impurities arises from isolated impurities or only from clusters of three or more Ni atoms [43-45]. This motivated us to perform calculations for Ti to Ni impurities in Pd to gain more insight into the behavior of such giant moment systems. Fig. 10 shows the calculated local moments inside atomic spheres of the impurities Cr to Ni which we find to be magnetic whereas Ti and V impurities are

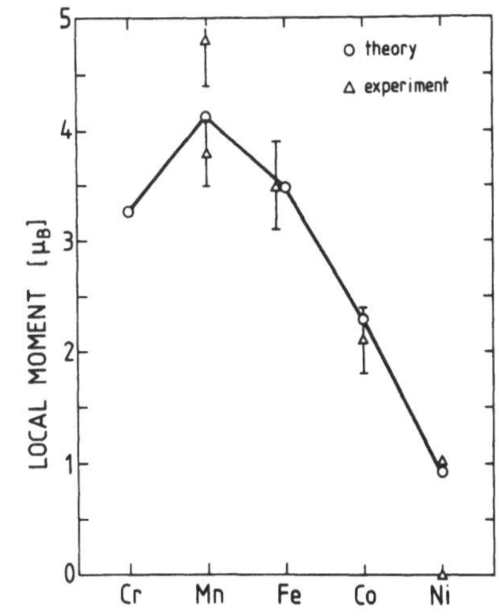

Fig.10 Local moments for 3d-impurities in Pd together with experimental values [42].

nonmagnetic [46]. From a comparison with the experimental information for Mn, Fe and Co impurities [42] we conclude that our results are reliable within the experimental error bar. For Ni we obtain a local moment of 0.9 μ_B which is in disagreement with the usual conclusion that an isolated Ni impurity in Pd is non-magnetic, however in agreement with a recent reanalysis of the experimental data by Loram and Mirza [45]. For a more detailed discussion we refer to [45,46].

It is interesting to compare the local moment results for Ag or Cu with the ones for Pd. Whereas a Ni-impurity in Ag or Cu is non-magnetic, it is magnetic in Pd. Contrary for V we obtain a local moment in Ag and Cu and for Ti in Ag, whereas both are non-magnetic in Pd. Thus for Pd the whole curve for the local moments versus the 3d-elements is shifted to the right. This is a consequence of the strong hybridization of the impurity 3d-level with the Pd d-band. In Cu and Ag the peak of the virtual bound state of Ni is well below E_F (see Fig. 3) so that the local moment criterion (28) is not met. Contrary for Ni in Pd, the Ni 3d-level, being energetically higher than the Pd one, is pushed by hybridization to higher energies, so that there is an appreciable LDOS at E_F in a paramagnetic calculation which leads to the local moment. Contrary for V and Ti we have rather large values for the LDOS at E_F in Ag and Cu, but a much smaller one in Pd due to the hybridization induced shift of the virtual bound states above E_F. Therefore the criterion (28) for a local moment is not met for V and Ti in Pd.

Most interesting is the host polarization of Pd, which leads to the observation of giant moments. Whereas in noble metals a RKKY-type behavior and a rapid decrease of the induced magnetization is found with increasing distance from the impurity, in Pd the induced moments do not change sign and slowly decrease over the region considered in the calculation, i.e. the impurity atom plus three shells of Pd atoms (43 atoms in total). Table 1 shows the calculated moments M_o on the impurity site, $M_{1,2,3}$ on a Pd atom in the first three shells around the impurity and the cluster moment M_{cl} of all 43 atoms. M_o^{exp} and M_t^{exp} are experimental values for the local moment and the total giant moment. From the slow decrease of the moments M_i we conclude that the polarization clouds extend over more than the 42 Pd atoms for which we allow perturbations. Because the use of a higher number of atoms is prohibitively expensive we have checked the sensitivity of our results varying the number of perturbed shells. A single site calculation where only the impurity potential is assumed to be perturbed and all Pd potentials are kept fixed to their values as in the ideal Pd host, already gives a good description of the impurity moments, the results differ a few percent from those of table 1. Similarly the polarization of the nearest Pd shell is already obtained within a few percent if only the potentials of these twelve atoms and the impurity are assumed to be perturbed. This indicates that the results in table 1 will not much change if more atoms are used. Because of the extended polarization it is clear that our calculations with only 43 perturbed atoms cannot reproduce the experimental values for the giant moments.

In [46] we give two methods to estimate the giant moments from our calculations. One is based on a remarkable ΔZ dependence of the host polarization, where ΔZ denotes the difference in valence electrons (ΔZ = -4, -3, -2, -1, 0 for Cr, Mn, Fe, Co, Ni). The dependence can be written as $M_i/M_o \cong \alpha_i\ (1+0.28\ \Delta Z)$ and is illustrated in Fig.11. Assuming that the formula is valid for all shells, we obtain for the total induced moment M_{ind} of the Pd host M_{ind}/M_o =

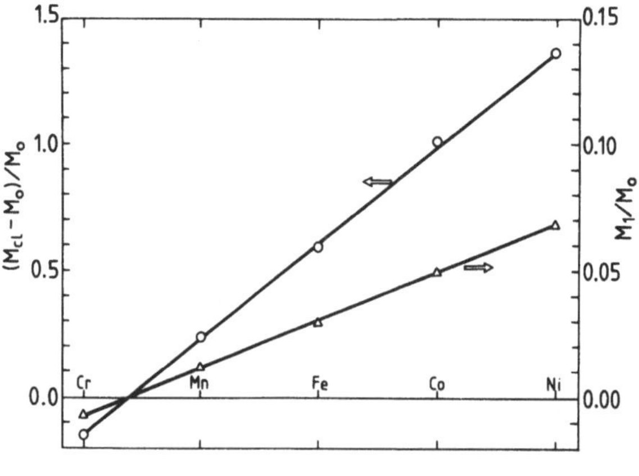

Fig. 11 Linear dependence of the ratio M_1/M_o (moment M_1 of a first shell atom divided by the impurity moment M_o) and of the total induced moment $(M_{cl}-M_o)/M_o$.

α (1+0.28 ΔZ). Physically this means that the spatial form of the polarization cloud is nearly the same for all impurities; only its amplitude varies. By determining the constant α using the value 4.6 μ_B for the value of the giant moment of Ni, we obtain the curve denoted by "method A" in fig. 12 for the giant moments of the other impurities. Both methods give giant moments in remarkable agreement with the experimental values. For Cr only a "dwarf" moment is obtained since the cloud moment partly cancels the local moment.

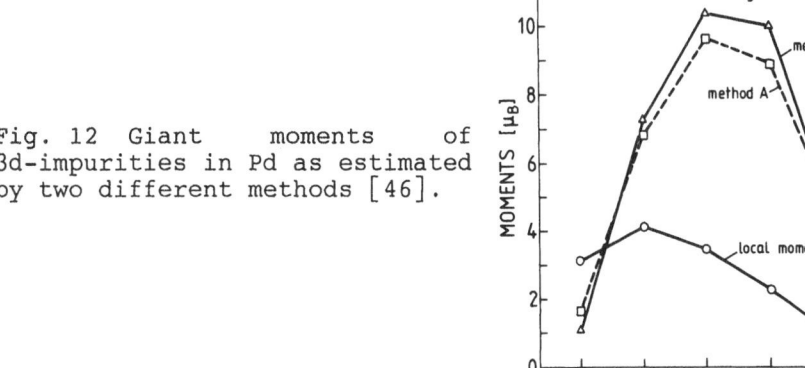

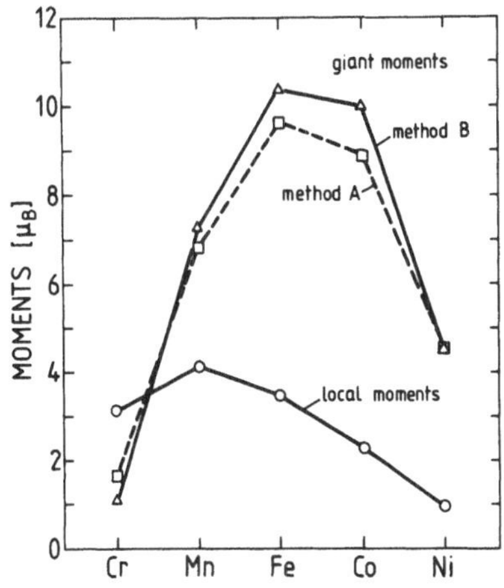

Fig. 12 Giant moments of 3d-impurities in Pd as estimated by two different methods [46].

3.5. Interaction of magnetic impurities

The interaction of magnetic impurities is of importance for the understanding of spin glasses, mictomagnets or ferromagnetic disordered alloys. The simplest model for the interaction is the Alexander-Anderson model [47], a generalization of the Anderson model [36] to two impurities on nearest neighbor sites. Here the direct overlap of the d-wave functions dominates the interaction and the RKKY interaction via the conduction electrons of the host is of minor importance. The essence of the Alexander-Anderson model is demonstrated in Fig. 13. For the ferromagnetic configuration (Fig. 13a) of the impurity pair we start with the same LDOS of the isolated impurities (dashed lines) on both sites. Since only states with the same spin direction hybridize, the virtual bound states for both spin directions split-up, but the centers remain unshifted. Contrary for the antiferromagnetic configuration (Fig. 13b) the virtual bound states for the same spin direction are at different energies. By hybridization they repel each other and small satellites appear at the "wrong" energies.

From the figure it becomes clear that for both configurations one can gain energy, in the antiferromagnetic case due to the covalent repulsion of the energy levels, in the ferromagnetic case due to the splitting of the energy levels. The latter process is most effective if the Fermi energy coincides with the peak position of the unperturbed virtual bound

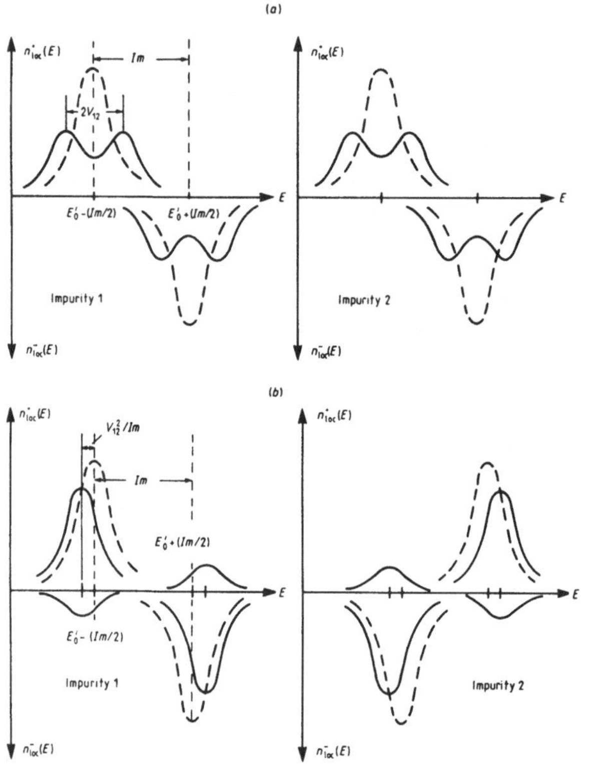

Fig.13 Schematic representation of the LDOS for two configurations of a pair of magnetic impurities: a) ferromagnetic configuration, b) antiferromagnetic configuration. Full curves: LDOS of the pair; broken curves: LDOS of the single impurity.

states. For more details we refer to [47,48].

As an example for a realistic calculation Fig. 14 shows the LDOS for a ferromagnetic and an antiferromagnetic pair of Mn impurities on nearest neighbor sites in Cu [48]. For the ferromagnetic configuration (Fig. 14a) the bonding-antibonding splitting of the minority virtual bound state is clearly seen, the corresponding splitting for the majority state being smaller due to the stronger localization of the corresponding wavefunctions. For the antiferromagnetic (Fig. 14b) configuration the changes of the LDOS from the isolated impurity behavior is quite small and qualitatively in agreement with the Alexander-Anderson model (Fig. 13). The peaks are practically unshifted.

The local moments are only slightly affected by the interaction [48]. For the ferromagnetic pair the moment is reduced by 6 %, versus a 5 % reduction for the antiferromagnetic pair. Similar small reductions are also obtained for Cr and Fe impurities in Cu. In the Ag host the interaction is even smaller due to the larger lattice constant of Ag and the moments are practically unchanged.

Of prime importance is the exchange interaction energy, i.e. the energy difference between the ferromagnetic and the antiferromagnetic configuration. We have calculated these energy difference using the frozen potential method [48], i.e. in first order perturbation theory. In this approximation the

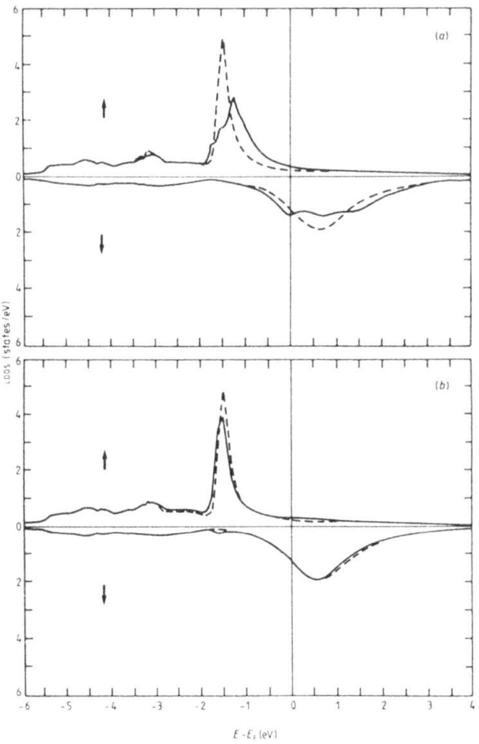

Fig.14 Pairs of Mn impurities on nearest neighbor sites in Cu: LDOS of (a) the ferromagnetic and (b) the antiferromagnetic configurations. The broken curve refers to a single Mn impurity.

difference of the total energies is given by the difference of the single particle energies, provided the same "frozen" trial potential is used for both configurations. The energy difference ΔE_n for various nearest neighbor distances of Cr, Mn and Fe impurities in Cu is listed in table 2. The calculations show that the antiferromagnetic configuration is

TABLE 1: The moments M_o on the impurity site, $M_{1,2,3}$ on the first-, second-, and third-nearest-neighbor Pd sites, and the cluster moment M_{cl} of all 43 atoms. M_o^{exp} and M_t^{exp} denote the experimental values for the local and giant moments.

Impurity	M_o	M_1	M_2	M_3	M_{cl}	M_o^{exp}	M_t^{exp}
Cr	3.14	-0.024	-0.012	-0.004	2.69	...	...
Mn	4.13	0.048	0.011	0.015	5.12	3.8±0.3 4.8±0.4	6.5-8
Fe	3.47	0.102	0.028	0.028	5.53	3.5±0.4	10-12 10
Co	2.28	0.114	0.032	0.031	4.60	2.1±0.3	9-10
Ni	0.92	0.063	0.022	0.015	2.18	0.1	4.6±1.8

more stable for Cr and Mn n.n. pairs, whereas for Fe n.n. pairs the ferromagnetic configuration is preferred. The results for Fe and Mn are in agreement with neutron scattering results, which point to a ferromagnetic n.n. configuration for Fe pairs [49] and to an antiferromagnetic n.n. pair in the case of Mn [50]. For Mn pairs the interaction is so strong that the frozen potential approximation becomes somewhat unreliable. Here the interaction has been recently calculated by the method of constraints [51,52] yielding however qualitatively the same results.

TABLE 2: Exchange interaction energy ΔE_n for the nth nearest neighbor separation of 3d impurity pairs in Cu [48]. Positive values mean that the ferromagnetic, negative values that the antiferromagnetic configuration is more stable. All values in meV.

	ΔE_1	ΔE_2	ΔE_3	ΔE_4
Cr	-364	42	-10	16
Mn	-141*	55	22	-22
Fe	130*	-6	23	-28

* calculated by the method of constraints [52]

4. TRANSITION METAL IMPURITIES IN Fe AND Ni

Fe (bcc) and Ni (fcc) are elemental ferromagnets with moments of $2.2\mu_B$ and $0.6\mu_B$ respectively. Whereas in Ni the majority band is well below E_F and completely filled, this is not the case in Fe, since a small tail of the majority DOS remains unoccupied. For this reason one refers to Ni as a strong ferromagnet and to Fe as a weak one. We will see that this rather small difference in the DOS is very important for the understanding of Fe and Ni alloys. Due to the filled majority band strong ferromagnetic alloys show an especially simple magnetic behavior, whereas weakly ferromagnetic alloys are more complicated.

4.1. 3d-impurities in Ni

Fig. 15 shows the calculated LDOS of the 3d-impurities Ti, V, ..., Cu in Ni [53,54]. For Co as well as for Fe impurities the majority band is practically unchanged, just one electron for Co and two electrons for Fe are missing locally in the minority band leading to sharp peaks above the band edge. Consequently the moment of Co ($1.7\mu_B$) is about $1\mu_B$ larger than the moment of pure Ni, whereas the Fe-moment ($2.7\mu_B$ is larger by $2\mu_B$. This simple behavior can be understood as follows. For Co and even more for Fe the electrostatic potential is weaker as in pure Ni. However, this decrease is effectively canceled

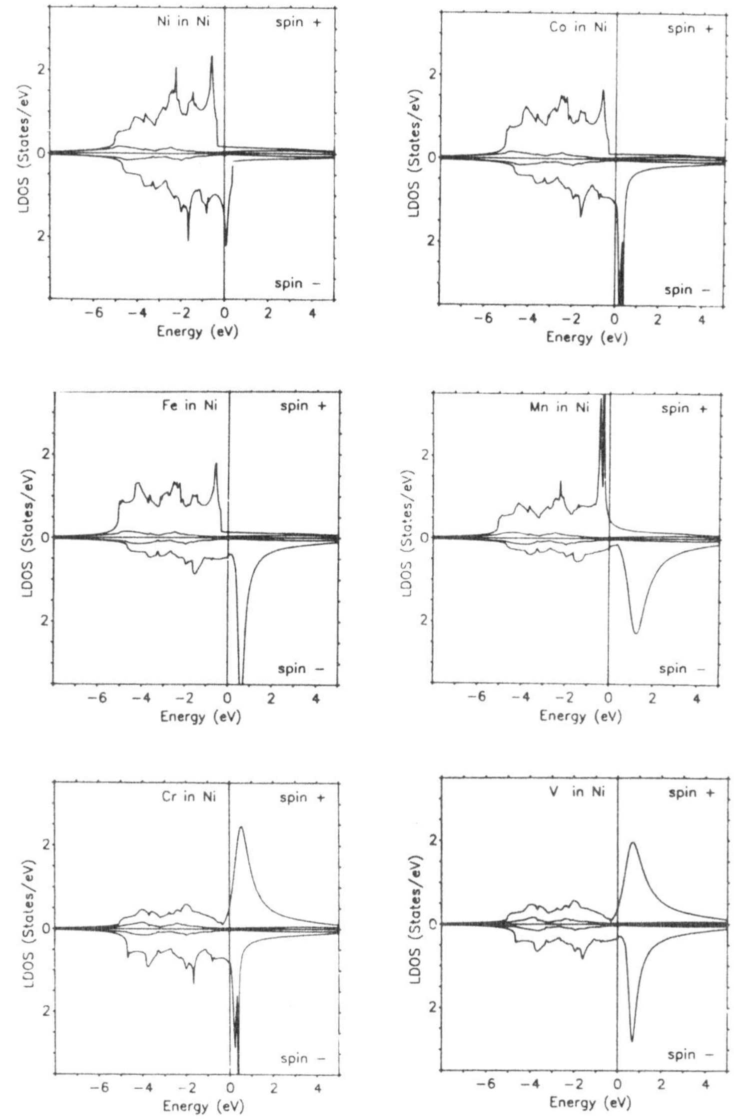

Fig.15 LDOS of 3d-impurities in Ni. The figure "Ni in Ni" refers to the DOS of pure Ni.

by an increase of the spin dependent exchange potential since the local moment increases in steps of $1\mu_B$. Consequently the majority band is effectively unchanged. In the minority band both potential contributions are added shifting the weight above E_F. In order to have a similar situation for Mn and Cr, one would need impurity moments of 3.7 and $4.7\mu_B$. However, this is not possible, since the potential needed for completely emptying the minority band has to be too repulsive.

With the calculated moment of $3\mu_B$ for Mn a virtual bound state develops between the upper band edge and the Fermi energy. Due to the small tail above E_F the majority band is no longer completely filled, but still nearly so. For Cr and also for the other early transition metal impurities (V, Ti, Sc) a completely different type of solution appears. In the majority band the virtual bound state shifts above the Fermi energy and the minority band contains slightly more electrons than the majority one leading to a negative moment which is antiparallel to the host moment.

Fig. 16a shows the calculated local moments for 3d-impurities in comparison with experimental values obtained from neutron scattering. With the exception of Cr the agreement is quite good. For Cr the calculation is very sensitive to the chosen exchange correlation potential as well as to details of the numerical procedures. Therefore the disagreement with the experimental value ($-0.2\pm0.6\mu_B$) should not be taken too seriously.

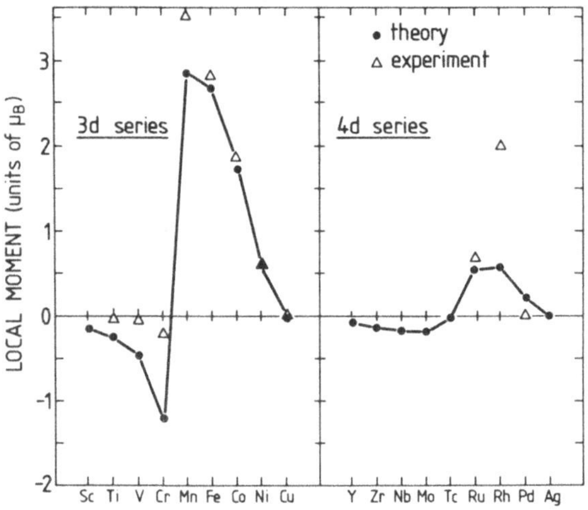

Fig.16 Local moments of 3d- and 4d-impurities in Ni together with experimental values (Δ) from neutron scattering.

The situation for the local moments is in fact more complicated than shown in Fig. 16. For Mn one obtains two solutions, one with a positive moment of $2.9\mu_B$ and another one with a negative moment being presumably less stable. By performing calculations for non-integer nuclear charges we find that in general two solutions exist for nuclear charges between 24.2 to 26.5, i.e. between Cr and Fe (Fig. 17). This behavior can be easily understood if one realizes that the moment of pure Ni ($0.6\mu_B$) is rather small. For a vanishing host moment, i.e. in a paramagnet such as Cu or Ag, the total energy E(M) as a function of the local moment M has two degenerate minima at the correct local moment values $\pm M_0$. If the host polarization is weak, it will slightly lower one energy minimum, but increase the other so that still two solutions M_o and $M'_o \tilde{=} -M_o$ exist , one being slightly more stable than the other. If however the host polarization becomes stronger or the local moment smaller then the second energetically unfavorable minimum will disappear and only a single solution remains.

Fig.17 Local moments for 3d-impurities in Ni. The thick line refers to calculations for non-integer nuclear charges Z between Cr and Co. For a large range of Z values two solutions exist.

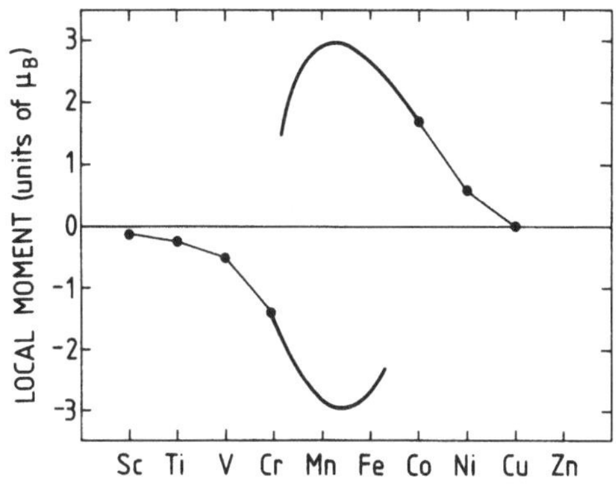

From macroscopic point of view the total change of the magnetization induced by the defect is more important than the local moment since it determines the concentration dependence of the alloy magnetization. From neutron scattering [55] it is known that the host perturbations are rather long ranged. For this reason we have recently performed calculations including four shells of Ni atoms in the self-consistency procedure [56]. The resulting total change ΔM of the magnetization for each impurity atom is shown in Fig. 18. The open circles represent results of magnetization measurements for the dilute alloys which are in excellent agreement with our calculations. The dashed curve shows the contribution from the impurity site, i.e. the difference between the local impurity moment and the host moment. Whereas for Co, Fe and Mn impurities the perturbation is well confined to the impurity cell, for the early transition metal impurities, which couple antiferromagnetically to the host, as well as for the sp-impurities there exist large and long ranged host perturbations.

The basic understanding of this, at first sight rather unusual behavior is due to Friedel [57], Kanamori [58] and others [59,60]. It can be most easily understood in a tight binding model by neglecting the sp-electrons. The essential point is that for strong ferromagnets, in our case for <u>Ni</u>Co, <u>Ni</u>Fe and <u>Ni</u>Mn, the majority band remains full (Fig.15) so that the integrated density of states $N_\uparrow(E_F)$ for majority spin does not change due to the addition of an impurity, i.e. $\Delta N_\uparrow(E_F) = 0$. Then the neutrality condition $\Delta N_\uparrow(E_F) + \Delta N_\downarrow(E_F) = \Delta Z$, where ΔZ is the valence difference, implies that the total magnetization of the alloy changes for each impurity by $\Delta M =$

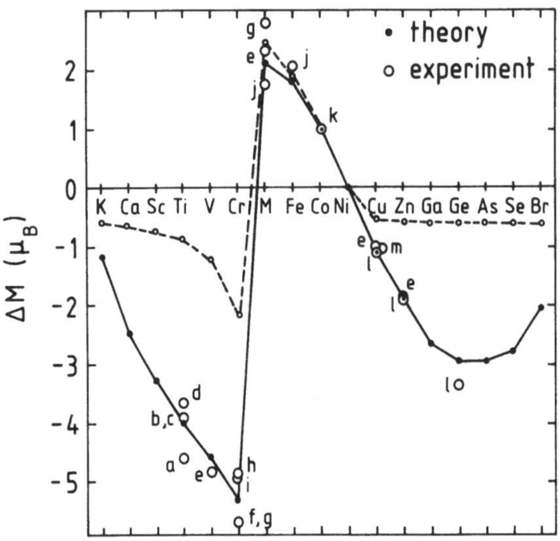

Fig.18 Change ΔM of the total magnetization for 3d and 4sp impurities in Ni. The full line connects the calculated values, the dashed line gives the contribution from the impurity cell. The large open circles denote results of magnetization measurements.

$\Delta N_{\uparrow}(E_F) - \Delta N_{\downarrow}(E_F) = -\Delta Z$. This directly explains the linear increase of ΔM for Co and Fe and to a lesser degree also for Mn. Contrary for the early transition metal impurities an empty d-state is located in the majority band above the Fermi energy. Due to the fivefold degeneracy we have $\Delta N_{\uparrow}(E_F) = -5$, since this δ-function like state has been split-off from the majority band. From the Friedel sum rule for the charge we obtain then $\Delta N_{\downarrow}(E_F) = \Delta Z+5$ so that the change ΔM of the magnetization is given by $\Delta M = -10 - \Delta Z$. The simplicity of these arguments is striking. All what is needed is the Friedel sum rule for charge neutrality and the notion that the majority band is either filled or that five states for each impurity are split-off above E_F.

Due to the neglect of sp-electrons one cannot expect the relation $\Delta N_{\uparrow}(E_F) = 0$ or -5 to be exactly satisfied. For instance in a tight-binding description the virtual bound states in Fig. 15 are infinitely sharp δ-functions. The calculated changes of the populations of both bands are shown in Fig. 19. Indeed one sees that $\Delta N_{\uparrow}(E_F)$ is practically zero for Co, Fe and Mn (as well as the early 4sp-impurities) and then jumps to a value of about -5 for the early 3d-impurities. The small but important deviations from this behavior are discussed in detail in [56] (as well as the more complicated behavior of the sp-impurities).

The above discussion remains also valid for concentrated Ni 3d alloys. In this case the virtual bound states for the 3d-impurities (Fig. 13) broaden into an impurity band, being separated from the genuine Ni-band by a hybridization minimum.

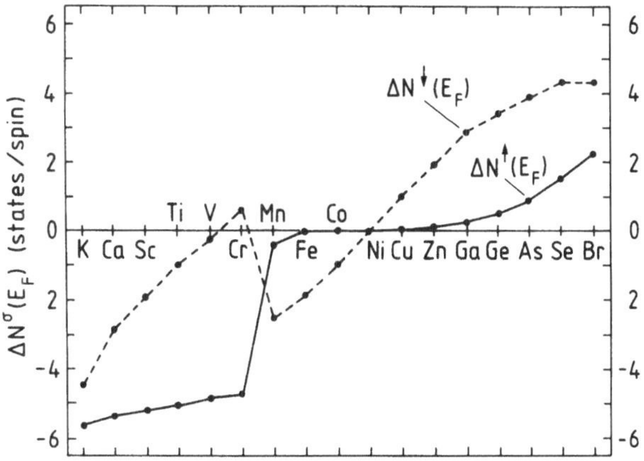

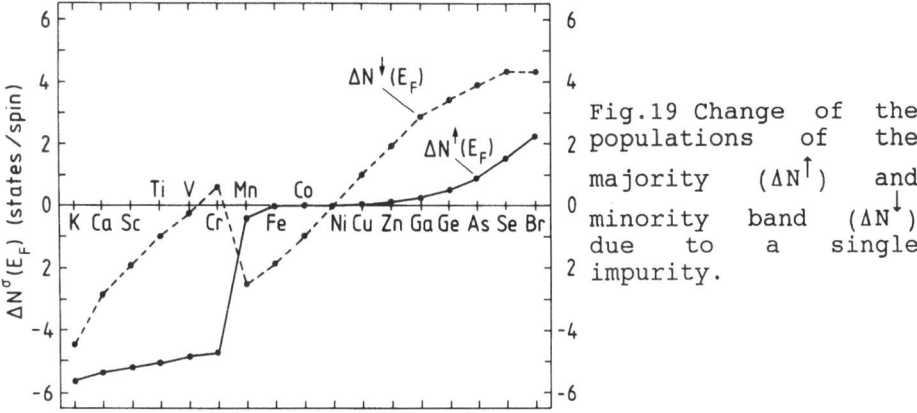

Fig.19 Change of the populations of the majority ($\Delta N^{\uparrow}$) and minority band ($\Delta N^{\downarrow}$) due to a single impurity.

Since this "band gap" is close to E_F the relation $\Delta N_{\uparrow}(E_F) \cong -5$ for each early 3d-impurity is well satisfied. Similarly concentrated NiCo and NiFe alloys remain strong ferromagnets, the latter however only for concentrations up to 60 % Fe. NiMn alloys are an interesting exception. While being strong in the dilute limit, they become weak for higher concentrations due to the broadening and emptying of the narrow majority virtual bound state of Mn (see Fig. 13). The simplicity and importance of the Friedel sum rule for the understanding of ferromagnetic alloys has recently been stressed and elaborated by Williams et al. [61-64]. We refer the reader to these articles for a more detailed discussion.

4.2 4d-Impurities in Ni

4d-impurities show a quite different behavior than 3d-ones since their wave functions are much more extended. As discussed previously these impurities therefore develop no local moment in paramagnetic hosts. However in ferromagnetic hosts a small moment is induced at the impurity site, being positive for the late and negative for the early 4d-impurities (see Fig. 14b). Whereas the experimental values for Ru and Pd are in good agreement with our calculations, the large experimental value of $2\mu_B$ for Rh [65] cannot be understood and is presumably not correct.

The LDOS of all 4d-impurities are characterized by a two peak structure [54]. Due to the strong hybridization with the Ni 3d-electrons two peaks are splitting off the band, one from the upper and one from the lower band edge, yielding similar LDOS for both spin directions. The changes $\Delta N_{\uparrow}(E_F)$ and $\Delta N_{\downarrow}(E_F)$ of the majority and minority band populations are plotted in Fig. 20, together with the change ΔM of the total moment. While the general behavior of the curves is similar to the ones for 3d-impurities (Fig. 16 and 17), the transition from strong ferromagnetism ($\Delta N_{\uparrow}(E_F) = 9$) to weak ferromagnetism ($\Delta N_{\uparrow}(E_F) = 9-5 = 4$) is much smoother. Strong ferromagnetism only occurs for dilute NiPd, and nearly so for NiRh. Tc, being isoelectronic to Mn, shows already all signs of weak ferromagnetism: $\Delta N_{\uparrow}(E_F)$ is off by -3.5 from the ideal value of 9 and the moment ΔM is strongly negative ($\Delta M = -3.9\mu_B$). This is a direct consequence of the stronger hybridization: The analog to the virtual bound state of Mn is much broader and mostly empty. The behavior of the early 4d-impurities (Mo, Nb, Zr, Y) is similar to their 3d-counter parts and characterized by a long ranged reduction of the host moments.

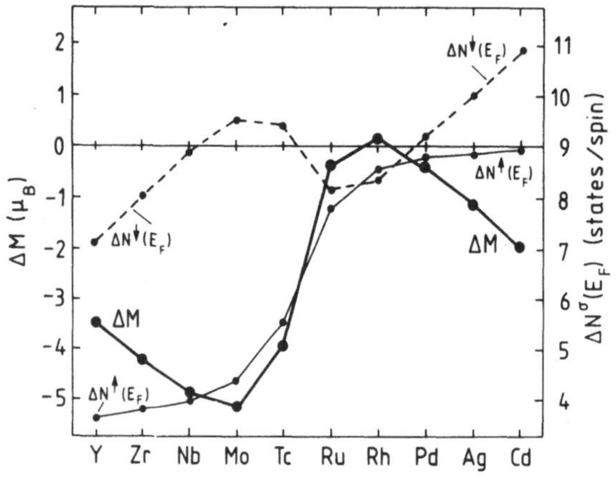

Fig.20 Change of the total magnetization (ΔM) and of the band populations for 4d-impurities in Ni.

4.3. Transition Metal Impurities in Fe

Selfconsistent calculations for impurities in Fe have been performed within the single site approximation by M. Akai et al. [21]. The present results are based on recent calculations [66] where the potentials of the impurity and five shells of host atoms (59 atoms in total) have been determined selfconsistently.

Fig. 21 shows the LDOS of 3d-impurities in Fe. For Co and Ni-impurities we observe an increase of the local minority population as well as the filling up of the local majority band. The behavior of the early transition metal impurities is similar to the behavior in Ni. The calculated local moments for 3d- and 4d-impurities are shown in Fig. 22. The general trend of the local moments is very similar to the Ni-case. The agreement with the experimental data is excellent.

Mn-impurities deserve some special comments. The results depend critically on the exchange correlation potential used in the calculation. For the approximation of Barth-Hedin (BH) we find a positive Mn-moment of 0.95 μ_B and, by varying the nuclear charge continuously around the Mn-value of 25, a continuous transition from positive to negative values. The same is true for the exchange correlation potential of Vosko-Wilk-Nusair (VWN) but the value for the Mn moment is negative, -0.42 μ_B. Contrary for the Barth-Hedin potential

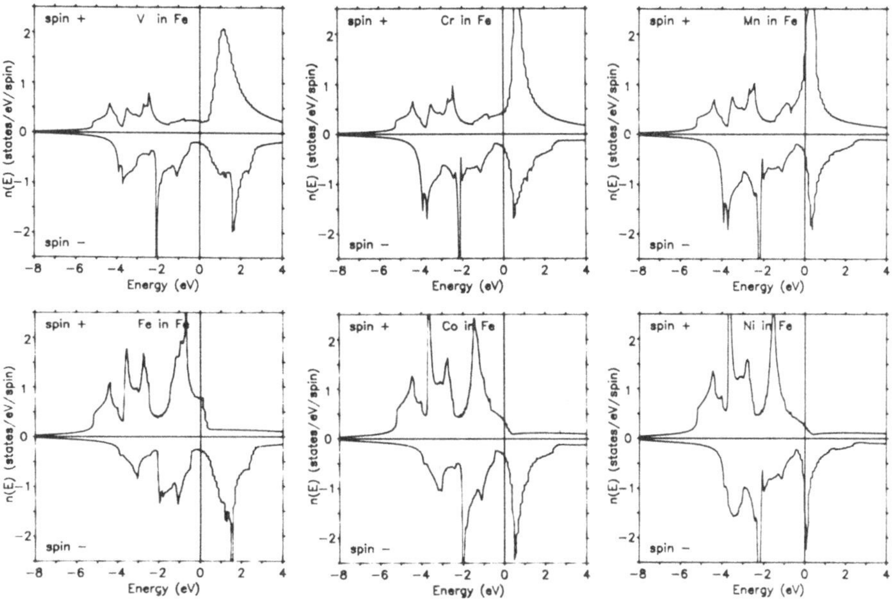

Fig.21 LDOS of 3d-impurities in Fe.

with the constants as determined by Moruzzi-Janak-Williams (MJW) we find a discontinuous transition from a positive moment state for $Z \gtrsim 25.04$ to a negative moment for $Z \lesssim 25.04$ with an extremely narrow region of two solutions around $Z \approx 25.04$. The resulting moment of Mn is then negative, $-1.15\ \mu_B$. The latter behavior is also found in the single site calculations of M. Akai et al [21] which gave a negative moment for $Z \lesssim 25.17$. Compared to Ni, where a broad coexistence region for two solution was found, in Fe practically only one solution survives which is due to the much larger host magnetization. Although conflicting experimental data exist, it is now generally accepted that Mn has a moment of about $0.6 - 0.8\mu_B$. Since our results depend so critically on the exchange correlation potential we cannot make any definite theoretical conclusion about the Mn moment. Thus Mn in Fe is a very critical case and also numerical approximations might affect the result. Note that for all other impurities the different exchange correlation potentials give practically the same results.

While the local behavior of the 3d-impurities in Fe and Ni is rather similar, this is not the case for the host perturbation around these impurities. To demonstrate this we have plotted in Fig. 23 the change ΔM of the macroscopic alloy magnetization due to a single impurity. ΔM_o is the change arising from the impurity cell alone. Note that the global properties (ΔM) and the local properties (ΔM_o) show a quite

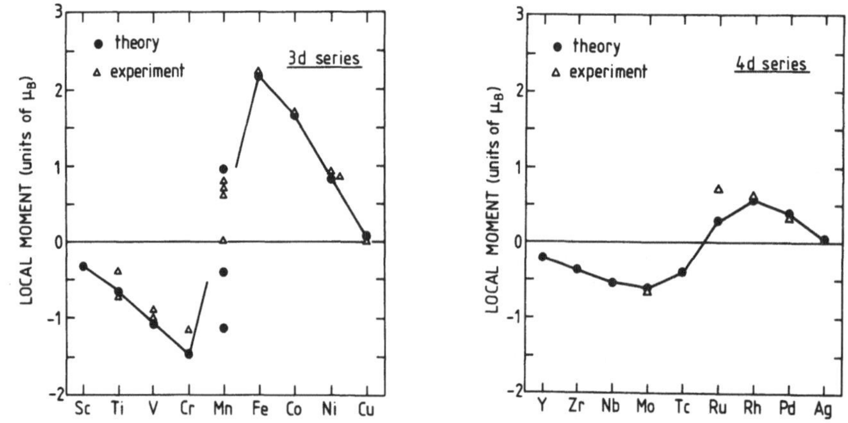

Fig.22 Local moments of 3d- and 4d-impurities in Fe together with experimental values (Δ) form neutron scattering.

different trend. For Co and Ni impurities the local moment is smaller than the host moment but the alloy magnetization increases due to positive host polarizations in several shells around the impurity. A similar behavior is also found for the 4d impurities Rh and Pd (see Fig. 24). Thus these alloys show a trend towards strong magnetism. Also for the early transition metal impurities the local and the global changes

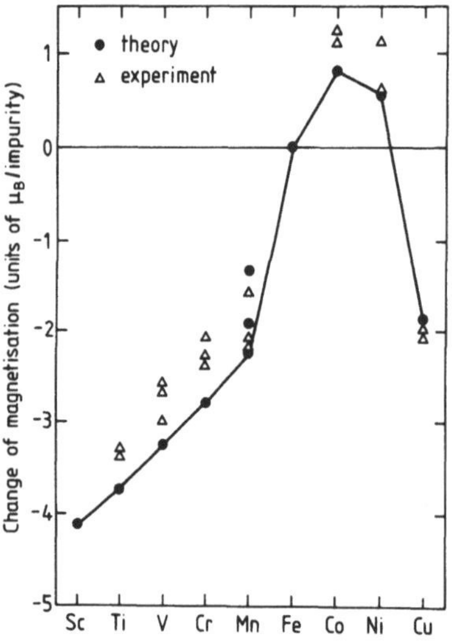

Fig.23 Change ΔM of the total magnetization due to a 3d-impurity in Fe. The Δ-symbols refer to the results of magnetization measurements.

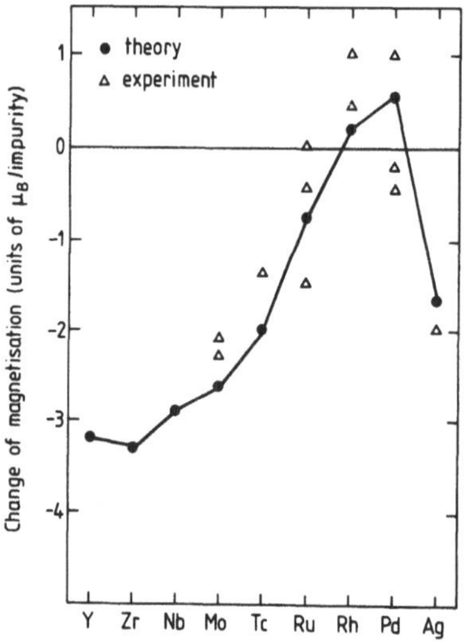

Fig.24 Change ΔM of the total magnetization for 4d-impurities in Fe.

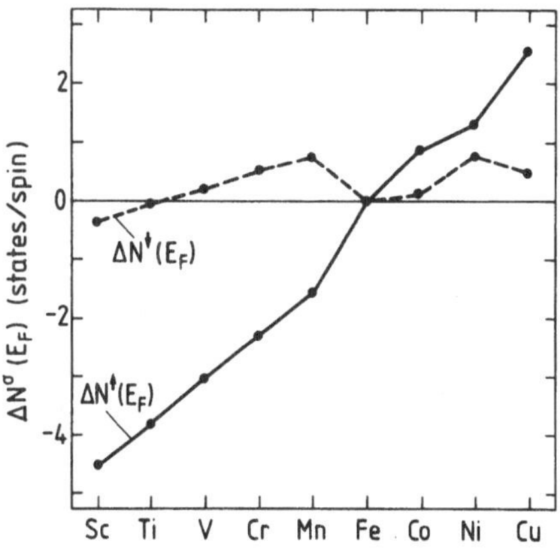

Fig. 25 Changes of the majority ($\Delta N^{\uparrow}$) and minority ($\Delta N^{\downarrow}$) band populations for 3d-impurities in Fe.

show a different behavior. Whereas the negative local moment decreases when progressing from Cr to Sc, from Mo to Y respectively, the change ΔM of the total magnetization becomes stronger negative, which is quite opposite to the behavior seen in Fig. 18 and 20 for the Ni-alloys.

The reason for this difference is the ferromagnetic weakness of Fe. The changes of the band population $\Delta N_{\uparrow}(E_F)$ and $\Delta N_{\downarrow}(E_F)$ as shown in Fig. 25 reveal that the majority population changes more or less linearly from -4.5 for Sc to +1.3 for Ni impurities, whereas the changes in the minority band are much smaller. This is in contrast to the situation for Ni alloys where except for the jump of -5 between Mn and Cr the majority population remains constant and the screening has to be performed by the minority band. Thus in a rough approximation one might characterize these Fe alloys by $\Delta N_{\downarrow}(E_F) \cong 0$ so that the moment changes as $\Delta M \cong \Delta Z$. By comparison with Fig. 21 this approximation, while quantitatively not very good, is qualitatively correct and accentuates the difference with respect to the Ni-alloys.

The shell dependence of the perturbed host magnetization is rather complicated. Whereas for Co and Ni impurities we observe enhanced moments in all five shells, for the early transition metal impurities we obtain positive as well as negative contributions. The perturbation of the first shell is strongly negative, the second one gives a weakly negative contribution whereas from the 3rd, 4th and 5th shells we obtain a net positive contribution. The systematic deviation from the experimental values for Cr, V and Ti seems to indicate that there are important positive contributions from shells further out.

5. HYPERFINE FIELDS AND ISOMER SHIFTS IN Fe- AND Ni-ALLOYS

Hyperfine fields and isomer shifts provide a unique microscopic information about dilute alloys. The dominant contribution to the hyperfine field results from the Fermi-contact interaction and is given by the magnetization density m(o) at the nucleus

$$H_{hf} = \frac{8\pi}{3} \mu_B \, m(o) \quad \text{with} \quad m(o) = \int^{E_F} d\epsilon \, (n_+(o,\epsilon) - n_-(o,\epsilon)) \qquad (29)$$

Here $n_\pm(o,\epsilon)$ are the local densities of states for spin up and spin down electrons at the nuclear position. There exists a huge amount of experimental data for impurity hyperfine fields in the elemental ferromagnets Fe, Co, Ni and Gd. Also the perturbed host hyperfine fields have been measured in many cases.

Isomer shifts of Mössbauer nuclei [67] are directly related to the local charge density m(o) at the nuclear position. More precisely the isomer shift ΔS is given by

$$\Delta S = \alpha \, \Delta n(o) \quad \text{with} \quad n(o) = \int^{E_F} d\epsilon \, (n_+(o,\epsilon) + n_-(o,\epsilon)) \qquad (30)$$

where $\Delta n(o)$ is the change of n(o) with respect to a reference medium. The isomer-shift calibration constant α depends only on nuclear properties and is essentially unknown. It can only be determined by comparison of measured ΔS data with calculated $\Delta n(o)$ values. For dilute Fe-alloys a large amount of information exists about the shifts of Fe-neighbors close to the impurities.

While hyperfine fields and isomer shifts give a very detailed and complimentary information about the magnetization and charge densities at the nuclear position, this information is of questionable value as long as the relation of these quantities at the nucleus to the general valence properties is not well understood. Note that only s-electrons have a finite density at the origin, which in transition metals play only a minor role for the binding. In the following we will represent calculations of both quantities for dilute Ni- and Fe-alloys. Especially we will address the question what valence properties determine the hyperfine fields and the isomer shifts and what such measurements can tell us about the local moments and the charge transfers in these alloys.

5.1. Hyperfine fields in Ni

The recent progress in the understanding of hyperfine fields in alloys was initiated by the work of Katayama-Yoshida et al. [20] which led to an understanding of the systematic trends of the hyperfine fields of sp-impurities in Ni and Fe. Selfconsistent calculations for transition metal impurities in Fe were performed by Akai et al [21] and in parallel by Blügel [68] for 3d and 4d impurities in Ni. The work presented here is based on a combined effort [69,70].

Fig. 26 shows the calculated hyperfine fields for 3d- and 4d-impurities in Ni in comparison with the experimental values. Given are both the non-relativistic results (NRL, dashed line) and the semirelativistic results (SRA, full lines). Relativistic corrections are of importance for 4d-impurities, but in the 3d-series they give only a small

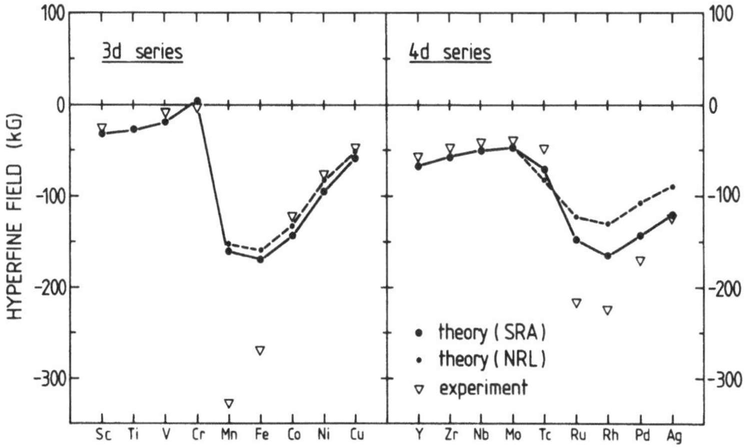

Fig.26 Calculated hyperfine fields for 3d- and 4d-impurities in Ni. The large dots refer to the semirelativisitc approximation, the small ones to the non-relativistic calculation. The experimental values are taken from [1].

enhancement of about 10 %. Whereas the results for the impurities with a negative moment agree very well with the experiments, this is not true for the ferromagnetically aligned impurities. Especially the values for Fe and Mn are considerably too small (A similar error also occurs for pure Fe). The disagreement is not due to errors of the local moments, which agree very well with neutron scattering data, nor due to the neglect of orbital contributions, which would give a positive contribution thus enhancing the discrepancy. Rather we believe that this is due to an error of the local density approximation in calculating the magnetization close to the nucleus.

For the relativistic calculation one has to use the proper relativistic Breit formula [71] for the hyperfine field. We have recently shown [70] that for the contact interaction this leads to a simple generalization of the contact formula where the spin density m(o) at the nucleus has to be replaced by an averaged spin density,

$$m_{av} = \int d\underline{r} \; \frac{1}{4\pi r^2} \; \frac{r_T/2}{(r+r_T/2)^2} \; m(\underline{r}) \tag{31}$$

being spherically averaged over a region the diameter of which is the Thomson radius $r_T = Ze^2/mc^2$. Here the relativistic spin density is defined by the expectation value of the large component Dirac wave function ϕ_1

$$m(\underline{r}') = \langle\phi_1|\sigma_z\delta(\underline{r}-\underline{r}')|\phi_1\rangle \qquad (32)$$

Since the relativistic wave function ϕ_1 diverges at the origin, the spin density behaves for small r' as

$$m(r') \sim r'^{2\lambda-2} \qquad \text{with} \qquad \lambda = \sqrt{1-(Z/137)^2} \qquad (33)$$

When averaging over the Thomson radius this leads to a modest enhancement of about 9-10 % for Ni-atoms. Very often the classical contact formula (29) is used, but with relativistic spin densities (32) averaged over the nuclear radius. This procedure is unjustified and leads to a relativistic enhancement being considerably too large.

Next we discuss the individual contributions to the hyperfine fields, i.e. the core and valence contributions. The **core polarization** is large for magnetic impurities. It is due to the exchange interaction of the polarized d-shell with the s-orbitals of the core. As a result a weak s-polarization is induced at the nuclear position, which is in general opposite to the local moment. Since the exchange interaction is weak the core polarization and thus the core hyperfine field H_{hf}^{C} is expected to scale with the local moment M_{loc}. Fig. 27 illustrates that this is indeed the case. Plotted are the local moments (left scale) and the core hyperfine fields (inverted right scale) for the 3d-series. Since both curves more or less coincide we get the following simple relation: $H_{hf}^{C} = C \cdot M_{loc}$ with $C \cong -100\ kG/\mu_B$ for the 3d-series and $C \cong -200\ kG/\mu_B$ for the 4d-series.

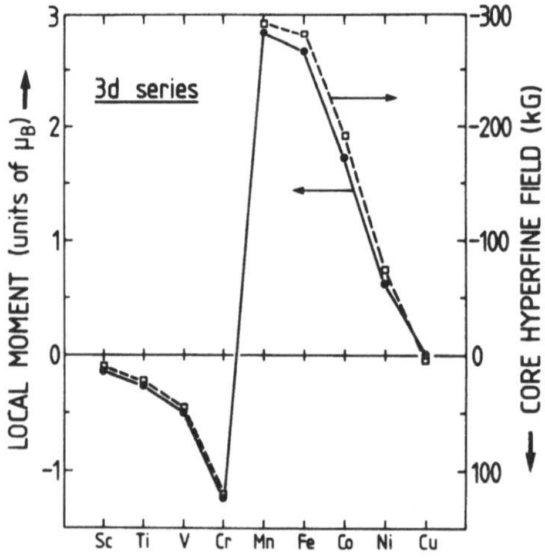

Fig.27 Core hyperfine fields (right scale) and local moments (left scale) for 3d-impurities in Ni. Note the inverted scale for the core hyperfine field.

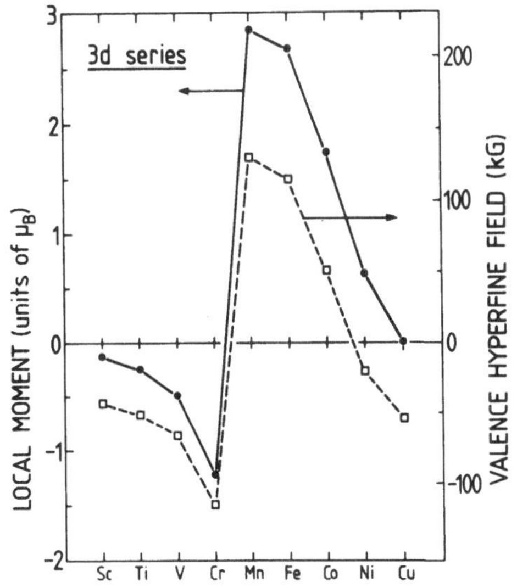

Fig.28 Valence hyperfine fields (right scale) and local moments (left scale) for 3d-impurities in Ni.

The **valence contributions** to the hyperfine fields are more complicated. They are plotted in Fig. 28 for the 3d-impurities (right scale) together with the corresponding local moments (left scale). It is seen that up to a more or less constant negative contribution the valence hyperfine fields follow closely the local moment curves. The constant negative contribution is quite analogous to the negative hyperfine field of the early sp-impurities [20]. The hybridization of the impurity s-orbitals with the spin-polarized d-orbitals of the neighboring Ni-atoms induces a weak s-polarization, which is negative for the early sp-impurities. This "**transferred hyperfine field**" is also negative for the transition metal impurities, since their potentials are weak in the sense of Katayama et al. [20]. One expects the transferred contribution to be essentially proportional to the local moments of the neighboring hosts atoms. Since these moments are strongly reduced for the impurities with negative moments (Sc ... Cr) the corresponding transferred fields are substantially smaller than the ones of the ferromagnetic impurities.

In addition to this negative transferred hyperfine field we obtain for magnetic impurities a "**local valence hyperfine field**" which dominates the behavior seen in Fig. 28. The figure demonstrates that this local contribution is proportional to the local moment; however contrary to the core hyperfine field the proportionality constant is positive. The physical origin of this local valence field is also different from the core polarization. The dominating mechanism is a repopulation effect, i.e. more majority and less minority s-states are occupied, leading to a net valence s-moment in the impurity Wigner-Seitz cell. Fig. 29 shows, that the valence hyperfine field is in a rather good approximation proportional to this local s-moment. Similar proportionalities as shown in Fig. 27, 28 and 29 for the 3d-series are also valid for the 4d-impurities in Ni [70].

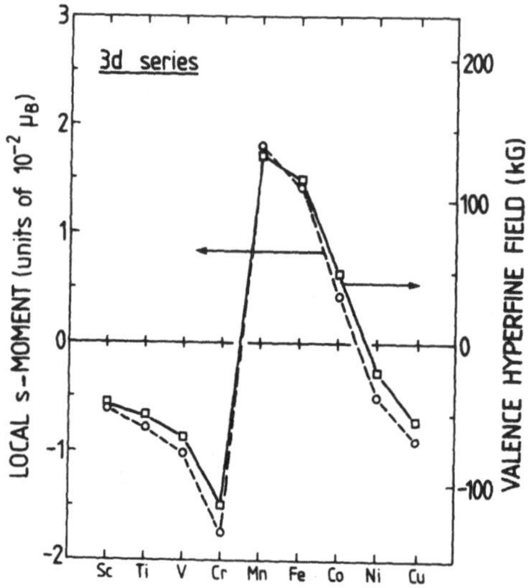

Fig.29 Valence hyperfine fields (right scale) and local s-moments in the impurity Wigner-Seitz sphere (left scale) for 3d-impurities in Ni.

The division of the hyperfine field into a local and a transferred contribution and the above discussion suggests a simple **interpolation formula** for the hyperfine field in terms of the moments of the considered atom and its neighbors. We therefore set

$$H_{hf} = a\, M_{loc} + b\, M_{host} + c \sum_{i=n.n.} \Delta M_i \qquad (34)$$

The first term $a \cdot M_{loc}$ represents the local hyperfine field, being the sum of the local core and valence contributions, which are both proportional to the local moment (see Fig. 27 and 28). From the behavior of the hyperfine field through the 3d- and 4d-series it is clear that the constant a is negative since the core contribution dominates. The second and third term represent the transferred hyperfine field consisting of a constant contribution $b \cdot M_{host}$ and a correction $c \sum_i \Delta M_i$ if the moments M_i on the neighboring sites are changed by the amount ΔM_i. Further ranging changes are neglected. Using the calculated values for the moments a reasonable fit to the hyperfine fields of both 3d-impurities and their nearest neighbors is obtained with $a \cong -37$, $b \cong -115$, $c \cong -10.4$ $[kG/\mu_B]$. It is clear that such a fit cannot accurately reproduce the calculated values. However it gives the correct trends and contains the most important physics, i.e. the relation to the local and the neighboring moments which determine the local and the transferred contributions of the hyperfine field.

5.2. Isomer shifts in dilute Fe alloys

We have recently made a systematic theoretical study of isomer shifts in dilute Fe-alloys [72]. Impurities in Fe with nuclear charges between Z=0 (vacancy) and Z=56 (Ba) are considered and the isomer shift on the first nearest neighbor (nn) Fe site is calculated. The neighboring atoms are fixed at their original lattice positions, so that lattice relaxations are not allowed. The potentials of the impurities and the nearest neighbors are calculated selfconsistently. Our present method of calculating the n.n. isomer shifts has the advantage that all impurities in Fe can be calculated with the same host Green's functions. The analogous calculation for Fe impurities in the equivalent number of hosts would be much more difficult and timeconsuming due to the many lattice structures and different lattice constant involved.

Fig. 30 shows the calculated isomer shift values for the n.n. Fe-atom and the comparison with the experimental data for dilute Fe-alloys as a function of the nuclear charge Z of the impurity. We have chosen a calibration constant α = -0.24

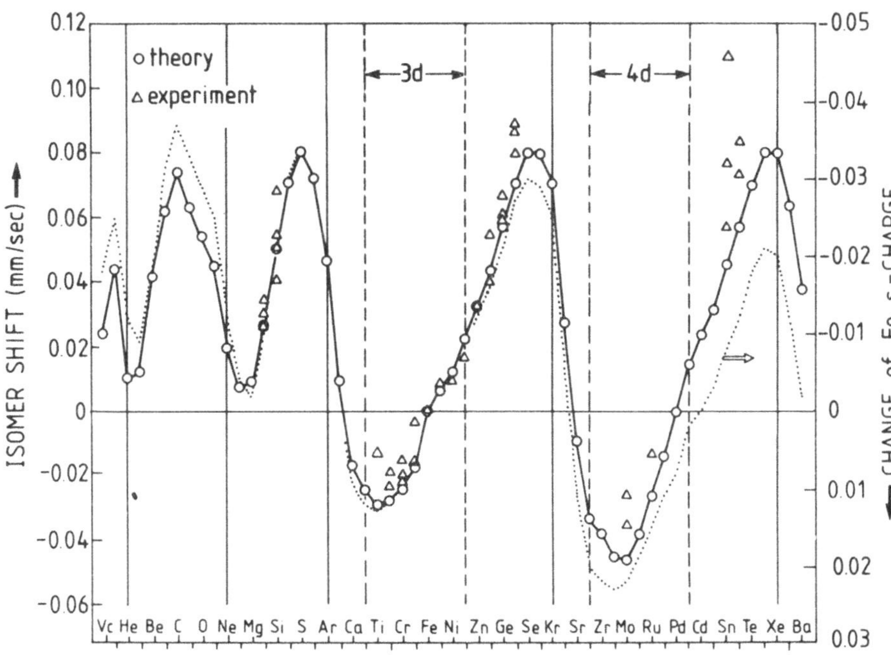

Fig.30 Calculated isomer shift of a nearest neighbor Fe atom as a function of the nuclear charge Z of the impurity (left scale). The triangles are experimental values for dilute Fe alloys. The dotted line gives the change Δn_s of the number of Fe s-electrons and refers to the right inverted scale.

a_o^3 mm s^{-1} being compatible with the experimental data for the analysis of which we refer to [73]. There and in ref. [67] part of the data shown in Fig. 30 are given. The general agreement is fairly good and the experimental trends are reproduced. The figure clearly demonstrates that the local density approximation represents a reliable basis to calculate isomer shifts. In general the shifts show a minimum at the beginning of each row of the periodic system and slowly increase to a maximum at the end of the row, in order to sharply drop to the minimum at the beginning of next row etc. However the periodicity is far from being perfect. For instance, with increasing row number the maxima systematically shift to larger Z-values and the minima get more pronounced. We come back to this later.

Since a negative (positive) isomer shift means that the charge $\rho(o)$ at the n.n. Fe nucleus has increased (decreased), the behavior seen in Fig. 30 follows simple electronegativity arguments that the charge flows from the element with the lower electronegativity to the one with the higher electronegativity. For instance in the 3d series the charge should flow from the impurity away to the n.n. Fe atoms if the impurity is on the left side of Fe in the periodic table, whereas the charge should flow to the impurity, if it is on the right side of Fe. However, as we will show later this is a too simplified description, since the isomer shift measures only the change of the charge density $\Delta\rho(o)$ at the Fe nucleus, which moreover is only of s-character. No direct information is obtained about p- or d-electrons. The question therefore arises how the changes $\Delta\rho(o)$ are related to the changes of the valence charges, i.e. the changes Δn_s, Δn_p and Δn_d of the number of s-, p- and d-valence electrons of the n.n. Fe atom.

The presently accepted view about isomer shifts in metallic alloys [75] is summarized in a linear relationship between $\Delta\rho(o)$ and the changes of the valence s- and d-electrons: $\Delta\rho(o) \cong C(\Delta n_s + R\Delta n_d)$. Whereas the first term describes the proportionality to the changes of the valence s-electrons, the second term describes the effectiveness of the valence d-electrons to "screen out" the s-electrons. Even for a constant total s-charge, $\Delta n_s=0$, the charge density at the origin should strongly decrease, if the total d-charge increases and vice versa. According to Watson and Bennett [75] typical R-values for metallic alloys are in the range $R \approx -0.5...-1.0$ and the term $R\Delta n_d$ is thought to describe the reduction of the valence s-charge density at the origin due to the increase of the d-electron number. Contrary Ingals [74,76] has pointed out the importance of the polarization of the core s-electrons and proposed a value $R \approx -5/8$.

Our calculations do not confirm these ideas. For instance we find that the core electrons are of secondary importance for the isomer shifts. Clearly the trends of the isomer shifts are determined by the valence contribution. Most important is that the valence polarization effect, emphasized by Watson and Bennett [75], is appreciably smaller then previously assumed. In order to demonstrate this we have plotted in Fig. 30 by the dotted line the change Δn_s of the total s-charge of the n.n. Fe-atom. Here Δn_s is defined as the change of the number of s-electrons in the Wigner-Seitz-sphere of the n.n. Fe-atom. Note that this curve refers to the right (inverted) scale which has been adjusted such that both curves have about the same amplitude. The close agreement suggests that the number Δn_s of valence s-electrons is the only relevant parameter determining the isomer shift. This excludes any major role of the d-electrons in screening the s-charge distribution.

The physical reason for the small R-values in alloys can be understood from renormalized atom calculations [77]. While in atoms of the transition metal series more than 70 % of the valence s-charge is located outside the metal Wigner-Seitz sphere, in the renormalized atom model the s-wave function is cut-off at the Wigner-Seitz radius and renormalized to one inside the sphere. While calculations for atomic Fe, which we performed, yield similar large R-values (R = -0.54) as found

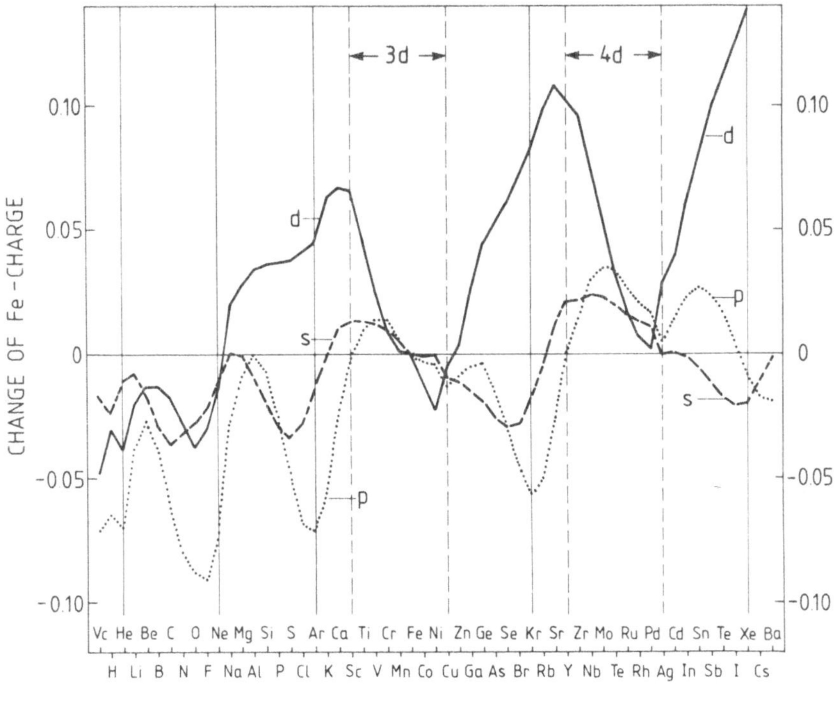

Fig. 31 Changes Δn_s, Δn_p and Δn_d of the number of s, p and d electrons in the n.n. Fe Wigner-Seitz sphere versus the nuclear charge of the impurity.

by other authors [74,78], calculations with the same, but renormalized wave functions yield a much smaller value (R = -0.11). Clearly this is due to the small spatial extent of the renormalized s-wave function, which makes the d-screening rather inefficient.

Fig. 31 shows the calculated changes of the s-, p- and d-electrons of the n.n. Fe-atom. Without going into details, one sees that for impurities within the 3d- and 4d-series the changes Δn_s, Δn_p and Δn_d show a similar trend and have the same sign. At the beginning of the 3d- and 4d-series the changes of the d-charge are much larger than the changes of the s-charge. Contrary for the 4sp- as well as the 5sp-impurities the d-charge shows a different trend. It increases and saturates towards the end of the period whereas the s-charge continuously decreases.

The p-charge shows an intermediate behavior between the d- and the s-charge. Fig. 31 clearly demonstrates that the isomer shift is not related to the total charge transfer, but only to its s-component. Otherwise a completely different trend would have been observed in Fig. 31.

At present we have not a complete understanding of the different trends of the s-, p- and d-components. Relatively simple is the behavior within the 3d-series which we will shortly discuss in the following. For the Fe d-charge the hybridization between the impurity d-electrons and the ones of Fe is most important, leading to the formation of bonding and antibonding states. We will consider here a very simple model for the charge transfer being based on the hybridization of two non-degenerate and orthogonal states ψ_A and ψ_B on neighboring sites. For the bonding and antibonding levels one has

$$E_{\pm} = \frac{1}{2}(E_A+E_B) \mp \frac{1}{2}\sqrt{(E_A-E_B)^2+V^2} \qquad (35)$$

where E_A and E_B are the atomic levels and V is the hybridization matrix element. The bonding and antibonding eigenfunctions $\psi_{\pm}$ are

$$\psi_{\pm} = \frac{1}{\sqrt{2}}\left(1 \pm \frac{\epsilon}{\sqrt{1+\epsilon^2}}\right)^{1/2}\psi_A \pm \frac{1}{\sqrt{2}}\left(1 \mp \frac{\epsilon}{\sqrt{1+\epsilon^2}}\right)^{1/2}\psi_B \qquad (36)$$

with $\epsilon = (E_B-E_A)/2V$

In the following we assume that the energy levels E_A and E_B change linearly with the atomic population $\overset{o}{N}_A$ and $\overset{o}{N}_B$, i.e. $E_A = E_o - \alpha \overset{o}{N}_A$ and similarly for the B-level. The A and B populations after hybridization are denoted by N_A and N_B. For $\overset{\circ}{N}_A+\overset{\circ}{N}_B \leq 2$ only the bonding level will be occupied so that (the + spin is valid for the A and - spin for the B-atom)

$$N_{A,B} = \frac{1}{2}\left(1 \pm \frac{\epsilon}{\sqrt{1+\epsilon^2}}\right)(\overset{\circ}{N}_A+\overset{\circ}{N}_B) \quad , \quad \epsilon = \frac{E_B-E_A}{2V} = \frac{\alpha}{2V}(\overset{\circ}{N}_A-\overset{\circ}{N}_B) \qquad (37)$$

whereas for $\overset{\circ}{N}_A+\overset{\circ}{N}_B \geq 2$ also the antibonding states have to be populated leading to

$$N_{A,B} = \left(1 \pm \frac{\epsilon}{\sqrt{1+\epsilon^2}}\right) + \left(1 \mp \frac{\epsilon}{\sqrt{1+\epsilon^2}}\right)(\overset{\circ}{N}_A+\overset{\circ}{N}_B-1) \qquad (38)$$

Two limiting cases are especially interesting. If both levels are nearly equal, i.e. $\overset{\circ}{N}_{A,B} = \bar{\overset{\circ}{N}} \pm \Delta\overset{\circ}{N}$, $\Delta\overset{\circ}{N} \ll 1$, one obtains up to linear terms in ΔN_o

$$\Delta N_{A,B} = N_{A,B}-\overset{\circ}{N}_{A,B} = \pm\, \epsilon\, \bar{N} = \pm \frac{\alpha}{4V}(\overset{\circ}{N}_A-\overset{\circ}{N}_B)(\overset{\circ}{N}_A+\overset{\circ}{N}_B) \qquad (39)$$

Thus the charge flows from the higher level to the lower level. This is the general trend shown Fig. 31 for the d-charge transfer. The charge flows to the Fe atom for the impurities listed to the left of Fe in the periodic table and from the Fe atom to the impurity for the atoms to the right of Fe. However an opposite behavior is obtained from eq. (37) and (38), if one of the levels is either nearly filled or nearly empty. For instance for $\overset{\circ}{N}_B = 0$ we obtain always $N_A < \overset{\circ}{N}_A$, since by hybridization charge is transferred to the B-atom. Similarly for a filled B-level ($\overset{\circ}{N}_B=2$) charge is transferred to the A-atoms since by hybridization there is some B admixture in the unoccupied antibonding states. A qualitatively similar, but more elaborate discussion of the d-charge transfer in transition metals has been given by Pettifor [79].

REFERENCES

1. G.N. Rao, Hyperfine Interactions 7, 141 (1979)
   K.S. Krane, Hyperfine Interactions 15/16, 1069 (1983)
2. C.O. Almbladh and V. von Barth, Phys. Rev. B13, 3307 (1976)
3. G. Ries and H. Winter, J. Phys. F9, 1589 (1979); F10, 1 (1980)
4. H. Winter, J. Phys. F11, 2283 (1981)
   J. Benkowitsch and H. Winter, J. Phys. F13, 991 (1983)
5. R.P. Gupta and R.W. Siegel, Phys. Rev. B22, 4572 (1980)
6. B. Chakraborty and R.W. Siegel, Phys. Rev. B27, 4535 (1983)
7. G.F. Koster and J.C. Slater, Phys. Rev. 94, 1392 (1954); 95, 1165 (1954)
8. R. Riedinger, J. Phys. F1, 392 (1971)
9. C.L. Cook and P.V. Smith, J. Phys. F4, 1344 (1974)
10. G.A. Baraff, M. Schlüter, Phys. Rev. Lett. 41, 892 (1978); Phy. Rev. B19, 4965 (1979); J. Bernholc, N.O. Lipari, S.T. Pantelides, Phys. Rev. Lett. 41, 895 (1978); Phys. Rev. B21, 3545 (1980)
11. S.P. Singhal and J. Callaway, Phys. Rev. B19, 5049 (1979)
12. T.H. Dupree, Ann. Phys. (N.Y.) 15, 63 (1961)
13. J.L. Beeby, Proc. Roy. Soy. London, Ser. A302, 113 (1967)
14. J. Morgan, Proc. Roy. Soc. London 16, 365 (1966)
15. N.A.W. Holzwarth, Phys. Rev. B11, 3718 (1975)
16. G. Lehmann, Phys. Stat. Sol. (b) 70, 735 (1975)
17. R. Harris, J. Phys. C3, 172 (1970)
18. M. Hamasaki, S. Asano and J. Yamashita, J. Phys. Soc. Jpn. 41, 378 (1976)
19. K. Terakura, J. Phys. Soc. Jpn. 40, 450 (1976); J. Phys. F6, 1385 (1976); Physica 91B, 162 (1977)
20. H. Katayama-Yoshida, K. Terakura and J. Kanamori, J. Phys. Soc. Jpn. 46, 822 (1979); 48, 1504 (1980); 49, 972 (1980)
21. M. Akai, H. Akai and J. Kanamori, J. Phys. Soc. Jpn. 54, 4246 (1985); 54, 4257 (1985)
22. R. Podloucky, R. Zeller and P.H. Dederichs, Phys. Rev. B22, 5777 (1980)
23. P.J. Braspenning, R. Zeller, A. Lodder and P.H. Dederichs, Phys. Rev. B29, 703 (1984)
24. H. Shiba, Progr. Theor. Phys. 46, 77 (1971)
25. P. Soven, Phys. Rev. B2, 4715 (1970)
26. C. Koenig and E. Daniel, J. de Phys. Lettres 42, L193 (1981)
    C. Koenig, N. Stefanou and J.M. Koch, Phys. Rev. B33, 5307 (1986)
27. O. Gunnarsson, O. Jepsen and O.K. Andersen, Phys. Rev. B27, 7144 (1983)
28. H. Akai and P.H. Dederichs, J. Phys. C18, 2455 (1985)
29. C.G. Broyden, Math. Comput. 19, 577 (1965); G.P. Srivastava, J. Phys. A17, L317 (1984)

30. R. Zeller, J. Deutz and P.H. Dederichs, Sol. State Commun. 44, 993 (1982)
31. A.R. Williams, P.J. Feibelman and N.D. Lang, Phys. Rev. B26, 5433 (1982)
32. R.M. Nieminen and M. Puska, J. Phys. F10, L123 (1980)
33. J. Deutz, P.H. Dederichs and R. Zeller, J. Phys. F11, 1787 (1981)
34. P. Steiner, H. Höchst, W. Steffen and S. Hüfner, Z. Physik 38, 191 (1980)
35. R. Zeller and P.H. Dederichs, Phys. Rev. Lett. 42, 1713 (1979)
36. P.W. Anderson, Phys. Rev. 124, 41 (1961)
37. H.S. Reehal and P.T. Andrews, J. Phys. F10, 1631 (1980)
38. P.T. Andrews and L.T. Brown, in: Physics of Transition Metals 1980, edited by P. Rhodes (IOP, Bristol, 1980), p. 141
39. D. van der Marel, G.A. Sawatzky and F.U. Hillebrecht, Phys. Rev. Lett. 53, 206 (1984)
40. D. van der Marel, C. Westra, G.A. Sawatzky and F.U. Hillebrecht, Phys. Rev. B31, 1936 (1985)
41. R.G. Jordan, W. Drube, D. Straub and F.J. Himpsel, Phys. Rev. B33, 5280 (1986)
42. G.J. Nieuwenhuys, Adv. Phys. 24, 515 (1975)
43. G. Chouteau, Physica (Amsterdam) 84B, 25 (1976)
44. T.D. Cheung, J.S. Kouvel and J.W. Garland, Phys. Rev. B23, 1245 (1981)
45. J.W. Loram and K.A. Mirza, J. Phys. F15, 2213 (1985)
46. A. Oswald, R. Zeller and P.H. Dederichs, Phys. Rev. Lett. 56, 1419 (1986)
47. S. Alexander and P.W. Anderson, Phys. Rev. 133, A1594 (1964)
48. A. Oswald, R. Zeller, P.J. Braspenning and P.H. Dederichs, J. Phys. F15, 193 (1985)
49. J.R. Davis and T.J. Hicks, J. Phys. F9, L7 (1979)
50. J.R. Davis, S.K. Burke and B.D. Rainford, J. Magn. Magn. Mater. 15-18, 151 (1980)
51. P.H. Dederichs, S. Blügel, R. Zeller and H. Akai, Phys. Rev. Lett. 26, 2512 (1984)
52. A. Oswald, R. Zeller and P.H. Dederichs, J. Magn. Magn. Mater. 54-57, 1247 (1986)
53. R. Zeller, in: Physics of Transition Metals 1980, p. 265 (ref. 38)
54. R. Zeller, J. Phys. F17, 2123 (1987)
55. G.G. Low, Adv. in Physics 18, 371 (1969)
56. N. Stefanou, A. Oswald, R. Zeller and P.H. Dederichs, Phys. Rev. B35, 6911, (1987)
57. J. Friedel, Nuovo Cimento 10, Suppl. No. 2, 287 (1958)
58. J. Kanamori, J. Appl Phys. 16, 929 (1965)
59. H. Hayakawa, Prog. Theor. Phys. 37, 213 (1967)
60. I.A. Campbell and A.A. Gomes, Proc. Phys. Soc. 91, 319 (1967)
61. A.R. Williams, A.P. Malozemoff, V.L. Moruzzi and M. Matsui, J. Appl. Phys. 55, 2353 (1984)
62. A.P. Malozemoff, A.R. Williams and V.L. Moruzzi, Phys.

Rev. B29, 1620 (1984)
63. A.R. Williams, V.L. Moruzzi, A.P. Malozemoff and K. Terakura, IEEE Trans. Magn. MAG-19, 1983 (1983)
64. A.P. Malozemoff, A.R. Williams, K. Terakura, V.L. Moruzzi and K. Fukamichi, J. Magn. Magn. Mater. 35, 192 (1983)
65. J.W. Cable, Phys. Rev. B15, 3477 (1977)
66. N. Stefanou et al, to be published.
67. Mößbauer Isomer Shifts, edited by G.K. Shenoy and F.E. Wagner, North Holland Publ. Co. (1978)
68. S. Blügel, diploma thesis, Technische Hochschule Aachen 1983
69. H. Akai, M. Akai, S. Blügel, R. Zeller and P.H. Dederichs, J. Magn. Magn. Mater. 45, 291 (1984)
70. S. Blügel, H. Akai, R. Zeller and P.H. Dederichs, Phys. Rev. B35, 3271 (1987)
71. G. Breit, Phys. Rev. 35, 1447 (1930)
72. H. Akai, S. Blügel, R. Zeller and P.H. Dederichs, Phys. Rev. Lett. 56, 2407 (1986)
73. I. Vincze and I.A. Campbell, J. Phys. F3, 647 (1973)
74. R. Ingals, F. van der Woude and G.A. Sawatzky, in Ref. 67, Chap. 7
75. R.E. Watson and L.H. Bennett, Phys. Rev. B15, 5136 (1977) and B17, 3714 (1978)
R.E. Watson, L.J. Swartzendruber and L.H. Bennett, Phys. Rev. B24, 6211 (1981)
76. R. Ingals, Phys. Rev. 155, 157 (1967) and 162, 18(E) (1967)
77. L. Hodges, R.E. Watson and H. Ehrenreich, Phys. Rev. B5, 3953 (1972)
78. A.J. Freeman and D.E. Ellis, in Ref. 67, Chap. 4
79. D.G. Pettifor, Sol. State Commun. 28, 621 (1978)

THE ELECTRONIC STRUCTURE AND THE STATE OF COMPOSITIONAL ORDER IN METALLIC ALLOYS

B.L. GYORFFY,+ D.D. JOHNSON,# F.J. PINSKI,† D.M. NICHOLSON,* AND G.M. STOCKS*

+ H.H. WILLS PHYSICS LABORATORY, UNIVERSITY OF BRISTOL, BRISTOL, U.K.
# NAVAL RESEARCH LABORATORY, WASHINGTON, D.C. 20375, U.S.A.
† UNIVERSITY OF CINCINNATI, CINCINNATI, OHIO 45221, U.S.A.
* OAK RIDGE NATIONAL LABORATORY, OAK RIDGE, TENNESSEE 37831, U.S.A.

1. INTRODUCTION

1.1. The nature of compositional order

Many two-component (A,B) systems crystallize into a random solid solution (1). In such a state the atoms occupy a more or less regular array of lattice sites but each site can be A or B in a random fashion. Then, on lowering the temperature, the system will either phase separate or order, starting at some transition temperature T_C. The aim of these lectures is to present a microscopic approach to the understanding of these scientifically interesting and technologically important processes.

To have a clear idea of the basic physics we have in mind, let us recall the salient features of both kinds of phase transformations.

In Fig. 1, we display one of the simplest phase diagrams in the book of Hansen (1). It shows a single coexistence curve along which regions of fcc solid solution of Pd and Rh with concentration on c_1 coexist with macroscopically large regions of concentration c_2. Except at the critical point,

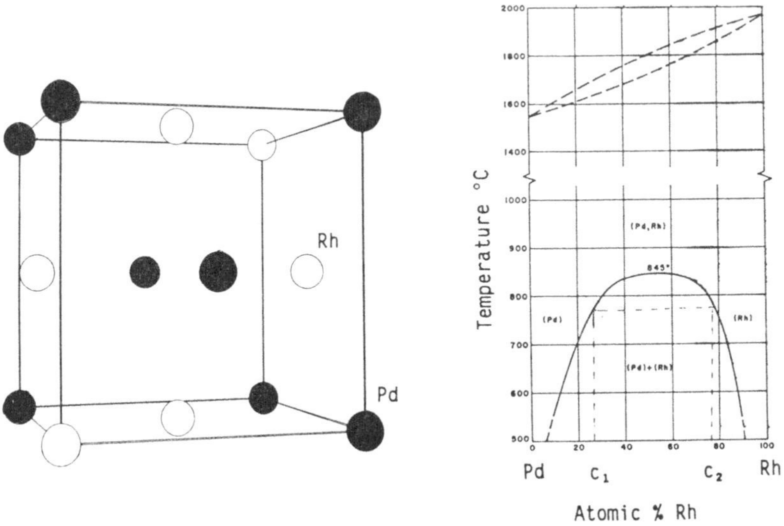

FIGURE 1. Crystal structure of substitutional solid solution Rh_CPd_{1-C} alloys (left). Phase diagram of fcc Rh_CPd_{1-C} alloys (right).

G. M. Stocks and A. Gonis (eds.), Alloy Phase Stability, 421–468.

c = 0.5, the phase-separation phase-transition is first order. At c = 0.5 the transition is continuous and the full range of the associated critical phenomena (2) should occur.

In Fig. 2, we show two well-known examples of compositional ordering. In the case of $Cu_{0.5}Zn_{0.5}$, the ordering takes place on a bcc lattice which naturally breaks up into two sublattices: the corners (I) and the body centers (II).

Above the transition temperature T_c both sublattices are equally likely to be occupied by both kinds of atoms. At T_c the concentration of the copper atoms on sublattice I, c_I, suddenly becomes different from that on sublattice II, c_{II}. Evidently, when this happens the translational invariance on the lattice, i.e., a symmetry of the high-temperature phase, is broken. The order parameter of the phase transition is

$$\eta(T) = c_I - c_{II} \, . \tag{1}$$

It is zero above T_c and increases from 0 to 1 as T decreases to T = 0 following the trajectory depicted in Fig. 2a. Thus, in this instance, the ordering transition is continuous.

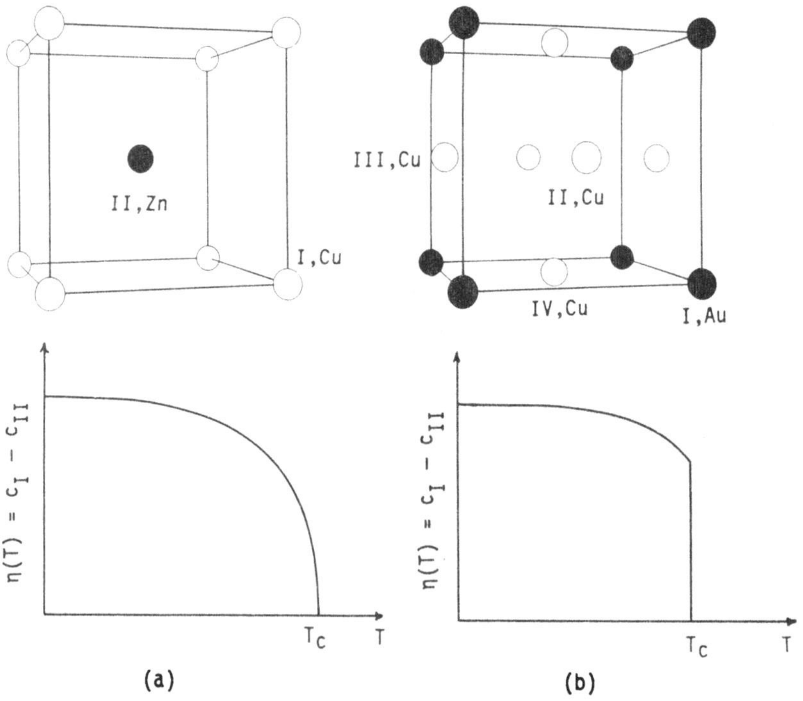

FIGURE 2. The crystal structure and the order parameter η(T) for (a) $Cu_{0.5}Zn_{0.5}$ and (b) $Cu_{0.75}Au_{0.25}$ alloys.

Another famous ordering process takes place in the Cu_cAu_{1-c} system (1), near c = 0.75, on an fcc lattice. This can be thought of as four interpenetrating simple cubic sublattices starting at the points labeled by 1, 2, 3, and 4, respectively. At the transition temperature the concentration of copper on the sublattice forming the corners of the picture shown, c_I, becomes less than that on the other three sublattices $c_{II} = c_{III} = c_{IV}$. The order parameter, η, is again $c_I - c_{II}$. However, as shown in Fig. 2b, η drops to zero at T_c discontinuously. Namely, the phase transition is first order.

To increase the precision of our discussion let us introduce the occupation variable

$$\xi_i = \begin{cases} 1 \text{ if there is an A atom at } \vec{R}_i \\ 0 \text{ if there is a B atom at } \vec{R}_i \end{cases} , \tag{2}$$

where $\vec{R}_i$ is the position vector of the ith site. The thermal average of ξ_i is the local concentration

$$c_i = \langle \xi_i \rangle . \tag{3}$$

In a homogeneous random solid solution c_i is the same, $\bar{c}$, for all sites. Symmetry is broken if some subset of $\{c_i\}$ is different from the others. This situation is conveniently described in the language of concentration waves (3). In short, one writes the deviation of c_i from the average concentration $\bar{c}$ as a linear superposition of static concentration waves

$$c_i = \bar{c} + \frac{1}{2} \sum_{\nu} (c_\nu e^{i\vec{k}_\nu \cdot \vec{R}_i} + c_\nu^* e^{-i\vec{k}_\nu \cdot \vec{R}_i}) , \tag{4}$$

with wave vectors $\vec{k}_\nu$ and amplitudes c_ν. Usually only a few wave vectors are needed. For instance, the Cu_3Au ($L1_2$) structure is described by

$$c_i = \bar{c} + \frac{1}{4} \eta \, (e^{i\vec{k}_1 \cdot \vec{R}_i} + e^{i\vec{k}_2 \cdot \vec{R}_i} + e^{i\vec{k}_3 \cdot \vec{R}_i}) , \tag{5}$$

where the wave vectors are $\vec{k}_1 = 2\pi/a(1,0,0)$, $\vec{k}_2 = 2\pi/a(0,1,0)$, $\vec{k}_3 = 2\pi/a(0,0,1)$ and *a* is the lattice parameter. It can be easily ascertained that Eq. (5) gives $c_I = \bar{c} - \frac{1}{4}\eta$ and $c_{II} = c_{III} = \bar{c}_{IV} = c + \frac{1}{4}\eta$.

A fact that makes the above description particularly attractive is that the concentration waves can be directly observed in X-ray, electron, and neutron diffraction experiments (4). If we denote $\{\vec{K}_j\}$, the reciprocal lattice vectors of the underlying lattice then, in the solid solution, one observes Bragg peaks at $\{\vec{K}_j\}$. These are called the fundamentals. When the symmetry is broken and one or more concentration waves acquire a finite amplitude, there will be some extra, superlattice spots in the diffraction pattern at $\{\vec{K}_j + \vec{k}_\nu\}$. The intensity of the superlattice peaks measure the amplitudes $\{c_\nu\}$.

In addition to the Bragg peaks, diffraction experiments can also measure the diffuse scattering intensity distributed all through the Brillouin zone (BZ). From the point-of-view of our present concern, this is interesting for it contains useful information about the occupational, two-point, correlation function

$$q(i,j) = \langle \xi_i \xi_j \rangle - \langle \xi_i \rangle \langle \xi_j \rangle \ , \tag{6}$$

in other words, short-range order.

In the disordered translational invariant, state q(i,j) depends only on the coordinate difference $\vec{R}_i - \vec{R}_j$ and hence may be written as

$$q(i,j) = \int_{BZ} d^3k \ q(\vec{k}) e^{i\vec{k}\cdot(\vec{R}_i - \vec{R}_j)} \ . \tag{7}$$

It turns out that the measured diffuse scattering intensity $I(\vec{k})$, where $\hbar\vec{k}$ is the momentum transfer in the scattering process, is proportional to $q(\vec{k})$ (Ref. 4).

Interestingly, for alloys with a tendency for the like elements to cluster $q(\vec{k})$ peaks around the fundamentals. For ordering, alloys $q(\vec{k})$ show peaks where the superlattice Bragg peak will be in the ordered state.

To summarize, the state of compositional order is well-described by the wave vectors $\{\vec{k}_\nu\}$ and amplitudes $\{c_\nu\}$ of the concentration waves, which spontaneously break the translational symmetry of the random solid solution, and the correlation function q(i,j). Moreover, these quantities can be directly measured.

The fundamental question these lectures aim to answer is:

1.2 What drives these phase transformations?

This question may be answered on three different levels. The first is the frankly phenomenological approach of Landau (5). The second is more microscopic and is based on effective pair-potential models. The third is the fully microscopic theory which describes the energies of various compositional configurations in electronic terms. We shall now outline all three approaches to the problem.

1.2.1 The Landau theory. Landau formulated his celebrated theory of second order (continuous) phase transition with the specific example of ordering in metallic alloys in mind (5). His basic ideas were elaborated at length by Lifschitz (6), Lyubarskii (7), and more recently by de Fontaine (8), Birman (9), and Khachaturyan (3). They consist of two quite seperate sets of notions (9). The first set of these comprise a group theoretic analysis of the ways the symmetry of the random solid solution may be broken and the energy cost of the symmetry breaking modulation of the concentration (8). Explicitly, denoting the deviation of the local concentration c_i from the average c by δc_i the free energy associated with a concentration fluctuation defined by the set $\{\delta c_i\}$ is assumed to be given by

$$\delta F(\{\delta c_i\}) = \frac{1}{2!} \sum_{ij} \gamma_{ij}^{(2)} \delta c_i \delta c_j + \frac{1}{3!} \sum_{ijk} \gamma_{ijk}^{(3)} \delta c_i \delta c_j \delta c_k$$

$$+ \frac{1}{4!} \sum_{ijk\ell} \gamma_{ijk\ell}^{(4)} \delta c_i \delta c_j \delta c_k \delta c_\ell \ . \qquad (8)$$

Then the algebra of symmetries, namely group theory, is used to classify the symmetry allowed values of the coefficients, $\gamma_{ij}^{(2)}$, $\gamma_{ijk}^{(3)}$, etc. They are assumed to be functions of the state variables such as temperature and pressure and are properties of the high-temperature symmetric phase.

The second set of notions refer to the use the above expression for δF can be put to. For example, in the original formulation the equilibrium state was given by the minimum of $\delta F(\{\delta c_i\})$. Namely, the equilibrium distribution of concentration given by $\{\delta \bar{c}_i\}$ is a solution of the Euler-Lagrange equation

$$\left.\frac{\partial(\delta F)}{\partial(\delta c_i)}\right|_{\{\delta \bar{c}_i\}} = 0 \ . \qquad (9)$$

Following this line of reasoning a second-order phase transition will occur when the high-temperature homogeneous state, $\delta c_i = 0$ for all i, becomes unstable to some specific distortion $\{\delta c_i^o\}$. The relevant modes of distortions $\{\delta c_i^\nu\}$ are eigen "vectors" of the matrix $\gamma_{ij}^{(2)}$:

$$\sum_{ij} \gamma_{ij}^{(2)} \delta c_i^\nu = \lambda_\nu \delta c_i^\nu \ . \qquad (10)$$

Evidently, the free-energy associated with a small amplitude concentration fluctuation, described by $\{\delta c_i^\nu\}$, is given by $\delta F_\nu = \lambda_\nu \sum_i (\delta c_i^\nu)^2$. Thus a fluctuation $\{\delta c_i\}$ for which the eigenvalue $\lambda_\nu < 0$ lowers the free energy. In short, the homogeneous state becomes unstable when the lowest eigenvalue λ_0 becomes zero and therefore the transition temperature T_c is given by $\lambda_0(T_c) = 0$.

The basis of the concentration wave approach of Khachaturyan (3) to compositional fluctuations is the observation that for $\gamma_{ij}^{(2)}$ which depends only on the coordinate difference $\vec{R}_i - \vec{R}_j$ between the positions of the ith and the jth site the eigenvectors are lattice waves: $\delta c_i^{\vec{k}} \sim e^{i\vec{k}\cdot\vec{R}_i}$. Then, the ordered state is described by finite amplitude concentration waves as in Eq. (4). It also follows that the eigenvalues which can be labeled by wave vectors, $\vec{k}$, in the first Brillouin zone of the disordered state, are the lattice Fourier transforms of $\gamma_{ij}^{(2)}$. Namely,

$$\lambda_{\vec{k}} = \gamma^{(2)}(\vec{k}) = \frac{1}{N}\sum_{i} e^{i\vec{k}\cdot(\vec{R}_i-\vec{R}_j)}\, \gamma_{ij}^{(2)} \;. \tag{11}$$

Since the instability occurs when the lowest eigenvalue goes through zero, the ordering wave vectors $\{\vec{k}_\nu\}$ are given by a set of degenerate minima of $\gamma^{(2)}(\vec{k})$. In other words they are the solutions of the equation

$$\vec{\nabla}_{\vec{k}}\, \gamma^{(2)}(\vec{k})\Big|_{\vec{k}_\nu} = 0 \;. \tag{12}$$

A surprising feature of the above theory is that although nothing specific is assumed about the physical forces which give rise to the coefficients $\gamma_{ij}^{(2)}$ some discrete set of characteristic wave vectors emerge as generic solutions of Eq. (12). In fact, as was shown in a remarkable analysis by Lifshitz (6), the interplay of the point group and space group of the disordered phase yields a set of Special Points, at which $\gamma^{(2)}(\vec{k})$ is a minimum, that are determined by symmetry alone.

For example, the Special Points of the fcc lattice are the

$$\begin{array}{l} (1\ 0\ 0),\ (0\ 1\ 0),\ (0\ 0\ 1) \\ (\tfrac12\ \tfrac12\ \tfrac12),\ (\bar{\tfrac12}\ \tfrac12\ \tfrac12),\ (\tfrac12\ \bar{\tfrac12}\ \tfrac12),\ (\tfrac12\ \tfrac12\ \bar{\tfrac12}) \\ (\tfrac12\ 1\ 0),\ (1\ 0\ \tfrac12),\ (0\ \tfrac12\ 1),\ (\bar{\tfrac12}\ \bar{1}\ 0),\ (\bar{1}\ 0\ \bar{\tfrac12}),\ (0\ \bar{\tfrac12}\ \bar{1})\ , \end{array} \tag{13}$$

The $L1_2$ structure described by Eq. (5) is generated by the Special Points (100), (010), and (001). As a further example, we note that the Special Points of the bcc lattice are

$$\begin{array}{l} (1\ 1\ 1) \\ (\tfrac12\ \tfrac12\ 0),\ (0\ \tfrac12\ \tfrac12),\ (\tfrac12\ 0\ \tfrac12),\ (\tfrac12\ \bar{\tfrac12}\ 0),\ (0\ \tfrac12\ \bar{\tfrac12}),\ (\tfrac12\ 0\ \bar{\tfrac12}) \\ (\tfrac12\ \tfrac12\ \tfrac12),\ (\bar{\tfrac12}\ \bar{\tfrac12}\ \bar{\tfrac12}) \end{array} \tag{14}$$

and the ordered $Cu_{0.5}Zn_{0.5}$ superstructure is generated by the Special Point $2\pi/a(1,1,1)$ where a is the lattice parameter.

In a way the above theory is spectacularly successful: the Special Points generate most of the observed ordered phases and there are representative systems corresponding to most Special Points (3,8). Even where it seems to fail, as in the case of incommensurate ordering, the failure is more apparent than real. Since these processes will be of interest later, we shall now comment on them, briefly, to stress their significance.

Landau's group theoretic arguments do no preclude the possibility that $\gamma^{(2)}(\vec{k})$ have minima other than those forced by symmetry. However, it is asserted that if there were some ordering wave vectors $\vec{Q}$, other than the Special Points, they would depend on the accidental, special features of the microscopic interactions and therefore would be extremely rare. Moreover, the periodicities corresponding to such wave vectors would most likely be incommensurate with the underlying lattice. Since at the time when the theory was developed (6) no incommensurate alloy phases were observed, the corresponding sector of the theory was not developed.

However, during the past two decades or so a variety of incommensurate phases have been discovered and the subject of their theoretical description has attracted considerable attention. Our current general understanding of incommensurate phases have been well reviewed by Levanyuk (10). In short, they are natural outcomes of the phenomenological Landau theory with many unexpected features (11,12).

What such phenomenological considerations cannot give are suggestions as to where the new periodicities, incommensurate with that of the disordered state, might come from. For insight in this matter we must turn to more microscopic theories. As we shall try to show in these lectures a very general mechanism which naturally drives incommensurate order has to do with the size and geometry of the Fermi Surface.

1.2.2. Pair potential models. This is a semi-microscopic approach which provides a more satisfactory answer to the question: "What drives the transition?" than the otherwise very powerful symmetry arguments above. Nevertheless, it is based on an effective, classical Hamiltonian and bypasses the essentially quantum mechanical nature of the electronic binding of solids.

The basic assumption of the model is that the configurational part of the energy of a two-component crystalline solid may be written as a sum over pairwise contributions:

$$H_{eff} = -\frac{1}{2}\sum_{ij}\left[\upsilon_{ij}^{AA}\xi_i\xi_j + \upsilon_{ij}^{AB}\xi_i(1-\xi_j) + \upsilon_{ij}^{BA}(1-\xi_i)\xi_j + \upsilon_{ij}^{BB}(1-\xi_i)(1-\xi_j)\right], \tag{15}$$

where $\upsilon_{ij}^{\alpha\beta}$ is the interaction energy of an α(=A,B) atom at the site i with β (=A,B) atom at the site j and the ξ_i's are the occupation variables defined previously. Although a great deal of physics has been neglected in writing down Eq. (15) (Ref. 8), the above Hamiltonian provides a very direct and plausible answer to our question. To see this we rearrange the terms according to the powers of the occupation variables as

$$H_{eff} = -\frac{1}{2}\sum_{ij}\upsilon_{ij}^{(2)}\xi_i\xi_j - \sum_i \upsilon_i^{(1)}\xi_i - \upsilon^{(0)}, \tag{16}$$

where

$$\upsilon_{ij}^{(2)} = \upsilon_{ij}^{AA} + \upsilon_{ij}^{BB} - 2\upsilon_{ij}^{AB},$$

$$\upsilon_i^{(1)} = \sum_j(\upsilon_{ij}^{AB} - \upsilon_{ij}^{BB}),$$

$$v^{(0)} = \frac{1}{2} \sum_{ij} v_{ij}^{BB} . \tag{17}$$

Evidently, if $v_{ij}^{(2)} > 0$ two *A* atoms at sites *i* and *j* ($\xi_i = 1$, $\xi_j = 1$) have lower energy than a pair of AB atoms ($\xi_i = 1$, $\xi_j = 0$) then like atoms cluster together leading to a lower energy and hence, at low enough temperature, to phase separation. On the other hand, for $v_{ij}^{(2)} < 0$ unlike pairs are energetically more favourable than like pairs. Clearly, this gives rise to an ordering tendency. Thus, $-v_{ij}^{AB}$ being more negative than $-v_{ij}^{AA}$ and $-v_{ij}^{BB}$ means $v_{ij}^{(2)} < 0$ and hence drives ordering processes. The opposite case, i.e., $-v_{ij}^{AA}$ and $-v_{ij}^{BB}$ being more negative than $-v_{ij}^{AB}$, implies $v_{ij}^{(2)} > 0$ and hence clustering.

From a technical point-of-view the problem defined by the above Hamiltonian is to find the grand partition function

$$Z(T,V,\nu) = \sum_{\{\xi_i\}} e^{-\beta[H_{eff}(\{\xi_i\}) - \nu \sum_i (\xi_i - ½)]} , \tag{18}$$

where *V* is the volume, ν is the chemical potential difference $\mu^A - \mu^B$ and hence the grand potential

$$\Omega = -\frac{1}{\beta} \ln Z . \tag{19}$$

Although we shall not make use of this formalism directly in these lectures, we mentioned the above details because there is one result we shall need. Allow, formally, the chemical potential difference ν_i to vary from site to site in Eq. (18). Namely, replace $\nu\sum_i \xi_i$ by $\sum_i \nu_i \xi_i$. Then it is straightforward to show that Ω becomes the generator of the correlation functions

$$\langle\xi_i\rangle = c_i = -\left(\frac{\partial\Omega}{\partial\nu_i}\right)_\nu \tag{20}$$

$$\langle\xi_i\xi_j\rangle - \langle\xi_i\rangle\langle\xi_j\rangle = q(i,j) = \frac{1}{\beta}\left(\frac{\partial c_i}{\partial\nu_j}\right)_\nu = -\frac{1}{\beta}\left(\frac{\partial^2\Omega}{\partial\nu_i\partial\nu_j}\right)_\nu . \tag{21}$$

As usual, after the various derivatives have been taken, ν_i is set equal to the thermodynamic value of ν at all sites. In what follows we shall make use of the above device repeatedly for generating the correlation functions of the occupation variables.

1.2.3. The electronic theory of compositional order. In the real world the positively charged atomic nuclei of solids are held together by the electron "glue" between them. This "glue" is a strongly interacting degenerate Fermi-liquid with deeply quantum mechanical properties.

Although it is tempting to attempt to apportion the electrons among the nuclei in such a way that the resulting roughly neutral, atomic-like, objects are only weakly, and therefore only pairwise, interacting; this cannot, in general, be done with impunity. The resulting attractively simple picture, which we have described in Sect. 1.2.2. will continue to break down in unpredictable fashion. Thus, there is a strong case for developing a theory of compositional fluctuations as well as that of crystal structures on the basis of an electronic description of the configurational energies. Obviously, this style of reasoning will provide an explanation of ordering and clustering in terms of the corresponding, presumably favorable, changes in the electronic structure in the spirit of the celebrated paragraph of Peierls' book (13). Closer to our specific concern, an example of how useful such an approach can be is provided by the work of da Silva, et al. (14).

Of course, the charged Fermi-liquid of electrons in an external potential, in the present case resulting from the atomic nuclei, is not fully understood by any means. Consequently, the following discussion will involve many approximations.

However, the burden of the above remarks is that even a very approximate treatment of the electrons may contain qualitative features which are central to the physics of the phenomena in question but have no counterpart in the essentially classical and local description by effective pair potential models.

The first step towards a sensible formulation of the problem is to make the adiabatic approximation. This can be safely done since the mechanism by which an *A* atom can be interchanged with a *B* atom is diffusion, and it is observed to be a very slow process compared with the fast motion of the electrons $\upsilon_{fermi} \sim 10^8$ cm/sec $< \upsilon_{diffusion} \sim 10^5$ cm/sec. Thus, the energy associated with a compositional configuration, specified by a set of occupation variables $\{\xi_i\}$ is the electronic grand potential $\Omega_e(T,\mu,\{\xi_i\})$. It depends on the temperature T, because it includes the entropy associated with thermal fluctuations of electron hole pairs, and the electronic chemical potential μ which determines the average number of electrons $\bar{N}$. This latter quantity has to be carefully related to the total number of *A* and *B* atoms, N^A and N^B, respectively, in such a way that the whole system is neutral. Clearly, $\Omega_e(\{\xi_i\})$, where to simplify the notation we supressed the dependence on T and μ, should be regarded as an effective Hamiltonian for the occupational degrees of freedoms. Hence, the problem at hand is to study the statistical mechanics summarized by the partition function

$$Z = \sum_{\{\xi_i\}} e^{-\beta[\Omega_e(\{\xi_i\}) - \sum_i \nu_i(\xi_i - \frac{1}{2})]} , \qquad (22)$$

and the free energy

$$\Omega = -\frac{1}{\beta} \ln Z . \qquad (23)$$

Although we have arrived at a formulation which is very close to that encapsulated by Eqs. (18) and (19), it is important to note the difference. $H_{eff}(\{\xi_i\})$ in Eq. (18) is a simple polynomial in the occupation variables while $\Omega_e(\{\xi_i\})$ in Eq. (22) is decidedly not. This difference reflects the long-ranged and nonlocal nature of the interactions between site occupations communicated by the electrons. Of the three different modes of analysis described above we shall follow the last in these lectures.

Although the problems have been simplified by the adiabatic approximation, it is still too difficult to be usefully tackled head on. Note that according to Eqs. (22) and (23) we have to solve the full many-electron problem for each compositional configuration $\{\xi_i\}$ and then the summation over each configuration has to be carried out. To render the problem tractable we shall make three closely related approximations, each in the spirit of mean field theory. Therefore, we shall call the collective results of these a *first principles mean field theory*. The logic of the whole scheme will be described in the next section. The way it can be implemented will be the subject of the third section. In the final section we shall make contact with the phenomenological Landau theory and derive a first principles version of the Cahn-Hilliard theories which are particularly useful for describing alloys with macroscopic inhomogeneities.

2. A FIRST PRINCIPLES MEAN-FIELD THEORY OF COMPOSITIONAL ORDER

2.1. A mean-field theory for the statistical mechanics of the concentration fluctuations

In this subsection we assume that $\Omega_e(\{\xi_i\})$ is known and investigate how it determines such physical quantities as c_i and $q(i,j)$ defined in Eqs. (3) and (6). Ways of calculating the relevant features of $\Omega_e(\{\xi_i\})$ by solving the quantum mechanical problem of the many electron system in the presence of the atomic nuclei will be dealt with in the next subsection.

The three most well-tried methods for solving classical statistical mechanical problems like the one defined in Eqs. (22) and (23) are the mean field approximation (2,3,8), the cluster variational method (15), and Monte Carlo simulations (16). In its general aim the first of these is the most modest and we shall follow this. We do this for three slightly different reasons. The first is the technical one that only for the mean field theory can we devise a tractable scheme for a consistent treatment of the electronic structure. The second, obviously less relevant, consideration is that of the three, only the mean field theory can be equally reliably implemented for short- and long-range interactions. Thirdly, we note that, by now, the mean field theory has a very special and well-understood status in statistical mechanics as the reference from which fluctuations are reckoned. Indeed, it is at this level that a microscopic theory can make contact with the Landau theory mentioned earlier. Also, even if the mean field theory phase diagrams have the wrong shape, as they invariably do, they represent a first systematic step toward understanding the forces which determine them. Hence, the mean field theory provides a useful level at which to compare different systems. In fact such comparisons are quite central to our material specific, nonuniversal concern which is to answer the question, "What drives the transitions?"

If the electronic grand potential $\Omega_e(\{\xi_i\})$ consisted of pairwise contributions, like H_{eff} in Eq. (16), the approximation procedure of the mean field theory would be entirely straightforward. One would merely replace products like $\xi_i\xi_j$ by $\langle\xi_i\rangle\xi_j$ and carry out the sums over all configurations in Eq. (22) without further difficulty. However, because $\Omega_e(\{\xi_i\})$ is not even a polynomial in the occupation variables, a more complicated argument has to be deployed. One way to proceed is the concentration functional approach of Gyorffy and Stocks (17). Here we follow an alternative root to the mean field theory for a general interaction function $\Omega_e(\{\xi_i\})$. The elegant formal device we shall use was invented by Peirls (18), Bogulubov (19), Feynman (20), and probably others, independently. It consists of taking a trial Hamiltonian H_0, which is a function of the occupation variables and depends on a number of parameters to be determined later, and expanding Ω in Eq. (23) in powers of $\Omega_e - H_0$. To first order we find

$$\Omega \cong \Omega^{(1)} = \Omega_0 + \langle \Omega_e - H_0 \rangle \ , \tag{24}$$

where $\Omega_0 = (1/\beta)\ell n Z_0$, with $Z_0 = \sum_{\{\xi_i\}} e^{-\beta[H_0(\{\xi_i\}) - \sum_i \nu_i(\xi_i - ½]}$ and $\langle \ \rangle_0$ means averaging with respect to the probability distribution

$$P(\{\xi_i\}) = \frac{1}{Z_0} e^{-\beta[H_0(\{\xi_i\}) - \sum_i \nu_i(\xi_i - ½)]} \ . \tag{25}$$

The key to the method is the proof, given in Feynman's book (20), that $\Omega^{(1)}$ is a variational upperbound for an arbitrary H_0, i.e.,

$$\Omega \leq \Omega^{(1)} \ . \tag{26}$$

The procedure to follow is now quite clear. We have to pick some reasonable trial Hamiltonian and then minimize $\Omega^{(1)}$ with respect to the parameters contained in H_0. For the minimal values of the parameters $\Omega^{(1)}$ is a variational approximate to Ω.

As it turns out the mean field theory corresponds to taking H_0 to be of the form

$$H_0 = \sum_i V_i \xi_i \ , \tag{27}$$

where the set $\{V_i\}$ are the variational coefficients. They are determined by the Euler-Lagrange equations

$$\frac{\partial \Omega^{(1)}}{\partial V_i} = 0 \ . \tag{28}$$

To put the above theory into a convenient form we note that in view of Eq. 27 $P(\{\xi_i\})$ factorizes into a product of local probabilities

$$P(\{\xi_i\}) = \prod_i P_i(\xi_i) \ , \tag{29}$$

where $P_i(\xi_i) = e^{-\beta(V_i - \nu_i)\xi_i}/[e^{-\beta(V_i-\nu_i)} + 1]$. Furthermore, the average of ξ_i is given by $\langle \xi_i \rangle = c_i = [e^{\beta(V_i-\nu_i)} + 1]^{-1}$. Let us now reparametrize $P_i(\xi_i)$ in terms of the local concentration c_i in place of V_i. It is easy to show that

$$P_i(\xi_i) = c_i\xi_i + (1 - c_i)(1 - \xi_i) \ . \tag{30}$$

Since the relation between V_i and c_i is monotonic we may change variables in $\Omega^{(1)}$ from V_i to c_i and minimize with respect to c_i. To find $\Omega^{(1)}$ in terms of $\{c_i\}$ we recall that $\Omega_0 - \langle H_0\rangle_0 = - TS_0$ where S_0 is the entropy of the trial ensemble corresponding to $P_0\{\xi_i\}) = \prod_i P_i(\{\xi_i\})$. It then follows that

$$\Omega^{(1)}(\{c_i\}) = k_BT \sum_i [c_i \ln c_i + (1 - c_i)\ln(1 - c_i)] + \langle\Omega_e\rangle_0 - \sum_i \nu_i(c_i - ½) \ . \tag{31}$$

Thus, our new Euler-Lagrange equations are $\partial\Omega^{(1)}/\partial c_i = 0$ and hence the equation of state is given by

$$k_BT \ln \frac{\bar{c}_i}{1 - \bar{c}_i} + S_i^{(1)}(\{c_i\}) - \nu_i = 0 \ , \tag{32}$$

where $\bar{c}_i$ is our approximation to the equilibrium concentration and

$$S_i^{(1)} = \left(\frac{\partial\langle\Omega_e\rangle_0}{\partial c_i}\right)_{\{\bar{c}_i\}} \ . \tag{33}$$

As we have mentioned in the introduction, a theory of compositional order must be, at least, a theory of c_i and the correlation function $q(i,j)$. To find the latter, within the above framework, we take the derivative of Eq. (32) with respect to c_j. This gives

$$\frac{k_BT}{\bar{c}_i(1 - \bar{c}_i)} \delta_{ij} + S_{ij}^{(2)} - \frac{\partial\nu_i}{\partial c_j} = 0 \ , \tag{34}$$

where

$$S_{ij}^{(2)} = \left(\frac{\partial S_i^{(1)}}{\partial c_j}\right)_{\{\bar{c}_i\}} = \left(\frac{\partial^2\langle\Omega_e\rangle_0}{\partial c_j \partial c_i}\right)_{\{\bar{c}_i\}} \ . \tag{35}$$

Then, recalling Eq. (21), we note that $\partial\nu_i/\partial c_j = k_BTq^{-1}(i,j)$ and hence Eq. (34) can be rearranged to read

$$q(i,j) = \beta\bar{c}_i(1 - \bar{c}_i)\delta_{ij} - \beta\bar{c}_i(1 - \bar{c}_i) \sum_\ell S_{i\ell}^{(2)} q(\ell,j) \ . \tag{36}$$

This is a rather general equation for $q(i,j)$ useful for studying fluctuations in a variety of states of long-ranged order. However, the most frequently studied problem is that of correlations in the disordered state. Then $\bar{c}_i = \bar{c}$ for all i and q(i,j) depends only on the coordinate difference $\vec{R}_i - \vec{R}_j$. Under these circumstances we can take the lattice Fourier transform of Eq. (36) and solve the resulting algebraic equation for $q(\vec{k})$ to obtain

$$q(\vec{k}) = \frac{\beta\bar{c}(1-\bar{c})}{1-\beta\bar{c}(1-\bar{c})S^{(2)}(\vec{k})} . \tag{37}$$

There are many versions of this formula in the literature. The oldest one is due to Krivgolaz (4) and Clapp and Moss (21). These authors were working with a pair-potential model, like the one described by Eq. (16), and used the random-phase approximation. They arrived at Eq. (37) with $S^{(2)}(\vec{k})$ replaced by the lattice Fourier transform of the interchange potential $\upsilon_{ij}^{(2)}$ namely, $\upsilon^{(2)}(\vec{k})$. However, as was stressed by Gyorffy and Stocks (17), Eq. (37) is an exact formula if $S_{ij}^{(2)}$ is the exact direct correlation function (22). In the present derivation $S_{ij}^{(2)}$ is the mean field theory approximation (equivalent to the random-phase approximation) to the direct correlation function. However, it is for an electronic description of the interchange energy rather than an effective pair-potential model. The first electronic theory of $q(\vec{k})$ in the form of Eq. (37) was proposed by Ducastelle et al. (23,24). Apart from minor technical details their work differs from those described in these lectures only in the technique of handling the electronic structure. Theirs are non-self-consistent tight-binding calculations while we shall make use of the local-density approximation within the context of density functional theory. Also there is a slight difference in the interpretation given to $S^{(2)}(\vec{k})$. They regard it as an effective pair potential to be used in a full statistical mechanical calculation while we prefer to view it as the mean field theory approximation to the direct correlation function. The reasons for this interpretation will be elaborated on in the next section. A recent discussion of these various views has been given by Turchi et al. (25).

To summarize this section, our principal results are the equation of state given in Eq. (32), the formula for the correlation function q(k) given in Eq. (37), and the definitions of the direct correlation functions $S_i^{(1)}$ and $S_{ij}^{(2)}$ given in Eqs. (33) and (35). The main point is that all this has been achieved on the basis of a fully electronic description of the free energy associated with thermal fluctuations of the composition. Evidently, this information is contained in the averaged electronic grand potential $\langle\Omega_e\rangle_0$ which is by construction, a function of the local concentrations $\{c_i\}$. The task for the rest of the theory is to evaluate the relevant derivatives of $\langle\Omega_e\rangle_0$.

2.2. The mean field theory for the crystal potential

We shall base our discussion of the electrons on the density functional theory of Hohenberg and Kohn (26). In particular we shall describe the states of the electron system using the local density approximation (LDA) (27). Clearly, given the success of the corresponding self-consistent band theory in accounting for cohesive and structural energies of ordered solids (28–30), a study of ordering energies on the same basis is called for. As we have stressed already, our aim here is to provide the foundation of just such analysis.

It would take us too far from the main line of our argument to recall the formal structure of the density functional theory or to derive the LDA. Let it suffice to summarize its implications for the problem at hand. These are that for a fixed configuration $\{\xi_i\}$ the equilibrium charge density $n(\vec{r};\{\xi_i\})$ is given by

$$n(\vec{r};\{\xi_i\}) = \sum_n |\psi_n(\vec{r};\{\xi_i\}|^2 f[\epsilon_n(\{\xi_i\})] \ , \tag{38}$$

where $f(\epsilon)$ is the usual Fermi function: $1/[e^{+\beta(\epsilon-\mu)} + 1]$ for the inverse temperature β and electron chemical potential μ, and the Kohn-Sham wave functions $\psi_n(\vec{r};\{\xi_i\})$ are solutions of the self-consistant Schrödinger equation

$$\{-\nabla^2 + \sum_i \upsilon_i(\vec{r};[n])\}\psi_n(\vec{r}) = \epsilon_n\psi_n(\vec{r}) \ . \tag{39}$$

Unlike in the usual applications of the theory for ordered systems, here the potential function in each unit cell $\upsilon_i(\vec{r};[n])$ is different from that in all the others. Nevertheless, this is only a consequence of the configuration dependence of $n(\vec{r};\{\xi_i\})$ and $\upsilon_i(\vec{r};[n])$ is the same LDA functional of $n(\vec{r};\{\xi_i\})$ as it is in the ordered case.

The above two equations, Eqs. (38) and (39), are a complete statement of the theory for the ground state charge density $n(\vec{r};\{\xi_i\})$. Moreover, since the electronic grand potential Ω_e is a unique functional of the charge density $n(\vec{r};\{\xi_i\})$, it is also the LDA theory of the sought-after quantity: $\Omega_e(\{\xi_i\})$. To find an explicit expression for $\Omega_e(\{\xi_i\})$ one merely has to integrate the Maxwell relation

$$\bar{N}(\mu,\{\xi_i\}) = -\left(\frac{\partial\Omega_e}{\partial\mu}\right) \ , \tag{40}$$

where the average number of electrons is given by

$$\bar{N}(\mu,\{\xi_i\}) = \int d^3r \ n(\vec{r};\{\xi_i\}) \ . \tag{41}$$

Thus, to find $\langle\Omega_e\rangle_0$ of Eqs. (33) and (34), we must solve Eq. (39) self-consistantly, using the LDA for the crystal potential, for each configuration $\{\xi_i\}$, evaluate the corresponding grand potential $\Omega_e(\{\xi_i\})$ and average the result with respect to the product distribution given in Eq. (29). Clearly, this is an impossible task which must be simplified if the theory is to be of practical use. The purpose of this section is to introduce such a simplification in the spirit of the mean field theory.

The way to make progress is not very difficult to see: when considering the site at $\vec{R}_i$ we should neglect the fluctuations at the other sites. Namely, in evaluating $\upsilon_i\{\vec{r};[n(\vec{r};\{\xi_i\})]\}$ we should replace all ξ_j $j\neq i$ by their thermal averages c_j:

$$\upsilon_i(\vec{r};n) \equiv \upsilon_i(\vec{r};\{\xi_i\}) \cong \bar{\upsilon}_i(\vec{r};\xi_i;\{c_j\}) \ . \tag{42}$$

More explicitly,

$$\bar{\upsilon}_i(\vec{r};\xi_i;\{c_j\}) = \xi_i\bar{\upsilon}^A(\vec{r} - \vec{R}_i;\{c_j\}) + (1 - \xi_i)\bar{\upsilon}^B(\vec{r} - \vec{R}_i;\{c_j\}) \ . \tag{43}$$

To highlight the physical meaning of this approximation, we note that in the disordered state, where most calculations to date have been carried out, $c_i = \bar{c}$ for all *i* and hence all the potential wells on the *A* or *B* sites are described by the same potential functions $\bar{\upsilon}^A(\vec{r})$ or $\bar{\upsilon}^B(\vec{r})$, respectively. In other words the fluctuations in the environment have been neglected. This is fully consistant with our guiding spirit of the mean field approximation.

When Eq. (43) is used in Eq. (39) the problem of self-consistant solution is still intractable. However, the mean-field-like approximation in the previous paragraph suggests another. In short, we may interchange the process of self-consistancy and averaging and demand self-consistancy only on the average.

This very fruitful idea was first proposed and implemented by Stocks and Winter (31). To put it into explicit form consider the equation for the Greens function of Eq. (39)

$$[\epsilon + \nabla^2 + \sum_i \upsilon_i(\vec{r} - \vec{R}_i)]G(\vec{r},\vec{r}\,';\epsilon) = \delta(\vec{r} - \vec{r}\,') \ , \tag{44}$$

where

$$\upsilon_i(\vec{r} - \vec{R}_i) = \xi_i\bar{\upsilon}_i^A(\vec{r} - \vec{R}_i) + (1 + \xi_i)\bar{\upsilon}_i^B(\vec{r} - \vec{R}_i) \ , \tag{45}$$

and, for simplicity, we denoted the full dependence of υ^A and υ^B defined in Eq. (43), on $\{c_i\}$ by the suffixes *i*.

Let us now assume that for a fixed set of potential wells we can solve for various averages of the solution to Eq. (44), $G(\vec{r},\vec{r}\,';\epsilon;\{\xi_i\})$, with respect to the distribution defined in Eq. (29). Clearly, if there is an α (*A* or *B*) type of atom at the site *i* the configurationally average charge density is given by

$$\bar{n}_i^\alpha(\vec{r} - \vec{R}_i) = -\frac{1}{\pi}\int d\epsilon f(\epsilon)\mathrm{Im}\langle G(\vec{r},\vec{r};\epsilon;\{\xi_i\})\rangle_{i,\alpha} \ , \tag{46}$$

where $\alpha = A$ or B and the partial average $\langle G \rangle_{i,\alpha}$ denotes the average of G over all members of the compositional ensemble which have an α-type atom at $\vec{R}_i$. It should be stressed that these partially averaged charge densities are defined in such a way that the fully averaged charge density is

$$\bar{n}_i(\vec{r} - \vec{R}_i) = c_i \bar{n}_i^A(\vec{r} - \vec{R}_i) + (1 - c_i)\bar{n}^B(\vec{r} - \vec{R}_i) . \qquad (47)$$

It is now fairly straightforward to say what is meant by a theory which is self-consistent on the average. Since $\bar{v}_i^\alpha(\vec{r} - \vec{R}_i)$ depends only on the charge density in the unit cell centered on $\vec{R}_i$ and the average charge density elsewhere, at the end of the above calculations we can use $\bar{n}_i^A(\vec{r} - \vec{R}_i)$, $\bar{n}_i^B(\vec{r} - \vec{R}_i)$, and $\bar{n}_i(\vec{r} - \vec{R}_i)$ to recalculate $\bar{v}_i^A(\vec{r} - \vec{R}_i)$ and $\bar{v}_i^B(\vec{r} - \vec{R}_i)$. With these new values of $\bar{v}_i^A$ and $\bar{v}_i^B$ one can, at least in principle, return to Eq. (44) and start the whole process over again. This iteration procedure must be continued until the output potentials are the same as the input potentials within some preset limit, i.e., until convergence in the charge density is obtained. This scheme is depicted in Fig. 3.

$$\leftarrow \leftarrow \leftarrow \leftarrow \leftarrow \leftarrow \leftarrow \leftarrow \leftarrow \leftarrow \quad \text{If } \bar{v}_{old}^\alpha \neq \bar{v}_{new}^\alpha \quad \leftarrow \leftarrow \leftarrow \leftarrow \leftarrow \leftarrow \leftarrow \leftarrow \leftarrow \leftarrow$$

$$\begin{pmatrix} \bar{v}_{old}^A(\vec{r}) \\ \\ \bar{v}_{old}^B(\vec{r}) \end{pmatrix} \Rightarrow \left\{ \begin{array}{c} \mathrm{Im}\langle G(\vec{r},\vec{r};\epsilon)\rangle_{iA} \\ \\ \mathrm{Im}\langle G(\vec{r},\vec{r};\epsilon)\rangle_{iB} \end{array} \right\} \Rightarrow \begin{pmatrix} \bar{n}_i^A(\vec{r}) \\ \\ \bar{n}_i^B(\vec{r}) \end{pmatrix} \Rightarrow \begin{pmatrix} \bar{v}_{new}^A(\vec{r}) \\ \\ \bar{v}_{new}^B(\vec{r}) \end{pmatrix}$$

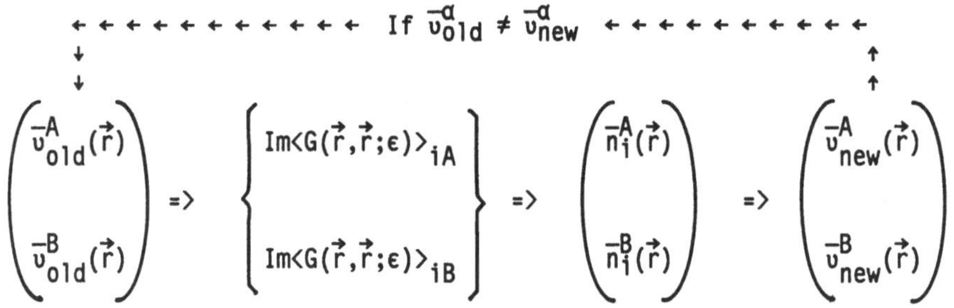

FIGURE 3. The density-functional scheme for self-consistency on the average as implemented in SCF-KKR-CPA calculations.

To summarize, Eq. (43), together with the above self-consistency procedure, constitute our mean field theory for the crystal potential. Of course, the theory is incomplete until a method for calculating the partial averages $\langle G \rangle_{i,A}$ and $\langle G \rangle_{i,B}$ are given. We shall provide the appropriate recipes in the next section.

2.3. The mean field theory for the disorder scattering of electrons - the KKR-CPA

Finally, in this section we complete the theory of $\langle \Omega_e \rangle_0$ by specifying an approximation scheme for calculating $\langle G(\vec{r},\vec{r}';\epsilon;\{\xi_i\})\rangle_0$ starting with Eqs. (44) and (45). The method is based on a multiple scattering version of the Korringa (32), Kohn, and Rostoker (33) (KKR) method of band theory and the coherent-potential approximation (CPA) idea (34,35). The combination of these two techniques is the only reliable first principles method for dealing with the electronic structure of disordered systems. It is called the KKR-CPA. Over the past decade or so it has been reviewed several times at length (31,36–38), and therefore we shall give only the bare outline of the argument here.

Nevertheless, before entering into a technical description of the KKR-CPA, a general remark concerning the CPA is in order. As is emphasized in Ref. 33, it is a very general idea for describing elementary excitations in disordered solids, and it has proved to be remarkably reliable in a variety of circumstances. Significantly, from the point of view of our present concern, it is sometimes referred to as the mean-field theory of disorder. Although it might sound strange to talk about a mean-field approximation in the present context, which is so different from statistical mechanics where the term originates, it is, nevertheless, a legitimate use of the phrase because for many models the CPA can be shown to be exact in the limit where the number of nearest-neighbors go to infinity (39). It is this mean-field-like feature which makes the CPA ideally suited as an approximation scheme for calculating $\langle\Omega_e\rangle_0$. In fact, we believe that this is the only approach which is consistent with the rest of the theory.

As we have already indicated, the KKR-CPA consist of two main parts. Firstly, it is a multiple scattering theory. It is this feature which enables the method to treat the crystal potential from first principles. Secondly, it is a way of accounting for the effects of disorderly arrangement of sites. We begin our discussion by focusing on the first part.

The multiple scattering approach to solving Schrödinger equations, like that in Eq. (44), is most powerful when the scattering centers are described by nonoverlapping potential wells separated by regions, however small, of a constant, interstitial, potential. Band theory calculations based on the density functional theory and LDA often assume that the crystal potential is of this form and maintain this constraint during the self-consistency iterations. For metals this procedure appears to work very well, and in what follows we shall adopt it without further ado. Moreover, we shall assume that each potential well is spherically symmetric. Thus, $\upsilon_i(\vec{r}-\vec{R}_i)$ in Eq. (44) depends only on the magnitude $|\vec{r}-\vec{R}_i|$ and only out to the radius of the inscribed sphere of the ith unit cell, i.e., the muffin-tin radius, r_{MT}. Between the cell boundary and the muffin-tin sphere, the potential is taken to be a constant, V_{MTZ}, which is assumed to be the same over the whole crystal.

The essence of multiple scattering theory is that one solves all the relevant single scattering problems first and then combines the results to find the energy eigenstates in the presence of many scatterers. When the scattering centers are not overlapping, as in the present case, it is sufficient to know the partial wave scattering amplitudes $f_{i,\ell}(\epsilon)$, where ℓ is the asymuthal quantum number and polar quantum number m does not appear because of spherical symmetry. As shown in Ref. 36, for example, the scattering amplitudes $f_{i,\ell}$ may be written in terms of the corresponding scattering phase shifts

$$f_{i,\ell}(\epsilon) = \frac{1}{2i}\left[e^{i2\delta_\ell^i(\epsilon)} - 1\right] . \tag{48}$$

For technical reasons we shall not make use of these scattering amplitudes directly but develop the theory in terms of the closely related t-matrix on the energy shell

$$t_{i,L}(\epsilon) = -\frac{1}{\sqrt{\epsilon}}\, f_{i,\ell}(\epsilon) , \tag{49}$$

where L stands for both ℓ and m.

The t-matrix $t_{i,L}(\epsilon)$ is the amplitude of the out-going spherical wave with angular momentum L due to an incident wave to i^{th} site of unit amplitude in the same angular momentum channel. The multiple scattering generalization of $t_{i,L}(\epsilon)$ is the scattering-path t-matrix $\tau_{LL'}^{ij}(\epsilon)$ which is the amplitude of the outgoing wave emanating from the site i in the angular momentum channel L if there was a unit amplitude scattering wave incident to the site j in the angular momentum channel L'. As is also shown in Ref. 36, this quantity is determined by the individual scattering matrices, $\{t_{i,L}(\epsilon)\}$, and the geometrical arrangement of the scattering centers according to the matrix equation

$$\sum_{\ell,L''} [t_{i,L}^{-1}(\epsilon)\delta_{i\ell}\delta_{LL''} - G_{LL''}(\vec{R}_i - \vec{R}_\ell;\epsilon)]\tau_{L''L'}^{\ell j}(\epsilon) = \delta_{ij}\delta_{LL'} , \qquad (50)$$

where the real space KKR structure constant $G_{LL'}(\vec{R}_i - \vec{R}_j;\epsilon)$ is given by

$$G_{LL'}(\vec{R}_i - \vec{R}_j;\epsilon) = 4\pi \sum_{L''} i^{\ell-\ell'} C_{LL'}^{L''} h_{L''}^{+}(\sqrt{\epsilon}\ |\vec{R}_i - \vec{R}_j|) Y_{L''}(\hat{R}_{ij}) \qquad (51)$$

for $i \neq j$. The quantities $C_{LL'}^{L''}$, $h_{L''}^{+}$ and Y_L are the Clebsch-Gordon coefficient-like Gaunt numbers, outgoing wave spherical Hankel functions and the real spherical harmonics, respectively.

Evidently, the matrix $\underset{\approx}{t}^{-1} - \underset{\approx}{G}$, where $\underset{\approx}{t}^{-1}$ stands for $t_{i,L}^{-1}\delta_{ij}\delta_{LL'}$ and $\underset{\approx}{G}$ for $G_{LL'}(\vec{R}_i - \vec{R}_j;\epsilon)$ is an infinite by infinite matrix in both the site, i, j, ···, etc. and angular momentum lables, L, L′, ···, etc. Because the relevant energy bands are all within a Ry or so of V_{MTZ} it is customary to make the low energy approximation of neglecting angular momentum labels beyond a certain ℓ_{max}. In our discussions ℓ_{max} will be mostly two, but occasionally the theory will have been implemented with $\ell_{max} = 3$.

There are two more general remarks about the matrix $\underset{\approx}{t}^{-1} - \underset{\approx}{G}$ which, as can be seen in Eq. (50), is the inverse of $\underset{\approx}{\tau} = \tau_{LL'}^{ij}$, that are in order. The first is the observation that the dependence on the potential functions is limited to the diagonal elements, and the off-diagonal elements are purely concerned with the propagation of free particle spherical waves between the sites, namely the geometrical arrangement of the sites. This separation of potential dependence and geometry of scattering centers is a unique feature of the multiple scattering approach to solving the Schrödinger equation for arrays of potential wells and is the main reason for the fact that the CPA idea can be readily implemented in this framework. The second point which may serve to orient the reader is that if, in addition to the sites forming a regular lattice, all the potential wells and, therefore, all the single-site t-matrices are the same, the matrix $\underset{\approx}{t}^{-1} - \underset{\approx}{G}$ can be diagonalized with respect to the site indicies by taking its lattice Fourier transform. Indeed, by taking the lattice Fourier transform of Eq. (50), we find, using an obvious notation, that

$$\tau_{LL'}(\vec{k};\epsilon) = [t_L^{-1}(\epsilon)\delta_{LL'} - G_{LL'}(\vec{k};\epsilon)]^{-1} \ . \tag{52}$$

A brief scrutiny of the meaning of $\tau_{LL'}(\vec{k};\epsilon)$ reveals that for the $\vec{k}$ and ϵ where it diverges there is a scattered wave even if there is no incident wave. This means that there is an energy eigenstate for the system of scatterers at that $\vec{k}$ and ϵ. Thus to find the spectrum of eigenvalues we must find for each $\vec{k}$ the ϵ's for which the determinant $||\underset{\approx}{t}^{-1} - \underset{\approx}{G}||$ vanishes. Since we have neglected $\ell > \ell_{max}$, $|\underset{\sim}{t}^{-1} - \underset{\sim}{G}(\vec{k};\epsilon)|$ is a finite matrix and hence its determinant can be evaluated with relative ease. In fact, this is the well-known KKR method. Because of that we shall refer to $\underset{\approx}{t}^{-1} - \underset{\approx}{G}$, whether in real space or in Fourier transform space, as the KKR matrix. Evidently, the scattering path t-matrix $\tau_{LL'}^{ij}$ and its lattice Fourier transform $\tau_{LL'}(\vec{k};\epsilon)$ are the inverses of the corresponding KKR matrices.

Assuming that the multiple scattering theory has been solved for $\tau_{LL'}^{ij}(\epsilon)$, one can calculate the Greens function $G(\vec{r},\vec{r};\epsilon)$. For $\vec{r}$ within the ith muffin-tin sphere and $\vec{r}'$ is the jth muffin-tin sphere, as shown in Fig. 4,

$$G(\vec{r},\vec{r}';\epsilon) = \sum_{LL'} [Z_L^i(\vec{r}_i;\epsilon)\tau_{LL'}^{ij}(\epsilon)Z_{L'}^i(\vec{r}_j;\epsilon) - Z_L^i(\vec{r}_i;\epsilon)J_L^j(\vec{r}_j;\epsilon)\delta_{ij}\delta_{LL'}], \tag{53}$$

where the functions $Z_L^i(\vec{r}_i;\epsilon)$ are the regular solutions of the Schrödinger equation in the ith muffin-tin sphere matching smoothly to $t_{i,L}^{-1}j_\ell + h_\ell^+ Y_L$ where j_ℓ and h_ℓ^+ are the spherical Bessel and Hunkel functions, respectively, $J_L^i(\vec{r}_i;\epsilon)$ is a solution which is irregular at the origin and matches smoothly to $j_\ell Y_L$ and, as before, $Y_L(\hat{r})$ is a real spherical harmonic.

The two main uses for the above Greens function are the calculations of the local density of states

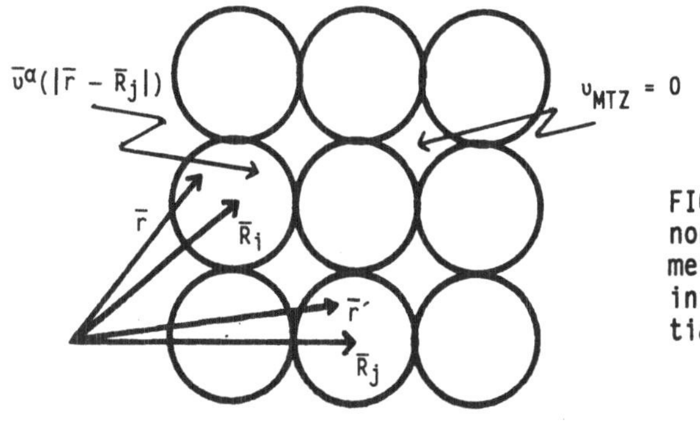

FIGURE 4. Schematic diagram of nonoverlapping spherically symmetric potential wells. In the interstitial region the potential is constant.

$$n_i(\epsilon) = -\frac{1}{\pi}\int_{V_i} d^3r \ \mathrm{Im}\ G(\vec{r},\vec{r};\epsilon) \ , \tag{54}$$

where the integral is over the ith unit cell and the local charge densities

$$n_i(\vec{r}) = -\frac{1}{\pi}\int d\epsilon f(\epsilon)\mathrm{Im}G(\vec{r},\vec{r};\epsilon) \ . \tag{55}$$

Clearly, Eq. (53) is the relevant expression to be used in Eq. (46). In particular

$$\bar{n}_i^{\alpha}(\vec{r}_i) = -\frac{1}{\pi}\ \mathrm{Im} \sum_{LL'} \int d\epsilon f(\epsilon)[Z_L^{i,\alpha}(\vec{r}_i;\epsilon)Z_{L'}^{i,\alpha}(\vec{r}_i;\epsilon)\langle\tau_{LL'}^{ii}(\epsilon)\rangle_{i\alpha}$$

$$- Z_L^{i,\alpha}(\vec{r}_i;\epsilon)J_L^{i,\alpha}(\vec{r}_i;\epsilon)] \ , \tag{56}$$

where $\alpha = A$ or B. Thus the purpose of our theory must be to calculate $\langle\tau_{LL'}^{ii}(\epsilon)\rangle_{i,\alpha}$.

Consider now the relatively simple, homogeneous case: $c_i = \bar{c}$ for all i. Then the CPA argument runs as follows: We assume that the average lattice is described by the effective scattering amplitude $t_{c,L}(\epsilon)$ in the sense that

$$\langle\tau_{LL'}^{ii}(\epsilon)\rangle = \tau_{LL'}^{c,ii}(\epsilon) \ , \tag{57}$$

where $\tau_{LL'}^{c,ii}$ is the solution to Eq. (50) for $t_{i,L} = t_{c,L}$ for all i. More explicitly, using Eq. (52), we find

$$\tau_{LL'}^{c,ii}(\epsilon) = \int_{V_{BZ}} d^3k[t_{c,L}^{-1}(\epsilon)\delta_{LL'} - G_{LL'}(\vec{k};\epsilon)]^{-1} \ . \tag{58}$$

We then study an α- (A or B) type impurity at i in this effective lattice. Such impurity is described by the solution of Eq. (50) for $t_{j,L}(\epsilon) = t_{j,c}(\epsilon)$ for all $j \neq i$ and $t_{i,L} = t_{\alpha,L}(\epsilon)$. Namely,

$$\underset{\approx}{\tau}^{\alpha,ii}(\epsilon) = [\underset{\approx}{1} + \underset{\approx}{\tau}^{c,ii}(\underset{\approx}{t}_A^{-1} - \underset{\approx}{t}_c^{-1})]^{-1}\underset{\approx}{\tau}^{c,ii} \ , \tag{59}$$

where, for simplicity of notation, we replaced the explicit angular momentum label by matrix notation. Up to now the effective t-matrix is not known. The CPA prescription for finding $t_{c,L}(\epsilon)$ is to require that the average of $\underset{\approx}{\tau}^{\alpha,ii}$ is equal to $\underset{\approx}{\tau}^{c,ii}$, i.e.,

$$\bar{c}\tau_{LL'}^{A,ii}(\epsilon) + (1-\bar{c})\tau_{LL'}^{B,ii}(\epsilon) = \tau_{LL'}^{c,ii}(\epsilon)\ . \tag{60}$$

Together Eqs. (58), (59), and (60) are the fundamental equations of the KKR-CPA. Their solution gives $t_{c,L}(\epsilon)$ and, obviously, determines $\tau_{LL'}^{c,ii}$ directly. Practical schemes for solving Eqs. (58), (59), and (60) are discussed at length in Refs. 30, 35, and 36.

Combining the theory presented in this section with the ideas of the previous one, we can formulate the following self-consistent scheme. Starting with the charge densities $\bar{n}^A(\vec{r})$ and $\bar{n}^B(\vec{r})$, we can calculate the partially averaged potential functions $\bar{\upsilon}^A(\vec{r})$ and $\bar{\upsilon}^B(\vec{r})$ using the LDA. Then placing these randomly on a lattice, we can solve for the partially averaged Greens functions $\langle G(\vec{r},\vec{r}';\epsilon)\rangle_{i,\alpha}$ using the KKR-CPA. From such Greens functions we can calculate the partially averaged charge densities $\bar{n}^A(\vec{r})$, $\bar{n}^B(\vec{r})$ and compare them with their starting values. If they disagree at a preset level, repeat the process until convergence. The first such self-consistent KKR-CPA calculation was performed by Stocks and Winter (40). A variety of interesting applications of the method are described in their review article (31).

However, the above homogeneous self-consistent KKR-CPA is not sufficient to solve the problem of calculating the direct correlation functions $S_i^{(1)}$, and $S_{ij}^{(2)}$ defined in Eqs. (33) and (35). Fortunately, the formal structure of the theory generalizes to the case of the inhomogeneous randomness described by the distribution function given in Eqs. (29) and (30). Clearly, for such an ensemble one should search for an effective lattice with the CPA t-matrix varying from site to site $t_{c,i,L}(\epsilon)$. This implies an infinite set of CPA equations. One for each site:

$$c_i\tau_{LL'}^{A,ii}(\epsilon) + (1-c_i)\tau_{LL'}^{B,ii}(\epsilon) = \tau_{LL'}^{c,ii}\ . \tag{61}$$

These are coupled since $\tau_{LL'}^{c,ii}$ depends on the $t_{c,i,L}(\epsilon)$ on every site. Nevertheless, if they are solved the $\tau_{LL'}^{\alpha,ii}$'s will lead, via Eq. (56) and the starting *A* and *B* potential functions, to different *A* and *B* charge densities $\bar{n}_i^{\alpha}(\vec{r})$ on each site. These differences reflect the differences in the local probabilities c_i that the sites *i* is occupied by an atom of the *A* species. Although one might start with the same charge density $\bar{n}^A(\vec{r})$ on all *A* sites and the same $\bar{n}^B(\vec{r})$ on the *B* sites at the end of the first iteration all the $\bar{n}_i^{\alpha}(\vec{r})$'s will be different and hence further iterations will converge to an inhomogeneous distribution of local charge densities.

The above self-consistent inhomogeneous KKR-CPA scheme is the foundation of our electronic theory of concentration fluctuations. However, as it stands, it is not fully complete. We still need an explicit expression for $\Omega_e^{CPA}(\{c_i\})$ which will be our approximation to $\langle\Omega_e\rangle_0$ in Eq. (31). To find such an expression we start with the appropriate version of Eq. (41). Namely,

$$\overline{N}^{CPA}(\mu;\{c_i\}) = \sum_i [c_i \int_{v_i} d^3r_i\overline{n}_i^A(\vec{r}_i) + (1 - c_i) \int_{v_i} d^3r_i\overline{n}^B(\vec{r}_i)] \,, \qquad (62)$$

and integrate it with respect to the electron chemical potential:

$$\Omega_e^{CPA}(\mu;\{c_i\}) = -\int_0^{\mu} d\mu' \overline{N}^{CPA}(\mu';\{c_i\}) \,. \qquad (63)$$

As shown in Ref. 41, Eq. (IV-7), this leads to

$$\Omega_e^{CPA}(\mu;\{c_i\}) = \mu Z - \int_{-\infty}^{\infty} d\epsilon \overline{N}^{CPA}(\epsilon;\{c_i\})f(\epsilon) - e^2 \sum_{i\neq j} \int_{v_i} d^3r \int_{v_j} d^3r' \frac{\overline{n}_i(\vec{r})\overline{n}_j(\vec{r}')}{|r - r'|}$$

$$- e^2 \sum_{i\alpha} c_{i,\alpha} \int_{v_i} d^3r \int d^3r' \frac{\overline{n}_i^{\alpha}(\vec{r})\overline{n}_i^{\alpha}(\vec{r}')}{|\vec{r} - \vec{r}'|}$$

$$+ \sum_{i\alpha} c_{i,\alpha} \int_{v_i} d^3r[\epsilon_{xc}^{LDA}(\overline{n}_i^{\alpha}) - V_{xc}^{LDA}(\overline{n}_i^{\alpha})]\overline{n}_i^{\alpha}(\vec{r}) \,, \qquad (64)$$

where the first two terms are the single particle contributions, Ω_{sp}^{CPA}, and the rest are usually referred to as the double-counting corrections. The averaged integrated density of states $\overline{N}^{CPA}(\epsilon;\{c_i\})$ is given by

$$\overline{N}^{CPA}(\epsilon;\{c_i\}) = N_0(\epsilon) - \frac{1}{\pi} \operatorname{Im}\ln||\underset{\approx}{t}_{c,i}^{-1}(\epsilon)\delta_{ij} - \underset{\approx}{G}(\vec{R}_i - \vec{R}_j;\epsilon||$$

$$- \frac{1}{\pi} \sum_{i\alpha} c_{i,\alpha} \operatorname{Im}\ln||\underset{\approx}{1} + (\underset{\approx}{t}_{i,\alpha}^{-1} - \underset{\approx}{t}_{i,c}^{-1})\underset{\approx}{\tau}^{c,ii}|| \,, \qquad (65)$$

where $N_0(\epsilon)$ is the free electron contribution.

Having arrived at the principal result of this section, Eq. (64), we pause to take note of its general features.

Clearly, the coupled equations of the inhomogeneous KKR-CPA, Eq. (61), cannot be solved in general. The cases where it has been solved or is conceivable that a numerical solution will be affordable are the homogeneous limit $c_i = \overline{c}$ for all i or circumstances where the c_i's form a regular pattern. Examples of the latter cases are superlattices and modulated alloys. Recall that such patterns, as the examples also indicate, correspond to long-range order which breaks the underlying symmetry of the lattice. However, crucial as such solutions will be to practical applications of the first principle's mean-field theory, they are not the main function of the above self-consistent inhomogeneous KKR-CPA scheme. That role is purely formal. The point is that because the scheme allows, explicitly, all the local concentrations to be different, it facilitates the formal evaluation of the derivatives $\partial\Omega^{CPA}/\partial c_i$, $\partial^2{}^{CPA}/\partial c_i\partial c_j$ which determine the direct correlation functions $S_i^{(1)}$ and $S_{ij}^{(2)}$. Thus, we have derived a method for generating explicit expressions for all the direct correlation functions $S_i^{(1)}$, $S_{ij}^{(2)}$, $S_{ijk}^{(3)}$, ··· . Moreover, once these expressions have been derived for an arbitrary concentration configuration $\{c_i\}$ they can be evaluated in all of the soluable arrangements discussed above. As will be seen presently this means that we have tractable theories for various states of long-range order, i.e., spontaneously formed concentration waves, and fluctuations about these states.

The second feature of the above general framework we wish to call attention to is its self-consistency in the density functional sense. The inhomogeneous CPA idea was first used by Gautier et al. (24) for more or less the same purpose as we have done here. However, they worked with a tight-binding model Hamiltonian description of the electrons and hence, their scope for charge self-consistency was limited to ad hoc procedures. The present theory is a synthesis of advances in band theory (KKR), density functional theory (LDA), and theory of disordered systems (CPA) at a level where all these elements fit together in a consistent fashion. It is this consistency which makes the present approach a fully first-principles one in the sense that the calculations are free of parameters. We shall return to a general discussion of the full significance of this fact toward the end of these lectures.

To summarize Sect. II and Subsects. 1, 2, and 3, we have developed the three principal steps which together constitute a general first principles mean-field theory of concentration fluctuations. This theory is general in the sense that it is applicable, for a fixed lattice, to arbitrary states of long-range order. We hasten to add that the full implication of this formal structure remains largely unexplored. In fact, all calculations to date are studies of the disordered state only. Thus, rather than developing the theory further in general terms, in the next chapter we explore some of its explicit consequences in the disordered state.

3. CONCENTRATION FLUCTUATIONS IN THE DISORDERED STATES

3.1 The averaged ground state energy

The simplest example where the rather elaborate framework of the previous section can be applied is that of a homogeneous random alloy, $c_i = \bar{c}$ for all i at zero temperature T=0. Of course, in accordance with the third law of thermodynamics, at T=0 such a state would not be an equilibrium state. Nevertheless, its configurationally averaged energy, $\bar{E}$, is of general interest as a reference energy for a variety of considerations. In this subsection, we report on the first ab initio total energy calculation of $\bar{E}$ for a random alloy. The particular system to be studied is Cu_cZn_{1-c}.

Because it is much easier to work at a fixed chemical potential μ than with a fixed number of electrons, we shall study the T→0 limit of the averaged grand potential Ω_e, namely, $\bar{E} - \mu\bar{N}$.

It follows from Eq. (64) and the CPA equations given in Eq. (61), as was first shown by Johnson et al. (42), that

$$\bar{E} = \sum_{\alpha} c_\alpha E^J[\bar{n}_\alpha, \bar{n}_o] \,, \tag{66}$$

where E^J is an expression of the same form as that obtained by Janak (43) [see Eq. (25) in Ref. 43] for pure metal muffin-tin and interstitial charge densities, $n(\vec{r})$ and n_0, respectively. Thus, having iterated the self-consistent KKR-CPA to convergence in the partially averaged charge densities $\bar{n}^{\alpha}(\vec{r})$ (α = A,B) and the average interstitial charge density $\bar{n}_0$ the configurationally average total energy can be calculated with relatively modest further effort.

The first such calculation was carried out by Johnson et al. (42) using the exchange correlation potential of Hedin and Lundqvist. They calculated $\bar{E}$ for Cu_cZn_{1-c} in the α-phase at a number of lattice parameters for each concentration $\bar{c}$. They then determined the equilibrium lattice parameter 'a' for which $\bar{E}$ is minimum. Their results for 'a' are shown in Fig. 5 as a function of concentration. Evidently, they find an almost linear relationship between 'a' and $\bar{c}$, in good agreement with Vegard's rule (44) and experiments. Given the ab initio nature of the calculation, this agreement may be taken as strong evidence that the self-consistent KKR-CPA is a sound basis on which to build a theory of compositional order.

An important feature of the above calculation is the consistency between the mean-field-like approximation of the crystal potential, namely, the neglecting of fluctuations in the environment when calculating the potential function at a specified site, and the single-site nature of the CPA way of averaging over all configurations. It is due to this consistency

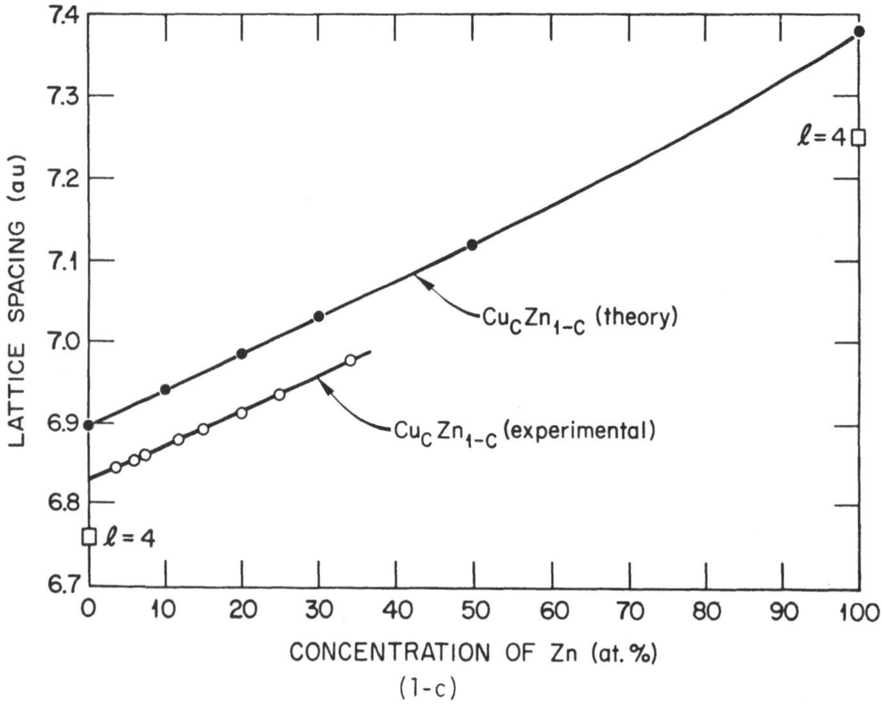

FIGURE 5. The calculated (solid circles) and experimentally observed (open circles) lattice parameter as a function of the concentration c for the fcc Cu_cZn_{1-c} alloys.

that the self-consistent KKR-CPA can be considered a proper, albeit approximate, density-functional theory for disordered systems. There are two formal consequences of Eq. (66) which are particularly significant from this point of view (45). The first is the fact that $\bar{E}$ is stationary with respect to variations in $\bar{n}_\alpha(\vec{r})$ for $\alpha = A$ or B, i.e.,

$$\frac{\delta}{\delta \bar{n}^\alpha(\vec{r})} \bar{E}[\bar{n}^A, \bar{n}^B, \bar{n}_\circ] = 0 \ . \tag{67}$$

The second is that $\bar{E}$ may be uniquely broken up into a kinetic energy T and potential energy, U, contributions and, as shown in Ref. 45

$$\frac{\delta}{\delta \bar{n}^\alpha(\vec{r})} \bar{U}[\bar{n}^A, \bar{n}^B, \bar{n}_\circ] = \bar{\upsilon}^\alpha(\vec{r}) \ . \tag{68}$$

It should be stressed that even though the order of charge self-consistency and averaging have been interchanged, the theory preserved these relations of the density functional theory because of the variational character of the CPA and the choice of the partially averaged local potentials.

3.2. Equation of state

It follows from Eqs. (32), (33), and the discussion in Sect. 2.1. that in the disordered state the equation of state is given by

$$k_B T \ \ell n \frac{\bar{c}}{1 - \bar{c}} + S^{(1)}(\bar{c}) - \nu = 0 \ , \tag{69}$$

where the local molecular field contribution to the chemical potential is given by

$$S^{(1)}(\bar{c}) = S_i^{(1)} \Big|_{c_i = \bar{c} \forall_i} \ . \tag{70}$$

Evidently, the finite temperature version of the homogeneous CPA in the previous section would not provide sufficient information to calculate the derivative with respect to c_i in Eq. (70). For this we must return to the inhomogeneous KKR-CPA of Sect. 2. Thus we proceed by taking the derivative of Eq. (64):

$$\frac{\partial \Omega_e^{CPA}}{\partial c_i} = - \int_{-\infty}^{\infty} d\epsilon \ \frac{\partial}{\partial c_i} \bar{N}^{CPA}(\epsilon;\{c_i\}) + \frac{\partial}{\partial c_i} (\delta \Omega_e^{CPA}) \ , \tag{71}$$

where we collected all the double-counting corrections to Ω_e^{CPA} into the symbol $\delta\Omega_e^{CPA}$. While it is fairly straightforward to continue our discussion retaining $\delta\Omega_e^{CPA}$, and it is frequently necessary in the interest of a more straightforward presentation, we shall drop this term in what follows.

The next step is to take the derivative of N^{CPA} in Eq. (65) with respect to c_i. This will have two contributions: the first is the result of taking the derivative with respect to c_i for fixed values of the effective t-matrices $\{t_{i,c,L}(\epsilon)\}$, and the second involves the derivatives of N^{CPA} with respect to the $t_{i,c,L}(\epsilon)$'s resulting from the chain rule. One of the striking properties of the KKR-CPA is that the averaged integrated density of states $\overline{N}^{CPA}$ is stationary with respect to variations in all the scattering amplitudes $t_{i,c,L}(\epsilon)$. Thus the second contributions to $\partial/\partial c_i$ $\overline{N}^{CPA}(\epsilon)$ will vanish identically and we obtain

$$S_{sp}^{(1)}(\bar{c}) = \int d\epsilon f(\epsilon)[\overline{N}^A(\epsilon) - \overline{N}^B(\epsilon)] \ , \tag{72}$$

where for $\alpha = A$ or B

$$\overline{N}^{\alpha}(\epsilon) = \frac{1}{\pi} \operatorname{Im} \ln || \underset{\approx}{1} + (\underset{\approx}{t}_{\alpha}^{-1} - \underset{\approx}{t}_{c}^{-1})\underset{\approx}{\tau}^{c,ii} || \ , \tag{73}$$

and we have taken $c_i = \bar{c}$ for all i after differentiating with respect to c_i. Consequently, all the quantities are to be evaluated in the same homogeneous self-consistent KKR-CPA we have studied in Subsect. 3.1. The subscript sp on $S_{sp}^{(1)}$ indicates that we have included only the single particle (sp) contributions, and we have neglected those coming from the double-counting corrections.

The notation $\overline{N}^{\alpha}(\epsilon)$ for the right-hand side of Eq. (73) is designed to suggest that the corresponding expression is closely related to the integrated density of states on an α-type site. This would be strictly true if $\underset{\approx}{t}_c(\epsilon)$ corresponded to an energy independent potential. However, the effective scattering matrix $t_{c,L}(\epsilon)$ corresponds to a potential function which is energy dependent, and therefore, the energy derivative of $N^A(\epsilon)$ and $N^B(\epsilon)$ defined in Eq. (73) is not the densities of states $n^A(\epsilon)$ and $n^B(\epsilon)$ defined by

$$\overline{n}_i^{\alpha}(\epsilon) = -\frac{1}{\pi} \int_{V_i} d^3r \ \operatorname{Im} \langle G(\vec{r},\vec{r};\epsilon)\rangle_{i,\alpha}^{CPA} \ . \tag{74}$$

In spite of this ambiguity it is useful to regard $N^A(\epsilon)$ and $N^B(\epsilon)$ as integrated densities of states on *A* and *B* sites because then we can interpret the "molecular field" contribution to the local chemical potential $S_{sp}^{(1)}(c)$ in Eq. (72) as the difference between the local grand potential on an *A* site, $\delta\Omega^A$, and on a *B* site, $\delta\Omega^B$. This is a simple and physically very appealing consequence of our theory.

To summarize, we may determine the equilibrium concentration at a given temperature and chemical potential by solving Eq. (69) using $S^{(1)}(c)$

calculated at a sequence of concentrations using the homogeneous, self-consistent KKR-CPA. A schematic construction of the graphical solution is shown in Fig. 6. Clearly when $S^{(1)}(c)$ is a decreasing function of concentration a van der Waals loop can develop. It signals a phase separation transition for which the coexistence curve is given by the usual Maxwell construction.

Although, as yet, no system has been analysed using the above theory, the coexistence curve of the Pd_cRh_{1-c} system has been calculated by an equivalent procedure using self-consistent KKR-CPA calculations (46). The results are shown in Fig. 7. Further details about this calculation will be provided in the next section. Here we merely wish to note that for this simple system we appear to have an even quantitatively adequate theory.

Clearly, Eq. (69) can be formally integrated with respect to c and we obtain for the grand potential per unit cell

$$\frac{1}{N}\,\Omega(\nu) = k_BT[\bar{c}\ell n\bar{c} + (1-\bar{c})\ell n(1-\bar{c}) + \Omega_e^{CPA}(\bar{c}) - \nu\bar{c}\ , \qquad (75)$$

where $\bar{\Omega}_e^{CPA}$ is to be calculated using a homogeneous self-consistent KKR-CPA. In fact, this latter computation would be just the finite temperature version of the average total energy calculation in the previous section. The temperature dependence of $\Omega^{CPA}(\bar{c})$ would come, principally, from the entropy associated with the production of electron-hole pairs by thermal fluctuations. Note that Eq. (75) might be thought of as a sensible alternative to Eq. (69), and hence, as the starting point for point for studying the equilibrium state.

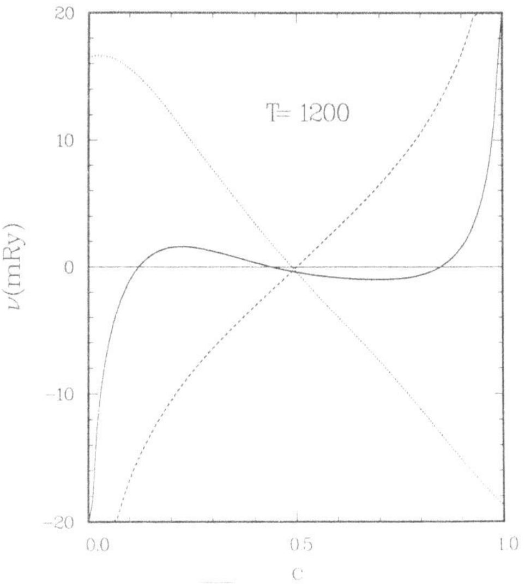

FIGURE 6. The chemical potential difference $\nu = \mu^A - \mu^B$ (solid line) versus concentration c for a generic phase separating binary alloys. The dotted line gives the concentration from $S^{(1)}$ and the dashed line that from the entropy. The actual data is that for $Rh_{0.5}Pd_{0.5}$ at a temperature of 1200 K obtained during the phase diagram calculation of Fig. 7.

Before drawing this section to a close it should be noted that the calculation of $S_i^{(1)}$ in various ordered states is a fairly straightforward generalization of those discussed above. As we have mentioned earlier such ordered states may be described by dividing up the basic lattice of all sites, fully invariant under lattice translations in the disordered state, into a finite set of sublattices with the concentration variable c_i being the same on all sites of a given sublattice but varying from sublattice to sublattice. Under these circumstances the basic CPA condition, Eq. (60), generalizes to a finite set of coupled equations, one for each crystallographically different site. The arguments leading to the charge self-consistency

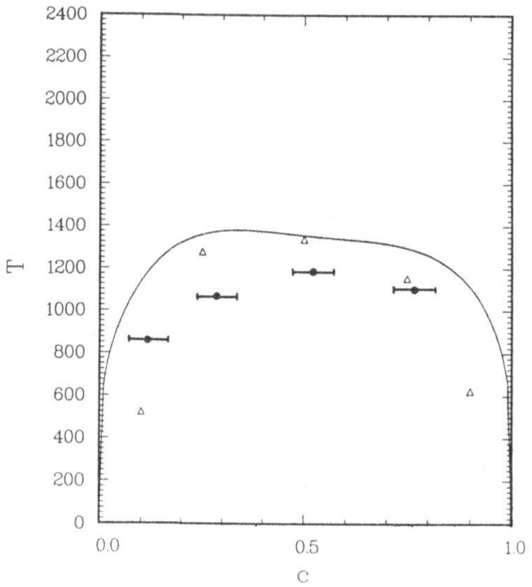

FIGURE 7. The phase boundary for Rh_cPd_{1-c} alloy as calculated by the first-principles KKR-CPA scheme (solid line), the calculated spinodal (triangles), and the experimentally determined coexistence line (dots with error bars). The experimental data is taken from J. E. Shield and R. K. Williams, Scripta Met. **21**, 1475 (1987).

procedure and the formula for $S_i^{(1)}$, Eq. (70), also generalizes without any difficulty. Clearly, the equation of state will be a set of coupled equations for the sublattice concentrations c_I, each of the form given in Eq. (32). The solutions of these will then give the mean-field theory prediction for the order parameters, $c_I - c_{I'}$, of which there may be several. Unlike Landau theories which are designed to work near the transition temperature T_c, the above calculations would yield the order parameters all the way to T=0. Moreover, together with the prescription of the Maxwell construction, the theory is equally applicable to first- and second-order phase transitions.

Unfortunately, as yet, no such full-blooded calculation of compositional order parameters has been performed. However, the time is ripe. The first multiple sublattice KKR-CPA calculations were reported by Pindor et al. (47). Since then a number of other workers (48) have also solved problems using that framework. Making these calculations self-consistent and pursuing them to the point where $S_i^{(1)}$, and therefore the order parameters are also calculated, is likely to be one of the more profitable fields in the first principles studies of energetics in solids.

3.3. The direct correlation functions

Using Eqs. (35) and (71), we may write the single particle contribution to the direct correlation function as

$$S_{ij}^{(2)} = \left(\frac{\partial^2}{\partial c_i \partial c_j} \Omega_e^{CPA}\right)\Big|_{c_i=\bar{c}\forall_i} = -\int_{-\infty}^{\infty} d\epsilon f(\epsilon)\left[\frac{\partial^2}{\partial c_i \partial c_j} \bar{N}^{CPA}(\epsilon;\{c_i\})\right]\Big|_{c_i=\bar{c}\forall_i} . \quad (75)$$

The first of the two derivatives in the above relations can be taken in the same way as in the case of calculating $S_i^{(1)}$ in the previous subsection. This step leads to

$$S_{ij}^{(2)} = -\int_{-\infty}^{\infty} d\epsilon f(\epsilon) \frac{\partial}{\partial c_j} \left[(\bar{N}_i^A(\epsilon) - \bar{N}_i^B(\epsilon)\right] , \quad (76)$$

where the partially averaged "integrated density of states," $\bar{N}_i^\alpha(\epsilon)$, are given by Eq. (73) with the suffix i and hence, the dependence on the whole inhomogeneous concentration field $\{c_i\}$ restored in a self-evident fashion.

When calculating $S_i^{(1)}$ in the disordered state, the $c_i = \bar{c}\forall_i$ limit was taken after a formal expression was derived for the first derivative $\partial/\partial c_i \bar{N}^{CPA}(\epsilon;\{c_i\})$. Here we retain the dependence of $\bar{N}_i^\alpha(\epsilon)$ on $\{c_i\}$ and proceed to derive a formal expression for its derivative with respect to $c_j \forall_j$ before the limit $c_i = \bar{c}\forall_i$ is taken.

With the help of the fully inhomogeneous version of Eq. (73), we find

$$\frac{\partial}{\partial c_j}[\bar{N}_i^A(\epsilon) - \bar{N}_i^B(\epsilon)] =$$

$$-\frac{1}{\pi}\,\mathrm{Im}\sum_{\ell\neq i}\mathrm{tr}\{(\underset{\approx}{D}_{i,A} - \underset{\approx}{D}_{i,B})(\underset{\approx}{\tau}^{c,ii})^{-1}[\underset{\approx}{\tau}^{c,i\ell}(\frac{\partial \underset{\approx}{t}_{\ell,c}^{-1}}{\partial c_j})\underset{\approx}{\tau}^{c,\ell i}] \qquad (77)$$

where "tr" denotes trace over the angular momentum indicies and

$$\underset{\approx}{D}_{i,\alpha}(\epsilon) = [\underset{\approx}{1} + (\underset{\approx}{t}_{i,\alpha}^{-1} - \underset{\approx}{t}_{i,c}^{-1})\underset{\approx}{\tau}^{c,ii}] \ . \qquad (78)$$

Let us denote the new quantity $\frac{\partial}{\partial c_j}\underset{\approx}{t}_{i,c}^{-1}(\epsilon)$ by

$$\Lambda_{LL'}^{ij}(\epsilon) = \frac{\partial}{\partial c_j}[t_{i,c;LL'}^{-1}(\epsilon;\{c_i\})] \ . \qquad (79)$$

Evidently, it is a response function which gives the change in $\underset{\approx}{t}_{i,c}^{-1}$ at $\vec{R}_i$ due to a change in the concentration c_j at $\vec{R}_j$, i.e.,

$$\delta t_{i,c;LL'}^{-1}(\epsilon) = \sum_j \Lambda_{LL'}^{ij}(\epsilon)\delta c_j \ . \qquad (80)$$

One may derive a Bethe-Saltpeter-like equation for this response function by taking derivatives of the basic inhomogeneous CPA equation given in Eq. (61). After some lengthy but straightforward algebra we find

$$\Lambda_{L_1L_2}^{ij}(\epsilon) = \Lambda_{L_1L_2}^{0;ij}(\epsilon) - \sum_{\ell\neq i}\sum_{L_3L_4} I_{L_1L_2;L_3L_4}^{i\ell}(\epsilon)\Lambda_{L_3L_4}^{\ell j}(\epsilon) \ , \qquad (81)$$

where

$$\Lambda_{L_1L_2}^{0;ij}(\epsilon) = \frac{1}{2}[\underset{\approx}{D}_{iA}(\underset{\approx}{t}_{i,A}^{-1} - \underset{\approx}{t}_{i,B}^{-1})\underset{\approx}{D}_{i,B}]_{L_1L_2}\delta_{ij} \ ,$$

and the vertex part

$$I^{i\ell}_{L_1L_2;L_3L_4}(\epsilon) =$$

$$\sum_{L'L''} \{\underset{\approx}{D}_{i,A}(\underset{\approx}{t}^{-1}_{i,A} - \underset{\approx}{t}^{-1}_{i,c})\gamma_{L_1L''}(\underset{\approx}{t}^{-1}_{i,B} - \underset{\approx}{t}^{-1}_{i,c})\underset{\approx}{D}_{i,B}\}_{L''L_2}X^{i\ell}_{L'L'';L_3L_4}(\epsilon)$$

with (82)

$$X^{i\ell}_{L'L'';L_3L_4}(\epsilon) = \tau^{c,i\ell}_{L'L_3}(\epsilon)\tau^{c,\ell i}_{L_4L''}(\epsilon) \qquad i \neq \ell .$$

These equations describe the response function $\underset{\approx}{\Lambda}^{ij}(\epsilon)$, and therefore, the direct correlation function $S^{(2)}_{ij}$ in an arbitrary inhomogeneous compositional state specified by $\{c_i\}$. As planned, we shall now focus on the disordered state $c_i = \bar{c}\forall_i$ where actual calculations become feasible.

In the disordered state $\Lambda^{o;ij}_{L_1L_2}$ becomes independent of i and j. Moreover, $\tau^{c,ij}_{L_1L_2}$ and hence $X^{i\ell}_{L_1L_2L_3L_4}$ depend only on the coordinate difference $R_i - R_\ell$. Under these circumstances we may take the lattice Fourier transform of Eq. (81). The result is a finite matrix equation (because the angular momentum cut off at $\ell = \ell_{max}$) at each k-point of the Brillouin zone and at each energy ϵ:

$$\Lambda_{L_1L_2}(\vec{k};\epsilon) = \Lambda^{o}_{L_1}(\epsilon)\delta_{L_1L_2} - \sum_{L_3L_4} I_{L_1L_2;L_3L_4}(\vec{k};\epsilon)\Lambda_{L_3L_4}(\vec{k},\epsilon)$$

$$\Lambda^{o}_{L}(\epsilon) = \frac{1}{2} D_{A,LL}(\epsilon)[t^{-1}_{A,L}(\epsilon) - t^{-1}_{B,L}(\epsilon)]D_{B,LL}(\epsilon)$$

(83)

$$I_{L_1L_2;L_3L_4}(\vec{k};\epsilon) = D_{A,L_1L_1}(t^{-1}_{A,L_1} - t^{-1}_{c,L_1})(t^{-1}_{B,L_2} - t^{-1}_{c,L_2})D_{B,L_2L_2}X_{L_1L_2;L_3L_4}(\vec{k};\epsilon)$$

$$X_{L_1L_2;L_3L_4}(\vec{k};\epsilon) = \frac{1}{\Omega_{BZ}} \int d^3k'[\tau^{C}_{L_1L_3}(\vec{k}' + \vec{k};\epsilon)\tau^{C}_{L_4L_2}(\vec{k}';\epsilon)] - \tau^{c,ii}_{L_1L_3}\tau^{c,ii}_{L_4L_2} .$$

To solve these is now a matter of matrix algebra at each $\vec{k}$ and ϵ point.

With these results we may now return to Eq. (77). In the disordered state, $S^{(2)}_{ij}$ is diagonal in reciprocal space. Indeed we find

$$S^{(2)}_{sp}(\vec{k}) = \frac{1}{\pi} \text{Im} \int d\epsilon f(\epsilon) \sum_{LL'L''} D_{A,LL}(t^{-1}_{A,L} - t^{-1}_{B,L})D_{B,LL}\Lambda_{L'L''}(\vec{k};\epsilon)$$

$$\cdot X_{LL;L'L''}(\vec{k};\epsilon) , \qquad (84)$$

which is the principle result of this section. Of course, as indicated by the subscript sp, this is only the single-particle contribution to $S^{(2)}(k)$. Contributions from the double-counting correction, $\partial/\partial c_i\ \partial/\partial c_j\ (\delta\Omega_e^{CPA})$, can be important and will have similar form to $S^{(2)}(\vec{k})$ but will not be dealt with here.

What has been achieved may be summarized as follows. At the end of a full self-consistant KKR-CPA calculation, in possession of $t_{A,L}(\epsilon)$, $t_{B,L}(\epsilon)$, and $t_{c,L}(\epsilon)$, we may proceed to calculate the "susceptibility" $\chi_{L_1L_2L_3L_4}(\vec{k};\epsilon)$, see Eq. (83), and then solve Eq. (83) for the response function $\Lambda_{L,L'}(\vec{k};\epsilon)$. Furthermore, we may evaluate the formula for $S^{(2)}(\vec{k})$ given in Eq. (84). Finally, using the direct correlation function $S^{(2)}(\vec{k})$, so obtained, we determine the sought-after-correlation function $q(\vec{k})$ by evaluating the right-hand side of Eq. (37). By now a number of such calculations have been performed for various systems (49). In the next section we review the particularly interesting example of studies of ordering in the Cu_cPd_{1-c} system.

4. FERMI SURFACE DRIVEN INCOMMENSURATE ORDERING

As we have mentioned in the introduction, the diffuse scattering intensity $I(\vec{k})$ of X-rays or electrons from a random binary alloy is proportional to the correlation function $q(\vec{k})$. In fact

$$I(\vec{k}) = (f_A - f_B)^2\ \bar{c}(1 - \bar{c})\alpha(\vec{k}) \ , \tag{85}$$

where f_A and f_B are the scattering amplitudes of the *A*- and *B*-type of atoms, respectively, and $\alpha(k)$ is the well-known Warren-Cowley order parameter (8) which is proportional to the correlation function: $q(\vec{k}) \equiv \bar{c}(1 - \bar{c})\alpha(\vec{k})$.

If a binary alloy, on an fcc lattice, near the AB_3 stochiometric composition orders into the $L1_2$ (Cu_3Au) structure below a transition temperature T_c then, as a precursor to the transition, near, but above T_c the scattering intensity $I(\vec{k})$ show diffuse scattering peaks at the superlattice Bragg peaks of the ordered state. In the case of the $L1_2$ structure, these peaks are at the face centers and at the center of edges of the fcc reciprocal space unit cell as shown in Fig. 8. In Fig. 9 we show a sequence of diffuse scattering electron-diffraction pictures for the Cu_cPd_{1-c} system for c = 0.126, 0.250, and 0.332, respectively taken from the work of Ohshima and Watanabe (50). Evidently, the precursor peak is not a single spot but four spots whose relative distance changes rapidly with concentration. There is a large class of alloys which behave in this interesting manner (11). They have attracted considerable attention for some thirty years. In particular, it was noted by Sato and Toth (11) that the distance of the four spots from the face center, where the peak would be in a normal $L1_2$ ordering process, is proportional to the electron per atom ratio e/a. In Fig. 10 we display their compilation of the relevant date. Furthermore, it was suggested by Moss (51) that e/a influences *m* through its effect on the geometry of the Fermi surface. However, the nature of the relevant Fermi surface and the precise mechanism by which it determined *m* remains obscure. In what follows we give a full microscopic

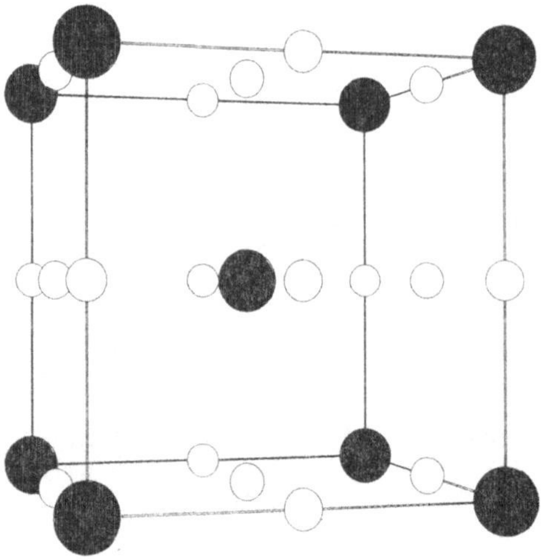

FIGURE 8. The large solid circles are the Bragg spots for the bcc reciprocal space corresponding to a fcc real-space lattice. The small open circles are the positions of the diffuse scattering maxima as the ordering transition into the $L1_2$ structure is approached from the high-temperature side.

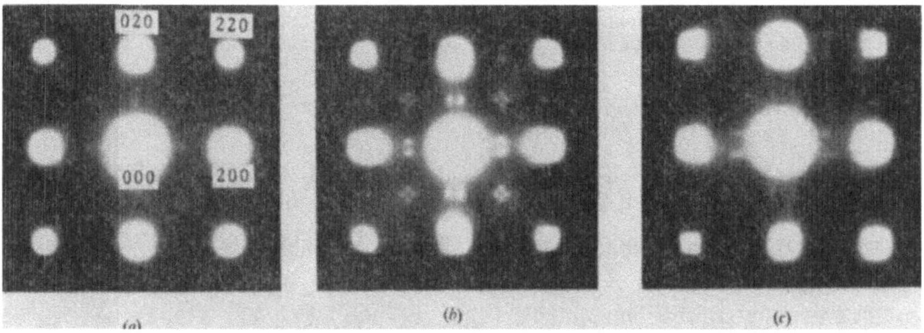

FIGURE 9. Electron diffraction patterns from various Cu_cPd_{1-c} alloys taken from the work of Oshima and Watanabe.

account of this phenomena using the theory developed in these lectures. The results presented here will bring up to date the arguments first presented by Gyorffy and Stocks (17).

The $L1_2$ superlattice points are Lifshitz special points in the sense of our introduction. Conventionally, they correspond to ordering by spontaneous appearance of concentration waves with wave vectors $Q_1^0 = 100$, $Q_2^0 = 010$, and $Q_3^0 = 001$. From the point-of-view of our present concern, it is more useful to regard the same concentration wave pattern as due to the equivalent three waves of wave vectors $Q_1 = 110$, $Q_2 = 101$, and $Q_3 = 011$. The way these produce the same pattern of superlattice spots are illustrated in Fig. 8. Evidently, in view of Eq. (85), the relation $c(1 - c)\alpha(\vec{k}) = q(\vec{k})$, and Eq. (37) to explain the experimental occurrence of

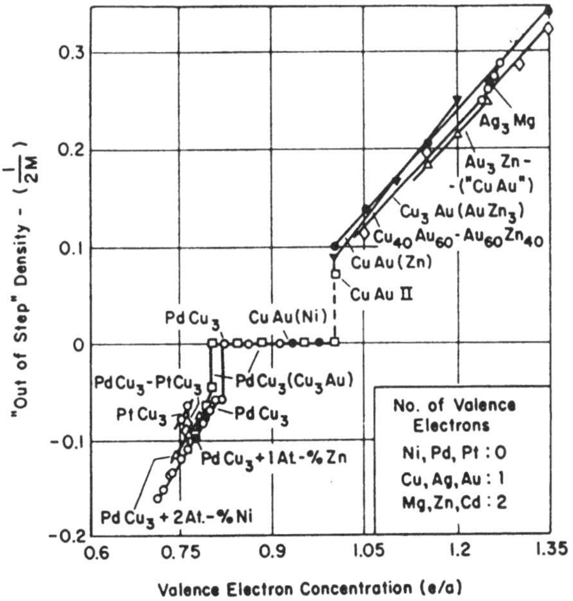

FIGURE 10. The deviation from the $L1_2$ commensurate wave of the modulation vectors of various incommensurately ordering systems. The data was compiled by Sato and Toth. The quantity (1/2M) defined by Sato and Toth and used in this figure is the same as the quantity *m* defined in the text.

these diffuse scattering intensity peaks we must look for peaks in $S^{(2)}(\vec{k})$ along the wave vectors Q_1, Q_2 and Q_3.

To see how such structure might arise let us simplify Eq. (84) by setting $\underline{\underline{D}}_A$, $\underline{\underline{t}}_A^{-1} - \underline{\underline{t}}_B^{-1}$, $\underline{\underline{D}}_B$ and $\underline{\underline{\Lambda}}$ equal to 1 on the grounds that the principal $\vec{k}$ dependence in this formula comes from the factor $X_{LL;L'L''}(\vec{k};\epsilon)$. Then, using Eq. (84) and a Krammers-Kronig relation for the scattering path matrices $\tau^{C}_{LL'}(\vec{k};\epsilon)$ we may write that

$$S^{(2)}(\vec{k}) \sim \sum_{L_1L_2} \int d\epsilon \int d\epsilon' \, \frac{f(\epsilon) - f(\epsilon')}{\epsilon - \epsilon'} \, \mathrm{Im}\tau_{L_1L_2}(\vec{k}',\epsilon')\mathrm{Im}\tau_{L_2L_1}(\vec{k}' + \vec{k};\epsilon) \; . \quad (86)$$

Next we note that the CPA approximation for the Bloch spectral function:

$$A(\vec{k};\epsilon) \equiv \sum_{ij} e^{i\vec{k}(\vec{R}_i - \vec{R}_j)} \int d^3r \mathrm{Im}\langle G(\vec{r} + \vec{R}_i;\vec{r} + \vec{R}_j;\epsilon)\rangle \; ,$$

can be similarly simplified to give

$$A(\vec{k};\epsilon) \sim \mathrm{Im}\tau^{C}_{LL'}(\vec{k};\epsilon) \; . \quad (87)$$

Hence, at least, schematically,

$$S^{(2)}(\vec{k}) \sim \int d\epsilon \int d\epsilon' \, \frac{f(\epsilon) - f(\epsilon')}{\epsilon - \epsilon'} \, A(\vec{k};\epsilon')A(\vec{k}' + \vec{k};\epsilon) \; .$$

The interpretation of the Bloch spectral function $A(k;\epsilon)$ is that it is the density of states at a k-point. Therefore, for an ordered system

$$A_B(\vec{k};\epsilon) = \sum_{\nu} \delta(\epsilon - \epsilon_{\vec{k},\nu}) , \tag{88}$$

where $\epsilon_{\vec{k},\nu}$ is the energy eigenvalue of a Bloch wave with wave vector $\vec{k}$ and band index ν. Consequently, for an ordered system

$$S^{(2)}(\vec{k}) \sim \sum_{\nu\nu'} \sum_{k'} \frac{f(\epsilon_{\vec{k}',\nu'}) - f(\epsilon_{\vec{k}'+\vec{k},\nu})}{\epsilon_{\vec{k}',\nu'} - \epsilon_{\vec{k}+\vec{k},\nu}} , \tag{89}$$

which is recognized as a generalized susceptibility factor.

As is well-known, the principle contributions to such factors comes from $\epsilon_{\vec{k},\nu}$'s close to the Fermi energy. Moreover, $S^{(2)}(k)$ will be particularly large at the spanning vector $\vec{Q}$ of two parallel sheets of the Fermi surface. The reason for this is the vanishing of the denominator $\epsilon_{\vec{k}',\nu'} - \epsilon_{\vec{k}'+\vec{k};\nu}$ at $\vec{k} = \vec{Q}$. Namely, $\epsilon_{\vec{k}',\nu} - \epsilon_{\vec{k}'+\vec{k},\nu} = 0$ if $\vec{k}'$ is on one of the sheets of the Fermi surface and $\vec{k}' + \vec{k}$ is on another. Evidently, the larger the surface area of the parallel sheets the larger the intensity of the peak in $S^{(2)}(\vec{k})$.

A large value of the response function at a particular $\vec{Q}$ implies large response to a small stimulus and therefore is an indication of an incipient instability. In physical terms, the above instability is due to the creation, by the stimulus (some external field), of a large number of electron-hole pairs with wave vector $\vec{Q}$ and essentially zero energy by exciting electrons from just below one of the Fermi surface sheets to previously empty states just above the other sheet.

This Fermi surface mechanism gives rise to a variety of celebrated instabilities of the metallic state. Usually, they are classified under the headings spin-density waves (SPD) (52) and charge-density waves (CDW) (53). We shall now add a third category to this list – concentration waves (CW).

The rough outline of the argument is now clear. Although the Bloch spectral function for an alloy describes smeared peaks in place of the delta-function peaks of the ordered systems, these may be quite well-defined both in $\vec{k}$ and ϵ. In short, $A(\vec{k},\epsilon)$ may describe only slightly broadened Fermi surfaces. That this is indeed the case is illustrated in Fig. 11 where we plot $A(k;\epsilon_F)$ at the Fermi energy ϵ_F along various directions in the Brillouin zone for a $Cu_{0.75}Pd_{0.25}$ alloy. Evidently, the Fermi surface is quite well-defined on the scale set by its size. Under these conditions, it may be sensible to look for parallel sheets of Fermi surfaces described by $A(k,\epsilon)$. Indeed, our picture in Fig. 11 suggests that the Fermi surface is pretty flat perpendicular to the Γ-K (110) direction. The evolution of this flat piece of Fermi surface with concentration is

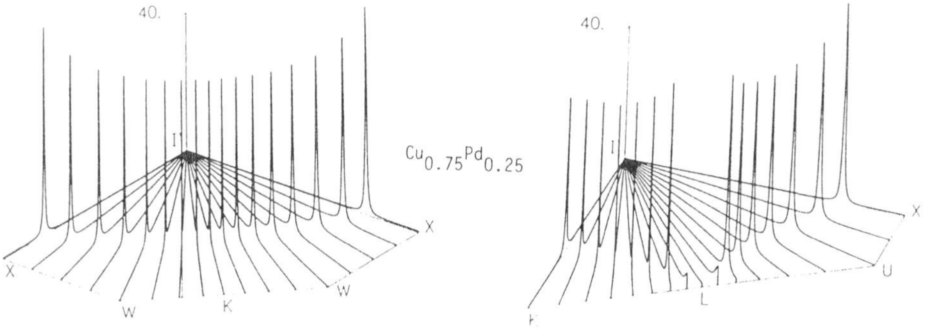

FIGURE 11. The Bloch spectral function $A_B(\vec{k},\epsilon_F)$ at the Fermi energy ϵ_F along various directions eminating from the center of the Brillouin zone for $Cu_{0.75}Pd_{0.25}$ alloy. Note the existence of a fairly well-defined Fermi surface.

shown in Fig. 12. Clearly, it gets flatter and fatter as the Pd concentration increases from 0 to 0.4 without being smeared very much. Moreover, there is another similar flat piece of Fermi surface centered on the nearest-neighbour Γ-point along the (110) direction. As we shall argue presently, these are the pairs of parallel sheets of Fermi surfaces which drive the incommensurate ordering seen in the experiments referred to in Fig. 9. Taking the spanning vectors $\vec{Q}$ from our calculations, we construct the expected diffuse scattering intensity maxima contours as shown in Fig. 13. Note that we expect peaks where the arcs from two different contributions cross.

Evidently, the arrangement of the predicted spots are in agreement with experiments. Moreover, they move with Pd concentration since the Fermi surface sheets move apart as the electron per atom ratio, e/a, decreases. This is also as found experimentally. Since the spanning vector $\vec{Q}$ is

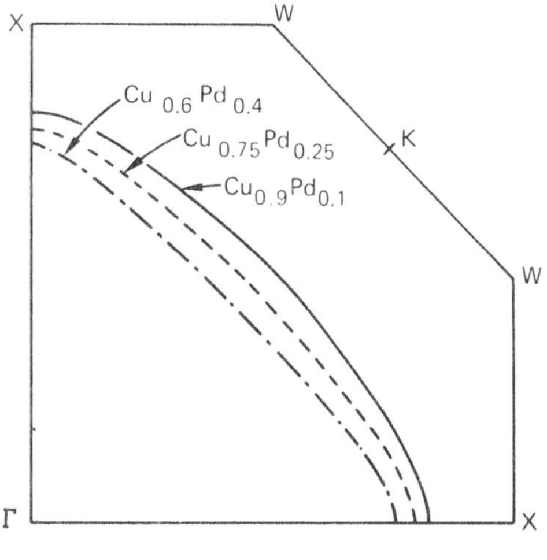

FIGURE 12. The evolution of the calculated Fermi surface of the Cu_cPd_{1-c} alloys with concentration. The focus of interest is the progressively flattened sheet perpendicular to the (110)(Γ-K) direction.

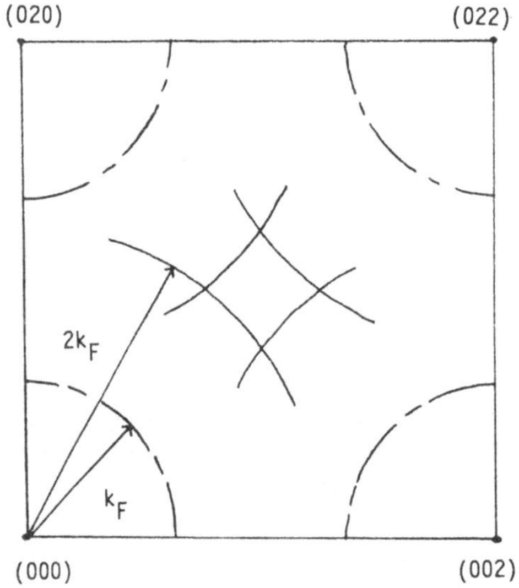

FIGURE 13. The loci of high-intensity diffuse scattering lines. The spots form where two such lines cross.

unlikely to be commensurate with the underlying lattice, the above mechanism provides a natural explanation of the four-way splitting of the $L1_2$ superlattice spots. Moreover, the changing size of the Fermi surface with changing *c* also furnishes a very natural cause for the relative movement of the spots. In fact, on the latter score, the agreement is even quantitative. To show this we plot the experimental and theoretical values of $m = 2[\sqrt{2} - 2k_F(011)]$ as a function of the Pd concentration in Fig. 14. Evidently, the agreement leaves little to be desired.

To summarize, random alloys with well-defined parallel sheets of Fermi surface will be unstable to formation of concentration waves whose wave vectors are the spanning vectors of the relevant Fermi surface sheets.

An appealing feature of the above mechanism is that it is quite robust in the sense that it does not depend on small delicate aspects of electronic structure. In fact, the Fermi surface of copper in the 110 direction is convex while that of palladium is concave. Thus, it is almost inevitable that, as the first changes into the second by alloying Pd into

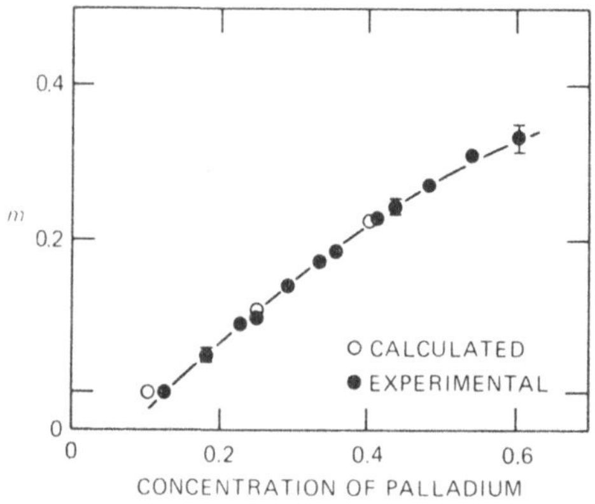

FIGURE 14. The calculated and measured concentration variation in the diffuse scattering spots separation for the Cu_cPd_{1-c} alloys.

Cu, at some concentration the alloy Fermi surface will get very flat. The only delicate question may be whether or not the Fermi surface is well defined when it becomes flat. Indeed, as will be seen later, severe smearing may render the Fermi surface mechanism ineffective. Thus, we may conclude that this mechanism may be relevant to all the compounds of Sato and Toth (11) where the ordering wave vector varies with e/a.

Although it is tempting to regard the above arguments as conclusive, there are good reasons for following up the detailed predictions of the theory by actually calculating $q(\vec{k})$. One of these is the caution prompted by past experience with other would-be Fermi surface driven instabilities (54). In short, on various occasions appropriately oriented parallel sheets of Fermi surfaces were found to explain an instability, but when the corresponding susceptibility was calculated, it showed either no structure or only some very minor features at the spanning vectors. We shall presently report on our calculations of $q(\vec{k})$ which shows that the expected features at the spanning vectors of the Fermi surface sheets identified above indeed occur.

However, before doing that we wish to mention another reason for studying $q(\vec{k})$ all through the Brillouin zone. It has to do with the fact that modern diffuse scattering intensity measurement show a fair amount of highly reproducible detailed structure. If we are able to reproduce these on the basis of our first principles calculations, in which all such structures are the direct consequence of the intricate details of the electronic energy bands, not just states near the Fermi energy, then we would significantly strengthen the case for believing that the present theory captures the physical essence of the forces which drive the ordering processes.

Having solved the self-consistant KKR-CPA equations for various concentrations of the $Cu_{1-c}Pd_c$ system at T = 0, we have evaluated the formula for $S^{(2)}(\vec{k})$ given in Eq. (84) for a number of temperatures. The computationally most demanding part of this calculation was the evaluation of the response function $X_{LL;L'L''}(k;\epsilon)$. This was done by integrating the product $\tau^{c}_{LL}(\vec{k}',\epsilon)\tau^{c}_{L'L''}(\vec{k}' + \vec{k};\epsilon)$ effectively over the whole Brillouin zone along directions emanating from the Γ point. In our most careful calculations we have used 136 directions in one irreducible forty-eighth of the Brillouin zone. The energy integral was performed by working at the complex energy points corresponding to the Matsubara frequencies, and we have summed over them for various temperatures. Interestingly, at $T \gg T_C$ instead of the Fermi surface driven structure at Q, a peak at the X point, one of the Lifshitz special points, was dominant. However, near to (~1000 K) $T = T_C$ the peak at Q emerged as the major feature reassuringly fulfilling our expectations.

On calculating $q(\vec{k})$ we found that $q(\vec{Q},T)$ diverged at $T = T_{sp}$ = 950 K. This is to be regarded as a spinodal temperature since the phase transition is expected to be first order. In fact, it is weakly first order experimentally. Unfortunately, we have not yet managed to do a sufficiently accurtate $S^{(1)}(\bar{c})$ calculation to warrant solving the equation of state. Nevertheless, it is gratifying that this spinodal temperature is not too far from the experimentally observed T_C = 927 K.

In Fig. 15 we show q(k,c), as calculated from Eq. (84) at c = 0.07, 0.25, and 0.40. Clearly, we find the split superlattice points as expected. Furthermore, they move with c in good agreement with the experiments. Thus, in this case, the Fermi surface mechanism dominates the

CALCULATED SHORT-RANGE ORDER DIFFUSE SCATTERING FROM Cu_cPd_{1-c} ALLOYS

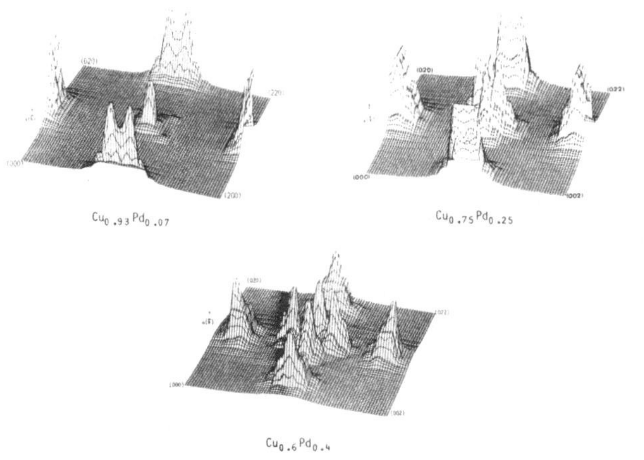

FIGURE 15. The correlation function $q(\vec{k},c)$, [proportional to the diffuse scattering intensity $I(\vec{k})$], in the plane containing the reciprocal lattice points 000, 020, 022, and 002 as calculated using the first-principles mean-field theory in the text for various Cu_cPd_{1-c} alloys.

symmetry-induced special points. As we remarked in the introduction this is allowed by the Landau argument. Indeed, in agreement with it, we found a material specific new length entering into the problem, Q^{-1}, which is not commensurate with the underlying lattice. The peak at $\vec{Q}$ dominating symmetry generated instabilities gives rise to an incommensurate modulation of the high-temperature symmetric phase at the temperature where it diverges.

In Fig. 16 we illustrate the temperature dependence of $q(\vec{k},t)$. Unfortunately, there are only limited experimental data on the fading out of the above short-range-order structure with rising temperature. Clearly, such a study would be a useful contribution to establishing where the theory stands on the energy scales governing the thermal fluctuations.

In conclusion, it must be stressed that the above discussion is incomplete in two important respects. Firstly, we have neglected the contribution to the direct correlation function $S_{ij}^{(2)}$ coming from the double-counting corrections $\delta\Omega_e^{CPA}$ to the free energy Ω_e^{CPA} in Eq. (64). These will play a particularly interesting role in systems where charge transfer, e.g., electronegativity, is a significant effect. We shall return to a discussion of this feature of the theory in a future publication. Secondly, we have studied only a small part of the phase diagram of the $Cu_{1-c}Pd_c$ system. Indeed, we have investigated the instability of the disordered state only, leaving the nature of the incommensurate state and how it locks into commensurate phases at low temperatures unexamined. The appropriate region of the phase diagram appears to display a number of novel ordering phenomena and is currently under intensive experimental investigation (55). While it is hoped that, eventually, the above first-principles mean-field theory may make contact with these interesting experimental features, for now these issues may be best addressed in the semi-phenomenological manner introduced in the next section.

The above remarks notwithstanding, the single particle formula for $S_{ij}^{(2)}$, given in Eq. (84), on its own is a useful tool for investigating ordering and clustering tendencies in the random phases. We like to stress that this is so even when the Fermi surface does not play a dominant role. As an example, we show our calculated $q(\vec{k})$ for the $Pd_{0.5}Rh_{0.5}$ alloy in Fig. 17. Clearly, the diffuse scattering intensity peaks are centered on

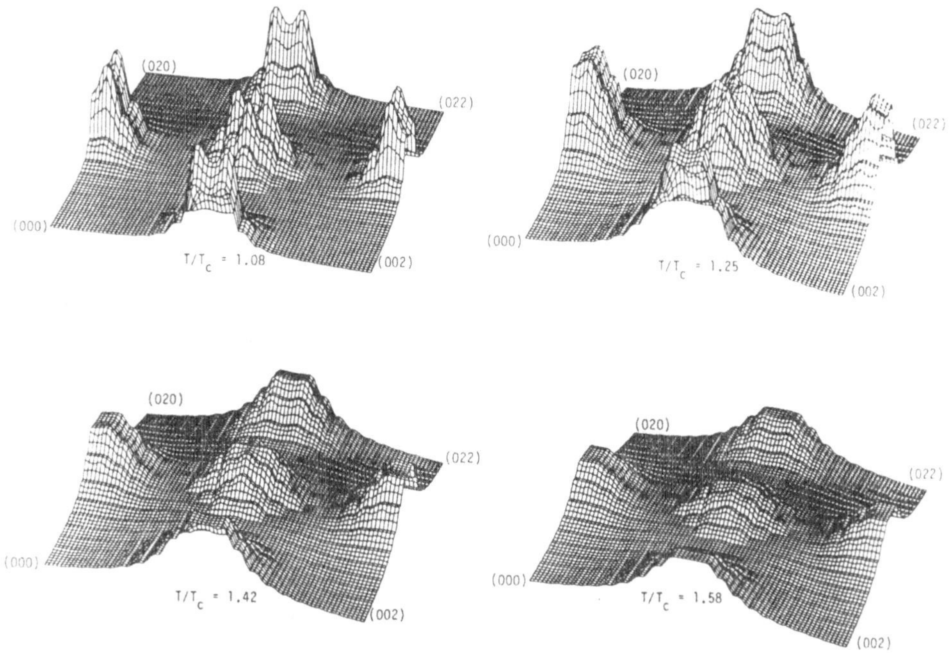

FIGURE 16. The evaluation of the correlation function $q(\vec{k},T)$ with temperature for $Cu_{0.75}Pd_{0.25}$ in the plane defined by the reciprocal lattice vectors 000, 020, 022, and 002.

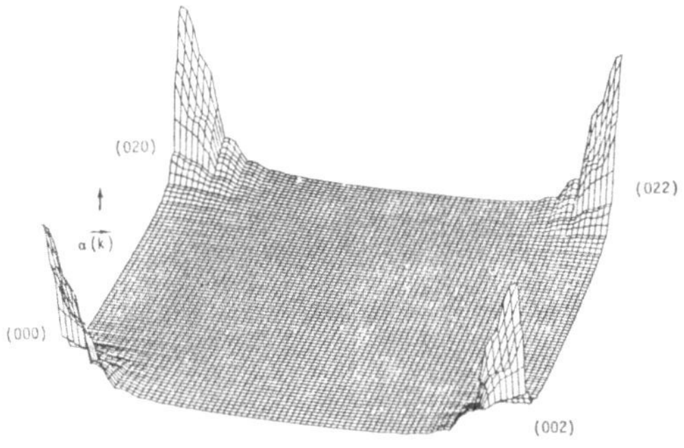

FIGURE 17. The correlation function $q(\vec{k})$ for $Pd_{0.25}Rh_{0.75}$ as calculated using the first-principles mean-field theory described in the text.

the fundamentals characteristic of the fully random alloys. Recall that this implies a clustering tendency and a phase-separation at low temperatures. It is gratifying to note that this is indeed what happens experimentally. The relevant phase separation curve is shown in Fig. 7. Given the parameter free nature of our calculation, the above qualitative success of the theory should be regarded as significant evidence that we have identified the electronic causes of the phase separation.

As a final example, we now comment on the possible origin of the spectacular sequences of long-period phases, discovered by van Tenderloo et al. in the $Al_{3-x}Ti_{1+x}$ system (56,57). As described in detail by Professors van Tenderloo and Selke in their contribution to this volume, these phases fit into a pattern usually associated with the axial next-nearest-neighbour Ising (ANNNI) model. Accordingly, it is argued that the sequences of commensurate phases are due to competing short-range interactions (58). On the other hand, given our experience with long-period structures in the $Cu_{1-c}Pd_c$ system, one might suspect that, again, the Fermi surface mechanism, which can be thought of as long-range oscillatory interaction, is involved. To check this hypothesis we have performed a self-consistant KKR-CPA calculation for the random alloy $Al_{0.75}Ti_{0.25}$ on an fcc lattice. The spectral function at ϵ_F is shown in Fig. 18. Evidently, the shape of the Fermi surface is very much like that of pure aluminum and there are few hints of parallel sheets. Moreover, whatever Fermi surface there is, it is significantly broadened on the scale of the Brillouin zone. Thus, it appears highly unlikely that this Fermi surface plays a role in the ordering process.

As we have noted earlier, even when the Fermi surface mechanism is not operating, $S^{(2)}_{sp}(\vec{k})$ may be a good guide as to the causes of ordering. In the hope that we may discover the short-range competing interactions required by the ANNNI model, we have calculated the single-particle contributions to $S^{(2)}_{ij}$ and find no indication of the (1 ½ 0) special point ordering of the DO_{22} structure. Moreover, calculations of the total energy of the $L1_2$ and DO_{22} ordered structure (59) on the same fcc lattice show that the $L1_2$ structure is favored.

As a means of investigating the matter further, Nicholson and Stocks have repeated the above calculation for a tetragonally distorted fcc lattice. Dramatically, it is found that the DO_{22} structure is favored over $L1_2$ for a c/a ratio close to that of experiment. This suggests that if we repeat our $S^{(2)}_{ij}$ calculation on a tetragonally distorted lattice we may find the relevant competing interactions. Thus, we appear to have discovered an example where lattice distortions and chemical ordering are critically coupled. Obviously, this calls for a theory where concentration and strain fluctuations are treated on an equal footing.

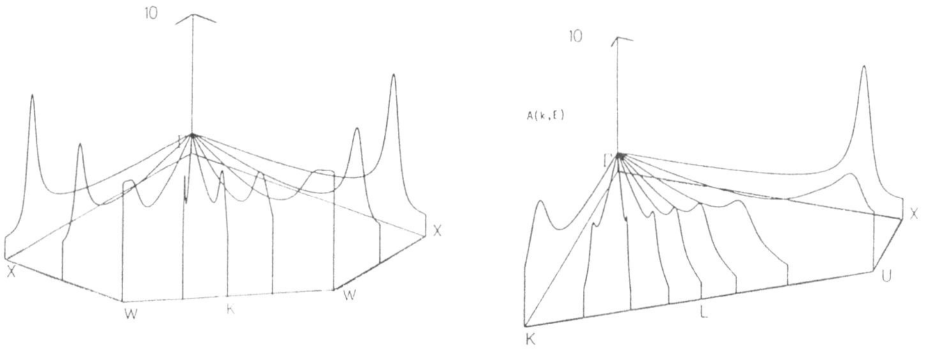

FIGURE 18. The Bloch spectral function at the Fermi energy $A_B(\vec{k},\epsilon_F)$, along lines eminating from the Γ point in the ΓXWKWXΓ (left) and ΓKLUXΓ (right) planes, for a fcc $TiAl_3$ random alloy.

5. WAYS OF USING THE FIRST PRINCIPLES MEAN FIELD THEORY

In addition to making use of the scheme we have presented in the previous sections directly, we can extend its usefulness by forging a bridge between it and various powerful phenomenological approaches to the problem of compositional variations in alloys. In what follows we shall discuss two such bridges.

5.1. First-principles coefficients for the Landau theory

Following Landau in his original paper, let us consider states for which the local concentration c_i deviates from the average $\bar{c}$ by a small amount δc_i. Then we may expand Eq. (64) about the homogeneous disordered state in powers of the deviations $\{\delta c_i\}$. This procedure is entirely analogous to the generalized perturbation theory of Ducastelle and Gautier (23). Comparing this expansion with the Landau theory given in Eq. (8), we find explicit microscopic expressions for the phenomenological coefficients as follows:

$$\gamma_{ij}^{(2)} = \frac{k_B T}{\bar{c}(1-\bar{c})}\delta_{ij} + S_{ij}^{(2)}$$

$$\gamma_{ijk}^{(3)} = \frac{k_B T(2\bar{c}-1)}{6\bar{c}^2(1-\bar{c})^2}\delta_{ij}\delta_{ik} + S_{ijk}^{(3)}$$

$$\gamma_{ijk\ell}^{(4)} = \frac{k_B T[1-3\bar{c}(1-\bar{c})]}{12\bar{c}^3(1-\bar{c})^3}\delta_{ij}\delta_{ik}\delta_{i\ell} + S_{ijk\ell}^{(4)}\,, \qquad (90)$$

where the higher-order direct-correlation functions $S_{ij}^{(3)}$ and $S_{ijk\ell}^{(4)}$ are straightforward generalization of $S_{ij}^{(2)}$ and are given by

$$S_{ijk}^{(3)} = \beta\left(\frac{\partial^3 \Omega_e^{CPA}}{\partial c_i \partial c_j \partial c_k}\right)_{c_i=\bar{c}\forall_i}$$

$$S_{ijk}^{(4)} = \beta\left(\frac{\partial^4 \Omega_e^{CPA}}{\partial c_i \partial c_j \partial c_k \partial c_\ell}\right)_{c_i=\bar{c}\forall_i}\,. \qquad (91)$$

Evidently, these microscopic coefficients $\gamma_{ij}^{(2)}$ and $\gamma_{ijk}^{(3)}$ depend on the temperature concentration and the volume as is usually assumed in the phenomenological theory. For $S_{ijk}^{(3)}$ and $S_{ijk\ell}^{(4)}$, we have not derived explicit KKR-CPA expressions. However, although laborious, it is a perfectly straightforward thing to do. As an indicator of the interesting results that can be obtained with such a powerful first-principles Landau theory, we mention the strong coupling charge density wave theory of Varma et al. (60). There important new features emerged when the analogies to the third- and fourth-order coefficients $S_{ijk}^{(3)}$ and $S_{ijk\ell}^{(4)}$ were calculated.

On the other hand, one may argue that $S^{(3)}$ and $S^{(4)}$ are small compared to $S^{(2)}$. This is the line taken in the generalized perturbation method (GPM) of Ducastelle and Gautier (23). However, while within their tight-binding model Hamiltonian framework, there appears to be evidence that the direct correlation functions fall off rapidly with their order. Our self-consistant density functional description of the electrons does not allow such conclusions. Nevertheless, it may frequently be so, and it would be very helpful if this hypothesis was tested by explicit calculations.

Once the coefficients in Eq. (8) have been established, one may proceed to studying the Euler-Lagrange equation corresponding to minimizing $F(\{c_i\})$. In this way one may examine, for example, the domain structure of alloys, as was done by McMillan (61) for Cu_3Au, on a first-principles basis. The theory may also be suitable for investigating anti-phase-boundary energies.

If for some reason the above mean-field theory proves insufficient, one may include the fluctuations by following the Ginzburg-Landau-Wilson prescription (62):

$$F = -\frac{1}{\beta}\ln Z$$

$$Z = \prod_i \int d(\delta c_i) e^{-\delta F(\{\delta c_i\})} \quad . \tag{92}$$

Clearly, this is a sensible description only when most of the contribution to Z comes from small values of δc_i. For instance, this is the case near second-order phase transitions. When the delta-function contribution to all the γ_{ij}'s are summed up to all orders (neglecting all the direct correlation functions but that for n = 2), this prescription becomes equivalent to using S_{ij} as a pair potential in Eq. (18). Again, this is an uncontrolled approximation which nevertheless may be useful when $S^{(2)}$ is a strong pair potential which gives rise to large concentration fluctuations, δc_i, but $S^{(n)}$ is small by comparison.

5.2. A first-principle Cahn-Hilliard theory - the local concentration approximation (LCA)

In this final section we shall develop a theory which, in contrast to the Landau theory, is valid for large deviations from the average concentration $\bar{c}$. Such a Cahn-Hilliard (63) or van der Waals (64) theory will be useful for studying surface and interface concentation profiles, or critical droplets of a new phase embedded in the old near-first-order transformations, to mention but a few applications.

The key idea which allows us to make progress with the above project is that we restrict our attention to cases where, while c_i changes a lot, it changes slowly. Namely, the change from one site to the next will be required to be small, but it may mount up over many lattice sites. In fact, the theory will be good for (a) small changes, whether fast or slow, and (b) large changes, if they are slow. These two conditions are depicted pictorially for the concentration function c(x), defined on a continuum, in Fig. 19. Obviously, this approximation is well-suited for taking the continuum limit.

We begin by breaking up the free energy $F(\{c_i\})$ into contributions from each site as follows

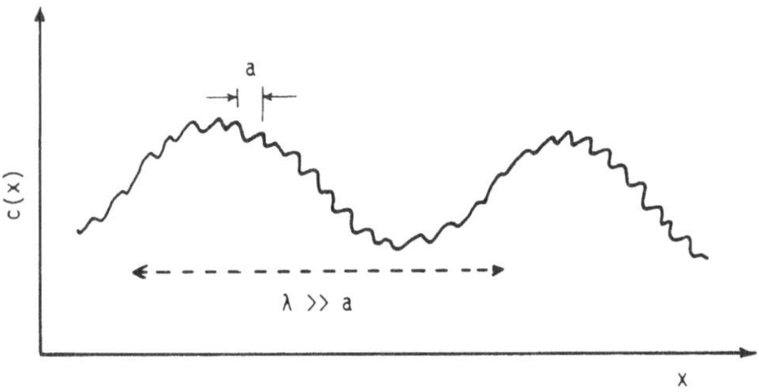

FIGURE 19. Schematic picture of compositional fluctuations for which the theory in the text is applicable.

$$\delta F(\{c_i\}) = \sum_i f_i(\{c_i\}) \ , \tag{93}$$

where f_i is a kind of internal Helmholtz free energy density which depends on the whole concentration configuration $\{c_i\}$. A completely local approximation would be if we assumed that f_i depends only on the local concentration c_i: $f_i(\{c_i\}) = f_i(c_i)$. Here we shall do better than that by expanding $f_i(\{c_i\})$ in the deviations $c_j - c_i$ for $j \neq i$. Note that we are expanding each free-energy contribution f_i about its own local concentration c_i.

$$f_i(\{c_i\}) = f_i^0 + \sum_j f_{ij}^{(1)}(c_i)(c_j - c_i)$$

$$+ \frac{1}{2} \sum_{j\ell} f_{ij\ell}^{(2)}(c_i)(c_j - c_i)(c_\ell - c_i) + \cdots \ . \tag{94}$$

This is like the gradient expansion in the Kohn-Sham theory of the electron gas.

Since δF is a unique function of {ci}, i.e., independent of the chemical potential, the above expansion must have the symmetry of the underlying lattice. Thus, as in the Landau theory, the coefficients obey symmetry rules. For instance, for cubic systems there can be no linear term in Eq. (94) and the quadratic term must have the form

$$\sum_j f_{ij}^{(2)}(c_i)(c_j - c_i)^2 \ . \tag{95}$$

It may be helpful if we recall that the above observation in the context of the gradient expansion of density functional theories for homogeneous systems corresponds to taking the quadratic term to have the form $|\vec{\nabla} n|^2$ where $n(\vec{r})$ is the charge density, on account of rotational invariance.

To continue, we note that in $\sum_i f_i$ there are two terms which are of the form of Eq. (95):

$$f_{ij}^{(2)}(c_i)(c_j - c_i)^2 \text{ and } f_{ji}^{(2)}(c_j)(c_i - c_j)^2 \ . \qquad (96)$$

We shall approximate these by $f_{ij}^{(2)}(\bar{c}_{ij})(c_j - c_i)^2$ where $\bar{c}_{ij} = \frac{1}{2}(c_i + c_j)$ and $f_{ij}(\bar{c}_{ij})$ depends only on $\vec{R}_i - \vec{R}_j$. Thus to quadratic order in the local deviations:

$$\delta F \cong \sum_i f_i^0(c_i) + \frac{1}{2}\sum_{ij} f_{ij}^{(2)}(\bar{c}_{ij})(c_i - c_j)^2 \ . \qquad (97)$$

Now, in the spirit of the previous section, we shall look for a microcscopic theory for the coefficients $f_i^0(c_i)$ and $f_{ij}^{(2)}(\bar{c}_{ij})$. The way to proceed is to Taylor expand both δF in Eq. (37) and $\Omega_e^{CPA}(\{c_i\})$ and compare coefficients. Note that the expansion of Ω_e^{CPA} is in powers of δc_i, not $c_i - c_j$. That is why Eq. (97) also has to be expanded in powers of the deviations of c_i from a common mean. The calculations are straightforward and we find

$$f_i^0(\bar{c}) = \Omega^{CPA}(\bar{c})$$

$$f_{ij}^{(2)}(\bar{c}) = -\frac{1}{2} S_{ij}^{(2)} \ . \qquad (98)$$

Consequently, according to the LCA

$$F(\{c_i\}) = \sum_i \Omega_e^{CPA}(c_i) - \frac{1}{4}\sum_{ij} S_{ij}^{(2)}(c_i - c_j) \ . \qquad (99)$$

It is useful to remark that for pair-potential models the mean-field theory already has the LCA form with $\Omega^{CPA}(\bar{c})$ replaced by $Z\tilde{v}\bar{c}(1 - \bar{c})$ in case of nearest-neighbour interactions. However, in the approach we are advocating $\Omega_e^{CPA}(\bar{c})$ and $S_{ij}^{(2)}$ are calculated from first principles and both may be quite complicated functions of $\bar{c}$.

Since Eq. (99) is only valid for slowly varying concentrations, we might as well take its continuum limit. Taking $c_i = c(\vec{R}_i)$ and for a nearby R_j $c(R_j) = c(R_i) + (\vec{R}_j - \vec{R}_i) \cdot \vec{\nabla}c(R_i)$, it is easy to derive the following expression for the free energy

$$F\{[c(\vec{r})]\} = \int d^3r\{\Omega_e^{CPA}[c(\vec{r})] + \vec{\nabla}[c(\vec{r}) \cdot \overset{\leftrightarrow}{S}{}^{(2)} \cdot \vec{\nabla}c(\vec{r})\} \ , \qquad (100)$$

where

$$S_{\alpha\beta}^{(2)} = \sum_j \vec{R}_{ij}^{\alpha}\vec{R}_{ij}^{\beta}S_{ij}^{(2)} \ . \qquad (101)$$

This is the principal result of this section. Evidently, Eq. (100) is of the form of the much used concentration functional of Cahn and Hilliard (3,8,63). The only difference is the anisotropic form of the gradient term. Such anistropy may often be useful in studying problems which involve interaction between compositional arrangements and the orientation of the lattice. However, when it can be neglected, Eq. (100) takes on the more usual form

$$F = F_0 \int d^3r[\bar{f}(\bar{c}) + \xi_0^2|\vec{\nabla}c|^2] ,$$

where

$$F_0 = -\frac{1}{2}\sum_i S_{ij}^{(2)} = \frac{k_BT}{\bar{c}(1-\bar{c})}$$

$$\bar{f}(c) = \Omega_e^{CPA}(c)/F_0 \qquad (102)$$

$$\xi_0^2 = \sum_j S_{ij}^{(2)}|\vec{R}_{ij}|^2/\sum_j S_{ij}^{(2)} .$$

Since f(c) and ξ_0 can now be calculated using the self-consistant KKR-CPA method, we have obtained a fully first-principles Cahn-Hilliard theory.

ACKNOWLEDGMENT

Research for G. M. Stocks and D. M. Nicholson was sponsored by the Division of Materials Sciences, U.S. Department of Energy, under contract DE-AC05-84OR21400 with Martin Marietta Energy Systems, Inc.

REFERENCES

1. Hansen M: Constitution of Binary Alloys. New York: McGraw Hill, 1958.
2. Stanley EH: Introduction to Phase Transitions and Critical Phenomena. Oxford: Clarendon Press, 1971.
3. Khachaturyan AG: Theory of Structural Transformations in Solids. New York: John Wiley & Sons, 1983.
4. Krivoglaz MA: Theory of X-ray and Thermal Neutron Scattering by Real Crystals. New York: Plenum Publishing, 1969.
5. Lifshitz EM and Pitaevskii LP: Statistical Physics, Part 1. New York: Pergamon Press, 3rd. edition, 1976.
6. Lifshitz EM: JETP **11**, 253-269 (1941); (J. Phys. USSSR **7**, 61 and 251 (1942).
7. Lyubarskii GY: Application of Group Theory in Physics. New York, Pergamon Press, 1960.
8. de Fontaine D: p. 73 in Solid State Physics, eds. H. Ehrenreich, F. Seitz and D. Turnbull, Vol. 34. New York: Academic Press, 1979.
9. Birman JL: Group Theoretical Methods in Physics (Lecture Notes in Physics), Vol. 79, eds. P. Kramer and A. Rieckers. New York: Springer-Verlag, 1980.
10. Levanyuk AP: Incommensurate Phases in Dielectrics, eds. R. Blinc and A. P. Levanyuk. Elsevier Science Publishers, B.V., 1986.
11. Sato H and Toth RS: Alloying Behavior and Effects in Concentrated Solid Solutions, ed. T. B. Massalski. New York: Gordon and Breach, 1965. Also *Phys. Rev.* **127**, 469 (1961).
12. Loiseau A, Van Tendeloo G, Portier R, and Ducastelle F: J. Physique **46**, 395 (1985).
13. Peierls RE: Quantum Theory of Solids. Oxford: Clarendon Press, 1955.
14. daSilva EZ, Strange P, Temmerman W, and Gyorffy BL: Phys. Rev. B **136**, 3015 (1987).
15. de Fontaine D and Kikuchi R: Applications of Phase Diagrams in Metallurgy and Ceramics **2**, 967 (1978). (NBS Special Publication 496)
16. Binder K: Festkorperproblem, ed. P. Grosse. New York: Springer-Verlag, XXVI, 1986.
17. Gyorffy BL and Stocks GM: Phys. Rev. Lett. **50**, 374 (1983).
18. Peierls R: Phys. Rev. **54**, 818 (1938).
19. Friedman LH: p. 185 in A Course in Structure Mechanics. Prentice Hall, Inc., 1985.
20. Feynman RP: Statistical Mechanics. Benjamin, 1972.
21. Clapp PC and Moss SC: Phys. Rev. **142**, 418 (1966); 171, 754 (1968).
22. Fisher ME: J. Math. Phys. **5**, 944 (1964).
23. Ducastelle F and Gautier F: J. Phys. F. Metal Physics **6**, 2039 (1976).
24. Gautier F, Ducastelle F, and Giner J: Phil. Mag. **31**, 1373 (1975).
25. Turchi PEA, Gonis A, Zhang XG, and Stocks GM: Configurational Energies in Terms of Effective Cluster Interactions in Binary Substitutional Alloys: Connection Between the Embedded Cluster Method and the Generalized Perturbation Method, this volume. Also, Sluiter M and Turchi PEA: Electronic Theory of Phase Stability in substitutional Alloys: A Comparison of the Connolly-Williams Scheme and the Generalized Perturbation Method, this volume.
26. Hohenberg P and Kohn W: Phys. Rev. **136**, B864 (1964).
27. Kohn W and Sham LJ: Phys. Rev. **140**, A1133 (1965).
28. Maruzzi VL, Janak JF, and Williams AR: Calculated Electronic Properties of Metals. New York, Pergamon Press, 1978.

29. Kohn W and Vashista P: Theory of the Inhomogeneous Electron Gas, eds. S. Lundquist and N.H. March. New York: Physics of Solids and Liquid Series, Plenum Press, 1983.
30. Williams AR and von Barth U: Theory of the Inhomogeneous Electron Gas, eds. S. Lundqvist and N.H. March. New York: Physics of Solids and Liquid Series, Plenum Press, 1983.
31. Stocks GM and Winter H: The Electronic Structure of Complex Systems, eds. P. Phariseau and W.M. Temmerman. New York: Plenum Press NATO ASI Series, Physics B113, 1984.
32. Korringa J: Physica **13**, 392 (1947).
33. Kohn W and Rostoker N: Phys. Rev. **34**, 1111 (1954).
34. Elliott RJ, Krumhansl JA, and Leath PL: Rev. Mod. Phys. **46**, 465 (1974).
35. Ehrenreich H and Schwartz L: Solid State Physics, eds. H. Ehrenreich, F. Seitz, and D. Turnbull. New York: Academic Press, 1976.
36. Gyorffy BL and Stocks GM: Electrons in Finite and Infinite Structures, eds. P. Phariseau and L. Scheire. New York: Plenum Press, NATO ASI Series, Physics B **24**, 1977.
37. Gyorffy BL and Stocks GM: Electrons in Disordered Metals and at Metallic Surfaces, eds. P. Phariseau, B.L. Gyorffy, and L. Scheire. New York: Plenum Press NATO ASI Series, Physics B **42**, 1979.
38. Faulkner JS: Progress in Materials Science **27**, 1 (1982).
39. Schwartz L and Siggia E: Phys. Rev. B **5**, 383 (1972).
40. Stocks GM and Winter H: Z. Phys. B. Condensed Matter **46**, 95 (1982).
41. Gyorffy BL, Kollar J, Pindor AJ, Stocks GM, Staunton J, and Winter H.: Eq. IV-7 in The Electronic Structure of Complex Systems, eds. P. Phariseau, W.M. Temmerman. New York: Plenum Press, 1984.
42. Johnson DD, Nicholson DM, Pinski FJ, Gyorffy BL, and Stocks GM: Phys. Rev. Lett. **56**, 2088 (1986).
43. Janak JF: Phys. Rev. B **9**, 3985 (1974).
44. Vegard L: Z. Phys. **5**, 17 (1921) and Z. Kristallogr. Kristallgeom. Krystallphys. Kristalchem. **67**, 239 (1928).
45. Johnson DD, Nicholson DM, Pinski FJ, Gyorffy BL, and Stocks GM: Phys. Rev. (submitted).
46. Nicholson DM, Stocks GM, Pinski FJ, Johnson DD, Gonis A, Gyorffy BL, and Wadsworth JS: Phys. Rev. Lett. (submitted).
47. Pindor AJ, Temmerman WM, and Gyorffy BL: J. Phys. F **13**, 1627 (1983).
48. Schandler G, Weinberger P, Gonis A, and Klima J: J. Phys. F **15**, (1985).
49. Stocks GM, Nicholson DM, Pinski FJ, Butler WH, Stern P, Temmerman WM, Gyorffy BL, Johnson DD, Gonis A, Chang X-G, and Turchi PEA: Mat. Res. Soc. Symp. Proc. **81**, 15 (1987).
50. Ohshima K and Watanabe D: Acta Cryst. A **33**, 520 (1977).
51. Moss SC: Phys. Rev. Lett. **22**, 1108 (1969).
52. Fawcett E: Rev. Mod. Phys. **60**, 209 (1988).
53. Wilson JA, Di Salvo FJ, and Mahajan F: Adv. Phys. **24**, 117 (1975).
54. Doran NJ: J. Phys. C **11**, L959 (1978).
55. Brodin D, Van Tenderloo G, Van Landwyt J, Amelinckx S, Portier R, Guymont M, Loiseau A: Phil. Mag. A **54**, 395 (1986).
56. Loiseau A, Van Tenderloo G, Portier R, and Ducastelle F: J. Phys. **46**, 595 (1985).
57. Miida R, Ajshimoto S, Watanabe D: Sci. Rep. RITU **A29** Suppl. 1 (1981).
58. de Fontaine D and Kulik J: Acta Met. **33**, 145 (1985).
59. Nicholson DM, Stocks GM, Temmerman WM, Sterne P, and Pettifor DG: (unpublished).
60. Varma CM and Simons AL: Phys. Rev. Lett. **51**, 138 (1983).

61. McMillan WL: Melting Localization and Chaos, eds. R.K. Kalia and P. Vashishta. The Netherlands: North Holland Publishing, 1982.
62. See page 471 of Reference 5.
63. Cahn JW and Hilliard JE: J. Chem. Phys. **28**, 258 (1958).
64. van der Waals JD: J. Stat. Phys. **20**, 197 (1979).

LOCAL DENSITY THEORY OF MAGNETISM AND ITS INTERRELATION WITH COMPOSITIONAL ORDER IN ALLOYS

J.B. STAUNTON
Department of Physics,
University of Warwick,
Coventry CV4 7AL, U.K.
B.L. GYORFFY
H.H.Wills' Physics Laboratory,
University of Bristol,
Tyndall Avenue, Bristol BS8 1TL, U.K.
D.D. JOHNSON
Condensed Matter Physics Branch,
Naval Research Laboratory,
Washington DC 20375, U.S.A.
F.J. PINSKI
Department of Physics,
University of Cincinnati,
Cincinnati OH 45221, U.S.A.
G.M. STOCKS
Metals and Ceramics Division,
Oak Ridge National Laboratory,
Oak Ridge TN 37830, U.S.A.

1 Introduction

It is well known that the most important source of magnetic interactions arises from electrostatic electron-electron interactions — the quantum mechanical exchange interactions. In magnetic insulators it is possible to associate electrons with particular atomic sites, define site 'spin' operators $\hat{s}_i$ and the famous Heisenberg -Dirac Hamiltonian can be used to describe the magnetic behaviour of such systems.

$$\hat{H} = \sum J_{ij} \hat{s}_i \cdot \hat{s}_j \tag{1}$$

in which J_{ij} is an exchange integral. In metallic systems, however, the situation is very different. It is not possible to allocate their itinerant electrons in this way and such pairwise interactions are inapplicable. In metals, magnetism is a complicated many electron phenomenon.

G. M. Stocks and A. Gonis (eds.), Alloy Phase Stability, 469–507.

In such systems magnetism is subtly connected with other properties. For example, some materials show a small thermal expansion coefficient below the Curie temperature, T_c, a large forced increase in volume when an external magnetic field is applied, a sharp decrease of spontaneous magnetism and of T_c when pressure is applied and large changes in the elastic constants on going through T_c. These are the famous invar effects, so called because they occur in the f.c.c. invar alloys Fe-Ni (65% Fe), Fe-Pd and Fe-Pt. Another example, of particular relevance to these lectures, is the connection between compositional ordering in an alloy and its magnetic state. When the states of magnetic and compositional order are strongly coupled, physically interesting and technologically important new phenomena may arise. Some alloys, such as Fe-Ni, develop directional chemical order when annealed in a magnetic field (1). The magnetic properties of many alloys are sensitive to the local environment. For example, ordered Ni-Pt (50%) is an anti-ferromagnetic alloy (2) whereas its disordered counterpart is ferromagnetic.

For several years it has generally been agreed that spin polarised band theory, established within the Spin Density Functional Formulism (3), is able to provide a reliable description of magnetic properties of transition metal systems at absolute zero temperature. This is essentially the modern version of the Stoner-Wohlfarth theory (4), (5), in which the magnetic moments are assumed to originate from itinerant d-electrons whose spins are aligned by exchange interactions. It provides a mechanism for non-integer moments together with a plausible account of the many body nature of magnetic moment formation at $T = 0K$ (6). This theory maps the many electron problem onto one of the independent electrons moving in the effective potentials and magnetic fields which themselves depend on all the other electrons. This is the self-consistent spin polarised band theory picture. A simple illustration uses the Hubbard Hamiltonian and Hartree -Fock approximation.

$$\hat{H} = \sum_{ij}(\epsilon_0\delta_{ij} + t_{ij})a^{\dagger}_{i,\sigma}a_{j,\sigma} + (1/2)I\sum_{i,\sigma} a^{\dagger}_{i,\sigma}a_{i,\sigma}a^{\dagger}_{i,-\sigma}a_{i,-\sigma} \tag{2}$$

$$a^{\dagger}_{i,\sigma}a_{i,\sigma}a^{\dagger}_{i,-\sigma}a_{i,-\sigma} \simeq a^{\dagger}_{i,\sigma}a_{i,\sigma}\langle a^{\dagger}_{i,-\sigma}a_{i,-\sigma}\rangle \tag{3}$$

$$\langle a^{\dagger}_{i,-\sigma}a_{i,-\sigma}\rangle = (1/2)\bar{n}_i - (1/2)\bar{\mu}_i\sigma \tag{4}$$

$$\bar{n}_i = \bar{n}_{i,+1} + \bar{n}_{i,-1} \tag{5}$$

$$\bar{\mu}_i = \bar{n}_{i,+1} - \bar{n}_{i,-1} \tag{6}$$

$$\hat{H}_{HF} = \sum_{ij}[(\epsilon_0 + (1/2)I\bar{n}_i - (1/2)I\bar{\mu}_i\sigma)\delta_{ij} + t_{ij}]a^{\dagger}_{i,\sigma}a_{j,\sigma} \tag{7}$$

in which $a^{\dagger}_i$, a_i are the creation and annihilation operators, ϵ_0 a site energy, t_{ij} a hopping parameter and I the many body Hubbard parameter representing the intrasite Coulomb integral. Equation 4 provides a picture of a $\sigma = +1$ electron 'seeing' a different potential on lattice site i from that seen by a $\sigma = -1$ electron, hence the picture of exchange split bands.

Whilst the modern version of the Stoner -Wohlfarth theory adequately describes the magnetic ground state of transition metals, its straightforward generalisation to finite temperatures fails badly. For example, the predicted Curie temperature, T_c, is too high typically by a factor of five and there are no moments and no Curie Weiss law above T_c. The theory seems to miss the dominant thermal fluctuation of the magnetisation. Its structure allows the thermally averaged magnetisation, $\bar{M}$, to vanish only as the exchange splitting between the spin polarised bands collapses together with the local magnetisation in each unit cell. Clearly an improved theory must allow thermal fluctuations of the orientations of these 'local moments', subtle entities set up by the collective behaviour of all the electrons. A detailed discussion of the implementation of this sort of theory will be left to a later lecture. For the time being we will be considering metallic systems in their magnetic states at temperatures well below their Curie temperatures.

2 Spin polarised electronic structure of disordered alloys.

We first discuss the parameter free, modern way of formulating the Stoner theory and start with a brief description of Spin Density Functional theory (for detailed reviews (3),(7)). Drs Dederichs, Freeman and Gyorffy also refer to this formulism in their lectures. Consider a many electron system in an external potential, $V^{ext.}$ and external magnetic field $\mathbf{B}^{ext.}$

$$
\begin{aligned}
V^{ext.} &= \sum_i v_i^{ext.}(\mathbf{r}_i) \\
\mathbf{B}^{ext.} &= \sum_i \mathbf{b}_i^{ext.}(\mathbf{r}_i) \\
\mathbf{r}_i &= \mathbf{r} - \mathbf{R}_i
\end{aligned}
$$

where $\mathbf{R}_i$ denotes the position of a lattice site. The many body Hamiltonian is

$$
\begin{aligned}
\hat{H} &= -(\hbar^2/2m)Tr. \int d\mathbf{r} \nabla \tilde{\psi}^\dagger(\mathbf{r},t) \cdot \nabla \tilde{\psi}(\mathbf{r},t) \\
&+ Tr. \int d\mathbf{r} \tilde{\psi}^\dagger(\mathbf{r},t)(V^{ext.}(\mathbf{r}) + \mathbf{B}^{ext.}(\mathbf{r}) \cdot \tilde{\sigma})\tilde{\psi}(\mathbf{r},t) \\
&+ (e^2/2)Tr. \int\int d\mathbf{r}\, d\mathbf{r}' \tilde{\psi}^\dagger(\mathbf{r},t)\tilde{\psi}^\dagger(\mathbf{r}',t)\tilde{\psi}(\mathbf{r}',t)\tilde{\psi}(\mathbf{r},t)/|\mathbf{r}-\mathbf{r}'| \qquad (8)
\end{aligned}
$$

in terms of creation and annihilation operators $\tilde{\psi}^\dagger$, $\tilde{\psi}$ ($Tr.$ denotes trace and $\tilde{}$ a 2×2 matrix). It is possible to prove that in the grand canonical ensemble at a given temperature, T, and chemical potential ν, the equilibrium charge $n(\mathbf{r})$ and magnetisation density $\mathbf{m}(\mathbf{r})$ are determined by the external potential and magnetic field. The correct equilibrium charge and magnetisation densities minimise the

Gibbs grand potential Ω

$$\begin{aligned}\Omega[n,\mathbf{m}] &= \int d\mathbf{r}V^{ext.}(\mathbf{r})n(\mathbf{r}) + \int d\mathbf{r}\mathbf{B}^{ext.}(\mathbf{r})\cdot\mathbf{m}(\mathbf{r}) \\ &+(e^2/2)\int\int d\mathbf{r}\,d\mathbf{r}'n(\mathbf{r})n(\mathbf{r}')/|\mathbf{r}-\mathbf{r}'| \\ &-\nu\int d\mathbf{r}n(\mathbf{r}) + G[n,\mathbf{m}] \end{aligned} \tag{9}$$

where G is a unique functional of charge and magnetisation densities at a given T, ν. The variational principle now states that Ω is a minimum for the equilibrium n, $\mathbf{m}$ and G can be written

$$G[n,\mathbf{m}] = T_s[n,\mathbf{m}] - TS_s[n,\mathbf{m}] + \Omega_{xc}[n,\mathbf{m}] \tag{10}$$

with T_s, S_s being respectively the kinetic energy and entropy of a system of non-interacting electrons with densities n, $\mathbf{m}$ at a temperature T. Ω_{xc} is the exchange and correlation contribution to the Gibbs Free Energy. The minimum principle can be shown to be identical to the corresponding equation for a system of non-interacting electrons moving in an effective potential $\tilde{W}^{eff.}$

$$\begin{aligned}\tilde{W}^{eff.}[n,\mathbf{m}] &= V^{ext.}\tilde{1} + \mathbf{B}^{ext.}\cdot\tilde{\sigma} + e^2\tilde{1}\int d\mathbf{r}'n(\mathbf{r}')/|\mathbf{r}-\mathbf{r}'| \\ &+\tilde{1}(\delta\Omega_{xc}/\delta n(\mathbf{r})) + (\delta\Omega_{xc}/\delta\mathbf{m}(\mathbf{r}))\cdot\tilde{\sigma}\end{aligned} \tag{11}$$

which satisfy the following set of equations

$$\{(\epsilon + (\hbar^2/2m)\nabla^2)\tilde{1} - \tilde{W}^{eff.}\}\tilde{G}(\mathbf{r},\mathbf{r}';\epsilon) = \tilde{1}\delta(\mathbf{r}-\mathbf{r}') \tag{12}$$

$$n(\mathbf{r}) = -Im(1/\pi)\int d\epsilon f(\epsilon-\nu)Tr.\tilde{G}(\mathbf{r},\mathbf{r};\epsilon) \tag{13}$$

$$\mathbf{m}(\mathbf{r}) = -Im(1/\pi)\int d\epsilon f(\epsilon-\nu)Tr.\tilde{\sigma}\tilde{G}(\mathbf{r},\mathbf{r};\epsilon) \tag{14}$$

$\tilde{G}$ represents an effective single electron Green function. Equation 9 can be re-written as

$$\begin{aligned}\Omega &= \int d\epsilon f(\epsilon-\nu)\epsilon n(\epsilon) - (e^2/2)\int\int d\mathbf{r}\,d\mathbf{r}'n(\mathbf{r})n(\mathbf{r}')/|\mathbf{r}-\mathbf{r}'| \\ &+\Omega_{xc} - \int d\mathbf{r}\{(\delta\Omega_{xc}/\delta n(\mathbf{r}))n(\mathbf{r}) + (\delta\Omega_{xc}/\delta\mathbf{m}(\mathbf{r}))\cdot\mathbf{m}(\mathbf{r})\}\end{aligned} \tag{15}$$

where the single particle density of states is

$$n(\epsilon) = -Im(1/\pi)\int d\mathbf{r}Tr.\tilde{G}(\mathbf{r},\mathbf{r};\epsilon) \tag{16}$$

Equations 12 to 14 demonstrate the spin polarised, self-consistent band structure basis to the many electron phenomenon. The scheme is exact but in practice approximations are made for the form of Ω_{xc} , the most common being

$$\Omega_{xc} \simeq \int d\mathbf{r}\Omega^0_{xc}(n,m)n(\mathbf{r}) \tag{17}$$

where Ω^0_{xc} is the exchange-correlation contribution to the Free Energy of a homogeneous electron gas. this is the 'local density' approximation and justification for its use are given in detail by Gunnarsson and Lundqvist (8) and others. This theory accurately describes a range of equilibrium properties of transition metals including whether a material is a magnet, the size of its magnetic moment per site, its cohesive energy etc. The underlying spin polarised bands are found, for example, in the book by Moruzzi, Janak and Williams (9).

Stocks and Winter (10) and Johnson et al. (11) have discussed in detail how this formulism may be applied to randomly disordered paramagnetic and ferromagnetic alloys A_cB_{1-c}. But,before summarising the salient points, it is necessary to consider how to treat accurately the problem of an electron moving through a lattice whose sites are randomly occupied by one atomic species or another. A reliable way of going about this is to use the Coherent Potential Approximation (C.P.A.) (12) and multiple scattering theory (KKR-CPA). Gyorffy and Stocks have provided a full description of this 'first principles' theory for electronic states in random alloys in two sets of lectures in previous NATO ASIs (13).

The effective potentials, of the form described by equation 11 appropriate to A and B atoms, are of spherically symmetric 'muffin-tin' form. In multiple scattering language, the single particle Green function $G(\mathbf{r},\mathbf{r}';\epsilon)$ of equations 12 to 14 is written (14)

$$\tilde{G}(\mathbf{r}_i,\mathbf{r}'_j;\epsilon) = \sum_{L,L'}\{Z_L(\mathbf{r}_i)\tau^{\tilde{i}\tilde{j}}_{L,L'}Z^{\dagger}_{L'}(\mathbf{r}'_j) - Z_L(\mathbf{r}_i)J^{\dagger}_L(\mathbf{r}'_i)\delta_{ij}\delta_{LL'}\} \tag{18}$$

in which $Z_L(\mathbf{r}_i)$ is the regular solution of the single site Schrödinger equation appropriate to the ith lattice site, J_L is the irregular solution. $\tau^{i\tilde{j}}$ represents the scattering path operator (15) and when it operates on the wave incident at the site at $\mathbf{R}_j$ it gives the scattered wave coming from the site at $\mathbf{R}_i$, including all the effects of scattering in between, i.e.

$$\tau^{\tilde{i}\tilde{j}}_{L,L'} = t^{\tilde{i}}_L\delta_{ij}\delta_{L,L'} + \sum_{k\neq i}\sum_{L''} t^{\tilde{i}}_L G^0_{L,L''}(\mathbf{R}_i - \mathbf{R}_k;\epsilon)\tilde{1}\tau^{k\tilde{j}}_{L'',L'} \tag{19}$$

where $t^{\tilde{i}}_L$ describes the Lth (l,m) channel single site scattering from site i and $G^0_{L,L'}$ the structure constants appropriate to the crystal lattice.

The idea behind the C.P.A. is to retrieve the Bloch theorem and construct a lattice of effective complex C.P.A. potentials such that the motion of an electron through this lattice approximates the motion of an electron, *on the average*, through the randomly disordered lattice. A C.P.A. potential described by a scattering t-matrix $t_{\tilde{c},L}$ is placed on every site and the multiple scattering in this ordered lattice specified by $\tau^{\tilde{c},ij}_{L,L'}$ (equation 19). An A atom impurity in this C.P.A. host can be described by an operator $\tau^{\tilde{A},ij}_{L,L'}$, similarly a B atom impurity by $\tau^{\tilde{B},ij}_{L,L'}$. The C.P.A. condition is then

$$c\tau^{\tilde{A},ij}_{L,L'} + (1-c)\tau^{\tilde{B},ij}_{L,L'} = \tau^{\tilde{c},ij}_{L,L'} \tag{20}$$

where c is the probability of any site being occupied by an A atom. This means that the optimum coherent potential is chosen by insisting that the substitution of a single site of the C.P.A. lattice by either an A or B site potential produces no further scattering, *on the average*. In practice $t_{\tilde{c},L}$ can be found by solving the site diagonal version of equation 20 only and for a ferro- or paramagnetic alloy in which the axis of spin polarisation is the same on every site this is written

$$c\tau^{A,00}_{L\sigma,L'\sigma} + (1-c)\tau^{B,00}_{L\sigma,L'\sigma} = \tau^{c,00}_{L\sigma,L'\sigma} \tag{21}$$

$$\tau^{c,00}_{L\sigma,L'\sigma} = (1/V_{BZ})\int d\mathbf{k}(t^{-1}_{c,\sigma} - G^0(\mathbf{k}))^{-1}_{LL'} \tag{22}$$

$$\tau^{A(B),00}_{L\sigma,L'\sigma} = \{[1 + (t^{-1}_{A(B),\sigma} - t^{-1}_{c,\sigma})\tau^{c,00}_{\sigma\sigma}]^{-1}\tau^{c,00}_{\sigma\sigma}\}_{LL'} = \sum_{L''} D^{A(B)}_{L\sigma,L''\sigma}\tau^{c,00}_{L''\sigma,L'\sigma} \tag{23}$$

(a quantity without L,L' labels denotes a matrix whose elements are labelled by angular momentum quantum numbers $L = l, m$). Equation 22 involves the inverse of the KKR-matrix. Observables are calculated by taking configurational averages estimated by considering fluctuations about this C.P.A. medium (16). For example, partially averaged charge and magnetisation densities $\langle n\rangle_{i,A(B)}$, $\langle\mu\rangle_{i,A(B)}$ on a site i in an A_cB_{1-c} binary alloy are given by

$$\langle n(\mathbf{r}_i)\rangle_{i,A(B)} = -(1/\pi)Im\int d\epsilon f(\epsilon-\nu)[\langle G_{++}(\mathbf{r}_i,\mathbf{r}_i;\epsilon)\rangle_{i,A(B)} + \langle G_{--}(\mathbf{r}_i,\mathbf{r}_i;\epsilon)\rangle_{i,A(B)}] \tag{24}$$

$$\langle \mu(\mathbf{r}_i)\rangle_{i,A(B)} = -(1/\pi)Im\int d\epsilon f(\epsilon-\nu)[\langle G_{++}(\mathbf{r}_i,\mathbf{r}_i;\epsilon)\rangle_{i,A(B)} - \langle G_{--}(\mathbf{r}_i,\mathbf{r}_i;\epsilon)\rangle_{i,A(B)}] \tag{25}$$

where

$$\langle G_{\sigma\sigma}(\mathbf{r}_i,\mathbf{r}_i;\epsilon)\rangle \simeq \sum_L \{Z^{A(B)}_{L\sigma}(\mathbf{r}_i)\tau^{A(B),00}_{L\sigma,L\sigma}Z^{A(B)}_{L\sigma}(\mathbf{r}_i) - Z^{A(B)}_{L\sigma}(\mathbf{r}_i)J^{A(B)}_{L\sigma}(\mathbf{r}_i\} \tag{26}$$

and $\langle\cdots\rangle_{i\alpha,j\beta,\ldots}$ represents the configurational average with the restriction that the ith site is occupied by an α atom, the jth site by a β atom etc. The Bloch spectral function (13),(14),$A^B_\sigma(\mathbf{k},\epsilon)$ is related to restricted averages of the non-site diagonal Green function which can be similarly expressed by considering fluctuations about the C.P.A. medium in terms of $Z^{A(B)}$,t_c and τ^c. It is a quantity through which it is possible to investigate the averaged single particle (quasiparticle) eigenvalue spectrum in both wave-vector and energy dependent detail.

A formal version of Spin Density Functional theory applied to an alloy of concentration c would require consideration of each possible configuration, the minimisation of the Gibbs Free Energy of each, finding the equilibrium charge and magnetisation densities, followed by the appropriate averaging. These steps are of course intractable. It is necessary to reverse the minimisation involved in the SDF theory with the configurational averaging in an approximate way which is consistent with the mean field spirit of the C.P.A. This is achieved by minimising the configurationally averaged Free Energy with respect to the partially averaged charge and magnetisation densities $\langle n\rangle_{i,A(B)} = \bar{n}_{iA(B)}$, $\langle \mu\rangle_{i,A(B)} = \bar{\mu}_{iA(B)}$ (11). The average Free Energy is written

$$\langle\Omega\rangle = c\langle\Omega\rangle_{i,A} + (1-c)\langle\Omega\rangle_{i,B} \tag{27}$$

$$\begin{aligned}
\langle\Omega\rangle_{i\alpha} = {} & Q_\alpha\nu - \int d\epsilon f(\epsilon-\nu)\sum_\sigma \langle N(\epsilon)\rangle_{i\alpha} \\
& -e^2/2\{\int\int d\mathbf{r}_i d\mathbf{r}_i' \bar{n}_{i\alpha}(\mathbf{r}_i)\bar{n}_{i\alpha}(\mathbf{r}_i')/|\mathbf{r}_i-\mathbf{r}_i'| \\
& +\sum_{j\neq i}\int\int d\mathbf{r}_i d\mathbf{r}_j' \bar{n}_{i\alpha}(\mathbf{r}_i)\bar{n}_j(\mathbf{r}_j')/|\mathbf{r}_i-\mathbf{r}_j'| \\
& +\sum_{j,k\neq i}\int\int d\mathbf{r}_j d\mathbf{r}_k' \bar{n}_j(\mathbf{r}_j)\bar{n}_k(\mathbf{r}_k')/|\mathbf{r}_j-\mathbf{r}_k'|\} \\
& -\int d\mathbf{r}_i\{(\delta\Omega_{xc}/\delta\bar{n}_{i\alpha}(\mathbf{r}_i) - \Omega_{xc}^0(\bar{n}_{i\alpha},\bar{\mu}_{i\alpha}))\bar{n}_{i\alpha}(\mathbf{r}_i) + (\delta\Omega_{xc}/\delta\bar{\mu}_{i\alpha}(\mathbf{r}_i))\bar{\mu}_{i\alpha}(\mathbf{r}_i)\} \\
& -\sum_{j\neq i}\sum_{\beta=A,B} c_\beta \int d\mathbf{r}_j\{(\delta\Omega_{xc}/\delta\bar{n}_{j\beta}(\mathbf{r}_j) - \Omega_{xc}^0(\bar{n}_{j\beta},\bar{\mu}_{j\beta}))\bar{n}_{j\beta}(\mathbf{r}_j) \\
& +(\delta\Omega_{xc}/\delta\bar{\mu}_{j\beta}(\mathbf{r}_j))\bar{\mu}_{j\beta}(\mathbf{r}_j)\}
\end{aligned} \tag{28}$$

in which $N_\sigma(\epsilon)$ is the integrated single particle density of states and may be written using the Lloyd determinant (13)

$$N_\sigma(\epsilon) = N_0(\epsilon) - Im(1/\pi)\ln\|t_\sigma^{-1} - G^0(\epsilon)\| \tag{29}$$

and Q_α is the charge appropriate to the $A(B)$ site. The minimisation

$$\delta\langle\Omega\rangle_{i\alpha}/\delta\bar{n}_{i\alpha} = \delta\langle\Omega\rangle_{i\alpha}/\delta\bar{\mu}_{i\alpha} = 0 \tag{30}$$

causes $\bar{n}_{i\alpha}$ and $\bar{\mu}_{i\alpha}$ to satisfy equations 24 to 26 and the self-consistently set up effective potential associated with an A ar B site is written

$$\begin{aligned}
\tilde{W}_{i\alpha}[\bar{n}_{i\alpha},\bar{\mu}_{i\alpha};\mathbf{r}_i] = {} & \{-(Q^\alpha e^2/r_i) \\
& +e^2\int d\mathbf{r}_i'\bar{n}_{i\alpha}(\mathbf{r}_i')/|\mathbf{r}_i-\mathbf{r}_i'| - \sum_{j\neq i}\bar{Q}e^2/|\mathbf{r}_i-\mathbf{R}_j+\mathbf{R}_i| \\
& +\sum_{j\neq i} e^2\int d\mathbf{r}_j'\bar{n}_j(\mathbf{r}_j')/|\mathbf{r}_i-\mathbf{r}_j'| \\
& +\delta\Omega_{xc}/\delta\bar{n}_{i\alpha}(\mathbf{r}_i)\}\tilde{1} + \delta\Omega_{xc}/\delta\bar{\mu}_{i\alpha}(\mathbf{r}_i)\tilde{\sigma}_z
\end{aligned} \tag{31}$$

where Q^{α} is the nuclear charge on an α site, $\bar{Q}$ the average.

To summarise, for an alloy of composition A_cB_{1-c}, the procedure begins with (i) guessed partially averaged charge and magnetisation densities for both A and B sites. (ii) Effective potentials are set up according to equation 31. (iii) The problem of an electron propagating through a random array of these potentials is solved for each spin polarisation, via the C.P.A. (iv) The partially averaged densities are recalculated according to equations 24 to 26 and (v) steps (ii) to (iv) repeated until consistency is achieved. (vi) Finally the converged quantities are used to calculate the Free Energy and other equilibrium properties.

Such self-consistent field SCF-KKR-CPA calculations have been carried out for several ferro- and paramagnetic alloys with good agreement with experimental measurements. For example, Pinski et al. (17) calculated magnetic moments for a range of b.c.c. $Fe-V$ and both b.c.c. and f.c.c. $Fe-Ni$ alloys and presented them on a Slater-Pauling curve with a comparison with experimental measurements.

As is well known, Stoner theory, applied to pure metals provides a picture of near rigidly exchange split bands (9). This picture, however, is not appropriate for alloys. The Stoner instability requires a large paramagnetic density of states and in a disordered paramagnetic alloy, electrons of either spin polarisation will be equally affected by the chemical disorder. In the spin polarised magnetic state it is possible for the effects of disorder to be confined to electrons of one 'spin'. (Later the importance of this for compositional ordering will be shown.) In this way the rigidly split bands of Stoner theory are lost. This is demonstrated by figures 1 to 3 of the spin polarised partially averaged densities of states of b.c.c. $Fe_{.5}V_{.5}$, f.c.c. $Ni_{.75}Fe_{.25}$ (17) and $Pd_{.75}Fe_{.25}$ alloys respectively.

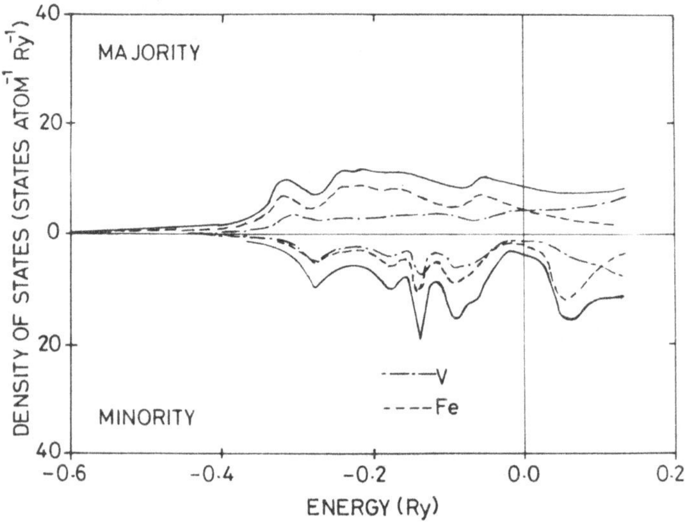

• Figure 1. Total and partially averaged densities of states of ferromagnetic $Fe_{.50}V_{.50}$ b.c.c. alloy.

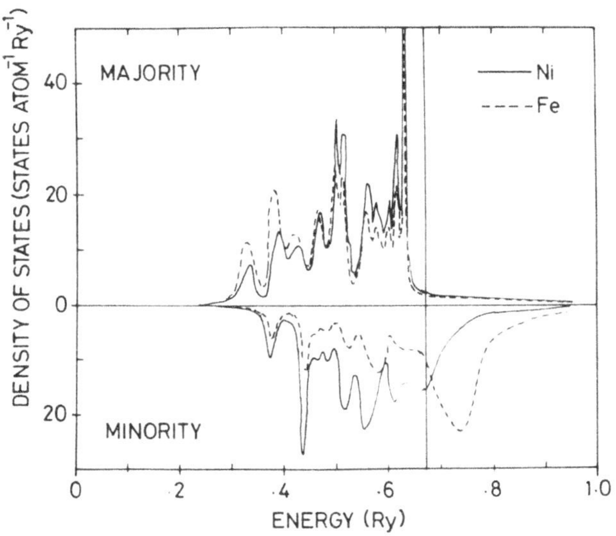

• **Figure 2a. Partially averaged densities of states of ferromagnetic $Ni_{.75}Fe_{.25}$ f.c.c. alloy.**

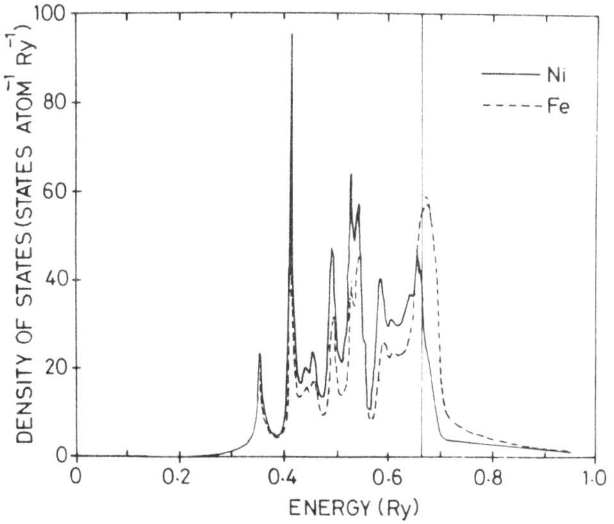

• **Figure 2b. Partially averaged densities of states of paramagnetic (Stoner) $Ni_{.75}Fe_{.25}$ f.c.c. alloy.**

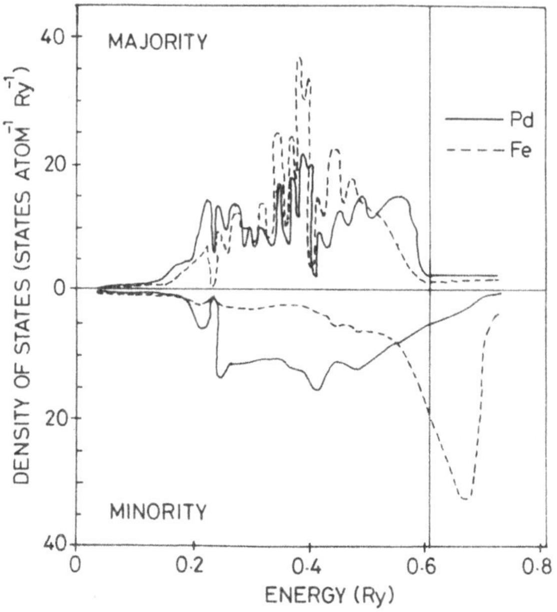

• Figure 3. Partially averaged densities of states of ferromagnetic $Pd_{.75}Fe_{.25}$ f.c.c. alloy.

Our $Fe - V$ calculations for a range of compositions indicate that the minority spin electrons 'see' little difference between Fe and V sites and it is the majority spin electrons that are most strongly affected by the disorder (figure 1). The Fermi energy is pinned in the bonding-antibonding valley of the minority density of states fixing a constant number of minority spin electrons for a range of iron-rich alloys. This produces a mechanism for the linear dependence of the average moment on the electron per atom ratio. (This has also been noted by Williams et al. (18) for ordered alloys.) An understanding of the more complicated dependence of the moments on Fe and V sites requires a more detailed analysis of the electronic structure. The moments on the V sites are formed anti-parallel to the larger ones on the Fe sites (a similar situation is found in $Fe - Mn$ alloys). When the concentration of Fe is less than 30% the alloys are found to be paramagnetic also in agreement with experiment.

The $Ni - Fe$ calculations indicate the opposite picture. Here it is the majority spin electrons which are scarcely affected by the chemical disorder in sharp distinction from the behaviour of the minority spin electrons. In the f.c.c. alloys the average magnetisation per site increases linearly with increase in Fe concentration. As more iron is added the partially averaged moment on a Fe site decreases whilst that on the nickel sites increases. In the Invar composition range, as the iron concentration approaches 70%, the size of moment is very sensitive to the lattice spacing. For a lattice parameter of 6.76a.u. an average moment of $2\mu_B$ exists per site whereas on a lattice spacing with a 4% smaller spacing no moment can be sustained (17). A small increase in Fe concentration in this range also causes a marked moment collapse. With volume decrease or a larger iron concentration the Fermi level is pushed into the majority d-bands precipitating the moment disappearance. A similar analysis is valid for interpreting calculations on pure f.c.c. Fe on a range of lattice spacings (19) and is useful in understanding the magnetic properties of many f.c.c. iron alloys.

As a summary of this section, it can be stated that at low temperatures (well below T_c, it is now accepted that the magnetic properties of metallic systems can be understood and predicted from the study of the behaviour of an electron moving through a lattice of effective potentials. These are set up self-consistently by all the electrons and the magnetic components are all aligned along one 'global' direction. Such SDF calculations have been carried out with success for pure transition metals (9), ordered alloys (20) and more recently, within the framework of the KKR-CPA (11),(13),(17), for disordered alloys. It is now therefore possible to investigate in a parameter-free way the wide range of magnetic properties exhibited by the transition metal alloys and the influence of magnetism upon other physical characteristics. The following discussion is concerned with the interrelation of these magnetic properties and compositional ordering.

3 Compositional order and magnetic structure of transition metal alloys

A 'first principles' theory of magnetic and compositional phase transitions in alloys should ideally treat both magnetic and compositional thermal fluctuations on an equal footing but here we will take the first step towards such a theory and look at the effect of the magnetic state upon compositional ordering. This is basically the generalisation to magnetic alloys of the theory that Dr. Gyorffy is discussing at this summer school for paramagnetic alloys (21).

We begin by writing down formally the Grand Potential for a system of electrons in a particular configuration of nuclei using the SDF theory. On a site i in the underlying lattice, the variable $\xi_i = 1$ if the site is occupied by a A atom, $= 0$ if occupied by a B atom, i.e. it is an Ising-like variable. A configuration is denoted $\{\xi_i\}$ and the Grand Potential $\Omega\{\xi_i\}$. Averaging over the compositional fluctuations with measure

$$P\{\xi_i\} = \exp(-\beta\Omega\{\xi_i\})/\prod_i \sum_{\xi_i=0,1} \exp(-\beta\Omega\{\xi_i\}) \tag{32}$$

gives an expression for the Free Energy

$$F = -(1/\beta)\ln\prod_i \sum_{\xi_i=0,1} \exp(-\beta\Omega\{\xi_i\}) \tag{33}$$

where $\beta = (k_B T)^{-1}$. $\Omega\{\xi_i\}$ is thus playing the role of a highly complicated concentration fluctuation Hamiltonian. By expanding about a suitable reference Hamiltonian, Ω_0, using the Feynman-Peierls Inequality (22)

$$F \leq F_0 + \langle \Omega - \Omega_0 \rangle^0 = \breve{F} \tag{34}$$

where

$$\langle X \rangle^0 = \prod_i \sum_{\xi_i=0,1} X\{\xi_i\}\exp(-\beta\Omega_0\{\xi_i\}) / \prod_i \sum_{\xi_i=0,1} \exp(-\beta\Omega_0\{\xi_i\}) \tag{35}$$

A mean field theory is produced with the choice

$$\Omega_0 = \sum_i f_i(\xi_i) \tag{36}$$

in which the best functions f_i are found following a functional minimisation of $\breve{F}$ and

$$f_i(\xi') = \langle \Omega \rangle^0_{\xi_i=\xi'} \tag{37}$$

where the restricted average is explicitly

$$\langle \Omega \rangle^0_{\xi_i=\xi'} = \prod_{j\neq i} \sum_{\xi_j=0,1} \Omega(\xi_1, \xi_2, \cdots, \xi_i = \xi', \cdots)\exp(-\beta f_j(\xi_j)) / \prod_{j\neq i} \sum_{\xi_j=0,1} \exp(-\beta f_j(\xi_j)) \tag{38}$$

i.e. the average of the Grand Potential is taken over all configurations with the restriction that on the ith site the occupation is fixed, $\xi_i = \xi'$. These partial averages are accessible from the SCF-KKR-CPA framework sketched in the last section and the mean field picture is consistent with that of the coherent potential approximation. The probability of finding an A atom on a particular site, i, at a temperature T, is

$$c_i = \exp(-\beta f_i(1)) / (\exp(-\beta f_i(1)) + \exp(-\beta f_i(0)) \tag{39}$$

Note that, in principle, the probability of occupation can vary from site to site but it is only the case of a homogeneous probability distribution, $c_i = c$, that can be tackled in practice with our KKR-CPA method.

Using response theory and the fluctuation dissipation theorem, it is possible to write down the various correlation functions appropriate to the system and hence investigate ordering. Firstly we will consider the response of the homogeneous system to the application of an inhomogeneous external field,

$$V^{ext.} = \sum_i (v_i - \eta) \tag{40}$$

which couples to the occupation variables $\{\xi_i\}$ such that a term

$$\sum_i (v_i - \eta)\xi_i = \sum_i u_i \xi_i \tag{41}$$

is added to the Grand Potential $\Omega\{\xi_i\}$ (η is the chemical potential difference such that the number of A and B sites remains constant). This applied field, different on every site, induces changes in the probabilities of occupation on every site $\{\delta c_i\}$. Furthermore, of crucial importance for this discussion, it also causes changes to the moments from site to site $\{\delta\mu_i^{A(B)}\}$. (We assume that the corresponding changes to the total charge densities are small and that the magnetisation changes occur in the muffin tin sphere only.) It is possible to formulate a theory through the formal framework of an inhomogeneous C.P.A. (21). Although the resulting equations, shown below, are not soluble in practice,

$$c_i \tau_{inh.}^{A,ii} + (1 - c_i)\tau_{inh.}^{B,ii} = \tau_{inh.}^{c,ii} \tag{42}$$

$$\tau_{inh.}^{c,ii} = t_{c,i,inh.} + \sum_{k \neq i} t_{c,i,inh.} G^0(\mathbf{R}_i - \mathbf{R}_k; \epsilon)\tau_{inh.}^{c,ki} \tag{43}$$

it is possible to find an approximate solution by perturbing about the homogeneous case. Clearly the single site C.P.A. t-matrix $t_{c,i,inh.}$ and the site diagonal C.P.A. scattering path operator $\tau_{inh.}^{c,ii}$ are different on every site, but to lowest order in the perturbation, they can be written

$$t_{c,i,inh.}^{-1} = t_c^{-1} + \delta t_{c,i}^{-1} \tag{44}$$

$$\tau_{inh.}^{c,ii} = \tau^{c,00} + \delta\tau^{c,ii} = \tau^{c,00} - \sum_j \tau^{c,ij}\delta t_{c,j}^{-1}\tau^{c,ji} \tag{45}$$

Equation 42, using equations 44 and 45 and the homogeneous solution $t_c, \tau^{c,ij}$ can now be solved to find $\{\delta t_{c,i}^{-1}\}$.

The induced change in probability of occupation on the ith site is written to low order in the applied field

$$\delta c_i \simeq -\beta c(1-c)(\langle\delta\Omega\rangle_{iA} - \langle\delta\Omega\rangle_{iB}) \tag{46}$$

which can be expressed, using the inhomogeneous version of equation 28, for $l \leq 2$ and assuming $\delta t_{c,i}^{-1}$ is a diagonal matrix in L -space

$$\begin{aligned} \delta c_i &= \beta c(1-c)(1/\pi) Im \int d\epsilon f(\epsilon - \nu) \sum_{L\sigma} D_{L\sigma}^A (t_{A,L\sigma}^{-1} - t_{B,L\sigma}^{-1}) D_{L\sigma}^B \\ & \sum_{L',j\neq i} \tau_{L\sigma,L'\sigma}^{c,ij} \tau_{L'\sigma,L\sigma}^{c,ji} \delta t_{c,j,L'\sigma}^{-1} \end{aligned} \tag{47}$$

The changes to the magnetisation on a site i which is either occupied by an A or B site can be expressed using the inhomogeneous version of equation 25.

$$\delta\mu_i^{A(B)} = K_i^{A(B)}/(1 - U^{A(B)}) \tag{48}$$

where

$$U^{A(B)} = (1/\pi)Im\int d\epsilon f(\epsilon-\nu)\sum_{L\sigma}(A_{L\sigma}^{A(B)}B_{L\sigma}^{A(B)}(D_{L\sigma}^{A(B)}\tau_{L\sigma,L\sigma}^{c,00})^2+\Delta_{l\sigma}^{A(B)}D_{L\sigma}^{A(B)}\tau_{L\sigma,L\sigma}^{c,00}) \tag{49}$$

$$K_i^{A(B)} = (1/\pi)\int d\epsilon f(\epsilon-\nu)\sum_{L\sigma}\sigma B_{L\sigma}^{A(B)}(D_{L\sigma}^{A(B)})^2\sum_{L',j\neq i}\tau_{L\sigma,L'\sigma}^{c,ij}\tau_{L'\sigma,L\sigma}^{c,ji}\delta t_{c,j,L'\sigma}^{-1} \tag{50}$$

The quantities A,B and Δ are the muffin tin integrals,($\sigma = \pm 1$).

$$A_{L\sigma}^{\alpha} = -\int d\mathbf{r}(Z_{L\sigma}^{\alpha}(\mathbf{r}))^2(\delta W_{\sigma}^{\alpha}(\mathbf{r})/\delta\mu_i^{\alpha})\sigma \tag{51}$$

$$B_{L\sigma}^{\alpha} = \int d\mathbf{r}(Z_{L\sigma}^{\alpha}(\mathbf{r}))^2 \tag{52}$$

$$\Delta_{L\sigma}^{\alpha} = -\int d\mathbf{r}(\delta B_{L\sigma}^{\alpha}/\delta W_{\sigma}^{\alpha}(\mathbf{r}))(\delta W_{\sigma}^{\alpha}(\mathbf{r})/\delta\mu_i^{\alpha})\sigma \tag{53}$$

For later use

$$\bar{U}_{L\sigma}^{\alpha} = A_{L\sigma}^{\alpha}/B_{L\sigma}^{\alpha} \tag{54}$$

The radial form of the magnetisation within a muffin tin sphere is assumed to be unchanged by the applied field (27). Equations 47 and 48 demonstrate how the induced compositional and magnetic changes are interrelated through alterations to the C.P.A. medium. By expanding about the homogeneous, unperturbed solution, as indicated by equations 44 and 45, it is possible to show

$$\begin{aligned}\delta t_{c,i,L\sigma}^{-1} = & [D_{L\sigma}^{A}(t_{A,L\sigma}^{-1} - t_{B,L\sigma}^{-1})D_{l\sigma}^{B}]\delta c_i \\ & +[c(D_{L\sigma}^{A})^2A_{L\sigma}^{A}]\delta\mu_i^{A}\sigma \\ & +[(1-c)(D_{L\sigma}^{B})^2A_{L\sigma}^{B}]\delta\mu_i^{B}\sigma \\ & -[D_{L\sigma}^{A}(t_{A,L\sigma}^{-1} - t_{c,L\sigma}^{-1})D_{L\sigma}^{B}(t_{B,L\sigma}^{-1} - t_{c,L\sigma}^{-1})] \\ & \sum_{j\neq i,L'}\tau_{L\sigma,L'\sigma}^{c,ij}\tau_{L'\sigma,L\sigma}^{c,ji}\delta t_{c,j,L'\sigma}^{-1}\end{aligned} \tag{55}$$

On combining equations 47,48 and 55, carrying out lattice Fourier transforms such that

$$\alpha(\mathbf{q}) = (1/N)\sum_j(\delta c_i/\delta u_j)\exp(i\mathbf{q}\cdot(\mathbf{R}_i - \mathbf{R}_j)) \tag{56}$$

and

$$\begin{aligned}\Upsilon(\mathbf{q}) &= (1/N)\sum_j(\delta\mu_i/\delta u_j)\exp(i\mathbf{q}\cdot(\mathbf{R}_i-\mathbf{R}_j)) \\ &= (1/N)\sum_{jk}(\delta\mu_i/\delta c_k)(\delta c_k/\delta u_j)\exp(i\mathbf{q}\cdot(\mathbf{R}_i-\mathbf{R}_j)) \\ &= \gamma(\mathbf{q})\alpha(\mathbf{q}) \qquad (57)\end{aligned}$$

in which

$$\alpha(\mathbf{q}) = 2\beta c(1-c)/(1-\beta c(1-c)S^{(2)\star}(\mathbf{q})) \qquad (58)$$

$$\begin{aligned}S^{(2)\star}(\mathbf{q}) &= S^{(2)}(\mathbf{q}) \\ &-\{[Q^A(\mathbf{q})L^A(\mathbf{q})(1-M^{AA}(\mathbf{q}))+Q^A(\mathbf{q})L^B(\mathbf{q})M^{AB}(\mathbf{q}) \\ &+Q^B(\mathbf{q})L^B(\mathbf{q})(1-M^{AA}(\mathbf{q}))+Q^B(\mathbf{q})L^A(\mathbf{q})M^{BA}(\mathbf{q})] \\ &/[(1-M^{AA}(\mathbf{q}))(1-M^{BB}(\mathbf{q}))-M^{AB}(\mathbf{q})M^{BA}(\mathbf{q})]\} \qquad (59)\end{aligned}$$

$$\gamma(\mathbf{q}) = \mu_A-\mu_B+c\gamma_A(\mathbf{q})+(1-c)\gamma_B(\mathbf{q}) \qquad (60)$$

$$\begin{aligned}\gamma_{A(B)}(\mathbf{q}) &= \{[L^{A(B)}(\mathbf{q})(1-M^{BB(AA)}(\mathbf{q}))+M^{AB(BA)}(\mathbf{q})L^{B(A)}(\mathbf{q})] \\ &/[(1-M^{AA}(\mathbf{q}))(1-M^{BB}(\mathbf{q}))-M^{AB}(\mathbf{q})M^{BA}(\mathbf{q})]\} \qquad (61)\end{aligned}$$

$$\begin{aligned}S^{(2)}(\mathbf{q}) &= (1/\pi)Im\int d\epsilon f(\epsilon-\nu)\sum_{L\sigma}D^A_{L\sigma}(t^{-1}_{A,L\sigma}-t^{-1}_{B,L\sigma})D^B_{L\sigma}\sum_{L'}X_{LL',\sigma}(\mathbf{q}) \\ &\sum_{L''}(1+I(\mathbf{q}))^{-1}_{L'L'',\sigma}D^A_{L''\sigma}(t^{-1}_{A,L''\sigma}-t^{-1}_{B,L''\sigma})D^B_{L''\sigma} \qquad (62)\end{aligned}$$

$$L^\alpha(\mathbf{q}) = (1/\pi)Im\int d\epsilon f(\epsilon-\nu)\sum_{L\sigma}\Lambda^\alpha_{L\sigma}(\mathbf{q}) \qquad (63)$$

$$\Lambda^\alpha_{L\sigma}(\mathbf{q}) = B^\alpha_{L\sigma}(D^\alpha_{L\sigma})^2\sigma\sum_{L'}X_{LL',\sigma}(\mathbf{q})\sum_{L''}(1+I(\mathbf{q}))^{-1}_{L'L'',\sigma}D^A_{L''\sigma}(t^{-1}_{A,L''\sigma}-t^{-1}_{B,L''\sigma})D^B_{L''\sigma} \qquad (64)$$

$$Q^\alpha(\mathbf{q}) = (1/\pi)Im\int d\epsilon f(\epsilon-\nu)\sum_{L\sigma}\bar{U}^\alpha_{L\sigma}\Lambda^\alpha_{L\sigma}(\mathbf{q}) \qquad (65)$$

$$M^{\alpha\beta}(\mathbf{q}) = (1/\pi)Im\int d\epsilon f(\epsilon-\nu)\sum_{L\sigma}\bar{U}^\alpha_{L\sigma}n^{\alpha\beta}_{L\sigma}c_\beta+U^\alpha\delta_{\alpha\beta} \qquad (66)$$

where $c_A = c$ and $c_B = 1-c$ and

$$n^{\alpha\beta}_{L\sigma} = B^\alpha_{L\sigma}(D^\alpha_{L\sigma})^2\sum_{L'}X_{LL',\sigma}(\mathbf{q})B^\beta_{L\sigma}(D^\beta_{L\sigma})^2c_\beta \qquad (67)$$

$$I_{LL',\sigma}(\mathbf{q}) = D^A_{L\sigma}(t^{-1}_{A,L\sigma}-t^{-1}_{c,L\sigma})D^B_{L\sigma}(t^{-1}_{B,L\sigma}-t^{-1}_{c,L\sigma})X_{LL',\sigma} \qquad (68)$$

$$X_{LL',\sigma}(\mathbf{q}) = (1/V_{BZ}) \int d\mathbf{k} \tau^{c}_{L\sigma,L'\sigma}(\mathbf{k}) \tau^{c}_{L'\sigma,L\sigma}(\mathbf{k}+\mathbf{q}) - (\tau^{c,00}_{L\sigma,L\sigma})^2 \delta_{LL'} \tag{69}$$

A key result is then that the compositional correlation function, related to $\alpha(\mathbf{q})$,i.e.

$$\alpha(\mathbf{q}) = (1/N) \sum_j \beta(\langle \xi_i \xi_j \rangle - \langle \xi_i \rangle \langle \xi_j \rangle) \exp(i\mathbf{q} \cdot (\mathbf{R}_i - \mathbf{R}_j)) \tag{70}$$

is given by equation 58 and its $\mathbf{q}$-dependent structure is given by $S^{(2)\star}$ of equation 59. The first term, $S^{(2)}$, is essentially the spin polarised version of the expression derived by Gyorffy and Stocks (21) and the second term, in brackets, arises from the change to the magnetic moments as the chemical environment alters. If there were no spins and the system were described by a pairwise interacting lattice gas model in the random phase approximation then equation 58 would represent the Krivoglaz-Moss-Clapp formula (23) and the real space version of $S^{(2)}(\mathbf{q})$, $S^{(2)}_{ij}$ would be the interchange energy

$$V_{ij} = V^{AA}_{ij} + V^{BB}_{ij} - 2V^{AB}_{ij} \tag{71}$$

with obvious notation. When the effect of magnetic moments interacting via phenomenological exchange interactions ,J_{ij}, in a moments-aligned case, is included then

$$S^{(2)}_{ij} = V_{ij} + J^{AA}_{ij} + J^{BB}_{ij} - 2J^{AB}_{ij} \tag{72}$$

Thus $S^{(2)}_{ij}$ can be interpreted as an effective pairwise ordering energy enhanced by the presence of rigid moments.

Another important result is that the magneto-compositional 'cross' correlation function is related to $\Upsilon(\mathbf{q})$ of equation 57 by

$$\Upsilon(\mathbf{q}) = (1/N) \sum_j \beta(\langle \mu_i \xi_j \rangle - \langle \mu_i \rangle \langle \xi_j \rangle) \exp(i\mathbf{q} \cdot (\mathbf{R}_i - \mathbf{R}_j)) \tag{73}$$

which itself is written as a product of $\alpha(\mathbf{q})$ and $\gamma(\mathbf{q})$. $\gamma(\mathbf{q})$ is expressed in terms of quantities γ_A and γ_B (equation 61) and the lattice Fourier transforms of these have a particularly revealing physical interpretation.

$$(1/V_{BZ}) \int d\mathbf{q} \gamma_{A(B)}(\mathbf{q}) \exp(-i\mathbf{q} \cdot (\mathbf{R}_i - \mathbf{R}_j)) = \gamma^{ij}_{A(B)} = \delta \mu^{A(B)}_i / \delta c_j \tag{74}$$

$\gamma^{ij}_{A(B)}$ describes the change in moment on a site i in the lattice if it is occupied by an $A(B)$ atom if the probability of occupation is altered on another site j. It measures the chemical environment effect on the moments' sizes. Thus

$$\gamma^{ij}_{A(B)} \Delta c_j = \gamma^{ij}_{A(B)}(1-c) \tag{75}$$

describes the moment change on the ith site occupied by an $A(B)$ atom if the site j in the random alloy is now *definitely* occupied by an A atom. Similarly,

$$\gamma^{ij}_{A(B)} \Delta c_j = \gamma^{ij}_{A(B)}(-c) \tag{76}$$

describes the change if site j is now *definitely* occupied by a B atom. Consequently once these quantities $\gamma^{ij}_{A(B)}$ have been calculated, it is possible, in a perturbative fashion, to investigate the magnetic structure of alloys of varying composition. Hence it provides a complementary approach to that supplied by explicit cluster calculations (24). For example the magnetic properties of modulated or clustering alloys can be examined and alloys designed theoretically to have specified magnetic properties.

In a similar vein to the preceding derivation, the calculation of the response to an external inhomogeneous magnetic field applied along the spontaneous magnetisation direction,

$$H^{ext.} = \sum_i h_i^{ext.} \tag{77}$$

delivers quantities $\chi(\mathbf{q})$, $\Upsilon(\mathbf{q})$ which are themselves related to the magnetic and magneto-compositional 'cross' correlation functions respectively.

$$\chi(\mathbf{q}) = (1/N)\sum_j \beta(\langle\mu_i\mu_j\rangle - \langle\mu_i\rangle\langle\mu_j\rangle)\exp(i\mathbf{q}\cdot(\mathbf{R}_i - \mathbf{R}_j)) \tag{78}$$

$$\chi(\mathbf{q}) = c\bar{\chi}^A(\mathbf{q}) + (1-c)\bar{\chi}^B(\mathbf{q}) + (\gamma(\mathbf{q}))^2\alpha(\mathbf{q}) \tag{79}$$

where

$$\begin{aligned}\bar{\chi}^{A(B)}(\mathbf{q}) &= [(1 - M^{BB(AA)}(\mathbf{q}))N^{A(B)}(\mathbf{q}) + M^{AB(BA)}(\mathbf{q})N^{B(A)}(\mathbf{q})] \\ &\quad /[(1 - M^{AA}(\mathbf{q}))(1 - M^{BB}(\mathbf{q})) - M^{AB}(\mathbf{q})M^{BA}(\mathbf{q})] \end{aligned} \tag{80}$$

where

$$\begin{aligned} N^{\alpha}(\mathbf{q}) &= (1/\pi)Im\int d\epsilon f(\epsilon-\nu)\sum_{L\sigma}\{n^{\alpha A}_{L\sigma}(\mathbf{q}) + n^{\alpha B}_{L\sigma}(\mathbf{q})\} \\ &\quad +(B^{\alpha}_{L\sigma}D^{\alpha}_{L\sigma}\tau^{c,00}_{L\sigma,L\sigma})^2 - (\delta B^{\alpha}_{L\sigma}/\delta h^{ext.})D^{\alpha}_{L\sigma}\tau^{c,00}_{L\sigma,L\sigma}\} \end{aligned} \tag{81}$$

Equation 79 once again shows the interrelation between the magnetic and compositional responses.

These theoretical quantities, correlation functions, determined fully by the underlying electronic structure of the homogeneous disordered alloy are closely related to information obtained from X-ray and neutron scattering, nuclear magnetic resonance and Mössbauer spectroscopic measurements. In particular, the cross-sections obtained from diffuse polarised neutron scattering data can be written

$$(d\sigma/d\omega)_{\zeta} = d\sigma^N/d\omega + \zeta(d\sigma^{NM}/d\omega) + d\sigma^M/d\omega \tag{82}$$

where $\zeta = +1(-1)$ if the neutrons are polarised (anti-) parallel to the magnetisation. The nuclear component $d\sigma^N/d\omega$ is proportional to the compositional correlation function (closely related to the Warren-Cowley short range order parameters) i.e. $\alpha(\mathbf{q})$ of equation 58.

$$d\sigma^N/d\omega = (\Delta b)^2\alpha(\mathbf{q}) \tag{83}$$

where Δb is the difference between the two nuclear scattering amplitudes. Similarly the magnetic component is proportional to $\chi(\mathbf{q})$ of equation 79. Finally $d\sigma^{NM}/d\omega$ describes the magnetic-compositional 'cross' correlation function, proportional to $\Upsilon(\mathbf{q})$ of equation 57. It is clear that the sum of the polarised cross-sections measures a combination of $\alpha(\mathbf{q})$ and $\chi(\mathbf{q})$ whilst the difference gives $\Upsilon(\mathbf{q})$. In order to separate $\alpha(\mathbf{q})$ and $\chi(\mathbf{q})$ experimentally the technique of isotope substitution can be used. Once $\alpha(\mathbf{q})$ and $\Upsilon(\mathbf{q})$ are measured, $\gamma(\mathbf{q})$,i.e. the moments' dependence on their chemical environment can be extracted. By considering a linear superposition of perturbations and phenomenological parameters, Marshall (25) gives formulae of the same form as equations 57 and 79 and the work of Cable and co-workers and others (26),(28) has continued from this and attempted to describe the experimental neutron scattering cross-sections in terms of a parameter for the range of perturbations and others to account for the chemical environment effect, a measure of the exchange interactions and also the type of lattice. It is the aim of this section to provide a parameter-free theory which can be tested against these experiments. We conclude this part of the lectures by illustrating the theory with studies of several iron alloys.

The first example that we discuss is an f.c.c. nickel rich nickel-iron alloy of composition $Ni(.75)Fe(.25)$. The work has also been described in reference 29. Figure 4 shows $S^{(2)}$ of equation 62 decomposed into its spin components together with the full quantity $S^{(2)\star}$ of equation 59. Now if such a quantity peaks at vanishing wavevectors $\mathbf{q} = 0$ then the system has a tendency to cluster or phase segregrate whereas if it peaks at a finite wavevector then an ordering tendency is suggested. Figure 4 shows that the majority spin contribution to $S^{(2)}$ is small.This is consistent with figure 2 a) which shows that a majority spin electron 'sees' very little difference between nickel and iron sites. Consequently these electrons contribute little to the compositional ordering mechanism. Figure 2 a) also indicates the contrary picture for the minority spin electrons and figure 4 now shows that these electrons provide the dominant contribution to $S^{(2)}$. This component $S_{-}^{(2)}$ peaks at a finite wavevector indicating an ordering tendency although our calculations are not accurate enough to predict absolute maxima at the (100) and (110) points which correspond to the experimentally observed L_{12} structure (30). Figure 4 also shows $S^{(2)}$ appropriate to paramagnetic $Ni(0.75)Fe(0.25)$ alloy, i.e.in the non-spin polarised Stoner paramagnetic state, (dashed line). It peaks at $\mathbf{q} = 0$ and therefore corresponds to a clustering tendency. As shown in figure 2 b) electrons of both spin polarisations contribute equally to the underlying mechanism, 'seeing' the same difference between nickel and iron sites. Exchange splitting the electronic structure in this system thus shifts the tendency from clustering to ordering.

By recalling some general features of the electronic forces responsible for order and disorder in metallic alloys we can understand the details described above. In transition metal alloys where the d-band is approximately half-filled ordering is expected. This is because here only bonding states are occupied and the bonding

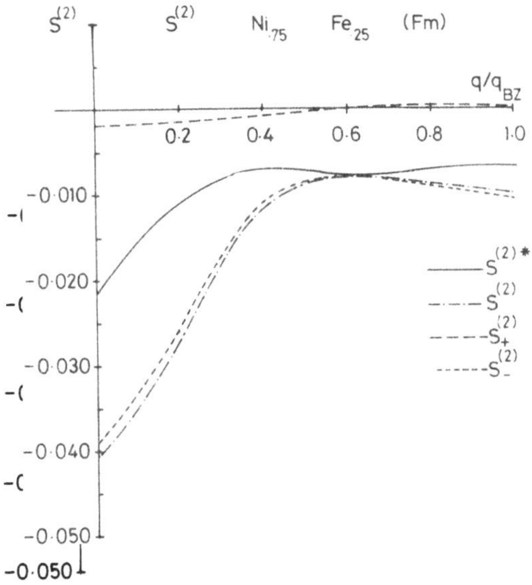

• **Figure 4a.** $S^{(2)}$ **components of ferromagnetic** $Ni_{.75}Fe_{.25}$ **in the (1,0,0) direction.**

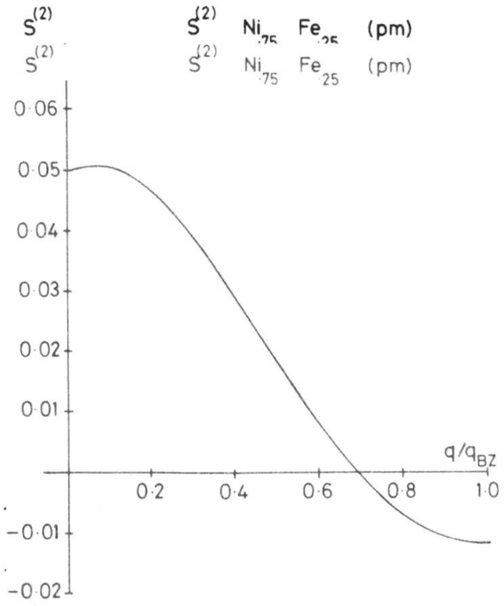

• **Figure 4b.** $S^{(2)}$ **of paramagnetic** $Ni_{.75}Fe_{.25}$ **in the (1,0,0) direction.**

states of the ordered state are lower than those of the disordered state due to level repulsion. On the other hand, almost completely filled bands can be expected to lead to clustering (31). Hence the half-filled bands of ferromagnetic $Ni(0.75)Fe(0.25)$ shown in fig.2 a) match the 'finite q' peak of $S_{-}^{(2)}$ of fig. 4 whereas the mostly full bands of paramagnetic nickel-iron shown in fig.2 b) correspond well with the 'zero q' peak of the dashed curve of $S^{(2)}$ also represented in fig.4.

We have also studied the f.c.c. ferromagnetic $Pd(0.75)Fe(0.25)$ alloy and its densities of states are shown in figure 3. Our calculation of $S_{+(-)}^{(2)}$ and $S^{(2)\star}$ appear in fig.5. An ordering tendency is suggested. In this case contributions from electrons of both spin types are important but the electronic mechanisms are different. Both $S_{+}^{(2)}$ and $S_{-}^{(2)}$ peak at finite wavevectors but only for the latter quantity can this be explained by half-filled d-bands (figure 3). The majority spin electrons ,although 'seeing' very little site diagonal disorder of the palladium and iron sites do experience substantial off-diagonal (band width) disorder. The structure of $S_{+}^{(2)}$ arises from a Fermi surface mechanism involving flat parallel sheets.This has been discussed by ref.(21) and in Dr. Gyorffy's lectures so no further details will be given here. The final example to be presented in these lectures refers to the $Fe(0.50)V(0.50)$ b.c.c. alloy and the relevant figures are fig.1. which shows the underlying electronic structure and fig. 6. which describes the appropriate $S^{(2)}$ calculation for this ferromagnetic alloy. The overall picture is of a complicated ordering tendency. The mostly full majority spin d-bands produce a 'clustering' $S_{+}^{(2)}$ whereas once again the half full minority spin states turn up an ordering $S_{-}^{(2)}$. In all three alloys the plots of the function $S^{(2)\star}$ which includes the contribution from the moments' response to their changing compositional environment show a greater weight towards small wavevectors. Figure 7 shows further details of this interrelation between the magnetisation and compositional order via the quantities γ_A and γ_B.

The lattice Fourier transforms of these quantities can be roughly estimated ($\gamma_{A(B)}^{ij}$ of eqn.74) and are found to be fairly short ranged. For the $Ni(.75)Fe(.25)$ alloy the moment on a nickel site will grow if surrounded by iron sites but decrease if enclosed by other nickel sites. On iron sites larger moments will form in local environments containing many nickel atoms but in iron rich environments the moments will be small. Similar more pronounced effects are found in our calculations for $Pd(.75)Fe(.25)$ alloy as shown in figure 7 b). Although part of these effects can be understood in terms of the polarisability of the nickel and palladium 'hosts' a useful interpretation arises from the behaviour of iron on an f.c.c. lattice. As shown in ref. 19 iron on a f.c.c. lattice with a spacing equivalent to that of this $Fe-Ni$ alloy would be in an antiferromagnetic state with a small magnetisation per site although a small expansion would indicate that a ferromagnetic state is stable with a larger moments. In these alloys surrounding the iron sites with nickel or palladium atoms causes the ferromagnetic state to be stable and the iron moments to grow. The work on iron-vanadium alloy, a b.c.c. system, has a different description. Moments on both iron and vanadium sites grow in magnitude if their environments contain

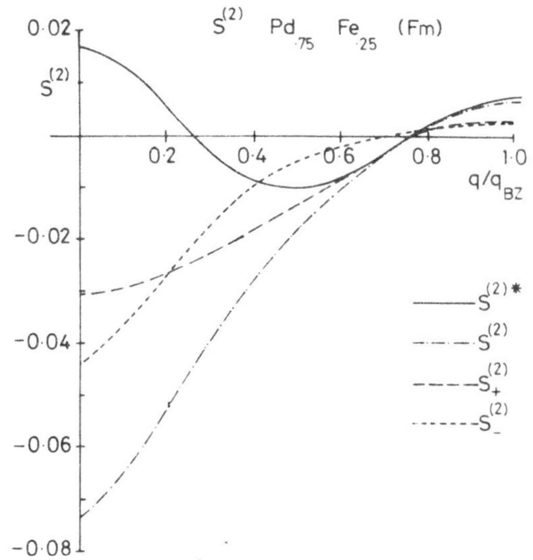

• **Figure 5.** $S^{(2)}$ **components of ferromagnetic** $Pd_{.75}Fe_{.25}$ **in the (1,0,0) direction.**

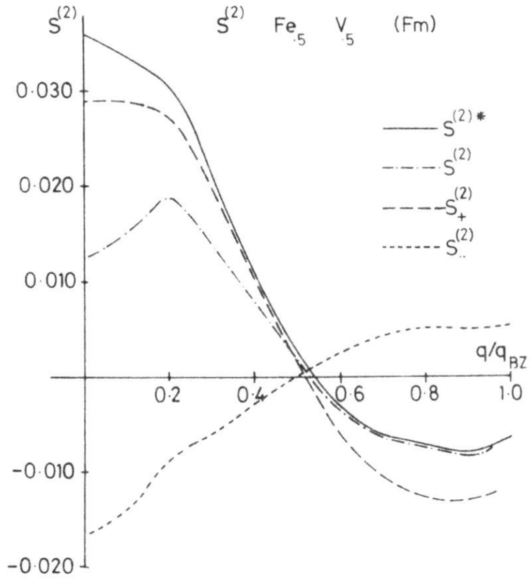

• **Figure 6.** $S^{(2)}$ **components of ferromagnetic** $Fe_{.50}V_{.50}$ **in the (1,0,0) direction.**

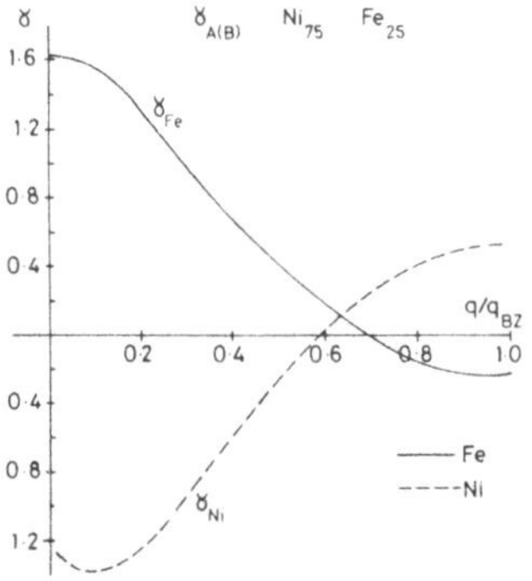

• Figure 7a. $\gamma_{A(B)}$ for alloy $A_cB_{1-c} = Ni_{.75}Fe_{.25}$ in the (1,0,0) direction.

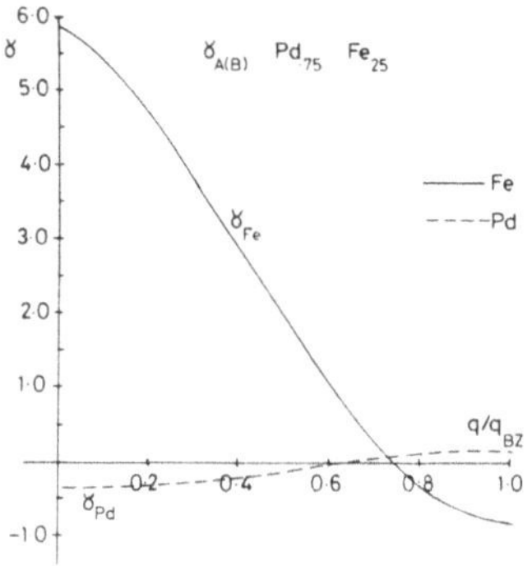

• Figure 7b. $\gamma_{A(B)}$ for alloy $A_cB_{1-c} = Pd_{.75}Fe_{.25}$ in the (1,0,0) direction

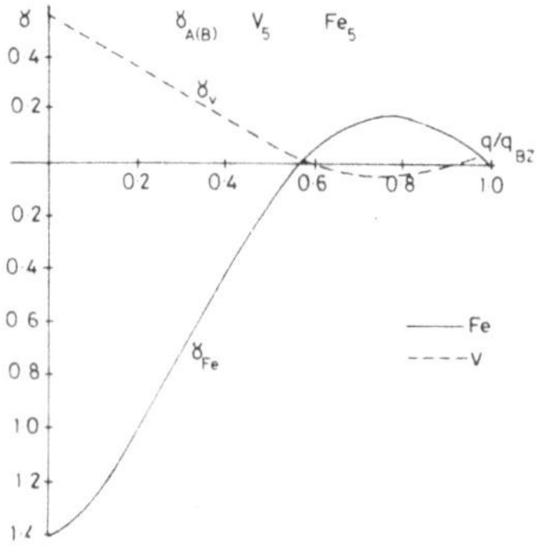

• Figure 7c. $\gamma_{A(B)}$ for alloy $A_cB_{1-c} = V_{.50}Fe_{.50}$ in the (1,0,0) direction.

larger numbers of iron atoms (figure 7c).

4 Metallic magnetism at finite temperatures

It has been emphasised in the previous two sections how the modern version of the Stoner-Wohlfarth theory accurately describes the magnetic ground state of metallic systems and can be usefully exploited. In this part, however, we will follow up the suggestion that its finite temperature extension usually seems to miss the dominant thermal fluctuation of the magnetisation. This neglected component comprises the orientational fluctuations of the 'local moments', which are the magnetisations within each unit cell of the crystal lattice set up by the collective behaviour of all the electrons. In the standard version of Stoner theory, these are constrained to be all aligned and consequently the thermally averaged magnetisation, $\bar{M}$, can only vanish along with the exchange splitting of the electronic bands. It is therefore believed to be necessary to construct a theory which allows the orientations to vary from site to site, enabling $\bar{M}$ to vanish as the orientational disorder of the 'local moments' grows. From this broad consensus (32), however, there remains a controversy concerning which orientational configurations are the most important.

The various approaches can be roughly partitioned into two. Firstly, there is the picture of the Fluctuating Local Band theory (FLB) (33),(34),(35) of a large amount of short range magnetic order even in the paramagnetic phase. This consists of large spatial regions in which the 'local moments' are nearly aligned, i.e. where the orientations vary gradually. In these regions Stoner theory can be applied and perturbations to it made. The quasi-elastic neutron scattering experiments of Ziebeck et al. (36), later confirmed by Shirane et al. (37) are given a simple though not uncontroversial (38) interpretation by this picture. In the case of inelastic neutron scattering, however, even the basic observations are controversial, (39),(40), let alone their interpretations in terms of 'spin waves' above T_c that may feature in such a model. A more rigorous test of this theory is required. It seems that it is difficult to carry out realistic calculations in which the magnetic and electronic structures are mutually consistent and consequently to examine the full implications of the FLB picture and to improve it systematically.

The second type of approach involves the 'disordered local moment' picture (41),(42),(43) (DLM). The 'local moment' entities of each site are commonly thought to fluctuate fairly independently. This picture is usually set up by assuming a Hubbard Hamiltonian, transforming the partition function Z for a system of interacting electrons into a functional integral over fluctuating fields to which non-interacting electrons are coupled and making static and single site approximations. The procedure is reasonably straightforward and more specific than in the case of FLB theory and there are many useful calculations with encouraging results. It does however have several unsatisfactory features connected with its simple parameter dependent basis. The Hubbard interaction I is not a canonical description of the electron-

electron interaction and therefore cannot safely be used as a fitting parameter and moreover the semi-phenomenological site energies,hopping integrals etc. are not necessarily consistent with the choice of I. Consequently this treatment does not provide a realistic description of the electronic structure which drives the important magnetic fluctuations. a simplified approach along these lines may lead to a mistaken identification of the dominant mechanisms and furthermore as they stand such theories are not open to systematic improvements. In this section we will discuss a 'first principles' formulation of this picture, based on the Spin Density Functional formulism which will attempt to rectify these drawbacks. (A full account of this work is provided in references (44) to (47)).

The basic physical insight underlying most recent work on metallic magnetism (32) assumes that it is possible to identify fast and slow motions. On a time scale τ long in comparison with the hopping time $\hbar/W$ (10^{-15} secs) where W represents a relevant band width but short when compared with an appropriate 'spin fluctuation' time ,($1/\omega^{SF}$ where ω^{SF} describes a typical spin fluctuation frequency), the spin orientations of electrons leaving a site are sufficiently correlated with those arriving such that a non-zero magnetisation exists when the appropriate quantity is averaged over a time, τ. These are the 'local moments' which can then change their orientations on the longer time scale. The many electron system which is ergodic is assumed not to cover its phase space uniformly in time and can be imagined as being restricted for long times τ near points in its phase space which can be labelled by particular orientational arrangements of the local moments and then moving rapidly to another similar point. We can borrow the notion of 'temporarily broken ergodicity' (48). Each orientational arrangement can be annotated by the set of unit vectors $\{\hat{e}_i\}$ picking out the orientations of the local moments

$$\hat{e}_i = \int_{V_i} d\mathbf{r}_i \bar{\mathbf{M}}_\tau(\mathbf{r}_i) / \left| \int_{V_i} d\mathbf{r}_i \bar{\mathbf{M}}_\tau(\mathbf{r}_i) \right| \tag{84}$$

where V_i is the volume of the ith unit cell. We now need to develop (i) a short time ($< \tau$) description of the system labelled by $\{\hat{e}_i\}$ in terms of a 'generalised' grand potential $\Omega\{\hat{e}_i\}$ followed by (ii) a prescription for the corresponding time evolution of the system in the reduced phase space of such configurations.

By generalising the finite temperature SDF theory (49),(50), one can write down formally that $\Omega\{\hat{e}_i\}$ is obtained as a result of a functional minimisation of the grand potential functional (equation 9) with respect to charge and magnetisation densities *subject to the constraint* that the magnetisation on every site is orientated consistently with $\{\hat{e}_i\}$ i.e.

$$\int_{V_j} d\mathbf{r}_j (\mathbf{m}(\mathbf{r}_j, \{\hat{e}_i\}) \times \hat{e}_i = 0 \tag{85}$$

appropriate to every site. This minimisation performed using Lagrange multipliers closely follows conventional SDF theory and $\Omega\{\hat{e}_i\}$, for a particular orientational configuration $\{\hat{e}_i\}$, which is a functional of charge and magnetisation densities $n(\mathbf{r},\{\hat{e}_i\})$,$\mathbf{m}(\mathbf{r},\{\hat{e}_i\})$, is found by the self-consistent solution of the 'Kohn-Sham' equations ($\hbar^2/2m = 1$).

$$\begin{aligned}\tilde{1}\delta(\mathbf{r}-\mathbf{r}') &= \{\epsilon+\nabla^2-\tilde{W}^{eff.}[n(\mathbf{r},\{\hat{e}_i\}),\mathbf{m}(\mathbf{r},\{\hat{e}_i\})] \\ &\quad -\sum_i \theta(\mathbf{r}-\mathbf{R}_i)\tilde{\sigma}\cdot(\mathbf{b}_i\times\hat{e}_i)\}\tilde{G}(\mathbf{r},\mathbf{r}';\epsilon) \end{aligned} \tag{86}$$

$$n(\mathbf{r},\{\hat{e}_i\}) = -(1/\pi)Im\int d\epsilon f(\epsilon-\nu)Tr\tilde{G}(\mathbf{r},\mathbf{r};\epsilon) \tag{87}$$

$$\mathbf{m}(\mathbf{r},\{\hat{e}_i\}) = -(1/\pi)Im\int d\epsilon f(\epsilon-\nu)Tr\tilde{\sigma}\tilde{G}(\mathbf{r},\mathbf{r};\epsilon) = \sum_i \mu_i(\mathbf{r},\{\hat{e}_i\})\hat{e}_i\theta(\mathbf{r}-\mathbf{R}_i) \tag{88}$$

$$\begin{aligned}\tilde{W}^{eff.} &= \sum_i\{-Ze^2/|\mathbf{r}-\mathbf{R}_i| \\ &\quad +e^2\int d\mathbf{r}'n(\mathbf{r}',\{\hat{e}_i\})/|\mathbf{r}-\mathbf{r}'| + \delta\Omega_{xc}[n,\mathbf{m};\{\hat{e}_i\}]/\delta n(\mathbf{r},\{\hat{e}_i\}) \\ &\quad +(\delta\Omega_{xc}[n,\mathbf{m};\{\hat{e}_i\}]/\delta\mathbf{m}(\mathbf{r},\{\hat{e}_i\}))\cdot\tilde{\sigma}\} \end{aligned} \tag{89}$$

In these equations, $\theta(\mathbf{r}-\mathbf{R}_i)$ equals unity if $\mathbf{r}$ lies within the ith unit cell, zero otherwise, and the Lagrange parameters $\{\mathbf{b}_i\}$ are adjusted such that the magnetisation is orientated along the set of directions $\{\hat{e}_i\}$.

The long time averages appropriate to the second stage of the scheme can be evaluated by taking averages over the ensemble of orientational configurations $\{\hat{e}_i\}$. The probability of finding one particular configuration is written

$$P\{\hat{e}_i\} = \exp(-\beta\Omega\{\hat{e}_i\})/\prod_j\int d\hat{e}_j\exp(-\beta\Omega\{\hat{e}_i\}) \tag{90}$$

and the Free Energy is found from

$$F = -(1/\beta)\ln\prod_i\int d\hat{e}_i\exp(-\beta\Omega\{\hat{e}_i\}) \tag{91}$$

The role of a classical 'spin' (local moment) Hamiltonian, albeit a highly complicated one, is played by $\Omega\{\hat{e}_i\}$. By choosing a suitable reference 'spin' Hamiltonian, expanding about it using the Feynman-Peierls inequality (22),an approximation to the Free Energy is obtained

$$F \leq F_0 + \langle\Omega-\Omega_0\rangle^0 = \breve{F} \tag{92}$$

with

$$F_0 = -(1/\beta)\ln\prod_i\int d\hat{e}_i\exp(-\beta\Omega_0) \tag{93}$$

$$\langle X\rangle^0 = \prod_i\int d\hat{e}_i X\exp(-\beta\Omega_0)/\prod_i\int d\hat{e}_i\exp(-\beta\Omega_0) = \prod_i\int d\hat{e}_i P_0\{\hat{e}_i\}X\{\hat{e}_i\} \tag{94}$$

By choosing Ω_0 to be of the form

$$\Omega_0 = \sum_i \omega_i^{(1)}(\hat{e}_i) + \sum_{i,j,i\neq j} \omega_{ij}^{(2)}(\hat{e}_i, \hat{e}_j) + \cdots \tag{95}$$

a flexible scheme is achieved which can be systematically improved. A functional minimisation of $\breve{F}$ of equation 92 gives the optimum functions $\omega_i^{(1)}$,$\omega_{ij}^{(2)}$ etc. which are expressed in terms of restricted averages

$$\omega_i^{(1)}(\hat{e}_i') = \langle\Omega\rangle^0_{\hat{e}_i=\hat{e}_i'} \tag{96}$$

$$\omega_{ij}^{(2)}(\hat{e}_i', \hat{e}_j') = \langle\Omega\rangle_{\hat{e}_i=\hat{e}_i',\hat{e}_j=\hat{e}_j'} \tag{97}$$

If only the first term of equation 95 , a set of single site functions, is used then we have established a mean field theory (this can easily be seen by replacing $\Omega\{\hat{e}_i\}$ with the Heisenberg classical Hamiltonian $\sum_{ij} J_{ij}\hat{e}_i \cdot \hat{e}_j$). Finally the whole scheme becomes feasible when $\langle\Omega\rangle_{\hat{e}_i}$ is minimised with respect to the partially averaged charge and magnetisation densities instead of minimising each $\Omega\{\hat{e}_i\}$ in turn and then carrying out the appropriate averages. Clearly the whole theory has strong formal similarities with that pertaining to compositional fluctuations but in this case the fluctuation space $\hat{e}_i$, connected with many electron spin fluctuations, is a continuous one rather than the simpler compositional two point space. Many of the numerical techniques developed for disordered alloys can be adapted for this magnetism problem. In summary this 'first principles' theory of magnetism involves a series of steps. (i) Firstly the assumption of a probability distribution

$$P\{\hat{e}_i\} = \prod_i P_i(\hat{e}_i) \tag{98}$$

(ii) then the solution of the 'Kohn-Sham' equations using the self-consistent field KKR-CPA method

$$\{\epsilon + \nabla^2 - \sum_i \tilde{W}_i^{eff.}(\mathbf{r}_i, \hat{e}_i)\}\tilde{G}(\mathbf{r}, \mathbf{r}'; \epsilon) = \tilde{1}\delta(\mathbf{r} - \mathbf{r}') \tag{99}$$

where

$$\begin{aligned}\tilde{W}^{eff.}(\mathbf{r}_i, \hat{e}_i) &= \{-Ze^2/r_i - \sum_{j\neq i} Ze^2/|\mathbf{r}_i - \mathbf{R}_j + \mathbf{R}_i| + e^2 \int d\mathbf{r}_i \bar{n}(\mathbf{r}_i', \hat{e}_i)/|\mathbf{r}_i - \mathbf{r}_i'| \\ &+ \sum_{j\neq i} e^2 \int d\mathbf{r}_j' \int d\hat{e}_j P_j(\hat{e}_j)\bar{n}_j(\mathbf{r}_j, \hat{e}_j)/|\mathbf{r}_i - \mathbf{r}_j'| \\ &+ \delta\Omega_{xc}/\delta\bar{n}_i(\mathbf{r}_i, \hat{e}_i)\}\tilde{1} + (\delta\Omega_{xc}/\delta\bar{\mu}_i(\mathbf{r}_i, \hat{e}_i))\tilde{\boldsymbol{\sigma}} \cdot \hat{e}_i \end{aligned} \tag{100}$$

and

$$\bar{n}_i(\mathbf{r}_i, \hat{e}_i) = \langle n(\mathbf{r}_i, \{\hat{e}_i\}\rangle_{\hat{e}_i} = -(1/\pi) Im \int d\epsilon f(\epsilon - \nu) Tr \sum_{LL'} \tilde{Z}_L^{i,\hat{e}_i}(\mathbf{r}_i)\langle\tilde{\tau}_{\hat{e}_i}^{ii}\rangle \tilde{Z}_{L'}^{i,\hat{e}_i\dagger}(\mathbf{r}_i) \tag{101}$$

$$\begin{aligned}\bar{\mu}(\mathbf{r}_i, \hat{e}_i) &= \langle \mu(\mathbf{r}_i, \{\hat{e}_i\})\rangle_{\hat{e}_i} \\ &= -(1/\pi) Im \int d\epsilon f(\epsilon-\nu) Tr(\tilde{\sigma}\cdot\hat{e}_i) \sum_{LL'} \tilde{Z}_L^{i,\hat{e}_i}(\mathbf{r}_i \langle \tilde{\tau}^{ii}_{LL'}\rangle_{\hat{e}_i} \tilde{Z}_{L'}^{i,\hat{e}_i\dagger}(\mathbf{r}_i)\end{aligned} \quad (102)$$

in which $\tilde{Z}_L^{i,\hat{e}_i}$ is the single site spinor wavefunction corresponding to $\tilde{W}_i^{eff.}(\mathbf{r}_i, \hat{e}_i)$ (equation 100). The partial averages $\langle \tilde{\tau}^{ii}_{LL'}\rangle_{\hat{e}_i}$ are carried out using the C.P.A.

$$\begin{aligned}\langle \tilde{\tau}^{ii}_{\hat{e}_i}\rangle &= \tilde{D}(\hat{e}_i)\tilde{\tau}^{c,ii} \\ &= (1 + (\tilde{t}_i^{-1}(\hat{e}_i) - \tilde{t}_{c,i}^{-1})\tilde{\tau}^{c,ii})^{-1}\tilde{\tau}^{c,ii}\end{aligned} \quad (103)$$

where $\tilde{t}_i(\hat{e}_i)$ is the 'on the energy shell' t-matrix describing scattering from a potential of the form of equation 100. The appropriate (inhomogeneous) C.P.A. condition is

$$\int d\hat{e}_i P_i(\hat{e}_i)\tilde{D}(\hat{e}_i) = \tilde{1} \quad (104)$$

and is analogous to equation 21. (iii)$\langle \Omega \rangle_{\hat{e}_i}$ is then reconstructed from the analogous generalisation of equation 28. (iv) The probability distribution $P_i(\hat{e}_i)$ is then reconstructed using equations 90 and 96 and the above steps repeated until a consistent result is achieved for a particular temperature T. (v) Finally quantities of interest e.g. the magnetisation, electronic structure as a function of temperature are calculated.

In the paramagnetic state there is an equal probability of finding a local moment on a site orientated in any direction such that the overall magnetisation is zero, i.e.

$$\begin{aligned}P_i(\hat{e}_i) &= 1/4\pi \\ \bar{\mathbf{M}} &= \int d\hat{e}_i P_i(\hat{e}_i)\hat{e}_i = 0\end{aligned} \quad (105)$$

and it is straightforward to show that the C.P.A. condition of equation 104 is identical to that appropriate to a disordered two component system of 'up' and 'down' pointing moments. The effective C.P.A. medium is described by (44)

$$\begin{aligned}\tilde{t}_{c,LL'} &= t_{c,L}\tilde{1}\delta_{LL'} \\ \tilde{\tau}^{c,00}_{LL'} &= \tau^{c,00}_{LL}\tilde{1}\delta_{LL'}\end{aligned} \quad (106)$$

and from the paramagnetic versions of equations 101 and 102 the charge and magnetisation densities are also independent of direction.

$$\begin{aligned}\bar{n}_i(\mathbf{r}_i, \hat{e}_i) &= \bar{n}(\mathbf{r}_i) \\ \bar{\mu}_i(\mathbf{r}_i, \hat{e}_i) &= \bar{\mu}(\mathbf{r}_i)\end{aligned} \quad (107)$$

and the single site t-matrix $\tilde{t}(\hat{e}_i)$ can now be written

$$\tilde{t}_L(\hat{e}_i) = 1/2(t_{L\uparrow} + t_{L\downarrow})\tilde{1} + 1/2(t_{L\uparrow} - t_{L\downarrow})\tilde{\sigma}\cdot\hat{e}_i \quad (108)$$

and

$$\tilde{D}_L^0(\hat{e}_i) = (1/2)(D_{L\uparrow}^0 + D_{L\downarrow}^0)\tilde{1} + (1/2)(D_{L\uparrow}^0 - D_{L\downarrow}^0)\tilde{\sigma} \cdot \hat{e}_i \tag{109}$$

$$D_{L\uparrow(\downarrow)}^0 = (1 + (t_{L\uparrow(\downarrow)}^{-1} - t_{c,L}^{-1})\tau_{LL}^{c,00})^{-1} \tag{110}$$

Consequently in the paramagnetic DLM state the techniques of alloy theory can be directly applied. In reference (44) the paramagnetic states of several transition metals were studied, the magnitudes of the local moments calculated together with the effects of Stoner excitations. Local moments would self-consistently form in b.c.c. and f.c.c. *Fe*, f.c.c. *Co* and *Ni* but not in b.c.c. *Cr*. The local moment in b.c.c. iron,1.85μ_B, was of nearly the same magnitude as the magnetisation per atom in the low temperature Stoner ferromagnetic state and increased a little with increase in lattice spacing whilst in nickel a small local moment 0.16μ_B could form which disappeared if the lattice was compressed slightly. Stoner excitations would also cause this moment to vanish at temperatures around 500K. We then went on to discuss the underlying electronic structure of these paramagnetic states (46). In b.c.c. *Fe* the electronic structure is characterised by a wave-vector and energy dependent local exchange splitting which collapses at some points in this space. This splitting is the cause of the establishment of the local moment. In nickel however the electronic structure resembles Stoner-Wohlfarth paramagnetic bands which are merely broadened by the disorder. The mechanism of local moment formation is now much harder to understand and presumably arises from assymetry of the 'smearing' of the bands which produces, in the appropriate energy and wavevector integrated quantity, a non-zero value for the local moment. Pinski et al. (51) studied he paramagnetic state of f.c.c. *Fe* finding that the local moment collapses as the lattice spacing is reduced and then analysed their results (52) in terms of the electronic structure.

A major advantage of this 'first principles' theory over its simpler parameter dependent counterparts is that it allows the details of the electronic structure underlying this 'disordered local moment' picture of the paramagnetic state to be compared with the results of spectroscopic experiments (53),(54),(55) which directly probe the electronic structure both as a stringent test for this picture and also as a guide for future improvements. Following such analyses of the electronic structure we now move on to consider the degree of magnetic order that it is capable of supporting and thus discover whether this electronic structure is consistent with the imposed DLM magnetic structure.

We consider a treatment of magnetic correlations in the paramagnetic state based upon a theory for the paramagnetic spin susceptibility of the system. In order to derive an expression for the wavevector dependent static susceptibility the response of the system to the application of a small external magnetic field is studied. The technical details are very similar to those used in setting up the theory for compositional ordering described in the previous section and are explained at length in reference (47) so here we present an outline only.

The applied magnetic field can vary in both orientation and magnitude from unit cell to unit cell in the crystal i.e.

$$\mathbf{H}^{ext.}(\mathbf{r}) = \sum_i \mathbf{h}_i \theta(\mathbf{r}_i) \tag{111}$$

and the magnetisation that it induces can be written

$$\mathbf{M}^{ind.} = \sum_i \mathbf{m}_i^{ind.} \theta(\mathbf{r}_i) \tag{112}$$

This arises from two coupled effects: (i) the 'local moments',orientationally disordered in the paramagnetic state, tend to 'flip' into alignment with the field causing a change in the chance of finding a moment orientated in a particular direction on a given site, $\{\delta P_i(\hat{e}_i)\}$,and (ii) the magnitudes of these moments alter depending on their orientation with respect to the field, $\{\delta\mu_i(\hat{e}_i)\}$. These effects are interrelated and in our inhomogeneous C.P.A. framework this occurs via changes to the C.P.A. medium. To lowest order the induced magnetisation on a site i is written

$$\mathbf{m}_i^{ind.} \simeq \bar{\mu} \int d\hat{e}_i \delta P_i(\hat{e}_i)\hat{e}_i + (1/4\pi) \int d\hat{e}_i \delta\mu_i(\hat{e}_i)\hat{e}_i \tag{113}$$

and

$$\delta P_i(\hat{e}_i) \simeq -(\beta/4\pi)\langle\delta\Omega\rangle_{\hat{e}_i} + (\beta/(4\pi)^2) \int d\hat{e}_i \langle\delta\Omega\rangle_{\hat{e}_i} \tag{114}$$

Expressions for both $\langle\delta\Omega\rangle_{\hat{e}_i}$ and $\delta\mu_i(\hat{e}_i)$ can be obtained by expanding equation 102 and that appropriate for $\langle\Omega\rangle_{\hat{e}_i}$ by considering low order deviations $\delta t^{-1}_{c,i,L}$ from the homogeneous C.P.A. medium so that

$$\tilde{t}^{-1}_{c,i,L,inh.} = t^{-1}_{c,L}\tilde{1} + \delta t^{-1}_{c,i,L}\tilde{\sigma} \cdot \hat{h}_i^{ext.} \tag{115}$$

and

$$\tilde{\tau}^{c,ii,inh.}_{LL} = \tau^{c,00}_{LL}\tilde{1} - \sum_{j,L''} \tau^{c,ij}_{LL''}\tau^{c,ji}_{L''L}\delta t^{-1}_{c,j,L''}\tilde{\sigma} \cdot \hat{h}_j^{ext.} \tag{116}$$

These deviations are obtained from the inhomogeneous C.P.A. equation

$$\int d\hat{e}_i P_i(\hat{e}_i)\tilde{D}_i(\hat{e}_i) = \tilde{1} \tag{117}$$

which is approximately written in terms of $\delta t^{-1}_{c,i,L}$ via

$$\int d\hat{e}_i \delta P_i(\hat{e}_i)\tilde{D}^0_L(\hat{e}_i)\delta_{LL'} + (1/4\pi) \int d\hat{e}_i \delta\tilde{D}_{i,LL'}(\hat{e}_i) = 0 \tag{118}$$

After considerable manipulation, including a lattice Fourier tranform, the susceptibility can be expressed as a sum of two contributions

$$\chi(\mathbf{q},T) = \chi^{flip}(\mathbf{q},T) + \chi^{mom}(\mathbf{q},T) \tag{119}$$

χ^{flip} represents the 'orientational' response of the system whilst χ^{mom} contains the moments' size change response. their interrelation can be seen from the following

$$\chi^{flip} = \chi_0^{flip}(1 + \Xi_1 \chi^{mom}) \tag{120}$$

$$\chi^{mom} = \chi_0^{mom}(1 + \Xi_2 \chi^{flip}) \tag{121}$$

where

$$\chi_0^{flip} = [\beta\bar{\mu}(\bar{\mu} + R)/3]/[1 - \beta J_0/3] \tag{122}$$

$$\chi_0^{mom} = N/[3(1-U) - M] \tag{123}$$

and

$$\Xi_1 = Q/(\bar{\mu} + R) \tag{124}$$

$$\Xi_2 = L/N\bar{\mu} \tag{125}$$

All these quantities depend on the electronic structure of the DLM state and are analogous to quantities given in the last section. The appropriate expressions are achieved by replacing the A and B site subscripts by $\uparrow$ and $\downarrow$ respectively and dropping the σ label. Then J_0 replaces $S^{(2)}$ of equation 62, L replaces $L^A + L^B$ with σ set equal to unity of equation 63, Q by $Q^A + Q^B$ with σ unity of equation 65. Furthermore M replaces $M^{AA} + M^{AB} + M^{BA} + M^{BB} - U^A - U^B$ and $c_A = c_B = 0.5$ in equation 66, U replaces $U^A + U^B$ of equation 49, N replaces $N^A + N^B$ of equation 81 and finally R replaces $L^A/6 + L^B/3$ of equation 63. The expressions are given clearly in reference 47.

The important results are equations 120 and 121 and they possess a clear physical interpretation. The first term of equation 120 is of the form of a classical 'Heisenberg' susceptibility and describes the response of the local moments to an external magnetic field under the constraint that they can only change orientation. Similarly the first term of equation 121 describes the analogous response under the restriction that the moments can alter their sizes and is an enhanced 'Stoner' paramagnetic susceptibility of a DLM environment. The remaining terms in these two expressions show the interdependence of the two types of response and underline the sensitivity of the magnitude of a 'local' moment to the orientational arrangement of the moments surrounding it. In conclusion these two equations can be combined and the total susceptibility $\chi(\mathbf{q}, T)$ has the form

$$\chi(\mathbf{q}, T) = [\beta C(\mathbf{q}, T)/3]/[1 - \beta J_{eff.}(\mathbf{q}, T)/3] + \chi_0^{mom}(\mathbf{q}, T) \tag{126}$$

with

$$C = (\bar{\mu} + L\chi_0^{mom}/N)(\bar{\mu} + R + Q\chi_0^{mom}) \tag{127}$$

$$J_{eff.} = J_0 - LQ\chi_0^{mom}/N \tag{128}$$

It has the physically transparent form of a classical Heisenberg susceptibility with an added itinerant part χ^{mom}. The Curie constant ($q = 0$ value of C in equation

127) is clearly enhanced by itinerant effects from that of strictly rigid 'spins' or local moments, $\bar{\mu}^2$. From the form of equation 126 a Curie-Weiss temperature behaviour can readily occur since the temperature dependence of C, $J_{eff.}$, χ_0^{mom}, all electronic structure quantities, arises from Fermi factors only.

We now illustrate the theory by briefly reviewing our calculations of the paramagnetic susceptibility for b.c.c. and f.c.c. Fe and also f.c.c. Ni.

(i)**b.c.c. Fe**

The results of these calculations are described in full in reference 47 so here we will just mention the important features. The uniform paramagnetic susceptibility ($q = 0$) was found to show a Curie-Weiss dependence upon temperature. The estimated Curie temperature was found to be 1280K which compares well with the experimental value of 1040K (58). The effective Curie constant of $(1.97\mu_B)^2$ although slightly enhanced from the strictly rigid moment case is somewhat small compared to the empirical value of $(3.12\mu_B)^2$. As a function of wavevector $\mathbf{q}$ we found that χ to have rather 'flat' large q 'tails' with peaks centred on $q = 0$ superimposed. Moreover this shows that the susceptibility is dominated by rigid moment effects i.e. χ_0^{flip} of equation 122. From the fluctuation dissipation theorem it is possible to infer that this is consistent with strong short range magnetic correlations with a weak effect from those of longer range. It is clear therefore that the magnetic correlations which can be supported by the disordered local moment (DLM) electronic structure are roughly consistent with the imposed magnetic structure of this mean field DLM picture.

A cautious comparison can be made with quasi-elastic polarised neutron scattering data (36) (37). Firstly the difference between scattering intensities obtained when the incident neutrons' polarisation is parallel, then perpendicular to the scattering vector (57) is proportional to the dynamical magnetic response function of the system $S(\mathbf{q},\omega)$. This in turn can be written

$$S(\mathbf{q},\omega) = |f(\mathbf{q})|^2 M(\mathbf{q},\omega) \tag{129}$$

where $f(\mathbf{q})$ is a magnetic atomic form factor and $M(\mathbf{q},\omega)$ represents the Fourier component of the magnetic response of the system to an applied magnetic field varying from site to site but uniform within each unit cell (of form of equation 111). $M(\mathbf{q},\omega)$ is written in terms of the dynamical spin susceptibility $\chi(\mathbf{q},\omega)$

$$M(\mathbf{q},\omega) = 3Im\chi(\mathbf{q},\omega)/(1-\exp(-\beta\omega)) \tag{130}$$

Taking the high temperature limit and integrating over all frequencies ω produces the relation involving the static susceptibility

$$\int_0^\infty d\omega M(\mathbf{q},\omega) = 3\chi(\mathbf{q})/\beta \tag{131}$$

Consequently our theory and calculations for the static susceptibility can be related to the quasi-elastic neutron scattering data provided the experimental 'energy window' is wide enough to capture all the magnetic scattering. If this assumption is made then our results appear to show too much weight at large wavevectors. A satisfactory comparison will really only be made with the development of a theory for the dynamical susceptibility.

In summary our mean field theory works fairly well for b.c.c. Fe, predicting a satisfactory Curie temperature and a Curie-Weiss law. The electronic structure underlying this DLM paramagnetic state predicted a wavevector and energy dependent *local* exchange splitting which collapses altogether at some points in this space. These features have been observed experimentally (54) (55) although there are indications that quantitative detailed agreement may be achieved by improving upon the mean field theory and including some effect of the correlations between the local moments at the outset. We will discuss a way of setting about this towards the end of these lectures.

(ii) **f.c.c. Fe**

We carried out a series of calculations for $\chi(\mathbf{q}, T)$ for iron on an f.c.c. lattice of varying spacing (reference 51). We found that the effective interaction between the local moments ($J_{eff.}$ from equation 128) to be ferromagnetic for the larger lattice spacings but to become anti-ferromagnetic for smaller spacings only slightly bigger than that for which no local moments can be sustained. These results complement the work of ref. 19 of f.c.c. Fe at low temperatures.

(iii) **f.c.c. Ni**

Our susceptibility calculations for nickel (reference 47) show the instability of the DLM paramagnetic state.The term χ_0^{mom} in the expression for the susceptibility in equation 123 dominates contrary to Fe and the sensitivity of the local moments to the orientations of those surrounding them is very strong. The Stoner ferromagnetic state is more stable for all temperatures for which local moments can be set up and not destroyed by Stoner excitations. In other words the DLM mean field theory of magnetism for nickel reduces to the Stoner-Wohlfarth theory. It is therefore necessary to go beyond this and incorporate some degree of correlation between the local moments.

Beyond the 'mean field' approximation in a 'first principles' theory of magnetism — the Onsager Reaction field.

A natural way of incorporating effects of correlation between the local moments' orientations from a mean field 'starting point' is via consideration of an Onsager reaction field. Before looking at our complicated 'first principles' theory it is useful to consider a simple Ising model (reference 56). Writing the Hamiltonian

$$\hat{H} = -\sum_{ij} J_{ij} e_i^z e_j^z + \sum_i h_i^{ext} e_i^z \tag{132}$$

in which $e_i^z = \pm 1$. The magnetisation m_i on a site i is given as $\langle e_i^z \rangle$ i.e. $\sum_{e_i^z=\pm 1} P_i(e_i^z) e_i^z$ so that

$$P_i(e_i^z) = (1 \pm m_i)/2 \tag{133}$$

Now in terms of a molecular field, h_i^{MF}

$$m_i = \tanh\beta(h_i^{MF} + h_i^{ext.}) \tag{134}$$

A mean field theory is provided by $h_i^{MF} = \sum_j J_{ij}m_j$. To improve upon this we need to subtract from the effective molecular field , set up on a particular site i, that contribution which derives from the magnetisation on the same site i i.e. the reaction field and obtain the Onsager cavity field h_i^c. The magnetisation is now

$$\begin{aligned} m_i &= \tanh\beta(h_i^c(m_i) + h_i^{ext.}) && (135)\\ h_i^c &= h_i^{MF} - h_i^{reaction} && (136)\\ &= \sum_j J_{ij}(m_j - \delta m_j^{(i)}) && (137) \end{aligned}$$

Now the contribution to the magnetisation on site j from that on site i, $\delta m_j^{(i)}$, can be written in terms of the susceptibility of the system as $\chi_{ij}(\chi_{ii})^{-1}m_i$ and now

$$m_i = \tanh\beta(\sum_{ij} J_{ij}m_j - \sum_i \lambda_i + h_i^{ext.}) \tag{138}$$

where

$$\begin{aligned} \lambda_i &= (\chi_{ii})^{-1}\sum_j J_{ij}\chi_{ji} && (139)\\ &= (\beta(1-m_i^2))^{-1}\int d\mathbf{q}\chi(\mathbf{q})J(\mathbf{q}) && (140) \end{aligned}$$

using the fluctuation dissipation theorem. A theory for the paramagnetic susceptibility can be completed by considering the moments that are induced by the external field $\chi_{ij} = \delta m_i/\delta h_j^{ext.}$ and

$$\chi(\mathbf{q}) = \beta/(1-\beta(J(\mathbf{q}) - \lambda)) \tag{141}$$

with $\lambda = (1/\beta)\int d\mathbf{q}\chi(\mathbf{q})J(\mathbf{q})$. The fluctuation dissipation theorem is contained naturally and in this paramagnetic limit the formulation turns out to be equivalent to the 'spherical model'.

We now turn to the question of implementing an analogous approach to our itinerant magnetism problem. In order to go beyond our 'mean field theory' (DLM) we need to consider correlations explicitly. Returning to the Feynman-Peierls Inequality, a reference 'local moment' Hamiltonian is written down (equation 95)

$$\begin{aligned} \Omega_0 &= \sum_i \omega_i^{(1)}(\hat{e}_i) + \sum_{ij,i\neq j} \omega_{ij}^{(2)}(\hat{e}_i,\hat{e}_j) \\ &= \sum_{ij} \tilde{\omega}_{ij}(\hat{e}_i,\hat{e}_j) && (142)\\ \tilde{\omega}_{ij}(\hat{e}_i,\hat{e}_j) &= \langle\Omega\{\hat{e}_i\}\rangle_{\hat{e}_i,\hat{e}_j} && (143) \end{aligned}$$

in which formally, using the generalised version of the SDF theory this implies that each $\Omega\{\hat{e}_i\}$ is minimised with respect to $n\{\hat{e}_i\}$ and $\mathbf{m}\{\hat{e}_i\}$ before the appropriate average is taken. The DLM theory is obtained once more from this way of writing things by using molecular field theory with the 'local moment' Hamiltonian written as above and then reversing the order of the SDF minimisation and the averaging consistent with the spirit of the molecular field approximation. We can consequently improve upon DLM theory by considering the subtraction of the Onsager reaction field and within the same philosophy again reversing the implied SDF minimisation and the averaging. It is possible to do this by considering $\delta t_{c,j,L}^{-1(i)}$ i.e. changes to the inhomogeneous C.P.A. medium on surrounding sites j caused by the magnetisation on a particular site i. The Onsager reaction field is thus incorporated via *changes to the C.P.A. medium.* With this idea we can now develop a theory for the paramagnetic spin susceptibility. The induced magnetisation caused by the application of an external magnetic field (equation 112) becomes

$$\begin{aligned}\mathbf{m}_i^{ind.} &\simeq \bar{\mu}\int d\hat{e}_i \delta P_i(\hat{e}_i;\{\delta t_{c,j}^{-1} - \delta t_{c,j}^{-1(i)}\})\hat{e}_i + (1/4\pi)\int d\hat{e}_i \delta\mu_i(\hat{e}_i,\{\delta t_{c,j}^{-1} - \delta t_{c,j}^{-1(i)}\})\hat{e}_i \\ &= \delta\mathbf{m}_i^{flip} + \delta\mathbf{m}_i^{mom} \qquad (144)\end{aligned}$$

in which a similar equation connects $\delta t_{c,j}^{-1(i)}$ as that connecting the $\delta t_{c,j}^{-1}$ (see equation 118 and reference 47)i.e.

$$\begin{aligned}\delta t_{c,j,L}^{-1(i)} &= (1/2\bar{\mu})D_{L\uparrow}^0(t_{L\uparrow}^{-1} - t_{L\downarrow}^{-1})D_{L\downarrow}^0\delta m_j^{flip(i)} \\ &+(1/2)(A_L^{\uparrow}(D_{L\uparrow}^0)^2 + A_L^{\downarrow}(D_{L\downarrow}^0)^2)\delta m_j^{mom(i)} \\ &+D_{L\uparrow}^0(t_{L\uparrow}^{-1} - t_{c,L}^{-1})D_{L\downarrow}^0(t_{L\downarrow}^{-1} - t_{c,L}^{-1}\sum_{L'})\sum_{k\neq j}\tau_{LL'}^{c,jk}\tau_{L'L}^{c,kj}\delta t_{c,k,L'}^{-1(i)} \qquad (145)\end{aligned}$$

The quantities $\delta m_j^{flip(i)}$ and $\delta m_j^{mom(i)}$ describe the changes to each component of the magnetisation on a site j induced by a change in magnetisation on a site i and are therefore related to correlation functions which are themselves related to susceptibilities χ^{flip} and χ^{mom}. It is clear that χ^{flip} corresponds to the correlation between the orientation of the magnetisation on a site j and the total magnetisation on a site i, $\langle \mathbf{m}_i\hat{e}_j\rangle$ whereas similarly χ^{mom} relates to the correlation between the size of the magnetisation on j and the total magnetisation on i, $\langle \mathbf{m}_i\mu_j\rangle$. By subtracting from the changes to the C.P.A. medium those portions $\delta t_j^{-1(i)}$ and then incorporating their effects into expressions for $\delta P_i(\hat{e}_i)$ and $\delta\mu_i(\hat{e}_i$ we can obtain the following expressions for the components of the paramagnetic spin susceptibility.

$$\chi^{flip} = \chi_0^{flip}(1 + \Xi_1\chi^{mom}) \qquad (146)$$
$$\chi^{mom} = \chi_0^{mom}(1 + \Xi_2\chi^{flip}) \qquad (147)$$

which are the same in form to the 'mean field' expressions of equations 120 and 121 but each quantity is different because of the reaction field. Instead of the terms shown in equations 122 to 125 we have

$$\chi_0^{flip}(\mathbf{q}) = (\beta/3)\bar{\mu}[\bar{\mu} + R(\mathbf{q})]/[1 - (\beta/3)(J_0(\mathbf{q}) - \Lambda_{J_0}^{flip} - \Lambda_Q^{mom})] \tag{148}$$

$$\chi_0^{mom}(\mathbf{q}) = N(\mathbf{q})/[3(1 - U + \lambda_U^{mom}) - M(\mathbf{q}) + \Lambda_L^{flip} + \Lambda_M^{mom}] \tag{149}$$

and

$$\Xi_1 = (Q(\mathbf{q}) - \Lambda_{J_0}^{flip} - \Lambda_Q^{mom})/(\bar{\mu} + R(\mathbf{q})) \tag{150}$$

$$\Xi_2 = (L(\mathbf{q}) - \Lambda(\mathbf{q}) - \Lambda_L^{flip} - \Lambda_M^{mom})/(N(\mathbf{q})\bar{\mu}) \tag{151}$$

The Λ quantities are appropriate Brillouin zone integrals e.g.

$$\Lambda_{J_0}^{flip} = (3/\beta)\int d\mathbf{q} J_0(\mathbf{q})\chi^{flip}(\mathbf{q}) \tag{152}$$

and

$$\lambda_U^{mom} = (3/\beta)U\int d\mathbf{q}\chi^{mom}(\mathbf{q}) \tag{153}$$

It remains to generate J_0,Q,R,N,M and L on a fairly dense mesh of **q**-points and to solve these coupled integral equations iteratively. There are several features that one might expect to emerge. (i) In the limit of vanishing local moments (for example as shown in the example of nickel) the theory reduces to a form similar to that proposed by Moriya (ref.59) (see also reference 60). It is a theory of weak itinerant magnetism with the Stoner excitations properly incorporated. It describes a non-local character of moment size fluctuations. As Moriya has shown, it is possible for such a theory to give a Curie-Weiss behaviour to the paramagnetic susceptibility with a Curie temperature reduced from that of conventional Stoner theory and an enhanced Curie constant. (ii) In the other limit of 'rigid' local moments (to which iron tends) the Onsager theory for the classical Heisenberg model is obtained. For iron however the theory may provide a larger enhancement of the Curie constant than that of mean field DLM theory once these correlation effects are applied to the interrelation between orientational and 'moments growth' effects. In summary and with emphasis, by coupling these effects by this Onsager reaction field framework we have achieved a unification of the theory of weak itinerant magnetism in which the spin fluctuations are non-local in nature and that of the 'local moment' picture.

5 Conclusion

In these lectures the spin polarised electronic structure of both magnetic metals and alloys has been focussed upon. This has been shown to provide the basis for a theory for a 'first principles' description of the mechanism driving the magnetic phase transition in pure metals at finite temperatures and in alloys it demonstrates how compositional ordering depends upon the state of magnetic order. It remains to combine these treatments of magnetic and compositional fluctuations and obtain a parameter-free microscopic theory of magnetic and compositional ordering in alloys. This work is in progress. A picture in which these magnetic and compositional fluctuations are subtly interrelated via the underlying electronic structure will emerge and the results of this theory will be able, step by step, to be compared with measurements from a wide range of experiments. The prospect of understanding a wide range of magnetic properties of alloys is in sight.

6 REFERENCES

1. Chikazurin,S., and Graham, C.D.,'Magnetism and Metallurgy' eds.A.E.Berkowitz and E.Kneller, vol.II,p.577,Academic Press, 1969.

2. Kuentzler,R.,'Physics of Transition Metals 1980' eds. P.Rhodes, Institute of Physics Conference Series no.55, 1980.

3. Kohn,W., and Vashishta, P.,'Physics of Solids and Liquids' eds. B.I.Lundqvist and N.March, Plenum,1982.

4. Stoner, E.C, Proc.Roy.Soc.**A139** ,339,1939.

5. Wohlfarth,E.P., Rev. Mod. Phys.,**25**,211,1953.

6. Herring, C.,'Magnetism' **IV** eds. G.T.Rado and H.Suhl,Academic Press,1966.

7. Rajagopal,A.K.,Adv.Chem.Phys. **41**,59,1980.

8. Gunnarsson, O., and Lundqvist,B.I.,Phys.Rev. **B13**,4274,1976.

9. Moruzzi, V.L.,Janak, J.F.,and Williams, A.R.,'Calculated Electronic Properties of Metals',Pergamon,1978.

10. Stocks,G.M., and Winter, H.,'Electronic Structure of Complex Systems' eds. P.Phariseau, W.M.Temmerman, NATO ASI series vol.B113,Plenum,1984.

11. Johnson,D.D.,Pinski,F.J., and Stocks,G.M., J.Appl.Phys. **57**,3018,1985.

12. Soven, P.,Phys.Rev.**156**,809,1967.

13. Gyorffy,B.L., and Stocks, G.M., 'Electrons in Finite and Infinite Structures' eds. P.Phariseau and L.Scheire,NATO ASI series B24,Plenum,1977. and 'Electrons in Disordered Metals and at Metallic Surfaces' eds. P.Phariseau,B.L.Gyorffy and L.Scheire,NATO ASI series B42,Plenum,1979.

14. Faulkner,J.S., and Stocks, G.M.,Phys.Rev. **B21**,3222,1980.

15. Gyorffy,B.L., and Stott,M.J.,'Band Structure Spectroscopy of Metals and Alloys' eds. D.J.Fabian and L.M.Watson,Academic Press,1973.

16. Durham,P.J.,Gyorffy,B.L., and Pindor,A.J.,J.Phys.**F10**,661,1980.

17. Johnson,D.D.,Pinski,F.J.,and Staunton,J.B.,J.Appl.Phys.**61**,3715,1987.

18. Williams,A.R.,Moruzzi,V.L.,Malozemoff,A.P., and Terakura, K.,IEE Trans. Magn. MAG-19,1983.

19. Wang, C.S.,Klein,B.M., and Krakauer, H.,Phys.Rev.Lett.**54**,1852,1985.

20. Williams,A.R.,Zeller,R.,Moruzzi,V.L.,Gelatt,C.D., and Kubler,J., J.Appl.Phys.**53**,2019,1982.

21. Gyorffy,B.L.,and Stocks,G.M.,Phys.Rev.Lett.**50**,374,1983.

22. Feynman,R.P.,Phys.Rev. **97**,660,1955.

23. Krivoglaz,M.A.,'Theory of X-ray and Thermal Neutron Scattering by Real Crystals',Plenum,1969. and de Fontaine,D., 'Solid State Physics' **34** eds. Ehrenreich,Seitz and Turnbull,1979.

24. Gonis,A.,Stocks,G.M.,Butler,W.H.,Winter,H.,Phys.Rev.**B29**,555,1984.

25. Marshall, W.,J.Phys.**C1**,88,1968.

26. Cable,J.W., and Medina,R.A.,Phys.Rev.**B13**,4868,1976.

27. Gunnarsson, O.,J.Phys.**F6**,587,1976.

28. Rainford,B.D.,J.Phys.Colloq.**43**,C-7,33,1982.

29. Staunton,J.B.,Johnson,D.D., and Gyorffy,B.L.,J.Appl.Phys.**61**,3693,1987.

30. Lefebvre,S.,Bley,F.,Fayard,M.,and Roth,M., Acta.Metall.**29**,749,1981.

31. Heine,V.,and Samson,J.H.,J.Phys.**F13**,2155,1983.

32. 'Electron Correlation and Magnetism in Narrow Band Systems' ed. T.Moriya,Berlin:Springer,1981.

33. Korenman,V.,Murray,J.L.,and Prange,R.E. ,Phys.Rev.**B16** a)4032, b)4048, c)4058,1977.

34. Capellmann,H.,Z.Phys.**B34**,29,1977.

35. Korenman,V.,J.Appl.Phys.**57**,3001,1985.

36. Ziebeck,K.R.A.,Brown,P.J.,Deportes,J.,Givord,D.,Webster,P.J., and Booth,J.G.,Helv.Phys.Acta**56**,117,1983.

37. Shirane, G.,Boni,P., and Wicksted,J.P.,Phys.Rev.**B33**,188,1986.

38. Edwards,D.M.,J.Mag.Magn.Mat.**45**,151,1984.

39. Lynn,J.W.,Phys.Rev.**B11**,2624,1975.

40. Steinsvoll, O.,Majkrazaak,C.F.,Shirane,G.,and Wicksted,J. ,Phys.Rev.Lett.**51**,300,1983.

41. Hubbard,J.,Phys.Rev.**B20**,4584,1979.

42. Hasegawa,H.,J.Phys.Soc.Jap.**46**,1504,1979.

43. Edwards,D.M.,J.Phys.**F12**,1789,1982.

44. Pindor,A.J.,Staunton,J.,Stocks,G.M.,and Winter,H.,J.Phys.**F13**,979,1983.

45. Gyorffy,B.L.,Pindor,A.J.,Staunton,J.,Stocks,G.M., and Winter,H., J.Phys.**F15**,1337,1985.

46. Staunton,J.,Gyorffy,B.L.,Pindor,A.J.,Stocks,G.M., and Winter,H., J.Phys.**F15**,1387,1985.

47. Staunton,J.,Gyorffy,B.L.,Stocks,G.M.,and Wadsworth, J., J.Phys.**F16**,1761,1986.

48. Palmer,R.G.,Adv.Phys.**31**,669,1982.

49. Mermin,N.D.,Phys.Rev.A137,1441,1965.

50. Gupta, U.,and Rajagopal,A.K.,Phys.Rep.**87**,259,1982.

51. Pinski,F.J.,Staunton,J.,Gyorffy,B.L.,Johnson,D.D.,and Stocks,G.M., Phys.Rev.Lett.**56**,2096,1986.

52. Shelton,W.A.,Pinski,F.J.,Stocks,G.M.,and Gyorffy,B.L., J.Appl.Phys.**61**,3712,1987.

53. Durham,P.J.,Staunton,J.,and Gyorffy,B.L.,J.Mag.Magn.**45**,38,1984.

54. Kisker,E.,J.Mag.Magn.Mat.**45**,23,1984.

55. Kirschner,J.,Globl,M.,Dose,V.,and Scheidt,H.,Phys.Rev.Lett.**53**,612,1984.

56. Brout,R.,and Thomas,H.,Physics,**3**,317,1967.

57. Ziebeck,K.R.A.,and Brown,P.J.,J.Phys.**F10**,2015,1980.

58. Crangle,J.,and Goodman,G.M.,Proc.Roy.Soc.**A321**,477,1971.

59. Moriya,T.,J.Mag.Magn.Mat.**45**,79,1984.

60. Cyrot,M., in 'Electron Correlation in Narrow Band Systems'
ed. T.Moriya,Berlin:Springer,1981.

(Research sponsored in part by the Science and Engineering Research Council of the U.K. and also by the Division of Materials Sciences, U.S. Department of Energy under Contract No. DE-AC05-840R21400 with the Martin Marietta Energy Systems Inc.)

CONFIGURATIONAL ENERGIES IN TERMS OF EFFECTIVE CLUSTER INTERACTIONS IN BINARY SUBSTITUTIONAL ALLOYS: CONNECTION BETWEEN THE EMBEDDED CLUSTER METHOD AND THE GENERALIZED PERTURBATION METHOD*

P. E. A. TURCHI, A. GONIS, X.-G. ZHANG** AND G. M. STOCKS#

LAWRENCE LIVERMORE NATIONAL LABORATORY LIVERMORE, CALIFORNIA 94550

ABSTRACT

Starting from the formal expansion of the configurational energy in terms of fully renormalized effective cluster interactions, it is shown that the Embedded Cluster Method (ECM) and the Generalized Perturbation Method (GPM) lead to identical expressions for the energy of a given alloy configuration within the Coherent Potential Approximation (CPA). Correction terms associated with fluctuations in the reference medium can be calculated in both methods. Numerical results are presented for a model tight binding Hamiltonian which clearly indicate the significance of multisite cluster interactions in the determination of the tendencies toward ordering or phase separation and phase stability in alloys.

In this paper we address the calculation of configurational energies in binary substitutional alloys in terms of uniquely defined concentration dependent effective cluster interactions (ECI). These ECI, in turn, can be used in conjunction with statistical models to study various alloy properties, e.g., stable ordered structures at low temperature and ultimately alloy phase diagrams.

The ECI can be naturally introduced to obtain an exact expression for the energy of an alloy in a given configuration specified by a set of occupation numbers $\left\{P^{\alpha_i^J}\right\}$ where $P^{\alpha_i^J}$ equals 1 (or 0) according to whether or not a particular site i of the alloy in configuration J is occupied by an atom of type α_i^J.

The thermodynamic potential, at T = OK, has the general form:

$$\Omega = \sum_J E^J P^J = \sum_J E^{\alpha_1^J \alpha_2^J \ldots \alpha_N^J} \prod_{i=1}^{N} P^{\alpha_i^J} \qquad (1)$$

where N is the number of sites in the alloy.

In the case of a binary alloy A_cB_{1-c}, the use of the relation $P_i^A + P_i^B = 1$ and the definitions $P_i^A = P_i$ and $\delta c_i = P_i - c$ allows us to express the configurational energy as follows:

G. M. Stocks and A. Gonis (eds.), Alloy Phase Stability, 509–514.

$$\Omega(\{\delta c_i\}) = E^{(0)} + \sum_{i=1}^{N} E_i^{(1)} \delta c_i + \frac{1}{2} \sum_{i,j}'' E_{ij}^{(2)} \delta c_i \delta c_j$$

$$+ \frac{1}{3} \sum_{i,j,k}'' E_{ijk}^{(3)} \delta c_i \delta c_j \delta c_k + \ . \ . \ . \qquad (2)$$

where the double prime denotes summations over sets of distinct sites. The coefficients $E^{(n)}_{i_i i_2 \ldots i_n}$ which occur in this expansion are interchange energies associated with clusters C_n of n sites and as the following definitions show explicitly, they are exact.

Let us first define an n-site interchange energy for a given configuration J_{C_n} outside cluster C_n:

$$E^{(n)}_{i_1 i_2 \ldots i_n}\left(J_{C_n}\right) = \prod_{j=1}^{n}\left(1 - Q^{AB}_{i_j}\right) E^{A\ A \ldots\ A}_{i_1 i_2 \ldots i_n}\left(J_{C_n}\right) \qquad (3)$$

where $Q^{AB}_{i_j}$ changes an A atom to a B atom on site i_j and $E^{A\ A \ldots\ A}_{i_i i_2 \ldots i_n}\ J_{C_N}$ denotes the configurational energy associated with A atoms on sites $i_1, \ldots$

Now, the expansion coefficients in (2), the fully renormalized effective interactions are given as the high temperature configurational average of $E^{(n)}_{i_1 i_2 \ldots i_n}\left(J_{C_n}\right)$ over all configurations J_{C_n} of the alloy surrounding cluster C_n:

$$E^{(n)}_{i_1 i_2 \ldots i_n} = < E^{(n)}_{i_1 i_2 \ldots i_n}\left(J_{C_n}\right) > = \sum_{J_{C_n}} P_{J_{C_n}} E^{(n)}_{i_1 i_2 \ldots i_n}\left(J_{C_n}\right) \qquad (4)$$

where $P_{J_{C_n}}$ is the probability of occurrence of J_{C_n}.

Since the expansion given in (2) is independent of the choice of a reference medium, it affords one considerable freedom in selecting a medium that yields an optimal rate of convergence. Let us, from now on, concentrate on the band contribution to the configurational energy (we emphasize that (4) holds for any configurationally dependent quantity of an alloy). In order to evaluate (2), we note that the (band) configurational energy E^J can be written in the form:

$$E^J = \frac{2}{\pi} \operatorname{Im} \int_{-\infty}^{E_F} d\varepsilon \ \operatorname{Tr} \ln G(\varepsilon) \qquad (5)$$

where $G(\varepsilon)$ denotes the Green function for the given configuration, E_F is the Fermi energy and the trace, Tr, is being taken over the orbital indices. Since the Green function obtained within the Coherent Potential Approximation (CPA) is exact to order t^4 in the single site scattering matrix, the CPA provides a particularly well suited medium for the summations in (2).

At the first level of this approximation, all sites outside a given cluster C_n are taken as being occupied by an effective medium of "CPA atoms." In other words, the true configurational averages of $E^{(n)}_{i_1 i_2 \ldots i_n}\left({}^J C_n\right)$, in (4), are replaced with their values assumed by the interchange energies of an n-site cluster embedded in a CPA medium. Calling this the coherent potential approximation for the configurational energy, we write:

$$E^J_{CPA} = E_{CPA} + \sum_i V^{(1)}_i \delta c_i + \frac{1}{2} \sum_{i,j}{}'' V^{(2)}_{ij} \delta c_i \delta c_j + \frac{1}{3} \sum_{i,j,k}{}'' V^{(3)}_{ijk} \delta c_i \delta c_j \delta c_k + \ldots \tag{6}$$

where E_{CPA} is the contribution of the CPA medium, i.e., it is the term arising from the site diagonal Green function, $\bar{G}$, of the uniform reference medium as defined within the CPA, and the various $V^{(n)}_{i_1 i_2 \ldots i_n}$ are the renormalized ECI as defined in the ECM [1]:

$$V^{(n)}_{i_1 i_2 \ldots i_n} = \left[V^{(n-1)}_{i_1 i_2 \ldots i_{n-1}}\right]^A_n - \left[V^{(n-1)}_{i_1 i_2 \ldots i_{n-1}}\right]^B_n$$

$$= -\frac{2}{\pi} \operatorname{Im} \int_{-\infty}^{E_F} d\varepsilon \operatorname{Tr} \ln \left[\prod_j Q^{j(\mathrm{even})}\right]\left[\prod_j Q^{j(\mathrm{odd})}\right]^{-1} \tag{7}$$

where j (even) and j (odd) denote cluster configurations with even and odd numbers of minority atoms (B), respectively, and Q is simply related to the full scattering matrix T by:

$$T = \delta\, D\, Q^{-1}$$

where δ and D^{-1} are site diagonal matrices with elements δ_i^{α} and $\left[D^{-1}\right]^{\alpha}_{ii}$ given in the well known tight binding formalism by $\varepsilon_i^{\alpha} - \sigma_i$ and $1 - \delta_i^{\alpha} \bar{G}_{ii}$ respectively (in the homogenous CPA: $\sigma_i \equiv \sigma_0$ is the self energy and the site diagonal average Green function $\bar{G}_{ii} \equiv \bar{G}_{00}$). As usual ε_i^{α} denotes the onsite energy for the atom of type α on site i.

Eq. (6) shows that, in the CPA, the configurational energy of an alloy can be calculated within the ECM as a sum of irreducible renormalized cluster interactions associated with clusters of atoms embedded in a CPA effective medium. In the case of monoatomic crystalline structures, the one body contribution vanishes identically since $\Sigma\, \delta c_i = 0$.

From the definition of Q, the expansion of Tr ℓn Q leads to:

$$\operatorname{Tr} \ln Q^J = \frac{1}{2} \sum_{i,j}{}' t^{\alpha_i^J} \bar{G}_{ij}\, t^{\alpha_j^J} \bar{G}_{ji} + \frac{1}{3} \sum_{ijk}{}' t^{\alpha_i^J} \bar{G}_{ij}\, t^{\alpha_j^J} \bar{G}_{jk} \times t^{\alpha_k^J} \bar{G}_{ki} + \ldots \tag{8}$$

where $\bar{G}$ is the site non diagonal Green function of the CPA reference medium and the prime refers to summations over sets of distinct consecutive sites. $t^{\alpha_i^J}$ is the scattering matrix characterizing an atom at site i, in configuration J of the alloy, embedded in the CPA medium.

By inserting (8) in (7) and then in (6), tedious but straightforward algebra shows that the ECI defined in the ECM correspond to the renormalized interactions as defined in the GPM [2] by taking into account higher order terms in the perturbative expansion of the configurational energy, given in (5). In other words, in the GPM approach, E^J is written as [3]:

$$E^J = E_{CPA} + \frac{1}{2}\sum_{ij}{}' V^{(2)}_{ij}\delta c_i\, \delta c_j + \frac{1}{3}\sum_{i,j,k}{}' V^{(3)}_{ijk}\, \delta c_i\, \delta c_j\, \delta c_k + \frac{1}{4}\sum_{i,j,k,l}{}' V^{(4)}_{ijkl}\, \delta c_i\, \delta c_j\, \delta c_k\, \delta c_l + \ldots \qquad (9)$$

where $$V^{(n)}_{i_1 i_2 \ldots i_n} = -\frac{2}{\pi}\,\mathrm{Im}\int_{-\infty}^{E_F} d\varepsilon\, \mathrm{Tr}\,(\Delta t\, \bar{G}_{i_1 i_2}\, \Delta t\, \bar{G}_{i_2 i_3} \ldots \Delta t\, \bar{G}_{i_n i_1})$$

the trace, Tr, being taken over the orbital indices. In this expression Δt denotes the difference in single site scattering matrices $t_A - t_B$. The major difference between (9) and (6) consists in the kind of summation which is performed: in (6) the summation is done on sets of distinct sites whereas in (9) this restriction applies to consecutive sites. Thus from (9) it is seen, for example, that the fourth order term $V^{(4)}_{ijij}$ will make contribution to the pair interaction $V^{(2)}_{ij}$ and also to E_{CPA}. Hereafter $V^{(2)}_{ij}$, $V^{(4)}_{ijij}$ . . . will be written as $V^{(2)}_s$, $V^{(4)}_s$ . . ., where $s = |\bar{R}_j - \bar{R}_i|$.

In order to illustrate the connection between ECM and GPM, numerical calculations of ECI were carried out with respect to a simple but realistic tight binding d-band Hamiltonian in the case of fcc paramagnetic transition metal alloys. The tight binding parameters chosen for these calculations are such that the Slater Koster hopping integrals between first neighbors only are given by $dd\sigma = -1.385$, $dd\pi = |dd\sigma|/2$, $dd\delta = 0$, and the onsite energies ε_d^A and ε_d^B take the values 0 and 3.12 respectively. These parameters lead to a common d-bandwidth for the fcc pure metals of $\bar{W} = 11.04$ in canonical units. Therefore for a typical d-bandwidth of 5 eV, the canonical energy unit, taking into account orbital and spin degeneracy, is roughly equal to 4.5 eV. The evaluation of the various Green functions entering the calculations was performed using the recursion method [4] with ten levels of continued fraction. Pair interactions $V_1^{(k)}$ (k = 2,4,6,8) and renormalized effective pair interaction as calculated in the GPM and ECM respectively, for nearest neighbor sites and concentration c = 0.75, are shown in Fig. 1 as a function of the Fermi energy E_F. The terms $V_1^{(2)}$, $V_1^{(4)}$, $V_1^{(6)}$, and $V_1^{(8)}$

which are expressed in terms of the corresponding powers of δt are depicted by the thin solid, dash-dot, dash and dot curves, respectively, whereas the thick solid curve shows the renormalized effective pair interaction calculated in the ECM. For a given distance s, it is easy to show that [1]:

$$V_s^{ECM} = \sum_{n=1}^{\infty} \frac{1}{n} \left[(1-c)^n - (-c)^n\right]^2 V_s^{(n),GPM} \tag{10}$$

It is seen that the various terms $V_1^{(k)}$, k>2 are several factors smaller than the leading term $V_1^{(2)}$ which in turn is extremely similar to the renormalized ECM result. Thus the lowest order term gives an extremely accurate representation of the sum (10) and indicates that the GPM is a valid, rapidly convergent perturbation expansion. Effective cluster interactions for 3-site clusters arranged linearly (along the (110) direction of the fcc crystalline structure) and in the form of an isoceles triangle (including two first neighbor and one second neighbor distances) are shown in Figs. 2a and b, respectively, as a function of the Fermi energy E_F. It is seen that the linear arrangement yields a renormalized effective interaction of the same order of magnitude as the nearest neighbor pair and about an order of magnitude larger than that of the triangular arrangement. Again the GPM expansion converges rapidly toward the renormalized ECM result as is easily inferred from a comparison of the thick solid (ECM) and thin solid (summation to 6th order in GPM) curves in Figs. 2a and b. It is worth noticing that around a zero of the first neighbor interaction, cluster interactions can possess large amplitudes, particularly for the triplet of atoms arranged along a line. Among all the clusters investigated in this study, this particular one appears to be one of the most important. Thus, when performing an analysis of the ground states of the Ising model or a study of alloy phase stability at $T \neq 0K$ such clusters cannot be ignored, contrary to a common consensus [3]. This remark also applies in the case of strongly concentration-dependent near-neighbor pair interactions. Hence in most cases, it is necessary to include the effects of multisite interactions in the statistical models commonly used in the calculation of alloy phase diagrams.

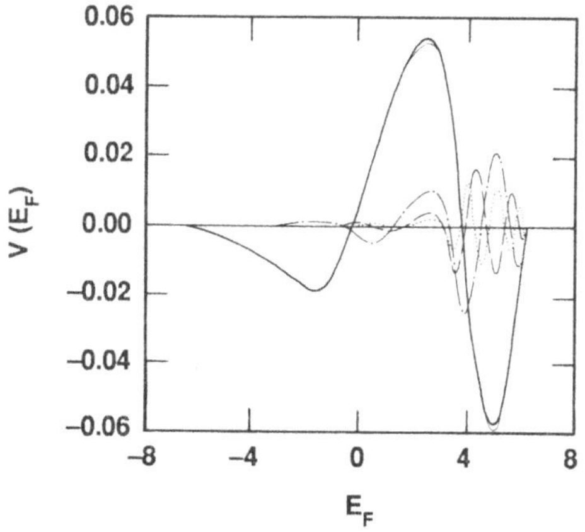

FIGURE 1. First neighbor effective pair interactions, for an $A_{0.75}B_{0.25}$ alloy, calculated in the GPM and the ECM, as a function of the Fermi energy E_F. The GPM contributions $V_1^{(k)}$ for k = 2, 4, 6, 8 are shown by the thin solid, dashed-dotted, dashed and dotted curves, respectively. The renormalized ECM first neighbor effective pair interaction is shown by the thick solid line.

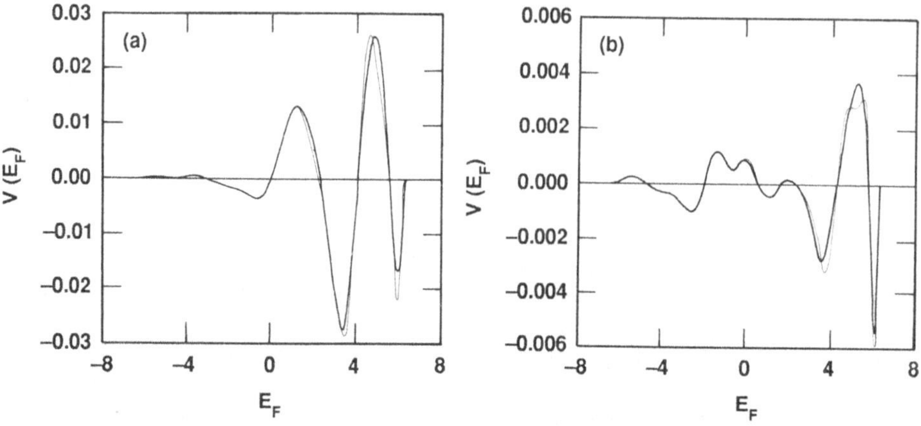

FIGURE 2. Cluster interactions for a) three sites arranged linearly along the (110) direction, b) an isoceles triangle-type of cluster in the fcc $A_{0.75}B_{0.25}$ alloy, as a function of the Fermi energy E_F. The thin solid curve represents the summation of the GPM expansion to 6th order, whereas the thick solid curve shows the renormalized ECM effective interaction.

ACKNOWLEDGMENT

The work of A.G. and X.-G.Z. was partly supported by the U.S. Dept. of Energy under Grant No. DE-FG02-84ER45 with Northwestern University. The work of G.M.S. and D.M.N. was sponsored by the Division of Materials Sciences, U.S. Dept. of Energy, under Contract DE-AC05-84OR21400 with the Martin Marietta Energy Systems, Inc.

*Work performed under the auspices of the U. S. Dept. of Energy by the Lawrence Livermore National Laboratory under contract number W-7405-ENG-48.

**Northwestern University, Dept. of Physics and Astronomy, Evanston, IL 60201

#Oak Ridge National Laboratory, Metals and Ceramics Division, P. O. Box X, Oak Ridge, TN 37831

REFERENCES

1. A. Gonis, X.G. Zhang, A.J. Freeman, P.E.A. Turchi, G.M. Stocks and D.M. Nicholson, Phys. Rev. B36, 4630, 1987.
2. F. Ducastelle and F. Gautier, J. Phys. F: Met. Phys., 6, 2039, 1976.
3. A. Bieber and F. Gautier, J. Phys. Soc. of Jap., 53, 2061, 1984; Z. Phys., B57, 335, 1984.
4. R. Haydock, Sol. St. Phys., 35, 215, 1980.

CLUSTER INTERACTIONS AND THERMODYNAMIC PROPERTIES OF Al-TRANSITION METAL ALLOYS

A. E. Carlsson

Department of Physics, Washington University, St. Louis, Missouri 63130

1. INTRODUCTION

Recent interest in Al-transition metal alloys is motivated both by their desirable technological properties[1] and by their basic scientific interest, including recently discovered phenomena such as icosahedral phase formation.[2] To perform calculations of the properties of these alloys at an atomistic level, it is necessary to ascertain the basic physical mechanisms underlying bonding and ordering. In this paper we study these mechanisms by calculating cluster interactions describing atomic rearrangements on a fixed underlying fcc lattice. We then use these interactions to analyze the systematics of heats of formation, solid solubilities, and structural preferences in Al–transition metal alloys.

2. METHOD

We obtain the cluster interactions using the matching method described in Ref. 3. One assumes a truncated expression for the alloy configurational energy at a lattice constant a of the form

$$E(a) = V_0 + V_1\langle\sigma\rangle + V_2\langle\sigma_i\sigma_j\rangle + V_3\langle\sigma_i\sigma_j\sigma_k\rangle + V_4\langle\sigma_i\sigma_j\sigma_k\sigma_l\rangle \ . \qquad (1)$$

The σ_i are spin-like variables which take on the values 1(−1) for transition metal (Al) atoms. The averages are taken over all sites, and nearest-neighbor pairs, triangles, and tetrahedra. The values of the interaction parameters for each alloy are obtained by matching to quantum-mechanical total energies for the two elemental fcc lattices, two A_3B compounds in the Cu_3Au structure, and the AB compound in the CuAu structure. The total energies are obtained via the augmented-spherical-wave (ASW) band structure method,[4] with an exchange-correlation functional of the Hedin-Lundquist form.[5] Magnetic and relativistic effects are neglected. This method has previously been used to obtain cluster interactions, effective-pair-interactions and bonding properties for Al–Ni[6,7,8] and Al–Cu[9] alloys.

3. RESULTS

Figure 1 shows the calculated values of V_2 ($V_2 > 0$ corresponds to ordering) for alloys of Al with 4d transition metals. We note that for all of the transition metals with partly filled d–bands, V_2 is large and positive. This is consistent with the pronounced ordering tendencies in these systems; in fact V_2 is sufficiently large that if confined to ϕ fcc lattice, most of the alloys would remain ordered up to the melting temperature. For Ag, the value of V_2 is considerably smaller. The dependence of V_2 on the number N_d of d–electrons can be interpreted as consisting of two contributions, one parabolic, and the other oscillating, with a minimum at a roughly half-filled band. The parabolic contribution

G. M. Stocks and A. Gonis (eds.), Alloy Phase Stability, 515–519.

would be expected if the transition metal d–bands were simply broadened by the presence of the Al. The oscillating contribution probably arises from changes in the band shape.[10]

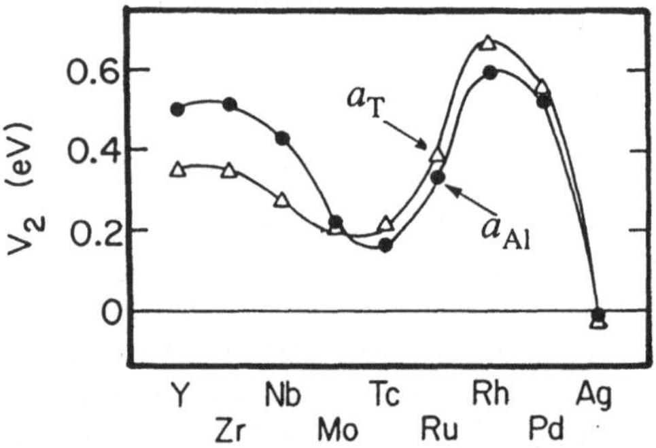

Figure 1. Calculated pair interactions for Al-4d transition metal alloys. Positive sign denotes ordering tendency. Solid circles and triangles denote values at Al and transition metal lattice constants, respectively.

Figure 2 shows the calculated heats of solution per solute atom, ΔH^{SS}, in the limit of small solute concentration. The heats of solution are obtained using the following model:

$$\Delta H^{SS} = \Delta H^{NN} + \Delta H^{LR} + \Delta H^{EL} \quad .$$

ΔH^{NN} and ΔH^{LH} are evaluated at the host lattice constant, and denote contributions from $V_0 - V_4$, and longer-ranged interactions respectively. ΔH^{LR} is evaluated from A_7B fcc supercell total energy calculations. ΔH^{EL} is simply the elastic energy (evaluated using the ASW method) required to compress or expand the solute to the host lattice constant. The alloys included in Fig. 2 are all of the fcc or hcp alloys for which we could find accurate heats of formation. Except for the Ag alloys, the values of ΔH^{NN} are large and exothermic. The values for Al–Ag solutions at both ends of the phase diagram are considerably smaller in magnitude, as expected from the small values of the interaction parameters. With the exception of this system, $|\Delta H^{LR}| << |\Delta H^{NN}|$, speaking favorably for the convergence of the cluster expansion. ΔH^{EL} is also considerably smaller than ΔH^{NN} for most of the systems; the largest contribution is in Cu in which it is almost half of ΔH^{NN}. The sum ΔH^{SS} of these contributions matches the chemical trends in the measured values[11] very well, and even fits the numerical values quite closely.

Al-transition metal phase diagrams consistently display larger solubilities at the transition metal end than at the Al end. To understand this overall trend, and also to elucidate differences between particular transition metals, we have calculated the energy difference ΔE associated with the solubility, defined by $\Delta E = \Delta H^{SS} - \Delta H^{ORD}$. Here ΔH^{ORD} is the heat of formation, per atom, for the ordered compound nearest in concentration to the pure host. If $\Delta E \gg k_B T$ and the configurational entropy of the ordered compound can be neglected, then the solid solubility is given by $C_{max} = (e^{-\Delta E/k_B T})$. Thus C_{max} provides an experimental measurement of ΔE. To evaluate ΔE theoretically we use our calculated values of ΔH^{SS}. ΔH^{ORD} is obtained from our calculations in cases where the ordered compound has the Cu_3Au structure; in other cases it is taken from experiment.[11]

The theoretical and experimental values of ΔE (solubilities obtained from Ref. 1), for an Al host, are shown in Fig. 3a. The values of ΔE correspond to small solubilities; for T = 700°K, corresponding to the temperature of a typical solubility measurement, $4 < \Delta E/k_B T < 20$ in all cases. The overall magnitudes and the chemical trends are consistent. Furthermore, the measured temperature dependence of ΔE in several systems suggests that the agreement would be further improved if the solubilities could be measured at lower temperatures.

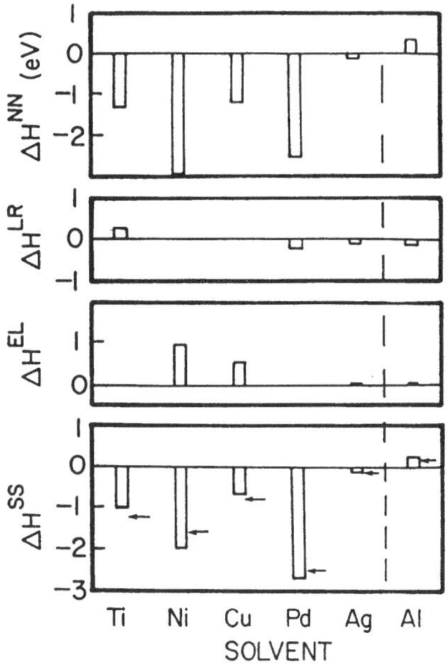

Figure 2. Contributions to heats of solution in Al-transition metal solid solutions. Negative sign denotes exothermic values. For case of Al denote experimental values.

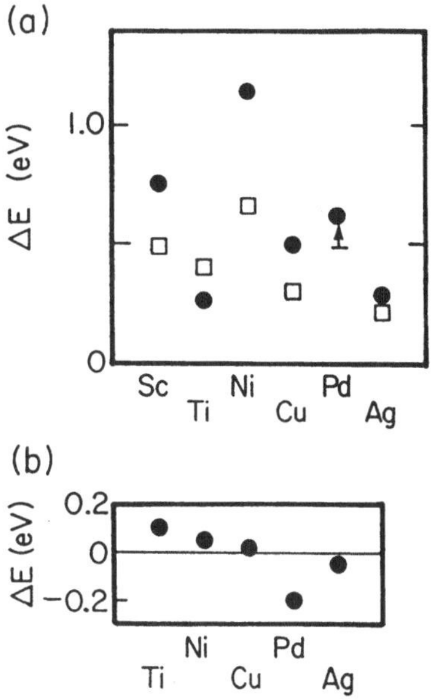

Figure 3. Solubility energy differences ΔE for Al host a) and transition metal host b). Solid circles: theoretical values. Open squares: experimental values for Pd in a) denotes estimate of lower bound from upper bound on solubility.

The theoretical values of ΔE for transition metal hosts, shown in Fig. 3b, are much smaller than those for Al hosts. For values this small one cannot obtain ΔE straightforwardly from the solubility data. However, the small values of ΔE are consistent with the observed phase diagrams, which for all the systems considered show solubilities greater than 10% even at temperatures far below the transition metal melting point. The simplest (although far from rigorous) explanation for the asymmetry between Al and transition metal hosts would be an asymmetry in the heat of formation.[12] For Ni, Cu, Pd, and Ag this asymmetry does appear to be the major contribution to that in ΔE. However, for Ti and Sc, the asymmetry in ΔH^{SS} has the opposite sign from that in ΔE. Thus in these latter two cases contributions from the symmetry in ΔH^{ORD} appear to dominate the contributions from ΔH^{SS}.

Finally, we point out a correlation between our cluster interactions and the local environments of transition metal atoms in Al–rich compounds.[13] Several heuristic arguments would suggest such a correlation. For example, the curvature of the ΔH^{SS} vs $<\sigma>$ curve for random solid solution at the Al end is $2V_2 + 6(-V_3 + 2V_4)$ if lattice strain effects are neglected. Thus if $V_3 - 2V_4$ is large and positive, one would expect a miscibility gap in the fcc phase diagram, which would likely be filled by an ordered compound on another underlying lattice.[7] Figure 4 shows that the predictions of such very qualitative arguments are confirmed by the observed crystal structures[14] for Al–4d transition metal alloys, at the Al–rich end of the phase diagram. Between Ru and Rh, $V_3 - 2V_4$ changes sign abruptly, remaining negative for the earlier transition metals and positive for the later ones. We would thus expect the later transition metals to have a coordination number other than twelve, but the earlier ones to be stable in a twelvefold structure. The Al_9Rh_2 structure has ninefold coordinated Rh; the Al_3Pd structure is probably similar to the Al_3Ni structure, in which Ni is ninefold coordinated. Al–Ag forms no compound at all at the Al–rich end. On the other hand, Ru and the earlier transition metals all form twelvefold coordinated compounds. We note that Ru and Rh have practically identical atomic sizes (as indicated in the figure). Thus the local transition metal coordination correlates much more closely with the cluster interactions than with atomic size differences.

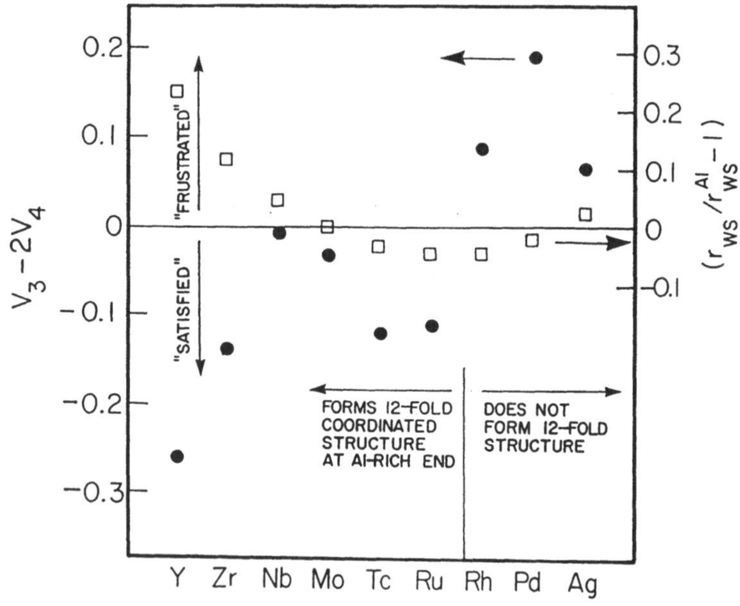

Figure 4. Cluster interactions entering ΔH for Cu_3Au structures and curvature of solid solution ΔH, for Al–4d transition metal alloys (solid circles). Open squares denote relative size difference between transition metal and Al atoms.

4. Acknowledgements

This work was supported by the Department of Energy Grant Number DE–FG02–84ER45130. The band-structure calculations were performed at the Production Supercomputer Facility of the Center for Theory and Simulation in Science and Engineering at Cornell University, under the National Science Foundation Grant No. DMR–8614232.

5. References

1. See, for example, L. F. Mandolfo, *Aluminum Alloys: Structure and Properties* (Butterworths, Boston, 1976).
2. D. S. Shectman, I. Blech, D. Gratias, and J. W. Cahn, Phys. Rev. Lett. **53**, 1951 (1984). For an overview, see C. L. Henley, Comm. Cond. Mat. Phys. **13**, 59 (1987).
3. J. W. D. Connolly and A. R. Williams, Phys. Rev. B **27**, 5168 (1983).
4. A. R. Williams, J. Kübler, and C. D. Gelatt, Phys. Rev. B **19**, 6094 (1979).
5. L. Hedin and B. I. Lundquist, J. Phys. C **4**, 2065 (1971).
6. A. E. Carlsson, Phys. Rev. B **35**, 4858 (1987).
7. A. E. Carlsson and J. M. Sanchez, Solid State Comm., in press.
8. D. Hackenbracht and J. Kübler, J. Phys. **F10** , 427 (1980).
9. A. E. Carlsson, Phys. Rev. Lett. **59**, 1108 (1987).
10. For example, contributions associated with changes in the fourth moment (with the lower moments fixed) must have at least two zero crossings as functions of the band filling. See F. Ducastelle and F. Cyrot-Lackmann, J. Phys. Chem. Solids **32**, 285 (1971); V. Heine and J. H. Sampson, J. Phys. **F13**, 2155 (1983).
11. R. Hultgren, R. D. Desai, D. T. Hawkins, M. Gleiser, and K. K. Kelley, *Selected Values of the Thermodynamic Properties of Binary Alloys* (American Society for Metals, Metals Park, Ohio, 1973).
12. R. E. Watson, L. H. Bennett, and D. A. Goodman, Acta Metall. **31**, 1285 (1983).
13. For another local approach to the relative stability of transition metal local environments, see A. C. Redfield and A. Zangwill, Phys. Rev. Lett. **58**, 2322 (1987).
14. P. Villars and L. D. Calvert, *Pearson's Handbook of Crystallographic Data for Intermetallic Phases* (American Society for Metals, Metals Park, Ohio, 1985).

ELECTRONIC THEORY OF PHASE STABILITY IN SUBSTITUTIONAL ALLOYS: A COMPARISON BETWEEN THE CONNOLLY-WILLIAMS SCHEME AND THE GENERALIZED PERTURBATION METHOD

M. SLUITER * and P. TURCHI **

* Department of Materials Science and Mineral Engineering, University of California - Berkeley, Berkeley, CA 94720, USA ** Department of Materials Science (L 280), Lawrence Livermore National Laboratory, P.O.Box 808, Livermore, CA 94550, USA

A detailed analysis of the tendencies toward ordering and phase separation and more generally the stability properties at T ≠ 0 K in substitutional alloys is carried out using the presciption proposed by Connolly and Williams and the Generalized Perturbation Method. This enables a discussion of the viability and the basic assumptions of both approaches. The effective cluster interactions which enter such a study are derived from a simple but realistic tight binding model.

The basic tools for studying ordering and phase stability in substitutional alloys and ultimately their phase diagrams, are based on the so called 3-D Ising model within various approximations, such as the Cluster Variation Method (CVM) [1] and Monte Carlo simulations [2]. In these models it is assumed that the internal energy can be written as a rapidly convergent sum of pair and multisite interactions. This statement has been proved in normal metals and their alloys [3] using the pseudo-potential theory, but in transition metals, where strong disorder effects can be important, such an expansion must be clearly justified. The success of phenomenological theories for the study of ordering processes has initiated extensive works in the field in order to provide a direct link between electronic band structure calculations and statistical models. Among other approaches the concentration functional theory of ordering [4], the Embedded Cluster Method [5] and the Generalized Perturbation Method (GPM) [6] can be mentioned.

In the GPM a perturbation treatment is derived by choosing a reference medium which is close to any particular configuration of the alloy. Hence, the intuitive idea to use the completely disordered state, as the one described in the Coherent Potential Approximation (CPA), as an appropiate reference medium. For each configuration of an alloy, specified by a set of occupation numbers $\{p_n^i\}$ (p_n^i takes the value 1 if site n is occupied by an atom of type i, and equals 0 otherwise), the total energy is written as a sum of two terms: the concentration dependent (but configuration independent) energy of the CPA medium, $E_{CPA}(c)$, and a configuration and concentration dependent ordering energy, $E_{ord}(\{p_n^i\})$, which can be written as an expansion in terms of concentration dependent l^{th} order effective cluster interactions $V^{(l)}_{n_1,n_2,\cdots,n_l}$ [7]. For example, by defining $\delta c_n = p_n^B$ for a binary alloy, the total energy for a given configuration can be expressed as:

$$E(\{\delta c_{n_i}\}) = E_{CPA}(c_B) + \frac{1}{2}\sum_{n_1,n_2} V^{(2)}_{n_1,n_2}(c_B)\delta c_{n_1}\delta c_{n_2} + \frac{1}{3}\sum_{n_1,n_2,n_3} V^{(3)}_{n_1,n_2,n_3}(c_B)\delta c_{n_1}\delta c_{n_2}\delta c_{n_3} + \cdots \quad (1)$$

Where the sums run over sets of distinct consecutive sites.

Studies in the framework of the tight binding approximation (TBA) concerning the phase stability of simple and complex phases at T=0 K [8] and for T≠0 K [9,10] have proved that the GPM is a viable method. Moreover, it has been shown that the expansion of the configurational contribution to the internal energy in terms of effective cluster interactions is, in most cases, rapidly convergent [7].

An alternative way to study phase transformations in alloys was proposed by Connolly and Williams [11]. In this approach (named hereafter the Connolly Williams Method, CWM) it is assumed that the total energy itself, for a given configuration c of the alloy, can be written as:

$$E^{(c)}(r) = \sum_{\gamma} V_{\gamma}(r)\, \xi_{\gamma}^{c} \quad (2)$$

G. M. Stocks and A. Gonis (eds.), Alloy Phase Stability, 521–526.

Where $V_\gamma(r)$ and ξ_γ^c are the configuration independent many body interaction potentials and the multisite correlation functions respectively, associated with cluster γ, r is the lattice parameter and the sum runs over all cluster types ("empty cluster" included) on a fixed lattice. The correlation functions are given by:

$$\xi_\gamma = \frac{1}{N_\gamma} \sum_{n_i} \sigma_{n_1} \sigma_{n_2} \ldots\ldots \sigma_{n_\gamma}$$

Where $\sigma_n = 1 - p_n$, takes the values +1 or -1 depending on the occupancy of the site n, N_γ is the total number of γ - site clusters and the sum runs over all γ^{th} order clusters of a given type on the lattice.

In practice, the set of equations (2) can only be solved if the existence of a maximum cluster α_{max} is assumed, beyond which the many body interactions are supposed to be negligible. Hence, once a choice of ordered structures has been made, this a priori arbitrary truncation allows the calculation of the remaining interactions by inverting (2). As a by-product, the energy of the totally disordered state can be estimated from:

$$E_{dis}(r) = \sum_{\gamma \le \alpha_{max}} V_\gamma(r)\, \xi_1^{n_\gamma} \tag{3}$$

Where ξ_1 is the point correlation function given by: $\xi_1 = 1 - 2c_B$, n_γ is the number of sites in the cluster γ and the sum extends from the empty cluster to the maximum cluster α_{max}.

Although the CWM has been applied recently [12], some crucial questions remain unanswered. In particular the rapid convergence of expansion (2) (a requirement for the truncation) has not been proved. Because the many body interactions, in this context, are the result of a fitting procedure; the question can be asked whether the V_γ are unambiguously defined. In other words, do the cluster interactions depend on the selection of the ordered structures, contrary to the implicit assumption made by truncation of (2)? In order to answer these questions, the CWM will be compared with the GPM and exact recursion results. For this purpose a simple tight binding model was designed to study phase stability from bandstructure calculations in binary substitutional paramagnetic transition metal alloys. The TBA was considered with the following assumptions: i) Only d-electrons are taken into account. Canonical Slater-Koster parameters involving first neighbor hopping integrals in the FCC crystal structure were used ($dd\sigma$=-1.385, $dd\pi = |dd\sigma|/2$, $dd\delta$=0, leading to a d-bandwidth of 11.08 for the pure metal); ii) The alloy effects come in through the difference in the atomic d-levels, also called the diagonal disorder parameter: $\delta_d = (\varepsilon_B - \varepsilon_A)/\overline{W}$ where $\overline{W}$ is the concentration weighted average of the half d-bandwidths of the pure elements. In the following δ_d takes the value 0.8. iii) No off-diagonal disorder (i.e. no variation of the hopping integrals with site occupancy) is assumed and charge transfer effects are neglected.

The density of states (dos) is obtained by the recursion method for each ordered structure (c) selected in the present study [13]. The one electron band energy (E_b), assumed to be the main contribution to the total energy, is calculated by integrating over the dos up to the Fermi energy E_F^c :

$$E_b^c = \int^{E_F^c} \varepsilon\, n^c(\varepsilon)\, d\varepsilon \tag{4}$$

Where the Fermi energy is determined so that integration of the dos up to the Fermi energy yields the d-band filling of the alloy.

In this description, most of the tendencies toward ordering and phase separation as well as the prediction of the most probable ordered states at T=0 K can be obtained [8]. Note, that because the recursion does not make any additional assumptions, its results can be regarded as exact in the context of the present model. As mentioned earlier, the CWM requires a rapid convergence of expansion (2) in order to justify its truncation. For this purpose the CWM many body interactions have been calculated considering a commonly chosen basis [11,12] of the [100] family of ordered phases, namely $L1_2$ (A_3B and AB_3) and $L1_0$ (AB), in addition to the pure metals, A and B.

In figure 1a the pair, triangle and tetrahedron cluster interactions (V_2^T, V_3^T, and V_4^T respectively) are shown as a function of the number of d-electrons of the B species for fixed $\Delta N = N_A - N_B$. Because of the proportionality between δ_d and ΔN, a value of $\Delta N = 6$ is chosen, in reasonable agreement with $\delta_d = 0.8$. As can be seen, V_3^T and V_4^T are much smaller than V_2^T, except for a large value of N_B. The CWM has been extended by including other ordered structures than the ones originally proposed and also by truncating expansion (2) at the tetrahedron-octahedron (TO) level (leading to 11 interactions V_γ^{TO}). Therefore the [1½0] family of ordered structures, namely DO_{22} (A_3B and AB_3), $MoPt_2$ type of order (A_2B and AB_2), A_2B_2 (phase 40 in Kanamori's notation), and CuPt type of order (AB) were selected (see ref [14] for a description of these ordered structures).

Figure 1b shows the V_γ^{TO} , γ=2,3,...10 computed with TO cluster truncation. Although the V_γ^{TO} γ=2,4,6 compare correctly with the corresponding V_γ^T γ=2,3,4, higher order (TO) cluster interactions are significant (note the isoceles triangle).

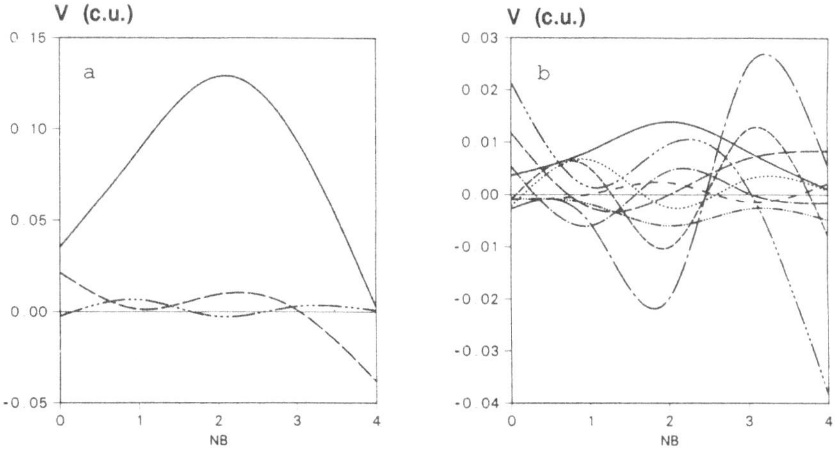

Figure 1. CWM cluster interactions in canonical units (c.u.): a) T-CWM _____ V_2 , _ _ _ V_3 , _..._ V_4 ; b) TO-CWM: _______ first neighbor pair divided by 10, ___ ___ _ second neighbor pair, ___...___... equilateral nearest neighbor triangle, ___.___. isoceles triangle, ... regular tetrahedron, ------- irregular tetrahedron, _..._... square, __.__.__ pyramid, _ _ _ _ _ octahedron.

Table 1 indicates the V_γ^T for various sets of ordered structures. Note that especially the values for the triangle and tetrahedron can vary significantly. Surprisingly, the [100] family gives the smallest values of V_3^T *and* V_4^T. We conclude that: i) the rapid convergence of expansion is not completely fulfilled and ii) that the V_γ , within the same cluster truncation, are not unambiguously defined. By contrast, in the GPM the concentration dependent effective cluster interactions do not suffer such an ambiguity [6,7].

Table 1. T-CWM cluster interactions (in c.u.) for various sets of ordered structures: a) $\{A, L1_2, L1_0, B\}$, b) $\{A, DO_{22}\text{–}A_3B, L1_0, L1_2\text{–}AB_3, B\}$, c) $\{A, DO_{22}, 40, B\}$.

V	a	b	c
0	-.13665	-.13953	-.13437
1	-.33290	-.33866	-.34127
2	.09455	.09455	.08647
3	.00118	.00694	.00956
4	.00299	.00587	.00841

A good test case for the the CWM and the GPM is to compare the predicted formation energy of a binary alloy in a given configuration c, ΔE_f^c , with the exact value obtained from the recursion:

$$\Delta E_f^c = E_b^c - (1-c_B)E_b^A - c_B E_b^B \tag{5}$$

Where E_b^A and E_b^B are the band energies of the pure elements.

From the V_γ^T depicted in figure 1a, one can calculate the formation energy of an ordered phase not included in the basis (the [100] family in this case). We choose the AB phase with CuPt type of order. The GPM formation energy is obtained by including only the zeroth term and the second order terms up to the fourth neighbor shell in expansion (1). A detailed description of the TBA-CPA-GPM can be found in refs [7,8,9]. Figure 2a shows that the GPM gives better agreement with the recursion results than the CWM does.

In figure 2b, the ordering energy ΔE_{ord}^c for the AB phase with CuPt structure is shown as a function of N_B. In the three methods, the ordering energy ΔE_{ord}^c, is calculated as follows:

$$\Delta E_{ord}^c = \begin{cases} E_b^c - E_{CPA} & \text{'exact'} \\ \frac{1}{2}\sum_{n \neq m} V_{nm}^{(2)} \delta c_n^c \delta c_m^c & GPM \\ \sum_{2 \le \gamma \le \alpha_{max}} V_\gamma \, (\xi_\gamma^c - \xi_1^{n_\gamma}) & CWM \end{cases}$$

The GPM predicts the possibility of CuPt type of order for $1.3 \le N_B \le 3.7$, in good agreement with the exact results and rather contrary to the CWM prediction. However, neither the CWM nor the GPM follows the exact results closely, for the GPM this may indicate that important multisite interactions have been neglected in truncating expansion (1) or that the configuration dependence of the Fermi energy is significant (an effect which is ignored in the GPM). Note, the difference in magnitude between the formation energy and the ordering energy of a compound (as a guideline, for a typical d-bandwidth of 5 eV, 1 canonical unit corresponds to ca. 4.5 eV).

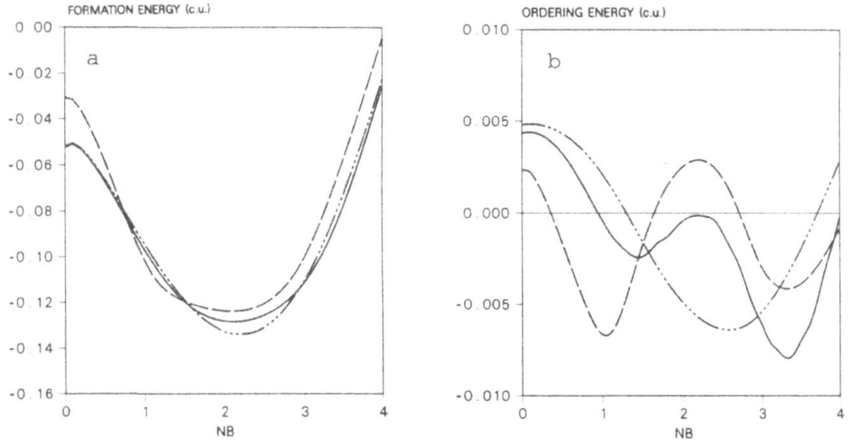

Figure 2: a) Formation energy of the CuPt type AB phase as a function of the number of d-electrons of element B N_B, as calculated by different methods: ------ T-CWM using [100] phases, ____... GPM, ______ exact recursion result. b) similar to a) for the ordering energy of the CuPt type AB phase, the solid line refers to CPA-recursion results.

Finally, the formation energy (ΔE_f^{dis}) of the random alloy (as calculated from eqns. (3) and (5) for various selections of ordered states in the T- and TO-cluster truncation of the CWM) has been compared with the CPA. The results are illustrated in figure 3, where ΔE_f^{dis} is plotted as a function of composition for an AB alloy with N_A=9 and N_B=3. In spite of the overall qualitative agreement, it is clear that within one level of the CWM approximation no unique mixing energy can be defined. The slow convergence in the CWM scheme is reflected in the disparity between the TO-CWM and the T-CWM curves.

Although at T=0 K qualitatively correct predictions can be made concerning the most probable ordered structures with both approaches. It is of crucial importance to improve the determination of the internal energy when addressing the calculation of phase diagrams because of the small value of the energy differences involved. The discrepancies between the GPM and the recursion results can be due to neglected multisite effects and higher order terms in expansion (1). Whereas the differences between the CWM and the recursion results are attributed to the slow convergence of expansion (2) and the non-uniqueness of the many body interactions. Fig. 4 and fig. 4b from ref. [9] show that these discrepancies can lead to very different phase diagrams. The configurational free energy was evaluated in the tetrahedron approximation of the CVM (for a detailed description, see ref [9]). Note that in the tetrahedron approximation of the CVM the $L1_2$ and DO_{22}, and the $L1_0$ and A_2B_2 ("40") phases are degenerate.

Going beyond a first neighbor pair analysis of phase stability of alloys is often needed to lift degeneracy between ordered states. Accurate and unambiguous cluster interactions are required to obtain quantitative answers in the field of alloy theory at $T\neq0$ K. Work along this line is currently under way.

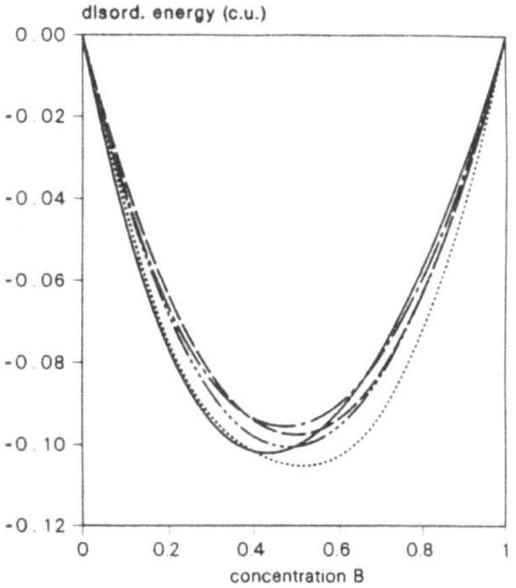

Figure 3. Formation energy of disordered alloy as a function of concentration of B species (N_A=9, N_B=3) as calculated with various methods: dashed line - T-CWM using [100] phases, triple dot dash line - T-CWM using {A, DO_{22}–A_3B, $L1_0$, $L1_2$–AB_3, B, dash dot line - T-CWM using {A, DO_{22}, 40, B} , solid line - TO-CWM and the dotted line represents the CPA.

One of the authors (P.T.) would like to thank P. Cenedese and A. Zunger for stimulating discussions. M.S. thanks Prof. D. de Fontaine for his interest and encouragement.

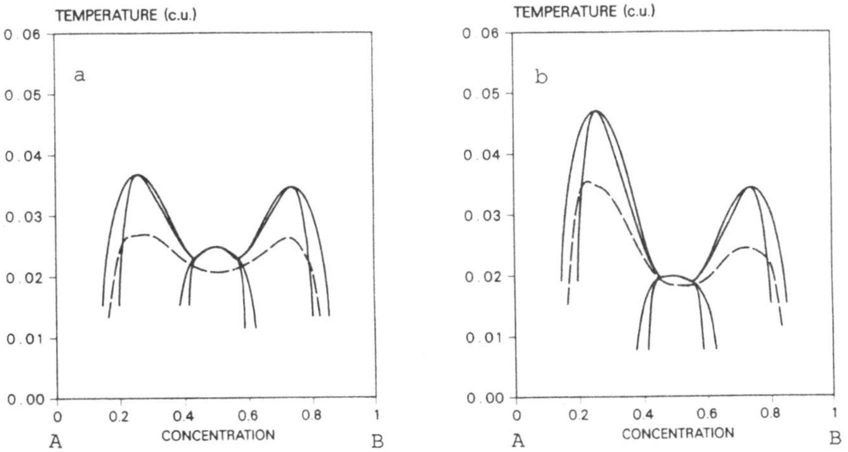

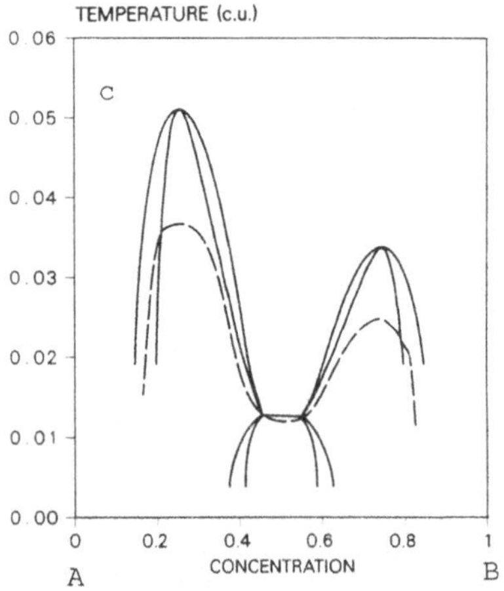

Figure 4. Equilibrium FCC order-disorder phase diagrams for an alloy with N_A=9, N_B=3 and δ_d=0.8, calculated with the tetrahedron approximation of the CVM using the cluster interactions from table 1. The dashed curve indicates the [100] ordering spinodal. The temperature is expressed in canonical units.

References:

1. R. Kikuchi, Phys. Rev. 81, 998 (1951).
   J.M. Sanchez, F. Ducastelle and D. Gratias, Physica 128A, 334 (1984).
   D. de Fontaine, Sol. St. Physics 34, 73 (1979).
2. K. Binder, J.L. Lebowitz, M.H. Phani and M.H. Kalos, Acta Metall. 29, 1655 (1981).
   K. Binder in 'Monte Carlo Methods in Statistical Physics', Springer-Verlag, Topics in Current Physics Vol.7, ed. K. Binder (1986).
3. V. Heine and D. Weaire, Sol. St. Physics 24, 249 (1970).
4. B.L. Gyorffy and G.M. Stocks, Phys. Rev. Lett. 50, 374 (1983).
5. A. Gonis, X.G. Zhang, A.J. Freeman, P. Turchi, G.M. Stocks and D.M. Nicholson, submitted to Phys.Rev. B.,and refs. therein.
6. F. Ducastelle and F. Gautier, J. of Phys. F 6, 2039 (1976).
7. A. Bieber and F. Gautier, J. of the Phys. Soc. of Jap. 53, 2061 (1984).
8. P. Turchi, Thesis, University Pierre et Marie Curie, Paris, France (1984), (unpublished).
   A. Bieber and F. Gautier, Acta Metall. 34, 2291 (1986) and refs therein.
9. P. Turchi, M. Sluiter and D. de Fontaine, Phys. Rev. B 36, 3161 (1987).
10. C. Sigli, M. Kosugi and J.M. Sanchez, Phys. Rev. Lett. 57, 253 (1986).
11. J.W.D. Connolly and A.R. Williams, Phys. Rev. B 27, 5169 (1983).
12. A.A. Mbaye, L.G. Ferreira and A. Zunger, Phys. Rev. Lett. 58, 49 (1987).
    K. Terakura, T. Oguchi, T. Mohri and K. Watanabe, Phys. Rev. B 35, 2169 (1987).
    T. Mohri, K. Terakura, T. Oguchi and K. Watanabe, submitted to Acta Metall.
    A.E. Carlsson, Phys. Rev. B 35, 4858 (1987).
13. R. Haydock, Sol .St. Phys. 35, 215 (1980).
14. J. Kanamori and Y Kakehashi, J. de Phys. 38, Colloq. C7-274 (1977).

SECTION 6

THE EFFECTS OF STRAIN AND MACROSCOPIC DEFECTS ON PHASE STABILITY

STRAIN CONTROLLED MORPHOLOGIES IN THE TWO-PHASE STATE

A. G. KHACHATURYAN

Center for Advanced Materials, Lawrence Berkeley Laboratory and Department of Materials Science and Mineral Engineering, University of California, Berkeley, CA 94720

I. INTRODUCTION

Phase transformation in solids usually involve crystal lattice rearrangement with the islands of the new phase inside the parent phase matrix. Crystal lattice mismatch produced by phase transformation is accomodated by elastic displacements generating the elastic strain field within the body. The elastic energy contained in the strain field may contribute considerably to the thermodynamics of the phase transformation, but the main effect of the elastic strain is far beyond the trivial renormalization of elastic energy. Unlike the "chemical" free energy depending only on the volume of phases, the elastic energy also depends on the morphology, shape, dispersion and mutual location of inclusions. In such a case the morphology of the alloy becomes an internal thermodynamic parameter that can be found from the free energy minimization. This, in fact, means that the conventional thermodynamics of phase transformations based on the free energy additivity should be questioned and validity of certain classical results has to be reexamined. To make more clear how far we can go in revising the theory of phase transformation when elastic energy is involved, it is noteworthy to look at the other cases when the bulk free energy proves to be dependent on morphology. The other cases where this situation takes place are ferromagnets and ferroelectrics whose magnetostatic and electrostatic energy also depend on shape, size and mutual location of domains. This dependence manifests itself, for example, in appearance of the so-called demagnetization factor, and it affects the ground state of ferromagnets. Indeed, the homogeneous single domain state that would be expected without magnetostatic energy transforms into an array of domains whose size tends to zero if the Bloch energy (surface energy of domain walls) vanishes. It is also known that the repulsive interaction between the similar domains results in the formation of so-called bubble domain structure.

The similar dramatic effects of the morphology on the thermodynamics of a phase transformation could be expected in the case of a transformation with large crystal lattice rearrangement. The specific examples of such effects are formation of platelet precipitates with the specific habit minimizing the strain energy as well as the formation of agglomerates of various orientational variants of the new phase accomodating crystal lattice mismatch and eliminating elastic strain. The latter effect is observed during martensitic transformation, ordering and decomposition and is a good example of the profound analogy with the magnetic and ferroelectric domain structure.

The intensive studies of the elastic strain effect caused by the other phase coherent inclusions were initiated by the classical works by Eshelby [1] in the fifties who calculated the elastic energy of an ellipsoidal inclusion in an isotropic case. The next step was made in our work [2] and the work by Roitburd [3] where the idea that the elastic strain energy minimization can be used for the habit plane determination was first proposed.

The general theory of elastic energy of an arbitrary distributed inclusions in elastically

G. M. Stocks and A. Gonis (eds.), Alloy Phase Stability, 529–555.

anisotropic medium in the homogeneous modulus case was formulated by Khachaturyan and Shatalov [4] and developed by Wen, Khachaturyan and Morris [5]. The theory was used for analyzing morphology of a single precipitate [6-13] and morphology transformations of a group of precipitates [4,5,9,14-18]. The exact solution of the elastic problem for an ellipsoidal inclusion in the heterogeneous modulus case and anisotropic crystals was obtained by Lee, Barnett and Aaronson [19].

The main topic of these lectures is the discussion of the application of the elastic theory to practical problems which arise in structural studies of the morphology of two-phase alloys. The consideration will be based on the theory developed for the homogeneous modulus case [2,4] because this approach enables one to treat arbitrary dispersoids using very simple mathematics. The cases where this approximation turns out to be insufficient will be discussed separately. The applied aspects of the theory will be especially emphasized. There are three groups of problems that deserve to be discussed in detail:

1. Morphology of a single coherent precipitate, its habit plane, equilibrium shape, orientational relations and crystal lattice parameters in the constraint state.
2. Shape transformations and morphology instabilities upon coarsening.
3. Strain-induced rearrangement of groups of precipitates upon coarsening.

2. ELASTIC ENERGY AND ELASTIC DISPLACEMENTS INDUCED BY ARBITRARY ARRAY OF COHERENT INCLUSIONS

Following [4] let us consider n types of inclusions that are produced by different crystal lattice rearrangements, for example, the rearrangements generating the different orientational variants of the same phase. These inclusions may be characterized by stress-free transformation strains, $\varepsilon^o_{ij}(1), ..., \varepsilon^o_{ij}(p), ..., \varepsilon^o_{ij}(n)$, describing the macroscopic shape change of the parent phase caused by the respective crystal lattice rearrangements. These inclusions can be produced by means of the five steps of the Eshelby cycle:

1. Cut inclusions from the matrix.
2. Let each inclusion be transformed to a new phase under the stress-free strains $\varepsilon^o_{ij}(1), ... , \varepsilon^o_{ij}(n)$.
3. Restore the initial shape applying the surface traction to create the opposite sign homogeneous elastic strains, $-\varepsilon^o_{ij}(1)$, change. The elastic energy required to induce this set of elastic strains is

$$E_{self} = (1/2)\sum_{p=1}^{n} v(p)\ \lambda_{ijkl}\ \varepsilon^o_{ij}(p)\ \varepsilon^o_{kl}(p) \tag{1}$$

   if the elastic moduli of all phases are the same , where λ_{ijkl} is the elastic modulus tensor, i, j, k, l are Cartesian indices, v(p) the volume of all inclusions of the pth type.
4. Reintroduce restored inclusions in their holes and weld them.
5. Remove the surface traction and allow the inclusions and matrix to relax. The relaxation energy, ΔE, by definition, should be a negative value reducing the energy (1).

The total elastic energy is then

$$E = (1/2)\sum_{p=1}^{n} v(p)\ \lambda_{ijkl}\ \varepsilon^o_{ij}(p)\ \varepsilon^o_{kl}(p)\ +\ \Delta E \tag{2}$$

Calculation of the relaxation energy ΔE requires solution of the elasticity problem. The elastic energy (2) is a functional of strain field $\varepsilon_{ij}(\mathbf{r})$ at point $\mathbf{r}$. In the approximation of linear elasticity this functional has the form

$$E = (1/2) \sum_{p=1}^{n} v(p)\ \lambda_{ijkl}\ \varepsilon^0{}_{ij}(p)\ \varepsilon^0{}_{kl}(q) \ + \ (1/2)\int f(\varepsilon_{ij})\ d^3r \tag{3}$$

where

$$f(\varepsilon_{ij}) \approx -\ \sigma^0{}_{ij}(\mathbf{r})\ \varepsilon_{ij} + \lambda_{ijkl}\ \varepsilon_{ij}\ \varepsilon_{kl} \tag{4}$$

where $\sigma^0{}_{ij}(\mathbf{r})$ and λ_{ijkl} are the first- and second-order expansion coefficients of the local elastic energy $f(\varepsilon_{ij})$ which are material constants. The unusual term in (4), linear in ε_{ij} , appears because in the system with inclusions the stress-free state is not a non-deformed state. The minimization of eq.(3) with respect to elastic displacements (finding mechanical equilibrium) requires solution of the equation of elasticity, $\partial\sigma_{ij} / \partial x_j = 0$, where $\sigma_{ij}(\mathbf{r})$ is stress at the point $\mathbf{r}=(x_1,x_2,x_3)$. Stress $\sigma_{ij}(\mathbf{r})$ is, by definition , the first variation of (3) with respect to ε_{ij}. Together with (4) , it gives

$$\sigma_{ij}(\mathbf{r}) = \frac{\delta E}{\delta\ \varepsilon_{ij}(\mathbf{r})} = -\ \sigma^0{}_{ij}(\mathbf{r}) + \lambda_{ijkl}\ \varepsilon_{kl}(\mathbf{r}) \tag{5}$$

Let us reveal the physical meaning of the material constants $\sigma^0{}_{ij}(\mathbf{r})$. At the stress-free state, $\sigma_{ij}(\mathbf{r})=0$, we have by definition of the stress-free state

$$\sigma^0{}_{ij}(\mathbf{r}) = \begin{cases} \lambda_{ijkl}\ \varepsilon^0{}_{kl}(p) & \text{if } \mathbf{r} \text{ is inside a particle of the type p} \\ 0 & \text{otherwise} \end{cases}$$

The latter condition can be rewritten in the condensed form

$$\sigma^0{}_{ij}(\mathbf{r}) = \sum_p \lambda_{ijkl}\ \varepsilon^0{}_{kl}(p)\ \tilde{\Theta}(p,\mathbf{r}) \tag{6}$$

where $\tilde{\Theta}(p,\mathbf{r})$ is the shape function of the precipitates of type (p); it is equal to unity if vector $\mathbf{r}$ corresponds to a point within an inclusion of the type p and is 0 otherwise. Introduction of shape function $\tilde{\Theta}(p,\mathbf{r})$ is very convenient because it describes spatial distribution of arbitrary inclusions. With definition (6) eq. (5) is

$$\sigma_{ij}(\mathbf{r}) = -\sum_p \sigma^0{}_{ij}(p)\ \tilde{\Theta}(p,\mathbf{r}) + \lambda_{ijkl}\ \varepsilon_{kl} \tag{7}$$

The elastic equilibrium equation $\partial\sigma_{ij} / \partial x_j = 0$ can then be rewritten in the form

$$\lambda_{ijkl} \frac{\partial^2 u_l}{\partial x_j\ \partial x_k} = \sum_p \sigma^0{}_{ij}(p) \frac{\partial \Theta(p,k)}{\partial x_j} \tag{8}$$

where the strain definition, $\varepsilon_{ij} = (1/2)(\partial u_i/\partial x_j + \partial u_j/\partial x_i)$, was used . Multiplying eq.(3) by the factor exp(-ikr) and integrating over **r** yields

$$\lambda_{ijkl}\, k_j\, k_k\, u(\mathbf{k})_l = -\, i \sum_p \sigma^0_{ij}(p)\, k_j\, \Theta(p,\mathbf{k}) \tag{9a}$$

or in the operator form

$$\hat{G}^{-1}(\mathbf{k})\, \mathbf{v}(\mathbf{k}) = -\, i \sum_p \hat{\sigma}_0(p)\;\; \mathbf{k}\; \Theta(p,\mathbf{k}) \tag{9b}$$

where $(\hat{\mathbf{G}}^{-1}(\mathbf{k}))_{ij} = \lambda_{ijkl}\, k_k\, k_l,\qquad (\hat{\sigma}_0(p))_{ij} = \sigma^0_{ij}(p)$

$$\mathbf{v}(\mathbf{k}) = \iiint_{-\infty}^{\infty} \mathbf{u}(\mathbf{r})\; \exp(-i\mathbf{k}\mathbf{r})\; d^3r$$

$$\Theta(p,\mathbf{k}) = \iiint_{-\infty}^{\infty} \tilde{\Theta}(p,r)\; \exp(-i\mathbf{k}\mathbf{r})\; d^3r \tag{10}$$

In transition from (8) to (9) , the boundary conditions on infinity $\mathbf{u}(\mathbf{r}) \longrightarrow 0$ and $\varepsilon_{ij}(\mathbf{r}) \longrightarrow 0$ at $\mathbf{r} \longrightarrow \infty$, where used.

The solution of eq. (9b) is

$$\mathbf{v}(\mathbf{k}) = -\, i \sum_p \hat{G}(\mathbf{k})\; \hat{\sigma}_0(p)\;\; \mathbf{k}\; \Theta(p,\mathbf{k})$$

or in indices

$$v_i(\mathbf{k}) = -\, i \sum_p G_{ij}(\mathbf{k})\, \sigma^0_{jk}(p)\;\, k_k\;\, \Theta(p,\mathbf{k}) \tag{11}$$

where $G_{ij}(\mathbf{k})=(\hat{\mathbf{G}}(\mathbf{k}))_{ij}$ is the matrix reverse to the matrix $(\hat{\mathbf{G}}(\mathbf{k})^{-1})_{ij} = \lambda_{ijkl}\, k_k\, k_l$, which is, in fact , the Fourier transform of the Green function of elasticity equation. Real displacements, $\mathbf{u}(\mathbf{r})$, can be found by the back Fourier transform

$$u(\mathbf{r})_i = -\, i \sum_p \int G(\mathbf{k})_{ij}\;\; \sigma^0_{jk}(p)\;\, k_k\;\; \Theta(p,\mathbf{k})\;\; \exp(i\mathbf{k}\mathbf{r})\;\; d^3k/(2\pi)^3 \tag{12}$$

Substituting (12) to (3) and integrating over **r** within infinite body results in the Fourier representation of the elastic energy

$$E = (1/2) \sum_p v(p)\, \lambda_{ijkl}\;\, \varepsilon^0_{ij}(p)\; \varepsilon^0_{kl}(p) \tag{13}$$

$$-\; (1/2) \sum_{p,q} \int < \mathbf{k} \mid \hat{\sigma}_0(p)\;\, \hat{G}(\mathbf{k})\;\;\, \hat{\sigma}_0(q) \mid \mathbf{k} > \Theta(p,\mathbf{k})\; \Theta(q,\mathbf{k})^*\; d^3k/(2\pi)^3$$

where $< \mathbf{k} \mid\; \hat{\sigma}_0(p)\; \hat{G}(\mathbf{k})\; \hat{\sigma}_0(q) \;\mid \mathbf{k} > \;=\; k_i\;\; \sigma^0_{ij}(p)\;\, G_{jk}(\mathbf{k})\;\, \sigma^0_{kl}(q)\;\, k_l$

The identities

$$\int \Theta(p,\mathbf{k})\,\Theta(q,\mathbf{k})^*\ d^3k/(2\pi)^3 = v(p)\,\delta_{pq}$$

where δ_{pq} is the Kronecker symbol simplify (13):

$$E = (1/2)\sum_{p,q} \int B(\mathbf{k}/k)_{pq}\ \Theta(p,\mathbf{k})\ \Theta(q,\mathbf{k})^*\ d^3k/(2\pi)^3 \tag{14}$$

where $B(\mathbf{n})_{pq} = \lambda_{ijkl}\,\varepsilon^0{}_{ij}(p)\,\varepsilon^0{}_{kl}(q) - \langle \mathbf{n} \,|\, \hat{\sigma}_0(p)\,\hat{\Omega}(\mathbf{n})\,\hat{\sigma}_0(q)\,|\,\mathbf{n}\rangle$ and $\hat{\Omega}(\mathbf{n})=k^2\,\hat{\mathbf{G}}(\mathbf{k})$. Since the shape function $\tilde{\Theta}(p,\mathbf{r})$ whose Fourier transform enters (14) can be a multiconnected function describing an array of inclusions of the type p eq.(14) may by used for calculation of elastic energy of both, an isolated arbitrary shape particle and groups of particles of different types. Therefore eq. (14) is , in fact, close equation for elastic energy of an arbitrary multiparticle system in an anisotropic matrix in the homogeneous modulus case. This energy is the sum of elastic energies of each isolated particle (self-energy) plus strain-induced pairwise interaction energies between particles.

3. A SINGLE PRECIPITATE IN AN INFINITE BODY

3.1. Elastic energy of a single precipitate

Closed equation for elastic energy of an isolated coherent particle in an infinite crystal body can be obtained from (14) as a particular case. The limit transition to a single particle may be readily done if we assume that the phase transition involves only one type of the crystal lattice rearrangement mode and if the shape function entering eq. (14) describes a simply-connected region enveloping the new phase particle. Then omitting summation over p in (14) we have a simple equation for the elastic energy:

$$E = (1/2)\int B(\mathbf{k}/k)\ |\Theta(\mathbf{k})|^2\, d^3k/(2\pi)^3 \tag{15}$$

where

$$B(\mathbf{n}) = \lambda_{ijkl}\ \varepsilon^0{}_{ij}\ \varepsilon^0{}_{kl} - n_i\ \sigma^0{}_{ij}\ \Omega_{jk}(\mathbf{n})\,\sigma^0{}_{kl}\ n_l \tag{16}$$

$\mathbf{n}=\mathbf{k}/k$, $\varepsilon^0{}_{ij}$ is stress free transformation strain, $\sigma^0{}_{ij} = \lambda_{ijkl}\ \varepsilon^0{}_{kl}$, $\Omega_{jl}(\mathbf{n}) = k^2\,G_{jl}(\mathbf{k})$ is the tensor inverse to $\Omega^{-1}{}_{jl}(\mathbf{n}) = \lambda_{ijkl}\,n_i\,n_k$, $B(\mathbf{n}) \geqslant 0$,

$$|\Theta(\mathbf{k})|^2 = \left|\int \tilde{\Theta}(\mathbf{r})\ \exp(-i\mathbf{kr})\ d^3r\right|^2 = \left|\int \exp(-i\mathbf{kr})\ d^3r\right|^2 \tag{17}$$

where integration is carried out over the particle volume V. Equation (17)yields , in fact, the Laue interference function describing diffraction on the particle. It is noteworthy that the first term in integrand of (15) , $B(\mathbf{n})$, depends on the elastic constants and crystal lattice mismatch only, being a material characteristic, while the second term, $|\Theta(\mathbf{k})|^2$ describes the geometry of inclusion only.

For simple shapes we have the following functions for $|\Theta(\mathbf{k})|^2$:

$$|\Theta(\mathbf{k})|^2 = \left(\frac{\sin(k_x a/2)}{k_x/2}\right)^2 \cdot \left(\frac{\sin(k_y b/2)}{k_y/2}\right)^2 \cdot \left(\frac{\sin(k_z c/2)}{k_z/2}\right)^2 \quad (18a)$$

for a parallelopiped with the edge lengths a, b, c, $\mathbf{k} = (k_x k_y k_z)$,

$$|\Theta(\mathbf{k})|^2 = V^2 \left| 3 \cdot \frac{\sin\Phi(\mathbf{k}) - \Phi(\mathbf{k})\cdot\cos\Phi(\mathbf{k})}{(\Phi(\mathbf{k}))^3} \right|^2 \quad (18b)$$

for an ellipsoid where $\Phi(\mathbf{k})^2 = L_{ij}\, k_i\, k_j$, L_{ij} is the tensor inverse to $L^{-1}{}_{ij}$ that determines the standard form of the ellipsoid surface, $L^{-1}{}_{ij}\, x_i\, x_j = 1$. The eigenvalues of L_{ij} are squares of the ellipsoid semiaxes, a^2, b^2, c^2.

The ellipsoid model is especially interesting since, as was shown by Eshelby [1] for isotropic elasticity and by Valpole [20] and Willis [21] for anisotropic elasticity, elastic strain inside an ellipsoidal inclusion (eigenstrain) is always homogeneous.

Equation (15) contains the Eshelby solution for an ellipsoid in the homogeneous modulus case as a particular case [22]. If the shape function (18b) is used and the limit transition to isotropic elasticity is made,

$$\lambda_{ijkl} \longrightarrow \delta_{ij}\,\delta_{kl}\; 2\mu\; v/(1-2v) \;+\; \mu(\delta_{ik}\,\delta_{jl} + \delta_{il}\,\delta_{jk})$$

where μ is the shear modulus, v is the Poission's ratio, eq.(15) is reduced to the Eshelby solution [1].

3.2. Optimal shape and habit at the low interphase energy limit

This problem can be solved minimizing elastic energy (15) at the fixed value of the precipitate volume V [2]. Since $B(\mathbf{n})$ and $|\Theta(\mathbf{k})|^2$ are always positive

$$E = (1/2)\int B(\mathbf{n})\, |\Theta(\mathbf{k})|^2\, d^3k/(2\pi)^3 \;\geqslant\; (1/2)(\min B(\mathbf{n}))\int |\Theta(\mathbf{k})|^2\, d^3k/(2\pi)^3 \quad (19)$$

where $\min B(\mathbf{n})$ is the minimum value of $B(\mathbf{n})$. With the identity

$$\int |\Theta(\mathbf{k})|^2\, d^3k/(2\pi)^3 = V$$

inequality (19) is

$$E = (1/2)\int B(\mathbf{k}/k)\, |\Theta(\mathbf{k})|^2\, d^3k/(2\pi)^3 \;\geqslant\; 1/2\,(\min B(\mathbf{n}))\, V \quad (20)$$

where the right side of (20) is the lowest possible limit for the elastic energy at a given volume.

Let us introduce the unit vector $\mathbf{n}_0$ providing the minimum of $B(\mathbf{n})$:

$$B(\mathbf{n}_0) = \min B(\mathbf{n}) \quad (21)$$

For an infinitely thin and infinitely extended platelike inclusion with the habit plane normal to $\mathbf{n}_0$, the function $|\Theta(\mathbf{k})|^2$ differs from zero only within infinitely thin and infinitely long rod in $\mathbf{k}$-space along $\mathbf{n}_0$ (this is a well known result from the diffraction theory; diffraction from plate yields rod in reciprocal space normal to the habit plane). In this case the inequality (20) becomes equality. Therefore the minimum elastic energy is attained if an inclusion is "rolled out" into the infinitely thin plate with the habit normal to the vector $\mathbf{n}_0$ minimizing $B(\mathbf{n})$. The elastic energy then is

$$E = E_{bulk} = (1/2)\,(\min B(\mathbf{n}))\,V = (1/2)\,B(\mathbf{n}_0)\,V \tag{22}$$

This strain energy (22) is proportional to the inclusion volume V.

3.3. <u>Bulk energy of an inclusion with invariant plane strain crystal lattice rearrangement</u>

The case of an invariant plane transformation strain plays an especially important role in the theory of phase transformations. For example, the idea of invariant plane strain is basic for the entire crystallographic theory of martensitic transformation which resulted in remarkable achievements in understanding the crystallography of this transformation. The theoretical results obtained above enable us to realize what is the reason behind this.

According to the crystallographic theory, the habit plane of a martensitic crystal is an invariant plane strain. It will be shown below that this directly follows from eq. (22) as a result of the minimization of the elastic energy (15).

The invariant plane strain has always a form of a diadic product

$$u^o_{ij} = \varepsilon_o\, l_i\, n^o_j \tag{23}$$

where $\mathbf{l}$ is a unit vector along the displacement direction, $\mathbf{n}_0$ is a unit vector normal to the invariant plane.

Substituting (23) for ε^o_{ij} to eq. (16) for $B(\mathbf{n})$ yields

$$B(\mathbf{n}) = \varepsilon_o^2 \left[\lambda_{ijkl}\, l_i\, l_k\, n^o_j\, n^o_l - \lambda_{ijsp}\, n^o_i\, n^o_p\, l_s\, \Omega_{jt}(\mathbf{n})\, \lambda_{tmqr}\, n^o_r\, n^o_m\, l_q \right] \tag{24}$$

One may see that at $\mathbf{n} = \mathbf{n}_0$, $B(\mathbf{n}_0) = 0$ since by definition of $\hat{\Omega}^{-1}(\mathbf{n})$

$$\lambda_{tmqr}\, n^o_r\, n^o_m = \Omega^{-1}(\mathbf{n}_0)_{tq}$$

Therefore the bulk energy (21) vanishes. We have the folowing simplification in (24)

$$\Omega_{jt}(\mathbf{n}_0)\, \lambda_{tmqr}\, n^o_r\, n^o_m = \Omega_{jt}(\mathbf{n}_0)\, \Omega^{-1}(\mathbf{n}_0)_{tq} = \delta_{jq}$$

Using the latter in (24) we get $B(\mathbf{n}_0)=0$. We have proved the result that in the case of an invariant plane transformation strain, the minimum of the bulk elastic energy equal to zero is attained when the inclusion is a plate whose habit plane coincides with the invariant plane. This exceptional situation when the choice of the optimal habit plane may eliminate the most substantial volume dependent positive elastic energy makes the case of the invariant plain strain so important. For example, one substantial conclusion can be immediately made:

if any group of new phase coherent precepitates may rearrange itself so that a plate-like aggregate of various orientational variants of the precipitate phase gives the macroscopic shape change described by an invariant plane strain, it will do this to eliminate the volume dependent elastic strain. This conclusion gives us the direction of strain-induced coarsening of such precipitate systems. The typical examples of such systems are tetragonal precipitates in a cubic phase matrix that ultimately form the martensite-type structure with the surface relief and habit plane determined by the conventional crystallographic theory of martensitic transformation.

3.4. Habit plane of tetragonal and hexagonal precipitates

Equation (21) for the habit plane was solved for the thin plate tetragonal plate-like inclusion in a cubic matrix in [12] and for a hexagonal inclusion in [11].

The solution for a tetragonal inclusion gives two types of the habit, (h0l) for the negative elastic anisotropy ($c_{11} - c_{12} - 2c_{44}$) < 0, and (hhl) for the positive anisotropy ($c_{11} - c_{12} - 2c_{44}$) > 0

(i) If ($c_{11} - c_{12} - 2c_{44}$) < 0, the normal to the habit plane, $\mathbf{n}_0$, is $\mathbf{n}_0 = (\sin\theta, 0, \cos\theta)$ where θ is an angle between $\mathbf{n}_0$ and the tetragonal axis given by equation

$$\cos^2\theta = \begin{cases} 0 & \text{if } -\infty < t < -[(c_{11}/c_{12})+1] \text{ and } 1 < t < \infty \\ 1 + \dfrac{c_{11}+2c_{12}}{c_{11}+c_{12}} \dfrac{t}{1-t} & \text{if } -[(c_{11}/c_{12})+1] < t < 0 \\ 1 & \text{if } 0 \leqslant t < 1 \end{cases} \tag{25a}$$

where $t = \varepsilon^0_{11}/\varepsilon^0_{33}$, ε^0_{11} and ε^0_{33} are non-zero components of the stress free transformation strain ε^0_{ij} (all other components are zero).

(ii) If ($c_{11} - c_{12} - 2c_{44}$) > 0, the normal to the habit plane is $\mathbf{n}_0 = \left((1/\sqrt{2})\sin\theta, (1/\sqrt{2}\sin\theta), \cos\theta\right)$ where

$$\cos^2\theta = \begin{cases} 0 & \text{if } -\infty < t < t_1 \text{ and } t_3 < t < \infty \\ 1 - 2\dfrac{(\xi+2)(c_{11}+2c_{12})\,t}{\xi(c_{11}+2c_{12})(2t-1)+4(c_{11}+c_{12})(t-1)} & \text{if } t_1 < t < 0 \\ 1 & \text{if } 0 \leqslant t \leqslant t_2 \\ \xi\dfrac{(c_{11}+2c_{12})+4c_{11}(1-t)}{\xi(c_{11}+2c_{12})(1+2t)} & \text{if } t_2 < t < t_3 \end{cases} \tag{25b}$$

where $t_1 = -(c_{11}/c_{12}) - 1 - \xi\dfrac{c_{11}+2c_{12}}{4c_{12}}$, $\quad t_2 = \dfrac{2c_{11}}{2c_1 + \xi(c_{11}+2c_{12})}$

$$t_3 = 1 + \xi\frac{c_{11}+2c_{12}}{4c_{11}}, \quad t_1 < t_2 < t_3, \quad \xi = \frac{c_{11}-c_{12}-2c_{44}}{c_{44}}$$

It follows from eq. (25a) and (25b) that for a cubic precipitate in a cubic matrix (when t=1) the habit plane is (001) if $(c_{11} - c_{12} - 2 c_{44}) < 0$ and is (111) if $(c_{11} - c_{12} - 2 c_{44}) > 0$.

Similar calculations were made by Mayo and Tsakalakos for precipitates of orthorhombic and hexagonal phase [11]. They derived explicit analytical expression for B(**n**) in terms of crystal lattice misfit and elastic constants of the hexagonal phase. Minimizing this equation with respect to **n** they were able to predict the habit of precipitates. This approach was applied to Al-Mg-Zn alloy with Zn/Mg ration between 2.5 and 7 , and the total Zn content less than 20 wt pct. The predicted $\{111\}_{fcc}$ habit plane of the η' phase in the fcc Al- based matrix is in agreement with electron microscopic observations [23, 24].

3.5. <u>Elastic energy of finite thin plate inclusions</u>

Finite thickness of a precipitate with the optimal habit normal to $\mathbf{n}_0$ should result in a positive correction to eq. (22), ΔE_{edge} . Mathematically it is associated with the fact that the rod in **k**-space where $|\Theta (\mathbf{k})|^2$ does not vanish for a finite thickness platelet has finite thickness and finite length. They are of the order of magnitude $2\pi/L$ and $2\pi/D$, respectively, where L and D are typical length and thickness of the plate-like precipitate. In this case, the energy correction to (22) is positive because integration over **k** in (15) is carried out over **k**-space region where B(**k**/k) does not assume its minimum value $B(\mathbf{n}_0)$. Therefore the correction ΔE_{edge} is of the order of $\Delta E_{edge}/E_{plate} \sim (D/L)$, i. e.,

$$\Delta E_{edge} \sim E_{plate} (D/L) = \lambda \varepsilon_0^2 (DL^2) (D/L) \sim \lambda (\varepsilon_0 D)^2 L \sim \lambda (\varepsilon_0 D)^2 P$$

where P is the platelet parameter. The physical meaning of the corection is quite clear. It is caused by the crystal lattice mismatch on the edges of the inclusion along its perimeter. The energy correction ΔE_{edge} can be interpreted as "string" energy with the line tensions $\sim \lambda (\varepsilon_0 D)^2$. In fact, this energy can be attributed to a dislocation loop with the Burgers vector $b = \sim \varepsilon_0 D$ enveloping the precipitate in its habit plane. Accurate calculations of the energy ΔE_{edge} for a tetragonal precipitate in the cubic matrix gives

$$\Delta E_{edge} = \beta (D^2/4\pi) \ln(L/D) P$$

where

$$\beta = \frac{(2 c_{12} \varepsilon^0_{11} + c_{11} \varepsilon^0_{33})^2}{c_{11}} \left[\xi \cdot \frac{c_{11}+c_{12}}{c_{11}} - \frac{c_{11}(1+\alpha_1^2) - 2 c_{12} \alpha_1 - 2(\alpha_1+1)}{c_{44}} \right]$$

ε^0_{33} and ε^0_{11} are crystal lattice mismatch along the tetragonalily axis and in the (001) plane, respectively, $\xi = (c_{11} - c_{12} - 2c_{44})/c_{44}$. is the anisotropy parameter [20]. In the case of a cubic precipitate in a cubic matrix $\varepsilon^0_{11} = \varepsilon^0_{33} = \varepsilon_0$,

$$\alpha_1 = \frac{(c_{11}+c_{12}) \varepsilon^0_{11} + c_{12} \varepsilon^0_{33}}{2 c_{12} \varepsilon^0_{11} + c_{11} \varepsilon^0_{11}} .$$

Then

$$\beta = -\frac{(c_{11} + 2\,c_{12})^2\ \varepsilon_0^2\ \xi\ (c_{11} - c_{12})}{c_{11}^2}$$

It was shown that in a general case of arbitrary symmetry phase the edge energy of a plate-like precipitate is

$$\Delta E_{edge} = (D^2/4\pi)\ \ln L/D \int \left[\beta_1 \frac{\left(\frac{dy}{dx}\right)^2}{1+\left(\frac{dy}{dx}\right)^2} + \beta_2 \frac{1}{1+\left(\frac{dy}{dx}\right)^2} \right] dl$$

where integration is taken over the contour $y = y(x)$ enveloping the precipitate in the habit plane, dl is the contour length element , β_1 and β_2 are second order expansion coefficinets with respect to $\mathbf{n}\text{-}\mathbf{n}_0$. The minimization of the ΔE_{edge} energy with respect to the shape in the habit plane described under the additional condition of conservation of its area in the habit plane, $S = \int y(x)dx$, gives the Lagrange equation for $y = y(x)$. The solution of this equation results in the forms presented in Fig.1.

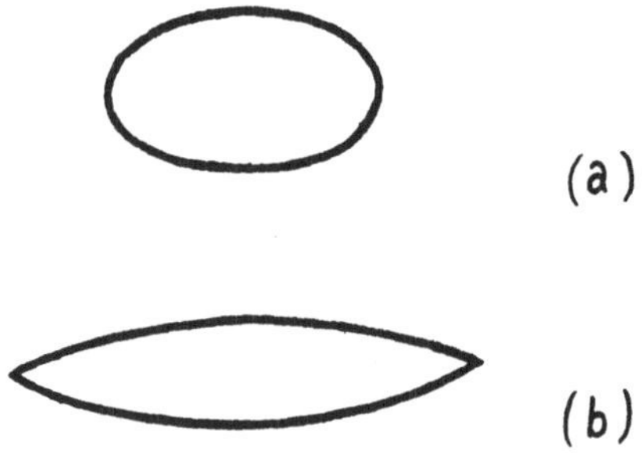

FIGURE 1. Calculated optimal shape of the platelet particle in the habit plane.
a) Slightly anisotropic case (oval shape). b) Strongly anisotropic case (lense shape).

3.6. Crystal lattice parameters and crystal lattice rotation in constraint plate-like inclusion

In the case of a single inclusion eq. (12) for elastic displacements yields

$$\mathbf{u}(\mathbf{r}) = -i \int \hat{\mathbf{G}}(\mathbf{k})\ \hat{\sigma}_0\ \mathbf{k}\ \Theta(\mathbf{k})\ \exp(i\mathbf{kr})\ d^3k/(2\pi)^3 \tag{26}$$

The coordinate derivative $\partial u_i/\partial x_j$ gives the distortion tensor, $u_{ij}(\mathbf{r})$,

$$u_{ij}(\mathbf{r}) = \partial u_i/\partial x_j = \int (\hat{\mathbf{G}}(\mathbf{k})\ \hat{\sigma}_0\ \mathbf{k})_i\ k_j\ \Theta(\mathbf{k})\ \exp(i\mathbf{kr})\ d^3k/(2\pi)^3$$

Since by definition of the Green function $\hat{\mathbf{G}}(\mathbf{k})$,

$$\hat{\mathbf{G}}(\mathbf{k}) = k^{-2}\,\hat{\Omega}(\mathbf{n}) \qquad \text{where } \mathbf{n} = \mathbf{k}/k$$

$$u_{ij}(\mathbf{r}) = \int n_j \left(\hat{\Omega}(\mathbf{n})\, \hat{\sigma}_0\ \mathbf{n} \right)_i \Theta(\mathbf{k})\ \exp(i\mathbf{kr})\ d^3k/(2\pi)^3 \tag{27}$$

In the case of plate-like inclusion eq.(27) is substantually simplified because the Fourier transform of its shape functions , $\Theta(\mathbf{k})$, does not vanish only within the thin and extended rod in $\mathbf{k}$-space emerging from the origin, $\mathbf{k}=0$, along the direction $\mathbf{n}_0$ normal to the habit. Then

$$u_{ij}(\mathbf{r}) \approx n^0_j \left(\hat{\Omega}(\mathbf{n}_0)\, \hat{\sigma}_0\ \mathbf{n}_0 \right)_i \int \Theta(\mathbf{k})\ \exp(i\mathbf{kr})\ d^3k/(2\pi)^3 \tag{28}$$

with accuracy of the ratio $D/L<<1$. The integral in the right hand part of (28) is the back Fourier transform of $\Theta(\mathbf{k})$ and therefore, by definition, is equal to $\tilde{\Theta}(\mathbf{r})$. Taking the latter into account we have

$$u_{ij}(\mathbf{r}) = S(\mathbf{n}_0)_i\, n^0_j\, \tilde{\Theta}(\mathbf{r}) = \begin{cases} u_{ij}^* = S(\mathbf{n}_0)_i\, n^0_j & \text{if } \mathbf{r} \text{ is inside the inclusion} \\ 0 & \text{otherwise} \end{cases} \tag{29a}$$

where

$$S(\mathbf{n}_0) = \hat{\Omega}(\mathbf{n}_0)\, \hat{\sigma}_0\, \mathbf{n}_0 \tag{29b}$$

For the cubic $\longrightarrow$ cubic phase transformation $\sigma^0_{ij} = (c_{11} + 2\,c_{12})\, \varepsilon_0\, \delta_{ij}$. It has been shown in section 3.4 that if $(c_{11} - c_{12} - 2\,c_{44}) < 0$, the habit is (001) and thus $\hat{\sigma}_0\, \mathbf{n}_0 = (c_{11} + 2\,c_{12})\, \varepsilon_0\, \mathbf{n}_0$. On the other hand $\hat{\Omega}(\mathbf{n}_0)\, \mathbf{n}_0 = (1/c_{11})\, \mathbf{n}_0$ if $\mathbf{n}_0 =(001)$. Using these relations gives $\mathbf{S}(\mathbf{n}_0) = \hat{\Omega}(\mathbf{n}_0)\, \hat{\sigma}_0\, \mathbf{n}_0 = \varepsilon_0\, \mathbf{n}_0\, (c_{11}+2\,c_{12})/c_{11}$. Then the eigenstrain u_{ij}^* in eq. (29) can be rewritten as

$$u_{ij}^* = S(\mathbf{n}_0)_i\, n^0_j = \varepsilon_0\, n^0_i\, n^0_j\, (c_{11} + 2\,c_{12})/c_{11}$$

$$= \varepsilon_0 \left((c_{11} + 2\,c_{12})/c_{11}\right) \begin{pmatrix} 000 \\ 000 \\ 001 \end{pmatrix} \tag{30}$$

Therefore a constraint coherent (001) platelet precipitate of a cubic phase has always a strain-induced tetragonality described by (30), the axial ration being $(c/a) = 1+\ \varepsilon_0\, (c_{11} + 2\,c_{12})/c_{11}$. If $(c_{11} - c_{12} - 2\,c_{44})> 0$, $\mathbf{n}_0 = 1/\sqrt{3}\ (111)$ and the similar calculation yields

$$u_{ij}^* = 1/3\ \frac{c_{11} + 2\,c_{12}}{c_{11} + 2\,c_{12} + 4\,c_{44}} \cdot \varepsilon_0 \begin{pmatrix} 111 \\ 111 \\ 111 \end{pmatrix}$$

Therefore a constraint coherent (111) platelet of a cubic phase has always strain-induced rhombohedricity.

Equation (29) leads us to the following important conclusions:

1. The distortion within a platelike inclusion, u_{ij}^*, is homogeneous. Distortion outside the inclusion in the matrix asymptotically vanishes when $D/L \longrightarrow 0$.

2. Since almost all elastic strain is concentrated within the platelet, the total elastic energy is not sensitive to the elastic moduli of the matrix. Therefore eq. (22) for elastic energy of a platelet derived in the homogeneous modulus case is nevertheless asymptotically correct also in the heterogeneous modulus case if the elastic moduli λ_{ijkl} in (22) are substituted by the elastic moduli of the precipitate.

3. The total distortion within a constraint platelet which transforms the matrix lattice to the constraint precipitate lattice is always an invariant plane strain, the invariant plain normal to $\mathbf{n}_0$ coinciding with the habit plane. Therefore any crystal lattice translation in the habit plane of a constraint precipitate exactly coincides with the corresponding translation in the matrix plane parallel to the habit.

4. The crystal lattice rotation caused by fitting two different lattices along the habit plane is described by asymmetric part of the distortion tensor u_{ij}^*:

$$\phi_{ij}^* = (1/2)\,[\,u_{ij}^* - u_{ji}^*] = (1/2)\,[\,S(\mathbf{n}_0)_i\, n^0_j - n^0_i\, S(\mathbf{n}_0)_j\,]$$

or by the rotation vector

$$\phi^* = 1/2\,\big(\,\mathbf{S}(\mathbf{n}_0)\times \mathbf{n}_0\big)$$

The direction of ϕ^* is the rotation axis direction, the absolute value of ϕ^* is the rotation angle.

3.7. Needle-like precipitates

As was shown above, the elastic energy assumes its minimum for a platelike precipitate whose habit is normal to the vector $\mathbf{n}_0$ minimizing $B(\mathbf{n})$. Mayo and Tsakalakos [11] have shown that this is not always the case . Needle-like precipitates may be more stable if the minimum of the function $B(\mathbf{n})$ is degenerate with respect to $\mathbf{n}$ lying in a plane. Such a situation may be expected if $B(\mathbf{n})$ has the cylindrical symmetry with respect to a certain direction.

If $B(\mathbf{n})$ has a cylindrical symmetry with respect to an axis directed along the direction $\mathbf{e}$, the function $B(\mathbf{n})$ depends on the scalar product $(\mathbf{ne})$: $B(\mathbf{n}) = B(\mathbf{ne})$. In the case of interest when $B(\mathbf{n})$ assumes its absolute minimum at $\mathbf{n}$ normal to the symmetry axis (at $(\mathbf{ne}) = 0$), one may expand $B(\mathbf{n})$ in a power series of $(\mathbf{ne})$:

$$B(\mathbf{n}) = \min B(\mathbf{n}) + \beta\,(\mathbf{ne})^2 + \ldots \qquad (31)$$

substituting (31) to (15) and using the identity

$$\int |\,\Theta(\mathbf{k})\,|^2\, d^3k/(2\pi)^3 = V$$

yields

$$E \cong (1/2) \min B(\mathbf{n})\cdot V + (1/2)\,\beta \int (\mathbf{ne})^2\,|\,\Theta(\mathbf{k})\,|^2\, d^3k/(2\pi)^3 \qquad (32)$$

The energy (32) is minimized for a needle along the direction $\mathbf{e}$ because the function

$|\Theta(\mathbf{n})|^2$ for a needle does not vanish within thin extended plate in the reciprocal space normal to the needle axis $\mathbf{e}$ where $(\mathbf{ne}) = 0$. Estimation of the integral in (32) for a needle along the axis $\mathbf{e}$ describing strain concentration near the needle tip yields [22]

$$\Delta E_{edge} = (1/2)\ \beta \int (\mathbf{ne})^2\ |\Theta(\mathbf{k})|^2\ d^3k/(2\pi)^3 \approx (1/2)\ \beta\ (4/3)\ \pi\ R_0^3$$

where $\beta \sim \lambda\, \varepsilon_0^2$, R_0 is a needle radius.

For a thin and long needle the correction to the bulk energy, ΔE_{edge}, is much smaller than the corresponding correction $\Delta E_{edge} \sim \lambda\, \varepsilon_0^2\, D^2 L$ for a thin plate. The bulk energy term (22) for a plate is the same as for a needle but the energy corrections ΔE_{edge} for a needle is much smaller than that for a plate (the first is proportional to R_0^3 while the second to D^2L). The calculations similar to that for a plate-like precipitate [22] give for a needle the following result:

1. Strain inside a coherent needle-like precipitate parallel to the direction $\mathbf{e}$ is homogeneous and has the form

$$\varepsilon_{ij}^* = (1/2)\ \alpha\ (\delta_{ij} - \varepsilon_i\, \varepsilon_j)$$

For a particular case of a tetragonal elastically isotropic precipitate $\alpha = (\varepsilon^0{}_{11} + \nu\, \varepsilon^0{}_{33})/(1 + \nu)$, where ν is the Poison ratio, $\varepsilon^0{}_{33}$ and $\varepsilon^0{}_{11}$ are the tetragonal stress-free strains along and perpendicular to the tetragonal axis.

2. Crystal lattice translations of constraint needle-like precipitate exactly coincide with the corresponding translations of the matrix phase. The latter can be seen from the equation

$$r'_i = [\,\delta_{ij} + \alpha\ (\delta_{ij} - e_i\, e_j)\,]\ r_0\, e_j = r_0\, e_i$$

which shows that the length $r_0\mathbf{e}$ along the direction $\mathbf{e}$ does not change.

Now a few words concerning cylindrical degeneration of $B(\mathbf{n})$. Analysis of equation (16) for the function $B(\mathbf{n})$ shows that the function $B(\mathbf{n})$ may be cylindrically degenerate with respect to $\mathbf{n}$ in the cases of the cubic-tetragonal phase transitions in an alloy based on almost isotropic cubic solvent (Al, Nb, Mo, W and so on) as well as for a cubic-hexagonal, cubic-trigonal and hexagonal-hexagonal phase transitions in anisotropic alloys. However, the cylindrical degeneracy is not sufficient for a needle to be formed. It occurs when the minimum of $B(\mathbf{n})$ is degenerate with respect to any $\mathbf{n}$ belonging to the plane normal to the cylinder axis. This puts a certain constraint on the transformation strain and elastic anisotropy.

Concluding this section two important points should be emphasized:

1. Formation of needles can be expected not only in the case of the cylindrical degeneracy of the function $B(\mathbf{n})$. Needles can also be stabilized by the interphase energy. Balance between elastic and interphase energy may produce the preferential needle-like shapes as an intermediate form during cooling a precipitate from a spheroid to platelet.

2. The function $B(\mathbf{n})$ is the Fourier transform of interaction energy of two precipitates in the long-distance limit. In the case of the cylindrical degeneracy when $B(\mathbf{n}) = \min B(\mathbf{n}) + \beta\ (\mathbf{ke})^2/k^2$, the back Fourier transform gives this interaction $V(\mathbf{r})$ in the form of the dipole-dipole interaction

$$V(r) = e_i e_j \beta \left(\delta_{ij}/r^3 - 3 (r_i r_j/r^5) \right) = \beta \left(1/r^3 - 3 (re)^2/ r^5\right)$$

where the coefficient β plays the role of the dipole moment magnitude. This fact will be discussed later in connection with the analogy between elastic strain and magnetostatic energy of magnets and electrostatic energy of ferroelectrics.

4. SHAPE TRANSFORMATIONS OF A CUBIC PHASE PRECIPITATE IN A CUBIC MATRIX UPON COARSENING

As was mentioned above, the equilibrium shape of a precipitate is a result of competition between the elastic and interphase energies. Equation (15) enables one to evaluate the elastic energy of an arbitrary shape precipitate while the interphase energy may be assumed to be equal to the product of surface tension coefficient, γ, and interphase area, S, if the interphase tension is isotropic, i. e.

$$E = E_{elast} + E_s \tag{33}$$

where

$$E_s = \gamma S$$

is the interphase energy, and the elastic energy E_{elast} is given by eq. (15). Let us consider the cubic-to-cubic phase transformation in the case of the negative elastic anisotropy, $(c_{11} - c_{12} - 2 c_{44}) < 0$.

Then integration in (15) yields the elastic energy in the form

$$E_{elast} \cong E_0 V + E_1 \eta V \tag{34}$$

where $E_0 = \varepsilon_0^2 (c_{11} + 2 c_{12}) (c_{11} - c_{12})/c_{11}$, E_0V is the elastic energy of an infinite thin plate of the volume V with the optimal (001) habit, and

$$E_1 = \frac{(c_{11}+ 2 c_{12})^2}{2} \cdot \frac{(2 c_{44} + c_{12} - c_{11})}{c_{11}(c_{11}+c_{12}+2 c_{44})} \varepsilon_0^2$$

The dimensionless coefficient η depends on the shape of the precipitate rather than its volume. The values of the coefficients η in eq. (34) for different shapes are given in Table 1.

TABLE 1

Shape	η
Sphere	0.7087
Cube	0.5580
Needle	0.4692
Thin Plate	0.0

The total energy (33), elastic and interphase, is thus,

$$E = E_0 V + E_1 V \eta + \gamma S$$

or in the reduced form

$$[(E - E_0 V)/(E_1 V)] = \eta + \frac{\gamma S}{E_1 V} = \eta + \frac{r_0}{a} \quad (35)$$

where

$$r_0 = \gamma / E_1 \quad (36)$$

is the material constant with dimension of length, volume-to-surface ratio, $a = V/S$, is a typical particle size, characterizing its degree of coarsening.

Comparing the reduced energy (35) for the various shapes versus the ratio r_0/a characterizing the degree of coarsening, the critical transition size from one shape to another can be found.The size a is related to the particle volume V by the following relations:

for a sphere	$a = (1/3)\,(3/4\pi)^{1/3}\,V^{1/3}$	$\approx 0.206\ V^{1/3}$
for a cube	$a = (1/6)\,V^{1/3}$	$\approx 0.1666\ V^{1/3}$
for an octahedron	$a = 2^{-2/3}\,3^{-7/6}\,V^{1/3}$	$\approx 0.1748\ V^{1/3}$
for a tetrahedron	$a = 2^{-1}\,3^{-7/6}\,V^{1/3}$	$\approx 0.1388\ V^{1/3}$

Therefore comparing the reduced energies (35) for a spheroid,

$$(E - E_0 V)/(E_1 V) = 0.7087 + 4.854\, r_0 / V^{1/3}$$

and for a cuboid

$$(E - E_0 V)/(E_1 V) = 0.5580 + 6\, r_0 / V^{1/3}$$

one can see that a spheroidal inclusion is more stable at $V^{1/3} \leqslant 7.64\, r_0$ and a cuboid precipitate becomes more stable upon coarsening when $V^{1/3} \gg 7.64\, r_0$.

In the case of positive anisotropy, $(c_{11} - c_{12} - 2c_{44}) > 0$, $B(\mathbf{n})$ assumes minimum value at $\mathbf{n} = \mathbf{n}_0 = (1/\sqrt{3}\,,\ 1/\sqrt{3}\,,\ 1/\sqrt{3})$ and thus the lowest elastic energy is attained for a platelet with the (111) habit. We can also assume that an optimal polihedron which is formed from a spheroid during its coarsening should be faceted by the optimal (111) planes only. Such a polihedron is either octahedron or tetrahedron. Elastic energy of an octahedron given by integrating eq. (15) was calculated by Tsakalakos.

5. APPLICATIONS

5.1. GP zones in Al-based alloys

X-ray and electron microscopic studies have shown that aging of some supersaturated alloys results in the formation of so-calley GP zones, small segregations of atoms that later develop into metastable or stable precipitate phase. A GP zone may be either equiaxial (Cu-Co, Al-Zn, Al-Ag etc.) or platelike (Al-Cu, Cu-Be) shapes.

Formation of GP zones can be well understood if they are regarded as new phase precipitates which are formed as a result of isomorphic decomposition occuring according to metastable diagrams with miscibility gap. If the initial solid solution has the cubic symmetry, such a decomposition results in atomic redistribution over crystal lattice sites of the solution and formation of cubic phase precipitates enriched by solute atoms within the cubic phase matrix. Then theoretical results formulated above may be applied to predict morphology and the structure of GP zones.

Since both phases have the same crystal lattice but different composition, the stress-free transformation strain is a pure dilatation :

$$\varepsilon^{o}_{ij} = \varepsilon_o \delta_{ij} = \left(\frac{da}{adc} \right) (c_p - c_m) \delta_{ij} \tag{37}$$

where $\frac{da}{adc}$ is the concentration coefficient of the crystal lattice expansion, c_p and c_m are atomic fractions of solute atoms in the precipitate and matrix, respectively.

5.1.1. GP zones in Al-Cu alloys. In Al-Cu alloys crystal lattice mismatch is very big. The concentration coefficient of the crystal lattice expansion is about 10% :

$$\frac{da}{adc_{Cu}} = -0.091$$

For such a large mismatch the plate-like morphology should be expected. Since $(c_{11} - c_{12} - 2c_{44}) < 0$ for Al $(c_{11} = 1.068 \cdot 10^{12},\ c_{12} = 0.607 \cdot 10^{12},\ c_{44} = 0.282 \cdot 10^{12}\ dn/cm^2)$, the minimum of $B(\mathbf{n})$ falls on the vector $\mathbf{n}_o = (001)$. This means that coherent precipitates should have {100} habit.

According to Gerold [25] a GP zone in Al-Cu alloy is a sole plane (001) of Cu atoms. This habit is in accordance with the above theoretical predictions. Recent electron microscopic observations seems to confirm this.

Let us estimate the distance between the Cu filled (001) plane and the nearest (001) planes. According to (29) and (30) for a plate-like precipitate of a cubic phase in the cubic matrix with the (001) habit,

$$\hat{\sigma}_o \mathbf{n}_o = (c_{11} + 2c_{12}) \varepsilon_o \mathbf{n}_o = (c_{11} + 2c_{12}) \left(\frac{da}{adc} \right) (c_p - c_m) \mathbf{n}_o$$

$$\hat{\Omega}(\mathbf{n}_o) = (1/c_{11}) \mathbf{n}_o$$

Therefore

$$S(\mathbf{n}_o) = \hat{\Omega}(\mathbf{n}_o) \hat{\sigma}_o \mathbf{n}_o = \frac{c_{11} + 2c_{12}}{c_{11}} \frac{da}{adc} (c_p - c_m) \mathbf{n}_o$$

and the transformation strain (30) within the constraint (001) plate-like precipitate is

$$u_{ij}^{*} = S(\mathbf{n}_o)_i n^o_j = \frac{c_{11} + 2c_{12}}{c_{11}} \frac{da}{adc} (c_p - c_m) n^o_i n^o_j =$$

$$= \frac{c_{11} + 2c_{12}}{c_{11}} \frac{da}{adc} (c_p - c_m) \begin{pmatrix} 000 \\ 000 \\ 001 \end{pmatrix} \tag{38}$$

It follows from (38) that the total strain within a constraint platelet precipitate with the (001) habit transforms its cubic lattice into a tetragonal one. This is the stress-induced tetragonality. The displacement of Al (001) plane nearest to the Cu (001) plane toward the Cu (001) plane produced by eignestrain u_{ij}^{*} given by (38) is

$$u = u_{33}^{*} a_{Al}/2 = (c_{11} + 2c_{12})/c_{11} \frac{da}{adc_{Cu}} (c_p - c_m) a_{Al}/2 \tag{39}$$

where a_{Al} is the Al crystal lattice parameter and $a_{Al}/2$ is the interplanar distance for a (001) plane.

The GP zone may be regarded as a plate-like precipitate whose thickness is equal to twice the interplanar distances (D = a). Half crystal lattice sites of such a precipitate are filled by Cu atoms and the other half by Al atoms. Therefore, we may assume that c_p = 1/2. Since matrix does not have Cu atoms, c_m=0. Using the latter in eq. (38) together with $\frac{da}{adc} \approx -0.091$ and Al lattice elastic constants we have

$$u = \frac{c_{11} + 2c_{12}}{c_{11}} \frac{da}{adc} (1/2) \frac{a_{Al}}{2} = \frac{1.068 + 2 \cdot 0.607}{1.068} (-0.091) \frac{4.041}{4} \approx -0.196 \text{ A}^o$$

The best fit between calculated and observed X-ray diffuse scattering has been obtained when displacement of the Al (001) plane toward the nearest (001) Cu plane is $u \approx -0.2$ A° [26]. The theoretically predicted value $u \approx -0.196$ A° is in the excellent agreement with that. It should also be mentioned that the calculation based on the crystal lattice static theory [22] gives the same equation for displacement if the phonon spectrum dispersion is neglected.

The matrix u_{ij}^* given by (38) also predicts that the crystal lattice parameteres of GP zone in (001) plane are exactly the same as in Al matrix.

5.1.2. <u>θ"-phase</u>. Aging above 100° C results in dissolution of GP zones and appearance of platelets of θ" metastable phase that is formed by alteration of Cu and Al (001) planes in the fcc lattice: Cu Al Al Al Cu Al Al Al....Therefore atomic fraction of Cu in θ" phase is c_p=1/4. The θ" phase formed due to such a sequence is a tetragonal phase with $c \approx 2a_{Al}$ and $a \approx a_{Al}$. The θ" phase being an ordered fcc-based superstructure enriched by Cu atoms has misfit described by eq. (38) and therefore should also be formed as plate-like precipitates with (100) habits.

Let us calculate the phase crystal lattice parameters using eq. (38). It predicts

$$c = (1 + u_{33}^*)\, 2a_{Al} = \left[1 + \frac{c_{11} + 2c_{12}}{c_{11}} \cdot \frac{da}{adc} \cdot (1/4)\right] 2 \cdot 4.041 =$$

$$= \left[1 + \frac{1.068 + 2 \cdot 0.607 \; (-0.091)\; 1/4}{1.068}\right] 2 \cdot 4.041 = 7.69 \text{ A}^o$$

Since the parameters of the constraint θ" phase in the (001) habit plane should be exactly the same as in the matrix we predict

$$a = a_{Al} = 4.041 \text{A}^o$$

The calculated values c = 7.69A° and a = 4.041A° are perfectly matched to the observed crystal lattice parameters c = 7.7A° and a = 4.04A°. Therefore both the habit plane orientation and crystal lattice paramaters of θ" fit very well the theoretical predictions.

5.1.3. <u>θ' phase.</u> The intermediate tetragonal θ' phase that succeeds the θ" phase in the course of aging has the fcc-faced lattice and is described by formula AlCu☐, where ☐

designates vacancy. The presence of built-in vacancies in the fcc lattice of the θ' phase introduces the additional contraction to the stress-free transformation strain. The theory predicts that θ' phase precipitates should also be platelets with the {001} habit. This conclusion is in agreement with electron microscopic observation.

Since crystal lattice mismatch for θ' phase in the Al matrix cannot be determined at the moment (we only know that its dilatational part is much larger than for the θ"phase), we cannot calculate parameter c using the same equation as with the θ" phase. However, the theory predicts that the crystal lattice parameter a which is situated in the habit plane must coincide exactly with the crystal lattice parameter of pure Al. Since $a = a_{\theta'}$, it means that $a_{\theta'} = a = a_{Al} = 4.041 A^o$. This prediction is also in excellent agreement with the observed results:

$$a_{\theta'} = 4.04\ A^o\ , \qquad c_{\theta'} \geqslant 5.8\ A^o$$

5.1.4. GP zones with small crystal lattice mismatch. GP zones were also observed in Cu-Co, Al-Zn, Al-Ag alloys. The difference between atomic diameters of solute and solvent atoms for them is less then about 3%. Since the GP zone volume is small , the theory predicts spherical shape of precipitates (see the end of section 4). This is in agreement with X-ray and electron microscopic observations. It is of interest to note that estimations of the elastic moduli of the precipitate phase in Al-Zn alloys gives $(c_{11}-c_{12} - 2\,c_{44}) > 0$ [27]. In this situation the theory developed above predicts spherical shape in the early stage of aging which should be transformed into octahedron (or tetrahedron) and later into {111} platelets.

5.1.5. GP zones when precipitate phase is hexagonal. Calculations by Mayo and Tsakalakos for GP zones and metastable η' hexagonal phase give the $\{111\}_{Al}$ habit [11] which is in agreement with electron microscopic observation. Precipitates of the η' metastable hexagonal phase in Al-Zn-Mg alloys give one more confirmation that a coherent plate-like precipitate and matrix have exactly the same crystal lattice parameters in the habit plane. Crystal lattice parameters of constraint hexagonal η' phase are

$$a_{\eta'} = 4.96 A^o\ , \qquad c_{\eta'} = 8.68 A^o$$

The Al-based matrix has the parameter $a_o = 4.054 A^o$. The $[\bar{1}/2\ \bar{1}/2\ 1]_{Al}$ and $[\bar{1}/2\ 1\ \bar{1}/2]_{Al}$ translations of the fcc matrix lying in the (111) plane which are transformed into the parameter $a_{\eta'}$ of the η' phase are equal to

$$T\left([\bar{1}/2\ \ \bar{1}/2\ \ 1] \right) = T\left([\bar{1}/2\ \ 1\ \ \bar{1}/2] \right) = a_o \sqrt{3/2} = 4.054 \sqrt{3/2} = 4.965 A^o$$

This value with acuracy of X-ray measurements coincides with the value $a = 4.96 A^o$ observed for η' phase and $a = 4.054 A^o$ for the cubic Al based matrix.

5.2. Precipitation of nitrides in Fe-N alloys

Elastic strain theory formulated above can be applied to determine morphology and crystal lattice correspondence of nitride precipitates in Fe-N martensite [9,22].

5.2.1. Precipitates of α" phase ($Fe_{16}N_2$) in Fe-N martensite. The decomposition reaction that occurs in tempered bct Fe-N martensite leads to the formation of ordered bcc-based tetragonal nitride, $Fe_{16}N_2$ (α") in the bcc α Fe matrix which later transforms into fcc-based γ' phase (Fe_4N). According to Jack [28], α" phase is a tetragonal phase with

$$a(\alpha'') = 2\,a_{Fe} = 2\cdot 2.86 = 5.72A^{o}$$
$$c(\alpha'') = 6.292A^{o} \sim 2\,a_{Fe}$$

The spacing (α'') is exactly equal to twice the crystal lattice parameters of the α Fe. This coincidence cannot be accidental. It would be explained if α'' phase precipitates are coherent platelets with the (001) habit. Then the theory predicts that parameter $a(\alpha'')$ situated in the habit plane (001) should be exactly equal to the corresponding parameter $2a_{Fe}$ of the (001) matrix plane. In other words, the crystal lattice parameters observed by Jack are the parameters of constraint precipitate. This conclusion which directly follows from the theory was proved by the crystal lattice parameter measurements of the single-phase ordered α'' phase solid solution which, by definition ,is stress-free . The measured crystal lattice parameters of Fe-8.56 at% Ni single phase alloy are:

$$a(\alpha'') = 5.692\ A^{o}$$
$$c(\alpha'') = 6.180\ A^{o} \qquad [29]$$

For this alloy $a(\alpha'') \neq 2a_{Fe}$ which proves that coincidence of $a(\alpha'')$ with $2a_{Fe}$ observed by Jack is a result of constraint. For the stoichiometric alloy the stress-free strain is $\varepsilon^{o}_{11} = -\,0.006537$, $\varepsilon^{o}_{33} = 0.107397$, $t = -\,0.006537/0.107397 = -\,0.0608$. These values enables one to calculate the crystal lattice parameters of the constraint (001) precipitate using eq. (29). The calculation gives

$$a(\alpha'') = 2\,a_{Fe} = 2\cdot 2.86 = 5.720\ A^{o}$$
$$c(\alpha'') = 6.289\ A^{o}$$

which is in excellent agreement with the Jack observations. Making use of the elastic constants of α Fe and value $t = -\,0.0608$ in eq. (25a) yields the vector $\mathbf{n}_0$ minimizing $B(\mathbf{n})$ in the form

$$\mathbf{n}_o = (\sin\theta,\ 0,\ \cos\theta) \quad \text{where } \theta = 16.3^{o} \quad \text{or} \quad \mathbf{n}_o = (0.279,\ 0,\ 0.960)$$

which deviates less than 1° from the normal to the $(207)_{bcc}$ plane. Therefore a large coherent precipitate of the α'' phase whose shape is predominantly dictated by the elastic energy relaxation should be produced in the form of a thin plate with habit close to $(207)_{bcc}$ [22]. When this result was first obtained there was the impression that it contradicts the electron microscopic observations of the α'' phase in the form of a thin plate with (001) habit. However, using the same theory Hong, et al. have demonstrated that for a small α'' phase precipitate whose equilibrium aspect ration L/D is less than 11, the habit plane is $(001)_{bcc}$ [9]. Only later with particle coarsening should it be transformed to the (207) habit. The electron microscopic observations seem to confirm this prediction. The (001) habit plane was observed to be transformed into puckered (001) plane composed of segments of planes close to the {207} planes [30].

5.2.2. Precipitates of γ' phase (Fe_4N) in Fe-N martensite. The γ' nitride is an interstitial cubic superstructure formed by ordering of N atoms over octahedral sites of the fcc Fe host lattice. Its crystal lattice parameter is

$$a_{\gamma'} = 3.791 A^o$$

while the crystal lattice parameter of the bcc-based Fe matrix is $a_0 = 2.860$ A^o. Since the stress-free transformation strain for bcc $\rightarrow$ fcc crystal lattice rearrangement is the tetragonal Bain strain, its components are

$$\varepsilon^0{}_{11} = \varepsilon^0{}_{22} = (a_{\gamma'} / a_0\sqrt{2}) - 1 = (3.791/2.86\sqrt{2}) - 1 = - 0.0627$$

$$\varepsilon^0{}_{33} = a_{\gamma'} / a_0 - 1 = (3.791/2.86) - 1 = 0.3255$$

$$t = \varepsilon^0{}_{11} / \varepsilon^0{}_{33} = - 0.1926$$

With this numerical value $t = - 0.1926$ using the elastic constants of pure iron in eq. (25a) we have

$$n_0 = (0.484, 0, 0.875)_{bcc}$$

This unit vector is normal to the predicted habit plane which deviates only by 2.4^o from the (102) habit observed [31].

5.3. Habit plane of β-phase in V-H alloys.

The β-phase in vanadium hydride is an interstitial bcc-based solid solution with H atom occupying the sole O_z octahedral sublatice of bcc V host lattice. Such an occupancy produces pseudotetragonal distortion. The β phase crystal lattice parameters are

$$a = 3.002 A^o , \qquad c = 3.311 A^o$$

The V matrix lattice has the parameter

$$a = 3.032 A^o$$

Therefore, the stress-free transformation strain is

$$\varepsilon^0{}_{11} = - 0.0099 , \quad \varepsilon^0{}_{33} = 0.0890 \quad \text{and} \quad t = - 0.111$$

With the V elastic constants eq. (25b) yields

$$n_0 = (0.277, 0.277, 0.920)$$

which is close to the normal to the $(227)_{bcc}$ habit. The normal to the observed habit plane of the β phase is

$$n_{obs} = (0.293, 0.236, 0.926) \quad [37,38]$$

Deviation of calculated habit from the observed one is about 0.9^o. This agreement can be regarded as very good because the theory does not have any fitting parameters.

6. MAGNETOSTATIC ENERGY AND ANOLOGY WITH ELASTIC STRAIN ENERGY

As was mentioned in the Introduction, there is the profound analogy between elastic strain energy of two-phase coherent dispersoid and magnetostatic energy of ferromagnets and electrostatic energy of ferroelectrics. The consequences of this analogy are so important that they deserve a special discussion. Below the equation for magnetostatic energy of the system of ferromagnetic domains will be described, and it will be shown that magnetostatic energy is analogous to one for the elastic strain energy. It will be demonstrated that the **k**-space technique developed above for the elastic energy can be with the same efficiency applied for magnetostatic energy of ferromagnets in the cases when the Bloch wall thickness is well below the typical siqe of domains [22].

As is known, the magnetostatic energy may always be represented as the sum of interacting magnetic dipoles

$$E_{mag} = 1/2\int\int m(\mathbf{r})_i \left[\frac{\delta_{ij}}{|\mathbf{r}-\mathbf{r}'|^3} - 3\cdot\frac{(\mathbf{r}-\mathbf{r}')_i\,(\mathbf{r}-\mathbf{r}')_j}{|\mathbf{r}-\mathbf{r}'|^3} \right] m(\mathbf{r}')_j \; d^3r \; d^3r' \qquad (40)$$

where $\mathbf{m}(\mathbf{r})$ is the magnetization density at the point $\mathbf{r}$, the integration in (40) is taken over the infinite crystal body. Using the Fourier representations :

$$\mathbf{m}(\mathbf{r}) = \int \mathbf{M}(\mathbf{k}) \; \exp(i\mathbf{k}\mathbf{r}) \; d^3k/(2\pi)^3$$

$$r^{-3}\,\delta_{ij} - r^{-5}\;3\;r_i\,r_j = 4\pi\int (\,k_i\,k_j\,/\,k^2\,)\;\exp(i\mathbf{k}\mathbf{r})\;d^3k/(2\pi)^3$$

in (40), one has

$$E_{mag} = (1/2)\iiint_{-\infty}^{\infty} 4\pi\,\frac{|M(k)k|^2}{k^2}\; d^3k/(2\pi)^3 \qquad (41)$$

where for simplicity the magnetic susceptibility is assumed to be equal to unity.

Let us consider an arbitrary system of ferromagnetic particles with various possible directions of magnetizations designated by the index p. Then the spatial distribution of the magnetization produced by the system of magnetic particles or magnetic domains is

$$\mathbf{m}\,(\mathbf{r}) = M_0 \sum_p \mathbf{e}(p)\;\tilde{\Theta}(p,\mathbf{r}) \qquad (42)$$

(compare with eq.(6)) , where $\tilde{\Theta}(p,\mathbf{r})$ is again the shape function of domains of the pth type, $\mathbf{e}(p)$ is the unit vector along the magnetization direction of pth domains, M_0 is the magnatization. The Fourier transform of (42) is

$$\mathbf{M}(\mathbf{k}) = M_0 \sum_p \mathbf{e}(p)\;\Theta(p,\mathbf{k})$$

Substituting this equation to (41) yields

$$E_{mag} = 2\pi\,(M_0)^2 \sum_{p,q} \int B(\mathbf{k}/k)^{mag}_{pq}\;\Theta(p,\mathbf{k})\;\Theta(q,\mathbf{k})^*\;d^3k/(2\pi)^3 \qquad (43)$$

where

$$B(\mathbf{n})^{mag}_{pq} = (\mathbf{e}(p)\,\mathbf{n})\;(\mathbf{e}(q)\,\mathbf{n})$$

is the angular function of the **k** vector direction, **n** = **k**/k. One may readily see that eq.(43) for the magnetostatic energy has absolutely the same form as eq. (14) for the elastic energy.

For a single domain particle eq.(43) gives the analog of eq.(15) for the elastic energy of a single coherent inclusion:

$$E_{mag} = 2\pi\,(M_0)^2 \int B(\mathbf{k}/k)^{mag}\;|\,\Theta(p,\mathbf{k})\,|^2\; d^3k/(2\pi)^3 \qquad (44)$$

where $B(\mathbf{n}) = (\mathbf{en})^2$. Equation (44) can also be rewritten as

$$E_{mag} = 2\pi\,(M_0)^2\,\alpha\,V$$

where

$$\alpha = V^{-1} \int \left[\,(\mathbf{ek})^2 / k^2\,\right]\;|\,\Theta(k)\,|^2\; d^3k/(2\pi)^3$$

is the **k**-space representation for the so-caled demagnetization factor, the dimensionless coefficient depending only on the shape of the particle.

It should be emphasized that eq.(43) gives the close solution for an arbitrary set of ferromagnetic domains whose size is well above the Bloch wall thickness. This equation can be efficiently used for calculation of the reverse magnetization and for analysis of morphologies of domain structures. To the authors knowledge, this **k**-space formulation of the magnetostatic energy is new and can be very useful in various applications because of its mathematical simplicity.

The formal analogy between the elastic energy (14) and (15) , and the magnetostatic energy (43) and (44) consists in the fact that both have the same mathematical form. The kernel function $B(\mathbf{k}/k)_{pq}$ in the elastic energy (14) as well as the corresponding kernel function $B(\mathbf{k}/k)^{mag}_{pq}$ depend on the direction of the wave vector **k** rather than on its absolute value. The kernel functions $B(\mathbf{k}/k)_{pq}$ are , in fact, the Fourier transform of the pairwise interaction between elements of volume of the domains (or coherent particles) of the type p and q. These energies can be found by the back Fourier transform which gives the singular function

$$V_{pq}(\mathbf{r}-\mathbf{r}') = 1/\,|\,\mathbf{r}-\mathbf{r}'| \cdot \psi\left(\,(\mathbf{r}-\mathbf{r}')\,/\,|\,\mathbf{r}-\mathbf{r}'\,|\,\right)$$

where $\psi(\mathbf{n})$ is the function of the direction, $(\mathbf{r}-\mathbf{r}')\,/\,|\,\mathbf{r}-\mathbf{r}'|$. This is the typical form of the dipole-dipole like interaction. The elastic interaction between elements of coherent precipitate volume has exactly the same form as the dipole-dipole interaction when $B(\mathbf{n})$ has the cylindrical symmetry about a certain axis **e**, i.e. when

$$B(\mathbf{n}) = B(\mathbf{ne}) \cong \min B(\mathbf{n}) + \beta\,(\mathbf{ne})^2 + \ldots$$

Then the back Fourier transform yields

$$V(\mathbf{r}-\mathbf{r}') = \beta\,|\,\mathbf{r}-\mathbf{r}'\,|^{-3}\left[\,1 - 3\left(\,(\mathbf{r}-\mathbf{r}')\,\mathbf{e}\,\right)^2 /\,|\,\mathbf{r}-\mathbf{r}'\,|^2\,\right]$$

which corresponds to interaction between two dipoles with dipole moments $\sqrt{\beta}\,\mathbf{e}$, separated by the distance $|\mathbf{r}-\mathbf{r}'|$.

The dependence of the magnetostatic energy on morphology of ferromagnetic particles is the well known effect. It results in the instability of a homogeneous state of the ferromagnetic phase with respect to decompositon into the system of domains. It is the major physical consequence of the fact that $B(\mathbf{k}/k)^{mag}_{pq}$ depends only on direction of $\mathbf{k}$ vector . In the case of a uniaxial ferromagnet film, a large domain with the opposite magnetization direction than the matrix also proves to be unstable with respect to splitting into the array of bubble domains. The reason for this is the same, the repulsion between volume elements of the domain which repel each other as parallel identical dipoles.

Summing up the foregoing one can see that instability of a homogeneous state of ferromagnet (and ferroelectric) is caused by the fact that the magnetostatic energy of a ferromagnetic phase unlike the exchange energy depends not only on the volume of the phase , but also on its morphology, shape and spatial distribution. It will be shown below that the same is true for the elastic energy of a coherent dispersoid.

7. STRAIN - INDUCED INSTABILITY OF COHERENT PARTICLES IN TWO - PHASE CUBIC ALLOYS

The elastic energy, unlike the "chemical " free energy of a two-phase alloy depends not only on the precipitate phase volume, but also on its shape and spatial distribution. The situation here is the same as with magnetostatic energy. Therefore, one could expect that dependence of elastic energy on morphology would produce the same effect, viz. instability of large coherent particles. This instability, splitting large coherent precipitates, analogous to the splitting instability resulting in formation of bubble domains, would seriously affect the traditional concepts of coarsenting in two-phase cubic alloys. The main result of the conventional theory of coarsening, that a two-phase alloy becomes more stable upon coarsening should be quenstioned.

First of all, all studies concerning evolution of alloy upon coarsening implicitly assume that precipitates remain intact and, if they coarsen, just monotonically increase their size. The theory [2,4] enables us to test this assumption. Following [32], we shall demonstrate that when a cuboidal particle of a cubic phase precipitate reaches a certain critical size, multiple of the typical length r_0 introduced above by eq. (36), the cuboid becomes unstable and decomposes into a doublet and later into an octet of subparticles. This phenomenon reflects repulsive interaction between elements of volume of a cuboid which , in fact, opposes the coarsening. Similarly we can predict that large plate should also be unstable with respect to splitting into several subplates and so on.

Splitting is not the only way to prevent the formation of too large overgrown precipitates. Elastic interaction between them may produce the same effect. This interaction would just oppose coarsening, the phenomenon which was really observed.

To analyze the elastic energy change upon transition of a monolytic precipitate into a group of subparticles we should compare the elastic energies of both states. To do this, let us represent the shape function, $\tilde{\theta}(\mathbf{r})$ of a group of identical subparticles as the sum of their shape functions, $\tilde{\theta}_0(\mathbf{r}-\mathbf{R}')$

$$\tilde{\theta}(\mathbf{r}) = \sum_j \tilde{\theta}_0(\mathbf{r}-\mathbf{R}_j)$$

where the index j labels all subparticles $\mathbf{R}_j$ describing the position of the center of gravity of the jth subpaticle. The Fourier transform of this function is

$$\Theta(\mathbf{k}) = \Theta_0(\mathbf{k}) \sum_j \exp(-i\,\mathbf{k}\mathbf{R}_j) \tag{45}$$

where the mutual localtion of precipitates is taken into account by the "structural factor", $\sum_j \exp(-i\mathbf{k}\mathbf{R}_j)$, $\Theta_0(\mathbf{k})$ is the Fourier transform of the shape function of a subparticle.

Substituting (45) to (15) gives the close equation for the elastic energy of this group of precipitates:

$$E = 1/2 \int B(\mathbf{k}/k)\, |\Theta_0(\mathbf{k})|^2 \left| \sum_j \exp(-i\,\mathbf{k}\mathbf{R}_j) \right|^2 d^3k/(2\pi)^3 \tag{46}$$

Using the expansion of the function B(**n**) in the series of cubic harmonics and terminating the corresponding series by two forms, we have

$$E = E_0 V + 4 E V_1 [\, I_1 + 27 \mu I_2 / 2]$$

where E_0 and E_1 are given in comments to eq. (34)

$$\mu = (c_{11} - c_{12} - c_{44}) / (c_{11} + c_{12} + 4c_{44})$$

The dimensionless coefficients I_1 and I_2 have the form

$$I_1 = V^{-1} \int \gamma_1(\mathbf{k}/k)\, |\Theta_0(\mathbf{k})|^2 \left| \sum_j \exp(-i\mathbf{k}\mathbf{R}_j) \right|^2 d^3k/(2\pi)^3$$

$$I_2 = V^{-1} \int \gamma_2(\mathbf{k}/k)\, |\Theta_0(\mathbf{k})|^2 \left| \sum_j \exp(-i\mathbf{k}\mathbf{R}_j) \right|^2 d^3k/(2\pi)^3 \tag{47}$$

where $\gamma_1(\mathbf{n}) = n_x^2 n_y^2 + n_x^2 n_z^2 + n_y^2 n_z^2$

$\gamma_2(\mathbf{n}) = n_x^2 n_y^2 n_z^2$,

(n_x, n_y, n_z) are Cartesian components of the unit vector **n**. The constants I_1 and I_2 are geometrical factors which depend on shape and mutual location of subparticles. Numerical calculations of the integrals (47) at $(c_{11} - c_{12} - 2c_{44}) < 0$ show that a cuboidal particle has greater elastic energy than an octet of cuboidal subparticles, the energy of the octet being the lowest when cuboidal subparticles are separated by the distance $u \simeq 0.4a$ where $2a = \sqrt[3]{V}$ is the edge length of the initial cuboid (Fig.2). The cuboid subparticle has also the greater elastic energy than the doublet of the identical parallelopiped subparticles formed due to splitting the cuboid. The lowest energy of the doublet is attained when separation distance between subparticles is 0.8a (Fig.3). The elastic energy of the octet is less than that of the doublet of the same volume (Fig 4). As for the interphase energy, its increase caused by the formation of new interphase because the splitting is less for the doublet than for the octet. Therefore, for a smaller precipitate when the interphase energy contribution dominates, a doublet should be expected. For the layer precipitate when the elastic energy prevails, the octet is favoured. The numerical calcuation and comparison of the elastic and surface energies of both morphologies show that the cuboid ⟶ doublet transformation may occur when $\sqrt[3]{V} \geqslant 27\, r_0$ where V is the cuboid volume, r_0 is given by eq.(36). Doublet ⟶ octet transformation may occur when $\sqrt[3]{V} \geqslant 82 r_0$. At greater volumes the octet ceases to be stable with respect to transformation to a platelet. These results naturally fit the result in section 4 concerning the shape transformation of a monolitic particle from a spheroid to a cuboid which occurs when $\sqrt[3]{V} \geqslant 7.6 r_0$. Together these results confirm our qualitative conclusions formulated above

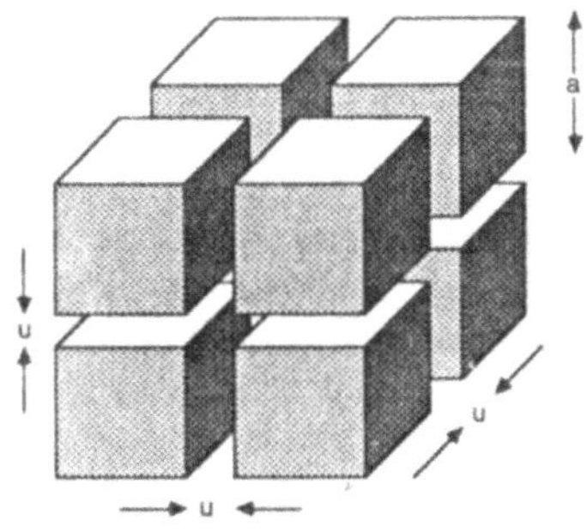

FIGURE 2. Schematic drawing of octet that results from the decomposition of a cuboidal particle.

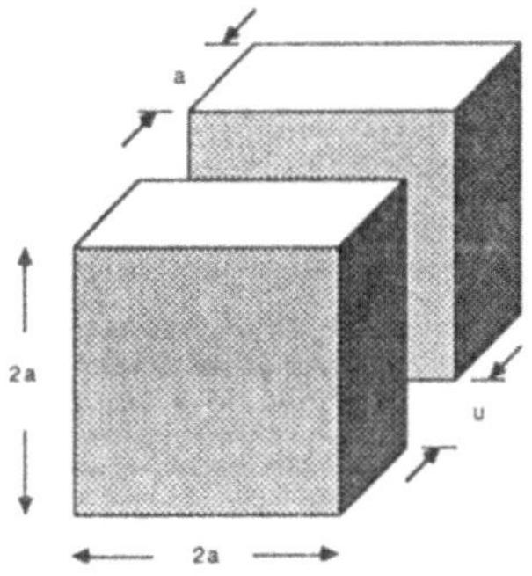

FIGURE 3. Schematic drawing of doublet of plates that results from the decomposition of a cuboidal particle.

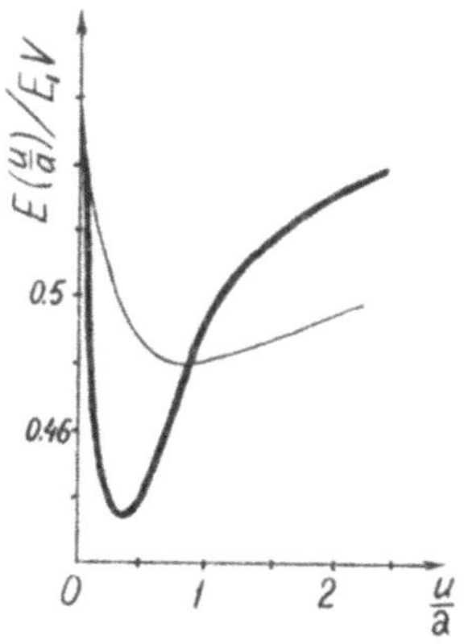

FIGURE 4. Configurational elastic energy, in dimensionless form, as a function of dimensionless particle spacing (u/a) for an octet of cubes (dark line) and a doublet of plates (light line).

that the morphology transformation is determined by the ratio between the efficient particle size $\sqrt[3]{V}$ and characteristic length r_0 depending on interphase energy, crystal lattice misfit and elastic moduli. The $\sqrt[3]{V}/r_0$ ratio is, in fact, the measure of contribution of the elastic energy with respect to the interphase energy to the coarsening process. The more is this ratio, the more the contribution of the elastic energy.

The stability limits found above characterize the conditions when one morphology becomes energetically more favourable than another. However, it should be emphasized that the "overgrown" microstructure is not automatically transformed into another. It may still be stable with respect to infinitesimal variations of the shape. In this situation the overgrown microstructure is metastable. It can be transformed into the stable one only by the finite shape transformation playing the same role as the critical nucleus fluctuation in the conventional phase transformation thermodynamics. Since the critical shape fluctuation required for the shape transformation is macroscopically large, all metastable morphologies should be very stable and transform into the stable morphology only near the metastability limit where a metastable particle also becomes unstable with respect to infinitesimal shape variations.

Simple qualitative interpretation of the microstructure transformation, spheroid $\longrightarrow$ cuboid $\longrightarrow$ doublet $\longrightarrow$ octet $\longrightarrow$ platelet, upon coarsening is the following. In the relevant case when $(c_{11} - c_{12} - c_{44}) < 0$, elements of the precipitate phase volume repel each other along the <111> directions. This repulsion transforms a spheroid into cuboid due to stretching of the spheroid volume along the <111> directions. The same effect results in splitting the cuboid along the same <111> direction transforming the cuboid into an octet. The only reason why new phase precipitates can exist as monolitic homogeneous particles is the interphase energy effect. The interphase energy opposes the splitting since it produces new interphase. The situation here is the same as in the case of ferromagnets, because both elastic energy and magnetostatic energies destroy the homogeneous state of a particle in the zero interphase energy limit, or in the limit of large particle volume when interphase energy plays the minor role. The example of such a behaviour gives a strip ferromagnetic domain which becomes unstable with respect to splitting into the set of bubble domains because of repulsion between the elements of the strip domain volume repelling each other as parallel magnetic dipoles.

The above results show that the mathematically simple theory based on the homogeneous modulus case approximation can be efficiently applied to the important technical alloys. For example, it has been shown above in section 3.6 that the homogeneous modulus approximation gives nevertheless asymptotically exact value of the elatic energy of plate-like precipitates if the elastic moduli of the precipitate phase are used. The reason for that is the concentration of elastic strain within the plate-like particle (the ratio of elastic energy concentrated outside and inside the particle tends to zero as the squared aspect ratio $(D/L)^2 \longrightarrow 0$, where D is the thickness, L is the length of the particle). On the other hand, equations for the edge energy of the plate in section 3.5 are also asymptotically correct if elastic moduli of the matrix are used. The reason for that is the same. The edge energy is the one that is concentrated in the matrix with the same asymptotic accuracy $(D/L) \longrightarrow 0$. Therefore, all results concerning the habit plane, eigenstrain and orientational relations obtained in section 3 in the homogeneous modulus approximation, proves to be valid if the moduli are different. In fact, the reduction of elastic energy upon strain-induced state of coarsening occurs mainly due to redistribution of elastic strain from the matrix to the precipitate phase so that ultimately all elastic strain proves to be contained in thin plate particles. In limit case of the martensite-like optimal structure at which bulk elastic energy completely vanishes, the value of elastic moduli does not matter at all. This is the reason why pure crystallographic theory of martensitic transformation proved to be so efficient.

ACKNOWLEDGMENTS

This work was supported by the Director, Office of Energy Research, Office of Basic Energy Science, Materials Sciences Division of the U.S. Department of Energy under Contract No. DE-AC03-76SF00098.

REFERENCES

1. J.D.Eshelby, Proc. Roy. Soc., A241, 376 (1957); A252, 56 (1959); Prog. Solid. Mech., 2, 89 (1961).
2. A.G.Khachaturyan, Soviet Physics-Solid State, 8, 2710 (1966)
3. A.L.Roitburd, Kristallography, 12, 567 (1967)
4. A.G.Khachaturyan and G.A.Shatalov, Soviet Physics-Solid State, 11, 118 (1969)
5. S.Wen, A.G.Khachaturyan, J.W.Morris,Jr., Proc. Int. Symp. Modulated Struct., AIP, N.Y. (1979) p. 168; Proc. Int. Conf. Martensitic Transformation, Canbridge, Mass, (1979) p. 94; Metallurgical Transactions 12A, 581 (1981).
6. J.W.Wert, Acta metallurgica, 24 , 65 (1976).
7. M.Hong, M.S. Thesis, University of California, Berkeley, CA (1978)
8. J.K.Lee and W.C.Johnson, Phys. Stat. Solidi 46(a), 375 (1976)
9. M.Hong, D.E.Wedge and J.W.Morris,Jr., Acta Metallurgica, 32, 279 (1984)
10. A.L.Roitburd and N.S.Kosenko, Phys. Stat. Solidi, 35(a), 735 (1976)
11. W.E.Mayo and T.Tsakalakos, Metallurgical Transactions, 11A, 1631 (1980)
12. S.H.Wen, E.Kostlan, M.Hong, A.G.Khachaturyan, J.W.Morris, Jr., Acta Metallurgica, 29, 124 (1981)
13. V.Lanteri, T.E.Mitchell and A.H.Heuer, Journal of American Ceramics Society, 69, 564 (1986)
14. A.G.Khachaturyan and G.A.Shatalov, Soviet Physics-JETP, 29, 557 (1969)
15. A.G.Khachaturyan, "The Theory of Phase Transformations and Structure of Solid Solutions", Moscow, Nauka (1974); Soviet Physics-JETP, 31, 98 (1970); Phys. Stat. Solidi, 35, 119 (1969)
16. A.G.Khachaturyan, V.N.Airapetyan, Phys. Stat. Solidi, (a) 26, 61 (1974).
17. E.Seitz and D.de Fontaine, Acta Metallurgica, 26, 1671 (1978).
18. H.Yamunchi and D.de Fontaine , Acta Metalurgica, 27, 763 (1979)
19. J.K.Lee, D.M.Barnett, and H.I.Aaronson, Metallurgical Transactions, 8A, 1447 (1977)
20. L.J.Valpole, Proc. Roy.Soc. (A) 300, 270 (1967).
21. J.R.Willis, "Asymmetric Problems of Elasticity", Adam Prize Essay, University of Cambridge, (1970)
22. A.G.Khachaturyan, "The Theory of Structural Transformations in Solids", Wiley & Sons, N.Y., (1983)
23. J.Gjonnes, Chr.J.Simensen, Acta Metallurgica, 18, 881 (1970)
24: G.Thomas and J.Nutting, Journal of Inst. Metals, 88, 81 (1959-60)
25. V.Gerold, Acta Cryst. 11, 236 (1958)
26. K.Doi, Acta Cryst. 13, 45 (1960)
27. V.Gerold, in Proc. of 116 TMS Meeting, Denver, Colorado, February 23-26, 1987
28. K.H.Jack, Proc. Roy. Soc. (A)208, 216 (1951)
29. A.V.Suyazov, M.P.Usikov and R.M.Mogutnov, Fiz. Met. Metalloved, 42, 755 (1976)
30. P.Ferguson, V.Dahemn and K.H.Westmacott, Scripta Metallurgica 18, 57 (1984)
31. M.P.Usikov and A.G.Khachaturyan, Phys. Met. Metallography, 30, 614 (1970)
32. A.G.Khachaturyan, S.V. Semenovskaya, and J.W.Morris,Jr.,(to be published in Acta Metallurgica).

THE INFLUENCE OF LATTICE DEFECTS ON ALLOY PHASE DIAGRAMS

F R N Nabarro

Condensed Matter Physics Research Unit, University of the Witwatersrand, Johannesburg, P O WITS 2050, and National Institute for Materials Research, CSIR, P O Box 395, Pretoria, 0001, South Africa.

ABSTRACT. An atom in the core of a lattice defect is effectively a foreign atom. Since the concentration of such core atoms is rarely greater than 10^{-4}, lattice defects usually have little influence on alloy phase diagrams, though they control the kinetics of phase changes. Static defects may affect the phase diagram if they are present in unusually high concentrations, or if the phase equilibrium is delicate. Defects such as phonons and magnons may be present in thermal equilibrium in atomic concentrations greater than or comparable with unity, and have a large influence on phase equilibria.

1. INTRODUCTION

In 1950, Seitz (1) offerred a synthesis of "imperfections in nearly perfect crystals" which is summarized in Fig. 1.

When the properties of a single imperfection such as a phonon or an electron are understood, one can begin to study pairwise interactions, both "diagonal" such as phonon-phonon scattering and "off-diagonal" such as phonon-electron scattering. Higher order "few-body" interactions would follow, but, although the Bardeen-Cooper-Schrieffer theory of superconductivity was mentioned as a current development, the class of truly "many-body" effects had not been generally recognized.

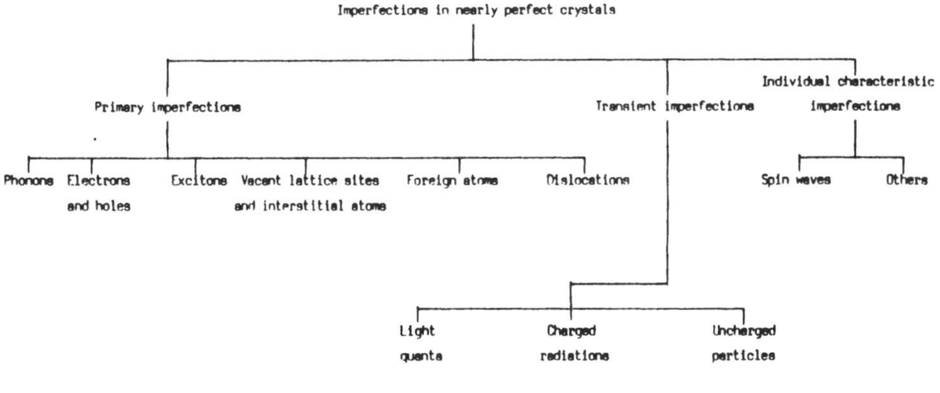

FIGURE 1. Seitz's classification of crystal defects.

G. M. Stocks and A. Gonis (eds.), Alloy Phase Stability, 557–584.

Nowadays we incline to use a classification based on the dimensionality of the defect, grouping vacant lattice sites, interstitial ions and foreign atoms as point defects, dislocations as 1-dimensional defects, stacking faults, grain boundaries and the free surface as 2-dimensional defects, and voids and particles of precipitate as 3-dimensional defects. Excitations such as phonons and magnons may fill the crystal and are essentially 3-dimensional. We regard them as transient, although they are normally present in thermal equilibrium, and Seitz classifies phonons as primary excitations. Seitz does not mention stacking faults, though they are discussed by Read and Shockley in the next article in the same volume, and he finds it "necessary to adopt an attitude towards the surface of the crystal". This attitude is that the surface is "part of the normal constitution of an ideal crystal". We shall see that there is much interesting physics associated with the surface.

An ordinary alloy phase diagram in the (c, T) plane may show about 5 phases in the whole range of concentrations $0 \leq c \leq 1$, and 3 in a temperature range $300 \leq T \leq 1500$ K. Each change of phase corresponds to a change in the electron/atom ratio e/a of order 0.2, or to an increase in the number of phonons/atom of order 5. The concentrations of lattice defects are usually far too low to produce changes in e/a or in the phonon spectrum of this order. For example, the most abundant point defect in a pure metal is usually the vacancy. The concentration of vacancies in thermal equilibrium even at the melting point is typically 10^{-4}. If the vacancy behaves like a substitutional alloying element of valency zero, this concentration of vacancies will alter e/a by a few parts in 10^4, and will change the phonon density of states by a few parts in 10^4. The mean density of dislocations in a heavily cold-worked metal is of the order of $10^{11}cm^{-2}$, the concentration of atoms which lie in dislocation cores and can therefore be expected to affect the electron/atom ratio or the phonon spectrum about as strongly as does a vacancy is therefore of order 10^{-4}, and their effect on phase equilibria will again be small. Plastic deformation also produces point defects. Annealing experiments (2, 3) show that the dislocations and the point defects make comparable contributions to the electrical resistivity, and so also to e/a and to the phonon spectrum. Small particles are expected to have phases different from those of the bulk if 1/5 - 1/10 of their atoms lie on or near the surface, which implies that they are 30 - 90 atoms in diameter.

Lattice defects can affect phase equilibria only in special circumstances. Their concentration may be anomalously high. The phase changes concerned may involve energy differences unusually small in comparison with the binding energy; these will be phase changes such as the transition from the superconducting to the normal state which occur at very low temperatures. They may occur at very low values of e/a; the metal/insulator transition occurs at e/a of the order of 8×10^{-5} in silicon and 4×10^{-6} in germanium (4), and these concentrations of electrons might be trapped by vacancies or dislocations.

A suggestion which has been discussed informally at this meeting but does not seem to have been worked out in detail, is that dislocations may have considerably more influence in systems which show long-range atomic or spin ordering. A dislocation introduces a sheet of antiphase boundary. If the mean separation of dislocations is n atomic spacings, the fraction of atoms in dislocation cores is $1/n^2$, but the fraction of wrong bonds is 1/n. If there is short-range order with a correlation distance of m interatomic spacings, the fraction of wrong bonds is m/n^2.

While defects can exert a substantial influence on alloy phase diagrams only in rather unusual circumstances, they usually provide the mechanism by which phase equilibria are achieved. Point defects provide the mechanism of diffusion by which phase separation, ordering and disordering can occur; dislocations are essential to the mechanisms of martensitic (shear) transformations (5).

2. POINT DEFECTS

Some types of point defect can occur in monatomic crystals, others only in crystals containing two or more distinct kinds of atom.

2.1 MONATOMIC CRYSTALS

The typical point defects of a monatomic crystal such as a pure metal are isolated lattice vacancies and isolated interstitial atoms. The vacancy normally occupies a lattice site. As a result of the nearest-neighbour repulsion between metal ions, the surrounding ions usually move inwards towards the vacancy, without destroying the local point-group symmetry. It is natural to believe that a self-interstitial, an additional atom of the same element inserted into the matrix, will occupy one of the largest interstices in the lattice. In the face-centred cubic lattice this would be the middle of a cube edge (Fig. 2(a)), which is also the centre of a cube (Fig. 2(b)). In practice, the configuration of lowest energy seems to be the dumb-bell, in which a single atom is replaced by a pair of atoms lying along < 100 >, either in a cube face (Fig. 2(c)) or, equivalently, perpendicular to a cube face (Fig. 2(d)). These defects are of lower local symmetry than the perfect lattice. Interstitials formed dynamically by irradiation may adopt the form of a crowdion, in which a < 110 > close-packed row of atoms is locally compressed to accommodate n + 1 atoms in the space normally occupied by n atoms (6) (Fig. 2(e)). Similarly, in the body-centred cubic lattice the obvious interstitial site is the centre of a cube face, but (9) a dumb-bell along < 111 > probably has a lower energy, while the dynamic crowdion again lies along the close-packed direction, here < 111 >.

A foreign "impurity atom" or "dopant" may either be in a substitutional site, so that if it were removed there would be a vacancy, or in an interstitial site. Usually, small atoms such as H, C, N occupy interstitial sites, while other impurities are substitutional. Presumably because of the large lattice strains which they introduce, interstitial atoms tend to form ordered superlattices even when they are present in very low concentrations, e.g. 0.1 atomic % of N in Ta (10) or 3 atomic % C in Ta (11).

Vacancies attract one another. This is most easily seen in the simple 2-dimensional nearest-neighbour bond metal of Fig. 3. The two isolated vacancies in Fig 3(a) have 8 broken bonds; the divacancy of Fig 3(b) has only 6. Interstitials also attract one another. To understand this we recognize that around an interstitial there are large distortions of the lattice. In general, these distortions will reduce the elastic

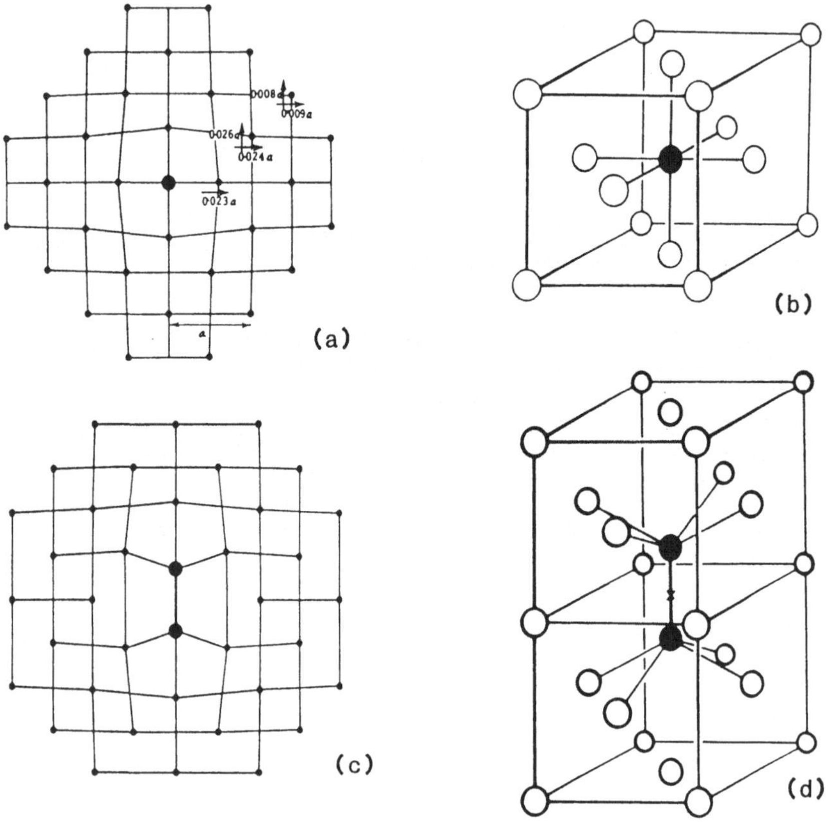

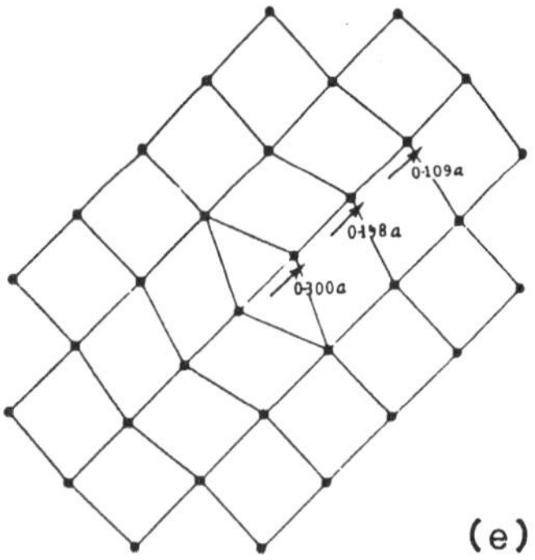

FIGURE 2. a) Octahedral interstitial in a face-centred cubic lattice (6). b) The same in a perspective representation (7). c) Dumb-bell interstitial in a face-centred cubic lattice (6). d) the same in a perspective representation (7). e) Crowdion in a face-centred cubic lattice (6).

constants, and the energy of a second interstitial is less if it is formed close to the first interstitial then if it is formed in the perfect lattice. If vacancies aggregate to form a platelet (Fig. 4(a)), the faces of this platelet may fall together to reform a region of perfect lattice surrounded by a dislocation line (Fig. 4(b)). There is thus an intimate connection between point defects and dislocations (12).

Alternatively, vacancies may aggregate to form a cavity in the crystal, usually called a void.

These interactions may lead to a striking effect in irradiated samples, the formation of a void lattice. The most satisfactory explanation of these structures (13) is as follows. Irradiation produces vacancies and interstitials in equal numbers. Dislocations preferentially attract and absorb interstitials, thus leading to an excess of vacancies which condense to form voids. These voids may themselves form a rather regular lattice (14), which is face-centred in the face-centred crystal lattice and body-centred in the body-centred crystal lattice (Fig. 5(a). The explanation is that the crystal under irradiation contains a high flux of interstitial crowdions which are mobile only in close-packed directions, and fill up voids when they strike them. Each void protects voids lying in a close-packed direction from itself up to a distance equal to the mean distance in which a crowdion decays to form a normal interstitial atom. Thus a void lattice is formed with a nearest-neighour distance equal to this mean free path, and with the same close-packed directions and so the same crystal structure as the crystal lattice. If this explanation is correct, the void lattice is not an equilibrium "phase", but a dissipative structure like a set of Taylor vortices or of Bénard cells. The void lattice may itself contain defects such as dislocations (Fig. 5(b)).

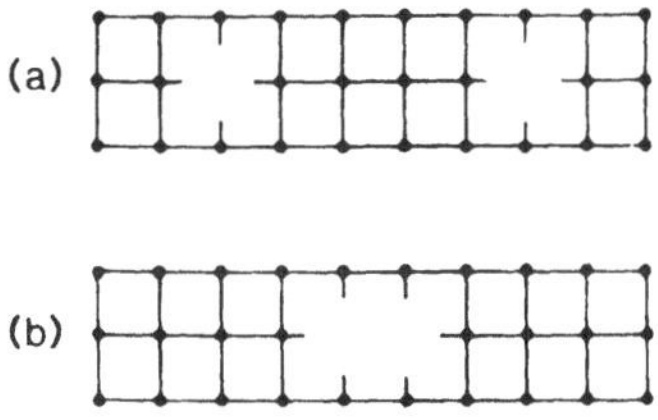

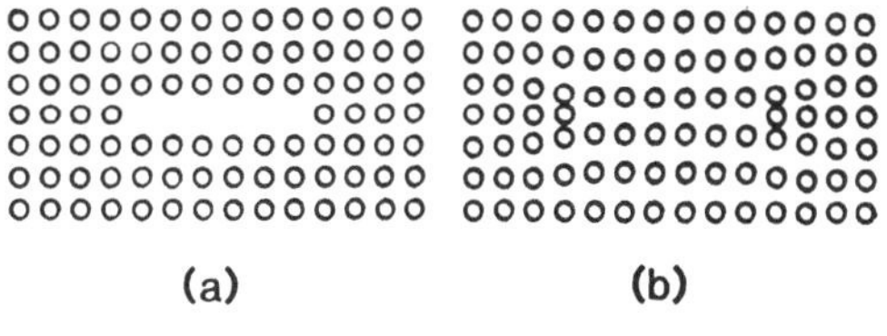

FIGURE 3

FIGURE 4

FIGURE 3. a) Two isolated vacancies in a simple square lattice have 8 dangling bonds. b) Two adjacent vacancies in a simple square lattice have 6 dangling bonds.

FIGURE 4. a) A Platelet of vacancies. b) The facs of the platelets fall together to form perfect crystal surrounded by a dislocation.

2.2 DIATOMIC CRYSTALS

In ionic crystals there are interstitial anions and cations which are negatively and positively charged defects respectively, and anion and

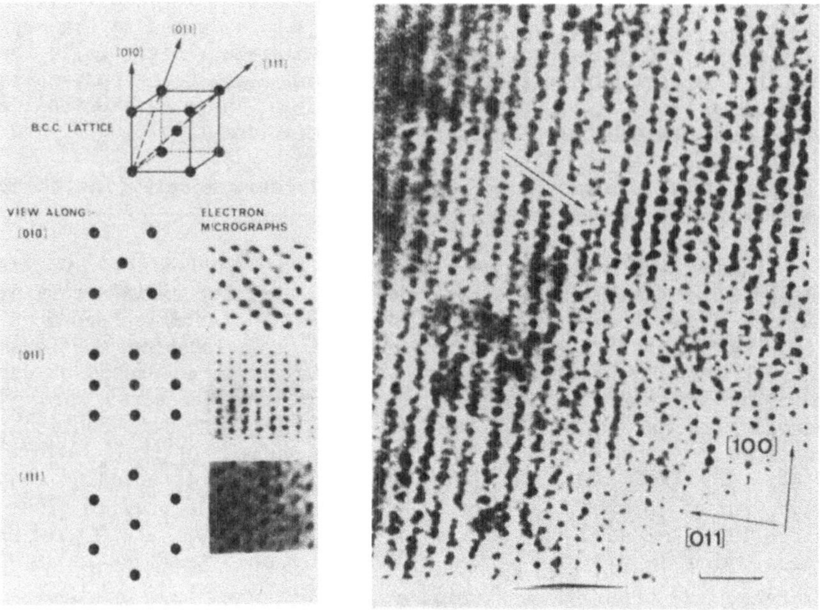

a) b)

FIGURE 5. a) The projections of a b.c.c. lattice along the (011), (010) and (111) directions, together with electron micrographs showing the projections of void lattices along the same directions (15). b) A dislocation in the void lattice in neutron-irradiated Mo-0.5% Ti (16).

cation vacancies which behave as positively and negatively charged defects respectively. The electrostatic energy would be extremely high unless the total defect charge in the bulk material was zero, so that at least two types of defects must be present. Usually there are only two types, of opposite sign, present in appreciable numbers. A cation vacancy may be compensated by a cation interstitial, in which case the disorder is described as of Frenkel type, or it may be compensated by an anion vacancy, in which case it is described as of Schottky type. A paradox appears when we recognize that the two compensating types of defects generally have different energies of formation, and so will be present in quite different concentrations in thermal equilibrium, while charge neutrality requires then to be present in equal numbers. The paradox was resolved by Frenkel (17), who showed that the concentrations would be different in the bulk and near the surface. There is a double layer of charge just inside the surface, which produces an electrostatic potential inside the crystal. This potential increases the free energy of one defect species, and decreases the free energy of the other. The thickness of the surface layer is a Debye electrostatic energy length. The corresponding length in an alloy would be less than one atomic diameter. As Frenkel pointed out (17, p.41), in an ionic crystal "an exchange of places between two adjacient ions of the opposite sign is excluded on the ground that it would require a vast increase of energy". This is not true in compound semiconductors such as GaAs, where the "antisite" defect of an arsenic atom on a gallium site seems to be present in as-grown crystals and to multiply on irradiation or plastic deformation, although the details are still far from clear (18, 19). In a disordered alloy, antisite defects are so common that the definition of a "correct" site has been lost. In such disordered alloys, one component tends to segregate to the surface in equilibrium. Hamilton (20) has explained this segregation very satisfactorily using an extension of the ideas of Miedema et al. (21-25). It would be interesting to relate his approach to that of Frenkel.

It sometimes happens that vacancies are present in a far higher equilibrium concentration than the 10^{-4} which can be achieved in pure metals. This is not uncommon in oxides, and can be of technological importance when the materials behave as fast-ion conductors (26, 27). It can also occur in alloys. The most famous of these is the series of Ni Al alloys studied by Bradley and Taylor (28) and the related Ni Al Cu alloys studied by Lipson and Taylor (29), and discussed by Raynor (30). The alloy Ni Al has an ordered body-centered cubic structure. The Al atoms contribute 3 electrons to the valence band. As a result of the high Fermi level, the effective valency of the smaller Ni atoms is zero. (The lectures of Dederichs in this volume discuss AlN, alloys in detail). With an excess of Ni, Al atoms are replaced by smaller and heavier Ni atoms, the lattice spacing falls and the density rises (Fig. 6). With excess of Al, the lattice spacing again falls, and the density falls very rapidly. This is explained by the assumption that the excess Al atoms remain on the Al sublattice, while the Ni sublattice contains vacancies. The concentration of vacancies can become so high that the number of atoms per unit cell falls from 2.00 to 1.67, while the number of electrons per unit cell remains within 2% of 3.00 (Table 1). The vacancies occupy Ni sites at random, but, when their concentration becomes high, a new phase Al_3Ni is formed in which the vacancies order in planes particular to a $\langle 111 \rangle$ cube axis, producing a trigonal structure. We shall see later that the ordering of defects to produce a new defect-free structure of lower symmetry is a common phenomenon. The behaviour of the binary alloy does not make it clear whether the structure is determined by the condition, or whether the large Al atoms form a framework into which small Ni atoms and effectively smaller vacancies can be inserted. The answer is given by the ternary alloy system Ni Al Cu, in which the Cu ion has much the same size as the Ni ion, but effective valency 1. The ternary phase diagram (Fig. 7) shows that vacancies appear that the number of valence electrons per cell is equal to 3.00 when e/a exceeds 1.50, and not at the composition $(NiCu)_{50}\,Al_{50}$.

TABLE 1: Electron and Atom Ratios for Nickel-Aluminium Alloys (30).

Atomic per cent Ni	Number of electrons/atom	Number of atoms/unit cell	Number of electrons/unit cell
49.6	1.51	2.00	3.02
48.9	1.53	2.00	3.06
48.4	1.55	1.96	3.04
46.6	1.60	1.89	3.02
45.25	1.64	1.84	3.02
40.0	1.80	1.67	3.00

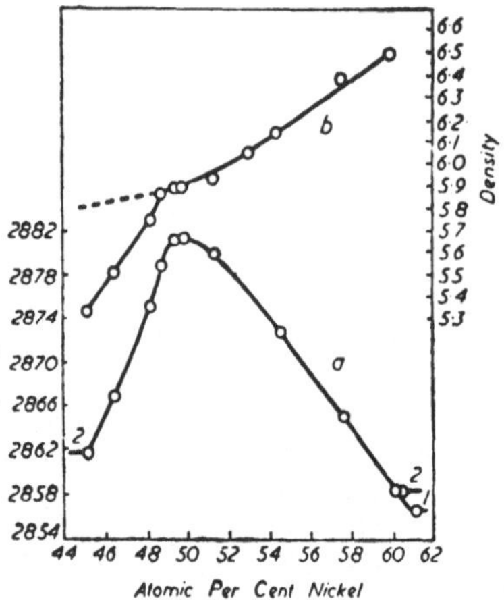

FIGURE 6

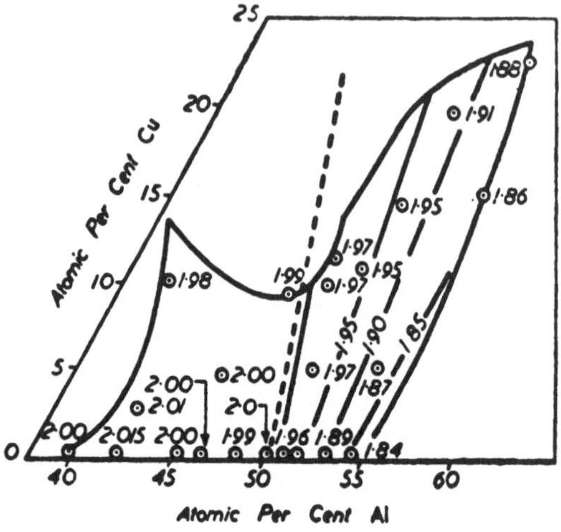

FIGURE 7

FIGURE 6. The lattice spacing (a) and the density (b) of nickel-aluminium alloys quenched from 900 °C. (The dashed curve is for an annealed two-phase structure) (28).

FIGURE 7. The number of atoms per unit cell in the ternary system nickel-aluminium-copper (29). The dashed lines is e/a = 1.15. The line $(CuNi)_{50}$ Al_{50} would be vertical.

If vacancies aggregate to form a vacancy superlattice in Al_3Ni, this ordering may be faulty, so that the new superlattice itself shows superlattice defects.

3. LINE DEFECTS

Three classes of line defect can occur in a crystal. The first, which might appear as a line of impurity atoms (not necessarily a straight line) or a line of vacancies, deforms the crystal as if it is a line of dilatation. Entropy considerations show that in equilibrium at a finite temperature such a line will never be unbroken, although short loops may be present. Needle-like precipitates, which are common, approximate to such line defects. The second class is the dislocation. These are abundant in most metal and alloy samples, and we shall discuss their effects. The third class is the disclination. If we take a closed circuit around a disclination (Fig. 8) and follow a vector which points along a lattice direction continuously around the circuit, the vector does not in general fall on its original lattice direction after it has been carried once around the circuit, but falls on another equivalent lattice direction. Such disclinations are found in 2-dimensional systems, in liquid crystals (32) and in the flux-line lattice in Type II superconductors (33), but their energy is too high for them to occur naturally in ordinary crystals. Tetrahedrally coordinated amorphous solids also contain linear geometric objects which are or resemble disclinations of strength $\pm\ ^1/_2$ (34), and their presence may be of importance in the theory of amorphous alloys.

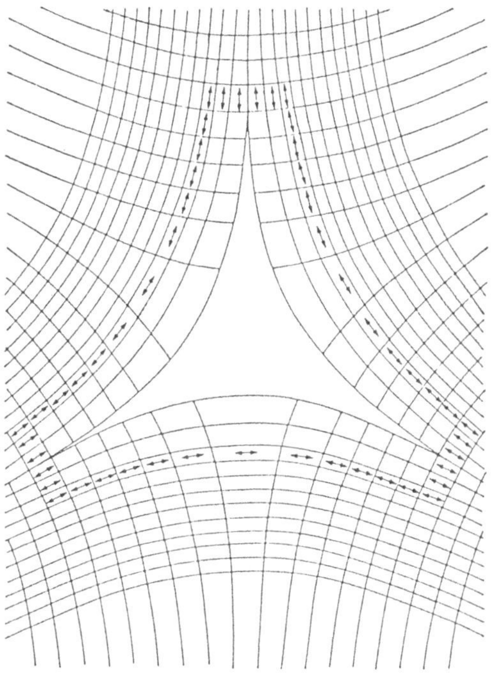

FIGURE 8. A wedge disclination of strength - $^1/_2$ (31).

While a circuit round a line of disclinations results in a change in orientation in the lattice, a circuit round a line of dislocation results in a change of location in the lattice.

A closed circuit surrounding the dislocation line does not contain equal numbers of lattice steps in opposite directions, up or down, left or right. For example, an anticlockwise circuit round the edge dislocation whose line meets Fig. 9 at O requires one more step to the left than there are steps to the right. This closure failure is the Burgers vector $\underset{\sim}{b}$ of the dislocation, and is constant along the line of the dislocation.

This dislocation can readily glide on its glide plane, which lies between the sheets of atoms marked A and B. If we look down on this plane, and imagine that there is a closed region over which the part of the crystal above the glide plane has slipped by one Burgers vector with respect to the part of the crystal below the glide plane (Fig. 10), we see that the boundary between the slipped and the unslipped regions is a closed dislocation line. All parts of the crystal more than a couple of lattice spacings from the dislocation line are topologically perfect, though they may be elastically strained. It is clear that there are regions of the dislocation line where its geometry will be particularly simple, the edge segments where the segment $d\underset{\sim}{\ell}$ is perpendicular to the Burgers vector $\underset{\sim}{b}$, and the screw segments where these two vectors are parallel.

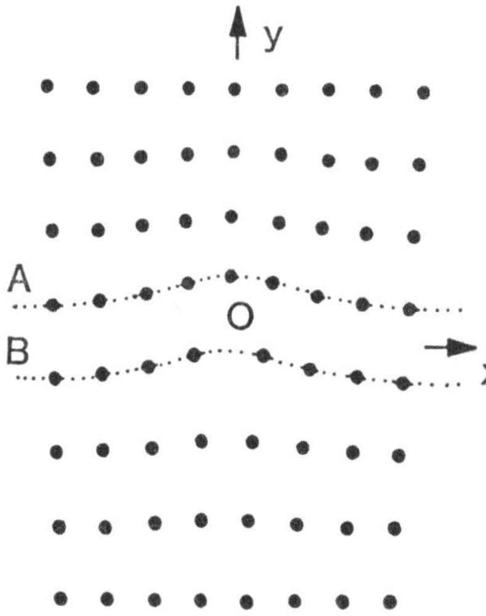

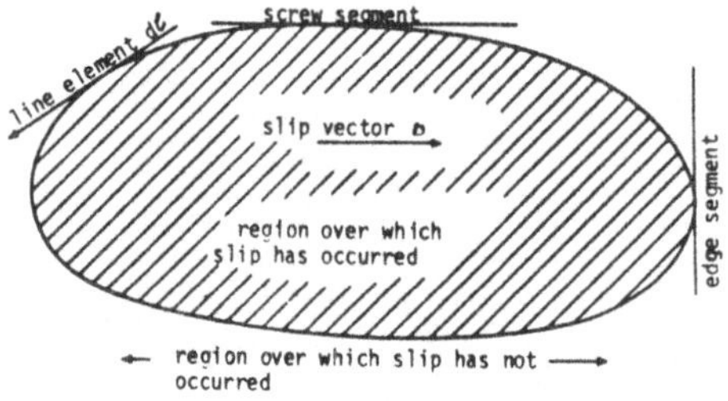

FIGURE 9. FIGURE 10.

FIGURE 9. Core structure of an edge dislocation in a simple lattice. The glide direction is Ox, the glide plane lies between the rows of atoms A and B (31).

FIGURE 10. A dislocation line, showing edge and screw segments. The slip plane is the plane of the figure, and the slip vector **b** lies in this plane. (35).

An additonal feature appears in close-packed lattices. Fig. 11 shows a close-packed plane of atoms with their centres at the points A. The next plane of atoms can have its centres at the points B or at the points C but not at both. A cubic crystal is built by using the stacking sequence ...ABCABCABC... or its twin ...ACBACBACB... A hexagonal crystal follows the sequence ...ABABAB... or ...ACACAC... A local error in this sequence, such as ...ABCABABCA... in the first cubic or ...ABABCBABA... in the first hexagonal stacking is called a stacking fault.

If we take a hard-sphere model of a close-packed structure, and try to slide one plane over another, we find that a sphere in a hollow such as B does not readily move in the direction B_1 B_2 into the nearest crystallographically equivalent position B_2. If we try to slide it in that direction, it moves over the col between spheres A_2 and A_5 into the non-equivalent hollow C_1. If the whole sheet of spheres does this, we have introduced a stacking fault over the whole slip plane. An elastic model behaves differently. If slip starts in a small region, the stacking fault will be bounded by a closed loop of partial dislocation similar to the dislocation loop in Fig. 10. A perfect dislocation leaves an area of perfect crystal behind it on the glide plane, but a partial dislocation leaves an area of stacking fault of surface energy γ. As a result, it is followed by another partial dislocation which removes the stacking fault by making the total displacement a lattice vector according to the scheme.

$$\underset{\sim}{b} = \underset{\sim}{b}_1 + \underset{\sim}{b}_2 \quad (1)$$

or

$$B_1B_2 = B_1C_1 + C_1B_2 \quad (2)$$

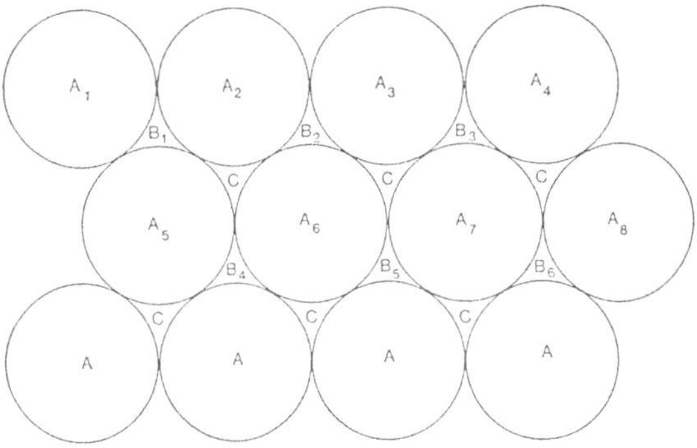

FIGURE 11

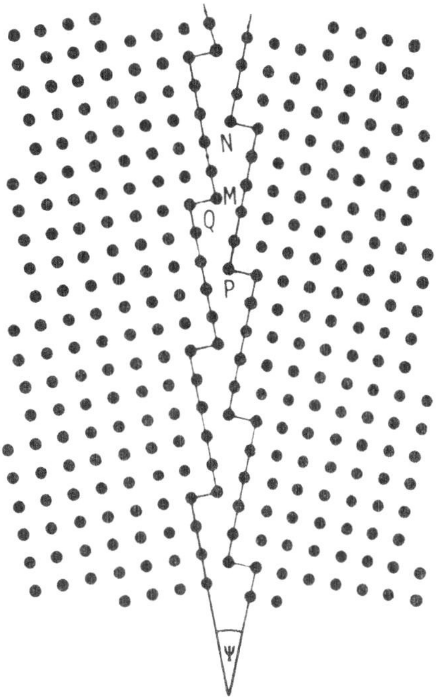

FIGURE 12

FIGURE 11. A close-packed plane of spherical atoms. The centres of the atoms define a hexagonal net A_1 A_2 A_3. The next sheet of atoms may lie above B_2 B_2 B_3, or above C_1 C_2 C_3 (35).

FIGURE 12. Two grains inclined at a small angle ψ. Edge dislocations appear near points N, M, P...

The distance d between the two partial dislocations which form this dissociated dislocation is given approximately by

$$d = 5\ \mu\ b^2/16\pi \quad (3)$$

where μ is the shear modulus of the material. It is possible that the presence of this stacking fault may explain the fact that the observed cross section for the scattering of conduction electrons by dislocations is 10-50 times greater than that calculated (36,37), but no real success has been achieved along these lines. Alternative approaches, such as those which attribute the scattering to the multivaluedness of the potential (38, 39) or to resonant scattering (40-42), do not yet have a firm physical basis. The fundamental difficulty in discussing the influence of dislocations on electron transport phenomena is that the displacements do not tend to zero at infinity, so that simple perturbation-theory treatments do not converge. Treatments based on the deformation potential of a quasi-continuum converge, but fail to take account of the periodic potential of the lattice.

A sheet of edge dislocations with Burgers vectors normal to the sheet forms a small-angle grain boundary (Fig. 12), so that planar defects can be constructed from line defects just as line defects can be constructed from point defects. However, when the misorientation between the grains is not small, their structure is better represented by other models (43). If the Burgers vectors of the edge dislocations lie in the sheet, it separates regions having different lattice parameters. The coarsening of lattices of precipitate particles such as are formed in Ni_3Al probably occurs by the expansion of loops of particle-lattice dislocations with Burgers vectors normal to their planes.

3.1 POLYTYPISM

The structure of some crystals may best be described by saying that they are built of layers of atoms which are stacked in a regular sequence interrupted by stacking faults. These faults do not occur at random, but in a repeating pattern which may have a very long period - up to 594 layers have been observed. Frank (44) showed that these structures, each of which is a distinct phase called a polytype (45), could be explained if the crystals grow around a dislocation or group of dislocations of large total Burgers vector. He has recently shown (46) that, if the crystal grows from a needle-shaped crystal surrounded by liquid or vapour, the Burgers vector of the included dislocation will tend to increase as growth proceeds, so that the formation of groups of dislocations of large total Burgers vector is a likely event. However, only one of these phases is in true thermodymanic equilibrium; the dislocation-assisted growth process determines which of many metastable phases will be formed.

3.2 INFLUENCE OF THE DISLOCATION CORE

Dislocations in semiconductors acquire electric charges. In a simple way one may say that the dislocation core contains broken "dangling" covalent bonds, each bond carrying a single electron. If this single electron is easily lost, the dislocation core acts as a line of donors; if the electron readily pairs with an electron in the conduction band, the dislocation core acts as a line of acceptors. Since the atomic concentration of carriers may be of the order of 10^{-6}, dislocations can change an n-type semiconductor into a p-type, effectively a change of phase. Since the trapped carriers act as recombination centres which downgrade the performance of semiconductor devices, their action is of practical as well as of theoretical interest, and has been widely studied (47, 48).

While both edge and screw dislocations acquire line charges, either because of their dangling bonds in a semiconductor or, because the mean density of the ions is, as a result of second-order elastic effects, less in the strained neighbourhood of a dislocation line than it is in the bulk, edge dislocations also acquire lines of electric dipole moment, because in first-order elasticity the lattice is expanded on one side of the glide plane and compressed on the other side (Fig. 9). In metals alloyed with atoms of a different valency, the alloying atoms will be attracted to lattice sites on one or other side of the glide plane (49). In minerals where the concentration of impurity changes during growth the resulting change of lattice parameter may be accommodated by arrays of edge dislocations whose temperature - dependent dipole moments can render a centrosymmetric crystal pyroelectric.

3.3 EFFECT OF PLASTIC DEFORMATION ON THE COLOURS OF ALLOYS

If the atomic concentration of current carriers (electrons or holes) is c_c, while the concentration of atoms which lie in the cores of defects is c_d, then in a free-electron band the fractional change in the Fermi level produced by the presence of the defects is given roughly by

$$\Delta E_F/E_F = c_d/c_c \qquad (4)$$

We have seen that in a metal c_c is of order unity, while c_d is at most about 10^{-4}, so that $\Delta E_F/E_F$ is very small, while in a semiconductor c_c is of order 10^{-6}, so that $\Delta E_F/E_F$ can be substantial even if c_d is only of order 10^{-6}.

Defects also scatter electrons, reducing the lifetime of their states and introducing an imaginary component $i\delta E_F$ into their energy. The scattering cross-section of a defect is of order b^2, and the defect-limited mean free path of a carrier is

$$\ell = b/c_d \qquad (5)$$

If the Fermi velocity is v_F, the defect-limited mean life of a state is

$$\tau = \ell/v_F = b/c_d v_F \qquad (6)$$

For a free-electron band, the Fermi velocity is given roughly by

$$v_F = hc_c^{1/3}/2mb \qquad (8)$$

where m is the carrier mass, so that

$$\delta E_F = h^2 c_c{}^{1/3} c_d/2mb^2 \qquad (9)$$

Since $$E_F = {}^1/_2\, mv_F{}^2 = h^2 c_c{}^{2/3}/8mb^2 \qquad (10)$$

we find $$\delta E_F/E_F = c_d/c_c{}^{1/3} \qquad (11)$$

Comparing equations (11) and (3), we see that

$$\delta E_F/\Delta E_F = c_c{}^{2/3} \qquad (12)$$

Thus, in a metal, $\delta E_F \simeq \Delta E_F < 10^{-4}$ E_F, while in a semiconductor with $c_c \simeq 10^{-6}$ we have $\delta E_F \simeq 10^{-4}$ ΔE_F.

The finite width δE_F of states at the Fermi level is likely to be associated with a shift of the mean energy of the order of $(\delta E_F)^2/2E_F$, which is negligible both in alloys and in semiconductors.

It is therefore surprising to find substantial evidence (50 - 53) that plastic deformation can change the colours of some ternary alloys of gold, silver and copper to such an extent that (54) the colour changes are of possible interest to manufacturers of jewellery. Several complicating factors could be present (53). While modern ternary phase diagrams have not yet been published, the best diagrams currently available (55, 56) suggest that some of the alloys which have been studied will not be in equilibrium as single phases at room temperature. Plastic deformation may accelerate phase separations from structures which are present in unstable form in annealed specimens. Some single - phase alloys are expected to be in equilibrium in a long-range ordered form at room temperature, and plastic deformation may encourage ordering in a disordered annealed sample. Both long-range (57) and short-range (58) ordering are known to effect the optical properties of alloys, and Jordan showed a striking example of such a colour change at this meeting.

Finally, we know (20) that one atomic species tends to segregate to the surface of an annealed binary alloy. Plastic deformation may simply be uncovering material of the bulk composition. All of these complicating factors can be examined experimentally, but the experiments have not yet been performed. Against these arguments which rest on the assumption of atomic reordering in alloys stand old experimental observations of substantial changes in the optical properties of "pure" Ag (59) and Au (60) on annealing.

3.4 THE TYPE I/TYPE II TRANSITION IN SUPERCONDUCTORS

Superconductors are divided into Type I, which may totally exclude or freely admit a magnetic field. and Type II, which totally exclude weak magnetic fields, freely admit strong magnetic fields, but show a partial field-dependent diamagnetism for a finite range of magnetic fields. The same material may in principle show a Type I phase and a Type II phase, separated by a phase boundary of the usual kind in the (p,T) plane. A material is of Type I if the Ginzburg-Landau parameter κ is less than 1/2 and of Type II if $\kappa > 1/2$. The parameter κ is defined by

$$\kappa = \lambda/\xi_0 \tag{13}$$

where λ is the penetration depth of a magnetic field and ξ_0 is the coherence length of the superconducting wave function in a pure metal. (We evade here the problems concerned with the temperature dependence of λ and ξ_0 and of the relation of λ to the theoretical (London) penetration depth).

In an imperfect metal, containing impurity atoms or lattice defects, ξ_0 is reduced to ξ roughly in accordance with the relation

$$\frac{1}{\xi} = \frac{1}{\xi_0} + \frac{1}{\ell} \tag{14}$$

where ℓ is the mean free path of normal electrons. The value of λ is hardly altered by these defects, and so the presence of defects will increase κ and may lead to the phase change Type I $\rightarrow$ Type II.

Typical values of ξ_0 are given in Table 2.

TABLE 2: Coherence Lengths ξ_0 for some pure Metals (61).

Metal	Sn	Al	Pb	Cd	Nb
ξ_0 (10^{-6}cm)	23	160	8.3	76	3.8
$1/\xi$ (10^3cm^{-1})	44	6	120	13	263

We have seen in equation (4) that $\ell = b/c_d$, and the estimates $b = 3 \times 10^{-8}$ cm and $c_d \leqslant 10^{-4}$ lead to $\ell \geqslant 3 \times 10^{-4}$cm, or $1/\ell \leqslant 3 \times 10^3$cm^{-1}. Comparison with the values of $1/\xi_0$ in Table 2 shows that, except in the case of Al, where λ is so small that κ is also very small, the contribution of $1/\ell$ to $1/\xi$ will always be much less than the contribution of $1/\xi_0$, so that the presence of defects will increase κ only slightly. It is only in the case of Pb that κ lies below, but close to, $1/\sqrt{2}$, so that defect scattering will induce the transition from Type I to Type II. It was first observed by Shaw and Mapother (62) in 1960, and confirmed by Druyvesteyn et al. in 1964 (63). Schenck and Shaw analysed it more thoroughly in 1969 (64), and showed that it did not occur in samples of In and Tl which were given similar plastic deformations. This is probably the only phase change which is caused (and not merely facilitated) by the presence of dislocations.

4. TWO-DIMENSIONAL DEFECTS

We classify two-dimensional defects into stacking faults, which can occur in monatomic crystals, and antiphase boundaries, which can only occur in ordered alloys. In both cases the two-dimensional defect separates two structures which are related to one another by a translation. A twin boundary separates two structures which are related to one another by a special rotation, while a grain boundary separates two structures which are related by an arbitrary rotation. Finally, the free surface separates the structure from an ideal or approximate vacuum.

4.1 STACKING FAULTS

We may define the perfect cubic crystal of section 3 by saying that the sequence ...ABC... is represented by the stacking operator Δ between each plane and the next, while its twin ...CBA... is represented by ∇ between each plane and the next. A stacking fault appears if Δ is replaced by ∇ or ∇ by Δ in a regular sequence. Stacking faults may appear at random in the cubic structure as in the sequence ...ΔΔΔ∇ΔΔΔΔΔΔ∇ΔΔ∇ΔΔΔ..., but if every stacking is faulted we obtain a different perfect structure, the twin ...∇∇∇∇... The twin may be generated by passing a partial dislocation between every two planes. If stacking faults are produced on alternate planes, we again have a perfect structure, the hexagonal close - packed structure of zinc, ...Δ∇Δ∇Δ∇..., while other regular arrays of stacking faults produce the double hexagonal close-packed structure

...ΔΔ∇∇ΔΔ∇∇... of several rare earth metals, or the structure ...ΔΔ∇ΔΔ∇... of samarium.

The change of notation from A,B,C to Δ,∇ allows structures such as ...ABC..., ...BCA... and ...CAB..., which can be derived from one another by a rigid-body translation, all to be represented by the common notation ...ΔΔΔ.... A further change of notation shows the equivalence of structures which can be derived from one another by a rigid-body rotation, such as the cube and its twin. A layer of atoms with stackings ΔΔ or ∇∇ on either side of it is cubically surrounded, and labelled c; one with Δ∇ or ∇Δ on its sides is hexagonally surrounded and labelled h. Then the face-centred cube and its twin are both ...ccc..., the double hexagonal close-packed is ...chch..., samarium is ...chhchh..., and hexagonal closed-packed is ...hhh.... A single stacking fault in f.c.c. is ...ccchhccc..., while ...cchccc... is a twin boundary. This analysis shows that the surface energy of a stacking fault should be close to twice that of a twin boundary, and experiment confirms this prediction rather well (65).

4.2 ANTIPHASE BOUNDARIES

The face-centred cubic lattic is composed of four symmetrically interpenetrating simple cubic lattices. Any one of these lattices may be arbitrarily chosen to define the corners of a cubic cell, and the atoms of the other sublattices then fall in the centres of x, y or z faces of those cubes. In a fully-ordered alloy of composition A_3B, the B atoms may occupy the cube corners and the A atoms the cube faces (Fig. 13(a)). This structure may continue indefinitely in the x direction (Fig. 13(b)), or the x faces of a sheet of cells may form an antiphase boundary, on the other side of which the B atoms no longer occupy cube corners, but occupy x (Fig. 13(c)), y (Fig. 13(d)) or z faces of the original cubic cells. If Fig. 13(c) is joined to Fig. 13(a), the composition A_3B is not disturbed, but the boundary between (Fig. 13(a) and Fig. 13(d) is enriched in B atoms. We call these conservative and non-conservative boundaries respectively. If a non-conservative boundary is repeated every n cells, the resulting structure has the composition $A_{3n+1}B_{n+1}$. If an alloy of this composition is prepared, it may have a tendency to form a super-superlattice with an antiphase boundary every n cells, producing a tetragonal structure with a cell approximately 1x1xn times the size of the original cubic cell. Similar structures with periodic conservative boundaries are also observed. Deviations from stoichiometry provide a clear driving force for the formation of non-conservative boundaries, but non-conservative boundaries in the structure illustrated in Fig 13 have a high antiphase- boundary energy, because they contain many B-B bonds. Conservative boundaries maintain the nearest neighbour ordering and are of low energy, but the mechanisms by which they reduce the free energy are subtler and weaker. The experimental observations are difficult, because the size of the original cell is at the limit of resolution of the electron microscope. They are described in refs. (66) and (67), and in the lectures of van Tendeloo in this meeting. Bak (68) has given a general survey of structures showing complex long-range order.

Complex structures of this kind are more prominent in metal oxides, where the ordering energy of metal and oxygen ions is much stronger. A typical case is that of tungsten trioxide WO_3 and related compounds such as ReO_3 and mixed oxides of tungsten, tantalum and molybdenum (69, 70).

They are composed of regular or slightly distorted octahedra of oxygen with a metal ion in the centre, joined at the corners to make a regular or slightly distorted cubic structure. Planar faults appear in these structures, in which a sheet of oxygen atoms is removed, and the adjacent octahedra join along edges instead of at corners. If the planes are of type (10m), and are separated by n cells of the original structure, the resulting oxide has the composition M_nO_{3n-m+1}. These structures have been observed with m = 2 and n = 8, 9, 10, 11, 12 or 14, with m = 3 and n = 18, 20, 26 and all values between 36 and 51, and with m = 4 and n in the range 55-70. A high resolution electron micrograph of a crystal with m = 3, n = 18 is shown in Fig. 14. Artificial super-lattices on much the same scale are becoming of technological importance (71,72).

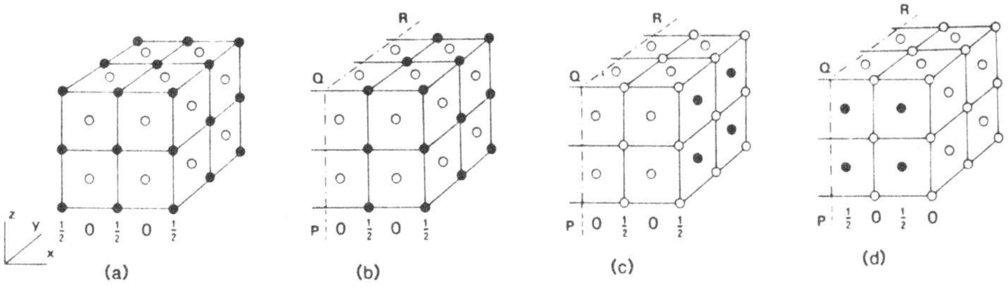

FIGURE 13. An ordered A_3B alloy. a) The A stoms (empty circles) are at cube faces, the B atoms (filled circles) at cube corners. b) The continuation of the same structure. c) To the right of PQR, B atoms are in the x faces of cubes. d) To the right of PQR, B atoms are in the y faces of cubes. The concentration of B atoms in each x plane is shown.

4.3 TWINS

Martensitic phase transformations are those which occur by a coherent shear process. Crystallographically equivalent shears may occur on several systems; for example, an f.c.c. crystal can transform into an h.c.p. crystal with any one of the f.c.c. $\langle 111 \rangle$ planes as base plane. If different portions of the parent crystal transform on different systems, large internal stresses are set up. The associated elastic energy can be greatly reduced if the martensitic product adopts a finely-twinned structure (73, 5), as Khachaturyan has explained in detail in his lectures. These multiply-twinned materials can be "superelastic", one twin of each pair growing reversibly at the expense of the other under the influence of an applied stress.

4.4 GRAIN BOUNDARIES

Atoms in a grain boundary have higher energies than those in the perfect crystal, but they also have higher configurational and vibrational entropies, so it is conceivable that grain boundaries could have negative free energies and be present in equilibrium. A crystal may be formed by assembling rodlike molecules in layers perpendicular to the rods, and then stacking the layers. If grain boundaries parallel to the rods, and formed by tilts about axes parallel to the rods, have negative free energies, while all other grain boundaries have positive free energies, the crystal undergoes the phase transformation solid → smectic A. Turbostratic graphite, in which layers of carbon atoms are rotated randomly about their normals, provides another example of a phase saturated with grain boundaries of a particular class.

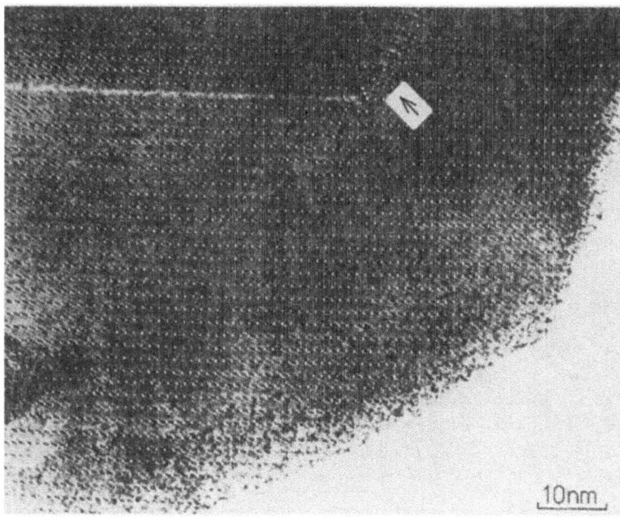

FIGURE 14. Electron micrograph of an almost perfectly ordered array of planar defects in WO_3, leading to the composition $W_{16}O_{52}$ (70).

4.5 THE FREE SURFACE

Every single-crystal sample posesses this defect. If the surface perturbation is noticeable to a depth of 3b, a spherical particle 2n atoms in diameter contains $^4/_3\ \pi\ n^3$ atoms, of which $3 \times 4\pi n^2$ are perturbed. If we again assume that a phase change may occur when $^1/_{10}$ of the atoms are perturbed, the largest sample for which the surface will induce a phase change is 2n atoms in diameter, where n = 90. Experiment shows that their small size appreciably affects the phase behaviour of metal particles as large as 40 nm in diameter, corresponding to n ≈ 70, of the order expected.

There is little doubt that the melting of pure metals is initiated at the surface, and much modern work (74, 75) assumes that a liquid sheath forms on the surface, and that melting of the sample requires this sheath to form and to propagate inwards. These processes have been observed over a temperature range of 100 K by ion scattering measurements on a {110} surface of lead (76), as many as 15-20 molten layers being observed just below the melting point. Measurements of the complex dielectric constant of KCl near the melting point (77) show pre- and post-melting phenomena over a range of 10 K. The thermodynamic order of any phase change at a surface is generally different from that in the bulk (eg 78). The effect has been observed experimentally in the case of the antiferromagnetic transition in NiO (79). One of the most interesting cases is that of a Type II superconductor in a tangential magnetic field lying between H_{c_2} and H_{c_3}. The normal material in the bulk is covered by a sheath of a gapless superconducting phase. A homogeneous topological sphere of a Type I superconductor is a perfect diamagnet, but if it is riddled with long channels it behaves as a permanent magnet.

Alloy particles with diameters of 40 nm or less show phase diagrams markedly different from those of the bulk (80). In the system SnBi the eutectic is depressed by about 80 K, and the mutual solubilities are greatly enhanced. In the system InSn, the two-phase regions appear to be replaced by amorphous regions which remain stable up to 250 $^{\circ}$C, or for over 100 hours at room temperature.

5. THREE-DIMENSIONAL DEFECTS

A structure undergoing a change from phase A to phase B, whether by shear or by diffusion, usually has to pass through an inhomogeneous transitional state I. While the free energy of B is less than that of A, it will often happen that the free energy of I is greater than that of A. If the system is controlled by varying a parameter such as temperature T, it is often important in practice to know not only the value of T for which phases A and B are in equilibrium, but also the values of T for which the changes A $\rightarrow$ B and B $\rightarrow$ A will occur at an observable rate. These problems are discussed in reference (73). A more recent treatment of shear transformations is to be found in reference (81). We outline in section 5.1 the considerations, treated more fully in reference (81), which govern diffusional transformations. Binder's lectures in this volume give a thorough account of recent developments.

The structure of some covalent crystals may conveniently be described as a piling of structural blocks, and this idea can be extended to complex alloy phases. Finally, there is evidence of a sort of anti-defect, which consists of the presence of a high concentration of almost perfectly crystalline regions in a material which is usually considered to have a typically amorphous structure.

5.1 INHOMOGENEOUS PHASE TRANSITIONS

Although intermediate cases are frequent, we concentrate our attention on phase transitions which occur either by the nucleation and growth of isolated particles of the new phase, or by the appearance and intensification of a concentration wave. We have already mentioned the elastic strain energy which appears if a small region of a new phase is formed by shear within an untransformed matrix. The transformation cannot occur if the chemical free energy of the unstrained product phase is only slightly less than that of the parent phase; the difference in chemical free energies must be large enough to compensate for the elastic energy.

Consider now a diffusional transformation, in which a homogeneous alloy decomposes into two phases of different compositions. This decomposition may occur, in our simplified picture, either by nucleation and growth or by the formation of a concentration wave. Suppose that, at a given temperature T, the free energy of an alloy with a given structure depends on its concentration in the way shown in Fig. 15. Then any alloy with a composition lying between A and B can reduce its free energy by separating into large unstrained regions, some of concentration A and some of concentration B. This might occur by the nucleation of small particles of concentration A, the concentration of the neighbouring regions of the matrix tending towards concentration B. The lattice parameters of phases A and B are usually different. If the particle of composition A is coherent with the matrix, that it to say, there is a 1:1 corr-

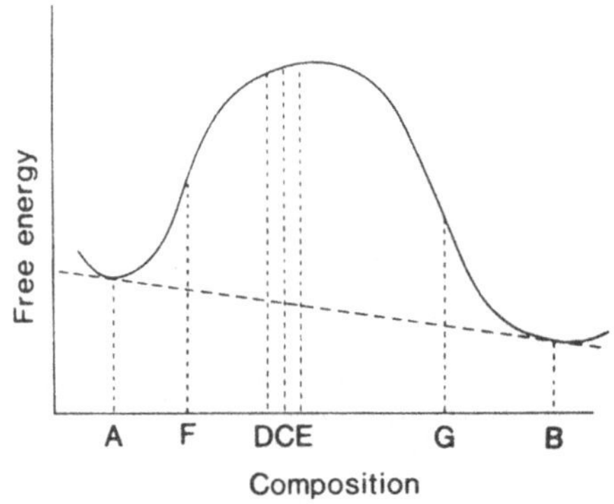

FIGURE 15. Free energy as a function of composition in an alloy. Alloys with compositions between A and B are unstable, but only alloys with compositions between F and G are unstable against small fluctuations in composition.

espondence between lattice sites in the matrix and in the precipitate, then elastic strain energy is present, and the free energy curve of Fig. 15 must be replaced by one in which account is taken of this elastic energy. The elastic energy need not be present if the precipitate is not coherent with the matrix, but there is then a surface energy at their interface. For a small nucleus of radius r, this surface energy, proportional to r^2, will exceed the chemical free-energy change, which is proportional to r^3, so that nucleation is possible only with the aid of thermal activation.

An alternative route to the phase separation is the formation of a concentration wave. A homogeneous solid solution of composition C may develop a concentration wave in which the minimum and maximum concentrations are D and E respectively. In the case shown, where the free-energy curve is concave downwards, this fluctuation decreases the free energy, is unstable, and leads to a phase separation. This remains true for all positions of C lying between F and G, which correspond to the inflections of the free-energy curve. The points F and G at temperature T are points on the spinodal for this phase. The concentration wave is a defect in the original structure, and in the concentration ranges AF and GB the energy required to form this defect acts as a barrier to the achievement of phase equilibrium.

5.2 STRUCTURES DETERMINED BY THE ASSEMBLY OF SUBUNITS

One of the most striking examples of structures determined by the assembly of subunits is provided by the solutions of PbS in Sb_2S_3 (82). The structure of Sb_2S_3 consists of long ribbons, seen end on in Fig. 16(a). The width of the ribbon is composed of four square pyramids, or half-octahedra, of sulphur atoms, with antimony atoms in the middle of each octahedron. By varying the width of the ribbon to 2n pyramids (Fig. 16(b), the composition can be altered to M_nS_{n+2}. More complicated stackings, such as that of Fig. 16(c), are also observed, in which intergrowths of ribbons of different widths occur.

Whereas the sulphide structures which have been considered consist of tightly-bonded units separated by sheets of weak bonding along which the crystal readily cleaves, every atom in an alloy structure is usually surrounded. For atoms of roughly the same size, each atom has Z = 12 neighbours, arranged roughly in cubic, hexagonal or dodecahedral configuration. We may call Z = 12 the ideal coordination, and we have already discussed how lattices are formed from the cubic and the hexagonal configurations. The dodecahedral configuration cannot be used alone to fill space, but it combines with other packing structures as readily as do the cubic and hexagonal configurations. Frank and Kasper (83, 84) showed that configurations with Z = 14, 15 and 16, and no others, are also likely to occur, and called these, together with Z = 12, normal coordinations.

We may regard the coordinations Z = 14, 15 and 16 as the characteristic defects of the ideal coordination Z = 12. Some authors (85, 86) consider these configurations as the cores of nodes of a disclination network which generates the actual structure from a simpler structure, but this approach seems to raise some conceptual difficulties. Frank and Kasper showed that atoms with normal coordinations would aggregate in a limited number of ways, mostly forming multilayer structures which themselves could be stacked according to simple rules. In this way they were able to explain the formation of very many complex metallic and alloy structures.

5.3 CRYSTALLINE REGIONS IN AMORPHOUS STRUCTURES

An amorphous region in a crystalline matrix is a defect in the state of complete order; a crystalline region in an amorphous matrix is a defect in the state of complete disorder. It is natural to expect that, if an amorphous material is prepared in a metastable state and then annealed, it will initially develop small regions of almost perfect crystalline order. Recent observations on a palladium-silicon alloy glass (87, 88) (Fig. 17) and on amorphous silicon (89) at atomic resolution have unexpectedly shown regions of crystalline order ten or more atoms in diameter in samples which would be regarded as typically amorphous. While the interpretation of these observations is somewhat controversial, the idea of a crystalline region as a defect in an ideally amorphous structure is established.

6. TRANSIENT DEFECTS

The important transient defects in alloys are phonons and magnons. At the Debye temperature, which is usually of the order of room temperature, each normal mode of the lattice has about one quantum of excitation. There are three modes per atom, and so at these temperatures the concentration of phonons is about three per atom. This high concentration scatters the conduction electrons, and renders the Fermi surface diffuse. The phonon-phonon interaction leads to thermal expansion, altering the number of electrons per unit volume and so the Fermi level. In a magnetic alloy, the destruction of ferromagnetism at the Curie temperature corresponds to the introduction of about one magnon per atom, and so it is not surprising that this transition also has a large influence on phase stability.

The earlier parts of this article have concentrated on the energies of phases. The high concentrations of phonons and magnons we are now considering are present in thermal equilibrium at finite temperatures, and the relative stability of two phases is now determined by their free energy rather than by their energy.

While excitons do not occur in metals there is a possibility that an excitonic phase containing a high concentration of excitons may appear close to metal-insulator transitions at low temperatures.

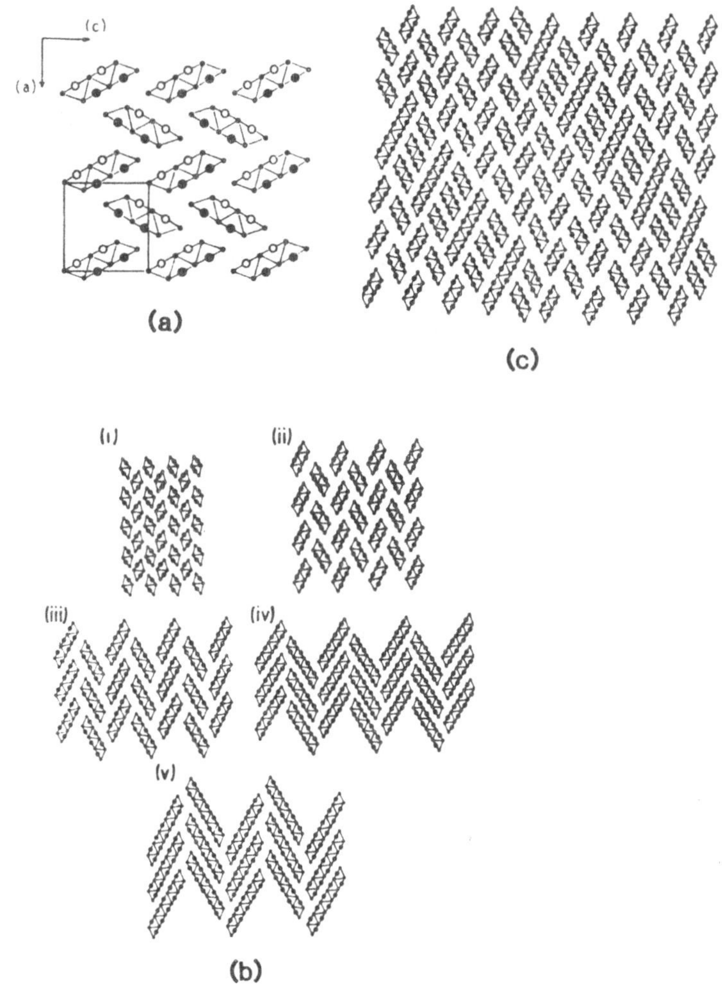

FIGURE 16. a) Projection of the structure of Sb_2S_3 on to (010), showing infinitely long ribbons, each four half-octahedra wide. Large circles: Sb, small circles: S. Filled circles: integral distances above plane of figure. Open circles: half-integral distances above plane of figure. b) Variants of a) with (i) two, (ii) four, (iii) six, (iv) eight and (v) ten half-octahedra in the width of a ribbon giving compositions (i) M_2S_4, (ii) M_4S_6, (iii) M_6S_8, (iv) M_8S_{10} and (v) $M_{10}S_{12}$. c) A close-packed array of ribbons of two different widths (82).

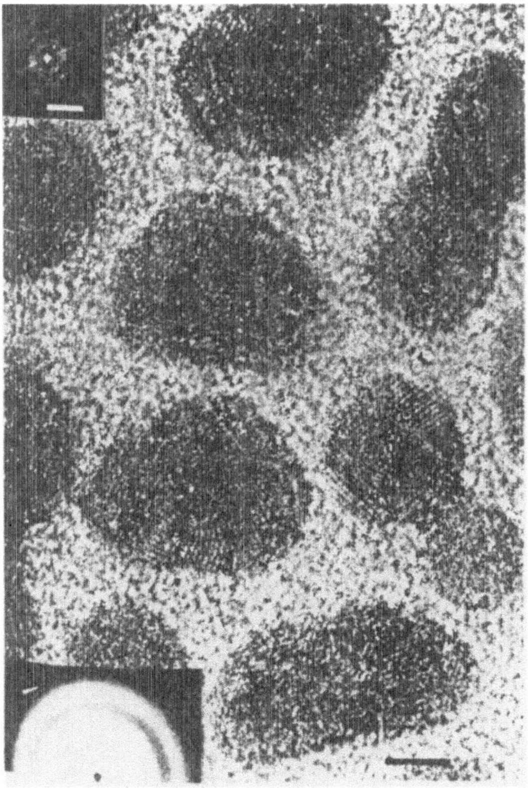

FIGURE 17. High-resolution electron micrograph of amorphous palladium-silicon alloy, showing domains of crystalline order (88).

6.1 PHONONS

According to Kittel (90), an atom of mass M vibrating at temperature T in a lattice of Debye temperature Θ scatters electrons with a scattering cross-section of order $\hbar^2T/Mk\Theta^2$. The number of phonons per atom is $3T/\Theta$, so the scattering cross-section $\sigma_{el\ ph}$ of a phonon for an electron is given by

$$\sigma_{el\ ph} = \hbar^2/3Mk\Theta \qquad (15)$$

We may compare this with the scattering cross-section for an electron of valency differing by unity from that of the matrix, which according to (91) is

$$\sigma_{el\ v} = (e^2/E_F)^2 \qquad (16)$$

where E_F is the Fermi energy, which in the free-electron approximation is given by

$$E_F = \frac{\hbar^2}{2m}(3\pi^2 N)^{2/3} \tag{17}$$

Here, m is the mass of the electron, and N is the number of electrons per unit volume. We take

$$N \approx (1/5a_0)^3 \tag{18}$$

where a_0 is the Bohr radius

$$a_0 = h^2/4\pi me^2 \tag{19}$$

From equations (15) - (19) we obtain

$$\frac{\sigma_{el\ ph}}{\sigma_{el\ v}} \approx 0.01\ \frac{m^2e^4}{h^2Mk\Theta} \tag{20}$$

The binding energy of the hydrogen atom is

$$E_H = \frac{2\pi^2 me^4}{h^2} \tag{21}$$

and this allows us to write

$$\frac{\sigma_{el\ Ph}}{\sigma_{el\ v}} \approx 5\times10^{-4}\ \frac{m}{M}\frac{E_H}{k\Theta} \tag{22}$$

Taking the values $m/M = 8.7 \times 10^{-6}$ and $E_H/k\Theta = 451$ for copper, we find

$$\frac{\sigma_{el\ ph}}{\sigma_{el\ v}} = 1.5 \times 10^{-6} \tag{23}$$

This means that the resistance produced in copper by 3 phonons per atom at room temperature is equal to that produced by 5 parts per million of zinc or nickel, while a resistance ratio of 1 000 between room temperature and liquid helium temperature implies an impurity content of only 5 parts in 10^9. This is compatible with the observation (92) by measurements of the Kondo effect that the concentration of transition-metal impurities in gold can be reduced to 1 part in 10^9. On the other hand, it suggests that high-resistance alloys could have 200 000 times the resistivity of copper at room temperature, whereas the actual ratio is only about 60. Ziman's estimate (91) that the phonon-limited free path of an electron at the melting point is about 50 lattice spacings, corresponding to about 250 lattice spacings at room temperature, suggests a ratio of about 250. It is in any case clear that the blurring of the Fermi surface produced by electron-phonon scattering is much less than that produced by the scattering of electrons by alloying atoms.

The scattering cross-section of $\hbar^2 T/Mk\Theta^2$ corresponds to an electron mean free path of

$$\Lambda_{el\ ph} = b^3 Mk\Theta^2/\hbar^2 T \tag{24}$$

a lifetime of $\Lambda_{el\ ph}/v_F$, and an uncertainty in energy of

$$\delta E_F = hv_F/\Lambda_{el\ ph}$$

$$= h^3 v_F T/4\pi^2 b^3 Mk\Theta^2 \tag{25}$$

Using equation (7) with $c_c = 1$, this becomes

$$\delta E_F = h^4 T/8\pi^2 b^4 Mmk\Theta^2 \tag{26}$$

The ratio of this to the thermal width of the Fermi surface is

$$\frac{\delta E_F}{kT} = \frac{h^4}{8\pi^2 b^4 Mmk^2\Theta^2} \tag{27}$$

Writing (17) in the form $E_F = h^2/8mb^2$ reduces this to the form

$$\frac{\delta E_F}{kT} = \frac{8}{\pi^2}\frac{m}{M}\left(\frac{E_F}{k\Theta}\right)^2 \tag{28}$$

Inserting the values for copper gives $\Delta E_F/kT = 1.4$, so that the broadening of the Fermi surface by phonon scattering is comparable to the thermal broadening.

We now turn to phonon-phonon interactions. In equilibrium, their principal effect is to cause thermal expansion. This expansion is conveniently expressed in terms of Grüneisen's constant Γ, which has a value of about 2 for most metals. The fractional increase in volume on heating from zero to temperature T is given by

$$\frac{\Delta V}{V} = \frac{\Gamma C_V T}{K_0 V} \tag{29}$$

where C_V is the thermal capacity, K_0 is the bulk modulus and V is the volume. At this temperature T the number of phonons per atom is $3T/\Theta$, and so the value of $\Delta V/V$ produced by one phonon per atom is

$$\frac{\Delta V}{V} = \frac{2C_V\Theta}{3\ K_0 V} \tag{30}$$

For 1 mole of copper, $C_V\Theta = 7.8 \times 10^{10}$ erg and $K_0V = 9.5 \times 10^{12}$ erg, so, for one phonon per atom,

$$\frac{\Delta V}{V} = 5.5 \times 10^{-3} \tag{31}$$

At room temperature, 3 phonons per atom lead to an expansion of $\Delta V/V = 1.6 \times 10^{-2}$. In a free-electron model, this reduces the Fermi energy by an amount equal to that produced by the addition of 1.6 percent of nickel, assumed zero-valent.

6.2 MAGNONS

As Hasegawa, Pettifor and others (93-95) have shown, the phase equilibria of the transition metals are largely controlled by their magnetic interactions. Magnons present in thermal equilibrium are defects in the ordered magnetic structure. When their concentration becomes high the magnetic order is destroyed and the destruction of magnetic order leads to changes in the crystal structure.

6.3 THE EXCITONIC PHASE

The semiconductor-semimetal transition at moderate temperatures is continuous, and not a phase change. However, at lower temperatures these two phases develop second-order transitions to "distorted" structures with either spin-density or charge-density waves. At very low temperatures there is probably a first-order phase transition between these two distorted structures. This transition has never been seen, but might be observable in pure ytterbium (96).

REFERENCES

1. F. Seitz, in Imperfections in Nearly Perfect Crystals, John Wiley, New York, 3, 1952.
2. J. Molenaar and W. H. Aarts, Nature, Lond., **166,** 690,1950.
3. T. H. Blewitt, R. R. Coltman and J. K. Redman, in Defects in Crystalline Solids, The Physical Society, London, 369, 1955.
4. K. F. Berggren, Phil Mag. **27,** 1027, 1973.
5. G. B. Olsen and M. Cohen in Dislocations in Solids **7,** North-Holland, Amsterdam, 295, 1986.
6. M. W. Thompson, Defects and Radiation Damage in Metals, Cambridge University Press, 1969.
7. P. H. Dederichs, in Vacancies '76, The Metals Society, London, 22, 1977.
8. H. R. Paneth, Phys. Rev. **80,** 708, 1950.
9. G. H. Vineyard, J. Phys. Soc. Japan Suppl. III **18,** 144, 1963.
10. D. P. Seraphim, N. R. Stemple and D. T. Novick, J. Appl. Phys. **33,** 136, 1962.
11. R. E. Villagrana and G. Thomas, Phys. Stat. Sol. **9,** 499, 1965.
12. F. R. N. Nabarro, in Strength of Solids, The Physical Society, London, 75, 1948.
13. C. H. Woo and W. Frank, J. Nucl. Mater. **137,** 7, 1985.
14. K. Krishan, Radiation Effects **66,** 121, 1982.
15. J. H. Evans, Radiation Effects **10,** 55, 1971.
16. V. K. Sikka and J. Moteff, Crystal Lattice Defects **3,** 113, 1972.
17. J. Frenkel, Kinetic Theory of Liquids, Clarendon Press, Oxford, 36-40, 1946.
18. R. Bray, Solid State Commun, **60,** 867, 1986.
19. E. R. Weber Solid State Commun, **60,** 871, 1986.
20. J.C. Hamilton, Phys. Rev. Lett **42,** 989, 1979.
21. A. R. Miedema, F. R. de Boer and P. F. de Chatel, J.Phys.F **3,** 1588, 1973.

22. A. R. Miedema, J. Less-Common Met. **32,** 117, 1973.
23. A. R. Miedema, R. Boom and F. R. de Boer, J. Less-Common Met. **41,** 283, 1975.
24. R. Boom, F. R. de Boer and A. R. Miedema, J. Less-Common Met. **46,** 271, 1976.
25. A. R. Miedema, Philips Tech. Rev. **36,** 217, 1976.
26. S. Geller, Solid Electrolytes, Springer, Berlin, 1977.
27. M. B. Salamon, Physics of Superionic Conductors, Springer, Berlin, 1979.
28. A. J. Bradley and A. Taylor, Proc. Roy. Soc. Lond., A **159,** 56, 1937.
29. H. Lipson and A. Taylor, Proc. Roy. Soc. Lond.,A **173,** 232, 1939.
30. G.V. Raynor, Progress in Metal Physics **1,** 1, 1949.
31. F. R. N. Nabarro, Theory of Crystal Dislocations, Clarendon Press, Oxford, 1967.
32. Y. Bouligand, Dislocations in Solids **5,** 299, 1980.
33. H. Träuble, Phys. Stat. Sol. **25,** 373, 1968.
34. N. Rivier, Phil. Mag. **40,** 859, 1979.
35. F. R. N. Nabarro in Mechanical and Thermal Behaviour of Metallic Materials, North-Holland, Amsterdam, 35, 1982.
36. J. M. Ziman, Electrons and Phonons, Clarendon Press, Oxford, §9.4, 1960.
37. Reference (31),§9.2.2.
38. Y. Yosida and K. Kawamura, Z. Physik B **32,** 355, 1979.
39. H.-J. Huang and K. Kawamura, Solid State Commun., **41,** 939, 1982.
40. R. A. Brown, J. Phys. F **7,** 1269, 1977.
41. R. A. Brown, J. Phys. F **7,** 1283, 1977.
42. R. A. Brown, J. Phys. F **8,** 1467, 1978.
43. V. Vitek, Dislocations 1984, CNRS, Paris, 435, 1984.
44. F. C. Frank, Adv. Physics **1,** 91, 1952.
45. A. R. Verma and P. Krishna, Polymorphism and Polytypism in Crystals, John Wiley, New York, 1966.
46. F. C. Frank, Phil. Mag., in the press.
47. R. Labusch and W. Schröter, Dislocations in Solids **5,** 127, 1980.
48. H. Alexander, Dislocations in Solids **7,** 113, 1986.
49. A.H. Cottrell, S. C. Hunter and F. R. N. Nabarro, Phil. Mag. **44,** 1064, 1953.
50. G. Tammann, Zeits. anorg. allg. Chemie **107,** 1,1919.
51. G. Tammann and C. Wilson, Zeits anorg. allg. Chemie **173,** 156, 1928.
52. W. Boas, in Dislocations and Mechanical Properties of Crystals, New York, John Wiley, 406, 1957.
53. E. L. Yates, Aust. J. Phys. **16,** 40, 1963.
54. Gold u. Silber u. Uhren u. Schmuck, 1975 No. 12, p.47.
55. R. Kikuchi, J. M. Sanchez, D. de Fontaine and H. Yamauchi, Acta Metall. **28,** 651, 1980.
56. H. Yamauchi, H.A. Yoshimatsu, A. R. Forouhi and D. de Fontaine, Precious Metals, Proc. Fourth Internat. Precious Metals Inst. Conf., Internat. Precious Metals Inst., New York p.241.
57. W. Scott and L. Muldawer, Phys. Rev. **39,** 1115, 1974.
58. J. B. Andrews, R. J. Nastasi-Andrews and R. E. Hummel, Phys. Rev. B **22,** 1837, 1980.
59. H. Margenau, Phys. Rev. **33,** 1035, 1929.
60. L. G. Schultz and F. R. Tangherlini, J. Opt. Soc. Amer. **44,** 362, 1954.
61. R. Meservey and B. B. Schwartz, in R. D. Parks (ed), Superconductivity **1,** 117, 1969.

62. R. W. Shaw and D. E. Mapother, Phys. Rev. **118,** 1474, 1960.
63. W. F. Druyvesteyn, D. J. van Ooijen and T. J. Berben, Phys. Rev. **36,** 58, 1964.
64. J. F. Schenck and R. W. Shaw, J. Appl. Phys. **40,** 5165, 1969.
65. J. P. Hirth and J. Lothe, Theory of Dislocations, 2nd edition, New York, John Wiley, 1982.
66. S. Amelinckx, Chemica Scripta **14,** 197, 1978-79.
67. A. Loiseau, G. van Tendeloo, R. Portier and F. Ducastelle, J.Phys. (Paris) **46** 595, 1985.
68. P. Bak, Physics Today, December 1986, p.38.
69. A. Magnéli, Acta Crystallogr. **6,** 495, 1953.
70. R. J. D. Tilley, Chemica Scripta **14,** 147, 1978-79.
71. G. H. Döhler, Scientific American **249,** (5), 118, Nov. 1983.
72. R. W. Cahn, Nature, Lond. **324,** 108, 1986.
73. J. W. Christian, The Theory of Transformations in Metals and Alloys, Oxford, Pergamon, 1965.
74. R. W. Cahn, Nature, Lond. **323,** 668, 1986.
75. G. L. Allen, R. A. Bayles, W. W. Gile and W. A. Jesser, Thin Solid Films **144,** 297, 1986.
76. J. W. M. Frenken and J. F. van der Veen, Phys. Rev. Lett, **54,** 134, 1985.
77. J. Tateno, Nature, Lond. **325,** 43, 1987.
78. R. Lipowsky and W. Speth, Phys. Rev. B **28,** 3983, 1983.
79. T. Wolfram, R. E. Dewames, W. F. Hall and P. W. Palmberg, Surface Science, **28,** 45, 1971.
80. G. L. Allen and W. A. Jesser, J. Crystal Growth, **70,** 546, 1984.
81. D. de Fontaine, Solid State Physics, **34,** 74, 1979.
82. R. J. D. Tilley, A. C. Wright and D. J. Smith, Proc. Roy. Soc. Lond. A **408,** 9, 1986.
83. F. C. Frank and J. S. Kasper, Acta Crystallogr. **11,** 184, 1958.
84. F. C. Frank and J. S. Kasper, Acta Crystallogr. **12,** 483, 1959.
85. D. R. Nelson, Phys. Rev. Lett. **50,** 982, 1983.
86. J. F. Sadoc, J. Phys. (Paris) Lett. **44,** L-707, 1983.
87. V. E. Cosslett and D. J. Smith, Chemica Scripta **14,** 39, 1978-79.
88. P. H. Gaskell, D. J. Smith, C. J. D. Catto and J. R. A. Cleaver, Nature, Lond. **281,** 465, 1979.
89. J. C. Phillips, J. C. Bean, B. A. Wilson and A. Ourmazd, Nature, Lond. **325,** 121, 1987.
90. C Kittel, Elementary Solid State Physics John Wiley, New York, p.169, 1962.
91. J. M. Ziman, Principles of the Theory of Solids, Cambridge University Press, 1964, p.190.
92. J. Kopp, J. Phys. F **5,** 1211, 1975.
93. D. G. Pettifor, CALPHAD **1,** 305, 1977.
94. H. Hasegawa and D. G. Pettifor, Phys. Rev. Lett. **50,** 130, 1983.
95. H. Hasegawa, M. W. Finnis and D. G. Pettifor, J. Phys. F **15,** 19, 1985.
96. B. I. Halperin and T. M. Rice, Solid State Physics **21,** 115, 1968.

PHASE STABILITY BY THE ARTIFICAL CONCENTRATION WAVE METHOD

A. F. Jankowski* and T. Tsakalakos**

*Lawrence Livermore National Laboratory, Livermore CA
**Rutgers University - College of Engineering, New Brunswick, NJ

1. INTRODUCTION

The effect of system fluctuations on the stability of alloy phases has been addressed by Khachaturyan.[1] A thermodynamic analysis of stability and transformations considers the dominance of either the thermal energy or interaction energy. At low temperatures, the interaction energy dominates. The phase state present may either be ordered, disordered, or some mixture of ordered and/or disordered. A redistribution of atoms over the crystal lattice sites accompanied by lattice rearrangements, further complicates the ordering or decomposition. The ordering and decomposition states may be interpreted by the loss of stability of a disordered solution with respect to static concentration waves (SCW). The loss of stability with respect to concentration waves characterized by asymptotically small wave vectors generates a, so called, spinodal miscibility. The loss of stability with respect to concentration waves characterized by finite wave vectors leads to the order-disorder transformation. The wave vector may be represented as $2\pi/\lambda$, where λ represents the repeat period of the wave, i.e. the scale of heterogeneity.

Ordering is a first-order phase transformation for finite concentrations at the transformation temperature. For infinitesimal amplitudes, a second-order phase transformation results. Therefore, a disordered state is unstable with respect to finite concentration waves at the temperature of the first-order phase. However, it is stable with respect to infinitesimal fluctuations. Supercooling of a disordered alloy is an example of such a metastable state.

Static concentration waves can be related to the equilibrium long range order (lro) parameters. The approach to the phase stability problem can therefore be made by means of SCW. Thus, the artifical SCW produced by evaporation/deposition techniques can be of great importance in determination of phase stability. In addition, the artifical SCW provides structural and physical parameters which determine the specific phase transformation.

2. ARTIFICAL CONCENTRATION WAVE APPROACH

2.1 Determination of Critical Temperatures

The composition-modulated foils are very well suited for characterization of the stability of a solid solution and for measurement of critical temperatures, using Khachaturyan's description for ordering (i.e. first- and second-order phase transformations) with respect to SCW. By preparing samples of different wavelengths and isothermally annealing, we can measure the decay or growth rate of the concentration waves. The $\theta/2\theta$ Bragg reflection satellite intensities $I_{\pm}$ are proportional to the square of the amplitude, that is

$$\ln \{I_{\pm}(t)/I_{\pm}(0)\} = 2R(k)t \qquad (1)$$

G. M. Stocks and A. Gonis (eds.), Alloy Phase Stability, 585–589.

where R(k) is the amplification (>0) or decay rate (<0) of the waves. Thus from a plot of ln $\{I_{\pm}(t)/I_{\pm}(0)\}$ vs. time, the slope 2R(k) can be computed. Consider the following examples of Cu/Ni and Cu/NiFe.

Slow kinetics of the Cu/Ni system at low temperatures have placed the critical temperature in doubt. Evidence from neutron scattering for a 43.5 Cu-56.5 Ni alloy suggest $320 < T_c < 340°C$.[2] To validate this suggestion a (111) Cu/Ni composition modulated foil with a 5-nm wavelength was fabricated for analysis.[3] To obtain an infinitesimal amplitude a high amplitude foil was annealed at 450°C for 2400 seconds. This reduced the amplitude to less than 8% of its initial value, whereas a 325°C - 72 hour anneal decreased the satellite intensity to only half its initial value. The foil was then annealed again at 325°C for 3 days. During this anneal there was an increase in intensity of 5%. Further annealing at 340°C produced a definite decay in amplitude. (The full aging treatment is shown in Fig. 1.). This aging behavior, therefore, suggests $325 < T_c < 340°C$. The large-amplitude wave decayed at 325°C because most of its spatial distribution lay outside the critical regime. The small-amplitude wave was confined within the critical region, however, so that a small growth was observed during aging for long periods.

The (111) Cu/NiFe composition modulated system has been recently examined for determination of the critical temperature[4] as well. Samples with wavelengths of 3.9, 4.3, and 7.1 nm and amplitudes of less than 10 at.% were isothermally annealed at 320, 345 and 400°C as shown in Fig. 2. The most rapid decay in amplitude was observed at 400°C with almost none at 345°C, whereas growth was evident at 320°C. The behavior is remarkably similar to that of Cu/Ni. These results suggest $320 < T_c < 345°C$. The contribution of short-circuit diffusion in the layered structures can be neglected as the kinetics are representative of bulk properties, even at the lowest annealing temperatures. For example, the composition amplitude was observed to grow with time for Cu/NiFe annealed at 320°C. Deviation from the linear diffusion equation is observed at short times for the following reasons. The enhanced diffusivity results primarily as recrystallization and grain growth drive the early stages of annealing. A second contributing factor is the nonlinearity of the diffusion equation, as formulated in detail.[5] A third factor influencing the enhanced diffusion at short times is the variation of modulus Y[111] with wave amplitude.[6]

2.2 Kinetics - Temperature Dependent Interdiffusion

A measure of the kinetics for a system is provided by the interdiffusion coefficients $\tilde{D}_B$. The interdiffusivity may be experimentally determined from the amplification factor R(k) as follows

$$R(k) = -B^2(h)\ \tilde{D}_B \tag{2}$$

$B^2(h)$, the dispersion relationship is given by

$$B^2(h) = a^{-2} \sum_r [1 - \cos(\underset{\sim}{k} \cdot \underset{\sim}{r})] \tag{3}$$

where a is the lattice parameter, h is the (h_1, h_2, h_3) triplet of the wave vector k, and the summation is taken over nearest neighbor sites r to the origin. Measurements of the satellite intensities ratioed to the Bragg reflection, following isothermal anneals, can then be used to calculate $\tilde{D}_B$ incorporating eqns. (1) and (2). The variation in $\tilde{D}_B$ versus $B^2(h)$ is

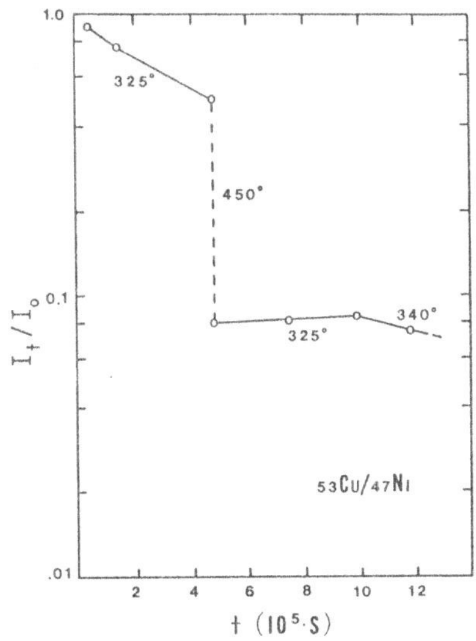

Fig. 1 Aging treatment of a 5 nm modulated Cu-Ni foil with 53 at.% Cu. Log (I/I_o) is plotted vs. annealing time.

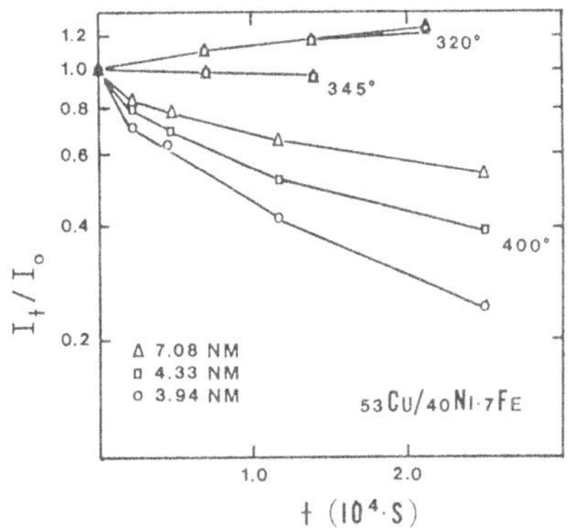

Fig. 2 The decay or growth of the (-) satellite intensity for 53 Cu/40Ni-7Fe foils of 3.94, 4.33, and 7.08 nm wavelength at isothermal annealing temperatures of 320, 345, and 400°C.

Fig. 3 Observed diffusivities $\tilde{D}_B$ vs B^2, for 53 at.% Cu/40 at% Ni-7 at.% Fe.

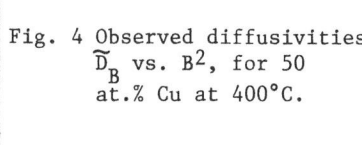

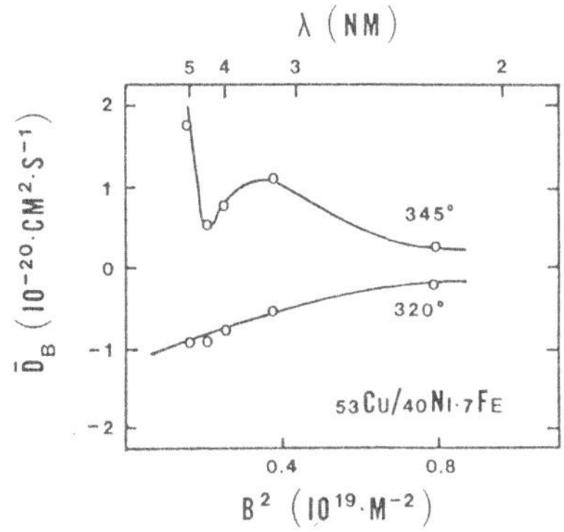

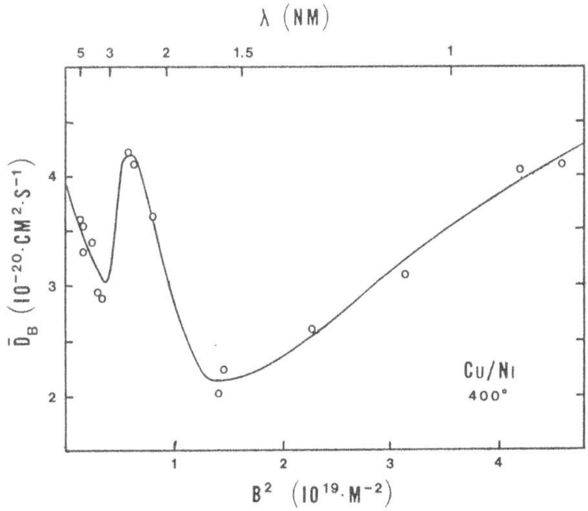

Fig. 4 Observed diffusivities $\tilde{D}_B$ vs. B^2, for 50 at.% Cu at 400°C.

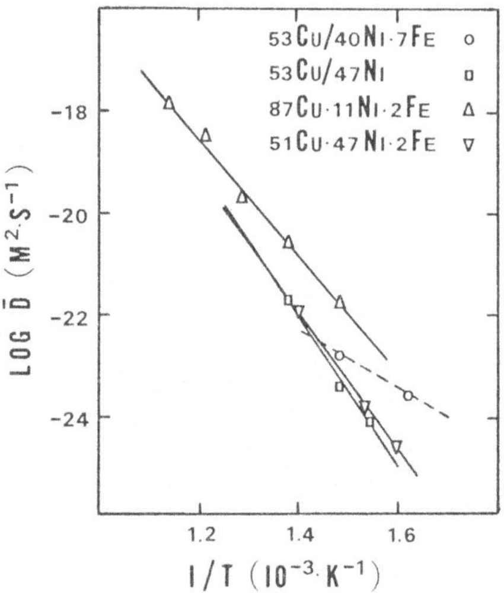

Fig. 5 Arrhenius plot of log $\tilde{D}$ for Cu/NiFe, Cu/Ni, and Cu-Ni-Fe alloys.

shown in Figs. 3 and 4 for Cu/NiFe at 320 and 345°C and Cu/Ni at 400°C, respectively. The decrease in $\tilde{D}_B$ at a composition wavelength of 2.5 nm is apparent for Cu/Ni at 400°C. Similar observations were also made for Cu/Pd,(7) Ag/Pd,(6) and Cu/NiFe.(4) The decrease is thought to coincide with loss of interfacial coherency. It is anticipated that the long wavelength modulations will be incoherent, i.e. the difference in lattice spacing between layers will be taken up by misfit dislocations. The reason for this is that the coherent strain energy is independent of the wavelength while the interfacial energy for an incoherent interface is inversely proportional to the wavelength. Thus an incoherent interface will be energetically favorable at wavelengths greater than some critical value. The decrease in $\tilde{D}_B$ is, therefore, due to the elimination from the driving force of the strain energy contribution.(3)

The interdiffusivities are not only dependent on the dispersion relationship but also upon temperature. It is customary to assume that there is a linear relationship between log $\tilde{D}$ and T^{-1} over a wide range of temperatures. The diffusion coefficient $\tilde{D}$, measured in a macroscopic couple, can be related to the interdiffusion coefficient $\tilde{D}_B$ by

$$\tilde{D}_B = \tilde{D}\,[1 + F_e(h)/f'' + (2/f'')\sum_{\mu=1}^{\infty} K_\mu\, B^{2\mu}(h)] \tag{4}$$

where $F_e(h)$ is the elastic Fourier energy,(8) f'' is the second derivative with respect to composition of the Helmholtz free energy per unit volume and K_μ are the gradient-energy coefficients.(9) In the long wavelength approximation - $B^2(h) = k^2$, and f'' and K_μ are identical with those expressions appearing in existing continuous theories(8,10,11) or the discrete theory.(12) They can be identified with the coefficients of an expansion of the free energy in composition gradients, often used to derive the diffusion equation of a nonuniform solid solution. They are important at short wavelengths and in systems where long-range interactions are not weak.

The diffusion coefficients and gradient-energy coefficients can be computed from eqn. (4) using a nonlinear regression fitting procedure developed for the experimental data of $\tilde{D}_B$ vs. $B^2(h)$, as that plotted in Figs. 3 and 4.(4,9) The Arrhenius plot of Fig. 5 can then be used to compute the activation energy Q. The slope of the log $\tilde{D}$ vs. T^{-1} plot is expressed as -Q/R, where R is the gas constant. The activation energies for 50Cu/50Ni and 53Cu/40Ni7Fe, from this plot, are 64.5 and ~50 Kcal/mole, respectively. For comparison, a value of 62 Kcal/mole has been reported for 50Cu-50Ni using interdiffusion measurements at high temperatures.(13) Data for the systems 51Cu-47Ni-2Fe(14) and 87.7Cu-10.7Ni-1.6Fe(15) are plotted as well.

3. DISCUSSION AND SUMMARY

Khachaturyan(1) has thoroughly addressed the effect of system fluctuations on the stability of alloy phase. The ordering and decomposition states can be understood by the loss of stability in a disordered solution through the use of static concentration waves (SCW). The approach to phase stability is therefore achievable by means of SCW. Experimentally, the stability phenomena can be studied using the method of artifical concentration waves. This has a twofold significance: (a) it eliminates many errors in the conventional diffuse scattering analysis; and (b) it points to the possibility of observing critical phenomena at certain composition wavelengths. The practical advantage of the technique is that the structure is controllable for studying alloys at, virtually, any composition and temperature. The composition modulation technique is an artifical perturbation in the crystal structure which can therefore be used to enhance our knowledge about the nature and properties of metallic crystal structures. In summary, the artifical concentration wave approach can be used for the determination of: (1) interatomic energies (not embellished upon in this paper); (2) critical temperatures; (3) the temperature dependence of interdiffusivities, i.e. the system kinetics; and (4) features which affect the structural stability, as the state of coherency and the presence of screening singularities (indicated by the appearance of sharp peaks/discontinuities in the effective diffusivity $\tilde{D}_B$.)(3)

Work performed in part under the auspices of the U.S. Department of Energy by Lawrence Livermore National Laboratory under Contract W-7405-Eng-48.

REFERENCES

1. Khachaturyan, A.G., <u>Theory of Structural Transformations in Solids</u> (J. Wiley, New York, 1983).
2. Vrijen, J. and Radelaar, S. 1978 Phys. Rev. B 17, 409.
3. Tsakalakos, T., 1981 Thin solid Films 86, 79.
4. Jankowski, A.F., 1984 PhD Thesis, Rutgers University.
5. Tsakalakos, T., 1977 PhD Thesis, Northwestern University.
6. Hénein, G., 1979 PhD Thesis, Northwestern University.
7. Philofsky, E.M. and Hilliard J.E., 1969 J. Appl. Phys. 40, 2198.
8. Cahn, J.W., 1962 Acta Metall. 10, 179.
9. Tsakalakos, T. and Hilliard, J.E., 1984 J. Appl. Phys. 55, 2885.
10. Cahn, J.W., 1961 Acta Metall. 9, 795.
11. Cahn, J.W. and Hilliard, J.E., 1958 J. Chem. Phys. 28, 258.
12. Cook, H.E., de Fontaine, D. and Hilliard, J.E., 1969 Acta metall. 17, 765.
13. Monma, K., Sutor, H., and Oikowa, H., 1964 Nippon Kinz. Gakk. 28, 188.
14. Poerschke, R., Wagner, W., and Wollenberger, H., 1986 J. Phys. F: Met. Phys 16, 421
15. Gust, W., Wachtel, E., Frühauf, B., and Predel, B., MRS Symp. Proc. <u>Phase Transformations in Solids</u> (North-Holland, Amsterdam, 1984) 21, 461.

Growth Modes and Metastability in Cu-Ni and Pd-Ni Compositionally Modulated Thin Films

N.K. Flevaris and Th. Karakostas
Department of Physics,
Aristotle University of Thessaloniki,
540 06, Thessaloniki, Greece

1. INTRODUCTION

More than half a century has elapsed since man devoted the first efforts in preparing materials that would possess artificial periodicities in composition. In recent times, these efforts have evolved in routine work mainly due to: a) the success in preparing such metallic thin films by vapour deposition [1,2], b) the advent of preparation methods such as the molecular beam epitaxy [3], sputtering techniques [4,5] etc and c) the very interesting (often unusual and technologically promising) physical behavior [6] of systems containing a one-dimensional periodic fluctuation in the composition.

Usually, this periodicity refers to a number of just a few to one hundred atomic planes. Novel mechanical [7-9], structural [10,11], magnetic [12-16] and transport [4,14,17-20] properties were reported for compositionally modulated metallic thin films with modulation wavelengths (λ) containing only 3 to 20 atomic planes. [The name compositionally modulated (CM) will be used for our metallic systems and we will reserve the often used term superlattices (SLs) for semiconducting such systems.]

The structure of metallic CM materials has been studied by x-ray diffraction [10,11] and transmission electron microscopy (TEM) techniques [21,22]. The widely accepted picture for the structure of an A_m-B_n CM system contains the following: First, the number of A-rich (m) and B-rich (n) atomic planes, in λ. Second, an approximation of the amplitute of the concentration wave. Third, estimates of the interplanar spacing (strain) modulation accompanying that of composition; the periodicity of this strain modulation does not necesarity scale with that of composition for all λ's [10]. Finally, accounts for a variety of defects such as dislocations, stacking faults and twins [22]

Several of the observed peculiarities in the properties of CM metallic films have been attributed to the structural uniqueness of these systems. Also, it has been recognized that at many cases the materials prepared by one group of workers exhibited different properties than those of similar systems prepared by others. Usually, these differences cannot be accounted for by the "macroscopic" structural characteristics as determined by the works mentioned above. In this contribution we will report on structural studies of CM Cu_m-Ni_n and Pd_m-Ni_n via reflection high-energy electron diffraction (RHEED) as well as planar (TEM), cross-sectional (XTEM) and scanning (SEM) electron microscopy, in an effort to improve the techniques of growth and justify (or predict) various physical properties.

G. M. Stocks and A. Gonis (eds.), Alloy Phase Stability, 591–598.

2. EXPERIMENTAL RESULTS AND DISCUSSION

2.1. Preparation of CM films

The samples were prepared in a dual e-gun evaporation unit. A first layer (~50-70 nm thick) of Cu or Pd was deposited onto heated mica substrates, prior to depositing the CM film. The background vacuum was 3-8x10^{-7} Torr. The substrate temperature during deposition of the CM film was high enough for crystallinity but not as high as needed for interdiffusion; varying with λ, usual values were between 400 and 600 K. Deposition rates were used in the range of 0.2-1.0 nms^{-1}. The two fluxes were periodically interrupted via a reciprocating shutter.

It was reported [11] that, for Cu_m-Ni_n films, it was possible to prepare short-λ modulations with sharp x-ray diffraction patterns when λ was commensurate with the constituent lattices (m and n being nearly integers).For more details see Refs. 6,11.

For the TEM work specimens were prepared with chemical etching or ion-beam thinning methods. For preparation for XTEM the foil was "sandwiched" between Si plates. Subsequently, strips of this construction were ion-beam thinned.

2.2. RHEED

The RHEED studies were performed at 100 kV with the specimens either on or stripped off of the mica.A typical reflection pattern, for CM Cu-Ni films, is shown in Fig. 1a. An obvious indication is the crystallinity of the surface layers. Taking into account the very small incidence angle of the beam it is apparent that the RHEED pattern refers to a few tens of surface atomic planes. To study any structural variations with depth, chemical etching of the surface was used. The results indicated that oxidation was not significant.

After rotating the specimen by 30° around the surface normal, [111], the pattern seen in Fig. 1b was obtained. This pattern indicates the existence of {111} twinning;such twinning would be invisible in observation along the [$\bar{2}$11] direction. This twinning could be studied in detail only by XTEM and this will be examined in a later Section. At this point it must be mentioned that the patterns of Fig. 1 were typical for both the Cu_m-Ni_n and Pd_m-Ni_n systems and for various m and n.

2.3. TEM

The TEM was performed at 120 kV. The general observation, for almost all specimens studied, was the absence of polycrystallinity rings from the diffraction patterns (DPs). However, the structure was far from being monocrystalline either. Figure 2 depicts DPs of the two twinned crystallites for Cu-Ni CM films. The DPs seen in Fig. 2 were obtained after the specimen was rotated 35° about the [1$\bar{1}$0] of the film plane.

Thus, one can conclude that it was typical for our CM Cu-Ni films to grow as crystallites between which a double-positioning twin relationship develops. The form of such crystallites is depicted in Fig. 3. Furthermore, it must be noticed that the crystallites tend to have the boundary planes nearly {211} from those planes belonging to the [111] zone axis. Finally, a more common observation about those twins is the formation of a "labyrinthic" ("puzzle-like") crystal extended over macroscopically large volumes. Bright- and dark-field micrographs of such a structure are shown in Fig. 4 where only a non-common reflection (the 111 of the [1$\bar{1}$0] zone axis) of the two crystals is excited. In micrographs, such as those of Fig. 4, for CM Cu-Ni another observation worth noticing is the existence of a variety of microtwins. Two more points must be noticed about the double-positioning twin boundaries: a) their continuous extension and the (apparently) curved pattern and b) their stepped growth across the thickness. A closer analysis of these boundaries indicates that they may be the result of coalescence of neighbouring crystallites. This would be controlable by adjusting the growth conditions and for this reason it will be adressed in the following where we will examine the effect of heating on the structure. This was performed in-situ. [Of course, when a CM Cu-Ni film is heated to temperatures in the range 650-850 K, interdiffusion is enhanced and the modulation is gradually destroyed. However, the behavior of CuNi alloys has been thoroughly studied [23] and one should not confuse disappearance of modulation with complete homogenization.]

After heating a $Cu_{8.8}$-$Ni_{7.0}$ sample to 850 K the DP seen in Fig. 5a was obtained. What is deduced from this DP is that the extra (1/3)422 spots were enhanced in intensity after the annealing. A dark-field micrograph (from the reflection indicated in the inset) is shown in Fig. 5b. It is clear that the annealing process resulted, via recrystallization, in the growth of the above mentioned large twinned crystallites at the expense of the smaller ones. Thus, the puzzle-like structure was further developed by the heating. It must be noted that these heating effects started at temperatures below those for interdiffusion. This suggests how it would possible to improve growth by appropriately selecting deposition rates and temperatures.

Finally, another form of defects was revealed by the planar TEM study. Namely, as seen in Fig. 6a, there is indication for the existence of crystallites having a rotation close to 30° with respect to the matrix. The dark- field micrograph taken from one of these extra reflections will be observed in Fig.6b. This observation, for a $Cu_{8.8}$-$Ni_{7.0}$ film, suggests that, although the system grows by developing the microtwins and the double-positioning twins, it also forms high-angle grains. Examination of the density of such crystallites in the matrix revealed an areal density of 10%. Further details about these grains will be presented in the Section of the XTEM studies.

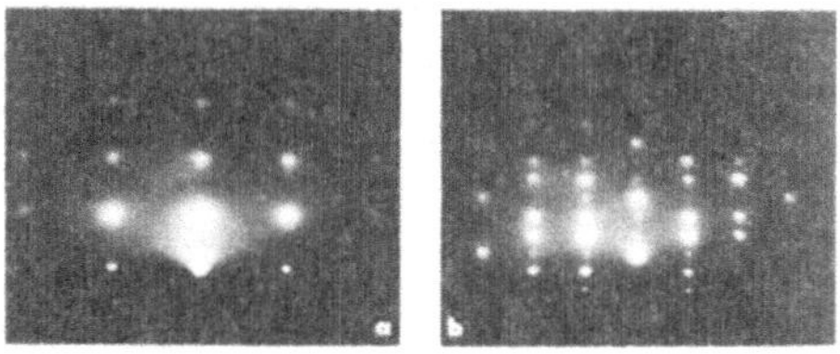

Fig. 1. Typical RHEED patterns for Cu-Ni along (a) [$\bar{2}$11] and (b) in 30° around the [111].

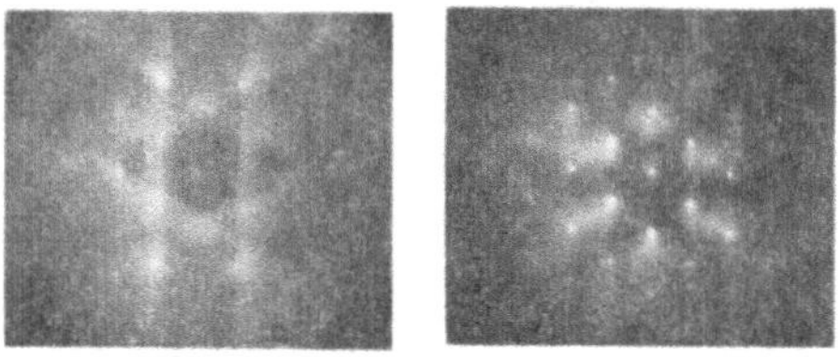

Fig. 2. TED patterns showing the twin relationship (of 35°) in modulated Cu-Ni films.

2.4. XTEM

The XTEM work was performed at 120 and 200 kV. The characteristic of this geometry is that it corresponds directly to the x-ray diffraction studies and to the RHEED ones.Thus, it is complementary to the TEM studies in planar geometries while it provides significant new information. The XTEM micrograph for a $Pd_{3.0}$-$Ni_{2.8}$ CM film is shown in Fig. 7. The corresponding DP is seen in the inset. The DP clearly depicts the satellite 111 reflections revealing a periodicity of ~6 atomic planes. The satellites of the transmitted beam will become distinctly visible in overfocused exposures. The twinning becomes clearly evident in the cross-sectional geometry when a rotation of ~30° around the [111] (from Fig. 7) renders the DP shown in Fig. 8a. (This diffraction of twinned crystallites corresponds directly to the case exposed by the RHEED experiments.) Consequently, the dark-field micrograph of the DP of Fig. 8a is seen in Fig. 8b. One can now see clearly the double-positioning and embeded twinned crystallites. Another observation is the appearance of the fringes in the modulation-sequence pattern. There is a distinct deviation from the perfectly planar form.Instead, a wavy (shuffles-like) pattern of the fringes is observed in Figs. 7 and 8b. This observation will be discussed together with those on Cu Ni.

The case of CM Cu-Ni for XTEM observations is not a desirable one because the difference in scattering amplitudes between the constituents is minimal and, thus, the composition fluctuation is not expected to give fringe patterns of considerable contrast. A micrograph, for a $Cu_{8.8}$-$Ni_{7.0}$ film, showing the modulation fringes will be seen in Fig. 9. Besides the fact that a mere composition variation in the sample would not be expected to produce such a fringe contrast, other features of this micrograph must be discussed. For example, the "variation" in λ in different areas as well as the "inclination" of the fringes.

For several reasons, one must not ignore the influence of the interplanar spacing fluctuation (strain) to these fringe patterns. The type and density of defects in the matrix will alter, locally, that modulation and the apparent periodicity will not be the same everywhere. As far as the "inclined" fringe sequence is concerned, the following must be noted. First, that it extends up to the outermost planes resulting in a tooth like surface morphology; this must be connected with the SEM observations to be presented in the next Section. Second, the rotational relationship between those columns which, also, relates to the RHEED and TEM studies in planar geometries. Third, the almost crystallographic-plane symmetry of the corresponding facets of the film surface. Fourth, the shuffling of the fringes. Fifth, the interpenetration of the bouble-positioning and embeded twins with these "inclined" fringes and, finally, the observation that those columns begin well over the Cu buffer layer which, often, is almost defect-free.

As mentioned in discussing Fig. 6 of the TEM work, there exist (in about 10% density) high-angle columnar grain boundaries with a 30° rotation to one another. Two such columns, as $2\bar{1}\bar{1}$ and $1\bar{1}0$ poles,would have the boundary as the $(0\bar{1}1)$ or $(11\bar{2})$ plane, respectively, and the surface facets (102) and (110). This latter outcome was a surprise to us but it seems to be a common feature for such columnar crystallites. Furthermore, the corresponding geometry appears to have set in for the periodicities of the "inclined" fringes. It must be noted that the embeded twins seem to leave unaffected the fringes, as can be seen from their interpenetration observed in Fig. 10a, while the principal interface for those twins, as Fig. 10b shows, was (111). For, we are inclined to consider these fringe patterns, the footprints of the two fluctuations, as direct observation of the anisotropy of the (artificially generated) internal strains resulting from the planar anisotropy of the corresponding biaxial elastic modulus, first introduced by Cahn [24].

2.5. SEM

The surface morphology of CM Cu-Ni and Pd-Ni films was examined by SEM. In Fig. 11 an example of these studies is shown revealing the "overgrowth" of a crystallite of hexagonal shape for a Cu_2-Ni_1 sample. The lateral size of this crystallite is of the order of the crystallites observed in TEM. The density of appearance of such surface anomalies was again of the same order with the high-angle grains observd in Fig. 6. Therefore, it appears that in the growth of CM short-λ Cu-Ni films, it is possible for the high-angle boundaries that form well above the Cu-layer to extend to the surface and result in an oriented surface anomaly. It was not possible to observe such surface anomalies in the CM films of Pd-Ni that we studied; it is worth noticing that those films did not exhibit a columnar growth mode nor a tooth-like surface section in the XTEM studies.

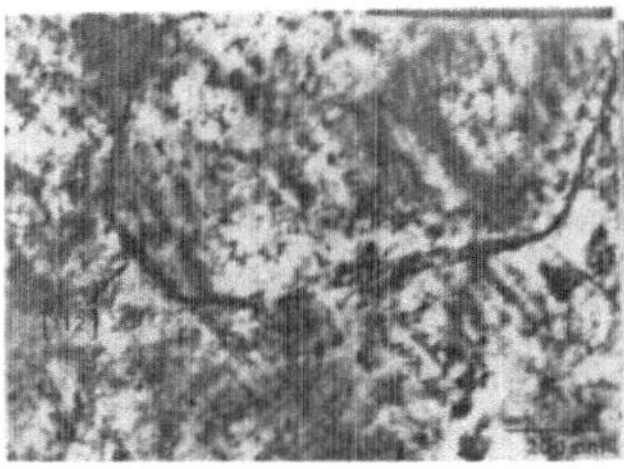

Fig. 3. A crystallite in Cu-Ni. Note symmetry of boundaries.

Fig. 4. a) Bright- and b) dark-field micrographs of a portion of a "puzzle-like" twinned crystallite.

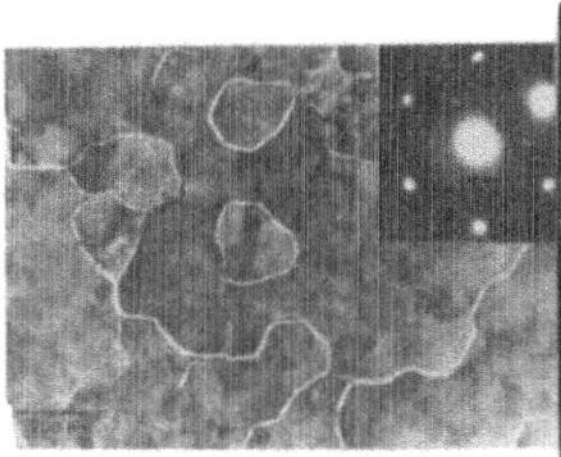

Fig. 5. Dark-field, as in the inset, for heated Cu-Ni.

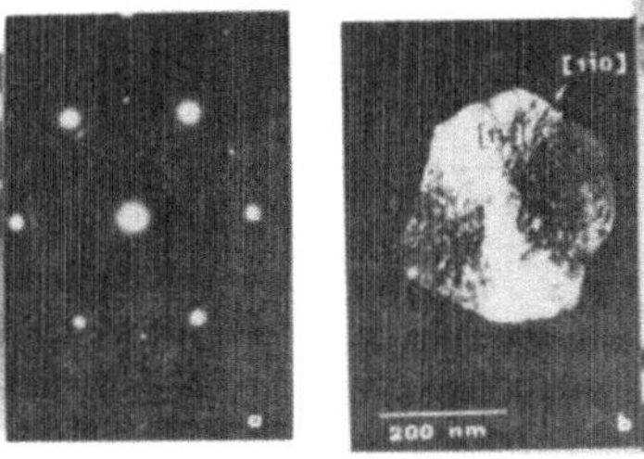

Fig. 6. The TED a) and the dark-field micrograph b).

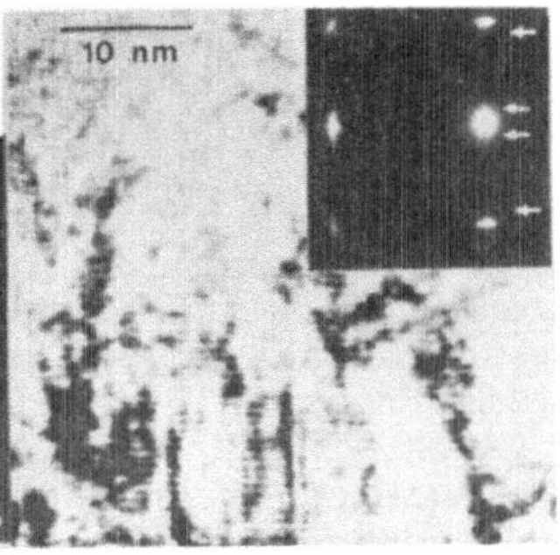

Fig. 7. XTEM micrograph for $Pd_{3.0}$-$Ni_{2.8}$ and its [2$\bar{1}\bar{1}$] XTED pattern in the inset.

3. CONCLUSIONS AND SUMMARY

Let us try to evaluate the observations reported here while combining the conclusions from the different electron microscopy techniques. The general observation was that CM Cu-Ni and Pd-Ni films, prepared by the method described in Section 2, did not grow in a polycrystalline (but well textured) manner as it was widely believed. However, departing from monocrystallinity, these systems developed a variety of defects with; twins were the primary ones. The crystallinity of the independent grains was maintained to the outer surface layers. Contrary to that, although the (Cu, especially) buffer layer was grown smootly and was followed by the formation of planar CM layers, a columnar growth mode would set in after some thickness. (This point point requires further examination so that it might be possible to improve the crystalline quality of the metallic CM films by appropriate alterations in the growth methodologies.)

The formation of various types of boundaries, such as twin boundaries and compositional planes, as well as high-angle boundaries, was examined and a brief account is given attempting a simple phenomenological modeling of CM metallic films. The modulation provides the means of tracking the growth, in XTEM studies, unlike the behavior of any homogeneous alloy. The coherency strains will be considered as the key parameter of influence in the growth of CM metallic systems. The growth mechanism must, therefore, be studied as an interplay of the coherency-strain-driven thermodynamics and the kinetics as determined primarily by the (external) growth conditions.

The substrate temperature has an upper limit, for each system, that is determined by the interdiffusion constant. Also, interface roughening would affect the choice of substrate temperature. However, one flexible parameter is the rate of deposition, at least for given external pressures. Thus, non-equilibrium growth of these systems would take the signature of the anisotropic internal (metastability) strains. (It must be borne in mind that we refer to CM system with appropriately [11] short λ's so that the structures would be -fully or, at least, partially- coherent.) The coherency strains, which result from the constituent lattice misfit, are determined by the corresponding [24] biaxial elastic modulus as also mentioned before. This "effective modulus", Y, can be determined by requiring relaxation of the stress along a particular direction while the constituent structures have been deformed to accommodate the misfit and achieve coherence. The morphology of the CM structure will be determined by this anisotropic modulus. A calculation of this quantity in a continuum (λ-independent) model has been given earlier for cubic [24-26], hexagonal [26,27], tetragonal [28], and orthorhombic [26] systems. Applying these model calculations in our case of Cu-Ni one can see that different anisotropy of Y results in different (e.g.{111},{112} or {110}) planes. Thus, the growth morphology can be studied thermodynamically on each plane via the "soft" directions of Y and, consequently,the energy balance for various types of boundaries would determine the form and the structure [29,30] of the surfaces and interfaces. Finally, the kinetics would provide means of perfecting the growth conditions of the modelled structures.

According to such a mechanism the observed columnar structure remains interesting but not so puzzling any longer. A detailed account of this work will be given elsewhere. What should be reminded at this point is that coherence was a requirement for such a treatment. As observed in Section 2, the columnar growth and the anisotropic fringe pattern were observed for Cu-Ni while Pd-Ni was free of such defects. The latter system comprises a misfit of about 10% while the former of only 2.5%. As a result the coherence is more easily destroyed by dislocations in Pd-Ni, as it was shown by XTEM.

In summary, studies of reflection, (planar and cross-sectional) transmission and surface-scanning electron microscopy were presented for metallic CM films. Those observations, firstly, suggest a possible mechanism of growth and, therefore, provide a model for improving preparation techniques and, secondly, impose a new frame for theoretical works on CM coherently strained systems towards several of the so far observed anisotropic properties. Also, in-situ heating was used in the TEM study to examine metastability phenomena.

Acknowledgements: We are indebted to Professor J. Stoemenos for his valuable contributions in the XTEM work. Also, Dr. D. Baral and Mr. G. Goldsmith deserve our thanks for their collaboration in preparing (at Professor J.E. Hilliard's laboratories at Northwestern University) some of the samples.

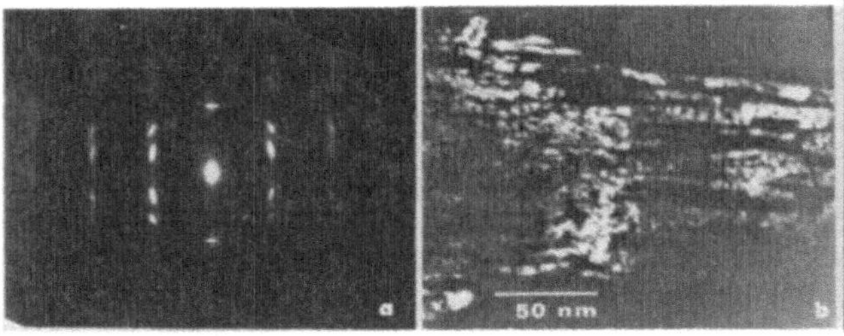

Fig. 8. a)XTED pattern showing twinning in $Pd_{3.0}$-$Ni_{2.8}$ and b) its dark-field micrograph with micro- and embeded twins.

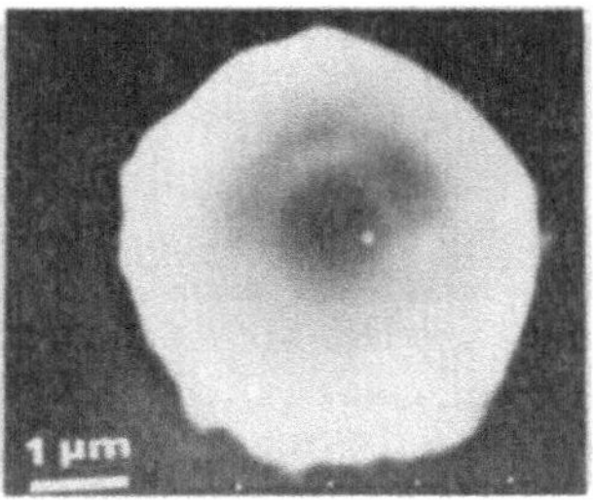

Fig. 11. Overgrowth on a Cu-Ni surface; by SEM.

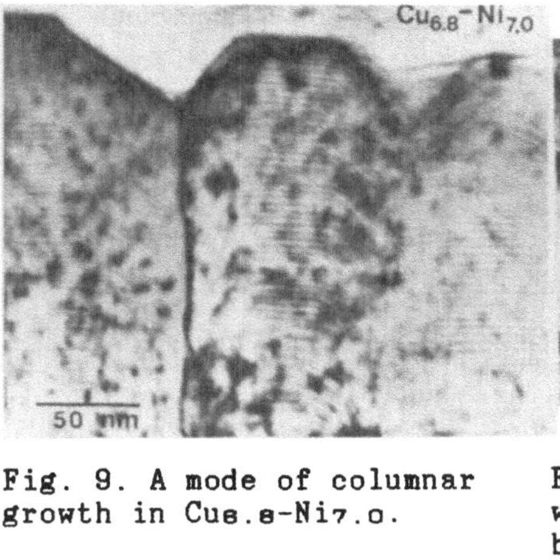

Fig. 9. A mode of columnar growth in $Cu_{6.8}$-$Ni_{7.0}$.

Fig. 10. a)Bright-field XTEM showing fringe-twins intersections. b)Dark-field XTEM showing sharp embeded twin boundaries.

REFERENCES

1. H.E.Cook and J.E.Hilliard,Appl. Phys. Lett. 8, 24 (1966).
2. L. Esaki, L.L. Chang and R. Tsu, Proc. 12th International Conference on Low Temp. Phys., Kyoto, Japan (1970).
3. A.Y. Cho, Appl. Phys. Letters 19, 467 (1971).
4. T. Barbee, Proc. NSF Workshop, Aurlie, Virginia, p. 94 (1978); S.T. Ruggerio, T.W. Barbee and M.R. Beasley, Phys. Rev. Letters 45, 1299 (1980).
5. I.K. Schuller, Phys. Rev. Letters 44, 1597 (1980).
6. See, for example: "Synthetic Modulated Structures", eds. L. L. Chang and B.C. Giessen (Academic Press, 1985).
7. W.M.C. Yang,T. Tsakalakos and J.E. Hilliard, J. Appl. Phys. 48, 876 (1977); T. Tsakalakos and J.E. Hilliard, J. Appl. Phys. 54, 734 (1983).
8. G. Henein and J.E.Hilliard, J. Appl. Phys.54, 728 (1983).
9. D. Baral, Ph.D. thesis, Northwestern Univ. Evanston (1983).
10. D.B. McWhan, in Ref. 6, p. 43.
11. N.K. Flevaris, D. Baral, J.E. Hilliard and J.B. Ketterson, Appl. Phys. Letters 38, 992 (1981).
12. E.M. Gyorgy, J.F. Dillon, Jr., D.B. McWhan, L.W. Rupp,Jr., L.R. Testardi, Phys. Rev. Lett. 45, 57 (1980).
13. N.K. Flevaris, J.B. Ketterson and J.E. Hilliard, J. Appl. Phys. 53, 8046 (1982).
14. L.H. Greene, W.L. Feldmann, J.M. Rowell, B. Batlogg, E.M. Gyorgy, W.P. Lowe and D.B. McWhan, Superlattices and Microstructures 1, 407 (1985); J. Kwo, E.M. Gyorgy, D. B. McWhan, M. Hong, F.J. DiSalvo, C. Vettier and J.E. Bower Phys. Rev. Letters 55, 1402 (1985).
15. M.B. Salamon, S. Sinha, J.J. Rhyne, J.E. Cunningham, R.W. Erwin, J. Borchers and C.P. Flynn, Phys. Rev. Letters 56, 259 (1986).
16. H. Sakakima, R. Krishnan and M. Tessier, J. Appl. Phys. 57, 3651 (1985); H.K. Wong, H.Q. Yang, J.E. Hilliard and J.B. Ketterson, J. Appl. Phys. 57, 3660 (1985).
17. D. Baral and J.E. Hilliard, Appl. Phys. Letters 41, 156 (1982).
18. I. Banerjee, Ph.D. thesis (unpublished), Northwestern Univ. Evanston (1982); I. Banerjee, Q.S. Yang, C.M. Falco and I. K. Schuller, Solid State Commun. 41, 805 (1982).
19. N.K. Flevaris, Ph.D. thesis, Northwestern Univ. Evanston (1983).
20. N.K. Flevaris and S. Logothetidis, Appl. Phys. Lett. 50, 1544 (1987).
21. A. Purdes, Ph.D. thesis, Northwestern Univ. Evanston (1976)
22. Th. Karakostas and N.K. Flevaris, J. Mater. Sci. Letters 5, 1235 (1986).
23. T. Tsakalakos, Scripta Metal. 15, 255 (1981).
24. J.W. Cahn, Acta Met. 9, 795 (1961);ibid 10, 179 (1962).
25. J.E. Hilliard, in "Phase Transformations", edited by H.I. Aaronson, ASM, Metals Park, Ohio (1970) p.497.
26. N.K. Flevaris, Scripta Metal. 15, 1111 (1981).
27. W.E. Mayo and T. Tsakalakos, Met. Trans. A11,1367 (1980).
28. N.K. Flevaris, J. Amer. Ceram. Soc. 70, 301 (1987).
29. C.S. Smith, Trans. Metall. Soc. 175, 5 (1948).
30. C. Herring, Phys. Rev. 82, 87 (1951).

PREMARTENSITIC MICROSTRUCTURES AS SEEN IN THE HIGH RESOLUTION ELECTRON MICROSCOPE: A STUDY OF A NI-AL ALLOY

D. SCHRYVERS[1,2], L.E. TANNER[1] , G. VAN TENDELOO[3]

1) Chemistry and Materials Science Department, Lawrence Livermore National Laboratory, Livermore, CA 94550, U.S.A.; 2) National Center for Electron Microscopy, Lawrence Berkeley Laboratory, University of California, Berkeley, CA 94720, U.S.A.; 3) University of Antwerp, RUCA, Groenenborgerlaan 171, B-2020 Antwerpen, Belgium.

1. INTRODUCTION

Recent investigations of displacive transformations in a wide range of metallic [1,2] and non-metallic materials [3,4] have been providing new insights into precursor behavior, viz., changes in parent phase properties and microstructure leading to these transformations. This behavior is viewed as evidence that the parent phase effectively "prepares itself" for the eventual transformation via related lattice displacements (usually incommensurate with the basic lattice), where the process is believed driven by anomalous temperature-dependent phonon effects [1-7]. The elucidation of this behavior is now aiding in the development of more effective models for displacive transformations, particularly those producing martensitic phases in metallic alloys [2,5-7].

Two common manifestations of the pretransformation regime in martensitic alloys are diffuse streaking in diffraction patterns and a related striated strain contrast ("tweed") in amplitude contrast TEM images [8-10], where both change continuously with temperature becoming more pronounced as the transformation temperature M_S is approached [11,12]. The diffuse streaks emanate from the Bragg reflections and are directed along $\langle 110\rangle^*$ of the cubic parent and are perpendicular to the quasi-periodic, non-continuous tweed striations lying parallel to {110} plane traces in the TEM image (e.g., see fig. 1 in ref. [13]). In essence, the foregoing are the signature of the lattice response to structural perturbations. This is a generic process with more than one origin and so, can be found in other contexts including phase transformations of distinctly different type (e.g., in the course of thermally activated diffusional transitions such as aging, ordering, etc. [8,14-16]). However, the nature of the underlying perturbations here and the associated temperature-dependence of the effects are unique to these athermal diffusionless transformations.

The high-temperature ordered β-phase (B2, CsCl-type) of Ni_xAl_{1-x} ($50<x<65$) is one of the more extensively studied parent phases exhibiting these precursor effects [10]. In this paper we describe an additional type of TEM contrast, much finer in scale than the tweed, which is seen in high-resolution phase contrast (HREM) images of $Ni_{62.5}Al_{37.5}$. This contrast is interpreted in terms of atomic displacements similar to those leading to martensite formation. The conclusions drawn from various experimental investigations of the tweed striations and related phenomena are taken as a general basis for this interpretation [8,10-12]. These studies indicate that the effects stem from static (or quasi-static) $\langle 110\rangle\langle 1\bar{1}0\rangle$ displacement waves having a range of wavelengths up to 75 nm [9,10]. The waves, no matter what their origin, are the lattice accomodation to perturbations concentrated along particular low-energy transverse displacement paths. From elasticity measurements [17] and inelastic neutron scattering [13] such low-energy modes, without being ideally "soft modes" [18,19], do indeed exist along $\langle 110\rangle^*$ directions in Ni-Al. Going a step further, Krumhansl [19] has recently examined

G. M. Stocks and A. Gonis (eds.), Alloy Phase Stability, 599–603.

the influence of temperature-dependent competing atomic interactions and severe anisotropy on phonon dispersion curves and finds that this may well be the source of modulated domain or embryo-like pretransformation microstructures.

2. DIFFUSE INTENSITY IN RECIPROCAL SPACE

In-situ heating and cooling electron diffraction experiments by Tanner et al. [12] documented that the distribution of intensity along the diffuse streaks at the Bragg reflections is temperature dependent. Microdensitometer traces reveal weak satellites at approximately ±0.16 <110>* which were just perceptable at room temperature and increased in amplitude as cooling approached M_S~ -100° C. The satellites were actually more easily observed by elastic neutron scattering experiments (see fig. 2 of ref. [13]). This indicates that, instead of all wavelengths being equally present before the transformation, a major fraction of these waves have an incommensurate wavelength of approximately 1.3 nm (i.e., an average of 6.2 {110} planes).

3. HIGH RESOLUTION IMAGES

When HREM images are obtained from the parent B2 phase at room temperature, the most informative viewing direction is along [001]. A suitable choice of the experimental parameters (e.g., defocus) will then result in a square pattern of white dots, resembling the configuration of the projected Ni (or Al) sublattice. In order to obtain such an image, the thin-foil specimen must be tilted into a symmetric zone axis orientation. Figure 1a shows such an image from a relatively thick region of a sample. In addition to the expected square white dot pattern, a diffuse contrast modulation along both [110] and [1$\bar{1}$0] directions is observed. The average period of this modulation (indicated in the figure) agrees with the ~1.3 nm determined from the satellite positions in the SAED pattern. No clear shift of the white dots can be observed between each {110} strip of differing contrast. It is emphasized once again that this modulation, as observed in these multi-beam phase contrast HREM images, is not to be confused with the two-beam amplitude contrast tweed pattern of the same orientation, but of much coarser spacing [8-10]. Since the basic white dot pattern is only slightly perturbed, one can look for a structural origin for these phase-contrast "micromodulations" in terms of small distortions of the ordered B2 parent lattice.

a b

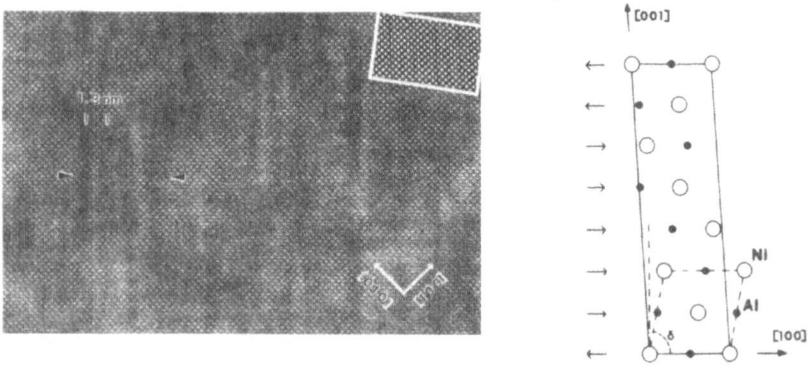

FIGURE 1a) [001] HR image showing the white dot lattice contrast together with the {110} "micromodulation". The inset is a calculated image (thickness = 5.7 nm, defocus = -80 nm) for the modulated parent B2 phase. b) [010] projection of the monoclinic unit cell of the final 7R martensitic phase. The arrows indicate the displacement with respect to the previous $(110)_{B2}$ plane.

4. INTERPRETATION OF THE IMAGES

In general, computer simulations calculating the expected HREM phase contrast images for a given type of model unit cell have to be performed in order to interpret the observed images of a real structure. These calculated images are then compared with the micrographs and, if necessary, the model unit cell is modified to obtain suitable agreement. Our calculated images show the same square pattern of white dots as the HREM images for the undistorted B2 structure. However, the simulation incorporating the introduction of transverse displacements is much more complicated owing to the large number of parameters involved and the large unit cell dimensions.

A $\langle 110\rangle\langle 1\bar{1}0\rangle$ transverse displacement wave may be applied to the basic crystal since the micromodulation contrast is related to the diffuse intensity distribution in reciprocal space. Although a number of different possibilities can be explored, due to space limitations we will confine the present discussion to the image simulations plus diffraction evidence that gave the most successful results. We use the approach suggested by Morii and Iizumi [20] in which the pretransformation state is viewed as a microstructural assembly "in which the stacking of the (110) planes is shuffled in a similar fashion as in the stacking structure of the (final martensite)". In the present case, the final structure is a 7-layer monoclinic cell denoted as 7R [21]. Our HREM images, as well as diffraction results [21], indicate that this phase has a $(5\bar{2})$ stacking sequence (Zhdanov notation), which can be seen as a long-period microtwinned superstructure based on the FCT $L1_0$ unit cell (3R-type) [22]. The [010] projection of the 7R structure is schematically represented in fig. 1b, including arrows indicating the relative displacements of the (110) planes of cubic B2 to produce this new close-packed monoclinic phase.

If HREM images of the asymmetric stacking arrangement are calculated using the multislice method [23], it becomes clear that the wide (5 plane) and narrow ($\bar{2}$ plane) portions of the modulation will produce differing contrast arising from the small spacings between successive twin planes in the latter. Using a model unit cell for the premartensitic state with an atomic shuffling distortion of only 30% of that required for 7R, but still resembling the displacements towards its $(5\bar{2})$ stacking, gives simulated images as shown in the inset of fig. 1a. The actual atomic displacements are too small to markedly affect the structure image of the basic B2 lattice (i.e., the square white dot pattern), but the change in local environment results in an apparent difference in background between the wide and narrow modulation subunits. Moreover, when viewing the simulated image at a grazing incidence along the [110] direction, the white dots are seen <u>not</u> to be perfectly aligned; an effect also discernable in the experimentally obtained image.

When comparing the above results with similar calculations using other distortional wave models we conclude that the shuffling of B2 {110} planes to mimic 7R gives the best agreement with the observed contrast. Since the periodicity of the 7R martensitic structure is retained in this model, the simulated modulation results in satellites at commensurate positions of $\pm 1/7\langle 110\rangle^*$. As indicated above, diffraction shows satellites at $\pm 0.16\langle 110\rangle^*$, hence, the distortion is actually incommensurate with the B2 lattice involving a non-integral or averaged modulation periodicity close to 6 planes with a domain or embryo-like pattern as discussed recently by Krumhansl [19] and others [7].

We are well aware that under normal circumstances, conclusions from image simulations of HREM micrographs should be based on the comparison of a set of images taken and calculated at different defocus and thickness values, rather than on the possible fortuitous correspondence of a single image. For our purposes here, however, we are more concerned with the disposition of the atomic displacements rather than the actual atomic coordinates and lattice parameters. A full comparison with other possible models and a discussion of different causes affecting the contrast modulation will be published in detail elsewhere.

5. DOMAIN STRUCTURE OR NOT?

From previous low-magnification and low-resolution amplitude contrast TEM experiments no firm evidence for a well-defined domain structure could be presented. However, since the tweed modulation is seen as a precursor effect to the martensitic transformation in Ni-Al, it is attractive to visualize an assembly of very small domains, with each domain only being distorted by one wave instead of six. At this point, it is important to note that no indication for fully martensitically transformed domains was found in the HREM images. Moreover, the [110] and $[1\bar{1}0]$ micromodulations do intersect with one another, indicating the possibilities that: (a) an atom can be displaced by more than one wave, or (b) that small domains overlap with one another.

From careful examination of the HREM images it is clear that, while the entire crystal is modulated, any set of the1.3 nm micromodulations does not extend over a very large area. This may be explained by the existence of numerous phase shifts (or perhaps discommensurations) [7] resulting in different wave packets arrayed along one particular lattice direction. From the observed images, the size of these packets, or domains, may be roughly estimated as a few wavelengths, i.e., 3 to 8 nm (see arrows in fig. 1). This is in reasonable agreement with the correlation length of local distortional fluctuations of about 4 nm determined from the breadth of the satellites observed with elastic neutron scattering by Shapiro et al. [13]. According to this view, the waves in different packets will then be out of phase. Since these regions are still relatively small, they can be considered as out-of-phase domains of a long-period structure, much as in the case of microdomains in short-range ordered alloys [24,25]. This ill-defined long-period will contribute to the streaking in regions between Bragg peaks and their satellites.

6. CONCLUDING REMARKS

The present study indicates that the B2 β-phase in quenched $Ni_{62.5}Al_{37.5}$ is distorted by displacement waves involving a planar shuffling of atoms resembling the final 7R martensite structure and with wavelengths of the order of 1.3 nm. The appearance of a $\langle 110\rangle\langle 1\bar{1}0\rangle$ type modulation with the indicated periodicity corresponds well with recent inelastic neutron scattering results which reveal nonlinear behavior in the TA$\langle 110\rangle$ phonon dispersion curve around the same wavelengths indicating a partial lattice softening for such waves [13]. In bulk material all six equivalent wave-vectors are equally present.

These distortional modulations are configured in some form of three-dimensional assembly. Following the interpretation given above, it can be concluded that a one-dimensional domain structure along one of six $\langle 110\rangle$ directions may exist. However, the beating of six displacement waves with apparently uncorrelated phase and wavelengths rules out the existence of a "conventional" three-dimensional domain structure [26]. For this reason it is uncertain whether much more information can be gained from image simulations as presented above.

There are now numerous indications that the underlying structure to the tweed contrast in this alloy is a precursor effect of the martensitic transformation. However, a detailed description of the effective correlation between the distorted parent phase and the martensitic product phase has yet to be developed. Recent HREM results reveal the existence of a sequence of different structures in the transition region between the modulated β-phase and the martensitic phase, depending on parameters such as the local composition and stress. Such transition structures include modulated β-phase in which only one [110] modulation is preferred or in which the periodicity differs from the above described 1.3 nm and the FCT $L1_0$ martensite with single shear defects. Details of these transformation microstructures and transitions will be described in a forthcoming paper.

ACKNOWLEDGEMENTS
The authors wish to thank the National Center for Electron Microscopy at the Lawrence Berkeley Laboratory, and the Center for Electron Microscopy for Materials Science at the University of Antwerp (RUCA) for the use of their facilities. Alloy single crystals were prepared by D. Pearson and co-workers at United Technologies Research Center. This work was carried out under the auspices U.S. Department of Energy, contract No. W-7405-ENG-48.

REFERENCES

1. Proc. Symp. on Pretransformation Behavior Related to Displacive Transformations, L. E. Tanner, W. A. Soffa, Eds., Met. Trans. A, 19A (1988) in press.
2. V.V. Kondrat'yev, V.G. Pushin, Fiz. Met. Metall., 60, 629 (1985).
3. Structural Phase Transitions 1, Topics in Current Physics, No. 23, K. A. Muller, H. Thomas, Eds. (Springer-Verlag, Berlin, 1981).
4. Incommensurate Phases in Dielectrics, 2, R. Blinc, A. P. Levanyuk, Eds. (Elsevier, Amsterdam, 1986).
5. A. Lasalmonie, Proc. Conf. on Solid State Phase Transitions in Metals and Alloys, (Les Editions de Physique, Orsay,1978) p.631.
6. P.-A. Lindgard, O. G.Mouritsen, Phys. Rev. Lett., 57, 2458 (1986); G. R. Barsch, J. A. Krumhansl, Met. Trans. A, 19A (1988) in press.
7. G.R. Barsch, J.A. Krumhansl, L.E.Tanner,M. Wuttig, Scripta Met. 21,1257 (1987)
8. L. E. Tanner, Phil. Mag. 14, 111 (1966).
9. L. E. Tanner, A. R. Pelton, R. Gronsky, J. de Phys., 43, Colloq. C4, C4-169 (1982).
10. I.M. Robertson, C.M. Wayman, Phil. Mag.A, 48, 421, 443, 629 (1983).
11. M. Sugiyama, R. Oshima, F.E. Fujita, J. Jpn. Inst. Met., 48, 881 (1984).
12. L.E. Tanner, A.R. Pelton, R. Gronsky, M.E. Wall, G. Van Tendeloo, submitted to Scripta Met., (1986).
13. S.M. Shapiro, J.Z. Larese, Y. Noda, S.C. Moss, L.E. Tanner, Phys. Rev. Lett., 57, 3199 (1986).
14. L. E. Tanner, H. J. Leamy, Order-Disorder Transformations in Alloys, H. Warlimont, Ed., (Springer, Berlin, 1974) p.180.
15. J.D.C. McConnell, Min. Mag., 38, 1 (1971).
16. P.K. Davies, J. Am. Ceram. Soc. 69, 796 (1986).
17. K. Enami, J. Hasunama, A. Nagasawa, S. Nenno, Scripta Met., 10, 879 (1976).
18. J.A. Krumhansl, Proc. Conf. on Nonlinearity in Condensed Matter, A.R. Bishop, D.K. Campbell,P.Kumar,S.E.Trullinger,Eds.(Springer- Verlag,Berlin,1987)p.255
19. J.A. Krumhansl, Proc. Conf. on Competing Interactions - Statics and Dynamics, R. LeSar, A. R. Bishop, Eds. (Springer-Verlag Berlin, 1988) in press.
20. Y. Morii, M. Iizumi, J. Phys. Soc. Jpn., 54, 2948 (1985).
21. V. V. Martynov, K. Enami, L. G. Khandros, S. Nenno, A. V. Tkachenko, Phys. Met. Metall., 55, 136 (1983).
22. M. Ahlers, Prog. Mat. Sci., 30, 135 (1986).
23. R. Kilaas, Proc. Microbeam Anal. Soc. Meeting 1987, (San Francisco Press, San Francisco, 1987) in press.
24. S. Hashimoto, Acta Cryst., A30, 792 (1974).
25. D. Van Dyck, G. Van Tendeloo, S. Amelinckx, Ultramicroscopy, 15, 357 (1984).
26. R. De Ridder, G. Van Tendeloo, D. Van Dyck, S. Amelinckx, J. de Phys., 38, Colloq. C7, C7-178 (1977).

SECTION 7

GENERAL TOPICS

ANNIHILATION MOMENTUM DENSITY OF POSITRONS TRAPPED AT VACANCY-TYPE DEFECTS IN METALS AND ALLOYS

A. Bansil

Physics Department, Northeastern University,
Boston, Massachusetts 02115

R. Prasad

Department of Physics, Indian Institute of Technology,
Kanpur, India 208016

R. Benedek

Materials Science Division, Argonne National Laboratory,
Argonne, Illinois 60439

I. INTRODUCTION

Positron annihilation, especially the angular correlation of annihilation radiation (ACAR), is a powerful tool for investigating the electronic spectra of ordered as well as defected materials. The tendency of positrons to trap at vacancy-type defects should enable this technique to provide a signature of the local environment of such defects; an understanding of the changes in the defect structure at phase transitions is relevant also in the context of this conference. In order to achieve this goal, however, we need to develop a theoretical basis for calculating the two-photon annihilation momentum density $\rho_{2\gamma}(\vec{p})$. With this motivation, we have recently formulated and implemented a theory of $\rho_{2\gamma}(\vec{p})$ from vacancy-type defects in metals and alloys.[1] This article gives an outline of our approach together with a few of our results. Within the constraints of space, the discussion is mostly illustrative; the citation of literature is minimal.

An outline of this article is as follows. Section II summarizes the basic equations for evaluating $\rho_{2\gamma}(\vec{p})$. Our Green's function-based approach is non-perturbative and employs a realistic (one-particle) muffin-tin Hamiltonian for treating electrons and positrons; we are therefore able to handle ordered as well as disordered d-band systems. Section III presents and discusses $\rho_{2\gamma}(\vec{p})$ results for a mono-vacancy in Cu.[1] These are the first such results in a d-band metal; previous work has been limited to Al.[2] For simplicity, we have neglected the effects of electron-positron correlations and of lattice distortion around the vacancy in our implementation of the theory; further work to delineate these effects is required. Section IV comments briefly on the question of treating defects such as divacancies and metal-impurity complexes in metals and alloys.[3] Finally, in Section V, we remark on the form of $\rho_{2\gamma}(\vec{p})$ for a mono-vacancy in jellium.[4]

G. M. Stocks and A. Gonis (eds.), Alloy Phase Stability, 607–612.

II. FORMAL CONSIDERATIONS

The starting point for the Green's function formulation of $\rho_{2\gamma}(\vec{p})$ is the equation:

$$\rho_{2\gamma}(\vec{p}) = \frac{1}{\pi^2}\int d\vec{r}\int d\vec{r}' exp[-i\vec{p}\cdot(\vec{r}-\vec{r}')]\int dE\, f(E)\int dE_+\, f_+(E_+) \times Im\, G(\vec{r},\vec{r}';E)\, Im\, G_+(\vec{r},\vec{r}';E_+), \qquad (1)$$

which expresses $\rho_{2\gamma}(\vec{p})$ in terms of the electron and positron Green's functions $G(E)$ and $G_+(E)$, and the associated Fermi-Dirac distribution functions $f(E)$ and $f_+(E)$ respectively. In the present application, it is useful to rewrite Eq. (1) in the form

$$\rho_{2\gamma}(\vec{p}) = \sum_{mn} M_{mn}(\vec{p})\, e^{i\vec{p}\cdot(\vec{R}_m-\vec{R}_n)}. \qquad (2)$$

Equation (2) decomposes $\rho_{2\gamma}(\vec{p})$ into a summation over the set $\{\vec{R}_m\}$ of direct lattice vectors with matrix elements $M_{mn}(\vec{p})$.

A computationally tractable equation for the matrix elements $M_{mn}(\vec{p})$ is obtained by using the angular momentum expansion for the Greens' function within the KKR scheme:[5]

$$Im\, G(\vec{r}_1,\vec{r}_2,E) = \sum_{LL'} Z_L^{(m)}(\vec{r}_1-\vec{R}_m,E)\, Im\, T_{LL'}^{mn}\, Z_{L'}^{(n)}(\vec{r}_2-\vec{R}_n,E), \qquad (3)$$

where $\vec{r}_1(\vec{r}_2)$ lies in the Wigner Seitz cell on the $m(n)$ site, $Z_L^{(i)}$ is the regular solution of the Schrödinger equation in the ith muffin-tin sphere, and

$$T(E) = [\tau^{-1} - B(E)]^{-1}, \qquad (4)$$

is the path operator matrix in the space of site indices n and the angular momenta L. B is similarly a matrix which in a perfect crystal is the Fourier transform of the KKR structure constants, and the atomic scattering matrix τ is related to the phase shifts of the individual scatterers. The use of expansion (3) for the electron and the positron Green's function straightforwardly yields the expression:

$$M_{mn}^{\alpha\beta}(\vec{p}) = \frac{1}{\pi^2}\int dE f(E)\int dE_+ f(E_+) \sum_{L_1L_1'}\sum_{L_2L_2'} \Delta_{L_1L_2}^{m\alpha}(\vec{p},E,E_+) \times\Delta_{L_1'L_2'}^{n\beta}(-\vec{p},E,E_+)[Im\, T_{L_1L_1'}^{e,mn}(E) Im\, T_{L_2L_2'}^{p,mn}(E_+)], \qquad (5)$$

where

$$\Delta_{LL'}^{i\alpha}(\vec{p},E,E_+) = 4\pi\sum_{L''}(-i)^{l''} Y_{L''}(\hat{p})\int_{\Omega_i} d^3r\, Z_L^{e,i\alpha}(\vec{r},E) Z_{L'}^{p,i\alpha}(\vec{r},E_+) Y_{L''}(\hat{r})\, j_{l''}(pr). \quad (6)$$

The superscript $\alpha(\beta)$ in Eqs. (5) and (6) indicates that the site m(n) is occupied by atom of type $\alpha(\beta)$; the superscript e(p) refers to electron(positron) quantities. The integration in Eq. (6) extends over the ith cell volume Ω_i.

The preceding formalism can, in principle, be used to calculate $\rho_{2\gamma}(\vec{p})$ for a general assembly of non-overlapping muffin-tin scatterers; we return to the treatment of specific defects below. It should be emphasized that the present real space approach is particularly suitable for evaluating $\rho_{2\gamma}(\vec{p})$ from a trapped positron, because the exponential decay of the positron wave-function from the trapping site will yield a rapid convergence of the sum in Eq. (2); by contrast, $\rho_{2\gamma}(\vec{p})$ for an extended Bloch-state positron is more naturally calculated by formally transforming these summations into reciprocal space.

III. MONO-VACANCY IN Cu

In any specific case, the main quantities which require further consideration are the electron and positron path operators T^{e} and T^{p}. For a mono-vacancy in an otherwise perfect crystal, the path operators can be obtained by using the exact solutions of the multiple scattering equations for the single impurity problem,[5] the impurity in the present case being the vacant site. For example, the (00) electron path operator is given by the equation:

$$T_{v}^{e,00}(E) = D_{v}(E)\, T^{e,c,00}(E), \tag{7}$$

where $T^{e,c,00}$ is the perfect crystal path operator, and

$$D_{v}(E) = [1 + (\tau_{v}^{-1}(E) - \tau_{c}^{-1}(E))\, T^{e,c,00}(E)]^{-1}. \tag{8}$$

Here, τ_{v} and τ_{c} refer to the electron muffin-tin potential for the vacant and the occupied crystalline sites respectively.

We have studied $\rho_{2\gamma}(\vec{p})$, and related physical quntities such as the electron and positron charge densities for a mono-vacancy trapped positron in Cu in some detail. Figure 1 shows illustrative results for $\rho_{2\gamma}(\vec{p})$. More specifically, partially summed contributions $\rho_{2\gamma}^{(j)}(\vec{p})$ upto the 4th shell are shown; the superscript (j) indicates the number of the most distant shell (surrounding the vacant site) upto which contributions to $\rho_{2\gamma}(\vec{p})$ in Eq. (2) are included. A comparison of the $j = 1$ through $j = 4$ curves (the $j = 3$ and $j = 4$ curves are indistinguishable on the scale of the figure) indicates that $\rho_{2\gamma}^{(1)}(\vec{p})$ is converged to within 10% of $\rho_{2\gamma}(\vec{p})$. We note that, $\rho_{2\gamma}^{(1)}(\vec{p})$ possesses all the characteristic features of the fully converged momentum density curve.

Figure 1 shows that the $j = 0$ (vacancy site) contribution to $\rho_{2\gamma}(\vec{p})$ is quite flat and featureless. The addition of the nearest neighbor (110)-shell(nn) contribution (the $j = 1$ curve) causes a dramatic change in $\rho_{2\gamma}(\vec{p})$ in the low momentum region. In fact, for $p \leq 5$ mrad, $\rho_{2\gamma}(\vec{p})$ is dominated by the nn-shell contribution. This may appear surprising in view of the exponential decay of the trapped positron wave function from the vacant site. We note however that $\rho_{2\gamma}(\vec{p})$ involves an overlap of the electron and positron wave functions. In the vacant site, the positron amplitude is large, but the electron density is quite small; on the other hand, in the nn-sites, a large electron density annihilates with the tail of the positron wave function. Therefore, the central site and nn-shell contributions should both be generally important. The exponential

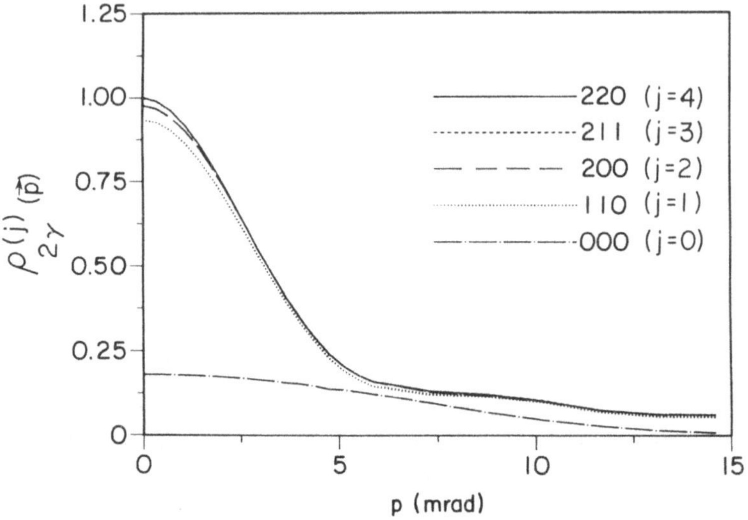

FIGURE 1: Partially summed momentum density $\rho^{(j)}_{2\gamma}(\vec{p})$ along [001] direction for various shells j for a mono-vacancy in Cu. After Ref. [1].

decay of the positron wave function makes contribution from more distant shells much smaller. Our calculations suggest that the nn-shell contributions would dominate in the low-p region more generally for mono-vacancies in metals and alloys.

IV. TREATMENT OF OTHER VACANCY-TYPE DEFECTS

We have in mind here situations which can be modelled by considering a relatively small (i. e. involving at most a few shells surrounding a central site) "defected region" placed in an otherwise perfect ordered medium. We consider a divacancy in Cu to illustrate our approach. If the system is modelled by placing two vacant sites in a perfect Cu crystal, we need to solve the "two impurity" path operator. By using matrix partitioning methods, the problem can be cast in terms of a (2×2) matrix equation (in angular momentum space) involving perfect Cu crystal path operators; the latter path operators are needed of course also for treating the mono-vacancy in Cu. The generalization to an impurity cluster of M-sites is quite straightforward, the solution is obtained then via an $(M \times M)$ matrix equation; see Ref. [3] for details of evaluating $\rho_{2\gamma}(\vec{p})$.

We comment finally on the treatment of vacancy type defects in disordered alloys. Here we need to ensemble average both sides of Eq. (2); this step formally yields:

$$< \rho_{2\gamma}(\vec{p}) > = \sum_{mn} < M_{mn}(\vec{p}) > e^{i\vec{p}\cdot(\vec{R}_m - \vec{R}_n)}, \tag{9}$$

where $< M >$ is seen from Eq. (5) to require the configuration average of the product $< Im T^e \, Im T^p >$ of electron and positron path operators; in Ref. [1], we propose to

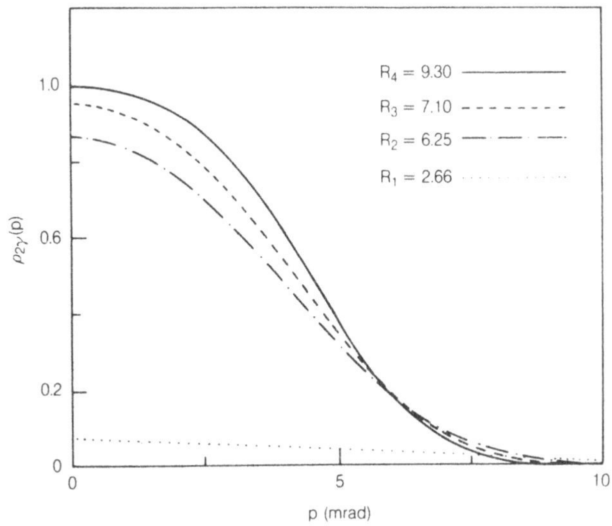

FIGURE 2: $\rho_{2\gamma}(\vec{p})$ for a mono-vacancy in jellium for various values of the radius R of the integration volume(see text). After Ref. [4].

decouple this average in terms of the electron and positron KKR-CPA medium path operators. Tractable expressions for $< \rho_{2\gamma}(\vec{p}) >$ are than straightforwardly obtained; see Ref. [1] for details.

V. MONO-VACANCY IN JELLIUM

A mono-vacancy in jellium is interesting because this model is amenable to an exact analytic solution at least for certain idealized potentials. We have used this model system to gain insights into the results obtained on the basis of the Green's function-based approach. Figure 2 shows $\rho_{2\gamma}(\vec{p})$ for a vacancy in jellium using a square well (square barrier) potential for positron (electrons).[4] The various parameters (well depth, barrier height, and Fermi energy) are chosen to mimic a mono-vacancy in Cu. The parameter R in the figure denotes the radius of the spherical volume to which the electron-positron overlap integral is restricted in calculating $\rho_{2\gamma}(\vec{p})$ (i.e. restrict $|\vec{r}|$, $|\vec{r}'| \leq R$ in the integrations in Eq. (1)); the values R_j correspond to an integration sphere of volume enclosed by the first j shells in Cu. Some of the characteristic features of Fig. 1 are seen to be reproduced by the model system. In particular, the magnitude of the R_1 curve is relatively small as is the case with the central site contribution in Fig. 1. Also, the R_2 curve shows the striking bell-shape of the $\rho^{(1)}_{2\gamma}(\vec{p})$ curve of Fig. 1; in contrast, as is well-known, $\rho_{2\gamma}(\vec{p})$ for the Bloch-state positron in Cu possesses an approximately constant value of unity up to the Fermi energy. It is clear that the jellium model can be useful in understanding the behavior of $\rho_{2\gamma}(\vec{p})$, although the model does not properly incorporate the Fermi surface and the d-bands in the electronic spectrum, in addition to neglecting the effects of nuclear repulsion of the positron from the host metal sites.

One of us (R.P.) acknowledges the hospitality of the Physics Department, Northeastern University, where part of this work was completed. This project was supported by the U. S. Department of Energy, BES-Materials Science, under contracts DE-FG02-85ER45223 and W-31-109-Eng- 38. It benefited from an allocation of supercomputer time on the ER-Cray at the MFE computer center.

REFERENCES

1. R. Prasad, R. Benedek, and A. Bansil in **Positron Annihilation**, edited by P. C. Jain, R. M. Singru, and K. P. Gopinathan (World Scientific, Singapore, 1985), p. 859; R. Prasad, R. Benedek, J. E. Robinson, and A. Bansil (to be published).
2. R.P. Gupta and R.W. Siegel, Phys. Rev. Letters **39**, 1212(1977); Phys. Rev. **B22**, 4572(1980); B.Chakraborty and R.W.Siegel, Phys. Rev.**B27**,4535(1983).
3. R. Prasad, A. Bansil, and R. Benedek (to be published).
4. J.E. Robinson, R. Benedek and R. Prasad, Phys. Rev. **B35**, 7392(1987).
5. For a recent review of the Green's function treatment of the electronic structure of ordered and disordered systems, see A. Bansil, in **Electronic Band Structure and its Applications**, Lecture Note Series, Vol. 283 (Springer-Verlag, Heidelberg, 1987), p. 273.

ELECTRONIC STRUCTURE, CHARGE TRANSFER EXCITATIONS AND HIGH T_c SUPERCONDUCTIVITY IN TRANSITION METAL OXIDES

A.J. FREEMAN AND JAEJUN YU

Materials Research Center and Department of Physics and Astronomy, Northwestern University, Evanston, Illinois 60208

ABSTRACT

This lecture describes and discusses the electronic structure and properties of the new high T_c transition metal oxides ($La_{2-x}M_xCuO_4$ and $YBa_2Cu_3O_{7-\delta}$) as determined from highly precise all-electron local density calculations. The important contributions of charge transfer excitations make excitons a likely candidate for their high T_c.

1. INTRODUCTION

Since the discovery of the high T_c superconductors $La_{2-x}M_xCuO_4$[1] and $YBa_2Cu_3O_{7-\delta}$[2], has set off an unprecendented effort to understand the mechanism giving rise to their superconductivity. It is now quite apparent that an understanding of the electronic structure and properties of the new high T_c superconductors, both $La_{2-x}M_xCuO_4$ and $YBa_2Cu_3O_{7-\delta}$, is emerging. This is an important step toward achieving an understanding of the origin of their high T_c. Detailed high resolution local density band structure results have served to demonstrate the close relation of the physics (band structure) and chemistry (bonds and valences) to the structural arrangements of the constituent atoms, and may provide insight into the basic mechanism of their superconductivity. In this paper, we provide a brief summary of the results on the detailed electronic structures of $La_{2-x}M_xCuO_4$ and $YBa_2Cu_3O_{7-\delta}$, compare them, and point out their relations to charge transfer excitations as a possible mechanism of superconductivity.

2. ELECTRONIC STRUCTURE AND PROPERTIES OF $La_{2-x}M_xCuO_4$.

In $La_{2-x}M_xCuO_4$ (M = Sr,Ba)[3], the results of highly precise all-electron local density full potential linearized augmented plane wave[4] (FLAPW) calculations of the energy band structure, charge densities, Fermi surface, etc., demonstrated: (i) that the material consisted of metallic Cu- O(1) planes separated by insulating (dielectric) La-O(2) planes and (ii) that this 2D character and alternating metal/insulator planes would have, as some of their most important consequences, strongly anisotropic (transport, magnetic, etc..) properties. Thus, the calculated band structure along high symmetry directions in the Brillouin zone shows only flat bands, i.e., almost no dispersion, along the c axis, demonstrating that the interactions between the Cu, O(2) and La atoms are quite weak. However, along the basal plane directions there are very strong interactions between the Cu-O(1) atoms leading to large dispersions and a very wide bandwidth (~ 9 eV).

The band structure near E_F has a number of interesting features. What is especially striking is that, in contrast to the complexity of its structure, only a single free electron-like band crosses E_F and gives rise to a simple Fermi surface (cf., Fig. 1). Since this band A in Fig. 1

G. M. Stocks and A. Gonis (eds.), Alloy Phase Stability, 613–620.

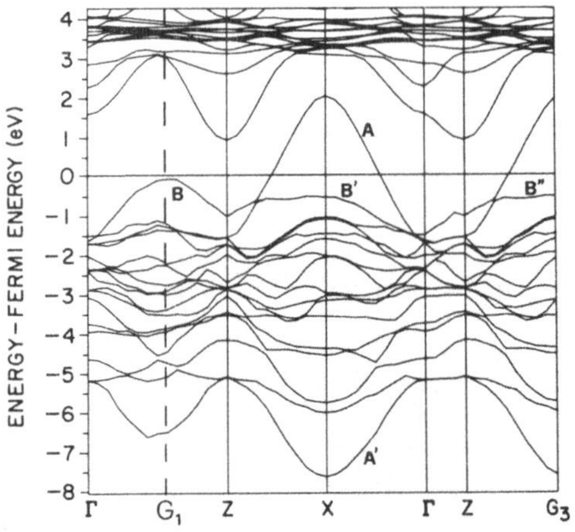

FIGURE 1. Band structure of La_2CuO_4 along symmetry lines in the extended Brillouin zone. (See the Ref. 3 for the notations used.)

originates from the Cu $d_{x^2-y^2}$-O(1) $p_{x,y}$ orbitals confined within the Cu-O(1) layer, it exhibits clearly all the characteristics of a two dimensional electron system. Particularly striking is the occurrence of a van Hove saddle point singularity (SPS). Such an SPS is expected, and found (cf. Fig. 2), to contribute strongly, via a singular feature, to the density of states (DOS). As we shall see, this dominance of the DOS near E_F by the SPS contribution is responsible for many of the striking properties of this material with M_x additions.

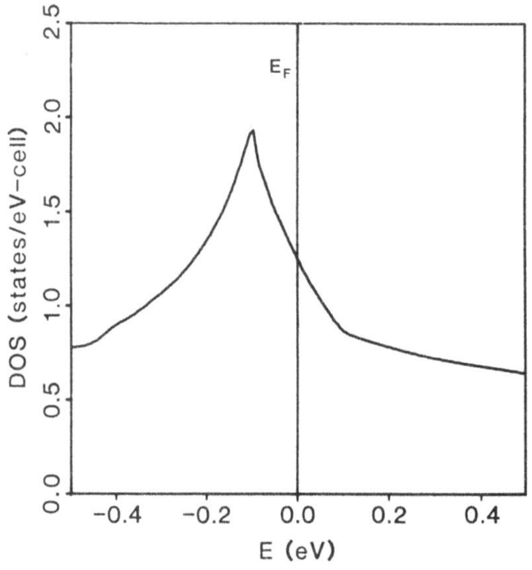

FIGURE 2. Blow-up of the density of states near E_F for La_2CuO_4.

The remarkable 2D nature of the electronic structure leads to a simple picture of the conductivity confined essentially to the metallic Cu-O(1) planes separated by insulating (ionic) planes of La-O(2). This picture is strongly confirmed by independent calculations[5] which model $La_{2-x}M_xCuO_4$ as a single slab consisting of a Cu-O(1) layer sandwiched by one La-O(2) layer on each side. (Note that such a slab has the correct stoichiometry and is charge neutral.) The electronic structure near E_F is dominated by the same single band of 2D p-d bonding character; the nesting feature[6] (with zone boundary spanning vector) and the van Hove SPS in the DOS are reproduced with this slab approach. In the band structure shown in Fig. 1, the strongly dispersed band A along the Γ-X (110) direction has only a Cu $d_{x^2-y^2}$-O(1) $p_{x,y}$ component, while band B along the Γ-G_1-Z (100) direction, especially at G_1 (i.e. the van Hove SPS point), is a mixture of Cu $d_{x^2-y^2}$-O(1) $p_{x,y}$ and Cu d_{z^2}- O(2)p_z orbitals. Another notable feature in the band structure of La_2CuO_4 is that the character of the bonding partner (band A') of the anti-bonding Cu $d_{x^2-y^2}$-O $p_{x,y}$ band A consists not only of the Cu $d_{x^2-y^2}$-O $p_{x,y}$ orbitals but has a large contribution from Cu sp orbitals, as also described by others[7]. (Note that these results are significantly different from the two dimensional tight binding model of Mattheiss[8], where the anti-bonding band A and B, as well as the bonding band A′, were regarded as having the same Cu $d_{x^2-y^2}$-O $p_{x,y}$ orbital character.)

The quasi-2D properties of the electronic structure are also supported by plots of the charge densities of electrons at E_F (cf. Fig. 3). This

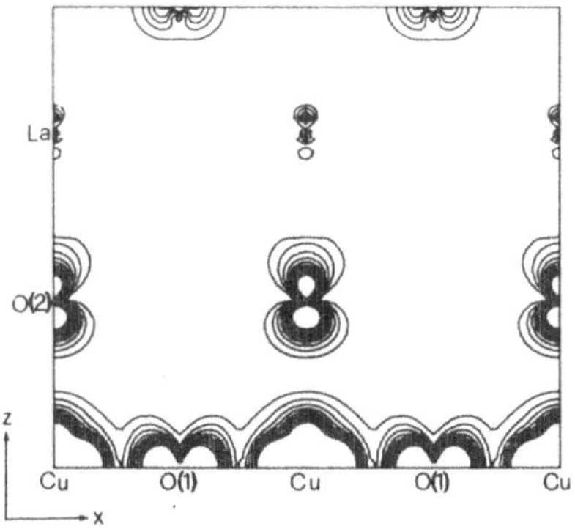

FIGURE 3. Contour plot in the xz vertical plane of valence charge density at E_F for La_2CuO_4.

charge density consists mainly of Cu $d_{x^2-y^2}$ and O(1) $p_{x,y}$ hybridized orbitals in the plane with some additional contribution of the Cu d_{z^2} and O(2) p_z components. There is essentially no electron density around the La site at E_F. This means that the La atoms do not contribute directly to the dynamical processes involving electrons near E_F. Further, an analysis of the band structure shows that the 5d level of La lies more than 1 eV above E_F; the 5p levels of La were found[3] to lie far below E_F (~ 15 eV). Thus, it is a fairly good approximation to consider the La atoms to be described in chemical terms as La^{3+} ions.

In view of the results presented above, we would expect - as a first approximation - that the introduction of divalent elements (e.g., M = Ba, Sr, etc..) as substitutional replacements for La would not change any major feature of the band structure, charge density, DOS, etc.. Thus, the use of a rigid band approximation to treat the case of alloys, $La_{2-x}M_xCuO_4$, may be considered as a quite good first approximation when x is small ($\leq$ 0.3). (This has been confirmed by independent virtual crystal approximation calculations[5]). Hence, in this spirit, the variation of composition x in $La_{2-x}M_xCuO_4$ can be taken into account simply as a change in the position of E_F, that is $E_F = E_F(x)$. Further, since E_F lies very close to the SPS, $N(E_F)$ is extremely sensitive to the position of E_F relative to the singular point (cf. Fig. 2). As a function of x, $N(E_F)$ varies from 1.2 states/eV-cell at x = 0 to 1.9 states/eV-cell at x = 0.16. This large variation in $N(E_F)$ will immediately affect a number of properties such as the magnetic susceptibility, specific heat, etc.

One major effect of the van Hove singularity on the properties of the system is the anomalous behavior of T_c with varying composition x in $La_{2-x}M_xCuO_4$. With increasing x, the superconducting critical temperature T_c increases rapidly from 14 K for $x \approx 0.05$ to a maximum of 37 K for $x \approx 0.2$ but then drops sharply for larger x values. Under the condition that the pairing potential V, in the pairing interaction parameter $\lambda = N(E_F)V$, is constant, the change in T_c with the composition x is associated with a change of $N(E_F)$. In fact, recent reports[9] show a large variation of T_c vs. x which rises sharply from 0 K at $x < 0.03$ to 22 K at x = 0.08, hits a maximum at $x \approx 0.15$ and then drops off sharply. These results are very consistent with our picture. Thus, it is clear that the strong variation in $N(E_F)$, derived from the quasi-2D van Hove singularity, plays a dominant role in the anomalous behavior of T_c as a function of x.

In total energy frozen phonon calculations[5] on $La_{2-x}M_xCu_4$, the role and effect of the optical breathing mode turned out to be significant. Since the breathing phonon mode involves the motion of oxygen atoms against the directional bonding of Cu d - O p in the plane, the in-plane Cu $d_{x^2-y^2}$-O(1) $p_{x,y}$ states of the 2D conduction bands are strongly affected by the breathing displacement. On the other hand, the out-of-plane Cu d_{z^2}-O(2) p_z orbitals, which are quite localized in the plane, are not much affected by the same breathing mode. But, because of the relative change of the Cu $d_{x^2-y^2}$-O(1) $p_{x,y}$ and Cu d_{z^2}-O(2) p_z, the charge fluctuations between Cu atoms, which can be as large as 0.3 electrons at the maximum of the O displacement, lead to transitions of the out-of-plane Cu d_{z^2}-O(2) p_z into the in-plane Cu $d_{x^2-y^2}$-O(1) $p_{x,y}$. Since the out-of-plane (anti-bonding) Cu d_{z^2}-O(2) p_z states near E_F are localized, we expect that the localized Cu d_{z^2}-O(2) p_z states, introduced by the charge fluctuation, may couple to the delocalized conduction electrons of the in-plane Cu $d_{x^2-y^2}$-O(1) $p_{x,y}$ orbital and possibly to form an excitonic state[5]. Thus, a key role in possible charge transfer excitations (CTE) is played by excitations between occupied localized Cu d_{z^2}-O(2) p_z and empty itinerant Cu $d_{x^2-y^2}$-O(1) $p_{x,y}$ states. We emphasized that these could couple resonantly with natural "Cu^{2+}-Cu^{3+}-like" charge fluctuations which exist in the $x > 0$ compounds, with important consequences for the superconductivity.

3. ELECTRONIC STRUCTURE AND PROPERTIES OF $YBa_2Cu_3O_7$.

For $YBa_2Cu_3O_{7-\delta}$, we presented[10,11] detailed high resolution results on the electronic band structure and density of states derived properties as obtained from the same highly precise state-of-the-art local density approach. These results demonstrated the close relation of the band structure to the structural arrangements of the constituent atoms and have helped to provide an integrated chemical and physical picture of the interactions.

The important structural features of the $YBa_2Cu_3O_{7-\delta}$ compounds arise from the fact that (2+δ) oxygen atoms are missing from the perfect triple perovskite, $YCuO_3(BaCuO_3)_2$. These vacancies arise from a total absence of O atoms in the Y-O planes (which seems to separate the Cu-Cu interactions across the Y plane) and an absence of O atoms in the Cu planes between the

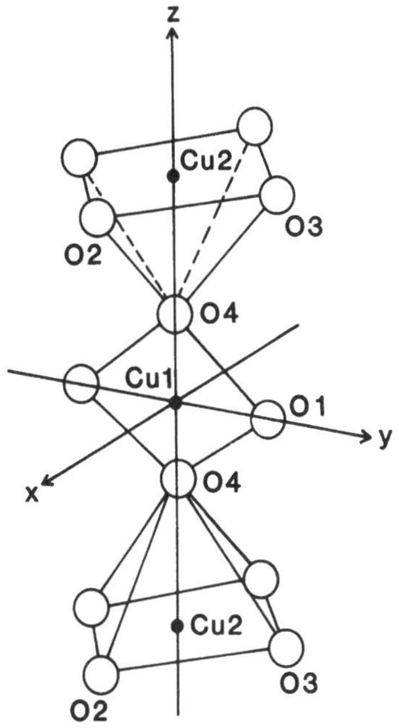

FIGURE 4. A local environment for the Cu1 and Cu2 atoms in $YBa_2Cu_3O_7$, following the Y-Cu2-Ba-Cu1-Ba-Cu2-Y ordering along z.

Ba-O planes (which leads to the formation of linear chains of Cu-O-Cu). As a result, there are two Cu ions (called Cu2) in five-coordinated positions - as shown in Fig. 4. Since the interatomic distance Cu2-O4 (2.303A) is much larger than Cu1-O4 (1.850A)[12], the Cu1 ions have a rather weak interaction with the Cu2 ions. The Cu2 ions are in the locally very strong tetragonal distortion and this yields a 2D structure for these planes similar to that of $La_{2-x}M_xCuO_4$. The additional distortions of the O2 and O3 ions (the so-called "dimpling") arises from the absence of O ions in the adjacent Y-O plane. From the contour plots of the valence charge density of $YBa_2Cu_3O_7$ on two vertical planes cutting the Cu-O bonds, it is apparent that there is an overall two-dimensionality to this system; the three horizontal Cu planes of the unit cell form, in fact, separated entities which do not interact with the neighboring entities along the c axis.

The calculated valence band structure of stoichiometric $YBa_2Cu_3O_7$ along high symmetry directions in the bottom (k_z = 0) plane of the orthorhombic BZ is shown in Fig. 5. The very close similarity in the band structure for the $k_z = 0$ and $k_z = \pi/c$ planes[10] indicates the highly 2D nature of the band structure.

It is seen from Fig. 5, that as in the case of La_2CuO_4, a remarkably simple band structure near E_F emerges from this complex set of 36 bands originating (from three Cu (3d) and seven O (2p) atoms). Four bands - two each consisting of Cu2(3d)-O2(p)-O3(p) orbitals and Cu1(d)-O1(p)-O4(p) orbitals - cross E_F. Two strongly dispersed bands C (S_1' and S_4 in Fig. 5; the labelling is given by their character at S) consist of Cu2($d_{x^2-y^2}$)-O2(p_x)- O3(p_y) combinations and have the 2D character which proved so important for the properties of $La_{2-x}M_xCuO_4$. The symmetry allowed interactions of the Cu2 bands with the Cu1 band results in a complicated disperson for the Cu2 bands (as occupied bands) along Γ-X and Γ-Y. Note that the S_1 and S_1' states have the same symmetry and so can interact (anticrossing) and disperse along the S-Y direction. Significantly, the Cu1 ($d_{z^2-y^2}$)-O1(p_y)-O4(p_z) anti-bonding band A (S_1 in Fig. 5) shows the (large) 1D dispersion expected from the Cu1-O1-Cu1 linear chains but is almost entirely unoccupied. This band is in sharp contrast to the π-bonding band B (formed from the Cu1 (d_{zy})-O1(p_z)-O4(p_y) orbitals) which is almost entirely occupied in the stoichiometric (δ = 0) compound and becomes fully occupied for the superconducting materials ($\delta \gg 0.1$). We will soon see that since this almost flat π-bonding band B (the state S_5 in Fig. 5) lies just below and crosses E_F (for δ = 0) along S-Y, it gives rise to peaks in the DOS near E_F making the DOS at E_F sensitive to the position of E_F (i.e., to δ).

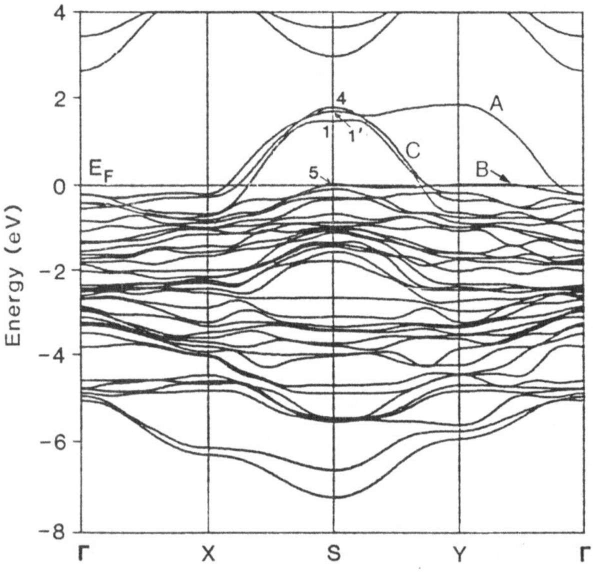

FIGURE 5. Band structure of $YBa_2Cu_3O_7$ along symmetry directions in the k_z = 0 plane of the orthorhombic Brillouin zone.

In our calculation for $\delta = 0$, the DOS at E_F, $N(E_F)$, is 1.13 states/eV Cu-atom, which is comparable to the 1.2 and 1.9 states/eV Cu-atom found earlier for $La_{2-x}M_xCuO_4$ at $x = 0$ and at the peak at $x = 0.16$, respectively. For increasing δ values (hence increasing E_F), $N(E_F)$ decreases sharply. Thus, for $\delta = 0.1$, $N(E_F) = 0.87$ states/eV Cu-atom while, for $\delta = 0.2$, $N(E_F) = 0.52$ states/eV Cu-atom (after which the DOS remains roughly constant). This means that the $N(E_F)$ per Cu atom values in the high T_c superconductor, $YBa_2Cu_3O_{7-\delta}$, are significantly lower than was found earlier (either experimentally or theoretically) for the (lower) high T_c superconductor, $La_{2-x}M_xCuO_4$. This result - which agrees with the conclusion of a recent experiment[13] - has a number of important possible consequences for DOS derived properties and superconductivity, including: reduced screening, an increased role for the polarization of ionic constituents, lowered conductivity (and reduced superconducting current carrying capacity), etc.

4. COMPARISON OF RESULTS FOR La_2CuO_4 AND $YBa_2Cu_3O_7$ AND EXCITONIC SUPERCONDUCTIVITY.

By comparing the electronic structure near E_F of La_2CuO_4 and $YBa_2Cu_3O_7$, we found similarities as well as different characteristics. First, the dominant electronic structure near E_F is the presence of very strongly dispersed conduction bands crossing E_F and almost "flat" bands. Both La_2CuO_4 and $YBa_2Cu_3O_7$ have 2D bands composed of Cu $d_{x^2-y^2}$-O(1) $p_{x,y}$ in La_2CuO_4 and Cu2 $d_{x^2-y^2}$-O(2) p_x - O(3) p_y in $YBa_2Cu_3O_7$. All of these 2D bands are strongly dispersed conduction bands. Second, however, the character of the localized states just below E_F is quite different. For La_2CuO_4, the strong tetragonal distortion of the CuO_6-octahedra leads to the splitting of the $d_{x^2-y^2}$-$p_{x,y}$ and d_{z^2}-p_z levels. Combined with the quasi-2D K_2NiF_4 structure, the in-plane $d_{x^2-y^2}$-$p_{x,y}$ states form a strongly dispersed conduction band; on the other hand, the out-of-plane d_{z^2}-p_z states remain localized. As shown in Fig. (1), the bands B´, B´´, which are formed from Cu d_{z^2}-O p_z orbitals, are located about 0.5 eV below E_F and show little dispersion. Also, as mentioned before, the band B is a mixture of two types of in-plane $d_{x^2-y^2}$-p_{xy} and out-of-plane d_{z^2}-p_z orbitals, because both belong to the same symmetry representation along Γ-G_1-Z.

As is now well-known, there are two types of Cu-atoms in the unit cell of $YBa_2Cu_3O_7$. The Cu2 in the 2D plane has a square pyramidal coordination of oxygens due to the missing oxygen in the Y-plane. From this structure, we expect that the Cu2 $d_{x^2-y^2}$ and Cu2 d_{z^2} energy levels are further apart than those in La_2CuO_4. Further, the distance between the Cu1 and O4 atoms in the chains is unusually short. Therefore, in $YBa_2Cu_3O_7$, the flat bands near E_F (band B in Fig. 5) dominantly have the character of the dpπ antibonding state of Cu(1)d_{zy}-O(1)p_z-O(4)p_y orbitals rather than of the Cu2d_{z^2}-O(4)p_z states in the 2D conduction plane. In addition to the flat dpπ bands, the 1D chain structure provides, as a partner, a strongly dispersed dpσ band composed of Cu(1) $d_{z^2-y^2}$-O(1)p_y-O(4)p_z orbitals, which is similar to the 2D $d_{x^2-y^2}$-$p_{x,y}$ in the case of La_2CuO_4.

We can draw an analogy between the electronic structures near E_F of La_2CuO_4 and $YBa_2Cu_3O_7$; we find the electronic structure of a well dispersed $d_{x^2-y^2}$-$p_{x,y}$ band and a fairly localized d_{z^2}-p_z state in the 2D Cu-plane of La_2CuO_4 corresponding to that of a strongly dispersed Cu1-O1-O4 dpσ band and a localized Cu1-O1-O4 dpπ state in the 1D chain of $YBa_2Cu_3O_7$. In fact, the 2D dpσ conduction bands in the Cu2-plane of $YBa_2Cu_3O_7$ are well separated at E_F from 1D bands, while the 2D conduction bands of the $d_{x^2-y^2}$-$p_{x,y}$ orbitals in La_2CuO_4 show a hybridization with the localized $d_{z^2}-p_z$ orbitals along the (100) direction.

For $La_{2-x}M_xCuO_4$, we discussed the importance of "Cu^{2+}-Cu^{3+}-like" charge fluctuations in connection with the possible role played by charge transfer excitations (CTE) in superconductivity. Concerning the superconductivity in $YBa_2Cu_3O_7$, it is expected that the Cu1-O1-O4 chain can play a critical role in the origin of superconductivity. A simple rigid band treatment of the band results[11] suggests that the Cu1(d_{zy})-O1(p_z)-O4(p_y) (dpπ anti-bonding) band becomes fully occupied for $\delta > 0.1$. Hence, excitations to its (almost) empty Cu1 ($d_{z^2-y^2}$)-O1(p_y)-O4(p_z) (dpσ anti-bonding) partner band can create strong polarization fields because the dpπ state is highly localized whereas the dpσ state is fairly itinerant along Cu1-O1 chains. As a result of the difference in the bonding character of the dpπ and dpσ states, a "local" charge transfer excitation (CTE) from the dpπ state to dpσ state may lead to significant electronic polarization. Incorporating the interactions between the 2D Cu d - Op conduction electrons and the charge transfer excitations (excitons), could produce the pairing interactions, via an exchange of excitons, which would enhance the T_c.

ACKNOWLEDGEMENTS

Work supported by NSF (through the Northwestern University Materials Research Center, Grant No. DMR85-20280). We thank our collaborators cited above (C.L. Fu, D.D. Koelling, S. Massidda).

REFERENCES

1. Bednorz, J.G. and K.A. Müller, Z. Phys. B 64, 189 (1986).
2. Wu, M.K., J.R. Ashburn, C.J. Torng, P.H. Hor, R.L. Meng, L. Gao, Z.J. Huang, Y.Q. Wang, and C.W. Chu, Phys. Rev. Lett. 58, 908 (1987).
3. Yu, J., A.J. Freeman, and J.-H. Xu, Phys. Rev. Lett. 58, 1035 (1987). Freeman, A.J., J. Yu, and C.L. Fu, Phys. Rev. B 36, Nov. 1 (1987).
4. Jansen, H.J.F. and A.J. Freeman, Phys. Rev. B 30, 561 (1984). Wimmer, E., H. Krakauer, M. Weinert and A.J. Freeman, Phys. Rev. B 24, 864 (1981).
5. Fu, C.L. and A.J. Freeman, Phys. Rev. B 35, 8861 (1987).
6. Xu, J.-H., T.J. Watson-Yang, J. Yu, and A.J. Freeman, Physics Lett. A 120, 489 (1987).
7. Takegahara, K., H. Harima, and A. Yanase, Jap. J. Appl. Phys. 26 L352 (1987). Oguchi, T., Jap. J. Appl. Phys. 26 L417 (1987).
8. Mattheiss, L.F., Phys. Rev. Lett. 58, 1028 (1987).
9. Kanbe,S., K. Kishio, K. Kitazawa, K. Fueki, H. Takagi, and S. Tanaka, Chem. Lett. 547 (1987). van Dover, R.B., R.J. cava, B. Batlogg, and E.A. Rietman, Phys. Rev. B 35, 5337 (1987).
10. Massidda, S., J. Yu, A.J. Freeman, and D.D. Koelling, Physics Lett. 122, 198 (1987).
11. Yu, J., S. Massidda, A.J. Freeman, and D.D. Koelling, Physics Lett. 122, 203 (1987).
12. Beno, M.A., L. Soderholm, D.W. Capone II, D.G. Hinks, J.D. Jorgensen, J.D. Grace, I.K. Schuller, C.U. Segre, and K. Zhang, Appl. Phys. Lett. 51, 57 (1987).
13. Cava, R.J., B. Batlogg, R.B. van Dover, D.W. Murphy, S. Sunshine, T. Siegrist, J.P. Remeika, E.A. Rietman, S. Zahurack, and G.P. Espinosa, Phys. Rev. Lett. 58, 1676 (1987). Hinks, D.G., G. Soderholm, D.W. Capone, II, J.D. Jorgensen, I.K. Schuller, C.U. Segre, K. Zhang, and J.D. Grace, Appl. Phys. Lett. 50, 1688 (1987).

ELECTRONIC AND STRUCTURAL PROPERTIES OF ORDERED III–V ALLOYS

Kathie E. Newman, Jun Shen, Dan Teng, Bing-Lin Gu,* Shang-Yuan Ren, and John D. Dow

Department of Physics, University of Notre Dame, Notre Dame, Indiana 46556
*Physics Department, Tsinghua University, Beijing, PRC

ABSTRACT

The strain energy of ordered phases of ternary III-V alloys has been estimated. A chalcopyrite structure is found to always have a lower strain energy than 1×1 superlattices oriented along the (0,0,1) or (1,1,1) directions. Ga-As and In-As bond lengths in ordered compounds $GaInAs_2$ are found to be in good agreement with the Mikkelsen-Boyce EXAFS measurements for the alloy. The effect of both ordering and strain on the bandstructure is also considered.

1. INTRODUCTION

Recent reports of new ordered structures in the ternary III-V alloys $A\ell_{1-x}Ga_xAs$, $GaAs_{1-x}Sb_x$ and $Ga_{1-x}In_xAs$ for $x \simeq \frac{1}{2}$ and $x \simeq \frac{1}{4}$[1] have kindled considerable interest in the semiconductor community. Remembering that a zinc-blende structure is composed of two face-centered cubic (fcc) sublattices, and ignoring the atoms located on the "passive" non-ordering sublattice, one finds that the newly discovered $x = \frac{1}{2}$ semiconductor structures, termed the (0,0,1) 1×1 superlattice, chalcopyrite, and (1,1,1) 1×1 superlattice (see Fig. 1), are all related to the fcc special-k point structures of metals: (0,0,1), (0,$\frac{1}{2}$,1), and (1,1,1), space groups P4/mmm, $I4_1/amd$, and $R\bar{3}m$.[2]

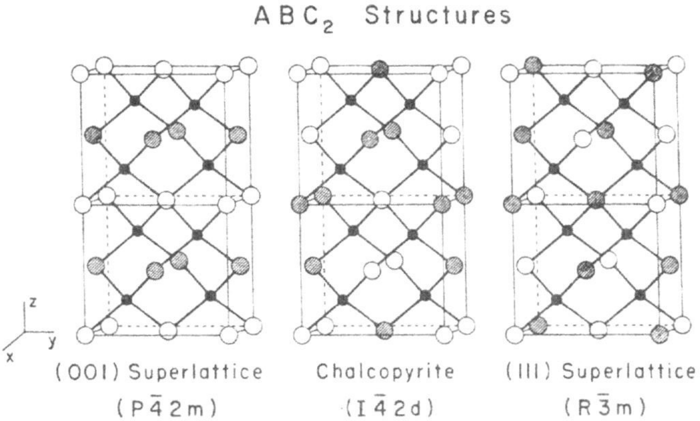

FIGURE 1. Special-k point structures for ordered compounds ABC_2: k = (0,0,1), 1×1 superlattice oriented along the z direction; k = (0,$\frac{1}{2}$,1), chalcopyrite; and k = (1,1,1), 1×1 superlattice oriented along the (1,1,1) direction. Also shown are the respective space-group designations of each crystal type. Note, each structure is shown undistorted from the parent zinc-blende form. Cations A and B are shown as large open and shaded circles, respectively, and the anions C are shown as smaller filled circles.

G. M. Stocks and A. Gonis (eds.), Alloy Phase Stability, 621–625.

The structure reported for $x = \frac{1}{4}$, famatinite, shown in Fig. 2, is also a $(1\ \frac{1}{2}\ 0)$ special-k structure.

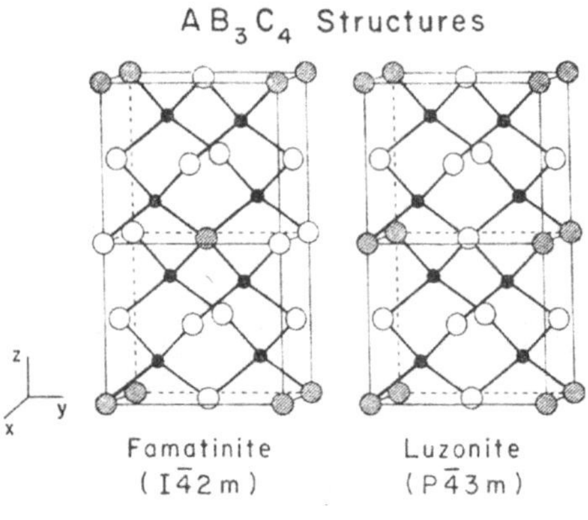

FIGURE 2. Special-k point structures for AB_3C_4 ordered compounds: k = (0,0,1), luzonite, and k = $(0,\frac{1}{2},1)$, famatinite.

Not known presently is the stability of the phases found. All experimental reports have used epitaxial growth, suggesting strongly the possibility of strain stabilization. Mbaye et al.,[3] who have calculated a phase diagram in a tetrahedron approximation, used strain arguments to explain why ordered phases are not normally seen.

The existence of sizable strains in the materials due to the tetrahedral sp^3 bond is clearly the feature that distinguishes the semiconductor ordering problem from that in metals. This paper investigates the effects of strain on the new ordered compounds. Strain energies for the five special-k structures are found. Finally, the effects of strain and order on the bandstructure of the new ordered forms is investigated, using as examples the (0,0,1)-oriented superlattice and chalcopyrite ordered structures.

2. THEORY

2.1 The Effects of Strain on Structure

We estimate the strain energy of a semiconductor from a simple phenomenological formula due to Harrison:[4]

$$E_{strain} = \frac{1}{2} C_0 (\delta d)^2 + \frac{1}{2} C_1 (\delta\theta_{ij})^2 \tag{1}$$

This formula has two contributions: A bond-stretching term, present because there is a natural bond length between any pair of atoms, and a bond-bending term, indicating that the energy should increase if angles deviate from those in a perfect tetrahedron. The parameters C_0 and C_1 are fit to elastic constants.

Our calculations start by specifying the atomic positions in the ordered structure. For each type of structure, there is an obvious symmetry-allowed distortion that best accommodates the bond-stretching and bond-bending forces. For example, for chalcopyrite, see Fig. 1, the positions of the 8 atoms in the basis are given in terms of just three parameters, the dimensions a and 2c of the body-centered tetragonal (bct) unit cell, with $c \simeq a \simeq a_0$, with a_0 the dimension of the parent zinc-blende cube, and an internal coordinate variable p ($p \simeq \frac{1}{4}a_0$).[5] Similarly, the superlattice oriented along the (0,0,1) direction has as its parameters the dimensions of the distorted cube, a^2c, and an internal coordinate h ($h \simeq \frac{1}{4}a_0$) that is the distance above the $z = 0$ plane of an anion C. [The superlattice oriented along the (1,1,1) direction is rhombohedral in symmetry and is characterized by 5 parameters.] Then, for each atom in the basis of the structure, we simply tabulate all contributions to Eq. (1), and then minimize the strain energy with respect to the free parameters. By evaluating Eq. (1) at its minima, we have found both the positions of all atoms as well as the size of the strain energy.

2.2 Bandstructure Calculations

In our calculations of bandstructure, we employ a modified empirical tight-binding Hamiltonian that includes the 5 atomic orbitals s, p_x, p_y, p_z and s*. Such a Hamiltonian is known to be accurate for the valence bands and the conduction band edge of zinc-blende semiconductors. The parameters of this Hamiltonian have been determined by fitting to experimental data for all III-V zinc-blende compounds by Vogl et al.[6]

Calculations use as input the positions of the atoms in each type of unit cell and follow the standard Slater-Koster method, e.g., strain forces a modification of the direction cosines used in the tight-binding Hamiltonian of the zinc-blende structure.[4] The required tight-binding parameters are found from the parent III-V compounds, e.g., we use Harrison's Law,[4] that the product of nearest-neighbor matrix elements V with the square of the bond length, Vd^2, should be nearly constant. Finally, we have included the effects of valence-band offsets on the ordered-compound's bandstructure,[7] and have concluded that these corrections are small.

3. RESULTS

We have computed the strain energies and effects of strain on five types of ternary III-V ordered compounds.[8] Here we simply quote results for a prototypical semiconductor alloy, $Ga_{1-x}In_xAs$.

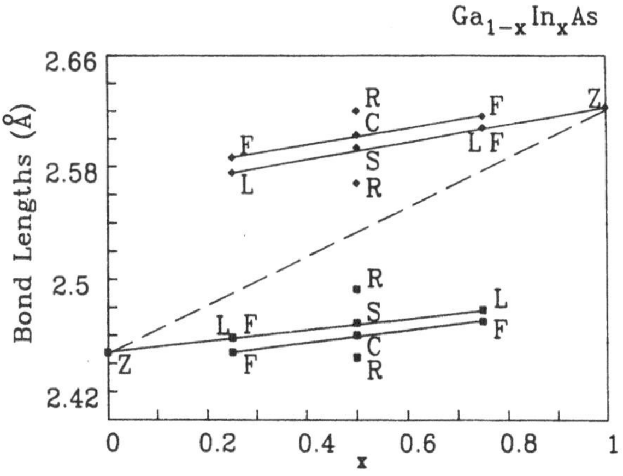

FIGURE 3. Calculated bond lengths in the $Ga_{1-x}In_xAs$ family of compounds.

In Fig. 3 we show results for the Ga-As bond lengths (squares) and In-As bond lengths (diamonds) in the $Ga_{1-x}In_xAs$ family of compounds. For $x = 0$ or 1, the bond length is that of the zinc-blende (Z) compounds GaAs or InAs. For $x = \frac{1}{4}$, luzonite compounds (L) contain three Ga-As bonds for every In-As bond, while famatinite compounds (F) have two long and one short Ga-As bonds for each In-As bond. A similar result holds for $x = 3/4$. For $x = \frac{1}{2}$, the -(0,0,1)-oriented superlattice (S) and chalcopyrite compounds have equal numbers of Ga-As and In-As bonds, while the rhombohedral (R) (1,1,1)-oriented superlattice contains three long and one short Ga-As bond and three short and one long In-As bond.

The dashed line shown in Fig. 3 indicates Vegard's Law expected for the average bond length measured in the alloy by x-ray diffraction. Mikkelsen and Boyce[9] have shown using EXAFS (extended x-ray absorption fine structure) measurements that there are two distinct bond lengths in $Ga_{1-x}In_xAs$ alloys, one for each type of bond. The solid lines in Fig. 3 that pass through the parent zinc-blende compounds are in good agreement with the slopes of the Mikkelsen-Boyce data for the alloy. These lines pass through compounds with either simple cubic or simple tetragonal symmetry. [Parallel to those lines are lines passing through the body-centered tetragonal compounds, famatinite, and chalcopyrite.] We interpret this agreement as follows. The alloy is to be thought of as a mixture of all possible orientations of tetrahedra $AsGa_nIn_{4-n}$. Portions of the alloy may show incipient order of the types shown in Figs.

1 and 2. The EXAFS data are statistical averages of the configurations found in the alloy; a line near the middle of the distribution of possible bond lengths is found. More importantly, the fact that the average results for the strain calculations alone reproduce well the EXAFS data imples that charge-transfer effects are probably only a small correction for the III-V family of compounds.[10]

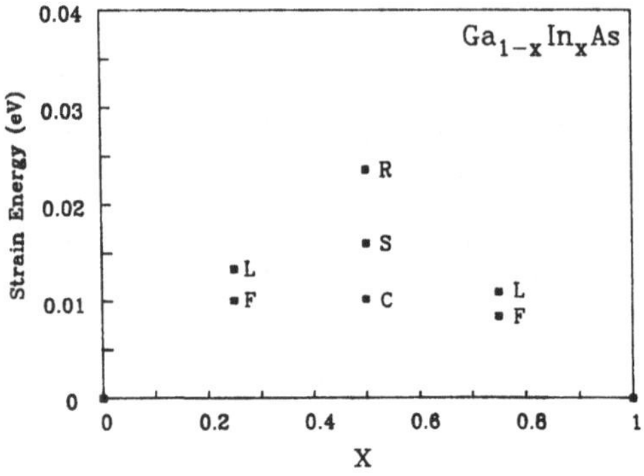

FIGURE 4. Calculated strain energies of the $Ga_{1-x}In_xAs$ family of compounds.

Shown in Fig. 4 are the calculated strain energies of the ordered structures. For x = 1/4 and 3/4, we find that the famatinite structure has the lowest energy, while for x = $\frac{1}{2}$, the chalcopyrite structure is lowest in energy, followed then by the (0,0,1) superlattice (S) and then by the rhombohedral-symmetry (R) superlattice oriented along the (1,1,1) direction. These results are consistent with the calculated bond lengths shown in Fig. 3. For example, for rhombohedral symmetry, there are two types of Ga-As (In-As bonds). One bond, found in a tetrahedron $GaIn_3As$, which is near the parent bond length, and the other one, found with a 3-fold degeneracy, in a tetrahedron $GaInAs_3$, at a distance quite far from the parent bond length. It is easily seen from Eq. (1) that those structures with large bond-length deviations from the parent zinc-blende compounds will tend to have a large strain energy. As a general trend, we find that structures with a body-centered tetragonal Bravais lattice most easily accommodate strain.

Turning finally to electronic properties, in Figs. 4 and 5 we show the low-temperature bandstructures of two $GaInAs_2$ ordered structures, the (0,0,1) superlattice and chalcopyrite structure. The band gap of the (0,0,1) superlattice is at 0.87 eV, compared with 0.81 eV in the alloy for x = 0.47,[11] and 0.83 eV in chalcopyrite. Notatable is the strain splitting of the top of the valence band. Because the local strain field of the tetrahedron has opposite orientations in the two structures, with the z direction singled out in the (0,0,1) superlattice, and an xy plane in chalcopyrite (see Fig. 1), the heavy hole is higher in the (0,0,1) superlattice than in chalcopyrite. Obviously, these two structures will have distinguishable experimental spectra.

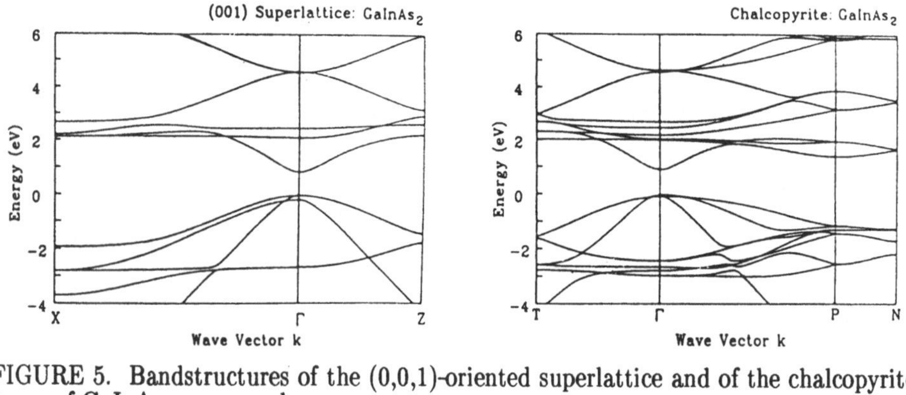

FIGURE 5. Bandstructures of the (0,0,1)-oriented superlattice and of the chalcopyrite forms of $GaInAs_2$ compounds.

4. CONCLUSIONS

We have investigated the relative stability for III-V compounds of the three possible ordered $x = \frac{1}{2}$ structures: the (0,0,1)- and (1,1,1)-oriented 1×1 superlattices and chalcopyrite. Without using a second-neighbor interaction on an fcc lattice,[2] we find that the strain energy due to the bonding of anions to cations alone is enough to distinguish the three phases. Our calculations predict that the chalcopyrite structure should have a lower energy than either type of superlattice. Similarly, we find that for $x = \frac{1}{4}$, the famatinite structure has lower strain energy than the luzonite structure.

Our strain calculations also allow us to predict structural parameters such as bond lengths for the new ordered III-V compounds. This in turn allows us to investigate the influence of order and strain on the bandstructure of these structures. We hope our calculations will aid experimentalists in the search for additional new ordered semiconductor compounds.

ACKNOWLEDGMENTS — We thank ONR and AFOSR for their financial support.

REFERENCES

1. T. S. Kuan, T. F. Kuech, W. I. Wang, and E. L. Wilkie, Phys. Rev. Lett. 54, 201 (1985); H. R. Jen, M. J. Cherng, and G. B. Stringfellow, Appl. Phys. Lett. 48, 1603 (1986); H. Nakayama and H. Fujita, in *Gallium Arsenide and Related Compounds -- 1985*, edited by M. Fujimoto, IOP Conference Proceedings No. 79 (Institute of Physics, Bristol and London, 1986), p. 289; and M. A. Shahid, S. Mahajan, D. E. Laughlin, and H. M. Cos, Phys. Rev. Lett. 58, 2567 (1987).
2. For a review, see D. de Fontaine, in *Solid State Physics*, edited by H. Ehrenreich, F. Seitz, and D. Turnbull (Academic, New York, 1979), Vol. 34, p. 73.
3. See A. A. Mbaye, L. G. Ferreira, and A. Zunger, Phys. Rev. Lett. 58, 49 (1987) and references quoted therein.
4. W. A. Harrison, *Electronic Structure and the Properties of Solids* (W. H. Freeman, San Francisco, 1980), p. 76, 196, and 481.
5. U. Kaufman and J. Schneider, *Festkörperprobleme XIV*, edited by H. J. Queisser (Pergamon, 1974) p. 229.
6. P. Vogl, H. P. Hjalmarson, and J. D. Dow, J. Phys. Chem. Solids 44, 365 (1983).
7. E. A. Kraut, J. Vac. Sci. Technol. B2, 488 (1984).
8. D. Teng, J. Shen, K. E. Newman, B.-L. Gu, S. Y. Ren, and J. D. Dow, unpublished.
9. J. C. Mikkelsen and J. B. Boyce, Phys. Rev. Lett. 49, 1412 (1982).
10. K. C. Hass and D. Vanderbilt, J. Vac. Sci. Technol. A5, 3019 (1987).
11. K. Alavi, R. L. Aggarwal, and S. H. Groves, Phys. Rev. B21, 1311 (1980); measurement is at 4.2 K.

LIQUID ALLOYS - NEW PERSPECTIVES AND CHALLENGES

Marie-Louise Saboungi, Susan R. Leonard,† Gerald K. Johnson, and David L. Price
Argonne National Laboratory
Argonne, Illinois 60439

1. INTRODUCTION

The opening address of S. C. Moss was effective in summarizing topics where progress has been achieved in the study of alloy phase stability; the long-term goals seemed to be within reasonable reach, e.g., calculation of phase diagrams from first principles, relations between structure and physicochemical properties. The purpose of this contribution is to add another dimension, and perhaps a challenge to theoreticians in this field, by showing complexities in the liquid state which must be included in the modeling of phase equilibria, such as liquid-solid coexistence and the influence of the liquid properties on the solidus curve.

In this paper, we will focus on one of many unusual liquid semiconducting alloys, K-Pb. The thermodynamic[1,2] and electrical[3] properties will be discussed and analyzed in terms of a disorder model introduced first by Wagner[4] for crystalline semiconductors. The structure of the equiatomic alloy will be presented; interpretation of the first sharp diffraction peak in the total structure factor is based on the local structures in the crystalline phase and should be viewed as an example of the interrelation between the solid and liquid properties.[5]

2. THERMODYNAMIC AND ELECTRICAL PROPERTIES

The thermodynamic properties of liquid K-Pb alloys have been measured by emf techniques.[1] The variations of the entropy, the activity coefficient, and the heat capacity with composition have characteristic features reminiscent of liquids with intermediate-range ordering. For example, the excess entropy of mixing has a "V" shape with the minimum located at the equiatomic composition. The partial Gibbs free energy of mixing has an inflection point at the equiatomic composition, which leads to a maximum in the Darken excess stability or a minimum in the concentration-concentration fluctuation function. The heat capacity of the liquid alloys exhibits a strong departure from additivity and goes through a pronounced sharp maximum at the equiatomic composition; furthermore, it has an anomalous temperature dependence with large magnitudes atypical of those observed for similar alloys[2] (Fig. 1). At this point, it should be noted that an analytical representation of the composition dependence of the thermodynamic properties in terms of power series is a meaningless exercise since such treatment cannot be used to extrapolate beyond the range of measurements.

The electrical resistivity and its temperature dependence have been measured by Meijer *et al.*[3] At the equiatomic composition, a maximum of about 870 $\mu\Omega$ cm is reported with a negative temperature coefficient of $-4.6\mu\Omega$ cm K^{-1}. According to the

†Present address: Department of Chemistry, University of California, Berkeley.

G. M. Stocks and A. Gonis (eds.), Alloy Phase Stability, 627–631.

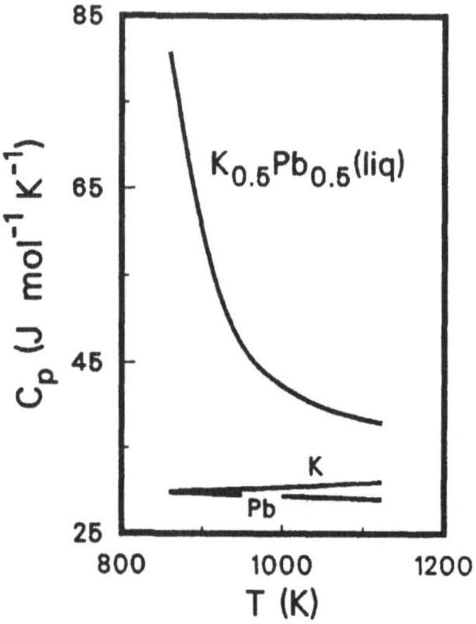

Fig. 1. Variations of C_p with temperature.[2]

well-known Mott classification, KPb is a semiconducting liquid. Thus, in order to correlate quantitatively the thermodynamic and transport properties, a model introduced by Wagner[4] for electronic disorder is used. At the 50-50 composition, one introduces the quantity δ, which is a measure of the deviation from the exact stoichiometry and is related to X_e and X_h by

$$\delta = X_e - X_h \tag{1}$$

where X_e (and X_h) is the ratio of the number of electrons (and electron holes) to the number of Pb atoms. The emfs, E, can be directly calculated by

$$E^o - E = \frac{RT}{F}\, sinh^{-1}\,(\delta/2X_e^o) \tag{2}$$

where the superscript ^{o} denotes the value at the exact stoichiometry. The thermodynamic functions can then be directly evaluated from the emfs. The electrical conductivity, σ, is given by

$$\sigma = \frac{2X_e^o(u_e u_h)}{V_m}\, F \cosh\left[(E^o - E)\frac{F}{RT} + \frac{1}{2} ln\left(\frac{u_e}{u_h}\right)\right] \tag{3}$$

where u_e and u_h are the mobilities of electrons and electron holes and V_m is the molar volume of KPb. The calculated and experimental values for the emfs and the electrical resistivity for an atom fraction of K varying from 0.40 to 0.60 are given in Figs. 2 and 3, respectively. A striking good representation by this model of different sets of measurements is obtained.

The interpretation of the thermodynamic properties, especially of the heat capacity anomalies, requires a model which allows for the presence of polynuclear species, such as K_nPb_n. The Wagner equations will have to be recalculated for such a model.

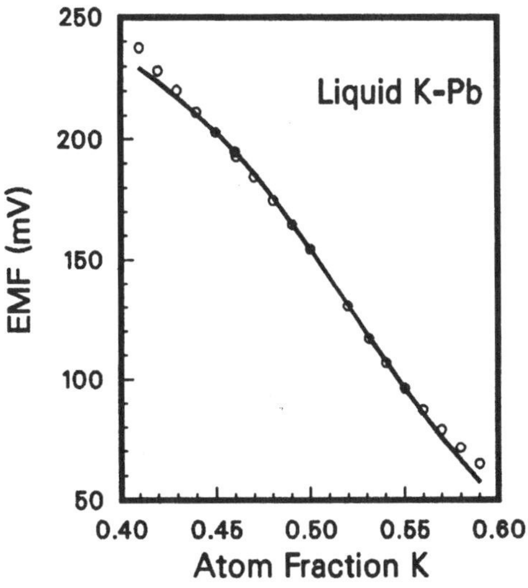

Fig. 2. Measured[1] (open circles) and calculated (solid line) emfs at T = 879 K and X_e^o =0.132 (Eq. 2).

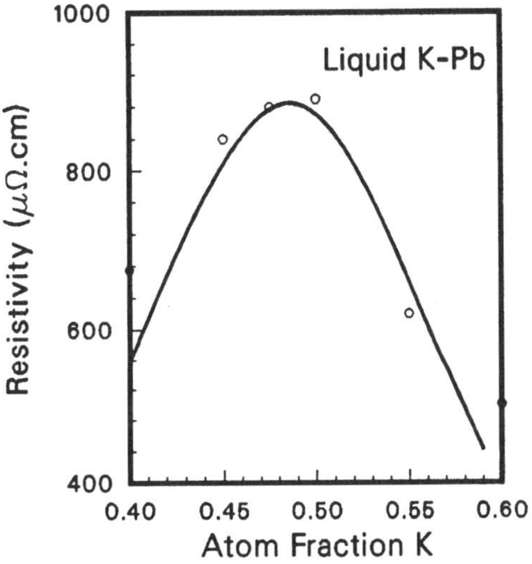

Fig. 3. Measured[3] (open circles) and calculated (solid line) electrical resistivities at T = 879 K and X_e^o = 0.132 (Eq. 2).

In order to check the necessity for such a treatment, an investigation of the structure of liquid K-Pb alloys was needed.

3. STRUCTURE OF LIQUID EQUIATOMIC KPb: EVIDENCE FOR THE PRESENCE OF MOLECULAR ENTITIES

Time-of-flight diffraction measurements were carried out on the Special Environment Powder Diffractometer at the Intense Pulsed Neutron Source at Argonne National Laboratory. The variations of the liquid structure factors with temperature were investigated. A first sharp diffraction peak (FSDP) was observed at Q~0.96 $\AA^{-1}$ with a magnitude which decreased with increasing temperature; the main peak remained unaltered. The presence of the FSDP is a distinctive signature in many amorphous materials. In the liquid, it is indicative of a degree of bonding and, in this case, should reflect the presence of molecular entities with intermediate-range order. The structure of solid KPb (a Zintl compound) has been investigated by x-ray diffraction and found to be similar to that of NaPb; the structural unit consists of nearly regular Pb_4 tetrahedra surrounded by a Na_4 antitetrahedron (i.e., Na atoms sit opposite the four faces of a Pb_4 tetrahedron). Using a model previously developed for chalcogenide glasses,[6] the total structure factor was calculated on the basis of structural units distributed according to a random packing of hard spheres with random uncorrelated orientations.[5] The structural units were constructed to be like the Na_4Pb_4 units in the solid, but with a size parameter expected of KPb. The agreement between the molecular model based on the presence of K_4Pb_4 in the liquid and the measurements is striking and lends credibility to the presence of K_4Pb_4 species in the liquid; with a reasonable choice of the two adjustable parameters, σ the hard sphere diameter and η the packing fraction of the structural units, the positions and magnitudes of the FSDP and the main peak are well reproduced (Fig. 4).

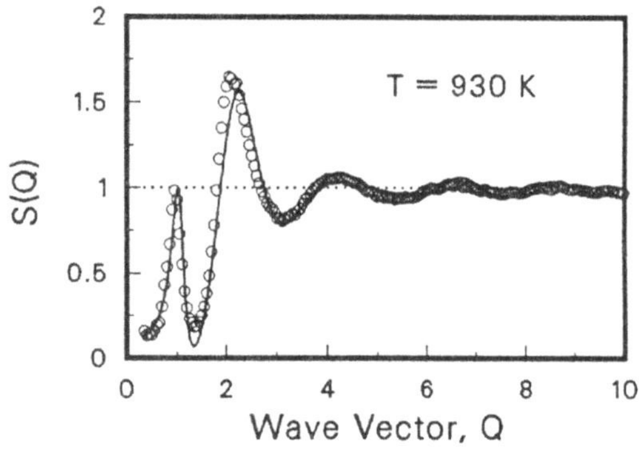

Fig. 4.Measured[5] (open cirles) and calculated (solid line) S(Q) for KPb(ℓ).

4. CONCLUSIONS

It is hoped that this short contribution on the physicochemical properties of an unusual class of liquids will serve as an incentive to explore the close relation between solids and liquids. First-principles calculations of the liquidus and of the coexistence of two phase equilibria must include the sometimes complex properties of the liquids and not assume ideal or regular behavior of the thermodynamic properties. Theoretical advances in the description of the solid properties will ultimately help to better define the properties of the liquid.

5. REFERENCES

1. Saboungi M.-L., Leonard S. R., and Ellefson J., J. Chem. Phys., Vol. **85(10)**, p. 6072, 1986.
2. Johnson G. K. and Saboungi M.-L., J. Chem. Phys., Vol. **86(11)**, p. 6376, 1987.
3. Meijer J. A., Geertsma W., and van der Lugt W., J. Phys. F, Vol. **15** , p. 899, 1985.
4. Wagner C. , in *Progress in Solid State Chemistry*, Vol 6, eds, Reiss H. and McCaldin J. O., Pergamon Press, 1977.
5. Saboungi M.-L., Blomquist R., Volin K. J., and Price D. L., J. Chem. Phys., Vol. **87(4)**, p. 2278, 1987.
6. Moss S. C. and Price D. L., in *Physics of Disordered Materials*, eds, Adler D., Fritzsche H., and Ovshinsky S. R., Plenum, New York, 1985.

ACKNOWLEDGMENT

This work was supported by the Division of Materials Science, Office of Basic Energy Sciences, U.S. Department of Energy under Contract W-31-109-ENG-38.

EXCITATION SPECTRA OF OPTICALLY EXCITED RARE-GAS ATOMS PHYSISORBED ON METAL SURFACES

CONSTANTINE MAVROYANNIS

Laser Chemistry Group, Division of Chemistry, National Research Council of Canada, Ottawa, Ontario, Canada K1A 0R6

1. INTRODUCTION

In the last decade, Flynn and coworkers (1-4) have extensively examined the optical excitation spectra of rare-gas atoms physisorbed on various metal surfaces as a function of coverage. The data were obtained by the use of differential reflectance methods with synchrotron radiation. The primary spectroscopic effort has focused on the way the substrate modifies levels of the free adsorbate, and on any newly formed excited configurations, which arise through the interaction of the adsorbate atom with the metal surface. The striking changes in the optical excitation spectra have been interpreted as due to the formation of a neutral excited adsorbate configuration on some surfaces and an ionized one on others (1-4). At high coverage (>1 monolayer), the spectra are clearly visible and resemble comparable excitations observed in absorption for both vapor phase and condensed rare gases. However, in the low coverage limit, the peak in the spectrum remains strong for some adatom-substrate combinations, while it completely vanishes for others. For instance, at low coverage the adsorption peaks of Xe on Al, Ti and Au, and Kr on Au are virtually eliminated while those of Xe and Ar on Mg, Kr and Ar on Al and Ar on Au persist (1-4). Arguments against the foregoing interpretation of the optical data have been recently made (5). No theoretical explanation has been formed (5,6) for the occasional disappearance of the excitonic component of the optical-absorption system at low coverage.

A microscopic theory has been recently developed (7), hereafter referred to as I, concerning the optical excitation spectra of neutral rare-gas atoms physisorbed on metal surfaces. In I a simple model has been used, where a neutral atom A and its image B are a distance 2R apart and interact through the dipole-dipole interaction (8-11). The metal surface separating the atom and its image is characterized by a dielectric function, which has been taken to be $\varepsilon(\omega)=1-\omega_p^2(\omega^2+i\omega\gamma_{sp})^{-1}$, where ω_p is the plasma frequency of the metallic electrons and γ_{sp} $(\gamma_{sp}\ll\omega)$ is the intrinsic damping of the surface plasmons due to various decay processes (12). The optical excitation spectra due to radiative and nonradiative processes have been considered in I in great detail. We shall describe here the excitation spectra due to the nonradiative processes.

2. NONRADIATIVE EXCITATION SPECTRA

In the limit when the surface plasmon mode is heavily damped, namely, when the nonradiative damping γ_{sp} is much greater than the radiative damping γ_0 $(\gamma_{sp}\gg\gamma_0)$, which is always valid, the radiative contributions to the damping can be neglected. In this limit the spectral function per atom takes the form (7)

G. M. Stocks and A. Gonis (eds.), Alloy Phase Stability, 633–637.

$$\gamma_{sp}\bar{I}_j(\omega) = \frac{N_j(\omega_{j+})\lambda_{sp}(\omega_{j+})-X^{sp}_{j+}M_j}{[X^{sp}_{j+}]^2 + \lambda^2_{sp}(\omega_{j+})} + \frac{N_j(\omega_{j-})\lambda_{sp}(\omega_{j-})+X^{sp}_{j-}M_j}{[X^{sp}_{j-}] + \lambda^2_{sp}(\omega_{j-})} , \quad (1)$$

The energies of excitation $\omega_{j\pm}$ are determined by

$$\omega^2_{j\pm} = \tfrac{1}{2}\{E^2_{\nu 0}+\omega^2_{sp} \pm [(E^2_{\nu 0}-\omega^2_{sp})^2 \pm 4E^2_{\nu 0}\omega^2_{sp}\alpha]^{\frac{1}{2}}\} , \quad (2)$$

where ω_{sp} is the frequency of the surface plasmons defined as $\omega_{sp}=\omega_p/\sqrt{2}$ and the coupling function $\alpha \equiv \alpha(\vec{R})=\alpha_s/2R^3$ with α_s being the static polarizability (11) of the atom $\alpha_s=(2/3)\Sigma_\nu|\vec{P}_{0\nu}|^2/E_{\nu 0}$. The atomic transition frequency $E_{\nu 0}=E_\nu - E_0$, where E_ν and E_0 are the frequencies of the excited state $|\nu\rangle$ and the ground state $|0\rangle$ of the atom, respectively; units in which $\hbar=1$ are used throughout. In Eqs. (1) and (2), j=s and j=a, as well as the + and - signs in the square root of Eq. (2), denote the symmetric and the antisymmetric modes, respectively. The functions $N_j(\omega_{j\pm})$, M_j and $\lambda_{sp}(\omega_{j\pm})$ are defined as

$$N_j(\omega_{j\pm}) = \pm\left(\frac{2E_{\nu 0}}{\omega_{j\pm}}\right)\frac{[\omega^2_{j\pm}-\omega^2_{sp}(1\mp\alpha)]}{(\omega^2_{j+}-\omega^2_{j-})} , \quad M_j = \frac{2E_{\nu 0}}{(\omega^2_{j+}-\omega^2_{j-})} , \quad (3)$$

$$\lambda_{sp}(\omega_{j\pm}) = \Gamma_{sp}(\omega_{j\pm})/\gamma_{sp} = \pm\tfrac{1}{2}\,\frac{(\omega^2_{j\pm}-E^2_{\nu 0})}{(\omega^2_{j+}-\omega^2_{j-})} , \quad (4)$$

while $X^{sp}_{j\pm}=(\omega-\omega_{j\pm})/\gamma_{sp}$ defines the reduced frequency.

The spectral function $\bar{I}_j(\omega)$ describes asymmetric Lorentzian lines, which are peaked at the frequencies $\omega=\omega_{j\pm}$ having spectral widths of the order of $\Gamma_{sp}(\omega_{j\pm})$; the extent of the asymmetry at $\omega\neq\omega_{j\pm}$ depends on the strength of the damping of the surface plasmons γ_{sp}. Inspection of Eq. (2) reveals that one of the energies of excitations is close to the value of the atomic frequency $E_{\nu 0}$ while the other is close to the value of the surface plasmon frequency ω_{sp}. Thus, when $E_{\nu 0}$ is greater than ω_{sp}, $E_{\nu 0}>\omega_{sp}$, the excitations at $\omega=\omega_{j+}(E_{\nu 0}<\omega_{j+})$ and $\omega=\omega_{j-}(\omega_{sp}>\omega_{j-})$ may be referred to as the atomic-like and surface plasmon-like excitations, respectively; the reverse is true for $E_{\nu 0}<\omega_{sp}$. The notation in Eqs. (1)-(4) is the same as in I, where all the expressions have been derived and discussed.

Using the values of R, α_s and ω_{sp} given in ref. (11), the computed spectra from Eq. (1) for the atomic-like excitations are illustrated in Figures 1 and 2, where the relative intensity $\gamma_{sp}\bar{I}_j(\omega)$ in units of γ_{sp} is plotted versus the relative frequency X^{sp}_j in units of eV/γ_{sp}. Because of lack of space, we report here only the numerical results for two frequencies of Xe and Kr, which are physisorbed on the metal surfaces of Al and Au, respectively. Computed values of the energies of excitation $\omega_{j\pm}$ from Eq. (2) reveal that when the frequencies of the symmetric modes are blue shifted the corresponding ones of the antisymmetric modes are red shifted relative to the atomic frequencies $E_{\nu 0}$ and vice versa. Numerical data for $N_j(\omega_{j\pm})$ and $\lambda_{sp}(\omega_{j\pm})$ for the atomic-like spectra imply that the relative intensities of the peaks for the symmetric and antisymmetric modes take positive and negative values indicating that the physical processes of absorption (attenuation) and stimulated emission (amplification) occur at the corresponding frequencies, respectively. Hence, the peaks of the

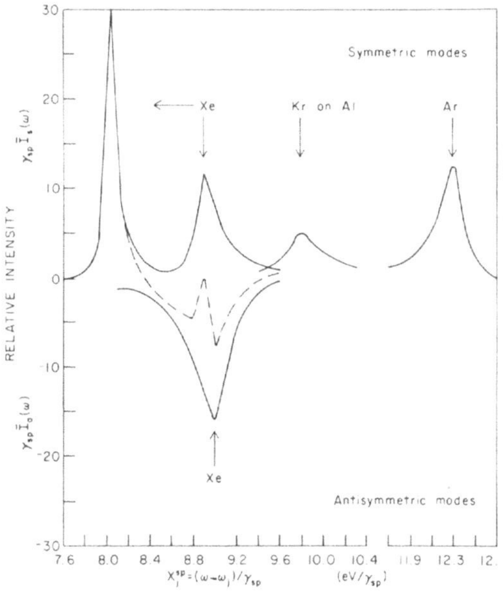

FIGURE 1. Spectra of the atomic-like excitations of excited rare-gas atoms physisorbed on the surface of Al. The relative intensities per atom $\gamma_{sp}\bar{I}_j(\omega)$ in units of γ_{sp} are plotted versus the relative frequencies $X_j^{sp} = (\omega-\omega_j)/\gamma_{sp}$ in eV/γ_{sp}, where $j=s\pm$ and $a\pm$ for the symmetric (positive intensities) and antisymmetric (negative intensities) modes, respectively. The numbering on the abscissa refers to the frequency ω in eV. The dashed line indicates the net lineshape of excited Xe on Al, which results from the combination of the red shifted peak of the symmetric mode at ω_{s-}= 8.9178 eV (atomic excitation 9.57 eV) and the blue shifted peak of the antisymmetric mode at ω_{a-}=8.9952 eV (atomic excitation 8.437 eV).

symmetric and antisymmetric modes may cancel each other out either completely or partially. Because of large values of $\lambda_{sp}(\omega_{s\pm})$ and $\lambda_{sp}(\omega_{a\pm})$ and very small values of $N_s(\omega_{s\pm})$ and $N_a(\omega_{a\pm})$, the spectra of the surface plasmon-like excitations of the symmetric and antisymmetric modes are broadened to the extent that it is impossible to present them graphically.

Figure 1 illustrates the spectra of the atomic like excitations of the symmetric modes of Xe, Kr and Ar on Al as well as one peak of the antisymmetric mode for Xe on Al. The dashed line in Fig. 1 indicates the net lineshape of excited Xe on Al, which originates from the combin-

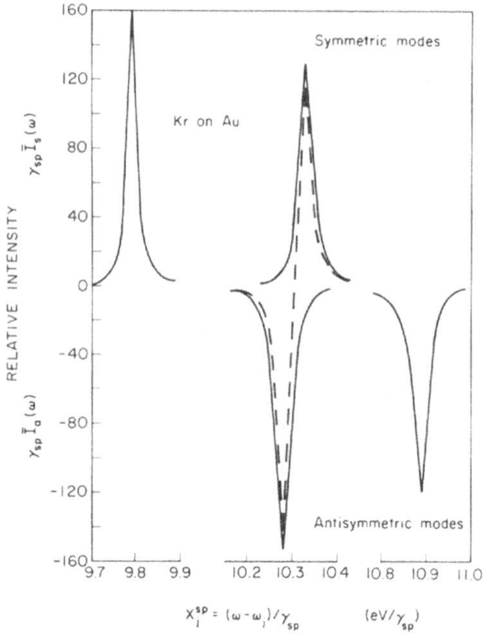

FIGURE 2. As in Figure 1 but for excited Kr at the frequencies $E_{\nu 0}$= 10.03 and 10.60 eV, respectively, physisorbed on the surface of Au. The dashed line illustrates the net lineshape arising from the combination of the red shifted peak of the symmetric mode at ω_{s-}=10.326 eV (atomic excitation 10.6 eV) and the blue shifted peak of the antisymmetric mode at ω_{a-}= 10.2815 eV (atomic excitation 10.03 eV).

ation of two lineshapes the red shifted peak of the symmetric mode at ω_{s-}= 8.9178 eV (atomic excitation $E_{\nu 0}$=9.57 eV) and the blue shifted peak of the antisymmetric mode at ω_{a-}=8.9952 eV (atomic excitation $E_{\nu 0}$=8.437 eV). These two peaks, whose frequency profiles are nearly overlapping, cancel each other out with the result that the peak at the atomic frequency $E_{\nu 0}$= 9.57 eV vanishes. A peak or a shoulder having a small negative intensity appears on each side of the vanished peak at ω_{s-}=8.9178 eV as indicated by the dashed line in Fig. 1.

The spectra of Kr on Au are illustrated by solid lines in Fig. 2 for the atomic-like excitations of the two symmetric and two antisymmetric modes, respectively. There is an overlap of the frequency profiles between the red shifted peak at ω_{s-}=10.33 eV (atomic excitation $E_{\nu 0}$=10.60 eV) of the symmetric mode and the blue shifted peak at ω_{a-}= 10.28 eV (atomic excitation $E_{\nu 0}$=10.03 eV) of the antisymmetric mode. The net result of the two lineshapes is illustrated by the dashed line in Fig. 2, which is of the absorption-amplification type, where the system absorbs and amplifies at the frequencies ω_{s-}=10.33 eV and ω_{a-}= 10.28 eV, respectively. Inspection of the lineshape of the dashed line in Fig. 2 reveals that the system at this condition is very unstable so that an infinitesimal frequency shift or a very small disturbance in the right direction either on the symmetric or on the antisymmetric mode will stabilize the system, where the positive and negative peaks will cancel each other out; the result will be the disappearance of the peak arising at the atomic excitation $E_{\nu 0}$=10.6 eV. For the remaining peaks at the frequencies ω_{s-}= 9.78 eV and ω_{a-}=10.88 eV of the symmetric and antisymmetric modes, respectively, other nearby excited states of the atom must be taken into consideration.

3. CONCLUDING REMARKS

The disappearance of the peaks of the spectral lines is due to the existence of the symmetric and antisymmetric states, whose relative intensities take positive and negative values, respectively; hence they may have a chance to cancel each other out whenever the frequency profiles of the peaks in question coincide. The relative intensities of the antisymmetric modes become negative only, at least for the cases under investigation, when the damping of the surface plasmons is much greater than the effective radiative damping (7). On the other hand, when the effective radiative damping is greater than the damping of the surface plasmons the relative intensities of the peaks of the antisymmetric modes take positive values (7). This implies that the occasional disappearance of the peaks of the spectral lines in the low coverage limit should only be observed in optical experiments provided that the appropriate conditions are fulfilled. This result is compatible with the observation that the disappearance of the peaks has been found only in optical experiments (1-5).

REFERENCES

1. Cunningham, J.E., Greenlaw, D.K., Erskine, J.L., Layton, R.P., and Flynn, C.P.: J. Phys. F: Metal Phys. 7, L281 (1977); Cunningham, J.E., Greenlaw, D.K., and Flynn, C.P.: Phys. Rev. Lett. 42, 328 (1979); Cunningham, J.E., Greenlaw, D.K., and Flynn, C.P.: Phys. Rev. B22, 717 (1980).
2. Cunningham, J.E., Gibbs, D., Chiu, T.H., and Flynn, C.P.: J. Phys. C: Solid State 14, L1113 (1981); Flynn, C.P., and Chen, Y.C.: Phys. Rev. Lett. 46, 447 (1981).
3. Flynn, C.P., and Cunningham, J.E.: J. Phys. C: Solid State 15, L1169 (1982); Gibbs, D., and Cunningham, J.E.: Phys. Rev. B29, 5292 (1984); Cunningham, J.E., Gibbs, D., and Flynn, C.P.: Phys. Rev. B29, 5304 (1984); Chen, Y.C., Cunningham, J.E., and Flynn, C.P.: Phys. Rev. B30, 7317 (1984).
4. Flynn, C.P.: Surface Sci. 158, 84 (1985).
5. Lang, N.D., Williams, A.R., Himpsel, F.J., Reihl, B., and Eastman, D.E.: Phys. Rev. B26, 1728 (1982); Chiang, T.C., Kaindl, G., and Eastman, D.E.: Solid State Comm. 41, 661 (1982); Eberhardt, W., and Zangwill, A.: Phys. Rev. B28, 5960 (1983); Demuth, J.E., Avouris, Ph., and Schmeisser, D.: Phys. Rev. Lett. 50, 600 (1983).
6. Bagchi, A., Barrera, R.G., and Dasgupta, B.B.: Phys Rev. Lett. 44, 1475 (1980); Bagchi, A., Barrera, R.G., and Fuchs, R.: Phys. Rev. B25, 7086 (1982).
7. Mavroyannis, C.: Mol. Phys. (in press).
8. Mavroyannis, C.: Mol. Phys. 6, 593 (1963).
9. Mavroyannis, C. and Hutchinson, D.A.: Solid State Comm. 23, 463 (1977); Mavroyannis, C.: Solid State Comm. 23, 467 (1977).
10. Lennard-Jones, J.E.: Trans. Faraday Soc. 28, 334 (1932).
11. Mavroyannis, C.: Mol. Phys. 36, 1565 (1978); Chem. Phys. Lett. 56, 263 (1978).
12. Endriz, J.G., and Spicer, W.: Phys. Rev. B4, 4144 (1971); Ritchie, R.H.: Surface Sci. 34, 1 (1973); Shinjo, K., Suyano, S., and Sasada, T.: Phys. Rev. B28, 5570 (1983).

DYNAMICS OF SPINODAL DECOMPOSITION IN POLYMER GELS

R. BANSIL, B. CARVALHO, and J. LAL

Department of Physics, Boston University, Boston, MA 02215

1. INTRODUCTION

The process of phase separation via spinodal decomposition has been the subject of extensive investigation in recent years(1). The dynamics of spinodal decomposition in binary phase separation processes has been shown to exhibit very similar behaviour in a wide variety of systems ranging from mixtures of simple fluids(2) to polymer blends(3-4). In all of these previous studies with polymers only linear, uncrosslinked polymers were investigated. However it is well known that linear polymers can crosslink and eventually form a gel. The process of gelation involves a percolation transition characterized by the sudden appearance of an infinite network at the gel-point which leads to a divergence of the shear viscosity from below the transition and the appearance of elasticity above it(5). Since the space inside a gel is compartmentalized into a mesh and since viscosity directly influences the diffusional movement it is possible that gelation may have significant effects on binary phase separation processes in polymers capable of forming a gel. To study this coupling between the thermodynamically driven binary phase separation process and the connectivity driven gelation transition we considered the system of gelatin in a poor solvent. Gelatin dissolved in a water-methanol mixture undergoes a binary phase-separation into a polymer rich phase and a polymer poor phase as the temperature is lowered and simultaneously undergoes a reversible sol-gel transition(6). In this paper we report the dynamics of spinodal decomposition in the gelatin-water-methanol (GWM) system and show that the occurence of gelation has a pronounced effect on the dynamics of phase separation and on the scaling laws describing the time evolution of domain growth.

2. EXPERIMENTAL METHODS

All the experiments reported in this paper were made on swine-skin gelatin (Sigma Chemical, approx. molecular weight 50,000 daltons) dissolved in a 40:60 mixture of methanol: water (by volume). It should be pointed out that although the gelatin-water-methanol system has 3 components, as far as phase separation is concerned the system can be treated as a binary system of polymer (gelatin) and solvent, since water and methanol are completely miscible and can be considered as the single solvent component.

2.1 ***Equilibrium phase diagram***. The cloud point curve and the spinodal line were determined from measurements of turbidity. Samples with different concentration of gelatin (ranging from 0.75% to 12% by weight in 100 ml solvent) were made in 1 cm path length cuvettes. For each sample the transmitted intensity from a 5 mw He-Ne laser was recorded as a function of temperature. Between each measurement

G. M. Stocks and A. Gonis (eds.), Alloy Phase Stability, 639–643.

the sample was heated to 40°C which is well above the upper consolute point for this system and thus the sample was remixed in the one phase regime. The transmitted light intensity decreases with temperature and the decrease is characterized by two very distinct slopes. Since the turbidity is proportional to the integral of the scattered intensity over the entire solid angle, we can interpret the temperature where the slope changes as the the cloud point. Following Tanaka(6) we define the spinodal temperature as the temperature where the transmitted light intensity extrapolates to zero. Both of these measures are just meant to be practical measures; in fact, the spinodal curve is sharply defined only in the limit of infinite range potentials(7). Nevertheless, the concept of a spinodal temperature has been used extensively in polymer systems and this may in part be due to the fact that mean-field theoretical description works very well for polymeric systems. Figure 1 shows the the cloud-point (T_{cl}) and the spinodal (T_s) temperatures as a function of gelatin concentration, i.e. the phase diagram of the GWM system.

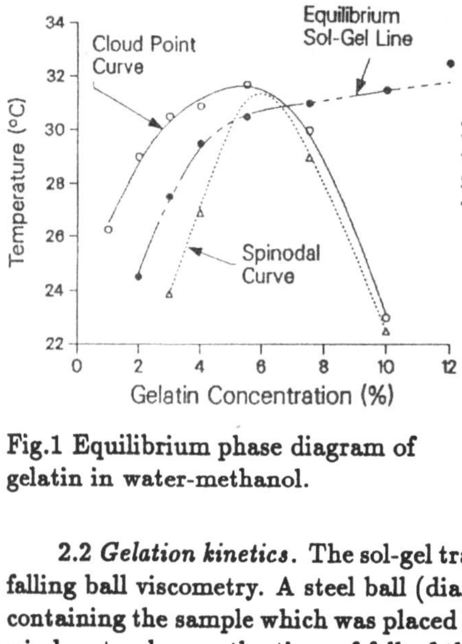

Fig.1 Equilibrium phase diagram of gelatin in water-methanol.

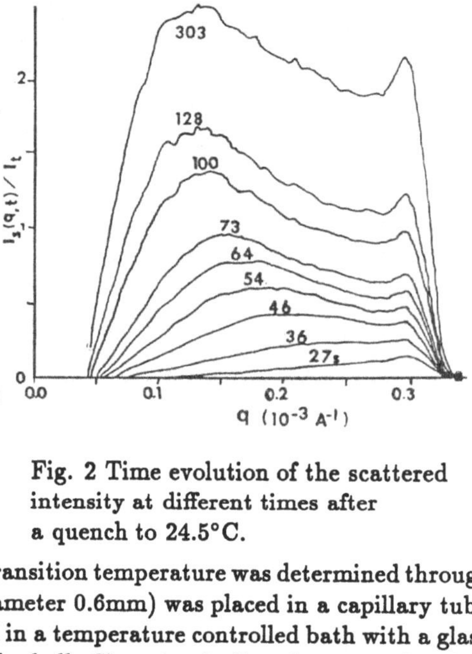

Fig. 2 Time evolution of the scattered intensity at different times after a quench to 24.5°C.

2.2 *Gelation kinetics*. The sol-gel transition temperature was determined through falling ball viscometry. A steel ball (diameter 0.6mm) was placed in a capillary tube containing the sample which was placed in a temperature controlled bath with a glass window to observe the time of fall of the ball. Since t_{fall} is directly proportional to the shear viscosity of the solution, the experiment shows the increase of the shear viscosity with time after a gelatin solution is quenched to a temperature below the equilibrium sol-gel transition temperature. By extrapolating the inverse of the time of fall, t_{fall} to zero we determine $\tau_{gel}(T)$, the time at which gelation occurs at the temperature T. We observed that the viscosity increases by more than two orders of magnitude as one approaches the gel point and that $\tau_{gel}(T)$ increases very rapidly with increasing temperature, varying from less than a minute at 24°C to over 12 hours at 31°C. Defining the equilibrium sol-gel transition temperature T_{gel}, as the temperature above which the sample never gels we find $\tau_{gel}(T) \sim (T_{gel} - T)^{-3.7}$ as the equilibrium sol-gel temperature approached. The equilibrium sol-gel transition curve shown in Fig. 1 was obtained by extrapolating these data. The main point to be noted from Fig. 1 is that for the GWM system a quench to a temperature T_q in the two phase region can also result in gelation if $T_q < T_{gel}$. However, the dynamics of phase separation will be influenced by gelation only if the sample gels in a time comparable to the characteristic time of phase separation . For the GWM system used in this study this condition is met for $T_q < 28C$.

2.3 *Spinodal decomposition dynamics.* The dynamics of phase separation was studied using the conical lens technique to measure the angular dependence of the scattered light intensity at various times after the sample was quenched to a temperature in the two phase region. The conical lens focusses light scattered at a given angle θ to a single point along the optical axis. Thus by moving a photodiode mounted on a computer-controlled linear stage we measure the angular dependence of the scattered light intensity, $I_s(q,t)$ which is proportional to the dynamic structure factor $S(q,t)$, at a fixed time t after the quench. Here $q = 4\pi n \sin(\theta/2)/\lambda$ is the scattering vector where $\lambda = 633nm$ and n is the refractive index. The sweep time (2 sec.) was small compared to the collapse time of the diffraction ring ($\sim$ 10 min.) The angular resolution was $\pm 0.35°$.

The experimental procedure involved mixing the gelatin solution thoroughly in a reentrant quartz cell (path length 0.1mm) at 40°C (far above the coexistence curve and sol/gel line). The sample was then quenched to the desired final temperature by quickly transferring to a water bath held at a temperature in the spinodal region. We found that for all cases the sample equilibrated to 95% of the quench temperature in at least 15 seconds. For each quench we recorded both the transmitted light intensity (I_t at q near 0) and the scattered light intensity ($I_s(q,t)$ at $0.05 < q < 0.35A \times 10^3$) as a function of time.

3. RESULTS AND DISCUSSION

We observe that upon quenching the sample to a temperature below the spinodal a diffuse, weak scattering occurs which sharpens into a well-defined diffraction ring as time proceeds. As the process of phase separation proceeds this ring grows more intense and its diameter decreases. These qualitative observations are characteristic of a spinodal process with the diameter of the ring related to the characteristic wave vector at which growth is most pronounced.

Figure 2 shows a typical plot of $I_s(q,t)$ the scattered intensity, versus the scattering vector q at different times after the sample was quenched to 24.5°C. The sharp feature at high q arises from the edge of the conical lens. The ring initially appears at large q. As time proceeds, the peak of the ring shifts in to smaller q (the domain size increases) and the intensity of the ring grows (the concentration gradient increases). This collapse of the ring continues till $\sim$ 100s. after the quench; beyond this time the peak position remains virtually unchanged, although the intensity still continues to increase. The increase of correlation length and concentration gradient are both signs of normal spinodal decomposition, however the occurence of a trapped ring implies that the domains are 'frozen' once the sample gels. However when the sample is quenched to $28.5°C$ the ring collapses completely in $\sim$ 20 minutes.

Figure 3 shows log-log plots of q_{max} vs. time for different quench temperatures; q_{max} is defined as the wavevector where the maximum intensity occurs. All of the quenches show some type of power law behavior but the exact nature of this behavior differs as the quench depth is changed. For the 24.5 °C data there are only two types of behavior. In the early stages, the ring collapses with an exponent of 0.5 but as time procceds the collapse of the ring stops. The 26.5°C data and the 27.5 °C data show three types of behavior. In the early stages, the ring follows a power law collapse; in the middle stages, we see a crossover region and the late stages also show a trapped ring. The 28.5 °C data shows power law collapse followed by a complete collapse — there is no crossover region.

Some other qualitative features of this phase transition are that: (a) the final wavevector at which the ring 'freezes' increases as the quenches become deeper and (b) the ring freezes at earlier times for deeper quenches. Since the rate of gelation increases at lower temperatures there will be more gel network at a given time t for a deep quench than for a shallow quench. If we postulate that the growth of the domains is limited by the gel network it follows that the system becomes trapped at earlier times and smaller characteristic domain size (i.e. larger $q_{\max}$) for deeper quenches.

Figure 4 shows plots of intensity at $q_{\max}$ (normalized by the forward intensity) as a function of time. One striking feature of these plots is that they all obey an equation of the form $I_s(q,t) = I_{\text{final}}(1 - e^{-\beta t})$. We believe this phenonmena is directly related to the the formation of the gel network. Another notable feature of these plots is the variation of β and I_{final} with quench depth: as the quenches are made shallower, β increases and I_{final} or the final concentration gradient decreases. The dependence of β on T_q is related to the relative magnitudes of the time scales on which phase separation and gelation occur. When the system is quenched far below the spinodal temperature (e.g., 24.5 C) the driving force for the phase transition is large; the transition therefore proceeds quickly (smaller β and a larger I_{final}). For these deep quenches the time scale on which gelation occurs is comparable to the time scale for phase separation. When the system is quenched just below the spinodal the driving force for the phase separation is smaller so that the phase separation proceeds slowly. However gelation is slowed down even more than phase separation for this case and thus the process of phase separation goes to completion before any significant gelation has occured.

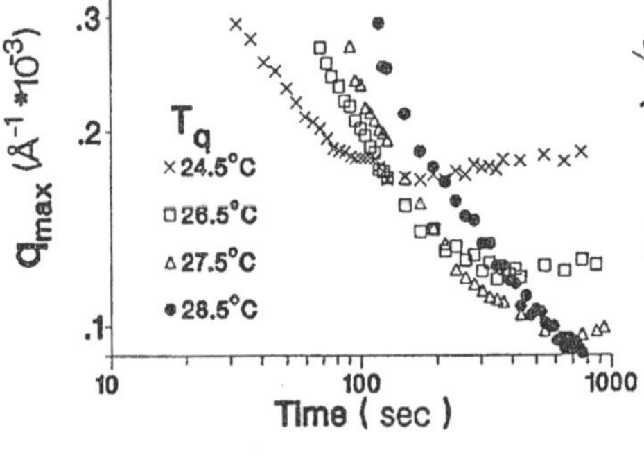

Fig.3 Time dependence of q_{max} at different T_q as indicated.

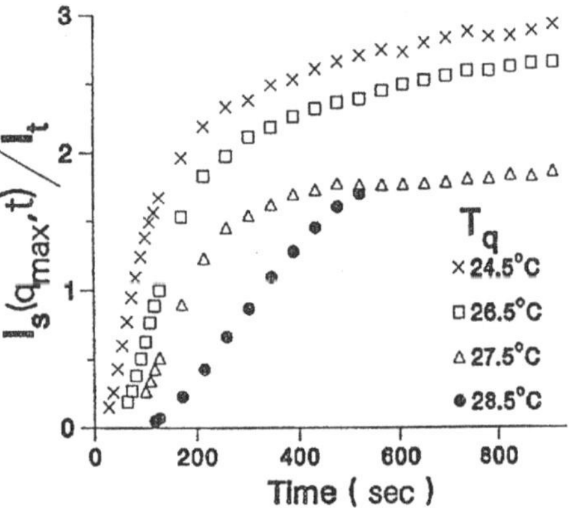

Fig.4 Time dependence of maximum scattered intensity at different T_q.

A comparison of Figs. 3 and 4 yields some interesting observations: for a deep quench, the final correlation length is greater than that for a shallow quench; yet, the final concentration gradient is greater in the deep quench case than for the shallow quench. This difference in behavior between the correlation length and concentration gradient indicates that the time scale on which domain growth occurs is different from the time scale on which the concentration gradient varies. For normal spinodal decomposition,without gelation $q_{max} = t^{-\alpha}$ and $I_s(q,t) = t^{\gamma}(1)$. In the GWM system we also see power law growth of the domain size. However, it is interesting to note that $I_s(q,t)$ differs drastically from the power law behavior seen in other systems. This difference is due to the role that gelation plays in the phase separation dynamics.

In conclusion we note that in spite of the qualitative similarity with other systems, the dynamical behavior in gelatin solution differs in several important ways from the dynamics in either small molecule mixtures or polymer-polymer mixtures. These are summarized below:

(i)The dynamical behavior depends on the location of the quench temperature T_q, relative to the equilibrium sol-gel transition temperature, T_{gel} and also on the relative magnitudes of the time taken for gelation (τ_{gel}) and the time taken for the phase separation to occur (τ_p). For the case, where $\tau_{gel} \gg \tau_p$, the dynamics is essentially the same as that observed in binary liquids with a power law relationship $q_{max} \sim t^{-0.5}$ as compared with a $t^{-1/3}$ or t^{-1} law observed in binary liquids. Interestingly, this behavior is observed even when T_q may be less than T_{gel}, the equilibrium sol-gel transition temperature, suggesting that if the gelation process is very slow compared to the phase separation process, then it has virtually no influence on phase separation dynamics. However, when the sample is quenched to $T_q < T_{gel}$ and $\tau_{gel} \sim \tau_p$ or $\tau_{gel} < \tau_p$ the spinodal decomposition process does not go to completion. The concentration fluctuation produces domains which grow to some final size and then get trapped. Experimentally, this is seen in the occurrence of a trapped diffraction ring which persists for days. The deeper the quench into the gel phase, the earlier in time is the onset of trapping and the larger the final wave vector (i.e., the smaller the domain size) at which the system becomes trapped. In this respect gelation may be similar to the behaviour of glasses.

(ii)A very unusual behavior is observed in the growth of the intensity, or, equivalently, the structure factor or the growth of the concentration fluctuations. In spinodal decomposition studies with other systems only two types of microscopic growth laws have been observed: (a) power law growth in late stages or (b) exponential growth in early stages. In the gelatin system we observed the intensity growing to a saturation limit.

REFERENCES

1 J. D. Gunton, M. San Miguel, and P. S. Sahni, *Phase Transitions and Critical Phenomena*, eds. Domb and Lebowitz (1983), Vol. 8.

2 W. E. Goldburg, in *Scattering Techniques Applied to Supramolecular and Nonequilibrium Systems*, eds. S. H. Chen, B. Chu, and R. Nossal (Plenum Press, New York, 1980), and references therein.

3 H. L. Snyder and P. Meakin, J. Chem. Phys. **79**, 5588 (1983).

4 C. C. Han, M. Okada, Y. Muroga, F. L. McCrackin, B. J. Bauer and Q. Tran-Cong, Polymer Engineering and Science **26**, 3 (1986).

5 D. Stauffer, A. Coniglio, and M. Adam, Adv. Polym. Sci. **44**, 103 (1982).

6 T. Tanaka, G. Swislow, and I. Ohmine, Phys. Rev. Lett. **42**, 1556 (1979).

7 K. Binder, Phys. Rev. A **29**, 341 (1984).

8 This research was supported by a grant from the NSF.

CLOSING ADDRESS

R.W. CAHN

Department of Materials Science and Metallurgy, University of Cambridge, Pembroke Street, Cambridge CB2 3QZ, England

Si Moss gave us a splendid Opening Address; by the time he had finished, the matter remaining for the rest of the speakers had already been substantially circumscribed. When the numerous theorists and occasional experimentalists who followed him had finished with Phase Stability, I kept asking myself insistently what there was left for me to say: indeed, I felt distinctly destabilized. There's only one thing for it: I'll percolate, and try to wind my way sinuously through the narrow gaps remaining between what others have presented.

Most people percolate by courtesy of a computer. That thought set me reflecting on the ways of computers, and (since after all I am giving a Closing Address) I was puzzling over the problem of removing an Opening Address from a computer's memory and replacing it by a Closing Address. I have been trying for more than a year to persuade a computer deep in the hidden recesses of Tennessee to admit the existence of the address you see at the top of this page. The fact that I failed dismally simply goes to show that an Opening Address is far more important, persistent and influential than a Closing Address. That is only fair in the light of the speech that Si Moss gave us...but makes it pretty tricky for me!

How, when you come down to it, does one set about cleansing a computer's memory? Following Pavlov, one might attempt a behaviourist's approach: electric shocks alternating with food pellets (bytes would please the computer best, I suppose). Another approach would be to have a go at psychoanalysis: everybody knows that a psychoanalyst's task is to extract information from deep within the unconscious. Shakespeare knew all about it:

Macbeth: Canst thou not minister to a mind diseased,

G. M. Stocks and A. Gonis (eds.), Alloy Phase Stability, 645–650.

Pluck from the memory a rooted sorrow,
Raze out the written troubles of the brain
And with some sweet oblivious antidote
Cleanse the stuffed bosom of that perilous stuff
That weighs upon the breast?
Physician: Therein the patient
Must minister to himself.
Macbeth: Throw physic[s] to the dogs,
I'll none of it!

There is some uncertainty, when one compares the different versions of the First Folio, whether Shakespeare wrote 'physic' (an old word for medicine) or 'physics'. Shakespeare was unfortunately born too early to use a spelling checker...he would have been a prime candidate...but personally I am convinced that he was an early member of the anti-science-and-computer brigade. Macbeth, certainly, would have been even more distraught if he had wanted to pluck a rooted sorrow from a computer's memory instead of his wife's.

Of course, computers have other uses besides creating opening data bases, correcting spelling and doing sums. People do computer simulations. Nancy Wright remarked that there is theory, there is experiment...and there is computer simulation. Is that a form of theory or a form of experiment? Maybe we should think of it as theory, because after all the simulationist can alter Nature as (s)he sees fit, and that's something no experimentalist can get away with.

Wigner once enunciated a splendid theoretician's principle:

'When theory and experiment agree,throw away the theory.'
'When theory and experiment *dis* agree, throw away the experiment.'

I wonder what he would have said about computer simulation? Perhaps, 'when simulation and experiment disagree, throw away the computer'?

In any case, I wonder whether people are too obsessed with achieving detailed agreement . I incline more to the outlook of Oscar Wilde:

'Ah don't say you agree with me! When people agree with me, I always feel that I must be wrong!'

And in another place, he put the seal on this philosophy:

'The truth is rarely pure, and never simple.'

Wigner and Wilde, alike, would have felt at home with the order-disorder transition, a topic I have loved for the past 32 years and understand less with every year that passes, as its true, diabolic complexity gradually emerges.

A strikingly large proportion of the speakers these past two weeks have addressed themselves to different aspects of phase *transitions*, and in particular the disorder-order transition (maybe we should write it that way round...after all, disorder is the primal state natural to anarchic nature). Following Si Moss, the heavy brigade (Ducastelle et al.) got to work on modern aspects of electron theory applied to the problem, and it really is remarkable what complexities can now be interpreted. Pettifor took it all a step further, by showing how things yet unknown can be confidently predicted, with the help of Mendeleev superstrings. De Fontaine appeared in his role as the prophet of CVM (Kikuchi, of course, is the father), Binder and Selke, in their different ways, achieved the apotheosis of statistical mechanics, Staunton shared with Freeman a mastery of 'advanced band structures'. We even heard from Newman how order-disorder concepts can be applied to semiconductors...and who ever previously thought of those as being potentially disordered?

All this was theory. Yet the experimentalists also put in their oar, notably van Tendeloo with his superb high-resolution electron micrographs, Liu, who showed us that the practical uses of ordered intermetallic compounds are exciting much activity, and Jordan, from whose measurements insights about order-disorder emerged as a sort of byproduct. Bansil's intriguing work on polymers provided a novel echo of what Binder had told us earlier, and Saboungi's work on metallic liquids provided fresh mysteries. Finally, a whole group of scientists, including Khachaturyan and Nabarro, high-lighted a variable which has been largely neglected up to now in considering phase transitions in general (and order-disorder transitions in particular): the role of stresses, both those applied externally and those internally generated by the transition itself and exerting a form of negative feedback: this part of the proceedings itself was a fascinating amalgam of theory and experiment., and is apposite in view of my namesake's researches on the role of stresses on phase equilibria (1) and diffusion (2), and the special emphasis by Williams (3) on stress effects in ordering alloys.

The complexities of both electron theory and statistical mechanics applied to ordering alloys have now reached a level, as we have seen these past days, when protagonists of different approaches experience some difficulties in fully appreciating each other's subtlety. This set me to thinking about adjustable parameters, orders of magnitude, rival models, and the insidious temptations of professional jargon, abbreviations in particular. That led me to three quotations, followed by a pictorial summary of all three together:

1) Wigner: 'If you give me two parameters, I can fit an elephant; if you give me three, I can make it wiggle its toes; and if you give me four, I can make a little bird sit on its back.'

2) de Fontaine: 'Ordering energy is like a little bird on the back of an elephant.'

3) Kipling: 'There are four-and-twenty ways
Of constructing tribal lays...
And every single one of them is right!'

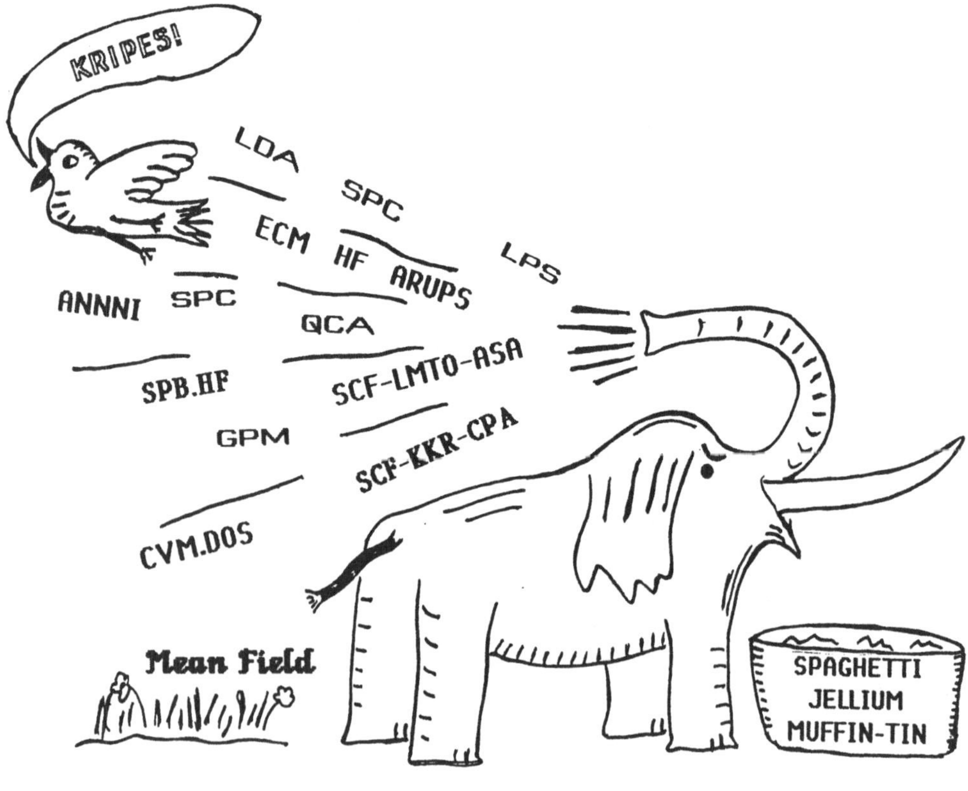

The little bird which initially, at least, perches on the elephant's back (before it becomes self-conscious about its precarious existence and flies off, squawking) represents, of course, the smallness of the ordering energy compared with a typical cohesive energy. In this it resembles the energy of ferromagnetic ordering, with which atomic ordering is indeed sometimes closely intertwined, as we have heard earlier in the week. Atomic and magnetic ordering have another intriguing feature in common: magnetic domains in ferromagnetic crystals and antiphase domains in atomically ordered crystals each have a powerful effect on the properties of their respective crystals (although no specific uses for antiphase domains, akin to the bubble domains in certain magnetic memories, have as yet surfaced). This year, spurred by an elephantine memory, we celebrate the fiftieth anniversary of the discovery of antiphase domains by Charles Sykes, who was working at the time in the British steel industry.

Antiphase domains, in my opinion, are the most unjustly neglected feature of atomically ordered alloys (although they do get their brief place in the sun in connection with long-period superlattices, about which we have learnt much this week). Their presence, size, crystallography and anisotropy affects various properties of ordered alloys, particularly the mechanical behaviour. Recently it has even transpired (3,4) that an antiphase domain boundary (APDB) in an alloy close to the Ni_3Al composition serves as a preferred heterogeneous nucleation site for the formation of the disordered phase above the transition temperature. This in turn seems to be linked with the theoretically predicted (6) localized changes in composition and order parameter not only at the boundary itself, but over a distance of some atomic parameters either side. In fact, like a magnetic domain wall, an APDB plainly has a finite effective thickness. The anomalous behaviour of APDBs, which constitute a kind of internal surface, matches that of free surfaces, where the ordering process is changed from first order to (nearly) second order (7-9). That, in turn, occasioned a magnificent application of formal critical theory, by Lipowsky and Speth (10), to the process of ordering at free surfaces, and that theory in turn has found application to recent experimental observations of preferential melting at free surfaces (11-13). There is a further consideration: as is now recognized (5), an intermetallic compound—i.e., one which freezes fully ordered— contains no antiphase domains, whereas an alloy which orders after it has solidified, does. Since APDs affect properties in various ways, this constitutes an important distinction between the two types of ordered alloy. Whether or not an alloy freezes in the ordered form depends on the ordering energy...the little bird on the elephant's back...and the ordering energy in turn is linked with other aspects of behaviour, such as the ordering kinetics (14).— In my view, theorists of the order-disorder transition would do well to devote more attention to the ways of antiphase domains.

I had hoped to avoid a relapse into seriousness at the end of two very serious, instructive and hard-working weeks, but it seems that my enthusiasm for a particular theme has got the better of me. An elephant never forgets what a little bird has once told him. You may not agree with me about the interest inherent in antiphase domains...but if so, I refer you to my first quotation from Oscar Wilde.

It remains for me to express, on behalf of all of us here, warm thanks to Tony Gonis and Malcolm Stocks for organizing a first-class scientific experience, and to NATO for making that possible in the first instance.

REFERENCES

1. Larché FC, Cahn JW, *Acta Metall.* **26** (1978) 53.
2. Larché FC, Cahn JW, *Acta Metall.* **30** (1982) 1835.
3. Williams RO, *Met. Trans.* **11A** (1980) 247.
4. Cahn RW, Siemers PA, Geiger JE, Bardhan P, *Acta Metall.* **35** (1987) 2737.
5. Cahn RW, Siemers, PA, Hall EL, *Acta Metall.* **35** (1987) 2753.
6. Sanchez JM, Eng S, Wu YP, Tien JK, *Mat. Res. Soc. Symp. Proc.* **81** (1987) 57.
7. Sundaram VS, Alben RS, Robertson WD, *Surf. Sci.* **46** (1974) 653.
8. Buck TM, Wheatley GH, Marchut L, *Phys. Rev. Lett.* **51** (1983) 43.
9. McRae EG, Malic RA, *Surf. Sci.* **148** (1984) 551.
10. Lipowsky R, Speth W, *Phys. Rev. B* **28** (1983) 3983.
11. Frenken JWM, van der Veen JF, *Phys. Rev. Lett.* **54** (1985) 134.
12. Cahn RW, *Nature* **323** (1986) 668.
13. Editorial, *Nature* **330** (1987) 599.
14. Cahn RW, *Mat. Res. Soc. Symp. Proc.* **57** (1987) 385.

LIST OF LECTURERS

Binder, K.	Universitat Meinz, Federal Republic of Germany
Cahn, R.W.	Keck Laboratory, United States
Dederichs, P.H.	KFA Institute, Federal Republic of Germany
deFontaine, D.	University of California, Berkeley, United States
Ducastelle, F.	ONERA, France
Freeman, A.J.	Northwestern University, United States
Gyorffy, B.L.	University of Bristol, United Kingdom
Jordan, R.G.	University of Birmingham, United Kingdom
Kaufman, L.	Mann Labs, Cambridge, MA, United States
Khachaturyan, A.G.	University of California, Berkeley, United States
Liu, C.T.	Oak Ridge National Laboratory, United States
Moss, S.C.	University of Houston, Texas, United States
Nabarro, F.R.N.	University of Witwatersrand, South Africa
Pettifor, D.D.	Imperial College, London, United Kingdom
Selke, W.	IBM Zurich, Switzerland
Staunton, J.B.	University of Warwick, Coventry, United Kingdom
Van Tendeloo, G.	Rijksuniversitair Centrum, Antwerpen, Belgium

LIST OF PARTICIPANTS

Alexandropoulos, N.G.	Physics Department, University of Ioannina, Greece
Alexandropoulos, T.	Physics Department, University of Ioannina, Greece
Andriotis, A.N.	Theoretical Physics Institute, Greece
Ardell, A.J.	University of California, United States
Badekas, H.	Technical University of Athens, Greece
Bacalis, N.	Physics Department, University of Crete, Greece
Banhart, J.	Universität München, Federal Republic of Germany
Bansil, A.K.	Northeastern University, United States
Bansil, R.	Boston University, United States
Bernard, C.	ENSEEG, Domaine Universitaire, France
Blander, M.	Argonne National Laboratory, United States
Broddin, D.	Universiteit Antwerpen, Belgium
Bunker, B.	Univ. of Notre Dame, United States
Carlsson, A.	Washington University, United States
Cenedese, P.	University of California, Berkeley, United States
Christides, C.	Demokritos Institute Aghia Paraskevi, Greece
Clapp, P.C.	University of Connecticut, United States
Coehoorn, R.C.	Philips Research Laboratories, The Netherlands
Connolly, J.W.D.	Office of Advanced Science, Computing-NSF, U.S.A.
Cressoni, J.C.	Imperial College, United Kindgom
daSilva, E.Z.	University of Campinas, Brazil
Dederichs, P.H	KFA Institute, Federal Republic of Germany
deGraef, M.	Katholieke Universiteit Leuven, Belgium
Dünbar. S.	Istanbul Technical University, Turkey
Faulkner, J.S.	Florida Atlantic University, United States
Finel, A.	ONERA, France
Flevaris, N.	University of Thessaloniki, Greece
Florio, G.M.	Instituto di Fisica dell, Univ. di Messina, Italy
Fontanet, R.	Universitat Autonoma de Barcelona, Spain
Ginatempo, B.	Instituto di Fisica dell, Univ. di Messina, Italy
Gautier, F.	LMSES, U. Louis Pasteur, France
Gonis, A.	Lawrence Livermore National Laboratory, U.S.A.
Guiliano, E.S.	Instituto di Fisica dell, Univ. di Messina, Italy
Harrison, N.	SERC, Daresbury Laboratory, United Kingdom
Hadjipanayis, G.	Kansas State University, United States
Jankowski, A.F.	Lawrence Livermore National Laboratory, U.S.A.
Johnson, D.D.	Naval Research Laboratory, United States
Karakostas, T.	University of Thessaloniki, Greece
Katsanos, D.	University of Ioannina, Greece
Kikuchi, R.	Max-Planck Inst. für Eisenforschung, F.R. Germany
Kiriakidis, G.	University of Crete, Greece
Kleppa, O.J.	James Franck Institute, Univ. of Chicago, U.S.A.
Komninou, P.	University of Thessaloniki, Greece
Kotsis, K.	University of Ioannina, Greece
König, U.	Technical University of Vienna, Austria
Koutsomihalis, A.M.	National Technical University of Athens, Greece
Leroux, C.	LMSES, Universite Louis Pasteur, France
Loiseau, A.	ONERA, France
Maclin, A.P.	National Research Council, United States
Madar, R.	CNRS, Domaine Universitaire, France

Marksteiner, P.	Technical University of Vienna, Austria
Mavroyannis, C.	National Research Council of Canada, Canada
Mishima, Y.	Tokyo Institute of Technology, Japan
Newman, K. E.	University of Notre Dame, United States
Nicholson, D.M.	Oak Ridge National Laboratory, United States
Ortin, J.	Katholieke Universiteit Leuven, Belgium
Papaconstantopoulos, D.	Naval Research Laboratory, United States
Papanicolaou, N.	University of Ioannina, Greece
Pfeiler, W.	Universität Wien, Austria
Pinski, F.J.	University of Cincinnati, United States
Polychroniadis, E.K.	University of Thessaloniki, Greece
Rand, M.H.	AERE, United Kingdom
Saboungi, M.L.	Argonne National Laboratory, United States
Sanchez, J.M.	Henry Krumb School of Mines, Columbia Univ., U.S.A.
Schadler, G.	Los Alamos National Laboratory, United States
Schryvers, D.	Lawrence Berkeley Lab., Univ. of California, U.S.A.
Schweika, W.	Kernforschungsanlage Jülich, Federal Rep. Germany
Shelton, W.A.	Oak Ridge National Laboratory, United States
Sluiter, M.	University of California, Berkeley, United States
Sohal, G.	University of Birmingham, United Kingdom
Solal, F.	ONERA, France
Stancanelli, A.	Instituto di Fisica dell Univ. di Messina, Italy
Stefanou, N.	Kernforschungsanlage Jülich, Federal Rep. Germany
Stocks, G.M.	Oak Ridge National Laboratory, United States
Takeda, M.	Rijksuniversitair Centrum, Antwerpen, Belgium
Tsakalakos, T.	State University of New York, United States
Tsagaraki, A.	University of Crete, Greece
Tsalapatis, P.A.	Technical University of Athens, Greece
Tsatis, D.	University of Patras, Greece
Turchi, P.E.A.	Lawrence Livermore National Laboratory, U.S.A.
Ucisik, A.H.	Istanbul Technical University, Turkey
Vedula, K.V.	Case Western Research University, Untied States
Virgji, H.	University of Bristol, United Kingdom
Weinberger, P.	Technical University of Vienna, Austria
Wille, L.T.	Univ. of California, Berkeley, United States
Wright, N.F.	Oak Ridge National Laboratory, United States
Zdetsis, A.D.	Research Center of Crete, Greece
Zhang, X.G.	Northwestern University, United States

AUTHOR INDEX

*Invited lecturer